FINANCIAL ENGINEERING AND COMPUTATION

During the past decade many sophisticated mathematical and computational techniques have been developed for analyzing financial markets. Students and professionals intending to work in any area of finance must not only master advanced concepts and mathematical models but must also learn how to implement these models computationally. This comprehensive text combines a thorough treatment of the theory and mathematics behind financial engineering with an emphasis on computation, in keeping with the way financial engineering is practiced in today's capital markets.

Unlike most books on investments, financial engineering, or derivative securities, the book starts from basic ideas in finance and gradually builds up the theory. The advanced mathematical concepts needed in modern finance are explained at accessible levels. Thus it offers a thorough grounding in the subject for MBAs in finance, students of engineering and sciences who are pursuing a career in finance, researchers in computational finance, system analysts, and financial engineers.

Building on the theory, the author presents algorithms for computational techniques in pricing, risk management, and portfolio management, together with analyses of their efficiency. Pricing financial and derivative securities is a central theme of the book. A broad range of instruments is treated: bonds, options, futures, forwards, interest rate derivatives, mortgage-backed securities, bonds with embedded options, and more. Each instrument is treated in a short, self-contained chapter for ready reference use.

Many of these algorithms are coded in Java as programs for the Web, available from the book's home page: www.csie.ntu.edu.tw/~lyuu/Capitals/capitals.htm. These programs can be executed on Windows, MacOS, or Unix platforms.

Yuh-Dauh Lyuu received his Ph.D. in computer science from Harvard University. His past positions include Member of Technical Staff at Bell Labs, Research Scientist at NEC Research Institute (Princeton), and Assistant Vice President at Citicorp Securities (New York). He is currently Professor of Computer Science and Information Engineering and Professor of Finance, National Taiwan University. His previous book is *Information Dispersal and Parallel Computation*.

Professor Lyuu has published works in both computer science and finance. He also holds a U.S. patent. Professor Lyuu received several awards for supervising outstanding graduate students' theses.

FINANCIAL ENGINEERING AND COMPUTATION

Principles, Mathematics, Algorithms

YUH-DAUH LYUU
National Taiwan University

CAMBRIDGE
UNIVERSITY PRESS

CAMBRIDGE
UNIVERSITY PRESS

32 Avenue of the Americas, New York NY 10013-2473, USA

Cambridge University Press is part of the University of Cambridge.

It furthers the University's mission by disseminating knowledge in the pursuit of education, learning and research at the highest international levels of excellence.

www.cambridge.org
Information on this title: www.cambridge.org/9780521781718

© Yuh-Dauh Lyuu 2002

First published 2002

A catalogue record for this publication is available from the British Library

ISBN 978-0-521-78171-8 Hardback

In Loving Memory of RACHEL *and* JOSHUA

Contents

Preface

> [A book] is a node within a network.
>
> Michel Foucault (1926–1984), *The Archaeology of Knowledge*

Intended Audience

As the title of this book suggests, a modern book on financial engineering has to cover investment theory, financial mathematics, and computer science evenly. This interdisciplinary emphasis is tuned more to the capital markets wherever quantitative analysis is being practiced. After all, even economics has moved away from a time when "the bulk of [Alfred Marshall's] potential readers were both unable and unwilling to read economics in mathematical form" according to Viner (1892–1970) [860] toward the new standard of which Markowitz wrote in 1987, "more than half my students cannot write down the formal definition of [the limit of a sequence]" [642].

This text is written mainly for students of engineering and the natural sciences who want to study quantitative finance for academic or professional reasons. No background in finance is assumed. Years of teaching students of business administration convince me that technically oriented MBA students will benefit from the book's emphasis on computation. With a sizable bibliography, the book can serve as a reference for researchers.

This text is also written for practitioners. System analysts will find many compact and useful algorithms. Portfolio managers and traders can obtain the quantitative underpinnings for their daily activities. This work also serves financial engineers in their design of financial instruments by expounding the underlying principles and the computational means to pricing them.

The marketplace has already offered several excellent books on derivatives (e.g., [236, 470, 514, 746, 878]), financial engineering (e.g., [369, 646, 647]), financial theory (e.g., [290, 492]), econometrics (e.g., [147]), numerical techniques (e.g., [62, 215]), and financial mathematics (e.g., [59, 575, 692, 725]). There are, however, few books that come near to integrating the wide-ranging disciplines. I hope this text succeeds at least partially in that direction and, as a result, one no longer has to buy four or five books to get good coverage of the topics.

Presentation

This book is self-contained. Technically sophisticated undergraduates and graduates should be able to read it on their own. Mathematical materials are added where they are needed. In many instances, they provide the coupling between earlier chapters and upcoming themes. Applications to finance are generally added to set the stage. Numerical techniques are presented algorithmically and clearly; programming them should therefore be straightforward. The underlying financial theory is adequately covered, as understanding the theory underlying the calculations is critical to financial innovations.

The large number of exercises is an integral part of the text. Exercises are placed right after the relevant materials. Hints are provided for the more challenging ones. There are also numerous programming assignments. Those readers who aspire to become software developers can learn a lot by implementing the programming assignments. Thoroughly test your programs. The famous adage of Hamming (1916–1998), "The purpose of computing is insight, not numbers," does not apply to erroneous codes. Answers to all nontrivial exercises and some programming assignments can be found near the end of the book.

Most of the graphics were produced with *Mathematica* [882]. The programs that generate the data for the plots have been written in various languages, including C, C++, Java, JavaScript, Basic, and Visual Basic. It is a remarkable fact that most – if not all – of the programming works could have been done with spreadsheet software [221, 708]. Some computing platforms admit the integration of the spreadsheet's familiar graphical user interface and programs written in more efficient high-level programming languages [265]. Although such a level of integration requires certain sophistication, it is a common industry practice. Freehand graphics were created with Canvas and Visio.

The manuscript was typeset in LaTeX [580], which is ideal for a work of this size and complexity. I thank Knuth and Lamport for their gifts to technical writers.

Software

Many algorithms in the book have been programmed. However, instead of being bundled with the book in disk, my software is Web-centric and platform-independent [412]. Any machine running a World Wide Web browser can serve as a host for those programs on *The Capitals* page at

www.csie.ntu.edu.tw/~lyuu/capitals.html

There is no more need for the (rare) author to mail the upgraded software to the reader because the one on the Web page is always up to date. This new way of software development and distribution, made possible by the Web, has turned software into an Internet service.

Organization

Here is a grand tour of the book:

Chapter 1 sets the stage and surveys the evolution of computer technology.

Chapter 2 introduces algorithm analysis and measures of complexity. My convention for expressing algorithms is outlined here.

Chapter 3 contains a relatively complete treatment of standard financial mathematics, starting from the time value of money.

Chapter 4 covers the important concepts of duration and convexity.

Chapter 5 goes over the static term structure of interest rates. The coverage of classic, static finance theory ends here.

Chapter 6 marks the transition to stochastic models with coverage of statistical inference.

Chapters 7–12 are about options and derivatives. Chapter 7 presents options and basic strategies with options. Chapter 8 introduces the arbitrage argument and derives general pricing relations. Chapter 9 is a key chapter. It covers option pricing under the discrete-time binomial option pricing model. The celebrated Black–Scholes formulas are derived here, and algorithms for pricing basic options are presented. Chapter 10 presents sensitivity measures for options. Chapter 11 covers the diverse applications and kinds of options. Additional derivative securities such as forwards and futures are treated in Chap. 12.

Chapters 13–15 introduce the essential ideas in continuous-time financial mathematics. Chapter 13 covers martingale pricing and Brownian motion, and Chap. 14 moves on to stochastic integration and the Ito process. Together they give a fairly complete treatment of the subjects at an accessible level. From time to time, we go back to discrete-time models and establish the linkage. Chapter 15 focuses on the partial differential equations that derivative securities obey.

Chapter 16 covers hedging by use of derivatives.

Chapters 17–20 probe deeper into various technical issues. Chapter 17 investigates binomial and trinomial trees. One of the motives here is to demonstrate the use of combinatorics in designing highly efficient algorithms. Chapter 18 covers numerical methods for partial differential equations, Monte Carlo simulation, and quasi–Monte Carlo methods. Chapter 19 treats computational linear algebra, least-squares problems, and splines. Factor models are presented as an application. Chapter 20 introduces financial time series analysis as well as popular time-series models.

Chapters 21–27 are related to interest-rate-sensitive securities. Chapter 21 surveys the wide varieties of interest rate derivatives. Chapter 22 discusses yield curve fitting. Chapter 23 introduces interest rate modeling and derivative pricing with the elementary, yet important, binomial interest rate tree. Chapter 24 lays the mathematical foundations for interest rate models, and Chaps. 25 and 26 sample models from the literature. Finally, Chap. 27 covers fixed-income securities, particularly those with embedded options.

Chapters 28–30 are concerned with mortgage-backed securities. Chapter 28 introduces the basic ideas, institutions, and challenging issues. Chapter 29 investigates the difficult problem of prepayment and pricing. Chapter 30 surveys collateralized mortgage obligations.

Chapter 31 discusses the theory and practice of portfolio management. In particular, it presents modern portfolio theory, the Capital Asset Pricing Model, the Arbitrage Pricing Theory, and value at risk.

Chapter 32 documents the Web software developed for this book.

Chapter 33 contains answers or pointers to all nontrivial exercises.

This book ends with an extensive index. There are two guiding principles behind its structure. First, related concepts should be grouped together. Second, the index should facilitate search. An entry containing parentheses indicates that the term within should be consulted instead, first at the current level and, if not found, at the outermost level.

Acknowledgments

Many people contributed to the writing of the book: George Andrews, Nelson Beebe, Edward Bender, Alesandro Bianchi, Tomas Björk, Peter Carr, Ren-Raw Chen, Shu-Heng Chen, Oren Cheyette, Jen-Diann Chiou, Mark Fisher, Ira Gessel, Mau-Wei Hung, Somesh Jha, Ming-Yang Kao, Gow-Hsing King, Timothy Klassen, Philip Liang, Steven Lin, Mu-Shieung Liu, Andrew Lo, Robert Lum, Chris McLean, Michael Rabin, Douglas Rogers, Masako Sato, Erik Schlögl, James Tilley, and Keith Weintraub.

Ex-colleagues at Citicorp Securities, New York, deserve my deep thanks for the intellectual stimuli: Mark Bourzutschky, Michael Chu, Burlie Jeng, Ben Lis, James Liu, and Frank Feikeh Sung. In particular, Andy Liao and Andy Sparks taught me a lot about the markets and quantitative skills.

Students at National Taiwan University, through research or course work, helped improve the quality of the book: Chih-Chung Chang, Ronald Yan-Cheng Chang, Kun-Yuan Chao [179], Wei-Jui Chen [189], Yuan-Wang Chen [191], Jing-Hong Chou, Tian-Shyr Dai, [257, 258, 259] Chi-Shang Draw, Hau-Ren Fang, Yuh-Yuan Fang, Jia-Hau Guo [405], Yon-Yi Hsiao, Guan-Shieng Huang [250], How-Ming Hwang, Heng-Yi Liu, Yu-Hong Liu [610], Min-Cheng Sun, Ruo-Ming Sung, Chen-Leh Wang [867], Huang-Wen Wang [868], Hsing-Kuo Wong [181], and Chao-Sheng Wu [885].

This book benefited greatly from the comments of several anonymous reviewers. As the first readers of the book, their critical eyes made a lasting impact on its evolution. As with my first book with Cambridge University Press, the editors at the Press were invaluable. In particular, I would like to thank Lauren Cowles, João da Costa, Caitlin Doggart, Scott Parris, Eleanor Umali, and the anonymous copy editor.

I want to thank my wife Chih-Lan and my son Raymond for their support during the project, which started in January 1995. This book, I hope, finally puts to rest their dreadful question, "When are you going to finish it?"

Useful Abbreviations

Acronyms

APT	Arbitrage Pricing Theory
AR	autoregressive (process)
ARCH	autoregressive conditional heteroskedastic (process)
ARM	adjustable-rate mortgage
ARMA	autoregressive moving average (process)
BDT	Black–Derman–Toy (model)
BEY	bond-equivalent yield
BOPM	binomial option pricing model
BPV	basis-point value
CAPM	Capital Asset Pricing Model
CB	convertible bond
CBOE	Chicago Board of Exchange
CBT	Chicago Board of Trade
CD	certificate of deposit
CIR	Cox–Ingersoll–Ross
CME	Chicago Mercantile Exchange
CMO	collateralized mortgage obligation
CMT	constant-maturity Treasury (rate)
COFI	Cost of Funds Index
CPR	conditional prepayment rate
DEM	German mark
DJIA	Dow Jones Industrial Average
FHA	Federal Housing Administration
FHLMC	Federal Home Loan Mortgage Corporation ("Freddie Mac")
FNMA	Federal National Mortgage Association ("Fannie Mae")
forex	foreign exchange
FRA	forward rate agreement
FV	future value

GARCH	generalized autoregressive conditional heteroskedastic
GLS	generalized least-squares
GMM	generalized method of moments
GNMA	Government National Mortgage Association ("Ginnie Mae")
HJM	Heath–Jarrow–Morton
HPR	holding period return
IAS	index-amortizing swap
IMM	International Monetary Market
IO	interest-only
IRR	internal rate of return
JPY	Japanese yen
LIBOR	London Interbank Offered Rate
LTCM	Long-Term Capital Management
MA	moving average
MBS	mortgage-backed security
MD	Macauley duration
ML	maximum likelihood
MPTS	mortgage pass-through security
MVP	minimum-variance point
NPV	net present value
NYSE	New York Stock Exchange
OAC	option-adjusted convexity
OAD	option-adjusted duration
OAS	option-adjusted spread
OLS	ordinary least-squares
PAC	Planned Amortization Class (bond)
P&I	principal and interest
PC	participation certificate
PO	principal-only
PSA	Public Securities Association
PV	present value
REMIC	Real Estate Mortgage Investment Conduit
RHS	Rural Housing Service
RS	Ritchken–Sankarasubramanian
S&P 500	Standard and Poor's 500 Index
SMBS	stripped mortgage-backed security
SMM	single monthly mortality
SSE	error sum of squares

SSR	regression sum of squares
SST	total sum of squares
SVD	singular value decomposition
TAC	Target Amortization Class (bond)
VA	Department of Veterans Affairs
VaR	value at risk
WAC	weighted average coupon
WAL	weighted average life
WAM	weighted average maturity
WWW	World Wide Web

Ticker Symbols

DJ	Dow Jones Industrial Average
IRX	thirteen-week T-bill
NDX	Nasdaq 100
NYA	New York Stock Exchange Composite Index
OEX	S&P 100
RUT	Russell 200
SPX	S&P 500
TYX	thirty-year T-bond
VLE	Value Line Index
WSX	Wilshire S-C
XMI	Major Market Index

Introduction

> But the age of chivalry is gone. That of sophisters, oeconomists, and
> calculators, has succeeded; and the glory of Europe is extinguished
> for ever.
>
> Edmund Burke (1729–1797), *Reflections on the Revolution
> in France*

1.1 Modern Finance: A Brief History

Modern finance began in the 1950s [659, 666]. The breakthroughs of Markowitz,
Treynor, Sharpe, Lintner (1916–1984), and Mossin led to the Capital Asset Pricing Model in the 1960s, which became the quantitative model for measuring risk.
Another important influence of research on investment practice in the 1960s was
the Samuelson–Fama efficient markets hypothesis, which roughly says that security
prices reflect information fully and immediately. The most important development in
terms of practical impact, however, was the Black–Scholes model for option pricing
in the 1970s. This theoretical framework was instantly adopted by practitioners. Option pricing theory is one of the pillars of finance and has wide-ranging applications
[622, 658]. The theory of option pricing can be traced to Louis Bachelier's Ph.D. thesis
in 1900, "Mathematical Theory of Speculation." Bachelier (1870–1946) developed
much of the mathematics underlying modern economic theories on efficient markets,
random-walk models, Brownian motion [ahead of Einstein (1879–1955) by 5 years],
and martingales [277, 342, 658, 776].[1]

1.2 Financial Engineering and Computation

Today, the wide varieties of financial instruments dazzle even the knowledgeable.
Individuals and corporations can trade, in addition to stocks and bonds, options,
futures, stock index options, and countless others. When it comes to diversification, one has thousands of mutual funds and **exchange-traded funds** to choose from.
Corporations and local governments increasingly use complex derivative securities
to manage their financial risks or even to speculate. **Derivative securities** are financial instruments whose values depend on those of other assets. All are the fruits of
financial engineering, which means structuring financial instruments to target investor preferences or to take advantage of arbitrage opportunities [646].

The innovations in the financial markets are paralleled by equally explosive progress in computer technology. In fact, one cannot think of modern financial systems without computers: automated trading, efficient bookkeeping, timely clearing and settlements, real-time data feed, online trading, day trading, large-scale databases, and tracking and monitoring of market conditions [647, 866]. These applications deal with *information*. Structural changes and increasing volatility in financial markets since the 1970s as well as the trend toward greater complexity in financial product design call for *quantitative* techniques. Today, most investment houses use sophisticated models and software on which their traders depend. Here, computers are used to model the behavior of financial securities and key indicators, price financial instruments, and find combinations of financial assets to achieve results consistent with risk exposures. The confidence in such models in turn leads to more financial innovations and deeper markets [659, 661]. These topics are the focus of **financial computation**.

One must keep in mind that every computation is based on input and assumptions made by the model. However, input might not be accurate enough or complete, and the assumptions are, at best, approximations.[2] Computer programs are also subject to errors ("bugs"). These factors easily defeat any computation. Despite these difficulties, the computer's capability of calculating with fine details and trying out vast numbers of scenarios is a tremendous advantage. Harnessing this power and a good understanding of the model's limitations should steer us clear of blind trust in numbers.

1.3 Financial Markets

A society improves its welfare through investments. Business owners need outside capital for investments because even projects of moderate sizes are beyond the reach of most wealthy individuals. Governments also need funds for public investments. Much of that money is channeled through the financial markets from savers to borrowers. In so doing, the financial markets provide a link between saving and investment,[3] and between the present and the future. As a consequence, savers can earn higher returns from their savings instead of hoarding them, borrowers can execute their investment plans to earn future profits, and both are better off. The economy also benefits by acquiring better productive capabilities as a result. Financial markets therefore facilitate real investments by acting as the sources of information.

A financial market typically takes its name from the borrower's side of the market: the government bond market, the municipal bond market, the mortgage market, the corporate bond market, the stock market, the commodity market, the foreign exchange (forex) market,[4] the futures market, and so on [95, 750]. Within financial markets, there are two basic types of financial instruments: **debt** and **equity**. Debt instruments are loans with a promise to repay the funds with interest, whereas equity securities are shares of stock in a company. As an example, Fig. 1.1 traces the U.S. markets of debt securities between 1985 and 1999. Financial markets are often divided into **money markets**, which concentrate on short-term debt instruments, and **capital markets**, which trade in long-term debt (bonds) and equity instruments (stocks) [767, 799, 828].

Outstanding U.S. Debt Market Securities (U.S. $ billions)

Year	Municipal	Treasury	Agency MBSs	U.S. corporate	Fed agencies	Money market	Asset-backed	Total
1985	859.5	1,360.2	372.1	719.8	293.9	847.0	2.4	4,454.9
1986	920.4	1,564.3	534.4	952.6	307.4	877.0	3.3	5,159.4
1987	1,010.4	1,724.7	672.1	1,061.9	341.4	979.8	5.1	5,795.4
1988	1,082.3	1,821.3	749.9	1,181.2	381.5	1,108.5	6.8	6,331.5
1989	1,135.2	1,945.4	876.3	1,277.1	411.8	1,192.3	59.5	6,897.6
1990	1,184.4	2,195.8	1,024.4	1,333.7	434.7	1,156.8	102.2	7,432.0
1991	1,272.2	2,471.6	1,160.5	1,440.0	442.8	1,054.3	133.6	7,975.0
1992	1,302.8	2,754.1	1,273.5	1,542.7	484.0	994.2	156.9	8,508.2
1993	1,377.5	2,989.5	1,349.6	1,662.1	570.7	971.8	179.0	9,100.2
1994	1,341.7	3,126.0	1,441.9	1,746.6	738.9	1,034.7	205.0	9,634.8
1995	1,293.5	3,307.2	1,570.4	1,912.6	844.6	1,177.2	297.9	10,403.5
1996	1,296.0	3,459.0	1,715.0	2,055.9	925.8	1,393.8	390.5	11,235.0
1997	1,367.5	3,456.8	1,825.8	2,213.6	1,022.6	1,692.8	518.1	12,097.2
1998	1,464.3	3,355.5	2,018.4	2,462.0	1,296.5	1,978.0	632.7	13,207.4
1999	1,532.5	3,281.0	2,292.0	3,022.9	1,616.5	2,338.2	746.3	14,829.4

Figure 1.1: U.S. debt markets 1985–1999. The Bond Market Association estimates. Sources: Federal Home Loan Mortgage Corporation, Federal National Mortgage Association, Federal Reserve System, Government National Mortgage Association, Securities Data Company, and U.S. Treasury. MBS, mortgage-backed security.

Borrowers and savers can trade directly with each other through the financial markets or direct loans. However, minimum-size requirements, transactions costs, and costly evaluation of the assets in question often prohibit direct trades. Such impediments are remedied by **financial intermediaries**. These are financial institutions that act as middlemen to transfer funds from lenders to borrowers; unlike most firms, they hold only financial assets [660]. Banks, savings banks, savings and loan associations, credit unions, pension funds, insurance companies, mutual funds, and money market funds are prominent examples. Financial intermediaries can lower the minimum investment as well as other costs for savers.

Financial markets can be divided further into primary markets and secondary markets. The **primary market** is often merely a fictional, not a physical, location. Governments and corporations initially sell securities – debt or equity – in the primary market. Such sales can be done by means of public offerings or private placements. A syndicate of investment banks underwrites the debt and the equity by buying them from the issuing entities and then reselling them to the public. Sometimes the investment bankers work on a best-effort basis to avoid the risk of not being able to sell all the securities. Subsequently people trade those instruments in the **secondary markets**, such as the New York Stock Exchange. Existing securities are exchanged in the secondary market.

The existence of the secondary market makes securities more attractive to investors by making them tradable after their purchases. It is the very idea that created the secondary market in mortgages in 1970 by asset securitization [54]. **Securitization** converts assets into traded securities with the assets pledged as collaterals, and these assets can often be removed from the balance sheet of the bank. In so doing, financial intermediaries transform illiquid assets into liquid liabilities [843]. By making mortgages more attractive to investors, the secondary market also makes them more affordable to home buyers. In addition to mortgages, auto loans, credit card receivables, senior bank loans, and leases have all been securitized [330]. Securitization has fundamentally changed the credit market by making the capital market a major supplier of credit, a role traditionally held exclusively by the banking system.

1.4 Computer Technology

Computer hardware has been progressing at an exponential rate. Measured by the widely accepted integer Standard Performance Evaluation Corporation (SPEC) benchmarks, the workstations improved their performance by 49% per year between 1987 and 1997. The memory technology is equally impressive. The dynamic random-access memory (DRAM) has quadrupled its capacity every 3 years since 1977. Relative performance per unit cost of technologies from vacuum tube to transistor to integrated circuit to very-large-scale-integrated (VLSI) circuit is a factor of 2,400,000 between 1951 and 1995 [717].

Some milestones in the industry include the IBM/360 **mainframe**, followed by Digital's **minicomputers**. (Digital was acquired by Compaq in 1998.) The year 1963 saw the first **supercomputer**, built by Cray (1926–1996) at the Control Data Corporation. Apple II of 1977 is generally considered to be the first **personal computer**. It was overtaken by the IBM Personal Computer in 1981, powered by Intel microprocessors and Microsoft's disk operating system (DOS) [638, 717]. The 1980s also witnessed the emergence of the so-called **massively parallel computers**, some of which had more than 65,000 processors [487]. Parallel computers have also been applied to database applications [247, 263] and pricing complex financial instruments [528, 794, 891]. Because commodity components offer the best performance/cost ratio, personal computers connected by fast networks have been uprooting niche parallel machines from most of their traditional markets [24, 200].

On the software side, high-level programming languages dominate [726]. Although they are easier to program with than low-level languages, it remains difficult to design and maintain complex software systems. In fact, in the 1960s, the software cost of the IBM/360 system already dominated its hardware cost [872]. The current trend has been to use the **object-oriented principles** to encapsulate as much information as possible into the so-called **objects** [101, 466]. This makes software easier to maintain and develop. Object-oriented software development systems are widely available [178].

The revolution fostered by the **graphical user interface** (**GUI**) brought computers to the masses. The omnipotence of personal computers armed with easy-to-use interfaces enabled employees to have access to information and to bypass several layers of management [140]. It also paved the way for the **client/server** concept [736].

Client/server systems consist of components that are logically distributed rather than centralized (see Fig. 1.2). Separate components therefore can be optimized based on their functions, boosting the overall performance/cost ratio. For instance, the **three-tier** client/server architecture contains three parts: user interface, computing (application) server, and data server [310]. Because the user interface demands fewer resources, it can run on lightly configured computers. Best of all, it can potentially be made *platform independent*, thus offering maximum availability of the

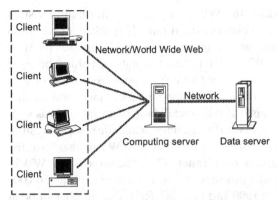

Figure 1.2: Client/server architecture. In a typical three-tier client/server architecture, client machines are connected to the computing server, which in turn is connected to the data server. As the bulk of the computation is with the computing server and the bulk of the data access is with the data server, the client computer can be lightly equipped.

server applications, thanks to Internet-induced developments in the mid-1990s. The server machines, on the other hand, can be powerful **multiprocessors** for the computing servers and machines with high disk throughputs for the data servers. The typical **World Wide Web** (**WWW**) architecture, for instance, is a three-tier client/server system consisting of the **browser**, **Web server**, and **database server**. The object-oriented methodology and client/server architecture can be profitably combined for financial computation [626, 867].

Database management systems are the backbone of information systems [497, 871]. With products from Computer Associates, IBM, Informix, Microsoft, Oracle, and Sybase, the database scenery is dominated by the **relational database** model invented by Codd at IBM in 1970 [216]. In a relational database, data are organized as two-dimensional tables. Consider the following table for storing daily interest rate data.

Attribute	Null?	Type
maturity	NOT NULL	CHAR(10)
ratedate	NOT NULL	DATE
rate	—	DECIMAL(15,8)

Name the table yieldcurve. The structured query language (SQL)[5] statement below can be used to retrieve the two-year U.S. Treasury yield as of December 1, 1994,

```
SELECT rate FROM yieldcurve
        WHERE maturity = '2YR' AND ratedate = '1994-12-01'
```

SQL can also be embedded into general-purpose programming languages. The advancement in the capability of low-cost personal computers and the release of truly multitasking operating systems for them (IBM's OS/2, Microsoft's Windows NT, and Linux) brought client/server database systems to the masses [1, 182, 688, 888]. However, by 1996, the relational database market started to be affected by the Internet momentum [311].

Prototyped in 1991 by Berners-Lee, the WWW is a global information system that provides easy access to Internet resources [63]. It quickly sparked a revolution in the use of the Internet for communications, information, and businesses [655]. A personal computer with access to the WWW – typically through a graphical browser from Microsoft or Netscape (part of America Online) – opens up a window to a world that can be described only as awesome: shopping, stock and bond quotes, online stock trading, up-to-date and historical financial data, financial analysis software, online versions of major newspapers and magazines, academic research results, journal archives and preprints, to mention just a few. The WWW can also form the information network *within* corporations, or **intranet** [733]. The surge of the WWW was one of the major reasons behind the Internet's growing from fewer than 500,000 hosts to more than 10 million between 1990 and 1996 [63, 655] (that number stood at 93 million as of July 2000). In 1998, 100 million people were using the Internet [852]. Even software development strategies were fundamentally changed [488]. These amazing developments are currently reshaping the business and the financial worlds [13, 338, 498, 831].

NOTES

1. Bachelier remained obscure until approximately 1960 when his major work was translated into English. His career problem seems to stem from some technical errors and the topic of his dissertation [637]. "The topic is somewhat remote from those our candidates are in the habit of treating," wrote his advisor, Poincaré (1854–1912) [277]. This is not the first time that ideas in economics have influenced other sciences [426, 660], the most celebrated being Malthus's simultaneous influence on Darwin and Wallace in 1838 [648].
2. Two Nobel laureates in economics, Merton and Scholes, helped found the hedge fund company, Long-Term Capital Management (LTCM). The firm's tools were "computers and powerful mathematics, not intuition nor inside information" [869]. The company underwent a U.S.\$3.6 billion forced bailout by 14 commercial and investment banks in September 1998.
3. Distinction is often made between *real* and *financial* investments. What economists mean by investment is the sort that produces real capital formation such as plants, land, and machinery [778]. Investments in this book will be of the financial kind as opposed to the real kind mentioned above. They involve only papers such as stocks and bonds [797].
4. The forex market is the world's largest financial market, in which an estimated U.S.\$1.5 trillion was traded in April 1998 [51]. Players are the major commercial and investment banks, with their traders connected by computers, telephones, and other telecommunication equipment [767].
5. The most widely used database language, SQL [315] is derived from SEQUEL (for **S**tructured **E**nglish **QUE**ry **L**anguage), which was designed and implemented at IBM.

CHAPTER TWO

Analysis of Algorithms

> In computer science there is no history of critical experiments that
> decide between the validity of various theories, as there are in phy-
> sical sciences.
>
> Juris Hartmanis [421]

Algorithms are precise procedures that can be turned into computer programs. A classical example is Euclid's algorithm, which specifies the exact steps toward computing the greatest common divisor. Problems such as the greatest common divisor are therefore said to be **computable**, whereas those that do not admit algorithms are **uncomputable**. A computable problem may have complexity so high that no efficient algorithms exist. In this case, it is said to be **intractable**. The difficulty of pricing certain financial instruments may be linked to their intrinsic complexity [169].

The hardest part of software implementation is developing the algorithm [264]. Algorithms in this book are expressed in an informal style called a **pseudocode**. A pseudocode conveys the algorithmic ideas without getting tied up in syntax. Pseudocode programs are specified in sufficient detail as to make their coding in a programming language straightforward. This chapter outlines the conventions used in pseudocode programs.

2.1 Complexity

Precisely predicting the performance of a program is difficult. It depends on such diverse factors as the machine it runs on, the programming language it is written in, the compiler used to generate the binary code, the workload of the computer, and so on. Although the actual running time is the only valid criterion for performance [717], we need measures of complexity that are machine independent in order to have a grip on the expected performance.

We start with a set of basic operations that are assumed to take one unit of time. Logical comparisons ($\leq$, $=$, $\geq$, and so on) and arithmetic operations of finite precision ($+$, $-$, $\times$, $/$, exponentiation, logarithm, and so on) are among them. The total number of these operations is then used as the total work done by an algorithm, called its **computational complexity**. Similarly, the **space complexity** is the amount of memory space used by an algorithm. The purpose here is to concentrate on the abstract complexity of an algorithm instead of its implementation, which involves so many details that we can never fully take them into account. Complexity serves

> **Algorithm for searching an element:**
>
> input: x, n, A_i ($1 \le i \le n$);
> integer k;
> for ($k = 1$ to n)
>   if $[x = A_k]$ return k;
> return not-found;

Figure 2.1: Sequential search algorithm.

as a good guide to an algorithm's actual running time. Because space complexity is seldom an issue in this book, the term complexity is used to refer exclusively to computational complexity.

The complexity of an algorithm is expressed as a function of the size of its input. Consider the search algorithm in Fig. 2.1. It looks for a given element by comparing it sequentially with every element in an array of length n. Apparently the worst-case complexity is n comparisons, which occurs when the matching element is the last element of the array or when there is no match. There are other operations to be sure. The for loop, for example, uses a loop variable k that has to be incremented for each execution of the loop and compared against the loop bound n. We do not need to count them because we care about the **asymptotic** growth rate, not the exact number of operations; the derivation of the latter can be quite involved, and its effects on real-world performance cannot be pinpointed anyway [37, 227]. The complexity from maintaining the loop is therefore subsumed by the complexity of the body of the loop.

2.2 Analysis of Algorithms

We are interested in worst-case measures. It is true that worst cases may not occur in practice. But an average-case analysis must assume a distribution on the input, whose validity is hard to certify. To further suppress unnecessary details, we are concerned with the rate of growth of the complexity only as the input gets larger, ignoring constant factors and small inputs. The focus is on the asymptotic growth rate, as mentioned in Section 2.1.

Let R denote the set of real numbers, R^+ the set of positive real numbers, and $N = \{0, 1, 2, \ldots, \}$. The following definition lays out the notation needed to formulate complexity.

DEFINITION 2.2.1 We say that $g = O(f)$ if $g(n) \le cf(n)$ for some nonnegative c and sufficiently large n, where $f, g : N \to R^+$.

EXAMPLE 2.2.2 The base of a logarithm is not important for asymptotic analysis because

$$\log_a x = \frac{\log_e x}{\log_e a} = O(\log_e x),$$

where $e = 2.71828\ldots$. We abbreviate $\log_e x$ as $\ln x$.

EXAMPLE 2.2.3 Let $f(n) = n^3$ and $g(n) = 3.5 \times n^2 + \ln n + \sin n$. Clearly, $g = O(f)$ because $g(n)$ is less than n^3 for sufficiently large n. On the other hand, $f \ne O(g)$.

Denote the input size by N. An algorithm runs in **logarithmic time** if its complexity is $O(\log N)$. An algorithm runs in **linear time** if its complexity is $O(N)$. The sequential search algorithm in Fig. 2.1, for example, has a complexity of $O(N)$

because it has $N = n + 2$ inputs and carries out $O(n)$ operations. A complexity of $O(N \log N)$ typifies sorting and various divide-and-conquer types of algorithms. An algorithm runs in **quadratic time** if its complexity is $O(N^2)$. Many elementary matrix computations such as matrix–vector multiplication have this complexity. An algorithm runs in **cubic time** if its complexity is $O(N^3)$. Typical examples are matrix–matrix multiplication and solving simultaneous linear equations. An algorithm runs in **exponential time** if its complexity is $O(2^N)$. Problems that *require* exponential time are clearly intractable. It is possible for an exponential-time algorithm to perform well on "typical" inputs, however. The foundations for computational complexity were laid in the 1960s [710].

▶ **Exercise 2.2.1** Show that $f + g = O(f)$ if $g = O(f)$.

▶ **Exercise 2.2.2** Prove the following relations: (1) $\sum_{i=1}^{n} i = O(n^2)$, (2) $\sum_{i=1}^{n} i^2 = O(n^3)$, (3) $\sum_{i=0}^{\log_2 n} 2^i = O(n)$, (4) $\sum_{i=0}^{\alpha \log_2 n} 2^i = O(n^\alpha)$, (5) $n \sum_{i=0}^{n} i^{-1} = O(n \ln n)$.

2.3 Description of Algorithms

Universally accepted mathematical symbols are respected. Therefore $+, -, \times, /, <, >, \leq, \geq$, and $=$ mean addition, subtraction, and so on. The symbol $:=$ denotes assignment. For example, $a := b$ assigns the value of b to the variable a. The statement **return** a says that a is returned by the algorithm.

The construct

> **for** $(i = a$ **to** $b)$ $\{ \cdots \}$

means that the statements enclosed in braces ({ and }) are executed $b - a + 1$ times, with i equal to $a, a + 1, \ldots, b$, in that order. The construct

> **for** $(i = a$ **down to** $b)$ $\{ \cdots \}$

means the statements enclosed in braces are executed $a - b + 1$ times, with i equal to $a, a - 1, \ldots, b$, in that order. The construct

> **while** $[S]$ $\{ \cdots \}$

executes the statements enclosed in braces until the condition S is violated. For example, **while** $[a = b]$ $\{ \cdots \}$ runs until a is not equal to b. The construct

> **if** $[S]$ $\{ T_1 \}$ **else** $\{ T_2 \}$

executes T_1 if the expression S is true and T_2 if the expression S is false. The statement **break** causes the current **for** loop to exit. The enclosing brackets can be dropped if there is only a single statement within.

The construct $a[n]$ allocates an array of n elements $a[0], \ldots, a[n-1]$. The construct $a[n][m]$ allocates the following $n \times m$ array (note that the indices start from *zero*, not *one*):

$$
\begin{matrix}
a[0][0] & \cdots & a[0][m-1] \\
\vdots & \ddots & \vdots \\
a[n-1][0] & \cdots & a[n-1][m-1]
\end{matrix}
$$

Although the zero-based indexing scheme is more convenient in many cases, the one-based indexing scheme may be preferred in others. So we use $a[1..n][1..m]$

to denote an array with the following $n \times m$ elements,

$$a[1][1], a[1][2], \ldots, a[n][m-1], a[n][m].$$

Symbols such as $a[\ ]$ and $a[\ ][\ ]$ are used to reference the entire array. Anything following // is treated as comment.

2.4 Software Implementation

Implementation turns an algorithm into a computer program on specific computer platforms. Design, coding, debugging, and module testing are all integral parts of implementation.[1] A key to a productive software project is the reuse of code, either from previous projects or commercial products [650]. The current trend toward object-oriented programming and standardization promises to promote software reuse.

The choice of algorithms in software projects has to be viewed within the context of a larger system. The overall system design might limit the choices to only a few alternatives [791]. This constraint usually arises from the requirements of other parts of the system and very often reflects the fact that most pieces of code are written for an existing system [714].

I now correct a common misconception about the importance of performance. People tend to think that a reduction of the running time from, say, 10 s to 5 s is not as significant as that from 10 min to 5 min. This view rests on the observation that a 5-s difference is not as critical as a 5-min difference. This is wrong. A 5-s difference can be easily turned into a 5-min difference if there are 60 such tasks to perform. A significant reduction in the running time for an important problem is always desirable.

Finally, a word of caution on the term **recursion**. Computer science usually reserves the word for the way of attacking a problem by solving smaller instances of the same problem. Take sorting a list of numbers as an example. One recursive strategy is to sort the first half of the list and the second half of the list separately before merging them. Note that the two sorting subproblems are indeed smaller in size than the original problem. Consistent with most books in finance, however, in this book the term "recursion" is used loosely to mean "iteration." Adhering to the strict computer science usage will usually result in problem formulations that lead to highly inefficient pricing algorithms.

NOTE

1. Software errors can be costly. For example, they were responsible for the crash of the maiden flight of the Ariane 5 that was launched on June 4, 1996, at a cost of half a billion U.S. dollars [606].

> Probably only a person with some mathematical knowledge would think of beginning with 0 instead of with 1.
>
> Bertrand Russell (1872–1970), *Introduction to Mathematical Philosophy*

CHAPTER THREE

Basic Financial Mathematics

> In the fifteenth century mathematics was mainly concerned with
> questions of commercial arithmetic and the problems of the architect.
>
> Joseph Alois Schumpeter (1883–1950), *Capitalism, Socialism and Democracy*

To put a value on any financial instrument, the first step is to look at its cash flow. As we are most interested in the present value of expected cash flows, three features stand out: magnitudes and directions of the cash flows, times when the cash flows occur, and an appropriate factor to discount the cash flows. This chapter deals with elementary financial mathematics. The following convenient time line will be adopted throughout the chapter:

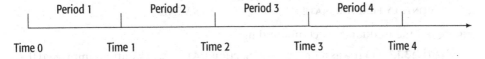

3.1 Time Value of Money

Interest is the cost of borrowing money [785, 787]. Let r be the annual interest rate. If the interest is **compounded** once per year, the **future value (FV)** of P dollars after n years is $FV = P(1+r)^n$. To look at it from another perspective, FV dollars n years from now is worth $P = FV \times (1+r)^{-n}$ today, its **present value (PV)**.[1] The process of obtaining the present value is called **discounting**.

In general, if interest is compounded m times per annum, the future value is

$$FV = P\left(1 + \frac{r}{m}\right)^{nm}. \tag{3.1}$$

Hence, $[1 + (r/m)]^m - 1$ is the equivalent annual rate compounded once per annum or simply the **effective** annual interest rate. In particular, we have *annual* compounding with $m = 1$, *semiannual* compounding with $m = 2$, *quarterly* compounding with $m = 4$, *monthly* compounding with $m = 12$, *weekly* compounding with $m = 52$, and *daily* compounding with $m = 365$. Two widely used yields are the **bond-equivalent yield (BEY)** (the annualized yield with semiannual compounding) and the **mortgage-equivalent yield (MEY)** (the annualized yield with monthly compounding).

An interest rate of r compounded m times a year is equivalent to an interest rate of r/m per $1/m$ year by definition. If a loan asks for a return of 1% per month, for example, the annual interest rate will be 12% with monthly compounding.

EXAMPLE 3.1.1 With an annual interest rate of 10% compounded twice per annum, each dollar will grow to be $[1 + (0.1/2)]^2 = 1.1025$ 1 year from now. The rate is therefore equivalent to an interest rate of 10.25% compounded once per annum.

EXAMPLE 3.1.2 An insurance company has to pay $20 million 4 years from now to pensioners. Suppose that it can invest money at an annual rate of 7% compounded semiannually. Because the effective annual rate is $[1 + (0.07/2)]^2 - 1 = 7.1225\%$, it should invest $20,000,000 \times (1.071225)^{-4} = 15,188,231$ dollars today.

As m approaches infinity and $[1 + (r/m)]^m \to e^r$, we obtain **continuous compounding**:

$$FV = Pe^{rn},$$

where $e = 2.71828$. We call scheme (3.1) **periodic compounding** to differentiate it from continuous compounding. Continuous compounding is easier to work with. For instance, if the annual interest rate is r_1 for n_1 years and r_2 for the following n_2 years, the future value of $1 will be $e^{r_1 n_1 + r_2 n_2}$.

▶ **Exercise 3.1.1** Verify that, given an annual rate, the effective annual rate is higher the higher the frequency of compounding.

▶ **Exercise 3.1.2** Below is a typical credit card statement:

NOMINAL ANNUAL PERCENTAGE RATE (%)	18.70
MONTHLY PERIODIC RATE (%)	1.5583

Figure out the frequency of compounding.

▶ **Exercise 3.1.3** (1) It was mentioned in Section 1.4 that workstations improved their performance by 54% per year between 1987 and 1992 and that the DRAM technology has quadrupled its capacity every 3 years since 1977. What are their respective annual growth rates with continuous compounding? (2) The number of requests received by the National Center for Supercomputing Applications (NCSA) WWW servers grew from ~300,000 per day in May 1994 to ~500,000 per day in September 1994. What is the growth rate per month (compounded monthly) during this period?

3.1.1 Efficient Algorithms for Present and Future Values

The PV of the cash flow $C_1, C_2, \ldots, C_n$ at times $1, 2, \ldots, n$ is

$$\frac{C_1}{1+y} + \frac{C_2}{(1+y)^2} + \cdots + \frac{C_n}{(1+y)^n}.$$

It can be computed by the algorithm in Fig. 3.1 in time $O(n)$, as the bulk of the computation lies in the four arithmetic operations during each execution of the loop that is executed n times. We can save one arithmetic operation within the loop by creating a new variable, say z, and assigning $1 + y$ to it before the loop. The statement $d := d \times (1 + y)$ can then be replaced with $d := d \times z$. Such optimization is often performed by modern compilers automatically behind the scene. This lends

Figure 3.1: Algorithm for PV. C_t are the cash flows, y is the interest rate, and n is the term of the investment. We can easily verify that the variable d is equal to $(1+y)^i$ at the beginning of the for loop. As a result, the variable x becomes the partial sum $\sum_{t=1}^{i} C_t (1+y)^{-t}$ at the end of each loop. This proves the correctness of the algorithm.

```
Algorithm for evaluating present value:

input:   y, n, C_t (1 ≤ t ≤ n);
real     x, d;
x := 0;
d := 1 + y;
for (i = 1 to n) {
          x := x + (C_i/d);
          d := d × (1 + y);
}
return x;
```

support to the earlier argument for asymptotic analysis: In a complex environment in which many manipulations are being done without our knowing them, the best we can do is often the asymptotics.

One further simplification is to replace the loop with the following statement:

for $(i = n$ down to $1) \ \{ \ x := (x + C_i)/d; \ \}.$

The above loop computes the PV by means of

$$\left\{ \cdots \left[\left(\frac{C_n}{1+y} + C_{n-1} \right) \frac{1}{1+y} + C_{n-2} \right] \frac{1}{1+y} + \cdots \right\} \frac{1}{1+y}.$$

This idea, which is due to Horner (1786–1837) in 1819 [582], is the most efficient possible in terms of the absolute number of arithmetic operations [103].

Computing the FV is almost identical to the algorithm in Fig. 3.1. The following changes to that algorithm are needed: (1) d is initialized to 1 instead of $1+y$, (2) i should start from n and run down to 1, and (3) $x := x + (C_i/d)$ is replaced with $x := x + (C_i \times d)$.

▶ **Exercise 3.1.4** Prove the correctness of the FV algorithm mentioned in the text.

3.1.2 Conversion between Compounding Methods

We can compare interest rates with different compounding methods by converting one into the other. Suppose that r_1 is the annual rate with continuous compounding and r_2 is the equivalent rate compounded m times per annum. Then $[1 + (r_2/m)]^m = e^{r_1}$. Therefore

$$r_1 = m \ln \left(1 + \frac{r_2}{m} \right), \tag{3.2}$$

$$r_2 = m \left(e^{r_1/m} - 1 \right). \tag{3.3}$$

EXAMPLE 3.1.3 Consider an interest rate of 10% with quarterly compounding. The equivalent rate with continuous compounding is

$$4 \times \ln \left(1 + \frac{0.1}{4} \right) = 0.09877, \quad \text{or} \quad 9.877\%,$$

derived from Eq. (3.2) with $m = 4$ and $r_2 = 0.1$.

For n compounding methods, there is a total of $n(n-1)$ possible pairwise conversions. Such potentially huge numbers of cases invite programming errors. To make that number manageable, we can fix a ground case, say continuous compounding, and then convert rates to their continuously compounded equivalents before any comparison. This cuts the number of possible conversions down to the more desirable $2(n-1)$.

3.1.3 Simple Compounding

Besides periodic compounding and continuous compounding (hence **compound interest**), there is a different scheme for computing interest called **simple compounding** (hence **simple interest**). Under this scheme, interest is computed on the original principal. Suppose that P dollars is borrowed at an annual rate of r. The simple interest each year is Pr.

3.2 Annuities

An **ordinary annuity** pays out the same C dollars at the end of each year for n years. With a rate of r, the FV at the end of the nth year is

$$\sum_{i=0}^{n-1} C(1+r)^i = C\frac{(1+r)^n - 1}{r}. \tag{3.4}$$

For the **annuity due**, cash flows are received at the beginning of each year. The FV is

$$\sum_{i=1}^{n} C(1+r)^i = C\frac{(1+r)^n - 1}{r}(1+r). \tag{3.5}$$

If m payments of C dollars each are received per year (the **general annuity**), then Eqs. (3.4) and (3.5) become

$$C\frac{\left(1+\frac{r}{m}\right)^{nm} - 1}{\frac{r}{m}}, \quad C\frac{\left(1+\frac{r}{m}\right)^{nm} - 1}{\frac{r}{m}}\left(1+\frac{r}{m}\right),$$

respectively. Unless stated otherwise, an ordinary annuity is assumed from now on. The PV of a general annuity is

$$\text{PV} = \sum_{i=1}^{nm} C\left(1+\frac{r}{m}\right)^{-i} = C\frac{1-\left(1+\frac{r}{m}\right)^{-nm}}{\frac{r}{m}}. \tag{3.6}$$

EXAMPLE 3.2.1 The PV of an annuity of \$100 per annum for 5 years at an annual interest rate of 6.25% is

$$100 \times \frac{1-(1.0625)^{-5}}{0.0625} = 418.387$$

based on Eq. (3.6) with $m=1$.

EXAMPLE 3.2.2 Suppose that an annuity pays $5,000 per month for 9 years with an interest rate of 7.125% compounded monthly. Its PV, $397,783, can be derived from Eq. (3.6) with $C = 5000$, $r = 0.07125$, $n = 9$, and $m = 12$.

An annuity that lasts forever is called a **perpetual annuity**. We can derive its PV from Eq. (3.6) by letting n go to infinity:

$$\text{PV} = \frac{mC}{r}. \tag{3.7}$$

This formula is useful for valuing perpetual fixed-coupon debts [646]. For example, consider a financial instrument promising to pay $100 once a year forever. If the interest rate is 10%, its PV is $100/0.1 = 1000$ dollars.

▶ **Exercise 3.2.1** Derive the PV formula for the general annuity due.

3.3 Amortization

Amortization is a method of repaying a loan through regular payments of interest *and* principal. The size of the loan – the **original balance** – is reduced by the principal part of the payment. The interest part of the payment pays the interest incurred on the **remaining principal balance**. As the principal gets paid down over the term of the loan,[2] the interest part of the payment diminishes.

Home mortgages are typically amortized. When the principal is paid down consistently, the risk to the lender is lowered. When the borrower sells the house, the remaining principal is due the lender. The rest of this section considers mainly the equal-payment case, i.e., fixed-rate **level-payment fully amortized mortgages**, commonly known as **traditional mortgages**.

EXAMPLE 3.3.1 A home buyer takes out a 15-year $250,000 loan at an 8.0% interest rate. Solving Eq. (3.6) with $\text{PV} = 250000$, $n = 15$, $m = 12$, and $r = 0.08$ gives a monthly payment of $C = 2389.13$. The amortization schedule is shown in Fig. 3.2. We can verify that in every month (1) the principal and the interest parts of the payment add up to $2,389.13, (2) the remaining principal is reduced by the amount indicated under the Principal heading, and (3) we compute the interest by multiplying the remaining balance of the previous month by 0.08/12.

Month	Payment	Interest	Principal	Remaining principal
				250,000.000
1	2,389.13	1,666.667	722.464	249,277.536
2	2,389.13	1,661.850	727.280	248,550.256
3	2,389.13	1,657.002	732.129	247,818.128
		. . .		
178	2,389.13	47.153	2,341.980	4,730.899
179	2,389.13	31.539	2,357.591	2,373.308
180	2,389.13	15.822	2,373.308	0.000
Total	430,043.438	180,043.438	250,000.000	

Figure 3.2: An amortization schedule. See Example 3.3.1.

Suppose that the amortization schedule lets the lender receive m payments a year for n years. The amount of each payment is C dollars, and the annual interest rate is r. Right after the kth payment, the remaining principal is the PV of the future $nm - k$ cash flows:

$$\sum_{i=1}^{nm-k} C\left(1+\frac{r}{m}\right)^{-i} = C\,\frac{1-\left(1+\frac{r}{m}\right)^{-nm+k}}{\frac{r}{m}}. \tag{3.8}$$

For example, Eq. (3.8) generates the same remaining principal as that in the amortization schedule of Example 3.3.1 for the third month with $C = 2389.13$, $n = 15$, $m = 12$, $r = 0.08$, and $k = 3$.

A popular mortgage is the **adjustable-rate mortgage** (**ARM**). The interest rate now is no longer fixed but is tied to some publicly available index such as the constant-maturity Treasury (CMT) rate or the Cost of Funds Index (COFI). For instance, a mortgage that calls for the interest rate to be **reset** every month requires that the monthly payment be recalculated every month based on the prevailing interest rate and the remaining principal at the beginning of the month. The attractiveness of ARMs arises from the typically lower initial rate, thus qualifying the home buyer for a bigger mortgage, and the fact that the interest rate adjustments are capped.

A common method of paying off a long-term loan is for the borrower to pay interest on the loan *and* to pay into a **sinking fund** so that the debt can be retired with proceeds from the fund. The sum of the interest payment and the sinking-fund deposit is called the **periodic expense** of the debt. In practice, sinking-fund provisions vary. Some start several years after the issuance of the debt, others allow a balloon payment at maturity, and still others use the fund to periodically purchase bonds in the market [767].

EXAMPLE 3.3.2 A company borrows \$100,000 at a semiannual interest rate of 10%. If the company pays into a sinking fund earning 8% to retire the debt in 7 years, the semiannual payment can be calculated by Eq. (3.6) as follows:

$$\frac{100000 \times 0.08/2}{1-(1+0.08/2)^{-14}} = 9466.9.$$

Interest on the loan is $100000 \times (0.1/2) = 5000$ semiannually. The periodic expense is thus $5000 + 9466.9 = 14466.9$ dollars.

▶ **Exercise 3.3.1** Explain why

$$\mathrm{PV}\left(1+\frac{r}{m}\right)^{k} - \sum_{i=1}^{k} C\left(1+\frac{r}{m}\right)^{i-1}$$

where the PV from Eq. (3.6) equals that of Eq. (3.8).

▶ **Exercise 3.3.2** Start with the cash flow of a level-payment mortgage with the lower monthly fixed interest rate $r - x$. From the monthly payment D, construct a cash flow that grows at a rate of x per month: $D, De^{x}, De^{2x}, De^{3x}, \ldots$. Both x and r are continuously compounded. Verify that this new cash flow, discounted at r, has the same PV as that of the original mortgage. (This identity forms the basis of the **graduated-payment mortgages** (**GPMs**) [330].)

➤ **Programming Assignment 3.3.3** Write a program that prints out the monthly amortization schedule. The inputs are the annual interest rate and the number of payments.

3.4 Yields

The term **yield** denotes the return of investment and has many variants [284]. The **nominal yield** is the **coupon rate** of the bond. In the *Wall Street Journal* of August 26, 1997, for instance, a corporate bond issued by AT&T is quoted as follows:

Company	Cur Yld.	Vol.	Close	Net chg.
ATT85/831	8.1	162	1061/2	−3/8

This bond matures in the year 2031 and has a nominal yield of $8\frac{5}{8}\%$, which is part of the identification of the bond. In the same paper, we can find other AT&T bonds: ATT43/498, ATT6s00, ATT51/801, and ATT63/404. The **current yield** is the annual coupon interest divided by the market price. In the preceding case, the annual interest is $8\frac{5}{8} \times 1000/100 = 86.25$, assuming a par value of \$1,000. The closing price is $106\frac{1}{2} \times 1000/100 = 1065$ dollars. (Corporate bonds are quoted as a percentage of par.) Therefore $86.25/1065 \approx 8.1\%$ is the current yield at market closing. The preceding two yield measures are of little use in comparing returns. For example, the nominal yield completely ignores the market condition, whereas the current yield fails to take the future into account, even though it does depend on the current market price.

Securities such as U.S. **Treasury bills (T-bills)** pay interest based on the **discount method** rather than on the more common **add-on method** [95]. With the discount method, interest is subtracted from the par value of a security to derive the purchase price, and the investor receives the par value at maturity. Such a security is said to be issued **on a discount basis** and is called a **discount security**. The **discount yield** or **discount rate** is defined as

$$\frac{\text{par value} - \text{purchase price}}{\text{par value}} \times \frac{360 \text{ days}}{\text{number of days to maturity}}. \tag{3.9}$$

This yield is also called the **yield on a bank discount basis**. When the discount yield is calculated for short-term securities, a year is assumed to have 360 days [698, 827].

EXAMPLE 3.4.1 T-bills are a short-term debt instrument with maturities of 3, 6, or 12 months. They are issued in U.S.\$10,000 denominations. If an investor buys a U.S.\$10,000, 6-month T-bill for U.S.\$9,521.45 with 182 days remaining to maturity, the discount yield is

$$\frac{10000 - 9521.45}{10000} \times \frac{360}{182} = 0.0947,$$

or 9.47%. It is this annualized yield that is quoted. The equivalent effective yield with continuous compounding is

$$\frac{365}{182} \times \ln\left(\frac{10000}{9521.45}\right) = 0.09835,$$

or 9.835%.

The **CD-equivalent yield** (also called the **money-market-equivalent yield**) is a simple annualized interest rate defined by

$$\frac{\text{par value} - \text{purchase price}}{\text{purchase price}} \times \frac{360}{\text{number of days to maturity}}.$$

To make the discount yield more comparable with yield quotes of other money market instruments, we can calculate its CD-equivalent yield as

$$\frac{360 \times \text{discount yield}}{360 - (\text{number of days to maturity} \times \text{discount yield})},$$

which we can derive by plugging in discount yield formula (3.9) and simplifying. To make the discount yield more comparable with the BEY, we compute

$$\frac{\text{par value} - \text{purchase price}}{\text{purchase price}} \times \frac{365}{\text{number of days to maturity}}.$$

For example, the discount yield in Example 3.4.1 (9.47%) now becomes

$$\frac{478.55}{9521.45} \times \frac{365}{182} = 0.1008, \quad \text{or} \quad 10.08\%. \tag{3.10}$$

The T-bill's **ask yield** is computed in precisely this way [510].

3.4.1 Internal Rate of Return

For the rest of this section, the yield we are concerned with, unless stated otherwise, is the **internal rate of return** (**IRR**). The IRR is the interest rate that equates an investment's PV with its price P:

$$P = \frac{C_1}{(1+y)} + \frac{C_2}{(1+y)^2} + \frac{C_3}{(1+y)^3} + \cdots + \frac{C_n}{(1+y)^n}. \tag{3.11}$$

The right-hand side of Eq. (3.11) is the PV of the cash flow $C_1, C_2, \ldots, C_n$ discounted at the IRR y. Equation (3.11) and its various generalizations form the foundation upon which pricing methodologies are built.

EXAMPLE 3.4.2 A bank lent a borrower $260,000 for 15 years to purchase a house. This 15-year mortgage has a monthly payment of $2,000. The annual yield is 4.583% because $\sum_{i=1}^{12 \times 15} 2000 \times [1 + (0.04583/12)]^{-i} \approx 260000$.

EXAMPLE 3.4.3 A financial instrument promises to pay $1,000 for the next 3 years and sells for $2,500. Its yield is 9.7%, which can be verified as follows. With 0.097 as the discounting rate, the PVs of the three cash flows are $1000/(1 + 0.097)^t$ for $t = 1, 2, 3$. The numbers – 911.577, 830.973, and 757.5 – sum to $2,500.

Example 3.4.3 shows that it is easy to *verify* if a number is the IRR. Finding it, however, generally requires numerical techniques because closed-form formulas in general do not exist. This issue will be picked up in Subsection 3.4.3.

EXAMPLE 3.4.4 A financial instrument can be bought for $1,000, and the investor will end up with $2,000 5 years from now. The yield is the y that equates 1000 with $2000 \times (1 + y)^{-5}$, the present value of $2,000. It is $(1000/2000)^{-1/5} - 1 \approx 14.87\%$.

Given the cash flow $C_1, C_2, \ldots, C_n$, its FV is

$$\text{FV} = \sum_{t=1}^{n} C_t (1+y)^{n-t}. \tag{3.12}$$

By Eq. (3.11), the yield y makes the preceding FV equal to $P(1+y)^n$. Hence, in principle, multiple cash flows can be reduced to a single cash flow $P(1+y)^n$ at maturity. In Example 3.4.4 the investor ends up with \$2,000 at the end of the fifth year one way or another. This brings us to an important point. Look at Eqs. (3.11) and (3.12) again. They mean the same thing because both implicitly assume that all cash flows are reinvested at the *same* rate as the IRR y.

Example 3.4.4 suggests a general yield measure: Calculate the FV and then find the yield that equates it with the PV. This is the **holding period return** (**HPR**) methodology.[3] With the HPR, it is no longer mandatory that all cash flows be reinvested at the same rate. Instead, explicit assumptions about the reinvestment rates must be made for the cash flows. Suppose that the reinvestment rate has been determined to be r_e. Then the FV is

$$\text{FV} = \sum_{t=1}^{n} C_t (1+r_e)^{n-t}.$$

We then solve for the **holding period yield** y such that $\text{FV} = P(1+y)^n$. Of course, if the reinvestment assumptions turn out to be wrong, the yield will not be realized. This is the **reinvestment risk**. Financial instruments without intermediate cash flows evidently do not have reinvestment risks.

EXAMPLE 3.4.5 A financial instrument promises to pay \$1,000 for the next 3 years and sells for \$2,500. If each cash flow can be put into a bank account that pays an effective rate of 5%, the FV of the security is $\sum_{t=1}^{3} 1000 \times (1+0.05)^{3-t} = 3152.5$, and the holding period yield is $(3152.5/2500)^{1/3} - 1 = 0.08037$, or 8.037%. This yield is considerably lower than the 9.7% in Example 3.4.3.

▶ **Exercise 3.4.1** A security selling for \$3,000 promises to pay \$1,000 for the next 2 years and \$1,500 for the third year. Verify that its annual yield is 7.55%.

▶ **Exercise 3.4.2** A financial instrument pays C dollars per year for n years. The investor interested in the instrument expects the cash flows to be reinvested at an annual rate of r and is asking for a yield of y. What should this instrument be selling for in order to be attractive to this investor?

3.4.2 Net Present Value

Consider an investment that has the cash flow $C_1, C_2, \ldots, C_n$ and is selling for P. For an investor who believes that this security should have a return rate of y^*, the **net present value** (**NPV**) is

$$\sum_{t=1}^{n} \frac{C_t}{(1+y^*)^t} - P.$$

The IRR is thus the return rate that nullifies the NPV. In general, the NPV is the difference between the PVs of cash inflow and cash outflow. Businesses are often assumed to maximize their assets' NPV.

EXAMPLE 3.4.6 The management is presented with the following proposals:

| | | Net Cash Flow at end of | | |
Proposal	Investment Now	Year 1	Year 2	Year 3
A	9,500	4,500	2,000	6,000
B	6,000	2,500	1,000	5,000

It believes that the company can earn 15% effective on projects of this kind. The NPV for Proposal A is

$$\frac{4500}{1.15} + \frac{2000}{(1.15)^2} + \frac{6000}{(1.15)^3} - 9500 = -129.57$$

and that for Proposal B is

$$\frac{2500}{1.15} + \frac{1000}{(1.15)^2} + \frac{5000}{(1.15)^3} - 6000 = 217.64.$$

Proposal A is therefore dropped in favor of Proposal B.

▶ **Exercise 3.4.3** Repeat the calculation for Example 3.4.6 for an expected return of 4%.

3.4.3 Numerical Methods for Finding Yields

Computing the yield amounts to solving $f(y) = 0$ for $y \geq -1$, where

$$f(y) \equiv \sum_{t=1}^{n} \frac{C_t}{(1+y)^t} - P \qquad (3.13)$$

and P is the market price. (The symbol $\equiv$ introduces definitions.) The function $f(y)$ is monotonic in y if the C_ts are all positive. In this case, a simple geometric argument shows that a unique solution exists (see Fig. 3.3). Even in the general case in which

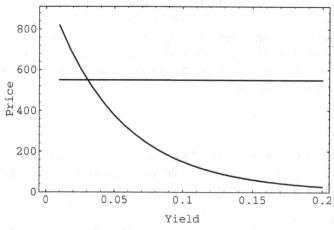

Figure 3.3: Computing yields. The current market price is represented by the horizontal line, and the PV of the future cash flow is represented by the downward-sloping curve. The desired yield is the value on the x axis at which the two curves intersect.

The bisection method for solving equations:

input: ϵ, a, and b ($b > a$ and $f(a) f(b) < 0$);
real length, c;
length $:= b - a$;
while [length $> \epsilon$] {
    $c := (b + a)/2$;
    if [$f(c) = 0$] return c;
    else if [$f(a) f(c) < 0$] $b := c$;
    else $a := c$;
}
return c;

Figure 3.4: Bisection method. The number ϵ is an upper bound on the absolute error of the returned value c: $|\xi - c| \le \epsilon$. The initial bracket $[a, b]$ guarantees the existence of a root with the $f(a) f(b) < 0$ condition. "if [$f(c) = 0$]" may be replaced with testing if $|f(c)|$ is a very small number.

all the C_ts do not have the same sign, usually only one value makes economic sense [547]. We now turn to the algorithmic problem of finding the solution to y.

The Bisection Method

One of the simplest and failure-free methods to solve equations such as Eq. (3.13) for any well-behaved function is the **bisection method**. Start with two numbers, a and b, where $a < b$ and $f(a) f(b) < 0$. Then $f(\xi)$ must be zero for some ξ between a and b, written as $\xi \in [a, b]$.[4] If we evaluate f at the midpoint $c \equiv (a + b)/2$, then (1) $f(c) = 0$, (2) $f(a) f(c) < 0$, or (3) $f(c) f(b) < 0$. In the first case we are done, in the second case we continue the process with the new bracket $[a, c]$, and in the third case we continue with $[c, b]$. Note that the bracket is halved in the latter two cases. After n steps, we will have confined ξ within a bracket of length $(b - a)/2^n$. Figure 3.4 implements the above idea.

The complexity of the bisection algorithm can be analyzed as follows. The while loop is executed, at most, $1 + \log_2[(b - a)/\epsilon]$ times. Within the loop, the number of arithmetic operations is dominated by the evaluation of f. Denote this number by C_f. The running time is $O(C_f \log_2[(b - a)/\epsilon])$. In particular, in computing the IRR, the running time is $O(n \log_2[(b - a)/\epsilon])$ because $C_f = O(n)$ by the algorithm in Fig. 3.1.

The Newton–Raphson Method

The iterative **Newton–Raphson method** converges faster than the bisection method. In **iterative methods**, we start with a first approximation x_0 to a root of $f(x) = 0$. Successive approximations are then computed by

$$x_0, F(x_0), F(F(x_0)), \ldots$$

for some function F. In other words, if we let x_k denote the kth approximation, then $x_k = F^{(k)}(x_0)$, where

$$F^{(k)}(x) \equiv \overbrace{F(F(\cdots(F(x))\cdots))}^{k}.$$

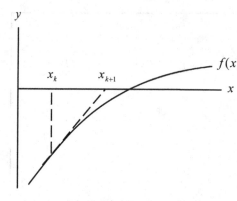

Figure 3.5: Newton–Raphson method.

In practice, we should put an upper bound on the number of iterations k. The necessary condition for the convergence of such a procedure to a root ξ is

$$|F'(\xi)| \leq 1, \tag{3.14}$$

where F' denotes the derivative of F [447].

The Newton–Raphson method picks $F(x) \equiv x - f(x)/f'(x)$; in other words,

$$x_{k+1} \equiv x_k - \frac{f(x_k)}{f'(x_k)}. \tag{3.15}$$

This is the method of choice when f' can be evaluated efficiently and is nonzero near the root [727]. See Fig. 3.5 for an illustration and Fig. 3.6 for the algorithm. When yields are being computed,

$$f'(x) = -\sum_{t=1}^{n} \frac{t C_t}{(1+x)^{t+1}}.$$

Assume that we start with an initial guess x_0 near a root ξ. It can be shown that

$$\xi - x_{k+1} \approx -(\xi - x_k)^2 \frac{f''(\xi)}{2 f'(\xi)}.$$

This means that the method converges *quadratically*: Near the root, each iteration roughly doubles the number of significant digits. To achieve $|x_{k+1} - x_k| \leq \epsilon$ required by the algorithm, $O(\log \log(1/\epsilon))$ iterations suffice. The running time is thus

$$O((\mathcal{C}_f + \mathcal{C}_{f'}) \log \log(1/\epsilon)).$$

The Newton–Raphson method for solving equations:

input: ϵ, x_{initial};
real x_{new}, x_{old};
$x_{\text{old}} := x_{\text{initial}}$;
$x_{\text{new}} := \infty$;
while $[\, |x_{\text{new}} - x_{\text{old}}| > \epsilon \,]$
          $x_{\text{new}} = x_{\text{old}} - f(x_{\text{old}})/f'(x_{\text{old}})$;
return x_{new};

Figure 3.6: Algorithm for the Newton–Raphson method. A good initial guess is important [727].

In particular, the running time is $O(n \log \log(1/\epsilon))$ for yields calculations. This bound compares favorably with the $O(n \log[(b-a)/\epsilon])$ bound of the bisection method.

A variant of the Newton–Raphson method that does not require differentiation is the **secant method** [35]. This method starts with two approximations, x_0 and x_1, and computes the $(k+1)$th approximation by

$$x_{k+1} = x_k - \frac{f(x_k)(x_k - x_{k-1})}{f(x_k) - f(x_{k-1})}.$$

The secant method may be preferred when the calculation of f' is to be avoided. Its convergence rate, 1.618, is slightly worse than that of the Newton–Raphson method, 2, but better than that of the bisection method, 1.

Unlike the bisection method, neither the Newton–Raphson method nor the secant method guarantees that the root remains bracketed; as a result, they may not converge at all. The **Ridders method**, in contrast, always brackets the root. It starts with x_0 and x_1 that bracket a root and sets $x_2 = (x_0 + x_1)/2$. In general,

$$x_{k+1} = x_k + \text{sign}[f(x_{k-2}) - f(x_{k-1})] \frac{f(x_k)(x_k - x_{k-2})}{\sqrt{f(x_k)^2 - f(x_{k-2})f(x_{k-1})}}.$$

The Ridders method has a convergence rate of $\sqrt{2}$ [727].

▶ **Exercise 3.4.4** Let $f(x) \equiv x^3 - x^2$ and start with the guess $x_0 = 2.0$ to the equation $f(x) = 0$. Iterate the Newton–Raphson method five times.

▶ **Exercise 3.4.5** Suppose that $f'(\xi) \neq 0$ and $f''(\xi)$ is bounded. Verify that condition (3.14) holds for the Newton–Raphson method.

▶ **Exercise 3.4.6** Let ξ be a root of f and J be an interval containing ξ. Suppose that $f'(x) \neq 0$ and $f''(x) \geq 0$ or $f''(x) \leq 0$ for $x \in J$. Explain why the Newton–Raphson method converges monotonically to ξ from any point $x_0 \in J$ such that $f(x_0) f''(x_0) \geq 0$.

3.4.4 Solving Systems of Nonlinear Equations

The Newton–Raphson method can be extended to higher dimensions. Consider the two-dimensional case. Let (x_k, y_k) be the kth approximation to the solution of the two simultaneous equations

$$f(x, y) = 0,$$
$$g(x, y) = 0.$$

The $(k+1)$th approximation (x_{k+1}, y_{k+1}) satisfies the following linear equations:

$$\begin{bmatrix} \partial f(x_k, y_k)/\partial x & \partial f(x_k, y_k)/\partial y \\ \partial g(x_k, y_k)/\partial x & \partial g(x_k, y_k)/\partial y \end{bmatrix} \begin{bmatrix} \Delta x_{k+1} \\ \Delta y_{k+1} \end{bmatrix} = -\begin{bmatrix} f(x_k, y_k) \\ g(x_k, y_k) \end{bmatrix}, \tag{3.16}$$

where $\Delta x_{k+1} \equiv x_{k+1} - x_k$ and $\Delta y_{k+1} \equiv y_{k+1} - y_k$. Equations (3.16) have a unique solution for $(\Delta x_{k+1}, \Delta y_{k+1})$ when the matrix is invertible. Note that the $(k+1)$th approximation is $(x_k + \Delta x_{k+1}, y_k + \Delta y_{k+1})$. Solving nonlinear equations has thus been reduced to solving a set of linear equations. Generalization to n dimensions is straightforward.

▶ **Exercise 3.4.7** Write the analogous n-dimensional formula for Eqs. (3.16).

▶ **Exercise 3.4.8** Describe a bisection method for solving systems of nonlinear equations in the two-dimensional case. (The bisection method may be applied in cases in which the Newton–Raphson method fails.)

3.5 Bonds

A bond is a contract between the issuer (borrower) and the bondholder (lender). The issuer promises to pay the bondholder interest, if any, and principal on the remaining balance. Bonds usually refer to long-term debts. A bond has a **par value**.[5] The **redemption date** or **maturity date** specifies the date on which the loan will be repaid. A bond pays interest at the coupon rate on its par value at regular time intervals until the maturity date. The payment is usually made semiannually in the United States. The **redemption value** is the amount to be paid at a redemption date. A bond is **redeemed at par** if the redemption value is the same as the par value. Redemption date and maturity date may differ.

There are several ways to **redeem** or **retire** a bond. A bond is redeemed at maturity if the principal is repaid at maturity. Most corporate bonds are **callable**, meaning that the issuer can retire some or all of the bonds before the stated maturity, usually at a price above the par value.[6]

Because this provision gives the issuer the advantage of calling a bond when the prevailing interest rate is much lower than the coupon rate, the bondholders usually demand a premium. A callable bond may also have **call protection** so that it is not callable for the first few years. **Refunding** involves using the proceeds from the issuance of new bonds to retire old ones. A corporation may deposit money into a sinking fund and use the funds to buy back some or all of the bonds. **Convertible bonds** can be converted into the issuer's common stock. A **consol** is a bond that pays interest forever. It can therefore be analyzed as a perpetual annuity whose value and yield satisfy the simple relation

$$P = c/r, \tag{3.17}$$

where c denotes the interest payout per annum.

The U.S. bond market is the largest in the world. It consists of U.S. Treasury securities, U.S. agency securities, corporate bonds, Yankee bonds, municipal securities, mortgages, and MBSs. **Agency securities** are those issued by either the U.S. Federal government agencies or U.S. Federal government-sponsored organizations. The mortgage market is usually the largest (U.S.$6,388 billion as of 1999), followed by the U.S. Treasury securities market (U.S.$3,281 billion as of 1999).

Treasury securities with maturities of 1 year or less are discount securities: the T-bills. Treasury securities with original maturities between 2 and 10 years are called **Treasury notes** (**T-notes**). Those with maturities greater than ten years are called **Treasury bonds** (**T-bonds**). Both T-notes and T-bonds are coupon securities, paying interest every 6 months.

Bonds are usually quoted as a percentage of par value. A quote of 95 therefore means 95% of par value. For T-notes and T-bonds, a quote of 100.05 means $100_{5/32}\%$ of par value, not 100.05%. It is typically written as 100-05.

▶ **Exercise 3.5.1** A consol paying out continuously at a rate of c dollars per annum has value $\int_0^\infty ce^{-rt}\, dt$, where r is the continuously compounded annual yield. Justify the preceding formula. (Consistent with Eq. (3.17), this integral evaluates to c/r.)

3.5.1 Valuation

Let us begin with **pure discount bonds**, also known as **zero-coupon bonds** or simply **zeros**. They promise a single payment in the future and are sold at a discount from the par value. The price of a zero-coupon bond that pays F dollars in n periods is $F/(1+r)^n$, where r is the interest rate per period. Zero-coupon bonds can be bought to meet future obligations without reinvestment risk. They are also an important theoretical tool in the analysis of coupon bonds, which can be thought of as a package of zero-coupon bonds. Although the U.S. Treasury does not issue such bonds with maturities over 1 year, there were companies that specialized in **coupon stripping** to create stripped Treasury securities. This financial innovation became redundant when the U.S. Treasury facilitated the creation of zeros by means of the Separate Trading of Registered Interest and Principal Securities program (STRIPS) in 1985 [799]. Prices and yields of stripped Treasury securities have been published daily in the *Wall Street Journal* since 1989.

EXAMPLE 3.5.1 Suppose that the interest rate is 8% compounded semiannually. A zero-coupon bond that pays the par value 20 years from now will be priced at $1/(1.04)^{40}$, or 20.83%, of its par value and will be quoted as 20.83. If the interest rate is 9% instead, the same bond will be priced at only 17.19. If the bond matures in 10 years instead of 20, its price would be 45.64 with an 8% interest rate. Clearly both the maturity and the market interest rate have a profound impact on price.

A **level-coupon bond** pays interest based on the coupon rate and the par value, which is paid at maturity. If F denotes the par value and C denotes the coupon, then the cash flow is as shown in Fig. 3.7. Its price is therefore

$$PV = \sum_{i=1}^{n} \frac{C}{\left(1+\frac{r}{m}\right)^i} + \frac{F}{\left(1+\frac{r}{m}\right)^n} = C\,\frac{1-\left(1+\frac{r}{m}\right)^{-n}}{\frac{r}{m}} + \frac{F}{\left(1+\frac{r}{m}\right)^n}, \qquad (3.18)$$

where n is the number of cash flows, m is the number of payments per year, and r is the annual interest rate compounded m times per annum. Note that $C = Fc/m$ when c is the annual coupon rate.

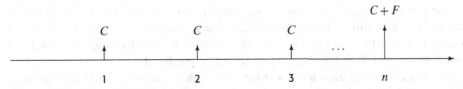

Figure 3.7: Cash flow of level-coupon bond.

EXAMPLE 3.5.2 Consider a 20-year 9% bond with the coupon paid semiannually. This means that a payment of $1000 \times 0.09/2 = 45$ dollars will be made every 6 months until maturity, and $1,000 will be paid at maturity. Its price can be computed from Eq. (3.18) with $n = 2 \times 20$, $r = 0.08$, $m = 2$, $F = 1$, and $C = 0.09/2$. The result is 1.09896, or 109.896% of par value. When the coupon rate is higher than the interest rate, as is the case here, a level-coupon bond will be selling above its par value.

The **yield to maturity** of a level-coupon bond is its IRR when the bond is held to maturity. In other words, it is the r that satisfies Eq. (3.18) with the PV being the bond price. For example, for an investor with a 15% BEY to maturity, a 10-year bond with a coupon rate of 10% paid semiannually should sell for

$$5 \times \frac{1 - [1 + (0.15/2)]^{-2 \times 10}}{0.15/2} + \frac{100}{[1 + (0.15/2)]^{2 \times 10}} = 74.5138$$

percent of par.

For a callable bond, the **yield to stated maturity** measures its yield to maturity as if it were not callable. The **yield to call** is the yield to maturity satisfied by Eq. (3.18), with n denoting the number of remaining coupon payments until the first call date and F replaced with the **call price**, the price at which the bond will be called. The related **yield to par call** assumes the call price is the par value. The **yield to effective maturity** replaces n with the **effective maturity date**, the redemption date when the bond is called. Of course, this date has to be estimated. The **yield to worst** is the minimum of the yields to call under all possible call dates.

▶ **Exercise 3.5.2** A company issues a 10-year bond with a coupon rate of 10%, paid semiannually. The bond is callable at par after 5 years. Find the price that guarantees a return of 12% compounded semiannually for the investor.

▶ **Exercise 3.5.3** How should pricing formula (3.18) be modified if the interest is taxed at a rate of T and capital gains are taxed at a rate of T_G?

▶ **Exercise 3.5.4** (1) Derive $\partial P / \partial n$ and $\partial P / \partial r$ for zero-coupon bonds. (2) For $r = 0.04$ and $n = 40$ as in Example 3.5.1, verify that the price will go down by approximately $d \times 8.011\%$ of par value for every $d\%$ increase in the period interest rate r for small d.

3.5.2 Price Behaviors

The price of a bond goes in the opposite direction from that of interest rate movements: Bond prices fall when interest rates rise, and vice versa. This is because the PV decreases as interest rates increase.[7] A good example is the loss of U.S.$1 trillion worldwide that was due to interest rate hikes in 1994 [312].

Equation (3.18) can be used to show that a level-coupon bond will be selling **at a premium** (above its par value) when its coupon rate is above the market interest rate, **at par** (at its par value) when its coupon rate is equal to the market interest rate, and **at a discount** (below its par value) when its coupon rate is below the market interest rate. The table in Fig. 3.8 shows the relation between the price of a bond and the required yield. Bonds selling at par are called **par bonds**.

The price/yield relation has a **convex** shape, as shown in Fig. 3.9. Convexity is attractive for bondholders because the price decrease per percent rate increase is

Yield (%)	Price (% of par)
7.5	113.37
8.0	108.65
8.5	104.19
9.0	100.00
9.5	96.04
10.0	92.31
10.5	88.79

Figure 3.8: Price/yield relations. A 15-year 9% coupon bond is assumed.

smaller than the price increase per percent rate decrease. This observation, however, may not hold for bonds with **embedded options** such as callable bonds. The convexity property has far-reaching implications for bonds and will be explored in Section 4.3.

As the maturity date draws near, a bond selling at a discount will see its price move up toward par, a bond selling at par will see its price remain at par, and a bond selling at a premium will see its price move down toward par. These phenomena are shown in Fig. 3.10. Besides the two reasons cited for causing bond prices to change (interest rate movements and a nonpar bond moving toward maturity), other reasons include changes in the yield spread to T-bonds for non-T-bonds, changes in the perceived credit quality of the issuer, and changes in the value of the embedded option.

▶ **Exercise 3.5.5** Prove that a level-coupon bond will be sold at par if its coupon rate is the same as the market interest rate.

3.5.3 Day Count Conventions

> Teach us to number our days aright,
> that we may gain a heart of wisdom.
> —Psalms 90:12

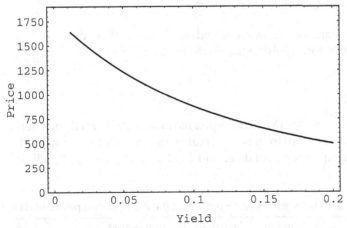

Figure 3.9: Price vs. yield. Plotted is a bond that pays 8% interest on a par value of $1,000, compounded annually. The term is 10 years.

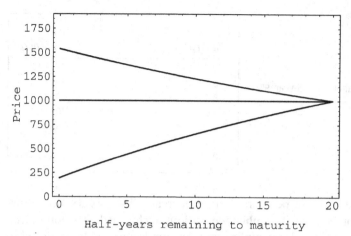

Figure 3.10: Relations between price and time to maturity. Plotted are three curves for bonds, from top to bottom, selling at a premium, at par, and at a discount, with coupon rates of 12%, 6%, and 2%, respectively. The coupons are paid semiannually. The par value is $1,000, and the required yield is 6%. The term is 10 years (the x axis is measured in half-years).

Handling the issue of dating correctly is critical to any financial software. In the so-called **actual/actual** day count convention, the first "actual" refers to the actual number of days in a month, and the second refers to the actual number of days in a coupon period. For example, for coupon-bearing Treasury securities, the number of days between June 17, 1992, and October 1, 1992, is 106: 13 days in June, 31 days in July, 31 days in August, 30 days in September, and 1 day in October.

A convention popular with corporate and municipal bonds and agency securities is **30/360**. Here each month is assumed to have 30 days and each year 360 days. The number of days between June 17, 1992, and October 1, 1992, is now 104: 13 days in June, 30 days in July, 30 days in August, 30 days in September, and 1 day in October. In general, the number of days from date $D_1 \equiv (y_1, m_1, d_1)$ to date $D_2 \equiv (y_1, m_1, d_1)$ under the 30/360 convention can be computed by

$$360 \times (y_2 - y_1) + 30 \times (m_2 - m_1) + (d_2 - d_1),$$

where y_i denote the years, m_i the months, and d_i the days. If d_1 or d_2 is 31, we need to change it to 30 before applying the above formula.

3.5.4 Accrued Interest

Up to now, we have assumed that the next coupon payment date is exactly one period (6 months for bonds, for instance) from now. In reality, the settlement date may fall on any day between two coupon payment dates and yield measures have to be adjusted accordingly. Let

$$\omega \equiv \frac{\text{number of days between the settlement and the next coupon payment date}}{\text{number of days in the coupon period}};$$

$$(3.19)$$

the day count is based on the convention applicable to the security in question. The price is now calculated by

$$PV = \sum_{i=0}^{n-1} \frac{C}{\left(1 + \frac{r}{m}\right)^{\omega+i}} + \frac{F}{\left(1 + \frac{r}{m}\right)^{\omega+n-1}}, \tag{3.20}$$

where n is the number of remaining coupon payments [328]. This price is called the **full price**, **dirty price**, or **invoice price**. Equation (3.20) reduces to Eq. (3.18) when $\omega = 1$.

As the issuer of the bond will not send the next coupon to the seller after the transaction, the buyer has to pay the seller part of the coupon during the time the bond was owned by the seller. The convention is that the buyer pays the quoted price plus the **accrued interest** calculated by

$$C \times \frac{\text{number of days from the last coupon payment to the settlement date}}{\text{number of days in the coupon period}}$$
$$= C \times (1 - \omega).$$

The yield to maturity is the r satisfying Eq. (3.20) when the PV is the invoice price, the sum of the quoted price and the accrued interest. As the quoted price in the United States does not include the accrued interest, it is also called the **clean price** or **flat price**.

EXAMPLE 3.5.3 Consider a bond with a 10% coupon rate and paying interest semi-annually. The maturity date is March 1, 1995, and the settlement date is July 1, 1993. The day count is 30/360. Because there are 60 days between July 1, 1993, and the next coupon date, September 1, 1993, the accrued interest is $(10/2) \times [\,(180 - 60)/180\,] = 3.3333$ per \$100 of par value. At the clean price of 111.2891, the yield to maturity is 3%. This can be verified by Eq. (3.20) with $\omega = 60/180$, $m = 2$, $C = 5$, PV$= 111.2891 + 3.3333$, and $r = 0.03$.

➤ **Exercise 3.5.6** It has been mentioned that a bond selling at par will continue to sell at par as long as the yield to maturity is equal to the coupon rate. This conclusion rests on the assumption that the settlement date is on a coupon payment date. Suppose that the settlement date for a bond selling at par (i.e., the *quoted price* is equal to the par value) falls between two coupon payment dates. Prove that its yield to maturity is less than the coupon rate.

➤ **Exercise 3.5.7** Consider a bond with a 10% coupon rate and paying interest semi-annually. The maturity date is March 1, 1995, and the settlement date is July 1, 1993. The day count used is actual/actual. Verify that there are 62 days between July 1, 1993, and the next coupon date, September 1, 1993, and that the accrued interest is 3.31522% of par value. Also verify that the yield to maturity is 3% when the bond is selling for 111.3.

➤ **Programming Assignment 3.5.8** Write a program that computes (1) the accrued interest as a percentage of par and (2) the BEY of coupon bonds. The inputs are the coupon rate as a percentage of par, the next coupon payment date, the coupon payment frequency per annum, the remaining number of coupon payments after the next coupon, and the day count convention.

3.5.5 Yield for a Portfolio of Bonds

Calculation for the yield to maturity for a portfolio of bonds is no different from that for a single bond. First, the cash flows of the individual bonds are combined. Then the yield is calculated based on the combined cash flow as if it were from a single bond.

EXAMPLE 3.5.4 A bond portfolio consists of two zero-coupon bonds. The bonds are selling at 50 and 20, respectively. The term is exactly 3 years from now. To calculate the yield, we solve

$$50 + 20 = \frac{100 + 100}{(1+y)^6}$$

for y. Because $y = 0.19121$, the annualized yield is 38.242%. The yields to maturity for the individual bonds are 24.4924% and 61.5321%. Neither a simple average (43.01225%) nor a weighted average (35.0752%) matches 38.242%.

3.5.6 Components of Return

Recall that a bond has a price

$$P = Fc \frac{1 - (1+y)^{-n}}{y} + \frac{F}{(1+y)^n},$$

where c is the period coupon rate and y is the period interest rate. Its **total monetary return** is $P(1+y)^n - P$, which is equal to

$$Fc \frac{(1+y)^n - 1}{y} + F - P$$

$$= Fc \frac{(1+y)^n - 1}{y} + F - Fc \frac{1 - (1+y)^{-n}}{y} - \frac{F}{(1+y)^n}.$$

This return can be broken down into three components: **capital gain/loss** $F - P$, **coupon interest** nFc, and **interest on interest** equal to

$$[P(1+y)^n - P] - (F - P) - nFc = P(1+y)^n - F - nFc$$

$$= Fc \frac{(1+y)^n - 1}{y} - nFc.$$

The interest on interest's percentage of the total monetary return can be shown to increase as c increases. This means that the higher the coupon rate, the more dependent is the total monetary return on the interest on interest. So bonds selling at a premium are more dependent on the interest-on-interest component, given the same maturity and yield to maturity. It can be verified that when the bond is selling at par ($c = y$), the longer the maturity n, the higher the proportion of the interest on interest among the total monetary return. The same claim also holds for bonds selling at a premium ($y < c$) or at a discount ($y > c$).

The above observations reveal the impact of reinvestment risk. Coupon bonds that obtain a higher percentage of their monetary return from the reinvestment of coupon interests are more vulnerable to changes in reinvestment opportunities. The yield to maturity, which assumes that all coupon payments can be reinvested at the yield to maturity, is problematic because this assumption is seldom realized in a changing environment.

Recall that the HPR measures the return by holding the security until the **horizon date**. This period of time is called the **holding period** or the **investment horizon**. The HPR is composed of (1) capital gain/loss on the horizon date, (2) cash flow income such as coupon and mortgage payments, and (3) reinvestment income from reinvesting the cash flows received between the settlement date and the horizon date. Apparently, one has to make explicit assumptions about the reinvestment rate during the holding period and the security's market price on the horizon date called the **horizon price**. Computing the HPR for each assumption is called **scenario analysis**. The scenarios may be analyzed to find the optimal solution [891]. The **value at risk** (**VaR**) methodology is a refinement of scenario analysis. It constructs a confidence interval for the dollar return at horizon based on some stochastic models (see Section 31.4).

EXAMPLE 3.5.5 Consider a 5-year bond paying semiannual interest at a coupon rate of 10%. Assume that the bond is bought for 90 and held to maturity with a reinvestment rate of 5%. The coupon interest plus the interest on interest amounts to

$$\sum_{i=1}^{2\times 5} \frac{10}{2} \times \left(1 + \frac{0.05}{2}\right)^{i-1} = 56.017 \text{ dollars.}$$

The capital gain is $100 - 90 = 10$. The HPR is therefore $56.017 + 10 = 66.017$ dollars. The holding period yield is $y = 12.767\%$ because

$$\left(1 + \frac{y}{2}\right)^{2\times 5} = \frac{100 + 56.017}{90}.$$

As a comparison, its BEY to maturity is 12.767%. Clearly, different HPRs obtain under different reinvestment rate assumptions. If the security is to be sold before it matures, its horizon price needs to be figured out as well.

▶ **Exercise 3.5.9** Prove that the holding period yield of a level-coupon bond is exactly y when the horizon is one period from now.

Additional Reading

Yield, day count, and accrued interest interact in complex ways [827]. See [244, 323, 325, 328, 895] for more information about the materials in the chapter. Consult [35, 224, 381, 417, 447, 727] for the numerical techniques on solving equations.

NOTES

1. The idea of PV is due to Irving Fisher (1867–1947) in 1896 [646].
2. There are arrangements whereby the remaining principal actually increases and then decreases over the term of the loan. The same principle applies (see Exercise 3.3.2).
3. Terms with identical connotation include **total return, horizon return, horizon total return**, and **investment horizon return** [646].
4. $[a, b]$ denotes the interval $a \le x \le b$, $[a, b)$ denotes the interval $a \le x < b$, $(a, b]$ denotes the interval $a < x \le b$, and (a, b) denotes the interval $a < x < b$.
5. Also called **denomination, face value, maturity value**, or **principal value**.
6. See [767] for the reasons why companies issue callable bonds. Callable bonds were not issued by the U.S. Treasury after February 1985 [325].
7. Reversing this basic relation is common. For example, it is written in [703] that "If Japanese banks are hit by a liquidity problem, they may have to sell U.S. Treasury bonds. A strong sell-off could have the effect of pushing down bond yields and rattling Wall Street."

CHAPTER FOUR

Bond Price Volatility

> Can anyone measure the ocean by handfuls or measure the sky with his hands?
>
> Isaiah 40:12

Understanding how interest rates affect bond prices is key to risk management of interest-rate-sensitive securities. This chapter focuses on bond price volatility or the extent of price movements when interest rates move. Two classic notions, **duration** and **convexity**, are introduced for this purpose with a few applications in risk management. Coupon bonds mean level-coupon bonds for the rest of the book.

4.1 Price Volatility

The sensitivity of the percentage price change to changes in interest rates measures price volatility. We define **price volatility** by $-(\partial P/P)/\partial y$. The price volatility of a coupon bond is

$$-\frac{\partial P/P}{\partial y} = -\frac{(C/y)\,n - (C/y^2)((1+y)^{n+1} - (1+y)) - nF}{(C/y)\,[\,(1+y)^{n+1} - (1+y)\,] + F(1+y)},\tag{4.1}$$

where n is the number of periods before maturity, y is the period yield, F is the par value, and C is the coupon payment per period. For bonds without embedded options, $-(\partial P/P)/\partial y > 0$ for obvious reasons.

Price volatility increases as the coupon rate decreases, other things being equal (see Exercise 4.1.2). Consequently zero-coupon bonds are the most volatile, and bonds selling at a deep discount are more volatile than those selling near or above par. Price volatility also increases as the required yield decreases, other things being equal (see Exercise 4.1.3). So bonds traded with higher yields are less volatile.

For bonds selling above or at par, price volatility increases, but at a decreasing rate, as the **term to maturity** lengthens (see Fig. 4.1). Bonds with a longer maturity are therefore more volatile. This is consistent with the preference for liquidity and with the empirical fact that long-term bond prices are more volatile than short-term ones. (The *yields* of long-term bonds, however, are less volatile than those of short-term bonds [217].) For bonds selling below par, price volatility first increases, then decreases, as shown in Fig. 4.2 [425]. Longer maturity here can no longer be equated with higher price volatility.

Percentage price change

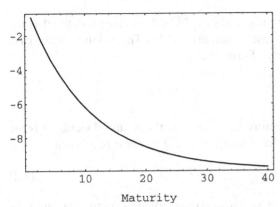

Maturity

Figure 4.1: Volatility with respect to terms to maturity: par bonds. Plotted is the percentage bond price change per percentage change in the required yield at various terms to maturity. The annual coupon rate is 10% with semiannual coupons. The yield to maturity is identical to the coupon rate.

➤ **Exercise 4.1.1** Verify Eq. (4.1).

➤ **Exercise 4.1.2** Show that price volatility never decreases as the coupon rate decreases when yields are positive.

➤ **Exercise 4.1.3** (1) Prove that price volatility always decreases as the yield increases when the yield equals the coupon rate. (2) Prove that price volatility always decreases as the yield increases, generalizing (1).

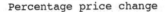

Percentage price change

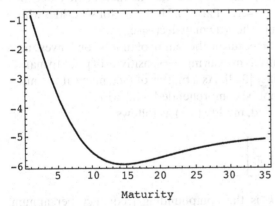

Maturity

Figure 4.2: Volatility with respect to terms to maturity: discount bonds. The annual coupon rate is 10% with semiannual coupons, and the yield to maturity is 40% (a deep discount bond). The terms to maturity are measured in half-years. The rest follows Fig. 4.1.

4.2 Duration

The **Macaulay duration (MD)**, first proposed in 1938 by Macaulay, is defined as the weighted average of the times to an asset's cash flows [627]. The weights are the cash flows' PVs divided by the asset's price. Formally,

$$\text{MD} \equiv \frac{1}{P} \sum_{i=1}^{n} \frac{i C_i}{(1+y)^i},$$

where n is the number of periods before maturity, y is the required yield, C_i is the cash flow at time i, and P is the price. Clearly, the MD, in periods, is equal to

$$\text{MD} = -(1+y) \frac{\partial P/P}{\partial y}. \tag{4.2}$$

This simple relation was discovered by Hicks (1904–1989) in 1939 [231, 496]. In particular, the MD of a coupon bond is

$$\text{MD} = \frac{1}{P} \left[\sum_{i=1}^{n} \frac{i C}{(1+y)^i} + \frac{n F}{(1+y)^n} \right]. \tag{4.3}$$

The above equation can be simplified to

$$\text{MD} = \frac{c(1+y)\left[(1+y)^n - 1\right] + ny(y-c)}{cy\left[(1+y)^n - 1\right] + y^2},$$

where c is the period coupon rate. The MD of a zero-coupon bond (corresponding to $c = 0$) is n, its term to maturity. In general, the Macaulay duration of a coupon bond is less than its maturity. The MD of a coupon bond approaches $(1+y)/y$ as the maturity increases, independent of the coupon rate.

Equations (4.2) and (4.3) hold only if the coupon C, the par value F, and the maturity n are all independent of the yield y, in other words, if the cash flow is independent of yields. When the cash flow is sensitive to interest rate movements, the MD is no longer inappropriate. To see this point, suppose that the market yield declines. The MD will be lengthened by Exercise 4.1.3, Part (2). However, for securities whose maturity actually decreases as a result, the MD may decrease.

Although the MD has its origin in measuring the length of time a bond investment is outstanding, it should be seen mainly as measuring the sensitivity of price to market yield changes, that is, as price volatility [348]. As a matter of fact, many, if not most, duration-related terminologies cannot be comprehended otherwise.

To convert the MD to be year based, modify (4.3) as follows:

$$\frac{1}{P} \left[\sum_{i=1}^{n} \frac{i}{k} \frac{C}{\left(1+\frac{y}{k}\right)^i} + \frac{n}{k} \frac{F}{\left(1+\frac{y}{k}\right)^n} \right],$$

where y is the *annual* yield and k is the compounding frequency per annum. Equation (4.2) also becomes

$$\text{MD} = -\left(1 + \frac{y}{k}\right) \frac{\partial P/P}{\partial y}.$$

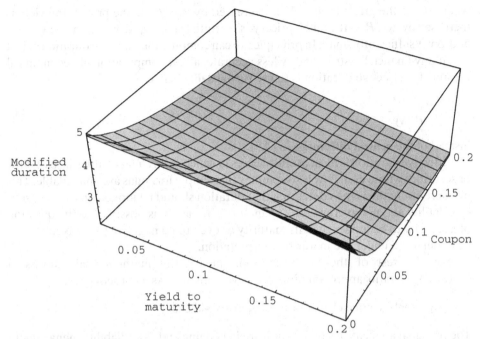

Figure 4.3: Modified duration with respect to coupon rate and yield. Bonds are assumed to pay semiannual coupon payments with a maturity date of September 15, 2000. The settlement date is September 15, 1995.

Note from the definition that

$$\text{MD (in years)} = \frac{\text{MD (in periods)}}{k}.$$

A related measure is the **modified duration**, defined as

$$\text{modified duration} \equiv -\frac{\partial P/P}{\partial y} = \frac{\text{MD}}{(1+y)}. \tag{4.4}$$

The modified duration of a coupon bond is Eq. (4.1), for example (see Fig. 4.3). By Taylor expansion,

percentage price change $\approx$ −modified duration $\times$ yield change.

The modified duration of a portfolio equals $\sum_i \omega_i D_i$, where D_i is the modified duration of the ith asset and ω_i is the market value of that asset expressed as a percentage of the market value of the portfolio. Modified duration equals MD (in periods)/$(1 + y)$ or MD (in years)/$(1 + y/k)$ if the cash flow is independent of changes in interest rates.

EXAMPLE 4.2.1 Consider a bond whose modified duration is 11.54 with a yield of 10%. This means if the yield increases instantaneously from 10% to 10.1%, the approximate percentage price change would be $-11.54 \times 0.001 = -0.01154$, or -1.154%.

A general numerical formula for volatility is the **effective duration**, defined as

$$\frac{P_- - P_+}{P_0(y_+ - y_-)}, \tag{4.5}$$

where P_- is the price if the yield is decreased by Δy, P_+ is the price if the yield is increased by Δy, P_0 is the initial price, y is the initial yield, $y_+ \equiv y + \Delta y$, $y_- \equiv y - \Delta y$, and Δy is sufficiently small. In principle, we can compute the effective duration of just about any financial instrument. A less accurate, albeit computationally economical formula for effective duration is to use **forward difference**,

$$\frac{P_0 - P_+}{P_0 \, \Delta y} \tag{4.6}$$

instead of the **central difference** in (4.5).

Effective duration is most useful in cases in which yield changes alter the cash flow or securities whose cash flow is so complex that simple formulas are unavailable. This measure strengthens the contention that duration should be looked on as a measure of volatility and not average term to maturity. In fact, it is possible for the duration of a security to be longer than its maturity or even to go negative [321]! Neither can be understood under the maturity interpretation.

For the rest of the book, duration means the mathematical expression $-(\partial P / P)/\partial y$ or its approximation, effective duration. As a consequence,

percentage price change $\approx$ $-$duration $\times$ yield change.

The principal applications of duration are in hedging and asset/liability management [55].

▶ **Exercise 4.2.1** Assume that 9% is the annual yield to maturity compounded semi-annually. Calculate the MD of a 3-year bond paying semiannual coupons at an annual coupon rate of 10%.

▶ **Exercise 4.2.2** Duration is usually expressed in percentage terms for quick mental calculation: Given duration $D_\%$, the percentage price change expressed in percentage terms is approximated by $-D_\% \times \Delta r$ when the yield increases instantaneously by $\Delta r \%$. For instance, the price will drop by 20% if $D_\% = 10$ and $\Delta r = 2$ because $10 \times 2 = 20$. Show that $D_\%$ equals modified duration.

▶ **Exercise 4.2.3** Consider a coupon bond and a traditional mortgage with the same maturity and payment frequency. Show that the mortgage has a smaller MD than the bond when both provide the same yield to maturity.[1] For simplicity, assume that both instruments have the same market price.

▶ **Exercise 4.2.4** Verify that the MD of a traditional mortgage is $(1 + y)/y - n/((1 + y)^n - 1)$.

4.2.1 Continuous Compounding

Under continuous compounding, the formula for duration is slightly changed. The price of a bond is now $P = \sum_i C_i e^{-y t_i}$, and

$$\text{duration (continuous compounding)} \equiv \frac{\sum_i t_i C_i e^{-y t_i}}{P} = -\frac{\partial P / P}{\partial y}. \tag{4.7}$$

Unlike the MD in Eq. (4.2), the extra $1 + y$ term disappears.

▶ **Exercise 4.2.5** Show that the duration of an n-period zero-coupon bond is n.

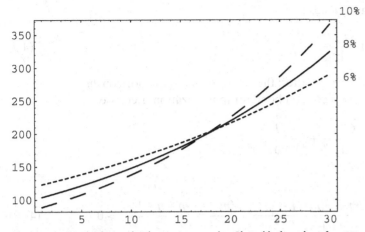

Figure 4.4: Bond value under three rate scenarios. Plotted is the value of an 8% 15-year bond from now to maturity if the interest rate is unchanged at 8% (solid curve), increased to 10% (dashed curve), and decreased to 6% (dotted curve). At the MD $m = 17.9837$ (half-years), the curves roughly meet [98].

4.2.2 Immunization

Buying coupon bonds to meet a future liability incurs some risks. Assume that we are at the horizon date when the liability is due. If interest rates rise subsequent to the bond purchase, the interest on interest from the reinvestment of the coupon payments will increase, and a capital loss will occur for the sale of the bonds. The reverse is true if interest rates fall. The results are uncertainties in meeting the liability.

A portfolio is said to **immunize** a liability if its value at the horizon date covers the liability for small rate changes now. How do we find such a bond portfolio? Amazingly, the answer is as elegant as it is simple: We construct a bond portfolio whose MD is equal to the horizon and whose PV is equal to the PV of the single future liability [350]. Then, at the horizon date, losses from the interest on interest will be compensated for by gains in the sale price when interest rates fall, and losses from the sale price will be compensated for by the gains in the interest on interest when interest rates rise (see Fig. 4.4). For example, a $100,000 liability 12 years from now should be matched by a portfolio with an MD of 12 years and a future value of $100,000.

The proof is straightforward. Assume that the liability is a certain L at time m and the current interest rate is y. We are looking for a portfolio such that

(1) its FV is L at the horizon m,
(2) $\partial FV/\partial y = 0$,
(3) FV is convex around y.

Condition (1) says the obligation is met. Conditions (2) and (3) together mean that L is the portfolio's minimum FV at the horizon for small rate changes.

Let $FV \equiv (1+y)^m P$, where P is the PV of the portfolio. Now,

$$\frac{\partial FV}{\partial y} = m(1+y)^{m-1}P + (1+y)^m \frac{\partial P}{\partial y}. \tag{4.8}$$

Imposing Condition (2) leads to

$$m = -(1+y)\frac{\partial P/P}{\partial y}.$$

(4.9)

This identity is what we were after: the MD is equal to the horizon m.

Suppose that we use a coupon bond for immunization. Because

$$FV = \sum_{i=1}^{n} \frac{C}{(1+y)^{i-m}} + \frac{F}{(1+y)^{n-m}},$$

it follows that

$$\frac{\partial^2 FV}{\partial y^2} = \sum_{i=1}^{n} \frac{(m-i)(m-i-1)\,C}{(1+y)^{i-m+2}} + \frac{(m-n)(m-n-1)\,F}{(1+y)^{n-m+2}} > 0$$

(4.10)

for $y > -1$ because $(m-i)(m-i+1)$ is either zero or positive. Because the FV is convex for $y > -1$, the minimum value of the FV is indeed L (see Fig. 4.5).

If there is no single bond whose MD matches the horizon, a portfolio of two bonds, A and B, can be assembled by the solution of

$$1 = \omega_A + \omega_B,$$
$$D = \omega_A D_A + \omega_B D_B$$

(4.11)

for ω_A and ω_B. Here, D_i is the MD of bond i and ω_i is the weight of bond i in the portfolio. Make sure that D falls between D_A and D_B to guarantee $\omega_A > 0$, $\omega_B > 0$, and positive portfolio convexity.

Although we have been dealing with immunizing a single liability, the extension to multiple liabilities can be carried out along the same line. Let there be a liability of size L_i at time i and a cash inflow A_i at time i. The NPV of these cash flows at

Horizon price

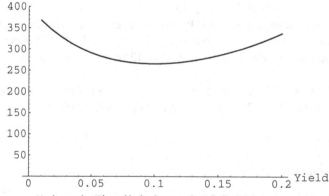

Figure 4.5: Horizon price. Plotted is the future value of a bond at the horizon. The yield at which the graph is minimized equates the bond's MD with the horizon. In this example, the bond pays semiannual coupons at an annual rate of 10% for 30 years and the horizon is 10 years from now. The FV is minimized at $y = 9.91\%$, and the MD at y is exactly 10 years. The bond's FV at the horizon will increase if the rate moves.

the horizon is

$$FV = \sum_i (A_i - L_i)(1 + y)^{m-i}.$$

Conditions (1)–(3) require that $FV = 0$, $\partial FV/\partial y = 0$, and $\partial^2 FV/\partial y^2 > 0$ around the current rate y. Together, they guarantee that the cash inflows suffice to cover the liabilities for small instantaneous rate movements now. In this more general setting, the distribution of individual assets' durations must have a wider range than that of the liabilities to achieve immunization (see Exercise 4.2.11).

Of course, a stream of liabilities can always be immunized with a matching stream of zero-coupon bonds. This is called **cash matching**, and the bond portfolio is called a **dedicated portfolio** [799]. Two problems with this approach are that (1) zero-coupon bonds may be missing for certain maturities and (2) they typically carry lower yields.

Immunization is a dynamic process. It has to be **rebalanced** constantly to ensure that the MD remains matched to the horizon for the following reasons. The MD decreases as time passes, and, except for zero-coupon bonds, the decrement is not identical to the decrement in the time to maturity [217]. This phenomenon is called **duration drift** [246]. This point can be easily confirmed by a coupon bond whose MD matches the horizon. Because the bond's maturity date lies beyond the horizon date, its duration will remain positive at the horizon instead of zero. Therefore immunization needs to be reestablished even if interest rates never change. Interest rates will fluctuate during the holding period, but it was assumed that interest rates change *instantaneously* after immunization has been established and then stay there. Finally, the durations of assets and liabilities may not change at the same rate [689].

When liabilities and assets are mismatched in terms of duration, adverse interest rate movements can quickly wipe out the equity. A bank that finances long-term mortgage investments with short-term credit from the savings accounts or certificates of deposit (CDs) runs such a risk. Other institutions that worry about duration matching are pension funds and life insurance companies [767].

▶ **Exercise 4.2.6** In setting up the two-bond immunization in Eqs. (4.11), we did not bother to check the convexity condition. Justify this omission.

▶ **Exercise 4.2.7** Show that, in the absence of interest rate changes, it suffices to match the PVs of the liability and the asset.

▶ **Exercise 4.2.8** Start with a bond whose PV is equal to the PV of a future liability and whose MD exceeds the horizon. Show that, at the horizon, the bond will fall short of the liability if interest rates rise and more than meet the goal if interest rates fall. The reverse is true if the MD falls short of the horizon.

▶ **Exercise 4.2.9** Consider a liability currently immunized by a coupon bond. Suppose that the interest rate changes instantaneously. Prove that profits will be generated when rebalancing is performed at time Δt from now (but before the maturity).

▶ **Exercise 4.2.10** The liability has an MD of 3 years, but the money manager has access to only two kinds of bonds with MDs of 1 year and 4 years. What is the right proportion of each bond in the portfolio in order to match the liability's MD?

▶ **Exercise 4.2.11** (1) To achieve **full immunization**, we set up cash inflows at more points in time than liabilities as follows. Consider a single-liability cash outflow L_t at time t. Assemble a portfolio with a cash inflow A_1 at time $t - a_1$ and a cash inflow A_2 at time $t + a_2$ with $a_1, a_2 > 0$ and $a_1 \leq t$. Conditions (1) and (2) demand that

$$P(y) = A_1 e^{a_1 y} + A_2 e^{-a_2 y} - L_t = 0,$$

$$\frac{dP(y)}{dy} = A_1 a_1 e^{a_1 y} - A_2 a_2 e^{-a_2 y} = 0$$

under continuous compounding. Solve the two equations for any two unknowns of your choice, say A_1 and A_2, and prove that it achieves immunization for any changes in y. (2) Generalize the result to more than two cash inflows.

4.2.3 Macaulay Duration of Floating-Rate Instruments

A **floating-rate instrument** makes interest rate payments based on some publicized index such as the prime rate, the London Interbank Offered Rate (LIBOR), the U.S. T-bill rate, the CMT rate, or the COFI [348]. Instead of being locked into a number, the coupon rate is reset periodically to reflect the prevailing interest rate.

Assume that the coupon rate c equals the market yield y and that the bond is priced at par. The first reset date is j periods from now, and resets will be performed thereafter. Let the principal be \$1 for simplicity. The cash flow of the floating-rate instrument is thus

$$\overbrace{c, c, \ldots, c}^{j}, \overbrace{y, \ldots, y}^{n-j}, y+1,$$

where c is a constant and $y = c$. So the coupon payment at time $j + 1$ starts to reflect the market yield. For example, when $j = 0$, every coupon payment reflects the prevailing market yield, and when $j = 1$, which is more typical, interest rate movements during the first period will not affect the first coupon payment. The MD is

$$-(1+y) \left. \frac{\partial P/P}{\partial y} \right|_{c=y}$$

$$= \sum_{i=1}^{j} i \frac{y}{(1+y)^i} + \sum_{i=j+1}^{n} \left[i \frac{y}{(1+y)^i} - \frac{1}{(1+y)^{i-1}} \right] + n \frac{1}{(1+y)^n}$$

$$= \text{MD} - \sum_{i=j+1}^{n} \frac{1}{(1+y)^{i-1}} = \frac{(1+y)[1-(1+y)^{-j}]}{y}, \tag{4.12}$$

where MD denotes the MD of an otherwise identical fixed-rate bond. Interestingly, the MD is independent of the maturity of the bond, n. Formulas for nonpar bonds are more complex but do not involve any new ideas [306, 348].

The attractiveness of floating-rate instruments is not hard to explain. Floating-rate instruments are typically less sensitive to interest rate changes than are fixed-rate instruments. In fact, the less distant the first reset date, the less volatile the instrument. And when every coupon is adjusted to reflect the market yield, there is no more interest rate risk. Indeed, the MD is zero when $j = 0$. In the typical case of $j = 1$, the MD is one period. By contrast, a bond that pays 5% per period for 30 periods has an MD of 16.14 periods.

▶ **Exercise 4.2.12** Show that the MD of a floating-rate instrument cannot exceed the first reset date.

4.2.4 Hedging

Hedging aims at offsetting the price fluctuations of the position to be hedged by the hedging instrument in the opposite direction, leaving the total wealth unchanged [222]. Define **dollar duration** as

$$\text{dollar duration} \equiv \text{modified duration} \times \text{price}\,(\%\text{ of par}) = -\frac{\partial P}{\partial y},$$

where P is the price as a percentage of par. It is the tangent on the price/yield curve such as the one in Fig. 3.9. The approximate dollar price change per \$100 of par value is

$$\text{price change} \approx -\text{dollar duration} \times \text{yield change}.$$

The related **price value of a basis point**, or simply **basis-point value** (**BPV**), defined as the dollar duration divided by 10,000, measures the price change for a one basis-point change in the interest rate. One **basis point** equals 0.01%.

Because securities may react to interest rate changes differently, we define **yield beta** as

$$\text{yield beta} \equiv \frac{\text{change in yield for the hedged security}}{\text{change in yield for the hedging security}},$$

which measures relative yield changes. If we let the **hedge ratio** be

$$h \equiv \frac{\text{dollar duration of the hedged security}}{\text{dollar duration of the hedging security}} \times \text{yield beta}, \tag{4.13}$$

then hedging is accomplished when the value of the hedging security is h times that of the hedged security because

$$\text{dollar price change of the hedged security}$$
$$= -h \times \text{dollar price change of the hedging security}.$$

EXAMPLE 4.2.2 Suppose we want to hedge bond A with a duration of seven by using bond B with a duration of eight. Under the assumption that the yield beta is one and both bonds are selling at par, the hedge ratio is 7/8. This means that an investor who is long \$1 million of bond A should short \$7/8 million of bond B.

4.3 Convexity

The important notion of **convexity** is defined as

$$\text{convexity (in periods)} \equiv \frac{\partial^2 P}{\partial y^2} \frac{1}{P}. \tag{4.14}$$

It measures the curvature of the price/yield relation. The convexity of a coupon bond is

$$\frac{1}{P}\left[\sum_{i=1}^{n} i(i+1)\frac{C}{(1+y)^{i+2}} + n(n+1)\frac{F}{(1+y)^{n+2}}\right]$$

$$= \frac{1}{P}\left\{\frac{2C}{y^3}\left[1 - \frac{1}{(1+y)^n}\right] - \frac{2Cn}{y^2(1+y)^{n+1}} + \frac{n(n+1)\left[F - (C/y)\right]}{(1+y)^{n+2}}\right\},$$

(4.15)

which is positive. For a bond with positive convexity, the price rises more for a rate decline than it falls for a rate increase of equal magnitude. Hence between two bonds with the same duration, the one with a higher convexity is more valuable, other things being equal. Convexity measured in periods and convexity measured in years are related by

$$\text{convexity (in years)} = \frac{\text{convexity (in periods)}}{k^2}$$

when there are k periods per annum. It can be shown that the convexity of a coupon bond increases as its coupon rate decreases (see Exercise 4.3.4). Furthermore, for a given yield and duration, the convexity decreases as the coupon decreases [325]. In analogy with Eq. (4.7), the convexity under continuous compounding is

$$\text{convexity (continuous compounding)} \equiv \frac{\sum_i t_i^2 C_i e^{-yt_i}}{P} = \frac{\partial^2 P/P}{\partial y^2}.$$

The approximation $\Delta P/P \approx -\text{duration} \times \text{yield change}$ we saw in Section 4.2 works for small yield changes. To improve on it for larger yield changes, second-order terms are helpful:

$$\frac{\Delta P}{P} \approx \frac{\partial P}{\partial y}\frac{1}{P}\Delta y + \frac{1}{2}\frac{\partial^2 P}{\partial y^2}\frac{1}{P}(\Delta y)^2$$

$$= -\text{duration} \times \Delta y + \frac{1}{2} \times \text{convexity} \times (\Delta y)^2.$$

See Fig. 4.6 for illustration.

A more general notion of convexity is the **effective convexity** defined as

$$\frac{P_+ + P_- - 2 \times P_0}{P_0\left[0.5 \times (y_+ - y_-)\right]^2},$$

(4.16)

where P_- is the price if the yield is decreased by Δy, P_+ is the price if the yield is increased by Δy, P_0 is the initial price, y is the initial yield, $y_+ \equiv y + \Delta y$, $y_- \equiv y - \Delta y$, and Δy is sufficiently small. Note that $\Delta y = (y_+ - y_-)/2$. Effective convexity is most relevant when a bond's cash flow is interest rate sensitive.

The two-bond immunization scheme in Subsection 4.2.2 shows that countless two-bond portfolios with varying duration pairs (D_A, D_B) can be assembled to satisfy Eqs. (4.11). However, which one is to be preferred? As convexity is a desirable feature, we phrase this question as one of maximizing the portfolio convexity among all the portfolios with identical duration. Let there be n kinds of bonds, with bond i having duration D_i and convexity C_i, where $D_1 < D_2 < \cdots < D_n$. Typically,

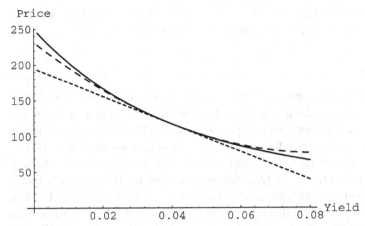

Figure 4.6: Linear and quadratic approximations to bond price changes. The dotted curve is the result of a duration-based approximation, whereas the dashed curve, which fits better, utilizes the convexity information. The bond in question has 30 periods to maturity with a period coupon rate of 5%. The current yield is 4% per period.

$D_1 = 0.25$ (3-month discount instruments) and $D_n = 30$ (30-year zeros). We then solve the following constrained optimization problem:

$$
\begin{aligned}
\text{maximize} \quad & \omega_1 C_1 + \omega_2 C_2 + \cdots + \omega_n C_n, \\
\text{subject to} \quad & 1 = \omega_1 + \omega_2 + \cdots + \omega_n, \\
& D = \omega_1 D_1 + \omega_2 D_2 + \cdots + \omega_n D_n, \\
& 0 \le \omega_i \le 1.
\end{aligned}
$$

The function to be optimized, $\omega_1 C_1 + \omega_2 C_2 + \cdots + \omega_n C_n$, is called the **objective function**. The equalities or inequalities make up the **constraints**. The preceding optimization problem is a **linear programming problem** because all the functions are linear. The solution usually implies a **barbell portfolio**, so called because the portfolio contains bonds at the two extreme ends of the duration spectrum (see Exercise 4.3.6).

▶ **Exercise 4.3.1** In practice, convexity should be expressed in percentage terms, call it $C_\%$, for quick mental calculation. The percentage price change in percentage terms is then approximated by $-D_\% \times \Delta r + C_\% \times (\Delta r)^2 / 2$ when the yield increases instantaneously by $\Delta r \%$. For example, if $D_\% = 10$, $C_\% = 1.5$, and $\Delta r = 2$, the price will drop by 17% because

$$
-D_\% \times \Delta r + \frac{1}{2} \times C_\% \times (\Delta r)^2 = -10 \times 2 + \frac{1}{2} \times 1.5 \times 2^2 = -17.
$$

Show that $C_\%$ equals convexity divided by 100.

▶ **Exercise 4.3.2** Prove that $\partial(\text{duration})/\partial y = (\text{duration})^2 - \text{convexity}$.

▶ **Exercise 4.3.3** Show that the convexity of a zero-coupon bond is $n(n+1)/(1+y)^2$.

▶ **Exercise 4.3.4** Verify that convexity (4.15) increases as the coupon rate decreases.

▶ **Exercise 4.3.5** Prove that the barbell portfolio has the highest convexity for $n = 3$.

> **Exercise 4.3.6** Generalize Exercise 4.3.5: Prove that a barbell portfolio achieves immunization with maximum convexity given $n > 3$ kinds of zero-coupon bonds.

Additional Reading

Duration and convexity measure only the risk of changes in interest rate levels. Other types of risks, such as the frequency of large movements in interest rates, are ignored [618]. They furthermore assume parallel shifts in the yield curve, whereas yield changes are not always parallel in reality (more is said about yield curves in Chap. 5). Closed-form formulas for duration and convexity can be found in [89, 209]. See [496] for a penetrating review. Additional immunization techniques can be found in [206, 325, 547]. The idea of immunization is due to Redington in 1952 [732]. Consult [213, 281, 545] for more information on linear programming. Many fundamental problems in finance and economics are best cast as optimization problems [247, 278, 281, 891].

NOTE

1. The bond was the standard design for mortgages, called **balloon mortgages**, before the Federal Housing Administration introduced fully amortized mortgages [330]. Balloon mortgages are more prone to default because the borrower may not have the funds for the balloon payment due. This exercise shows that fully amortized mortgages are less volatile than balloon mortgages if prepayments are nonexisting.

CHAPTER FIVE

Term Structure of Interest Rates

He pays least [...] who pays latest.
Charles de Montesquieu (1689–1755), *The Spirit of Laws*

The term structure of interest rates is concerned with how the interest rates change with maturity and how they may evolve in time. It is fundamental to the valuation of fixed-income securities. This subject is important also because the term structure is the starting point of any stochastic theory of interest rate movements. Interest rates in this chapter are period based unless stated otherwise. This simplifies the presentation by eliminating references to the compounding frequency per annum.

5.1 Introduction

The set of yields to maturity for bonds of equal quality and differing solely in their terms to maturity[1] forms the **term structure**. This term often refers exclusively to the yields of zero-coupon bonds. Term to maturity is the time period during which the issuer has promised to meet the conditions of the obligation. A **yield curve** plots yields to maturity against maturity and represents the prevailing interest rates for various terms. See Fig. 5.1 for a sample Treasury yield curve. A **par yield curve** is constructed from bonds trading near their par value.

At least four yield-curve shapes can be identified. A **normal** yield curve is upward sloping, an **inverted** yield curve is downward sloping, a **flat** yield curve is flat (see Fig. 5.2), and a **humped** yield curve is upward sloping at first but then turns downward sloping. We will survey the theories advanced to explain the shapes of the yield curve in Section 5.1.

The U.S. Treasury yield curve is the most widely followed yield curve for the following reasons. First, it spans a full range of maturities, from 3 months to 30 years. Second, the prices are representative because the Treasuries are extremely liquid and their market deep. Finally, as the Treasuries are backed by the full faith and credit of the U.S. government, they are perceived as having no credit risk [95]. The most recent Treasury issues for each maturity are known as the **on-the-run** or **current coupon** issues in the secondary market (see Fig. 5.3). Issues auctioned before the current coupon issues are referred to as **off-the-run** issues. On-the-run and off-the-run yield curves are based on their respective issues [325, 489].

Yield (%)

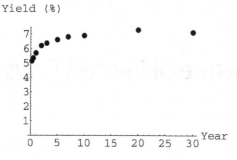

Figure 5.1: Treasury yield curve. The Treasury yield curve as of May 3, 1996, published by the U.S. Treasury and based on bid quotations on the most actively traded Treasury securities as of 3:30 PM with information from the Federal Reserve Bank of New York.

The yield on a non-Treasury security must exceed the base interest rate offered by an on-the-run Treasury security of comparable maturity by a positive spread called the **yield spread** [326]. This spread reflects the risk premium of holding securities not issued by the government. The base interest rate is also known as the **benchmark interest rate**.

5.2 Spot Rates

The i-period **spot rate** $S(i)$ is the yield to maturity of an i-period zero-coupon bond. The PV of $1 i periods from now is therefore $[1 + S(i)]^{-i}$. The one-period spot rate – the **short rate** – will play an important role in modeling interest rate dynamics later in the book. A **spot rate curve** is a plot of spot rates against maturity. Its other names include **spot yield curve** and **zero-coupon yield curve**.

In the familiar bond price formula,

$$\sum_{i=1}^{n} \frac{C}{(1+y)^i} + \frac{F}{(1+y)^n},$$

every cash flow is discounted at the same yield to maturity, y. To see the inconsistency, consider two riskless bonds with different yields to maturity because of their different cash flow streams. The yield-to-maturity methodology discounts their contemporaneous cash flows with different rates, but common sense dictates that cash flows occurring at the same time should be discounted at the same rate. The spot-rate methodology does exactly that.

A fixed-rate bond with cash flow $C_1, C_2, \ldots, C_n$ is equivalent to a package of zero-coupon bonds, with the ith bond paying C_i dollars at time i. For example, a

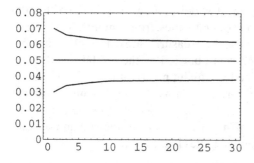

Figure 5.2: Three types of yield curves. Depicted from top to bottom are inverted, flat, and normal yield curves.

	Curr	Securities	Prev Close		9:28	
3	–	11/13/97	5.10	5.24	5.11	5.25
6	–	2/12/98	5.13	5.34	5.12	5.33
1	–	8/20/98	5.20	5.49	5.19	5.48
2	5.875	7/31/99	100-03+	5.81	100-04+	5.80
3	6.000	8/15/00	100-03+	5.96	100-04+	5.95
5	6.000	7/31/02	99-23+	6.06	99-24	6.06
10	6.125	8/15/07	99-07	6.23	99-09	6.22
30	6.375	8/15/27	97-25+	6.54	97-27+	6.54

Figure 5.3: On-the-run U.S. Treasury yield curve (Aug. 18, 1997, 9:28 AM EDT). Source: Bloomberg.

level-coupon bond has the price

$$P = \sum_{i=1}^{n} \frac{C}{[1+S(i)]^i} + \frac{F}{[1+S(n)]^n}. \tag{5.1}$$

This pricing method incorporates information from the term structure by discounting each cash flow at the corresponding spot rate. In general, any riskless security having a predetermined cash flow $C_1, C_2, \ldots, C_n$ should have a market price of

$$P = \sum_{i=1}^{n} C_i d(i),$$

where

$$d(i) \equiv [1+S(i)]^{-i}, \quad i = 1, 2, \ldots, n,$$

are called the **discount factors**. The discount factor $d(i)$ denotes the PV of $\$1$ i periods from now, in other words, the price of the zero-coupon bond maturing i periods from now. If the market price is less than P, it is said to be **undervalued** or **cheap**. It is said to be **overvalued** or **rich** otherwise. The discount factors are often interpolated to form a continuous function called the **discount function**. It is the discount factors, not the spot rates, that are directly observable in the market.

▶ **Exercise 5.2.1** Prove that the yield to maturity y is approximately

$$\frac{\sum_i [\partial C_i(y)/\partial y] S(i)}{\partial P/\partial y}$$

to the first order, where $C_i(y) \equiv C_i/(1+y)^i$ denotes the ith cash flow discounted at the rate y. Note that $\partial C_i(y)/\partial y$ is the dollar duration of the i-period zero-coupon bond. (The yield to maturity is thus roughly a weighted sum of the spot rates, with each weight proportional to the dollar duration of the cash flow.)

5.3 Extracting Spot Rates from Yield Curves

Spot rates can be extracted from the yields of coupon bonds. Start with the short rate $S(1)$, which is available because short-term Treasuries are zero-coupon bonds. Now $S(2)$ can be computed from the two-period coupon bond price P by use of Eq. (5.1),

$$P = \frac{C}{1 + S(1)} + \frac{C + 100}{[\,1 + S(2)\,]^2}.$$

EXAMPLE 5.3.1 Suppose the 1-year T-bill has a yield of 8%. Because this security is a zero-coupon bond, the 1-year spot rate is 8%. When the 2-year 10% T-note is trading at 90, the 2-year spot rate satisfies

$$90 = \frac{10}{1.08} + \frac{110}{[\,1 + S(2)\,]^2}.$$

Therefore $S(2) = 0.1672$, or 16.72%.

In general, $S(n)$ can be computed from Eq. (5.1), given the market price of the n-period coupon bond and $S(1), S(2), \ldots, S(n-1)$. The complete algorithm is given in Fig. 5.4. The correctness of the algorithm is easy to see. The initialization steps and step 3 ensure that

$$p = \sum_{j=1}^{i-1} \frac{1}{[\,1 + S(j)\,]^j}$$

at the beginning of each loop. Step 1 solves for x such that

$$P_i = \sum_{j=1}^{i-1} \frac{C_i}{[\,1 + S(j)\,]^j} + \frac{C_i + 100}{(1 + x)^i},$$

where C_i is the level-coupon payment of bond i and P_i is its price.

Each execution of step 1 requires $O(1)$ arithmetic operations because $x = [\,(C_i + 100)/(P_i - C_i p)\,]^{1/i} - 1$ and expressions like y^z can be computed by $\exp[\,z \ln y\,]$ (note that $\exp[\,x\,] \equiv e^x$). Similarly, step 3 runs in $O(1)$ time. The total running time is hence $O(n)$.

Algorithm for extracting spot rates from coupon bonds:

```
input:   n, C[1..n], P[1..n];
real     S[1..n], p, x;
S[1] := (100/P[1]) − 1;
p := P[1]/100;
for (i = 2 to n) {
        1. Solve P[i] = C[i] × p + (C[i] + 100)/(1 + x)^i for x;
        2. S[i] := x;
        3. p := p + (1 + x)^−i;
}
return S[ ];
```

Figure 5.4: Algorithm for extracting spot rates from a yield curve. $P[\,i\,]$ is the price (as a percentage of par) of the coupon bond maturing i periods from now, $C[\,i\,]$ is the coupon of the i-period bond expressed as a percentage of par, and n is the term of the longest maturity bond. The first bond is a zero-coupon bond. The i-period spot rate is computed and stored in $S[\,i\,]$.

In reality, computing the spot rates is not as clean-cut as the above **bootstrap-ping** procedure. Treasuries of the same maturity might be selling at different yields (the **multiple cash flow problem**), some maturities might be missing from the data points (the **incompleteness problem**), Treasuries might not be of the same quality, and so on. Interpolation and fitting techniques are needed in practice to create a smooth spot rate curve (see Chap. 22). Such schemes, however, usually lack economic justifications.

➤ **Exercise 5.3.1** Suppose that $S(i) = 0.10$ for $1 \le i < 20$ and a 20-period coupon bond is selling at par, with a coupon rate of 8% paid semiannually. Calculate $S(20)$.

➤ **Programming Assignment 5.3.2** Implement the algorithm in Fig. 5.4 plus an option to return the annualized spot rates by using the user-supplied annual compounding frequency.

5.4 Static Spread

Consider a *risky* bond with the cash flow $C_1, C_2, \ldots, C_n$ and selling for P. Were this bond riskless, it would fetch

$$P^* = \sum_{t=1}^{n} \frac{C_t}{[1 + S(t)]^t}.$$

Because riskiness must be compensated for, $P < P^*$. The **static spread** is the amount s by which the spot rate curve has to shift *in parallel* in order to price the bond correctly:

$$P = \sum_{t=1}^{n} \frac{C_t}{[1 + s + S(t)]^t}.$$

It measures the spread that a risky bond would realize over the entire Treasury spot rate curve if the bond is held to maturity. Unlike the yield spread, which is the difference between the yield to maturity of the risky bond and that of a Treasury security with comparable maturity, the static spread incorporates information from the term structure. The static spread can be computed by the Newton–Raphson method.

➤ **Programming Assignment 5.4.1** Write a program to compute the static spread. The inputs are the payment frequency per annum, the annual coupon rate as a percentage of par, the market price as a percentage of par, the number of remaining coupon payments, and the discount factors. Some numerical examples are tabulated below:

Price (% of par)	98	98.5	99	99.5	100	100.5	101
Static spread (%)	0.435	0.375	0.316	0.258	0.200	0.142	0.085

(A 5% 15-year bond paying semiannual interest under a flat 7.8% spot rate curve is assumed.)

5.5 Spot Rate Curve and Yield Curve

Many interesting relations hold between spot rate and yield to maturity. Let y_k denote the yield to maturity for the k-period coupon bond. The spot rate dominates the yield to maturity if the yield curve is normal; in other words, $S(k) \geq y_k$ if $y_1 < y_2 < \cdots$ (see Exercise 5.5.1, statement (1)). Analogously, the spot rate is dominated by the yield to maturity if the yield curve is inverted. Moreover, the spot rate dominates the yield to maturity if the spot rate curve is normal ($S(1) < S(2) < \cdots$) and is dominated when the spot rate curve is inverted (see Exercise 5.5.1, statement (2)). Of course, if the yield curve is flat, the spot rate curve coincides with the yield curve.

These results illustrate the **coupon effect** on the yield to maturity [848]. For instance, under a normal spot rate curve, a coupon bond has a lower yield than a zero-coupon bond of equal maturity. Picking a zero-coupon bond over a coupon bond based purely on the zero's higher yield to maturity is therefore flawed.

The spot rate curve often has the same shape as the yield curve. That is, if the spot rate curve is inverted (normal, respectively), then the yield curve is inverted (normal, respectively). However, this is only a trend, not a mathematical truth. Consider a three-period coupon bond that pays $1 per period and repays the principal of $100 at the end of the third period. With the spot rates $S(1) = 0.1$, $S(2) = 0.9$, and $S(3) = 0.901$, the yields to maturity can be calculated as $y_1 = 0.1$, $y_2 = 0.8873$, and $y_3 = 0.8851$, clearly not strictly increasing. However, when the final principal payment is relatively insignificant, the spot rate curve and the yield curve do share the same shape. Such is the case with bonds of high coupon rates and long maturities (see Exercise 5.5.3). When we refer to the typical agreement in shape later, the above proviso will be implicit.

▶ **Exercise 5.5.1** Prove the following statements: (1) The spot rate dominates the yield to maturity when the yield curve is normal, and (2) the spot rate dominates the yield to maturity if the spot rate curve is normal, and it is smaller than the yield to maturity if the spot rate curve is inverted.

▶ **Exercise 5.5.2** Contrive an example of a normal yield curve that implies a spot rate curve that is not normal.

▶ **Exercise 5.5.3** Suppose that the bonds making up the yield curve are ordinary annuities instead of coupon bonds. (1) Prove that a yield curve is normal if the spot rate curve is normal. (2) Still, a normal yield curve does not guarantee a normal spot rate curve. Verify this claim with this normal yield curve: $y_1 = 0.1$, $y_2 = 0.43$, $y_3 = 0.456$.

5.6 Forward Rates

The yield curve contains not only the prevailing interest rates but also information regarding future interest rates currently "expected" by the market, the **forward rates**. By definition, investing $1 for j periods will end up with $[1 + S(j)]^j$ dollars at time j. Call it the **maturity strategy**. Alternatively, suppose we invest $1 in bonds for i periods and at time i invest the proceeds in bonds for another $j - i$ periods, where $j > i$. Clearly we will have $[1 + S(i)]^i [1 + S(i, j)]^{j-i}$ dollars at time j, where $S(i, j)$ denotes the $(j - i)$-period spot rate i periods from now, which is unknown today.

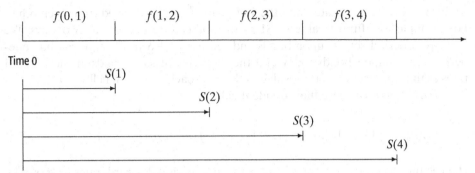

Figure 5.5: Time line for spot and forward rates.

Call it the **rollover strategy**. When $S(i, j)$ equals

$$f(i, j) \equiv \left\{ \frac{[1 + S(j)]^j}{[1 + S(i)]^i} \right\}^{1/(j-i)} - 1, \tag{5.2}$$

we will end up with $[1 + S(j)]^j$ dollars again. (By definition, $f(0, j) = S(j)$.) The rates computed by Eq. (5.2) are called the (**implied**) **forward rates** or, more precisely, the $(j - i)$-**period forward rate** i **periods from now**. Figure 5.5 illustrates the time lines for spot rates and forward rates.

In the above argument, we were not assuming any a priori relation between the implied forward rate $f(i, j)$ and the future spot rate $S(i, j)$. This is the subject of the term structure theories to which we will turn shortly. Rather, we were merely looking for the future spot rate that, *if realized*, would equate the two investment strategies. Forward rates with a duration of a single period are called **instantaneous forward rates** or **one-period forward rates**.

When the spot rate curve is normal, the forward rate dominates the spot rates:

$$f(i, j) > S(j) > \cdots > S(i). \tag{5.3}$$

This claim can be easily extracted from Eq. (5.2). When the spot rate curve is inverted, the forward rate is in turn dominated by the spot rates:

$$f(i, j) < S(j) < \cdots < S(i). \tag{5.4}$$

See Fig. 5.6 for illustration.

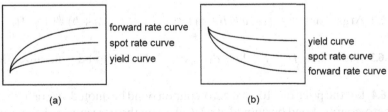

(a) (b)

Figure 5.6: Yield curve, spot rate curve, and forward rate curve. When the yield curve is normal, it is dominated by the spot rate curve, which in turn is dominated by the forward rate curve (if the spot rate curve is also normal). When the yield curve is inverted, on the other hand, it dominates the spot rate curve, which in turn dominates the forward rate curve (if the spot rate curve is also inverted). The forward rate curve here is a plot of one-period forward rates.

Forward rates, spot rates, and the yield curve[2] can be derived from each other. For example, the future value of $1 at time n can be derived in two ways. We can buy n-period zero-coupon bonds and receive $[1 + S(n)]^n$ or we can buy one-period zero-coupon bonds today and then a series of such bonds at the forward rates as they mature. The future value of this approach is $[1 + S(1)][1 + f(1, 2)] \cdots [1 + f(n - 1, n)]$. Because they are identical,

$$S(n) = \{[1 + S(1)][1 + f(1, 2)] \cdots [1 + f(n - 1, n)]\}^{1/n} - 1. \tag{5.5}$$

Hence, the forward rates, specifically the one-period forward rates $f(s, s + 1)$, determine the spot rate curve.

EXAMPLE 5.6.1 Suppose that the following 10 spot rates are extracted from the yield curve:

Period	1	2	3	4	5	6	7	8	9	10
Rate (%)	4.00	4.20	4.30	4.50	4.70	4.85	5.00	5.25	5.40	5.50

The following are the 9 one-period forward rates, starting one period from now.

Period	1	2	3	4	5	6	7	8	9
Rate (%)	4.40	4.50	5.10	5.50	5.60	5.91	7.02	6.61	6.40

If $1 is invested in a 10-period zero-coupon bond, it will grow to be $(1 + 0.055)^{10} = 1.708$. An alternative strategy is to invest $1 in one-period zero-coupon bonds at 4% and reinvest at the one-period forward rates. The final result,

$$1.04 \times 1.044 \times 1.045 \times 1.051 \times 1.055 \times 1.056$$
$$\times 1.0591 \times 1.0702 \times 1.0661 \times 1.064 = 1.708,$$

is exactly the same as expected.

▶ **Exercise 5.6.1** Assume that all coupon bonds are par bonds. Extract the spot rates and the forward rates from the following yields to maturity: $y_1 = 0.03$, $y_2 = 0.04$, and $y_3 = 0.045$.

▶ **Exercise 5.6.2** Argue that $[1 + f(a, a + b + c)]^{b+c} = [1 + f(a, a + b)]^b [1 + f(a + b, a + b + c)]^c$.

▶ **Exercise 5.6.3** Show that $f(T, T + 1) = d(T)/d(T + 1) - 1$ (to be generalized in Eq. (24.2)).

▶ **Exercise 5.6.4** Let the price of a 10-year zero-coupon bond be quoted at 60 and that of a 9.5-year zero-coupon bond be quoted at 62. Calculate the percentage changes in the 10-year spot rate and the 9.5-year forward rate if the 10-year bond price moves up by 1%. (All rates are bond equivalent.)

▶ **Exercise 5.6.5** Prove that the forward rate curve lies above the spot rate curve when the spot rate curve is normal, below it when the spot rate curve is inverted, and that they cross where the spot rate curve is instantaneously flat (see Fig. 5.7).

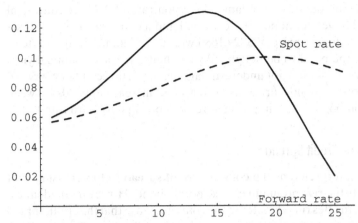

Figure 5.7: Spot rate curve and forward rate curve. The forward rate curve is built by one-period forward rates.

5.6.1 Locking in the Forward Rate

Although forward rates may or may not be realized in the future, we can lock in any forward rate $f(n, m)$ today by buying one n-period zero-coupon bond for $1/[1 + S(n)]^n$ and selling $[1 + S(m)]^m/[1 + S(n)]^n$ m-year zero-coupon bonds. Here is the analysis. There is no net initial investment because the cash inflow and the cash outflow, both at $1/[1 + S(n)]^n$ dollars, cancel out. At time n there will be a cash inflow of \$1, and at time m there will be a cash outflow of $[1 + S(m)]^m/[1 + S(n)]^n$ dollars. This cash flow stream implies the rate $f(n, m)$ between times n and m (see Fig. 5.8).

The above transactions generate the cash flow of an important kind of financial instrument called a **forward contract**. In our particular case, this forward contract, agreed on *today*, enables us to borrow money at time n *in the future* and repay the loan at time $m > n$ with an interest rate equal to the forward rate $f(n, m)$.

Now that forward rates can be locked in, clearly they should not be negative. However, forward rates derived by Eq. (5.2) may be negative if the spot rate curve is steeply downward sloping. It must be concluded that the spot rate curve cannot be arbitrarily specified.

▶ **Exercise 5.6.6** (1) The fact that the forward rate can be locked in today means that future spot rates must equal today's forward rates, or $S(a, b) = f(a, b)$, in a certain economy. Why? How about an uncertain economy? (2) Verify that forward rates covering the same time period will not change over time in a certain economy.

▶ **Exercise 5.6.7** (1) Confirm that a 50-year bond selling at par (\$1,000) with a semi-annual coupon rate of 2.55% is equivalent to a 50-year bond selling for \$1,000 with a semiannual coupon rate of 2.7% and a par value of \$329.1686. (2) Argue that a

Figure 5.8: Locking in the forward rate. By trading zero-coupon bonds of maturities n and m in the right proportion, the forward rate $f(n, m)$ can be locked in today.

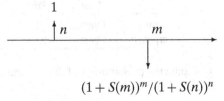

100-year bond selling at par with a semiannual coupon rate of 2.7% is equivalent to a portfolio of the above-mentioned 50-year bond and a contract to buy 50 years from now a 50-year bond at a price of $329.1686 with a semiannual coupon rate of 2.7%. (3) Verify that the bond to be bought 50 years hence has a semiannual yield of 8.209%. (Therefore we should not underestimate the importance of later forward rates on long-term coupon bonds' prices as even a small increase in yields between two long-term coupon bonds could imply an unreasonably high forward rate.)

5.6.2 Term Structure of Credit Spreads

Static spread can be interpreted as the constant **credit spread** to the Treasury spot rate curve that reflects the risk premium of a corporate bond. However, an identical credit spread at all maturities runs counter to the common sense that the credit spread should rise with maturity; a corporation is more likely to fail in, say, 10 years rather than in 1.

One theory of term structure of credit spreads postulates that the price of a corporate bond equals that of the Treasury times the probability of solvency. Furthermore, once default occurs, a corporation remains in that state and pays zero dollar. Because the probability of default is one minus the probability of solvency,

$$1 - \text{probability of default (1 period)} = \frac{\text{price of 1-period corporate zero}}{\text{price of 1-period Treasury zero}}.$$

After using the above equation to compute the probability of default for corporate bonds with one period to maturity, we can calculate the **forward probability of default**, the conditional probability of default in the second period given that default has not occurred in the first period. This forward probability of default clearly satisfies

$$[\,1 - \text{probability of default (1 period)}\,]$$
$$\times [\,1 - \text{forward probability of default (period 2)}\,]$$
$$= \frac{\text{price of 2-period corporate zero}}{\text{price of 2-period Treasury zero}}.$$

In general, the equation satisfied by the forward probability is

$$[\,1 - \text{probability of default } (i-1 \text{ periods})\,]$$
$$\times [\,1 - \text{forward probability of default (period } i)\,]$$
$$= \text{probability the corporate bond survives past time } i$$
$$= \frac{\text{price of } i\text{-period corporate zero}}{\text{price of } i\text{-period Treasury zero}}. \tag{5.6}$$

The algorithm for computing the forward probabilities of default is shown in Fig. 5.9.

▶ **Exercise 5.6.8** Consider the following four zero-coupon bonds:

Type	Maturity	Price	Yield	Type	Maturity	Price	Yield
Treasury	1 year	94	6.28%	Treasury	2 year	87	7.09%
Corporate	1 year	92	8.51%	Corporate	2 year	84	8.91%

Compute the probabilities of default and the forward probabilities of default.

Algorithm for forward probabilities of default:

input: n, $P[1..n]$, $Q[1..n]$;
real $f[1..n]$, $p[1..n]$;
$p[1] := 1 - (Q[1]/P[1])$;
$f[1] := p[1]$;
for $(i = 2 \text{ to } n)$ {
        $f[i] := 1 - (1 - p[i-1])^{-1} \times (Q[i]/P[i])$;
        $p[i] := p[i-1] \times f[i]$;
}
return $f[\,]$;

Figure 5.9: Algorithm for forward probabilities of default. $P[\,i\,]$ is the price of the riskless i-period zero, $Q[\,i\,]$ is the price of the risky i-period zero, $p[\,i\,]$ stores the probability of default during period one to i, and the forward probability of default for the ith period is calculated in $f[\,i\,]$.

▶ **Exercise 5.6.9** (1) Prove Eq. (5.6). (2) Define the **forward spread** for period i, $s(i)$, as the difference between the instantaneous period-i forward rate $f(i-1, i)$ obtained by riskless bonds and the instantaneous period-i forward rate $f_c(i-1, i)$ obtained by corporate bonds. Prove that $s(i)$ roughly equals the forward probability of default for period i.

5.6.3 Spot and Forward Rates under Continuous Compounding

Under continuous compounding, the pricing formula becomes

$$P = \sum_{i=1}^{n} Ce^{-iS(i)} + Fe^{-nS(n)}.$$

In particular, the market discount function is

$$d(n) = e^{-nS(n)}. \tag{5.7}$$

A bootstrapping procedure similar to the one in Fig. 5.4 can be used to calculate the spot rates under continuous compounding. The spot rate is now an arithmetic average of forward rates:

$$S(n) = \frac{f(0, 1) + f(1, 2) + \cdots + f(n-1, n)}{n}. \tag{5.8}$$

The formula for the forward rate is also very simple:

$$f(i, j) = \frac{jS(j) - iS(i)}{j - i}. \tag{5.9}$$

In particular, the one-period forward rate equals

$$f(j, j+1) = (j+1)\,S(j+1) - j\,S(j) = -\ln \frac{d(j+1)}{d(j)} \tag{5.10}$$

(compare it with Exercise 5.6.3).

Rewrite Eq. (5.9) as

$$f(i, j) = S(j) + [S(j) - S(i)] \frac{i}{j-i}.$$

Then under continuous time instead of discrete time, Eq. (5.10) becomes

$$f(T, T + \Delta T) = S(T + \Delta T) + [S(T + \Delta T) - S(T)] \frac{T}{\Delta T},$$

and the instantaneous forward rate at time T equals

$$f(T) \equiv \lim_{\Delta T \to 0} f(T, T + \Delta T) = S(T) + T \frac{\partial S}{\partial T}. \tag{5.11}$$

Note that $f(T) > S(T)$ if and only if $\partial S / \partial T > 0$.

▶ **Exercise 5.6.10** Derive Eqs. (5.8) and (5.9).

▶ **Exercise 5.6.11** Compute the one-period forward rates from this spot rate curve: $S(1) = 2.0\%$, $S(2) = 2.5\%$, $S(3) = 3.0\%$, $S(4) = 3.5\%$, and $S(5) = 4.0\%$.

▶ **Exercise 5.6.12** (1) Figure out a case in which a change in the spot rate curve leaves all forward rates unaffected. (2) Derive the duration $-(\partial P / \partial y)/P$ under the shape change in (1), where y is the short rate $S(1)$.

5.6.4 Spot and Forward Rates under Simple Compounding

This is just a brief subsection because the basic principles are similar. The pricing formula becomes

$$P = \sum_{i=1}^{n} \frac{C}{1 + i S(i)} + \frac{F}{1 + n S(n)}.$$

The market discount function is

$$d(n) = [1 + n S(n)]^{-1}, \tag{5.12}$$

and the $(i - j)$-period forward rate j periods from now is

$$f(i, j) = \frac{[1 + j S(j)][1 + i S(i)]^{-1} - 1}{j - i}. \tag{5.13}$$

To annualize the rates, multiply them by the number of periods per annum.

▶ **Exercise 5.6.13** Derive Eq. (5.13).

5.7 Term Structure Theories

Term structure theories attempt to explain the relations among interest rates of various maturities. As the spot rate curve is most critical for the purpose of valuation, the term structure theories discussed below will be about the spot rate curve.

5.7.1 Expectations Theory

Unbiased Expectations Theory

According to the **unbiased expectations theory** attributed to Irving Fisher, forward rate equals the average future spot rate:

$$f(a, b) = E[\, S(a, b)\,],\qquad\qquad(5.14)$$

where $E[\,\cdot\,]$ denotes mathematical expectation [653, 799]. Note that this theory does not imply that the forward rate is an accurate predictor for the future spot rate. It merely asserts that it does not deviate from the future spot rate systematically. The theory also implies that the maturity strategy and the rollover strategy produce the same result at the horizon on the average (see Exercise 5.7.2). A normal spot rate curve, according to the theory, is due to the fact that the market expects the future spot rate to rise. Formally, because $f(j, j+1) > S(j+1)$ if and only if $S(j+1) > S(j)$ from Eq. (5.2), it follows that

$$E[\, S(j, j+1)\,] > S(j+1) \quad \text{if and only if } S(j+1) > S(j)$$

when the theory holds. Conversely, the theory implies that the spot rate is expected to fall if and only if the spot rate curve is inverted [750].

The unbiased expectations theory, however, has been rejected by most empirical studies dating back at least to Macaulay [627, 633, 767], with the possible exception of the period before the founding of the Federal Reserve System in 1915 [639, 751]. Because the term structure has been upward sloping ∼80% of the time, the unbiased expectations theory would imply that investors have expected interest rates to rise 80% of the time. This does not seem plausible. It also implies that riskless bonds, regardless of their different maturities, earn the same return on the average (see Exercise 5.7.1) [489, 568]. This is not credible either, because that would mean investors are indifferent to risk.

▶ **Exercise 5.7.1** Prove that an n-period zero-coupon bond sold at time $k < n$ has a holding period return of exactly $S(k)$ if the forward rates are realized.

▶ **Exercise 5.7.2** Show that

$$[1 + S(n)]^n = E[1 + S(1)]\, E[1 + S(1, 2)] \cdots E[1 + S(n-1, n)]$$

under the unbiased expectations theory.

Other Versions of the Expectations Theory

At least four other versions of the expectations theory have been proposed, but they are inconsistent with each other for subtle reasons [232]. Expectation also plays a critical role in other theories, which differ by how risks are treated [492].

Consider a theory that says the expected returns on all possible riskless bond strategies are equal for all holding periods. In particular,

$$[1 + S(2)]^2 = [1 + S(1)]\, E[1 + S(1, 2)]\qquad\qquad(5.15)$$

because of the equivalence between buying a two-period bond and rolling over one-period bonds. After rearrangement, $E[1 + S(1, 2)] = [1 + S(2)]^2/[1 + S(1)]$. Now consider the following two one-period strategies. The first strategy buys a two-period bond and sells it after one period. The expected return is $E[\{1+S(1,2)\}^{-1}][1+S(2)]^2$. The second strategy buys a one-period bond with a return of $1+S(1)$. The theory says they are equal: $E[\{1+S(1,2)\}^{-1}][1+S(2)]^2 = 1+S(1)$, which implies that

$$\frac{[1+S(2)]^2}{1+S(1)} = \frac{1}{E[\{1+S(1,2)\}^{-1}]}.$$

Combining this equation with Eq. (5.15), we conclude that

$$E\left[\frac{1}{1+S(1,2)}\right] = \frac{1}{E[1+S(1,2)]}.$$

However, this is impossible, save for a certain economy. The reason is **Jensen's inequality**, which states that $E[g(X)] > g(E[X])$ for any nondegenerate **random variable** X and strictly convex function g (i.e., $g''(x) > 0$). Use $g(x) \equiv (1+x)^{-1}$ to prove our point. So this version of the expectations theory is untenable.

Another version of the expectations theory is the **local expectations theory** [232, 385]. It postulates that the expected rate of return of any bond over a single period equals the prevailing one-period spot rate:

$$\frac{E[\{1+S(1,n)\}^{-(n-1)}]}{[1+S(n)]^{-n}} = 1+S(1) \quad \text{for } all\ n > 1. \tag{5.16}$$

This theory will form the basis of many stochastic interest rate models later. We call

$$\frac{E[\{1+S(1,n)\}^{-(n-1)}]}{[1+S(n)]^{-n}} - [1+S(1)]$$

the **holding premium**, which is zero under the local expectations theory.

Each version of the expectations theory postulates that a certain expected difference, called the **liquidity premium** or the **term premium**, is zero. For instance, the liquidity premium is $f(a,b) - E[S(a,b)]$ under the unbiased expectations theory and it is the holding premium under the local expectations theory [694]. The incompatibility between versions of the expectations theory alluded to earlier would disappear, had they postulated *nonzero* liquidity premiums [143]. For example, the **biased expectations theory** says that

$$f(a,b) - E[S(a,b)] = p(a,b),$$

where the liquidity premium p is not zero [39, 653]. A nonzero liquidity premium is reasonably supported by evidence. There is also evidence that p is neither constant nor time-independent [43, 335].

▶ **Exercise 5.7.3** (1) Prove that

$$E\left[\frac{1}{\{1+S(1)\}\{1+S(1,2)\}\cdots\{1+S(n-1,n)\}}\right] = \frac{1}{[1+S(n)]^n}$$

under the local expectations theory. (2) Show that the local expectations theory is inconsistent with the unbiased expectations theory.

▶ **Exercise 5.7.4** The **return-to-maturity expectations theory** postulates that the maturity strategy earns the same return as the rollover strategy with one-period bonds, i.e.,

$$[1 + S(n)]^n = E[\{1 + S(1)\}\{1 + S(1, 2)\} \cdots \{1 + S(n - 1, n)\}], \quad n > 1.$$

Show that it is inconsistent with the local expectations theory.

5.7.2 Liquidity Preference Theory

The **liquidity preference theory** holds that investors demand a risk premium for holding long-term bonds [492]. The liquidity preference theory is attributed to Hicks [445]. Consider an investor with a holding period of two. If the investor chooses the maturity strategy and is forced to sell the two-period bonds because of an unexpected need for cash, he would face the **interest rate risk** and the ensuing **price risk** because bond prices depend on the prevailing interest rates at the time of the sale. This risk is absent from the rollover strategy. As a consequence, the investor demands a higher return for longer-term bonds. This implies that $f(a, b) > E[S(a, b)]$. When the spot rate curve is inverted,

$$[1 + S(i)]^{1/(i+1)}\{1 + E[S(1, i+1)]\}^{i/(i+1)}$$
$$< [1 + S(1)]^{1/(i+1)}\{1 + E[S(1, i+1)]\}^{i/(i+1)}$$
$$< 1 + S(i+1)$$
$$< 1 + S(i).$$

Thus $E[S(1, i+1)] < S(i)$. The market therefore has to expect the interest rate to decline in order for an inverted spot rate curve to be observed.

The liquidity preference theory seems to be consistent with the typically upward-sloping yield curve. Even if people expect the interest rate to decline and rise equally frequently, the theory asserts that the curve is upward sloping more often. This is because a rising expected interest rate is associated with only a normal spot rate curve, and a declining expected interest rate can sometimes be associated with a normal spot rate curve. Only when the interest rate is expected to fall below a threshold does the spot rate curve become inverted. The unbiased expectations theory, we recall, is not consistent with this case.

▶ **Exercise 5.7.5** Show that the market has to expect the interest rate to decline in order for a flat spot rate curve to occur under the liquidity preference theory.

5.7.3 Market Segmentation Theory

The **market segmentation theory** holds that investors are restricted to bonds of certain maturities either by law, preferences, or customs. For instance, life insurance companies generally prefer long-term bonds, whereas commercial banks favor shorter-term ones. The spot rates are determined within each maturity sector separately [653, 799].

The market segmentation theory is closely related to the **preferred habitats theory** of Culbertson, Modigliani, and Sutch [674]. This theory holds that the investor's horizon determines the riskiness of bonds. A horizon of 5 years will prefer a 5-year zero-coupon bond, demanding higher returns from both 2- and 7-year bonds, for example, because the former choice has reinvestment risk and the latter has price risk. Hence, in contrast to the liquidity preference theory, $f(a, b) < E[S(a, b)]$ can happen if the market is dominated by long-term investors.

5.8 Duration and Immunization Revisited

Rate changes considered before for duration were parallel shifts under flat spot rate curves. We now study duration and immunization under more general spot rate curves and movements.

5.8.1 Duration Measures

Let $S(1), S(2), \ldots,$ be the spot rate curve and $P(y) \equiv \sum_i C_i/[1+S(i)+y]^i$ be the price associated with the cash flow $C_1, C_2 \ldots$ Define duration as

$$-\frac{\partial P(y)/P(0)}{\partial y}\bigg|_{y=0} = \frac{\sum_i \frac{iC_i}{[1+S(i)]^{i+1}}}{\sum_i \frac{C_i}{[1+S(i)]^i}}.$$

Note that the curve is shifted in parallel to $S(1)+\Delta y, S(2)+\Delta y, \ldots,$ before letting Δy go to zero. As before, the percentage price change roughly equals duration multiplied by the size of the parallel shift in the spot rate curve. But the simple linear relation between duration and MD (4.4) breaks down. One way to regain it is to resort to a different kind of shift, the **proportional shift**, defined as

$$\frac{\Delta[1+S(i)]}{1+S(i)} = \frac{\Delta[1+S(1)]}{1+S(1)}$$

for all i [317]. Here, Δx denotes the change in x when the short-term rate is shifted by Δy. Duration now becomes

$$\frac{1}{1+S(1)}\left\{\frac{\sum_i \frac{iC_i}{[1+S(i)]^i}}{\sum_i \frac{C_i}{[1+S(i)]^i}}\right\}. \tag{5.17}$$

If we define **Macaulay's second duration** to be the number within the braces in Eq. (5.17):

$$\text{duration} = \frac{\text{Macaulay's second duration}}{[1+S(1)]}.$$

This measure is also called **Bierwag's duration** [71, 496].

Parallel shift does not reflect market reality. For example, long-term rates do not correlate perfectly with short-term rates; in fact, the two rates often move in opposite directions. Short-term rates are also historically more volatile. Practitioners sometimes break the spot rate curve into segments and measure the duration in each segment [470].

Duration can also be defined under custom changes of the yield curve. For example, we may define the **short-end duration** as the effective duration under the following shifts. The 1-year yield is changed by ± 50 basis points ($\pm 0.5\%$). The amounts of yield changes for maturities $1 \le i \le 10$ are $\pm 50 \times (11-i)/10$ basis points. Yields of maturities longer than 10 remain intact. If the yield curve is normal, the $+50$ basis-point change corresponds to **flattening** of the yield curve, whereas the -50 basis-point change corresponds to **steepening** of the yield curve. **Long-end duration** can be specified similarly. Two custom shifts are behind nonproportional shifts (see Exercise 5.8.3) and Ho's **key rate durations** (see Section 27.5).

Although durations have many variants, the one feature that all share is that the term structure can shift in only a fixed pattern. Despite its theoretical limitations, duration seems to provide as good an estimate for price volatility as more sophisticated measures [348]. Furthermore, immunization with the MD, still widely used [91], is as effective as alternative duration measures [424]. One explanation is that, although long-term rates and short-term rates do not in general move by the same amount or even in the same direction, roughly parallel shifts in the spot rate curve are responsible for more than 80% of the movements in interest rates [607].

▶ **Exercise 5.8.1** Assume continuous compounding. Show that if the yields to maturity of all fixed-rate bonds change by the same amount, then (1) the spot rate curve must be flat and (2) the spot rate curve shift must be parallel. (Hint: The yields of zero-coupon bonds of various maturities change by the same amount.)

▶ **Exercise 5.8.2** Verify duration (5.17).

▶ **Exercise 5.8.3** Empirically, long-term rates change less than short-term ones. To incorporate this fact into duration, we may postulate **nonproportional shifts** as

$$\frac{\Delta[1+S(i)]}{1+S(i)} = K^{i-1}\frac{\Delta[1+S(1)]}{1+S(1)} \quad \text{for some } K < 1.$$

Show that a t-period zero-coupon bond's price sensitivity satisfies

$$\frac{\Delta P}{P} = -tK^{t-1}\frac{\Delta[1+S(1)]}{1+S(1)}$$

under nonproportional shifts.

5.8.2 Immunization

The Case of NO Rate Changes

Recall that in the absence of interest rate changes and assuming a flat spot rate curve, it suffices to match the PVs of the future liability and the asset to achieve immunization (see Exercise 4.2.7). This conclusion continues to hold even if the spot rate curve is not flat, as long as the future spot rates equal the forward rates. Here is the analysis. Let L be the liability at time m. Then

$$P = \sum_{i=1}^{n} \frac{C}{[1+S(i)]^i} + \frac{F}{[1+S(n)]^n} = \frac{L}{[1+S(m)]^m}.$$

The PV of the liability at any time $k \leq m$ is hence

$$\frac{L}{[1+S(k,m)]^{m-k}} = P[1+S(k)]^k$$

by Eq. (5.2) and the premise that $f(a,b) = S(a,b)$. The PV of the bond plus the reinvestments of the coupon payments at the same time is

$$\sum_{i=1}^{k} C[1+S(i,k)]^{k-i} + \sum_{i=1}^{n-k} \frac{C}{[1+S(k,i+k)]^i} + \frac{F}{[1+S(k,n)]^{n-k}}$$

$$= \sum_{i=1}^{k} \frac{C[1+S(k)]^k}{[1+S(i)]^i} + \sum_{i=1}^{n-k} \frac{C[1+S(k)]^k}{[1+S(i+k)]^{i+k}} + \frac{F[1+S(k)]^k}{[1+S(n)]^n}$$

$$= P[1+S(k)]^k,$$

which matches the liability precisely. Therefore, in the absence of unpredictable interest rate changes, duration matching and rebalancing are not needed for immunization.

The Case of Certain Rate Movements

Recall that a future liability can be immunized by a portfolio of bonds with the same PV and MD under flat spot rate curves (see Subsection 4.2.2). If only parallel shifts are allowed, this conclusion can be extended to general spot rate curves. Here is the analysis. We are working with continuous compounding. The liability L is T periods from now. Without loss of generality, assume that the portfolio consists of only zero-coupon bonds maturing at t_1 and t_2 with $t_1 < T < t_2$. Let there be n_i bonds maturing at time $t_i, i = 1, 2$. Assume that $L = 1$ for simplicity. The portfolio's PV is

$$ V \equiv n_1 e^{-S(t_1) t_1} + n_2 e^{-S(t_2) t_2} = e^{-S(T) T}, $$

and its MD is

$$ \frac{n_1 t_1 e^{-S(t_1) t_1} + n_2 t_2 e^{-S(t_2) t_2}}{V} = T. $$

These two equations imply that

$$ n_1 e^{-S(t_1) t_1} = \frac{V(t_2 - T)}{t_2 - t_1}, \quad n_2 e^{-S(t_2) t_2} = \frac{V(t_1 - T)}{t_1 - t_2}. $$

Now shift the spot rate curve uniformly by $\delta \neq 0$. The portfolio's PV becomes

$$ n_1 e^{-[S(t_1)+\delta] t_1} + n_2 e^{-[S(t_2)+\delta] t_2} = e^{-\delta t_1} \frac{V(t_2 - T)}{t_2 - t_1} + e^{-\delta t_2} \frac{V(t_1 - T)}{t_1 - t_2} $$

$$ = \frac{V}{t_2 - t_1} [e^{-\delta t_1} (t_2 - T) + e^{-\delta t_2} (T - t_1)], $$

whereas the liability's PV after the parallel shift is $e^{-[S(T)+\delta] T} = e^{-\delta T} V$. As

$$ \frac{V}{t_2 - t_1} [e^{-\delta t_1} (t_2 - T) + e^{-\delta t_2} (T - t_1)] > e^{-\delta T} V, $$

immunization is established. See Fig. 5.10 for illustration.

Intriguingly, we just demonstrated that (1) a duration-matched position under parallel shifts in the spot rate curve implies a free lunch as any interest rate change generates profits and (2) no investors would hold the T-period bond because a portfolio of t_1- and t_2-period bonds has a higher return for any interest rate shock (in fact, they would own bonds of only the shortest and the longest maturities). Implausible as the assertions may be, the reasoning seems impeccable. The way to resolve the conundrum lies in observing that rate changes were assumed to be *instantaneous*. The problem disappears when price changes occur *after* rate changes [207, 848].[3]

A barbell portfolio often arises from maximizing the portfolio convexity, as argued in Section 4.3. Higher convexity may be undesirable, however, when it comes to immunization. Recall that convexity assumes parallel shifts in the term structure. The moment this condition is compromised, as is often the case in reality, the more

Asset/liability ratio

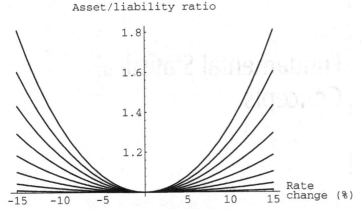

Figure 5.10: Asset/liability ratios under parallel shifts in the spot rate curve. Each curve is the result of a pair of zero-coupon bonds with maturities (t_1, t_2) to immunize a liability 10 periods away. All curves have a minimum value of one when there are no shifts. Interest rate changes move the portfolio value ahead of the liability, and the effects are more pronounced the more t_1 and t_2 are away from 10.

dispersed the cash flows, the more exposed the portfolio is to the **shape risk** (or the **twist risk**) [206, 246].

▶ **Exercise 5.8.4** Repeat the above two-bond argument to prove that the claims in Exercise 4.2.8 remain valid under the more general setting here.

Additional Reading

Consult [325, 514, 583, 629] for more information on the term structure of credit spread. Pointers to empirical studies of the expectations theory can be found in [144, 147]. Also called the **Fisher–Weil duration** [424], Macaulay's second duration is proposed in [350]. See [239, 245, 267, 318] for alternative approaches to immunization and [367] for immunization under stochastic interest rates.

NOTES

1. **"Maturity"** and **"term"** are usually used in place of "term to maturity."
2. The coupon rates of the coupon bonds making up the yield curve need to be specified.
3. We return to this issue in Exercises 14.4.4 and 24.6.8.

CHAPTER SIX

Fundamental Statistical Concepts

There are three kinds of lies: lies, damn lies, and statistics.
Benjamin Disraeli (1804–1881)

Statistics is vital to any scientific discipline that is confronted with the task of summarizing data and making inferences from them. This elementary chapter presents notations and results in probability and statistics that will be useful or extended later.

6.1 Basics

Several definitions are related to expressing the variability of a random variable. The **variance** σ_X^2 of a random variable X is defined as

$$\mathrm{Var}[\,X\,] \equiv E[\,(X - E[\,X\,])^2\,].$$

The **standard deviation** σ_X is the square root of the variance, $\sqrt{\mathrm{Var}[\,X\,]}$. The **skewness** of X with mean μ is $E[\,(X-\mu)^3/\sigma^3\,]$, and the **kurtosis** is $E[\,(X-\mu)^4/\sigma^4\,]$.

The **sample mean** of a random sample $X_1, X_2, \ldots, X_n$ is

$$\overline{X} \equiv \frac{1}{n}\sum_{i=1}^{n} X_i.$$

A measure of a random sample's variability is its **sample variance**, defined as

$$\overline{\sigma^2} \equiv \frac{1}{n-1}\sum_{i=1}^{n}(X_i - \overline{X})^2. \tag{6.1}$$

The **sample standard deviation** $\overline{\sigma}$ is the square root of the sample variance:

$$\overline{\sigma} \equiv \sqrt{\overline{\sigma^2}} = \sqrt{\frac{1}{n-1}\sum_{i=1}^{n}(X_i - \overline{X})^2}. \tag{6.2}$$

An estimator for a parameter θ is said to be **unbiased** if its expected value equals θ. Sample variance (6.1) is an unbiased estimator of σ_X^2 when each random sample X_i has the same distribution as X. Although the sample standard deviation is a biased estimator of the standard deviation, it converges to the unbiased one.

The **covariance** between two random variables X and Y is defined by

$$\text{Cov}[X, Y] \equiv E[(X - \mu_X)(Y - \mu_Y)],$$

where μ_X and μ_Y are the means of X and Y, respectively. If X and Y tend to move in the same direction, their covariance will be positive, whereas if they tend to move in opposite directions, their covariance will be negative. Call X and Y **uncorrelated** random variables if $\text{Cov}[X, Y] = 0$. A computational shortcut for covariance is

$$\text{Cov}[X, Y] = E[XY] - \mu_X \mu_Y. \tag{6.3}$$

The **correlation** (or **correlation coefficient**) between X and Y is

$$\rho_{X,Y} \equiv \frac{\text{Cov}[X, Y]}{\sigma_X \sigma_Y}, \tag{6.4}$$

provided that both have nonzero standard deviations. The variance of a weighted sum of random variables satisfies

$$\text{Var}\left[\sum_{i=1}^{n} a_i X_i\right] = \sum_{i=1}^{n} \sum_{j=1}^{n} a_i a_j \text{Cov}[X_i, X_j]. \tag{6.5}$$

The above becomes $\sum_{i=1}^{n} \sum_{j=1}^{n} a_i^2 \text{Var}[X_i]$ when X_i are uncorrelated.

Let $X \mid I$ denote X conditional on the **information set** I. The information set can be another random variable's value or the past values of X, for example. The **conditional expectation** $E[X \mid I]$ is the expected value of X conditional on I. Note that it is a random variable. The extremely useful **law of iterated conditional expectations** says that

$$E[X] = E[E[X \mid I]].$$

More generally, if I_2 contains at least as much information as I_1, then

$$E[X \mid I_1] = E[E[X \mid I_2] \mid I_1]. \tag{6.6}$$

A typical example is for I_1 to contain price information up to time t_1 and for I_2 to contain price information up to a later time t_2.

▶ **Exercise 6.1.1** Prove Eq. (6.3) by using the well-known identity $E[\sum_i a_i X_i] = \sum_i a_i E[X_i]$.

▶ **Exercise 6.1.2** Prove that if $E[X \mid Y = y] = E[X]$ for all realizations y, then X and Y are uncorrelated. (Hint: Use the law of iterated conditional expectations.)

6.1.1 Generalization to Higher Dimensions

It is straightforward to generalize the above concepts to higher dimensions. Let $X \equiv [X_1, X_2, \ldots, X_n]^{\text{T}}$ be a vector random variable (A^{T} means the transpose of A). Its **mean vector** and the $n \times n$ **covariance matrix** are defined, respectively, as

$$E[X] \equiv [E[X_1], E[X_2], \ldots, E[X_n]]^{\text{T}},$$
$$\text{Cov}[X] \equiv [\text{Cov}[X_i, X_j]]_{1 \leq i, j \leq n}.$$

In analogy with Eq. (6.3), $\mathrm{Cov}[\,X\,] = E[\,XX^{\mathrm{T}}\,] - E[\,X\,]\,E[\,X\,]^{\mathrm{T}}$. The **correlation matrix** is defined as $[\,\rho_{X_i,X_j}\,]_{1\le i,j\le n}$. Let $X_1, X_2, \ldots, X_N$ be N observations on X. The **sample mean vector** and the **sample covariance matrix** are defined, respectively, as

$$\overline{X} \equiv \frac{1}{N} \sum_{i=1}^{N} X_i, \quad \frac{1}{N-1} \sum_{i=1}^{N} (X_i - \overline{X})(X_i - \overline{X})^{\mathrm{T}}.$$

The sample covariance matrix is an unbiased estimator of the covariance matrix.

▶ **Exercise 6.1.3** Prove that $E[\,AX\,] = AE[\,X\,]$ and that $\mathrm{Cov}[\,AX\,] = A\,\mathrm{Cov}[\,X\,]\,A^{\mathrm{T}}$.

6.1.2 The Normal Distribution

A random variable X is said to have **normal distribution** with mean μ and variance σ^2 if its probability density function is $e^{-(x-\mu)^2/(2\sigma^2)}/(\sigma\sqrt{2\pi})$. This fact is expressed by $X \sim N(\mu, \sigma^2)$ where $\sim$ means equality in distribution. The **standard normal distribution** has zero mean, unit variance, and the distribution function

$$N(z) \equiv \frac{1}{\sqrt{2\pi}} \int_{-\infty}^{z} e^{-x^2/2}\,dx.$$

(The **distribution function** of a random variable X is defined as $F(z) \equiv \mathrm{Prob}[\,X \le z\,]$ for any real number z.) The normal distribution is due to de Moivre (1667–1754).

There are fast and accurate approximations to $N(z)$ [5, 423]. An example is

$$N(z) \approx 1 - \frac{1}{\sqrt{2\pi}}\, e^{-z^2/2}(a_1 x + a_2 x^2 + a_3 x^3 + a_4 x^4 + a_5 x^5)$$

for $z \ge 0$, where $x \equiv 1/(1 + 0.2316419z)$ and

$$a_1 = 0.319381530, \quad a_3 = 1.781477937, \quad a_5 = 1.330274429,$$
$$a_2 = -0.356563782, \quad a_4 = -1.821255978.$$

As for $z < 0$, use $N(z) = 1 - N(-z)$.

The **central moments** of the normal random variable X are

$$E[\,(X-\mu)^{2n}\,] = \frac{(2n)!}{2^n n!}\,\sigma^{2n}, \quad E[\,(X-\mu)^{2n+1}\,] = 0, \tag{6.7}$$

where $n = 0, 1, 2, \ldots$. For example, the skewness and the kurtosis of the standard normal distribution are zero and three, respectively. The **moment generating function** of a random variable X is defined as $\theta_X(t) \equiv E[\,e^{tX}\,]$. The moment generating function of $X \sim N(\mu, \sigma^2)$ is known to be

$$\theta_X(t) = \exp\left[\mu t + \frac{\sigma^2 t^2}{2}\right]. \tag{6.8}$$

If $X_i \sim N(\mu_i, \sigma_i^2)$ are independent (or, equivalently for normal distributions, uncorrelated), then $\sum_i X_i$ has a normal distribution with mean $\sum_i \mu_i$ and variance $\sum_i \sigma_i^2$. In general, let $X_i \sim N(\mu_i, \sigma_i^2)$, which may not be independent. Then $\sum_{i=1}^{n} t_i X_i$ is normally distributed with mean $\sum_{i=1}^{n} t_i \mu_i$ and variance $\sum_{i=1}^{n} \sum_{j=1}^{n} t_i t_j \mathrm{Cov}[\,X_i, X_j\,]$ [343]. The joint distribution of $X_1, X_2, \ldots, X_n$ has this joint moment generating function:

$$E\left[\exp\left[\sum_i^n t_i X_i\right]\right] = \exp\left[\sum_{i=1}^n t_i \mu_i + \frac{1}{2}\sum_{i=1}^n\sum_{j=1}^n t_i t_j \operatorname{Cov}[X_i, X_j]\right].$$

These X_i are said to have a **multivariate normal distribution**. We use $X \sim N(\mu, C)$ to denote that $X \equiv [X_1, X_2, \ldots, X_n]^{\mathrm{T}}$ has a multivariate normal distribution with mean $\mu \equiv [\mu_1, \mu_2, \ldots, \mu_n]^{\mathrm{T}}$ and covariance matrix $C \equiv [\operatorname{Cov}[X_i, X_j]]_{1 \leq i, j \leq n}$. With $M \equiv C^{-1}$ and the (i, j)th entry of the matrix M being $M_{i,j}$, the X's probability density function is

$$\frac{1}{\sqrt{(2\pi)^n \det(C)}} \exp\left[-\frac{1}{2}\sum_{i=1}^n\sum_{j=1}^n (X_i - \mu_i) M_{ij} (X_j - \mu_j)\right],$$

where $\det(C)$ denotes the determinant of C [23].

In particular, if X_1 and X_2 have a **bivariate normal distribution** with correlation ρ, their joint probability density function is

$$\frac{1}{2\pi\sigma_1\sigma_2\sqrt{1-\rho^2}} \exp\left[-\frac{\left(\frac{X_1-\mu_1}{\sigma_1}\right)^2 - 2\rho\left(\frac{X_1-\mu_1}{\sigma_1}\right)\left(\frac{X_2-\mu_2}{\sigma_2}\right) + \left(\frac{X_2-\mu_2}{\sigma_2}\right)^2}{2(1-\rho^2)}\right].$$

The sum $\omega_1 X_1 + \omega_2 X_2$ is normally distributed with mean $\omega_1\mu_1 + \omega_2\mu_2$ and variance

$$\omega_1^2\sigma_1^2 + 2\omega_1\omega_2\rho\sigma_1\sigma_2 + \omega_2^2\sigma_2^2. \tag{6.9}$$

Fast and accurate approximations to the bivariate normal random variable's distribution function are available [470].

If $X_i \sim N(\mu_i, \sigma^2)$ are independent, then $Y \equiv \sum_{i=1}^n X_i^2/\sigma^2$ has a **noncentral chi-square distribution** with n **degrees of freedom** and **noncentrality parameter** $\theta \equiv (\sum_{i=1}^n \mu_i^2)/\sigma^2 > 0$, denoted by $Y \sim \chi(n, \theta)$. The mean and the variance are $n + \theta$ and $2n + 4\theta$, respectively [463]. When μ_i are zero, Y has the **central chi-square distribution**.

The **central limit theorem**, which is due to Laplace (1749–1827), is a cornerstone for probability and statistics. It says that, if X_i are independent with mean μ and variance σ^2, then

$$\frac{\sum_{i=1}^n X_i - n\mu}{\sigma\sqrt{n}} \to N(0, 1).$$

Conditions for the theorem's applicability are rather mild [343].

▶ **Exercise 6.1.4** Prove that central moments (6.7) are equivalent to

$$E[(X-\mu)^n] = \begin{cases} 0, & \text{if } n \geq 1 \text{ is odd} \\ 1 \cdot 3 \cdot 5 \cdots (n-1)\sigma^n, & \text{if } n \geq 2 \text{ is even} \end{cases},$$

where $n = 1, 2, \ldots$.

6.1.3 Generation of Univariate and Bivariate Normal Distributions

Let X be uniformly distributed over $(0, 1]$ so that $\operatorname{Prob}[X \leq x] = x$ for $0 < x \leq 1$. Repeatedly draw two samples x_1 and x_2 from X until $\omega \equiv (2x_1 - 1)^2 + (2x_2 - 1)^2 < 1$. Then $c(2x_1 - 1)$ and $c(2x_2 - 1)$ are independent standard normal variables, where

Normal distribution

Lognormal distribution

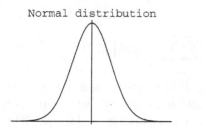
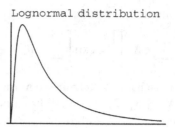

Figure 6.1: Normal and lognormal distributions: the standard normal distribution X and the lognormal distribution e^X.

$c \equiv \sqrt{-2(\ln \omega)/\omega}$. This is called the **polar rejection method** [727]. Pairs of normally distributed variables with correlation ρ can be generated as follows. Let X_1 and X_2 be independent standard normal variables. Then

$$
\begin{aligned}
U &\equiv a X_1, \\
V &\equiv \rho U + \sqrt{1 - \rho^2} \, a X_2
\end{aligned}
$$

are the desired random variables because $\mathrm{Var}[\,U\,] = \mathrm{Var}[\,V\,] = a^2$ and $\mathrm{Cov}[\,U, V\,] = \rho a^2$.

6.1.4 The Lognormal Distribution

A random variable Y is said to have a **lognormal distribution** if $\ln Y$ has a normal distribution (see Fig. 6.1). This distribution is due to Bachelier [147].

If X is normally distributed with mean μ and variance σ^2, then the density function of the lognormally distributed random variable $Y \equiv e^X$ is

$$
f(y) \equiv
\begin{cases}
\dfrac{1}{\sigma y \sqrt{2\pi}} \, e^{-(\ln y - \mu)^2/(2\sigma^2)}, & \text{if } y > 0 \\
0, & \text{if } y \le 0
\end{cases}.
\tag{6.10}
$$

The mean and the variance of Y are

$$
\mu_Y = e^{\mu + \sigma^2/2}, \quad \sigma_Y^2 = e^{2\mu + \sigma^2}\left(e^{\sigma^2} - 1\right),
\tag{6.11}
$$

respectively. Furthermore,

$$
\mathrm{Prob}[\,Y \le y\,] = \mathrm{Prob}[\,X \le \ln y\,] = N\left(\frac{\ln y - \mu}{\sigma}\right).
\tag{6.12}
$$

The nth **moment** about the origin, defined as $\int_{-\infty}^{\infty} x^n f(x)\, dx$ for a random variable x with density function $f(x)$, is $e^{n\mu + n^2\sigma^2/2}$ for Y. A version of the central limit theorem states that the product of n independent positive random variables approaches a lognormal distribution as n goes to infinity.

▶ **Exercise 6.1.5** Let Y be lognormally distributed with mean μ and variance σ^2. Show that $\ln Y$ has mean $\ln[\,\mu/\sqrt{1 + (\sigma/\mu)^2}\,]$ and variance $\ln[\,1 + (\sigma/\mu)^2\,]$.

▶ **Exercise 6.1.6** Let X be a lognormal random variable such that $\ln X$ has mean μ and variance σ^2. Prove the identity $\int_a^{\infty} x f(x)\, dx = e^{\mu + \sigma^2/2}\, N(\frac{\mu - \ln a}{\sigma} + \sigma)$.

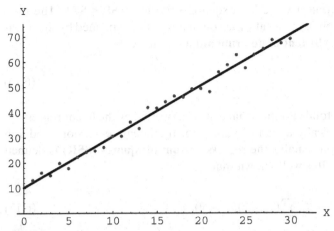

Figure 6.2: Linear regression. The linear function $Y = 10.1402 + 2.0238X$ is fit to the data under the least-squares criterion.

6.2 Regression

Suppose we are presented with the data $(x_1, y_1), (x_2, y_2), \ldots, (x_n, y_n)$. The data can be plotted on a rectangular coordinate system, resulting in the so-called **scatter diagram**, such as the dots in Fig. 6.2. If the scatter diagram suggests a linear relation between the variables, we can fit a simple straight-line model $y = \beta_0 + \beta_1 x$ to the data. The problem of finding such a fit is called **linear regression**.[1] To estimate the model parameters β_0 and β_1 with the **least-squares principle**, we find $\widehat{\beta}_0$ and $\widehat{\beta}_1$ that minimize

$$\sum_{i=1}^{n} [\, y_i - (\widehat{\beta}_0 + \widehat{\beta}_1 x)\,]^2. \tag{6.13}$$

This line is called the linear regression of y on x [632]. It is well known that

$$\widehat{\beta}_1 = \frac{\sum_i (x_i - \overline{x})(y_i - \overline{y})}{\sum_i (x_i - \overline{x})^2} = \frac{n \sum_i x_i y_i - (\sum_i x_i)(\sum_i y_i)}{n \sum_i x_i^2 - (\sum_i x_i)^2}, \tag{6.14}$$

$$\widehat{\beta}_0 = \frac{\sum_i y_i - \widehat{\beta}_1 \sum_i x_i}{n} = \overline{y} - \widehat{\beta}_1 \overline{x}. \tag{6.15}$$

The resulting line $y = \widehat{\beta}_0 + \widehat{\beta}_1 x$ is called the **estimated regression line** or the **least-squares line**. The ith **fitted value** is $\widehat{y}_i \equiv \widehat{\beta}_0 + \widehat{\beta}_1 x_i$. Note that $(\overline{x}, \overline{y})$ is on the estimated regression line by virtue of Eq. (6.15).

A few statistics are commonly used. The **error sum of squares** (**SSE**) is the sum of the squared deviation about the estimated regression line:

$$\text{SSE} \equiv \sum_i (y_i - \widehat{y}_i)^2 = \sum_i (y_i - \widehat{\beta}_0 - \widehat{\beta}_1 x_i)^2.$$

Because the SSE measures how much variation in y is *not* explained by the linear regression model, it is also called the **residual sum of squares** or the **unexplained variation**. The **total sum of squares** (**SST**) is defined as $\text{SST} \equiv \sum_i (y_i - \overline{y})^2$, which measures the total amount of variation in observed y values. This value is also known as the **total variation**. By the least-squares criterion, SSE ≤ SST. The ratio SSE/SST is the proportion of the total variation that is left unexplained by the linear regression model. The **coefficient of determination** is defined as

$$R^2 \equiv 1 - \frac{\text{SSE}}{\text{SST}};$$ (6.16)

it is the proportion of the total variation that can be explained by the linear regression model. A high R^2 is typically a sign of success of the linear regression model in explaining the y variation. Finally, the **regression sum of squares (SSR)** is defined as $\text{SSR} \equiv \sum_{i=1}^{n}(\widehat{y}_i - \overline{y})^2$. It is well known that

$$\text{SSR} = \text{SST} - \text{SSE} = \widehat{\beta}_1 \sum_{i=1}^{n}(x_i - \overline{x})(y_i - \overline{y}).$$ (6.17)

Thus $R^2 = \text{SSR}/\text{SST}$. Because the SSR is large when the estimated regression line fits the data closely (as SSE is small), it is interpreted as the amount of total variation that is explained by the linear regression model. For this reason it is sometimes called the **explained variation**.

The more general linear regression, also known as **multiple regression**, fits

$$y = \beta_0 + \beta_1 x_1 + \beta_2 x_2 + \cdots + \beta_k x_k$$

to the data. Equation (6.17) holds for multiple regression as well [422, 523]. Nonlinear regression uses nonlinear regression functions. In **polynomial regression**, for example, the problem is to fit

$$y = \beta_0 + \beta_1 x + \beta_2 x^2 + \cdots + \beta_k x^k$$

to the data. See Fig. 6.3 for the $k = 2$ case.

▶ **Exercise 6.2.1** Prove that $\text{SSE} = \sum_i y_i^2 - \widehat{\beta}_0 \sum_i y_i - \widehat{\beta}_1 \sum_i x_i y_i$.

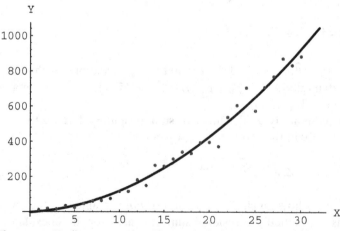

Figure 6.3: Nonlinear regression. The quadratic function $Y = -1.28204 + 2.52945X + 0.945518X^2$ is fit to the data under the least-squares criterion.

6.3 Correlation

Given n pairs of observations $(x_1, y_1), (x_2, y_2), \ldots, (x_n, y_n)$ on (X, Y), their **sample correlation coefficient** or **Pearson's** r is defined as

$$r \equiv \frac{\sum_i (x_i - \overline{x})(y_i - \overline{y})}{\sqrt{\sum_i (x_i - \overline{x})^2}\sqrt{\sum_i (y_i - \overline{y})^2}}. \tag{6.18}$$

The sample correlation coefficient is a point estimator for $\rho_{X,Y}$ and is traditionally used to summarize the strength of correlation. It can be shown that $-1 \le r \le 1$. In particular, $r = 1$ when the data lie on a straight line with positive slope and $r = -1$ when the data lie on a straight line with negative slope. In some sense r measures the *linear* relation between the variables.

In regression, one variable is considered dependent and the others independent; the purpose is to predict. Correlation analysis, in contrast, studies how strongly two or more variables are related, and the variables are treated symmetrically; it does not matter which of the two variables is called x and which y.

We used the symbol r deliberately: Squaring r gives exactly the coefficient of determination R^2. Indeed, from Eqs. (6.14) and (6.17),

$$r^2 = \widehat{\beta}_1^2 \frac{\sum_i (x_i - \overline{x})^2}{\sum_i (y_i - \overline{y})^2} = \widehat{\beta}_1 \frac{\sum_i (x_i - \overline{x})(y_i - \overline{y})}{\sum_i (y_i - \overline{y})^2} = \frac{\text{SSR}}{\text{SST}} = R^2. \tag{6.19}$$

Interestingly, Eq. (6.16) implies that $\text{SSE} = \text{SST} \times (1 - r^2)$.

EXAMPLE 6.3.1 Figure 6.4 plots the stock prices of Intel, Silicon Graphics, Inc. (SGI), VLSI Technology, and Wal-Mart from August 30, 1993, to August 30, 1995. The sample correlation coefficient between VLSI Technology and Intel is extremely high at 0.950376. The sample correlation coefficient between Intel and SGI is also high at 0.883291. Technology stocks seem to move together. In contrast, the sample correlation coefficient between Intel and Wal-Mart is low at 0.14917. From these numbers and Eq. (6.19), we can deduce, for example, that 90.3215% of the total variations between Intel's and VLSI Technology's stock prices can be explained by a linear regression model.

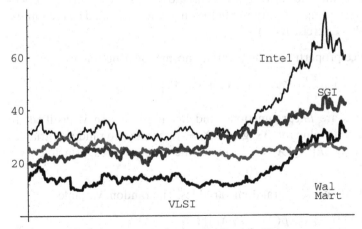

Figure 6.4: Correlation among stock prices. See Example 6.3.1.

▶ **Exercise 6.3.1** Find the estimated regression line for $\{(1, 1.0), (2, 1.5), (3, 1.7), (4, 2.0)\}$. Check that the coefficient of determination indeed equals the sample correlation coefficient.

6.4 Parameter Estimation

After a family of stochastic models has been chosen to capture the reality, the values of their parameters must be found to completely specify the distribution. Inferring those parameters constitutes the major goal of financial econometrics [147]. Three estimation techniques are mentioned below.

6.4.1 The Least-Squares Method

This method is due to Legendre (1752–1833) in 1806 and Gauss (1777–1855) in 1809 [582].[2] It works by minimizing the sum of squares of the deviations, in other words, the SSE. For example, the least-squares estimate of X, given the measurements x_i on it, is the number $\widehat{X}$ that minimizes

$$f(\widehat{X}) \equiv \sum_{i=1}^{n}(x_i - \widehat{X})^2. \tag{6.20}$$

This method was also used in the derivation of the estimated regression line in Section 6.2 by the minimization of (6.13). Recall that no stochastic models were assumed there.

Suppose that the following **linear regression model** is postulated between x and y:

$$y = \beta_0 + \beta_1 x + \epsilon,$$

where ϵ is a random variable with zero mean and finite variance. In other words, added to each observation of y is some uncorrelated noise ϵ. Then the estimated parameters of the estimated regression line, which are now random variables, have the smallest variances among all unbiased *linear* estimators. This is the **Gauss–Markov theorem**, which is due to Gauss in 1821 and Markov (1856–1922) in 1912 [75, 632]. It is interesting to observe that the least-squares estimate of β_1 – the $\widehat{\beta}_1$ in Eq. (6.14) – can be interpreted as the sample covariance between x and y divided by the sample variance of x (see also Exercise 6.4.1).

EXAMPLE 6.4.1 Two nice properties of the bivariate normal distribution are

$$E[X_2 \mid X_1] = \mu_2 + \rho\frac{\sigma_2}{\sigma_1}(X_1 - \mu_1), \quad \text{Var}[X_2 \mid X_1] = (1 - \rho^2)\sigma_2^2.$$

Hence the regressions are *linear* functions, and linear regression is justified. In fact, the fitted (predicted) value for X_2, given $X_1 = x$ for any two random variables X_1 and X_2, is exactly $E[X_2 \mid X_1 = x]$ under the least-squares principle (see Exercise 6.4.3) [846].

▶ **Exercise 6.4.1** Let X_1 and X_2 be random variables. The random variable

$$Y \equiv (X_2 - E[X_2]) - \{\alpha + \beta(X_1 - E[X_1])\}$$

is the prediction error of the linear prediction $\alpha + \beta(X_1 - E[X_1])$ of X_2 based on X_1. Show that (1) $\text{Var}[Y] = E[Y^2]$ is minimized at $\alpha \equiv 0$ and $\beta \equiv (\text{Cov}[X_1, X_2])/(\text{Var}[X_1])$, which is called **beta**, and (2) X_1 and Y are uncorrelated if the optimal linear prediction is used.

▶ **Exercise 6.4.2** Verify that the f in Eq. (6.20) is minimized at $\widehat{X} = (1/n)\sum_{i=1}^{n} x_i$.

▶ **Exercise 6.4.3** (1) Prove that a minimizes the **mean-square error** $E[(X-a)^2]$ when $a = E[X]$. (2) Show that the best predictor a of X_k based on $X_1, X_2, \ldots, X_{k-1}$ in the mean-square-error sense, that is, with minimum $E[(X_k - a)^2 \mid X_1, X_2, \ldots, X_{k-1}]$, is the **conditional least-squares estimator** $E[X_k \mid X_1, X_2, \ldots, X_{k-1}]$.

6.4.2 The Maximum Likelihood Estimator

Suppose that the sample has the probability density function $p(z \mid \theta)$. If Z is observed, $p(Z \mid \theta)$ is called the **likelihood** of θ.[3] The **maximum likelihood (ML) method** estimates θ by the number $\widehat{\theta}$ that maximizes the likelihood. Formally the **likelihood function** as the joint probability of the event $X_1 = x_1, X_2 = x_2, \ldots, X_n = x_n$ is

$$L(\theta) \equiv \text{Prob}[X_1 = x_1, X_2 = x_2, \ldots, X_n = x_n \mid \theta],$$

where $\theta \equiv (\theta_1, \theta_2, \ldots, \theta_k)$ is the vector of parameters to be estimated. The likelihood function product equals $\prod_{i=1}^{n} p_{X_i}(x_i \mid \theta)$, where $p_{X_i}(x_i \mid \theta)$ is the probability density function of $X_i = x_i$ when the samples are drawn independently. The ML method estimates θ with $\widehat{\theta} \equiv (\widehat{\theta}_1, \widehat{\theta}_2, \ldots, \widehat{\theta}_m)$ such that $L(\widehat{\theta}) \geq L(\theta)$ for all θ. It may be biased, however.

An estimator is **consistent** if it converges in probability to the true parameter as the sample size increases. The ML method, among consistent estimators, enjoys such optimality properties as minimum asymptotic variance and asymptotic normality under certain regularity conditions. These properties carry over to samples from a stochastic process [413, 422]. Unlike the least-squares method, which uses only the first two moments of the observations, the ML method utilizes the complete distribution of the observations.

Under certain regularity conditions, the ML estimate of θ is the solution to the simultaneous equations $\partial L(\theta)/\partial \theta_i = 0$. Often it is the logarithm of $L(\theta)$, called the **log-likelihood function**, that is more convenient to work with. Numerical techniques are needed when a closed-form solution for θ is not available.

EXAMPLE 6.4.2 Based on n independent observations $x_1, x_2, \ldots, x_n$ from $N(\mu, \sigma^2)$, the log-likelihood function is

$$\ln L(\mu, \sigma^2) = -\frac{n}{2} \ln(2\pi) - \frac{n}{2} \ln \sigma^2 - \frac{1}{2\sigma^2} \sum_{i=1}^{n} (x_i - \mu)^2.$$

After setting $\partial \ln L/\partial \mu$ and $\partial \ln L/\partial \sigma^2$ to zero, we obtain $\widehat{\mu} = (1/n)\sum_i x_i$, the sample mean, and $\widehat{\sigma^2} = (1/n)\sum_i (y_i - \widehat{\mu})$. The ML estimator of variance is biased because it differs from Eq. (6.1). It is consistent, however.

6.4.3 The Method of Moments

The **method of moments** estimates the parameters of a distribution by equating the population moments with their sample moments. Let $X_1, X_2, \ldots, X_n$ be random samples from a distribution characterized by k parameters $\theta_1, \theta_2, \ldots, \theta_k$. The method of moments estimates these parameters by solving k of the following equations:

$$M_1 = \frac{1}{n} \sum_{i=1}^{n} X_i, \quad M_2 = \frac{1}{n} \sum_{i=1}^{n} X_i^2, \quad M_3 = \frac{1}{n} \sum_{i=1}^{n} X_i^3, \ldots,$$

where the moments $M_i \equiv E[X^i]$ are functions of the parameters.

The name method of moments comes from the notion that parameters should be estimated by using moments. Also called the **analog method**, the method of moments requires no knowledge of the likelihood function. Although only certain moments of the observations instead of the full probability density function are used, this method is convenient and usually leads to simple calculations as well as to consistent estimators. Furthermore, it is the only approach of wide applicability in some situations.

Additional Reading

This chapter draws on [12, 23, 195, 273, 343, 463, 802, 816, 846] for probability theory, statistics, and statistical inferences. A very accurate approximation to the normal distribution appears in [678]. Regression analysis is covered by many books [317, 422, 632, 799]. See [273, 522, 584, 846] for more information about the lognormal distribution. The method of moments was introduced by Pearson (1857–1936) in 1894 [415].

NOTES

1. The idea of regression is due to Galton (1822–1911) [65].
2. Gauss claimed to have made the discovery in 1795 [75, 339].
3. The idea of likelihood is due to Ronald Fisher (1890–1962) [671].

CHAPTER
SEVEN

Option Basics

The shift toward options as the center of gravity of finance [...]
Merton H. Miller (1923–2000) [666]

Options grant their holder the right to buy or sell some **underlying asset**. Options are therefore **contingent claims** or **derivative securities** because their value depends on that of the underlying asset. Besides being one of the most important classes of financial instruments, options have wide-ranging applications in finance and beyond. As far as explaining empirical data goes, the option pricing theory ranks as the most successful theory in finance and economics [766].

7.1 Introduction

There are two basic types of options: **calls** and **puts**. More complex instruments can often be decomposed into a package of calls and puts. A call option gives its holder the right to buy a specified number of the underlying asset by paying a specified **exercise price** or **strike price**. A put option gives its holder the right to sell a specified number of the underlying asset by paying a specified strike price. The underlying asset may be stocks, stock indices, options, foreign currencies, futures contracts, interest rates, fixed-income securities, mortgages, winter temperatures, and countless others [54, 346, 698]. When an option is embedded, it has to be traded along with the underlying asset.

The one who issues an option is called a **writer**. To acquire the option, the holder pays the writer a **premium**. When a call is **exercised**, the holder pays the writer the strike price in exchange for the stock, and the option ceases to exist. When a put is exercised, the holder receives from the writer the strike price in exchange for the stock, and the option ceases to exist. An option can be exercised before the expiration date, which is called **early exercise**. It can also be sold at any trading date before the expiration date.

American and **European options** differ in when they can be exercised. American options can be exercised at any time up to the expiration date, whereas European options can be exercised only at expiration.[1] An American option is worth at least as much as an otherwise identical European option because of the early exercise feature.

Many strategies and analysis in the book depend on taking a **short position**. In stocks, short sales involve borrowing stock certificates and buying them back later; in short, selling what one does not own precedes buying. The short seller is apparently betting that the stock price will decline. Note that borrowed shares have to be paid back with shares, not cash. The short seller does not receive cash dividends; in fact, the short seller must make matching dividend payments to the person to whom the shares were sold. Every dividend payout hence reduces a short seller's return.

It is easier to take a short position in derivatives. All one has to do is to find an investor who is willing to buy them, that is, to be **long** the derivatives. For derivatives that do not deliver the underlying asset or those that are mostly settled by taking offset positions, their outstanding contracts may cover many times the underlying asset [60].

For the rest of this chapter, C denotes the call value, P the put value, X the strike price, S the stock price, and D the dividend. Subscripts are used to differentiate or emphasize times to expiration, stock prices, or strike prices. The notation $\mathrm{PV}(x)$ indicates the PV of x dollars at expiration.

7.2 Basics

An option does not oblige its holder to exercise the right. An option will hence be exercised only when it is in the best interest of its holder to do so. Clearly a call will be exercised only if the stock price is higher than the strike price. Similarly, a put will be exercised only if the stock price is less than the strike price. The value or **payoff** of a call at expiration is therefore $C = \max(0, S - X)$, and that of a put at expiration is $P = \max(0, X - S)$ (see Fig. 7.1). Payoff, unlike profit, does not account for the initial cost. For example, the payoff of a long position in stock is S, and the payoff of a short position in stock is $-S$ (see Fig. 7.2). At any time t before the expiration date,

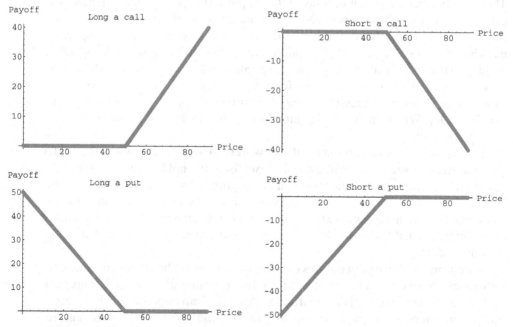

Figure 7.1: Option payoffs: the option payoffs at expiration with a strike price of 50.

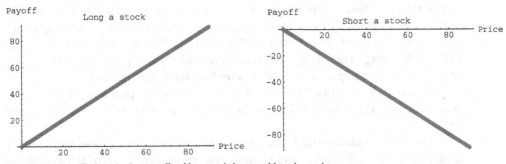

Figure 7.2: Payoff of stock: the payoffs of long and short positions in stock.

we call $\max(0, S_t - X)$ the **intrinsic value** of a call and $\max(0, X - S_t)$ the intrinsic value of a put. The part of an option's value above its intrinsic value is called its **time value** and represents the possibility of becoming more valuable before the option expires. The option premium thus consists of the intrinsic value and the time value. A call is said to be **in the money** if $S > X$, **at the money** if $S = X$, and **out of the money** if $S > X$. Similarly, a put is said to be **in the money** if $S < X$, **at the money** if $S = X$, and **out of the money** if $S > X$. Options that are in the money at expiration should be exercised. Surprisingly, more than 10% of option holders let in-the-money options expire worthless [340]. Although an option's terminal payoff is obvious, finding its value at any time before expiration is a major intellectual breakthrough. Figure 7.3 plots the values of put and call before expiration.

7.3 Exchange-Traded Options

Puts and calls first appeared in 1790. (Aristotle described a kind of call in *Politics* [29, Book 2, Chapter 11].) However, before 1973, options were traded in **over-the-counter markets** in which financial institutions and corporations traded directly with one another. The main distinction of over-the-counter options is that they are customized. Today, over-the-counter options are most popular in the area of foreign currencies and interest rates.

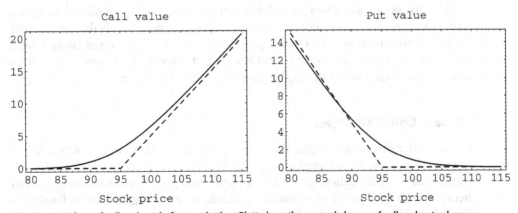

Figure 7.3: Values of call and put before expiration. Plotted are the general shapes of call and put values as functions of the stock price before expiration. Dashed lines denote the option values at expiration.

The Chicago Board Options Exchange (CBOE) started the options trading on April 26, 1973. Since then options have been traded in many exchanges such as the American Stock Exchange (AMEX) and the Philadelphia Stock Exchange (PHLX). **Exchange-traded options** standardize the terms of option contracts, create centralized trading and price dissemination facilities, and introduce the Options Clearing Corporation (OCC), all of which serve to promote an active secondary market. The term **listed option** is also used to refer to an exchange-traded option.

Terms on the exchange-traded stock options govern the expiration dates and the strike prices. The strike prices are centered on the current price of the underlying stock with fixed increments that depend on the price of the stock. Typical increments are $2_{1/2}$ for a stock price less than $25 per share, $5 for a stock price between $25 and $200 per share, and $10 for a stock price over $200 per share. A stock typically has options outstanding expiring at three expiration dates. The exchange also limits the maximum number of options an individual can take on one side of the market. Exchange-traded stock options are American.

Exchange-traded stock options are not **cash dividend protected** (or simply **protected**). This means that the option contract is not adjusted for cash dividends. As the stock price typically falls by the amount roughly equal to the amount of the cash dividend as it goes ex-dividend, dividends are detrimental for calls. The converse is true for puts. However, options are adjusted for stock splits. After an n-for-m stock split, the strike price is only m/n times its previous value, and the number of shares covered by one contract becomes n/m times its previous value. Exchange-traded stock options are also adjusted for stock dividends. Unless stated otherwise, options are assumed to be unprotected. Figure 7.4 shows a small sample of listed stock options.

EXAMPLE 7.3.1 For an option to buy 100 shares of a company for $50 per share, a 2-for-1 split would change the term to a strike price of $25 per share for 200 shares.

A contract normally covers 100 shares of stock. Option prices are quoted per unit of the underlying asset. For instance, the Merck July 35 call closed at $9_{1/2}$ on March 20, 1995, by Fig. 7.4. The total cost of the call was $950.

For exchange-traded options, an option holder can **close out** or **liquidate** the position by issuing an **offsetting order** to sell the same option. Similarly, an option writer can close out the position by issuing an offsetting order to buy the same option. This is called **settled by offset**, made possible by the OCC. The **open interest** is the total number of contracts that have not been offset, exercised, or allowed to expire – in short, the total number of long (equivalently, short) positions.

7.4 Basic Option Strategies

Option strategies involve taking positions in options, the underlying assets, and borrowing or lending. For example, six positions were mentioned before: long stock, short stock, long call, short call, long put, and short put. A strategy can be **bullish**, **bearish**, or **neutral** in terms of market outlook, it can be **aggressive**, **defensive**, or virtually **riskless** in terms of risk posture, and it can be designed to profit in volatile or calm markets. For example, buying a stock is a bullish and aggressive strategy, bullish because it profits when the stock price goes up and aggressive because

Option	Strike	Exp.	-Call- Vol.	-Call- Last	-Put- Vol.	-Put- Last
			...			
Exxon	60	Apr	1053	5 1/2	1000	3/16
65	65	Apr	951	15/16	830	11/16
65	65	May	53	17/16	10	1 1/16
65	65	Oct	32	2 3/4	...	...
65	70	Jul	2	1/4	40	5 1/4
			...			
Merck	30	Jul	328	15 1/4	...	...
44 1/2	35	Jul	150	9 1/2	10	1/16
44 1/2	40	Apr	887	4 3/4	136	1/16
44 1/2	40	Jul	220	5 1/2	297	1/4
44 1/2	40	Oct	58	6	10	1/2
44 1/2	45	Apr	3050	7/8	100	1 1/8
44 1/2	45	May	462	1 3/8	50	1 3/8
44 1/2	45	Jul	883	1 15/16	147	1 3/4
44 1/2	45	Oct	367	2 3/4	188	2 1/16
			...			
Microsoft	55	Apr	65	16 3/4	52	1/8
71 1/8	60	Apr	556	11 3/4	39	1/8
71 1/8	65	Apr	302	7	137	3/8
71 1/8	65	Jul	93	9	15	1 1/2
71 1/8	65	Oct	34	10 5/8	9	2 1/4
71 1/8	70	Apr	1543	3 1/8	162	1 1/2
71 1/8	70	May	42	4 1/4	2	2 1/8
71 1/8	70	Jul	190	5 3/4	61	3
71 1/8	70	Oct	94	7 1/2	1	4
			...			

Figure 7.4: Options quotations. In August 2000, the *Wall Street Journal* started quoting stocks traded on the New York Stock Exchange, the Nasdaq National Market, and the AMEX in decimals. All three exchanges are expected to convert to the decimal system by April 2001. Source: *Wall Street Journal*, March 21, 1995.

the investor runs the risk of maximum loss, dollar for dollar, if the stock goes down. More aggressive strategies include buying stocks on margin and buying options. For instance, the Exxon April 60 call allows the holder to control a $65 stock for a mere $5.5 (see Fig. 7.4). Selling stocks short, on the other hand, is aggressive but bearish. In **covered** positions, some securities protect the returns of others. There are three basic kinds of covered positions: **hedge**, **spread**, and **combination**.

> **Exercise 7.4.1** How would you characterize buying a call in terms of market outlook and risk posture?

7.4.1 Hedge

A **hedge** combines an option with its underlying stock in such a way that one protects the other against loss. A hedge that combines a long position in stock with a long put

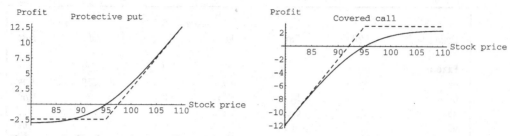

Figure 7.5: Profits of protective put and covered call. The strike price and the current stock price are both $95. The dashed lines represent the positions' profits at expiration. A **profit diagram** does not take into account the time value of the money used in setting up the position.

is called a **protective put**. A hedge that combines a long position in stock with a short call is called a **covered call** (see Fig. 7.5). Covered calls may be the most common option strategy used by institutional investors to generate extra income in a flat market. Because both strategies break even only if the stock price rises, their market outlook is bullish. They are also defensive: The investor owns the stock anyway in a covered call, and the protective put guarantees a minimum value for the portfolio. A **reverse hedge** is a hedge in the opposite direction such as a short position in stock combined with a short put or a long call.

Writing a **cash-secured put** means writing a put while putting aside enough money to cover the strike price if the put is exercised. The payoff is similar to that of a covered call. The maximum profit is $X - [\text{PV}(X) - P]$, and the maximum loss is $P - \text{PV}(X)$, which occurs when the stock becomes worthless. A **ratio hedge** combines two short calls against each share of stock. It profits as long as the stock price does not move far in either direction. See Fig. 7.6 for illustration.

▶ **Exercise 7.4.2** Verify the maximum profit of the cash-secured put.

▶ **Exercise 7.4.3** Both a protective put on a diversified portfolio and a fire insurance policy provide insurance. What is the essential difference between them?

▶ **Exercise 7.4.4** Start with $100 and put $100/(1+r)$ in the money market earning an annual yield of r. The rest of the money is used to purchase calls. (1) Figure out

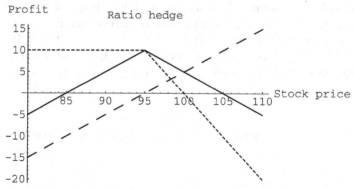

Figure 7.6: Profit of ratio hedge. The solid line is the profit diagram of a ratio hedge at expiration with a strike price of $95 and a current stock price of $95. The dashed line represents the profit diagram of the stock, and the dotted line represents the profit diagram of the option.

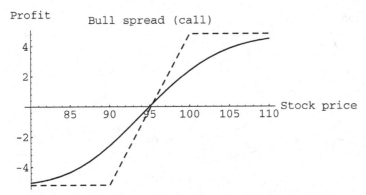

Figure 7.7: Profit of bull call spread. Plotted is the profit diagrams of a bull call spread at expiration (dashed line) and at 1 month before expiration (solid curve). Both the strike price and the current stock price are $95.

the payoff of this strategy when the option expires 1 year from now. (2) What is the r that makes the strategy a "90/10" one, meaning putting 90% in the money market today and earning just enough to exercise the option at expiration? (This strategy is called the **90/10 strategy**.)

7.4.2 Spread

A **spread** consists of options of the same type and on the same underlying asset but with different strike prices or expiration dates. They are of great interest to options market makers. We use X_L, X_M, and X_H to denote the strike prices with $X_L < X_M < X_H$.

A **bull call spread** consists of a long X_L call and a short X_H call with the same expiration date. The initial investment is $C_L - C_H$. The maximum profit is $(X_H - X_L) - (C_L - C_H)$, and the maximum loss is $C_H - C_L$. The risk posture is defensive. See Fig. 7.7 for illustration. This spread is also known as **price spread**, **money spread**, or **vertical spread** (vertical, because it involves options on different rows

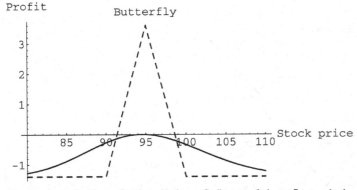

Figure 7.8: Profit of butterfly. Plotted is the profit diagram of a butterfly at expiration (dashed line) and at 1 month before expiration when it is initially set up (solid curve). The strike prices are $90, $95, and $100, and the current stock price is $95.

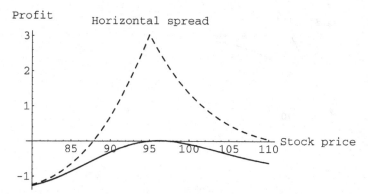

Figure 7.9: Profit of horizontal spread. Plotted is the profit diagram of a horizontal spread at expiration of the near call (dashed curve) and at the time when it is initially set up (solid curve). Both the strike price and the current stock price are $95. There is one month to the first expiration date and two months to the second expiration date.

of the same vertical column as is obvious from Fig. 7.4). Writing an X_H put and buying an X_L put with identical expiration dates will create the so-called **bull put spread**. A **bear spread** amounts to selling a bull spread. It profits from declining stock prices.

EXAMPLE 7.4.1 An investor bought a call. Afterwards, the market moved in her favor, and she was able to write a call for the same premium but at a higher strike price. She ended up with a bull spread and a terminal payoff that could never be negative.

Three calls or three puts with different strike prices and the same expiration date create a **butterfly spread**. Specifically, the spread is long one X_L call, long one X_H call, and short two X_M calls. The first two calls form the **wings**. See Fig. 7.8 for illustration. A butterfly spread pays off a positive amount at expiration only if the asset price falls between X_L and X_H. A butterfly spread with a small $X_H - X_L$ thus approximates a **state contingent claim**, which pays $1 only when a particular price results [346].[2]

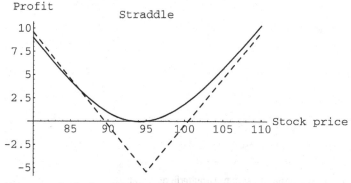

Figure 7.10: Profit of straddle. Plotted is the profit diagram of a straddle at expiration (dashed line) and at 1 month before expiration when it is initially set up (solid curve). The strike price and the current stock price are both $95.

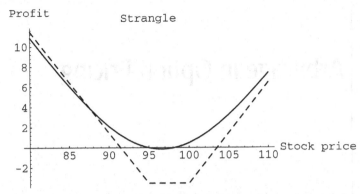

Figure 7.11: Profit of strangle. Plotted is the profit diagram of a strangle at expiration (dashed line) and at 1 month before expiration when it is set up (solid curve). Here the strike prices are $95 (for the put) and $100 (for the call), and the current stock price is $95.

A **horizontal spread** (also called **time spread** or **calendar spread**) involves two options with the same strike price but different expiration dates. A typical horizontal spread consists of a long call with a far expiration date and a short call with a near expiration date. Its profit pattern arises from the difference in the rate of time decay between options expiring at different dates. See Fig. 7.9 for illustration. A **diagonal spread** involves two options with different strike prices and different expiration dates.

▶ **Exercise 7.4.5** A state contingent claim has a payoff function f such that $f(x) = 0$ for all $x \neq X$ and $\int_{-\infty}^{\infty} f(x)\, dx = 1$. Mathematically, f is called the **Dirac delta function**. Argue that the value of a state contingent claim equals $\partial^2 C / \partial X^2$.

7.4.3 Combination

A **combination** consists of options of different types on the same underlying asset, and they are either all bought or all written. A **straddle** is created by a long call and a long put with the same strike price and expiration date. A straddle is neutral on the direction of price movements and has limited risk. Because it profits from high volatility, a person who buys a straddle is said to be **long volatility** [646]. See Fig. 7.10 for illustration. In contrast, selling a straddle benefits from low volatility with a maximum profit of $C + P$. A **strangle** is identical to a straddle except that the call's strike price is higher than the put's. Figure 7.11 illustrates the profit pattern of a strangle.

A **strip** consists of a long call and two long puts with the same strike price and expiration date. A **strap** consists of a long put and two long calls with the same strike price and expiration date. Their profit patterns are very much like that of a straddle except that they are not symmetrical around the strike price. Hence, although strips and straps also bet on volatile price movements, one direction is deemed more likely than the other.

NOTES

1. Like the Holy Roman Empire, the terms American and European have nothing to do with geography.
2. State contingent claims are also called **Arrow securities** in recognition of Arrow's contribution [836].

CHAPTER EIGHT

Arbitrage in Option Pricing

All general laws are attended with inconveniences, when applied to
particular cases.

David Hume, *"Of the Rise and Progress of the Arts and
Sciences"*

The **no-arbitrage principle** says there should be no free lunch. Simple as it is,
this principle supplies the essential argument for option pricing. After the argument is presented in Section 8.1, several important option pricing relations will be
derived.

8.1 The Arbitrage Argument

A riskless arbitrage opportunity is one that, without any initial investment, generates nonnegative returns under all circumstances and positive returns under some
circumstances. In an efficient market, such opportunities should not exist. This no-
arbitrage principle is behind modern theories of option pricing if not a concept that
unifies all of finance [87, 303]. The related **portfolio dominance principle** says that
portfolio A should be more valuable than portfolio B if A's payoff is at least as good
under all circumstances and better under some circumstances.

A simple corollary of the no-arbitrage principle is that a portfolio yielding a zero
return in every possible scenario must have a zero PV. Any other value would imply
arbitrage opportunities, which one can realize by shorting the portfolio if its value is
positive and buying it if its value is negative. The no-arbitrage principle also justifies
the PV formula $P = \sum_{i=1}^{n} C_i d(i)$ for a security with known cash flow $C_1, C_2, \ldots, C_n$
(recall that $d(i)$ is the price of the i-period zero-coupon bond with \$1 par value). Any
price other than P will lead to riskless gains by means of trading the security and
the zeros. Specifically, if the price P^* is lower than P, we short the zeros that match
the security's n cash flows in both maturity and amount and use P^* of the proceeds
P to buy the security. Because the cash inflows of the security will offset exactly the
obligations of the zeros, a riskless profit of $P - P^*$ dollars has been realized now. See
Fig. 8.1. On the other hand, if the security is priced higher than P, one can realize a
riskless profit by reversing the trades.

Here are two more examples. First, an American option cannot be worth less
than the intrinsic value for, otherwise, one can buy the option, promptly exercise
it, and sell the stock with a profit. Second, a put or a call must have a nonnegative

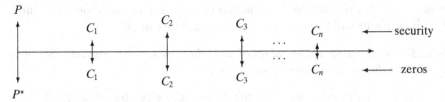

Figure 8.1: Price of fixed cash flow. Consider a security with cash flow $C_1, C_2, \ldots, C_n$ and price P^*. Assemble a portfolio of zero-coupon bonds with matching principals $C_1, C_2, \ldots, C_n$ and maturities $1, 2, \ldots, n$. Let its total price be P. Then $P = P^*$ to preclude arbitrage opportunities.

value for, otherwise, one can buy it for a positive cash flow now and end up with a nonnegative amount at expiration.

▶ **Exercise 8.1.1** Give an arbitrage argument for $d(1) \geq d(2) \geq \cdots$.

▶ **Exercise 8.1.2 (Arbitrage Theorem).** Consider a world with m states and n securities. Denote the payoff of security j in state i by D_{ij}. Let D be the $m \times n$ matrix whose (i, j)th entry is D_{ij}. Formulate necessary conditions for arbitrage freedom.

8.2 Relative Option Prices

We derive arbitrage-free relations that option values must satisfy. These relations hold regardless of the probabilistic model for stock prices. We only assume, among other things, that there are no transactions costs or margin requirements, borrowing and lending are available at the riskless interest rate, interest rates are nonnegative, and there are no arbitrage opportunities. To simplify the presentation, let the current time be time zero. $\text{PV}(x)$ stands for the PV of x dollars at expiration; hence $\text{PV}(x) = x d(\tau)$, where τ is the time to expiration.

The following lemma shows that American option values rise with the time to expiration. This proposition is consistent with the quotations in Fig. 7.4; however, it may not hold for European options.

LEMMA 8.2.1 *An American call (put) with a longer time to expiration cannot be worth less than an otherwise identical call (put) with a shorter time to expiration.*

Proof: We prove the lemma for the call only. Suppose instead that $C_{t_1} > C_{t_2}$, where $t_1 < t_2$. We buy C_{t_2} and sell C_{t_1} to generate a net cash flow of $C_{t_1} - C_{t_2}$ at time zero. Up to the moment when the time to t_2 is τ and the short call either expires or is exercised, the position is worth $C_\tau - \max(S_\tau - X, 0)$. If this value is positive, close out the position with a profit by selling the remaining call. Otherwise, $\max(S_\tau - X, 0) > C_\tau \geq 0$, and the short call is exercised. In this case, we exercise the remaining call and have a net cash flow of zero. In both cases, the total payoff is positive without any initial investment.

LEMMA 8.2.2 *A call (put) option with a higher (lower) strike price cannot be worth more than an otherwise identical call (put) with a lower (higher) strike price.*

Proof: We prove the lemma for the call only. This proposition certainly holds at expiration; hence it is valid for European calls. Let the two strike prices be $X_1 < X_2$. Suppose that $C_{X_1} < C_{X_2}$ instead. We buy the low-priced C_{X_1} and write the high-priced

C_{X_2}, generating a positive return. If the holder of C_{X_2} exercises it before expiration, just exercise the long call to generate a positive cash flow of $X_2 - X_1$.

LEMMA 8.2.3 *The difference in the values of two otherwise identical options cannot be greater than the difference in their strike prices.*

Proof: We consider the calls only. Let the two strike prices be $X_1 < X_2$. Assume that $C_{X_1} - C_{X_2} > X_2 - X_1$ instead. We buy the lower-priced C_{X_2}, write the higher-priced C_{X_1}, generating a positive return, and deposit $X_2 - X_1$ in a riskless bank account.

Suppose that the holder of C_{X_1} exercises the option before expiration. There are two cases. If $C_{X_2} > S - X_1$, then sell C_{X_2} to realize a cash flow of $C_{X_2} - (S - X_1) > 0$. Otherwise, exercise C_{X_2} and realize a cash flow of $X_1 - X_2 < 0$. In both cases, close out the position with the money in the bank to realize a nonnegative net cash flow.

Suppose the holder of C_{X_1} does not exercise the option early. At the expiration date, our cash flow is 0, $X_1 - S < 0$, and $X_1 - X_2 < 0$, respectively, if $S \le X_1$, $X_1 < S < X_2$, and $X_2 \le S$. The net cash flow remains nonnegative after the money in the bank account is added, which is at least $X_2 - X_1$.

LEMMA 8.2.4 *A call is never worth more than the stock price, an American put is never worth more than the strike price, and a European put is never worth more than the present value of the strike price.*

Proof: If the call value exceeded the stock price, a covered call position could earn arbitrage profits. If the put value exceeded the strike price, writing a cash-secured put would earn arbitrage profits. The tighter bound holds for European puts because the cash can earn riskless interest until expiration.

▶ **Exercise 8.2.1** Show that Lemma 8.2.3 can be strengthened for European calls as follows: The difference in the values of two otherwise identical options cannot be greater than the present value of the difference in their strike prices.

▶ **Exercise 8.2.2** Derive a bound similar to that of Lemma 8.2.4 for European puts under negative interest rates. (This case might be relevant when inflation makes the real interest rate negative.)

8.3 Put–Call Parity and Its Consequences

Assume that either the stock pays no cash dividends or that the options are protected so that the option values are insensitive to cash dividends. Note that analysis for options on a non-dividend-paying stock holds for protected options on a dividend-paying stock by definition. Results for protected options therefore are not listed separately.

Consider the portfolio of one short European call, one long European put, one share of stock, and a loan of $\mathrm{PV}(X)$. All options are assumed to carry the same strike price and time to expiration τ. The initial cash flow is therefore $C - P - S + \mathrm{PV}(X)$. At expiration, if the stock price S_τ is at most X, the put will be worth $X - S_\tau$ and the call will expire worthless. On the other hand, if $S_\tau > X$, the call will be worth $S_\tau - X$ and the put will expire worthless. After the loan, now X, is repaid, the net future cash flow is zero in either case. The no-arbitrage principle implies that the initial investment to set up the portfolio must be nil as well. We have proved the following **put–call parity** for European options:

$$C = P + S - \text{PV}(X). \tag{8.1}$$

This identity seems to be due to Castelli in 1877 and thence has been rediscovered many times [156].

The put–call parity implies that there is essentially only one kind of European option because the other can be **replicated** from it in combination with the underlying stock and riskless lending or borrowing. Combinations such as this create **synthetic securities**. For example, rearranging Eq. (8.1) as $S = C - P + \text{PV}(X)$, we see that a stock is equivalent to a portfolio containing a long call, a short put, and lending $\text{PV}(X)$. Other uses of the put–call parity are also possible. Consider $C - P = S - \text{PV}(X)$, which implies that a long call and a short put amount to a long position in stock and borrowing the PV of the strike price – in other words, buying the stock on margin. This might be preferred to taking a levered long position in stock as buying stock on margin is subject to strict margin requirements.

The put–call parity implies that $C = (S - X) + [X - \text{PV}(X)] + P \geq S - X$. Because $C \geq 0$, it follows that $C \geq \max(S - X, 0)$, the intrinsic value. An American call also cannot be worth less than its intrinsic value. Hence we have the following lemma.

LEMMA 8.3.1 *An American call or a European call on a non-dividend-paying stock is never worth less than its intrinsic value.*

A European put may sell below its intrinsic value. In Fig. 7.3, for example, the put value is less than its intrinsic value when the option is deep in the money. This can be verified more formally, as follows. The put–call parity implies that

$$P = (X - S) + [\text{PV}(X) - X + C].$$

As the put goes deeper in the money, the call value drops toward zero and $P \approx (X - S) + \text{PV}(X) - X < X - S$, its intrinsic value under positive interest rates. By the put–call parity, the following lower bound holds for European puts.

LEMMA 8.3.2 *For European puts, $P \geq \max(\text{PV}(X) - S, 0)$.*

Suppose that the PV of the dividends whose ex-dividend dates occur before the expiration date is D. Then the put–call parity becomes

$$C = P + S - D - \text{PV}(X). \tag{8.2}$$

▶ **Exercise 8.3.1** (1) Suppose that the time to expiration is 4 months, the strike price is $95, the call premium is $6, the put premium is $3, the current stock price is $94, and the continuously compounded annual interest rate is 10%. How to earn a riskless arbitrage profit? (2) An options market maker writes calls to a client, then immediately buys puts and the underlying stock. Argue that this portfolio, called **conversion**, should earn a riskless profit.

▶ **Exercise 8.3.2** Strengthen Lemma 8.3.1 to $C \geq \max(S - \text{PV}(X), 0)$.

> **Exercise 8.3.3** In a certain world in which a non-dividend-paying stock's price at any time is known, a European call is worthless if its strike price is higher than the known stock price at expiration. However, Exercise 8.3.2 says that $C \geq S - PV(X)$, which is positive when $S > PV(X)$. Try to resolve the contradiction when $X > S > PV(X)$.

> **Exercise 8.3.4** Prove put–call parity (8.2) for a single dividend of size D^* at some time t_1 before expiration: $C = P + S - PV(X) - D^* d(t_1)$.

> **Exercise 8.3.5** A European **capped call option** is like a European call option except that the payoff is $H - X$ instead of $S - X$ when the terminal stock price S exceeds H. Construct a portfolio of European options with an identical payoff.

> **Exercise 8.3.6** Consider a European-style derivative whose payoff is a piecewise linear function passing through the origin. A security with this payoff is called a **generalized option**. Show that it can be replicated by a portfolio of European calls.

8.4 Early Exercise of American Options

Assume that interest rates are positive in this section. It turns out that it never pays to exercise an American call before expiration if the underlying stock does not pay dividends; selling is better than exercising. Here is the argument. By Exercise 8.3.2, $C \geq \max(S - PV(X), 0)$. If the call is exercised, the value is the smaller $S - X$. The disparity comes from two sources: (1) the loss of the insurance against subsequent stock price once the call is exercised and (2) the time value of money as X is paid on exercise. As a consequence, every pricing relation for European calls holds for American calls when the underlying stock pays no dividends. This somewhat surprising result is due to Merton [660].

THEOREM 8.4.1 *An American call on a non-dividend-paying stock should not be exercised before expiration.*

The above theorem does not mean American calls should be kept until maturity. What it does imply is that when early exercise is being considered, a *better* alternative is to sell it. Early exercise may become optimal for American calls on a dividend-paying stock, however. The reason has to do with the fact that the stock price declines as the stock goes ex-dividend. Surprisingly, an American call should be exercised at only a few dates.

THEOREM 8.4.2 *An American call will be exercised only at expiration or just before an ex-dividend date.*

Proof: We first show that $C > S - X$ at any time other than the expiration date or just before an ex-dividend date. Assume otherwise: $C \leq S - X$. Now, buy the call, short the stock, and lend $Xd(\tau)$, where τ is time to the next dividend date. The initial cash flow is positive because $X > Xd(\tau)$. We subsequently close out the position just before the next ex-dividend date by calling the loan, worth X, and selling the call, worth at least $\max(S_\tau - X, 0)$ by Lemma 8.3.1. The proceeds are sufficient to buy the stock at S_τ. The initial cash flow thus represents an arbitrage profit. Now that the value of a call exceeds its intrinsic value between ex-dividend dates, selling it is better than exercising it.

Unlike American calls on a non-dividend-paying stock, it might be optimal to exercise an American put even if the underlying stock does not pay dividends. Part of the reason lies in the fact that the time value of money now favors early exercise: Exercising a put generates an immediate cash income X. One consequence is that early exercise becomes more profitable as the interest rate increases, other things being equal.

The existence of dividends tends to offset the benefits of early exercise in the case of American puts. Consider a stock that is currently worthless, $S = 0$. If the holder of a put exercises the option, X is tendered. If the holder sells the option, he receives $P \le X$ by Lemma 8.2.4 and keeps the stock. Doing nothing generates no income. If the stock will remain worthless till expiration, exercising the put now is optimal. It is therefore no longer true that we consider only a few points for early exercise of the put. Consequently, concrete results regarding early exercise of American puts are scarcer and weaker.

The put–call parity holds for European options only; for American options,

$$P \ge C + \mathrm{PV}(X) - S \tag{8.3}$$

because an American call has the same value as a European call by Theorem 8.4.1 and an American put is at least as valuable as its European counterpart.

▶ **Exercise 8.4.1** Consider an investor with an American call on a stock currently trading at $45 per share. The option's expiration date is exactly 2 months away, the strike price is $40, and the continuously compounded rate of interest is 8%. Suppose the stock is deemed overpriced and it pays no dividends. Should the option be exercised?

▶ **Exercise 8.4.2** Prove that if at all times before expiration the PV of the interest from the strike price exceeds the PV of future dividends before the expiration date, the call should not be exercised before expiration.

▶ **Exercise 8.4.3** Why is it not optimal to exercise an American put immediately before an ex-dividend date?

▶ **Exercise 8.4.4** Argue that an American put should be exercised when $X - S > \mathrm{PV}(X)$.

▶ **Exercise 8.4.5** Assume that the underlying stock does not pay dividends. Supply arbitrage arguments for the following claims. (1) The value of a call, be it European or American, cannot exceed the price of the underlying stock. (2) The value of a European put is $\mathrm{PV}(X)$ when $S = 0$. (3) The value of an American put is X when $S = 0$.

▶ **Exercise 8.4.6** Prove that American options on a non-dividend-paying stock satisfy $C - P \ge S - X$. (This and relation (8.3) imply that American options on a non-dividend-paying stock satisfy $C - S + X \ge P \ge C - S + \mathrm{PV}(X)$.)

8.5 Convexity of Option Prices

The convexity of option price is stated and proved below.

LEMMA 8.5.1 *For three otherwise identical calls with strike prices* $X_1 < X_2 < X_3$,

$$C_{X_2} \leq \omega C_{X_1} + (1 - \omega) C_{X_3},$$
$$P_{X_2} \leq \omega P_{X_1} + (1 - \omega) P_{X_3}.$$

Here $\omega \equiv (X_3 - X_2)/(X_3 - X_1)$. *(Equivalently,* $X_2 = \omega X_1 + (1 - \omega) X_3$.)

Proof: We prove the lemma for the calls only. Suppose the lemma were wrong. Write C_{X_2}, buy ωC_{X_1}, and buy $(1 - \omega) C_{X_3}$ to generate a positive cash flow now. If the short call is not exercised before expiration, hold the calls until expiration. The cash flow is described by

	$S \leq X_1$	$X_1 < S \leq X_2$	$X_2 < S < X_3$	$X_3 \leq S$
Call written at X_2	0	0	$X_2 - S$	$X_2 - S$
ω calls bought at X_1	0	$\omega(S - X_1)$	$\omega(S - X_1)$	$\omega(S - X_1)$
$1 - \omega$ calls bought at X_3	0	0	0	$(1-\omega)(S - X_3)$
Net cash flow	0	$\omega(S - X_1)$	$\omega(S - X_1) + (X_2 - S)$	0

Because the net cash flows are either nonnegative or positive, there is an arbitrage profit.

Suppose that the short call is exercised early when the stock price is S. If $\omega C_{X_1} + (1 - \omega) C_{X_3} > S - X_2$, sell the long calls to generate a net cash flow of $\omega C_{X_1} + (1 - \omega) C_{X_3} - (S - X_2) > 0$. Otherwise, exercise the long calls and deliver the stock. The net cash flow is $-\omega X_1 - (1 - \omega) X_3 + X_2 = 0$. Again, there is an arbitrage profit.

By Lemma 8.2.3, we know the slope of the call (put) value, when plotted against the strike price, is at most one (minus one, respectively). Lemma 8.5.1 adds that the shape is convex.

EXAMPLE 8.5.2 The prices of the Merck July 30 call, July 35 call, and July 40 call are $15.25, $9.5, and $5.5, respectively, from Fig. 7.4. These prices satisfy the convexity property because $9.5 \times 2 < 15.25 + 5.5$. Look up the prices of the Microsoft April 60 put, April 65 put, and April 70 put. The prices are $0.125, $0.375, and $1.5, respectively, which again satisfy the convexity property.

8.6 The Option Portfolio Property

Stock index options are fundamentally options on a stock portfolio. The American option on the Standard & Poor's 100 (S&P 100) Composite Stock Price Index is currently the most actively traded option contract in the United States [150, 746, 865]. Options on the Standard & Poor's 500 (S&P 500) Composite Stock Price Index are also available. They are European, however. Options on the Dow Jones Industrial Average (DJIA) were introduced in 1997. The underlying index, DJX, is DJIA divided by 100. Other popular stock market indices include the Russell 2000 Index for small company stocks and the broadest based Wilshire 5000 Index. Figure 8.2 tabulates some indices as of February 7, 2000.

As the following theorem shows, an option on a portfolio of stocks is cheaper than a portfolio of options. Hence it is cheaper to hedge against market movements as a whole with index options than with options on individual stocks.

	High	Low	Close	Net Chg.	From Dec. 31	%Chg.
DJ Indus (DJX)	109.71	108.46	109.06	−0.58	−5.91	−5.1
S&P 100 (OEX)	778.01	768.45	774.19	−1.32	−18.64	−2.4
S&P 500 -A.M.(SPX)	1427.23	1413.33	1424.24	−0.13	−45.01	−3.1
Nasdaq 100 (NDX)	3933.75	3858.89	3933.34	+58.97	+225.51	+6.1
NYSE (NYA)	627.03	621.14	623.84	−3.06	−26.46	−4.1
Russell 2000 (RUT)	532.40	525.52	532.39	+6.87	+27.64	+5.5
Major Mkt (XMI) .	1110.00	1096.63	1098.09	−11.64	−67.89	−5.8
Value Line (VLE) . .	1011.29	1003.91	1006.99	−2.13	−18.81	−1.8

Figure 8.2: Stock index quotations. Source: *Wall Street Journal*, February 8, 2000.

THEOREM 8.6.1 *Consider a portfolio of non-dividend-paying assets with weights ω_i. Let C_i denote the price of a European call on asset i with strike price X_i. Then the index call on the portfolio with a strike price $X \equiv \sum_i \omega_i X_i$ has a value of, at most, $\sum_i \omega_i C_i$. The same result holds for European puts as well. All options expire on the same date.*

The theorem in the case of calls follows from the following inequality:

$$\max\left(\sum_{i=1}^{n} \omega_i (S_i - X_i), 0\right) \geq \sum_{i=1}^{n} \max(\omega_i (S_i - X_i), 0),$$

where S_i denote the price of asset i. It is clear that a portfolio of options and an option on a portfolio have the same payoff if the underlying stocks either all finish in the money or out of the money. Their payoffs diverge only when the underlying stocks are not perfectly correlated with each other. The degree of the divergence tends to increase the more the underlying stocks are uncorrelated.

▶ **Exercise 8.6.1** Consider the portfolio of puts and put on the portfolio in Theorem 8.6.1. Because both provide a floor of $\sum_i \omega_i X_i$, why do they not fetch the same price?

Concluding Remarks and Additional Reading

The no-arbitrage principle can be traced to Pascal (1623–1662), philosopher, theologian, and founder of probability and decision theories [409, 410]. In the 1950s, Miller and Modigliani made it a pillar of financial theory [64, 853]. Bounds in this chapter are model free and should be satisfied by any proposed model [236, 346, 470]. Observe that they are all *relative* price bounds. The next chapter presents absolute option prices based on plausible models of stock prices. Justifications for the index options can be found in [236, Section 8.3].

CHAPTER NINE

Option Pricing Models

Life can only be understood backwards; but it must be lived forwards.
Søren Kierkegaard (1813–1855)

Although it is rather easy to price an option at expiration, pricing it at any prior moment is anything but. The no-arbitrage principle, albeit valuable in deriving various bounds, is insufficient to pin down the exact option value without further assumptions on the probabilistic behavior of stock prices. The major task of this chapter is to develop option pricing formulas and algorithms under reasonable models of stock prices. The powerful binomial option pricing model is the focus of this chapter, and the celebrated Black–Scholes formula is derived.

9.1 Introduction

The major obstacle toward an option pricing model is that it seems to depend on the probability distribution of the underlying asset's price and the risk-adjusted interest rate used to discount the option's payoff. Neither factor can be observed directly. After many attempts, some of which were very close to solving the problem, the breakthrough came in 1973 when Black (1938–1995) and Scholes, with help from Merton, published their celebrated option pricing model now universally known as the **Black–Scholes option pricing model** [87].[1] One of the crown jewels of finance theory, this research has far-reaching implications. It also contributed to the success of the CBOE [660]. In 1997 the Nobel Prize in Economic Sciences was awarded to Merton and Scholes for their work on "the valuation of stock options."

The mathematics of the Black–Scholes model is formidable because the price can move to any one of an infinite number of prices in any finite amount of time. The alternative **binomial option pricing model (BOPM)** limits the price movement to two choices in a period, simplifying the mathematics tremendously at some expense of realism. All is not lost, however, because the binomial model converges to the Black–Scholes model as the period length goes to zero. More importantly, the binomial model leads to efficient numerical algorithms for option pricing. The BOPM is the main focus of this chapter.

Throughout this chapter, C denotes the call value, P the put value, X the strike price, S the stock price, and D the dividend amount. Subscripts are used to emphasize or differentiate different times to expiration, stock prices, or strike prices. The

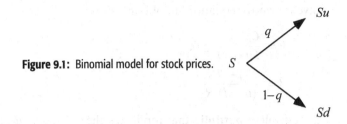

Figure 9.1: Binomial model for stock prices.

symbol PV(x) stands for the PV of x at expiration unless stated otherwise. Let $\hat{r} > 0$ denote the continuously compounded riskless rate per period and $R \equiv e^{\hat{r}}$ its **gross return**.

9.2 The Binomial Option Pricing Model

In the BOPM, time is discrete and measured in periods. The model assumes that if the current stock price is S, it can go to Su with probability q and Sd with probability $1 - q$, where $0 < q < 1$ and $d < u$ (see Fig. 9.1). In fact, $d < R < u$ must hold to rule out arbitrage profits (see Exercise 9.2.1). It turns out that six pieces of information suffice to determine the option value based on arbitrage considerations: S, u, d, X, $\hat{r}$, and the number of periods to expiration.

> **Exercise 9.2.1** Prove that $d < R < u$ must hold to rule out arbitrage profits.

9.2.1 Options on a Non-Dividend-Paying Stock: Single Period

Suppose that the expiration date is only one period from now. Let C_u be the price at time one if the stock price moves to Su and C_d be the price at time one if the stock price moves to Sd. Clearly,

$$C_u = \max(0, Su - X), \qquad C_d = \max(0, Sd - X).$$

See Fig. 9.2 for illustration.

   Now set up a portfolio of h shares of stock and B dollars in riskless bonds. This costs $hS + B$. We call h the **hedge ratio** or **delta**. The value of this portfolio at time one is either $hSu + RB$ or $hSd + RB$. The key step is to choose h and B such that the portfolio replicates the payoff of the call:

$$hSu + RB = C_u,$$
$$hSd + RB = C_d.$$

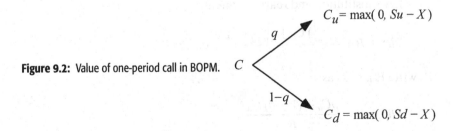

Figure 9.2: Value of one-period call in BOPM.

Solve the above equations to obtain

$$h = \frac{C_u - C_d}{Su - Sd} \geq 0, \tag{9.1}$$

$$B = \frac{uC_d - dC_u}{(u - d) R}. \tag{9.2}$$

An **equivalent portfolio** that replicates the call *synthetically* has been created. An equivalent portfolio is also called a **replicating portfolio** or a **hedging portfolio**. By the no-arbitrage principle, the European call should cost the same as the equivalent portfolio, or $C = hS + B$. As it is easy to verify that

$$uC_d - dC_u = \max(0, Sud - Xu) - \max(0, Sud - Xd) < 0,$$

the equivalent portfolio is a levered long position in stocks.

For American calls, we have to consider immediate exercise. When $hS + B \geq S - X$, the call should not be exercised immediately; so $C = hS + B$. When $hS + B < S - X$, on the other hand, the option should be exercised immediately for we can take the proceeds $S - X$ to buy the equivalent portfolio plus some more bonds; so $C = S - X$. We conclude that $C = \max(hS + B, S - X)$. For non-dividend-paying stocks, early exercise is not optimal by Theorem 8.4.1 (see also Exercise 9.2.6). Again, $C = hS + B$.

Puts can be similarly priced. The delta for the put is $(P_u - P_d)/(Su - Sd) \leq 0$, where $P_u = \max(0, X - Su)$ and $P_d = \max(0, X - Sd)$. The European put is worth $hS + B$, and the American put is worth $\max(hS + B, X - S)$, where $B = \{(uP_d - dP_u)/[(u - d) R]\}$.

▶ **Exercise 9.2.2** Consider two securities, A and B. In a period, security A's price can go from \$100 to either (a) \$160 or (b) \$80, whereas security B's price can move to \$50 in case (a) or \$60 in case (b). Price security B when the interest rate per period is 10%.

9.2.2 Risk-Neutral Valuation

Surprisingly, the option value is independent of q, the probability of an upward movement in price, and hence the expected gross return of the stock, $qSu + (1 - q) Sd$, as well. It therefore does not directly depend on investors' **risk preferences** and will be priced the same regardless of how risk-averse an investor is. The arbitrage argument assumes only that more deterministic wealth is preferred to less. The option value does depend on the sizes of price changes, u and d, the magnitudes of which the investors must agree on.

After substitution and rearrangement,

$$hS + B = \frac{\left(\frac{R-d}{u-d}\right) C_u + \left(\frac{u-R}{u-d}\right) C_d}{R} > 0. \tag{9.3}$$

Rewrite Eq. (9.3) as

$$hS + B = \frac{pC_u + (1 - p) C_d}{R}, \tag{9.4}$$

where

$$p \equiv \frac{R-d}{u-d}. \tag{9.5}$$

As $0 < p < 1$, it may be interpreted as a probability. Under the binomial model, the expected rate of return for the stock is equal to the riskless rate $\hat{r}$ under $q = p$ because $pSu + (1-p)Sd = RS$.

An investor is said to be **risk-neutral** if that person is indifferent between a certain return and an uncertain return with the same expected value. Risk-neutral investors care about only expected returns. The expected rates of return of all securities must be the riskless rate when investors are risk-neutral. For this reason, p is called the **risk-neutral probability**. Because risk preferences and q are not directly involved in pricing options, any risk attitude, including risk neutrality, should give the same result. The value of an option can therefore be interpreted as the expectation of its discounted future payoff in a **risk-neutral economy**. So it turns out that the rate used for discounting the FV is the riskless rate in a risk-neutral economy. Risk-neutral valuation is perhaps the most important tool for the analysis of derivative securities.

We will need the following definitions shortly. Denote the **binomial distribution** with parameters n and p by

$$b(j; n, p) \equiv \binom{n}{j} p^j (1-p)^{n-j} = \frac{n!}{j!\,(n-j)!}\, p^j (1-p)^{n-j}.$$

Recall that $n! = n \times (n-1) \cdots 2 \times 1$ with the convention $0! = 1$. Hence $b(j; n, p)$ is the probability of getting j heads when tossing a coin n times, where p is the probability of getting heads. The probability of getting at least k heads when tossing a coin n times is this **complementary binomial distribution function** with parameters n and p:

$$\Phi(k; n, p) \equiv \sum_{j=k}^{n} b(j; n, p).$$

Because getting fewer than k heads is equivalent to getting at least $n - k + 1$ tails,

$$1 - \Phi(k; n, p) = \Phi(n - k + 1; n, 1 - p). \tag{9.6}$$

▶ **Exercise 9.2.3** Prove that the call's expected gross return in a risk-neutral economy is R.

▶ **Exercise 9.2.4** Suppose that a call costs $hS + B + k$ for some $k \neq 0$ instead of $hS + B$. How does one make an arbitrage profit of M dollars?

▶ **Exercise 9.2.5** The standard arbitrage argument was used in deriving the call value. Use the risk-neutral argument to reach the same value.

9.2.3 Options on a Non-Dividend-Paying Stock: Multiperiod

Consider a call with two periods remaining before expiration. Under the binomial model, the stock can take on three possible prices at time two: Suu, Sud, and Sdd (see Fig. 9.2.3). Note that, at any node, the next two stock prices depend on only the

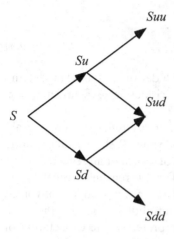

Figure 9.3: Stock prices in two periods. This graph is called a **binomial tree**, although **binomial lattice** is a better term as real tree branches do not merge.

current price, not on the prices of earlier times. This memoryless property is a key feature of an efficient market.[2]

Let C_{uu} be the call's value at time two if the stock price is Suu. Thus

$$C_{uu} = \max(0,\, Suu - X).$$

C_{ud} and C_{dd} can be calculated analogously:

$$C_{ud} = \max(0,\, Sud - X), \qquad C_{dd} = \max(0,\, Sdd - X).$$

See Fig. 9.4 for illustration. We can obtain the call values at time one by applying the same logic as that in Subsection 9.2.2 as follows:

$$C_u = \frac{pC_{uu} + (1-p)\,C_{ud}}{R}, \qquad C_d = \frac{pC_{ud} + (1-p)\,C_{dd}}{R}. \tag{9.7}$$

Deltas can be derived from Eq. (9.1). For example, the delta at C_u is $(C_{uu} - C_{ud})/(Suu - Sud)$.

We now reach the current period. An equivalent portfolio of h shares of stock and \$$B$ riskless bonds can be set up for the call that costs C_u (C_d) if the stock price goes to Su (Sd, respectively). The values of h and B can be derived from Eqs. (9.1) and (9.2).

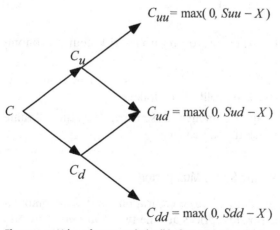

Figure 9.4: Value of a two-period call before expiration.

Because the call will not be exercised at time one even if it is American, $C_u >$ $Su - X$ and $C_d > Sd - X$. Therefore,

$$hS + B = \frac{pC_u + (1-p)\,C_d}{R} > \frac{(pu + (1-p)\,d)\,S - X}{R} = S - \frac{X}{R} > S - X.$$

So the call again will not be exercised at present, and

$$C = hS + B = \frac{pC_u + (1-p)\,C_d}{R}.$$

The above expression calculates C from the two successor nodes C_u and C_d and none beyond. The same computation happens at C_u and C_d, too, as demonstrated in Eqs. (9.7). This recursive procedure is called **backward induction** because it works backward in time [27, 66]. Now,

$$\begin{aligned} C &= \frac{p^2 C_{uu} + 2p(1-p)\,C_{ud} + (1-p)^2 C_{dd}}{R^2} \\ &= \frac{p^2 \times \max\,(0, Su^2 - X) + 2p(1-p) \times \max\,(0, Sud - X) + (1-p)^2 \times \max\,(0, Sd^2 - X)}{R^2}. \end{aligned}$$

The general case is straightforward: Simply carry out the same calculation at every node while moving backward in time. In the n-period case,

$$C = \frac{\sum_{j=0}^{n} \binom{n}{j} p^j (1-p)^{n-j} \times \max(0, Su^j d^{n-j} - X)}{R^n}. \tag{9.8}$$

It says that the value of a call on a non-dividend-paying stock is the expected discounted value of the payoff at expiration in a risk-neutral economy. As this C is the only option value consistent with no arbitrage opportunities, it is called an **arbitrage value**. Note that the option value depends on S, X, $\hat{r}$, u, d, and n. Similarly, the value of a European put is

$$P = \frac{\sum_{j=0}^{n} \binom{n}{j} p^j (1-p)^{n-j} \times \max(0, X - Su^j d^{n-j})}{R^n}.$$

The findings are summarized below.

LEMMA 9.2.1 *The value of a European option equals the expected discounted payoff at expiration in a risk-neutral economy.*

In fact, every derivative can be priced as if the economy were risk-neutral [420]. For a European-style derivative with the terminal payoff function $\mathcal{D}$, its value is

$$e^{-\hat{r}n} E^{\pi}[\mathcal{D}],$$

where E^{π} means that the expectation is taken under the risk-neutral probability.

Because the value of delta changes over time, the maintenance of an equivalent portfolio is a dynamic process. The dynamic maintaining of an equivalent portfolio does not depend on our correctly predicting future stock prices. By construction, the portfolio's value at the end of the current period, which can be either C_u or C_d, is precisely the amount needed to set up the next portfolio. The trading strategy is hence **self-financing** because there is neither injection nor withdrawal of funds throughout and changes in value are due entirely to capital gains.

Let a be the minimum number of upward price moves for the call to finish in the money. Obviously a is the smallest nonnegative integer such that $Su^a d^{n-a} \geq X$, or

$$a = \left\lceil \frac{\ln(X/Sd^n)}{\ln(u/d)} \right\rceil. \tag{9.9}$$

Hence,

$$
\begin{aligned}
C &= \frac{\sum_{j=a}^{n} \binom{n}{j} p^j (1-p)^{n-j} (Su^j d^{n-j} - X)}{R^n} \\
&= S \sum_{j=a}^{n} \binom{n}{j} \frac{(pu)^j [(1-p) d]^{n-j}}{R^n} - \frac{X}{R^n} \sum_{j=a}^{n} \binom{n}{j} p^j (1-p)^{n-j} \\
&= S \sum_{j=a}^{n} b(j; n, pue^{-\hat{r}}) - Xe^{-\hat{r}n} \sum_{j=a}^{n} b(j; n, p) \tag{9.10}
\end{aligned}
$$

The findings are summarized below.

THEOREM 9.2.2 *The value of a European call and the value of a European put are*

$$C = S\Phi(a; n, pue^{-\hat{r}}) - Xe^{-\hat{r}n}\Phi(a; n, p),$$

$$P = Xe^{-\hat{r}n}\Phi(n - a + 1; n, 1 - p) - S\Phi(n - a + 1; n, 1 - pue^{-\hat{r}}),$$

respectively, where $p \equiv (e^{\hat{r}} - d)/(u - d)$ and a is the minimum number of upward price moves for the option to finish in the money.

The option value for the put above can be obtained with the help of the put–call parity and Eq. (9.6). It can also be derived from the same logic as underlies the steps for the call but with $\max(0, S - X)$ replaced with $\max(0, X - S)$ at expiration. It is noteworthy that with the random variable S denoting the stock price at expiration, the options' values are

$$C = S \times \text{Prob}_1[S \geq X] - Xe^{-\hat{r}n} \times \text{Prob}_2[S \geq X], \tag{9.11}$$

$$P = Xe^{-\hat{r}n} \times \text{Prob}_2[S \leq X] - S \times \text{Prob}_1[S \leq X], \tag{9.11'}$$

where Prob_1 uses pu/R and Prob_2 uses p for the probability that the stock price moves from S to Su. Prob_2 expresses the probability that the option will be exercised in a risk-neutral world. Exercise 13.2.12 will offer an interpretation for Prob_1.

A market is **complete** if every derivative security is attainable [420]. There are $n + 1$ possible states of the world at expiration corresponding to the $n + 1$ stock prices $Su^i d^{n-i}, 0 \leq i \leq n$. Consider $n + 1$ state contingent claims, the ith of which pays \$1 at expiration if the stock price is $Su^i d^{n-i}$ and zero otherwise. These claims make the market complete for European-style derivatives that expire at time n. The reason is that a European-style derivative that pays p_i dollars when the stock price finishes at $Su^i d^{n-i}$ can be replicated by a portfolio consisting of p_i units of the ith state contingent claim for $0 \leq i \leq n$. In the case of **continuous trading** in which trading is allowed for each period, two securities suffice to replicate every possible derivative and make the market complete (see Exercise 9.2.10) [289, 434].

The existence of risk-neutral valuation is usually taken to *define* arbitrage freedom in a model in that no self-financing trading strategies can earn arbitrage profits. In fact, the existence of risk-neutral valuation does imply arbitrage freedom for discrete-time models such as the BOPM. The converse proposition, that arbitrage freedom implies the existence of a risk-neutral probability, can be rigorously proved; besides, this probability measure is unique for complete markets. The "equivalence" between arbitrage freedom in a model and the existence of a risk-neutral probability is called the (first) **fundamental theorem of asset pricing** .

▶ **Exercise 9.2.6** Prove that early exercise is not optimal for American calls.

▶ **Exercise 9.2.7** Show that the call's delta is always nonnegative.

▶ **Exercise 9.2.8** Inspect Eq. (9.10) under $u \to d$, that is, zero volatility in stock prices.

▶ **Exercise 9.2.9** Prove the put–call parity for European options under the BOPM.

▶ **Exercise 9.2.10** Assume the BOPM. (1) Show that a state contingent claim that pays \$1 when the stock price reaches $Su^i d^{n-i}$ and \$0 otherwise at time n can be replicated by a portfolio of calls. (2) Argue that continuous trading with bonds and stocks can replicate any state contingent claim.

▶ **Exercise 9.2.11** Consider a single-period binomial model with two risky assets S_1 and S_2 and a riskless bond. In the next step, there are only two states for the risky assets, $(S_1 u_1, S_2 u_2)$ and $(S_1 d_1, S_2 d_2)$. Show that this model does not admit a risk-neutral probability for certain u_1, u_2, d_1, d_2, and R. (Hence it is not arbitrage free.)

A Numerical Example

A non-dividend-paying stock is selling for \$160 per share. From every price S, the stock price can go to either $S \times 1.5$ or $S \times 0.5$. There also exists a riskless bond with a continuously compounded interest rate of 18.232% per period. Consider a European call on this stock with a strike price of \$150 and three periods to expiration. The price movements for the stock price and the call value are shown in Fig. 9.5. The call value is found to be \$85.069 by backward induction. The same value can also be found as the PV of the expected payoff at expiration:

$$\frac{(390 \times 0.343) + (30 \times 0.441)}{(1.2)^3} = 85.069.$$

Observe that the delta value changes with the stock price and time.

Any mispricing leads to arbitrage profits. Suppose that the option is selling for \$90 instead. We sell the call for \$90 and invest \$85.069 in the replicating portfolio with 0.82031 shares of stock as required by delta. To set it up, we need to borrow $(0.82031 \times 160) - 85.069 = 46.1806$ dollars. The fund that remains, $90 - 85.069 = 4.931$ dollars, is the arbitrage profit, as we will see shortly.

> *Time 1.* Suppose that the stock price moves to \$240. The new delta is 0.90625. Buy $0.90625 - 0.82031 = 0.08594$ more shares at the cost of $0.08594 \times 240 = 20.6256$ dollars financed by borrowing. Our debt now totals $20.6256 + (46.1806 \times 1.2) = 76.04232$ dollars.
>
> *Time 2.* Suppose the stock price plunges to \$120. The new delta is 0.25. Sell $0.90625 - 0.25 = 0.65625$ shares for an income of $0.65625 \times 120 = 78.75$ dollars. Use this income to reduce the debt to $(76.04232 \times 1.2) - 78.75 = 12.5$ dollars.

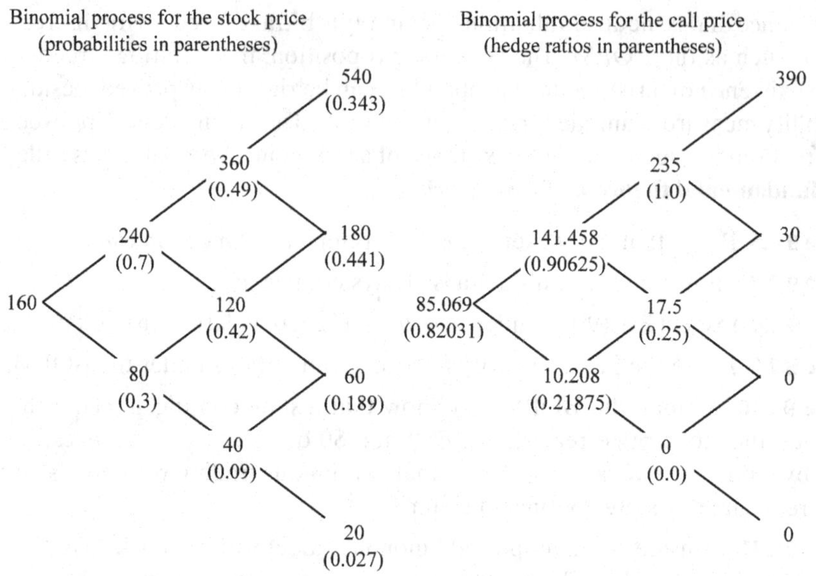

Binomial process for the stock price
(probabilities in parentheses)

Binomial process for the call price
(hedge ratios in parentheses)

Figure 9.5: Stock prices and European call prices. The parameters are $S = 160$, $X = 150$, $n = 3$, $u = 1.5$, $d = 0.5$, $R = e^{0.18232} = 1.2$, $p = (R - d)/(u - d) = 0.7$, $h = (C_u - C_d)/(Su - Sd) = (C_u - C_d)/S$, and $C = [pC_u + (1 - p)C_d]/R = (0.7 \times C_u + 0.3 \times C_d)/1.2$.

Time 3 (The case of rising price). The stock price moves to \$180, and the call we wrote finishes in the money. For a loss of $180 - 150 = 30$ dollars, we close out the position by either buying back the call or buying a share of stock for delivery. Financing this loss with borrowing brings the total debt to $(12.5 \times 1.2) + 30 = 45$ dollars, which we repay by selling the 0.25 shares of stock for $0.25 \times 180 = 45$ dollars.

Time 4 (The case of declining price). The stock price moves to \$60. The call we wrote is worthless. Sell the 0.25 shares of stock for a total of $0.25 \times 60 = 15$ dollars to repay the debt of $12.5 \times 12 = 15$ dollars.

9.2.4 Numerical Algorithms for European Options

Binomial Tree Algorithms

An immediate consequence of the BOPM is the **binomial tree algorithm** that applies backward induction. The algorithm in Fig. 9.6 prices calls on a non-dividend-paying stock with the idea illustrated in Fig. 9.7. This algorithm is easy to analyze. The first loop can be made to take $O(n)$ steps, and the ensuing double loop takes $O(n^2)$ steps. The total running time is therefore quadratic. The memory requirement is also quadratic. To adapt the algorithm in Fig. 9.6 to price European puts, simply replace $\max(0, Su^{n-i}d^i - X)$ in Step 1 with $\max(0, X - Su^{n-i}d^i)$.

The binomial tree algorithm starts from the last period and works its way toward the current period. This suggests that the memory requirement can be reduced if the space is reused. Specifically, replace $C[n+1][n+1]$ in Fig. 9.6 with a one-dimensional array of size $n+1$, $C[n+1]$. Then replace step 1 with

$$C[i] := \max(0, Su^{n-i}d^i - X);$$

Binomial tree algorithm for pricing calls on a non-dividend-paying stock:

input: $S, u, d, X, n, \widehat{r}\, (u > e^{\widehat{r}} > d, \widehat{r} > 0)$;
real $R, p, C[n+1][n+1]$;
integer i, j;
$R := e^{\widehat{r}}$;
$p := (R-d)/(u-d)$;
for $(i = 0$ to $n)$
        1. $C[n][i] := \max(0, Su^{n-i}d^i - X)$;
for $(j = n-1$ down to $0)$
        for $(i = 0$ to $j)$
                2.1. $C[j][i] := (p \times C[j+1][i] +$
                        $(1-p) \times C[j+1][i+1])/R$;
return $C[0][0]$;

Figure 9.6: Binomial tree algorithm for calls on a non-dividend-paying stock. $C[j][i]$ represents the call value at time j if the stock price makes i downward movements out of a total of j movements.

Step 2.1 should now be modified as follows:

$$C[i] := \{ p \times C[i] + (1-p) \times C[i+1] \}/R;$$

Finally, $C[0]$ is returned instead of $C[0][0]$. The memory size is now linear. The one-dimensional array captures the strip in Fig. 9.7 and will be used throughout the book.

We can make further improvements by observing that if $C[j+1][i]$ and $C[j+1]$ $[i+1]$ are both zeros, then $C[j][i]$ is zero, too. We need to let the i loop within the double loop run only from zero to $\min(n-a, j)$ instead of j, where a is defined in Eq. (9.9). This makes the algorithm run in $O(n(n-a))$ steps, which may be substantially smaller than $O(n^2)$ when a is large. The space requirement can be similarly reduced to $O(n-a)$ with a smaller one-dimensional array $C[n-a+1]$. See Fig. 9.8, in which the one-dimensional array implements the strip in that figure.

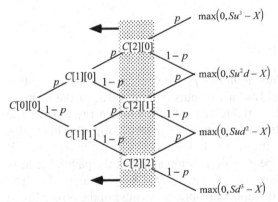

Figure 9.7: Backward induction on binomial trees. Binomial tree algorithms start with terminal values computed in step 1 of the algorithm in Fig. 9.6. They then sweep a strip backward in time to compute values at intermediate nodes until the root is reached.

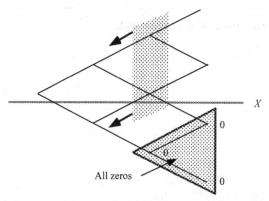

Figure 9.8: Skipping zero-valued nodes to improve efficiency. The stock expires worthless if it finishes below the horizontal line. Zeros at the terminal nodes propagate through the tree depicted here for a call. Such nodes can be skipped by binomial tree algorithms. Note that nodes at the same horizontal level have an identical stock price if $ud = 1$.

➤ **Programming Assignment 9.2.12** Implement the binomial tree algorithms for calls and puts.

An Optimal Algorithm

To reduce the running time to $O(n-a)$ and the memory requirement to $O(1)$, note that

$$b(j; n, p) = \frac{p(n-j+1)}{(1-p)j} b(j-1; n, p).$$

The following program computes $b(j; n, p)$ in $b[j]$ for $a \le j \le n$:

$$b[a] := \binom{n}{a} p^a (1-p)^{n-a};$$

```
for  (j = a + 1 to n)
      b[ j ] := b[ j − 1 ] × p × (n − j + 1)/[ (1 − p) × j ];
```

It clearly runs in $O(n-a)$ steps. With the $b(j; n, p)$ available, risk-neutral valuation formula (9.10) is trivial to compute. The case of puts is similar. As for the memory requirement, we need only a single variable instead of a whole array to store the $b(j; n, p)$s as they are being sequentially computed. The algorithm appears in Fig. 9.9. This linear-time algorithm computes the discounted expected value of $\max(S - X, 0)$. It can be adapted to price any European option. For example, if the payoff function is $\max(\sqrt{S-X}, 0)$, we simply replace $D - X$ with $\sqrt{D-X}$ in the algorithm. The above technique cannot be applied to American options because of the possibility of early exercise. As a result, algorithms for American options usually run in quadratic time instead of in linear time. The performance gap between pricing American and European options seems inherent in general.

➤ **Exercise 9.2.13** Modify the linear-time algorithm in Fig. 9.9 to price puts.

Linear-time, constant-space algorithm for pricing calls on a non-dividend-paying stock:

input: $S, u, d, X, n, \hat{r}$ ($u > e^{\hat{r}} > d$ and $\hat{r} > 0$);
real R, p, b, D, C;
integer j, a;
$a := \lceil \ln(X/Sd^n)/\ln(u/d) \rceil$;
$p := (e^{\hat{r}} - d)/(u - d)$;
$R := e^{n\hat{r}}$;
$b := p^a(1 - p)^{n-a}$; // $b(a; n, p)$ is computed.
$D := S \times u^a d^{n-a}$;
$C := b \times (D - X)/R$;
for $(j = a + 1$ to $n)$ {
      $b := b \times p \times (n - j + 1)/((1 - p) \times j)$;
      $D := D \times u/d$;
      $C := C + b \times (D - X)/R$;
}
return C;

Figure 9.9: Optimal algorithm for European calls on a stock that does not pay dividends. Variable b stores $b(j; n, p)$ for $j = a, a + 1, \ldots, n$, in that order, and variable C accumulates the summands in Eq. (9.10) by adding up $b(j; n, p) \times (Su^j d^{n-j} - X)/e^{n\hat{r}}, j = a, a + 1, \ldots, n$.

➤ **Programming Assignment 9.2.14** Implement the algorithm in Fig. 9.9 and benchmark its speed. Because variables such as b and D can take on extreme values, they should be represented in logarithms to maintain precision.

The Monte Carlo Method

Now is a good time to introduce the **Monte Carlo method**. Equation (9.8) can be interpreted as the expected value of the random variable Z defined by

$$Z = \max(0, Su^j d^{n-j} - X)/R^n \quad \text{with probability } b(j; n, p), \quad 0 \le j \le n.$$

To approximate the expectation, throw n coins, with p being the probability of getting heads, and assign

$$\max(0, Su^j d^{n-j} - X)/R^n$$

to the experiment if it generates j heads. Repeat the procedure m times and take the average. This average clearly has the right expected value $E[Z]$. Furthermore, its variance $\text{Var}[Z]/m$ converges to zero as m increases.

Pricing European options may be too trivial a problem to which to apply the Monte Carlo method. We will see in Section 18.2 that the Monte Carlo method is an invaluable tool in pricing European-style derivative securities and MBSs.

➤ **Programming Assignment 9.2.15** Implement the Monte Carlo method. Observe its convergence rate as the sampling size m increases.

The Recursive Formulation and Its Algorithms

Most derivative pricing problems have a concise and natural *recursive* expression familiar to programmers. Yet a brute-force implementation should be resisted. For example, the recursive implementation of the binomial option pricing problem for the call is as follows:

> Price($S, u, d, X, n, \widehat{r}$)   { // Pricing European calls recursively.
> $p := (e^{\widehat{r}} - d)/(u - d)$;
>   if $[n = 0]$ return max$(S - X, 0)$;
>   else return $[\, p \times$ Price$(Su, u, d, X, n - 1, \widehat{r}) + (1 - p)$
> $\times$ Price$(Sd, u, d, X, n - 1, \widehat{r})\,]/e^{\widehat{r}}$;
> }

If every possible stock price sequence of length n is traced, the algorithm's running time is $O(n2^n)$, which is not practical.

9.3 The Black–Scholes Formula

On the surface, the binomial model suffers from two unrealistic assumptions: (1) The stock price takes on only two values in a period and (2) trading occurs at discrete points in time. These shortcomings are more apparent than real. As the number of periods increases, the stock price ranges over ever-larger numbers of possible values, and trading takes place nearly continuously. What needs to be done is proper calibration of the model parameters so that the model converges to the continuous-time model in the limit.

9.3.1 Distribution of the Rate of Return

Let τ denote the time to expiration of the option measured in years and r be the continuously compounded annual rate. With n periods during the option's life, each period therefore represents a time interval of τ/n. Our job is to adjust the period-based u, d, and interest rate represented by $\widehat{r}$ to match the empirical results as n goes to infinity. Clearly $\widehat{r} = r\tau/n$. As before, let R denote the period gross return $e^{\widehat{r}}$.

We proceed to derive u and d. Under the binomial model, $\ln u$ and $\ln d$ denote the stock's two possible continuously compounded rates of return per period. The rate of return in each period is characterized by the following **Bernoulli random variable**:

$$B = \begin{cases} \ln u, & \text{with probability } q \\ \ln d, & \text{with probability } 1 - q \end{cases}.$$

Let S_τ denote the stock price at expiration. The stock's continuously compounded rate of return, $\ln(S_\tau/S)$, is the sum of n independent Bernoulli random variables above, and

$$\ln \frac{S_\tau}{S} = \ln \frac{Su^j d^{n-j}}{S} = j \ln(u/d) + n \ln d, \tag{9.12}$$

where the stock price makes j upward movements in n periods. Because each upward price movement occurs with probability q, the expected number of upward price movements in n periods is $E[\,j\,] = nq$ with variance

$$\text{Var}[\,j\,] = n[\,q(1-q)^2 + (1-q)(0-q)^2\,] = nq(1-q).$$

We use

$$\widehat{\mu} \equiv \frac{1}{n} E\left[\ln \frac{S_\tau}{S}\right], \qquad \widehat{\sigma}^2 \equiv \frac{1}{n} \operatorname{Var}\left[\ln \frac{S_\tau}{S}\right]$$

to denote, respectively, the expected value and the variance of the period continuously compounded rate of return. From the above,

$$\widehat{\mu} = \frac{E[\,j\,] \times \ln(u/d) + n\ln d}{n} = q\ln(u/d) + \ln d,$$

$$\widehat{\sigma}^2 = \frac{\operatorname{Var}[\,j\,] \times \ln^2(u/d)}{n} = q(1-q)\ln^2(u/d).$$

For the binomial model to converge to the expectation $\mu\tau$ and variance $\sigma^2\tau$ of the stock's true continuously compounded rate of return over τ years, the requirements are

$$n\widehat{\mu} = n(q\ln(u/d) + \ln d) \to \mu\tau, \tag{9.13}$$

$$n\widehat{\sigma}^2 = nq(1-q)\ln^2(u/d) \to \sigma^2\tau. \tag{9.14}$$

We call σ the stock's (annualized) **volatility**. Add $ud = 1$, which makes the nodes at the same horizontal level of the tree have an identical price (review Fig. 9.8). Then the above requirements can be satisfied by

$$u = e^{\sigma\sqrt{\tau/n}}, \qquad d = e^{-\sigma\sqrt{\tau/n}}, \qquad q = \frac{1}{2} + \frac{1}{2}\frac{\mu}{\sigma}\sqrt{\frac{\tau}{n}}. \tag{9.15}$$

(See Exercises 9.3.1 and 9.3.8 for alternative choices of u, d, and q.) With Eqs. (9.15),

$$n\widehat{\mu} = \mu\tau,$$

$$n\widehat{\sigma}^2 = \left[1 - \left(\frac{\mu}{\sigma}\right)^2 \frac{\tau}{n}\right]\sigma^2\tau \to \sigma^2\tau.$$

We remark that the no-arbitrage inequalities $u > R > d$ may not hold under Eqs. (9.15), and the risk-neutral probability may lie outside $[0, 1]$. One solution can be found in Exercise 9.3.1 and another in Subsection 12.4.3. In any case, the problems disappear when n is suitably large.

What emerges as the limiting probabilistic distribution of the continuously compounded rate of return $\ln(S_\tau/S)$? The central limit theorem says that, under certain weak conditions, sums of independent random variables such as $\ln(S_\tau/S)$ converge to the normal distribution, i.e.,

$$\operatorname{Prob}\left[\frac{\ln(S_\tau/S) - n\widehat{\mu}}{\sqrt{n}\,\widehat{\sigma}} \le z\right] \to N(z).$$

A simple condition for the central limit theorem to hold is the **Lyapunov condition** [100],

$$\frac{q\,|\ln u - \widehat{\mu}\,|^3 + (1-q)\,|\ln d - \widehat{\mu}\,|^3}{n\widehat{\sigma}^3} \to 0.$$

After substitutions, the condition becomes

$$\frac{(1-q)^2 + q^2}{n\sqrt{q(1-q)}} \to 0,$$

which is true. So the continuously compounded rate of return approaches the normal distribution with mean $\mu\tau$ and variance $\sigma^2\tau$. As a result, $\ln S_\tau$ approaches the normal distribution with mean $\mu\tau + \ln S$ and variance $\sigma^2\tau$. S_τ thus has a lognormal distribution in the limit. The significance of using the continuously compounded rate is now clear: to make the rate of return normally distributed.

The lognormality of a stock price has several consequences. It implies that the stock price stays positive if it starts positive. Furthermore, although there is no upper bound on the stock price, large increases or decreases are unlikely. Finally, equal movements in the rate of return about the mean are equally likely because of the symmetry of the normal distribution: S_1 and S_2 are equally likely if $S_1/S = S/S_2$.

▶ **Exercise 9.3.1** The price volatility of the binomial model should match that of the actual stock in the limit. As q does not play a direct role in the BOPM, there is more than one way to assign u and d. Suppose we require that $q = 0.5$ instead of $ud = 1$. (1) Show that

$$u = \exp\left[\frac{\mu\tau}{n} + \sigma\sqrt{\frac{\tau}{n}}\right], \qquad d = \exp\left[\frac{\mu\tau}{n} - \sigma\sqrt{\frac{\tau}{n}}\right]$$

satisfy requirements (9.13) and (9.14) as *equalities*. (2) Is it valid to use the probability 0.5 during backward induction under these new assignments?

Comment 9.3.1 Recall that the Monte Carlo method in Subsection 9.2.4 used a biased coin. The scheme in Exercise 9.3.1, in contrast, used a *fair* coin, which may be easier to program. The choice in Eqs. (9.15) nevertheless has the advantage that $ud = 1$, which is often easier to work with algorithmically. Alternative choices of u and d are expected to have only slight, if any, impacts on the convergence of binomial tree algorithms [110].

▶ **Exercise 9.3.2** Show that

$$\frac{E[(S_{\Delta t} - S)/S]}{\Delta t} \to \mu + \frac{\sigma^2}{2}, \tag{9.16}$$

where $\Delta t \equiv \tau/n$.

Comment 9.3.2 Note the distinction between Eq. (9.13) and convergence (9.16). The former says that the annual continuously compounded rate of return over τ years, $\ln(S_\tau/S)/\tau$, has mean μ, whereas the latter says that the instantaneous rate of return, $\lim_{\Delta t \to 0}(S_{\Delta t} - S)/S)/\Delta t$, has a larger mean of $\mu + \sigma^2/2$.

9.3.2 Toward the Black–Scholes Formula

We now take the final steps toward the Black–Scholes formula as $n \to \infty$ and q equals the risk-neutral probability $p \equiv (e^{r\tau/n} - d)/(u - d)$.

LEMMA 9.3.3 *The continuously compounded rate of return* $\ln(S_\tau/S)$ *approaches the normal distribution with mean* $(r - \sigma^2/2)\tau$ *and variance* $\sigma^2\tau$ *in a risk-neutral economy.*

Proof: Applying $e^y = 1 + y + (y^2/2!) + \cdots$ to p, we obtain

$$p \to \frac{1}{2} + \frac{1}{2} \frac{r - \sigma^2/2}{\sigma} \sqrt{\frac{\tau}{n}}. \tag{9.17}$$

So the q in Eq. (9.15) implies that $\mu = r - \sigma^2/2$ and

$$n\widehat{\mu} = \left(r - \frac{\sigma^2}{2}\right)\tau$$

$$n\widehat{\sigma}^2 = \left[1 - \left(\frac{r - \sigma^2/2}{\sigma}\right)^2 \frac{\tau}{n}\right]\sigma^2\tau \to \sigma^2\tau.$$

Because

$$\frac{(1-p)^2 + p^2}{n\sqrt{p(1-p)}} \to 0,$$

the Lyapunov condition is satisfied and the central limit theorem is applicable.

Lemma 9.3.3 and Eqs. (6.11) imply that the expected stock price at expiration in a risk-neutral economy is $Se^{r\tau}$. The stock's expected annual rate of return is thus the riskless rate r.

THEOREM 9.3.4 (The Black–Scholes Formula):

$$C = SN(x) - Xe^{-r\tau}N(x - \sigma\sqrt{\tau}),$$

$$P = Xe^{-r\tau}N(-x + \sigma\sqrt{\tau}) - SN(-x),$$

where

$$x \equiv \frac{\ln(S/X) + \left(r + \sigma^2/2\right)\tau}{\sigma\sqrt{\tau}}.$$

Proof: As the put–call parity can be used to prove the formula for a European put from that for a call, we prove the formula for the call only. The binomial option pricing formula in Theorem 9.2.2 is similar to the Black–Scholes formula. Clearly, we are done if

$$\Phi(a; n, pue^{-\widehat{r}}) \to N(x), \qquad \Phi(a; n, p) \to N(x - \sigma\sqrt{\tau}). \tag{9.18}$$

We prove only $\Phi(a; n, p) \to N(x - \sigma\sqrt{\tau})$; the other part can be verified analogously.

Recall that $\Phi(a; n, p)$ is the probability of at least a successes in n independent trials with success probability p for each trial. Let j denote the number of successes (upward price movements) in n such trials. This random variable, a sum of n Bernoulli variables, has mean np and variance $np(1-p)$ and satisfies

$$1 - \Phi(a; n, p) = \text{Prob}[\, j \le a - 1\,] = \text{Prob}\left[\frac{j - np}{\sqrt{np(1-p)}} \le \frac{a - 1 - np}{\sqrt{np(1-p)}}\right]. \tag{9.19}$$

It is easy to verify that

$$\frac{j - np}{\sqrt{np(1-p)}} = \frac{\ln(S_\tau/S) - n\widehat{\mu}}{\sqrt{n}\,\widehat{\sigma}}.$$

Now,

$$a - 1 = \frac{\ln(X/Sd^n)}{\ln(u/d)} - \epsilon$$

for some $0 < \epsilon \leq 1$. Combine the preceding equality with the definitions for $\widehat{\mu}$ and $\widehat{\sigma}$ to obtain

$$\frac{a - 1 - np}{\sqrt{np(1-p)}} = \frac{\ln(X/S) - n\widehat{\mu} - \epsilon \ln(u/d)}{\sqrt{n}\,\widehat{\sigma}}.$$

So Eq. (9.19) becomes

$$1 - \Phi(a; n, p) = \text{Prob}\left[\frac{\ln(S_\tau/S) - n\widehat{\mu}}{\sqrt{n}\,\widehat{\sigma}} \leq \frac{\ln(X/S) - n\widehat{\mu} - \epsilon \ln(u/d)}{\sqrt{n}\,\widehat{\sigma}} \right].$$

Because $\ln(u/d) = 2\sigma \sqrt{\tau/n} \to 0$,

$$\frac{\ln(X/S) - n\widehat{\mu} - \epsilon \ln(u/d)}{\sqrt{n}\,\widehat{\sigma}} \to z \equiv \frac{\ln(X/S) - \tau\left(r - \sigma^2/2\right)}{\sigma\sqrt{\tau}}.$$

Hence $1 - \Phi(a; n, p) \to N(z)$, which implies that

$$\Phi(a; n, p) \to N(-z) = N\left(\frac{\ln(S/X) + r\tau}{\sigma\sqrt{\tau}} - \frac{1}{2}\sigma\sqrt{\tau} \right) = N(x - \sigma\sqrt{\tau}),$$

as desired.

We plot the call and put values as a function of the current stock price, time to expiration, volatility, and interest rate in Fig. 9.10. Note particularly that the option value for at-the-money options is essentially a linear function of volatility.

▶ **Exercise 9.3.3** Verify the following with the Black–Scholes formula and give heuristic arguments as to why they should hold without invoking the formula. (1) $C \approx S - Xe^{-r\tau}$ if $S \gg X$. (2) $C \to S$ as $\tau \to \infty$. (3) $C \to 0$ as $\sigma \to 0$ if $S < Xe^{-r\tau}$. (4) $C \to S - Xe^{-r\tau}$ as $\sigma \to 0$ if $S > Xe^{-r\tau}$. (5) $C \to S$ as $r \to \infty$.

▶ **Exercise 9.3.4** Verify convergence (9.17).

▶ **Exercise 9.3.5** A **binary call** pays off $1 if the underlying asset finishes above the strike price and nothing otherwise.[3] Show that its price equals $e^{-r\tau} N(x - \sigma\sqrt{\tau})$.

▶ **Exercise 9.3.6** Prove $\partial^2 P / \partial X^2 = \partial^2 C / \partial X^2$ (see Fig. 9.11 for illustration).

▶ **Exercise 9.3.7** Derive Theorem 9.3.4 from Lemma 9.3.3 and Exercise 6.1.6.

Tabulating Option Values

Rewrite the Black–Scholes formula for the European call as follows:

$$C = Xe^{-r\tau}\left[\frac{S}{Xe^{-r\tau}} N(x) - N(x - \sigma\sqrt{\tau}) \right],$$

where

$$x \equiv \frac{\ln(S/(Xe^{-r\tau}))}{\sigma\sqrt{\tau}} + \frac{\sigma\sqrt{\tau}}{2}.$$

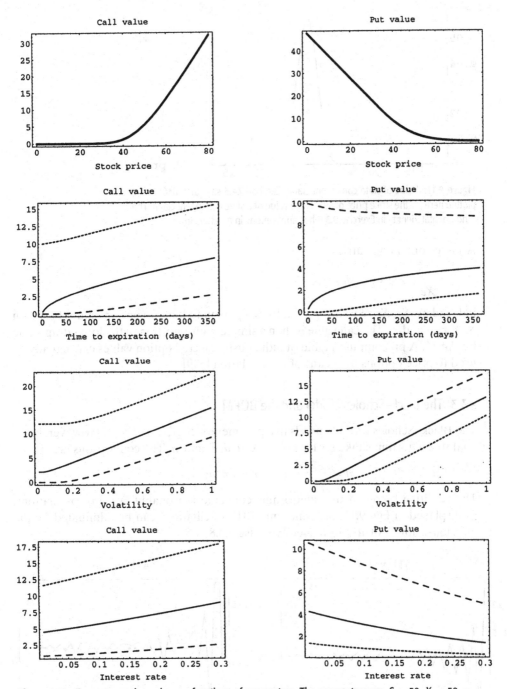

Figure 9.10: European option values as functions of parameters. The parameters are $S = 50$, $X = 50$, $\sigma = 0.3$, $\tau = 201$ (days), and $r = 8\%$. When three curves are plotted together, the dashed curve uses $S = 40$ (out-of-the-money call or in-the-money put), the solid curve uses $S = 50$ (at the money), and the dotted curve uses $S = 60$ (in-the-money call or out-of-the-money put).

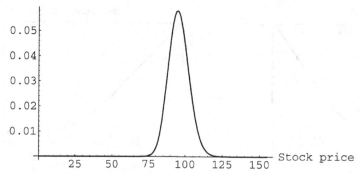

Figure 9.11: Value of state contingent claim. Exercise 7.4.5 says that $\partial^2 C / \partial X^2$, plotted here for the strike price of $95, is the value of a state contingent claim. The fundamental identity in Exercise 9.3.6 has applications in asset pricing.

A table containing entries

$$\frac{S}{Xe^{-r\tau}}\, N(x) - N(x - \sigma \sqrt{\tau})$$

indexed by $S/(Xe^{-r\tau})$ and $\sigma \sqrt{\tau}$ allows a person to look up option values based on S, X, r, τ, and σ. The call value is then a simple multiplication of the looked-up value by $Xe^{-r\tau}$. A precomputed table of judiciously selected option values can actually be used to price options by means of interpolation [529].

9.3.3 The Black–Scholes Model and the BOPM

The Black–Scholes formula needs five parameters: S, X, σ, τ, and r. However, binomial tree algorithms take six inputs: $S, X, u, d, \widehat{r}$, and n. The connections are

$$u = e^{\sigma \sqrt{\tau/n}}, \qquad d = e^{-\sigma \sqrt{\tau/n}}, \qquad \widehat{r} = r\tau/n.$$

The resulting binomial tree algorithms converge reasonably fast, but oscillations, as displayed in Fig. 9.12, are inherent [704]. Oscillations can be eliminated by the judicious choices of u and d (see Exercise 9.3.8).

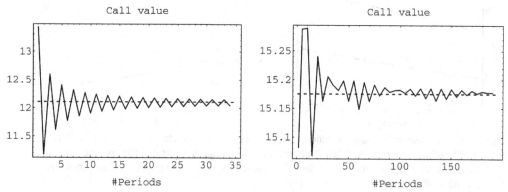

Figure 9.12: Convergence of binomial tree algorithms. Plotted are the European call values as computed by the binomial tree algorithm against the number of time partitions, n. The parameters used are $S = 100$, $X = 100$ (left) and 95 (right), $r = 8\%$, $\sigma = 0.2$, and $\tau = 1$. The analytical values, 12.1058 (left) and 15.1749 (right), are displayed for reference.

EXAMPLE 9.3.5 Consider a 3-month option when the interest rate is 8% per annum and the volatility is 30% per annum. This means that $\tau = 0.25, r = 0.08$, and $\sigma = 0.3$. If the binomial tree algorithm uses $n = 5$, it should use $u = e^{0.3\sqrt{0.25/5}} = 1.0694$ and $d = e^{-0.3\sqrt{0.25/5}} = 0.9351$.

▶ **Exercise 9.3.8** Here is yet another way to assign u and d:

$$u = e^{\sigma\sqrt{\tau/n}+(1/n)\ln(X/S)}, \qquad d = e^{-\sigma\sqrt{\tau/n}+(1/n)\ln(X/S)}, \qquad q = \frac{e^{r\tau/n} - d}{u - d}.$$

(1) Show that it works. (2) What is special about this choice?

9.4 Using the Black–Scholes Formula

9.4.1 Interest Rate

The riskless rate r should be the spot rate with a maturity near the option's expiration date (in practice, the specific rate depends on the investor [228]). The choice can be justified as follows. Let r_i denote the continuously compounded one-period interest rate measured in periods for period i. The bond maturing at the option's expiration date is worth $\exp[-\sum_{i=1}^{n} r_i]$ per dollar of face value. This implies that $r\tau = \sum_{i=1}^{n} r_i$. Hence a single discount bond price with maturity at time n (equivalently, the n-period spot rate) encompasses all the information needed for interest rates. In the limit, $\sum_{i=0}^{n-1} r_i \to \int_0^\tau r(t)\,dt$, where $r(t)$ is the short rate at time t. The relevant annualized interest rate is thus $r = (1/\tau)\int_0^\tau r(t)\,dt$.

Interest rate uncertainty may not be very critical for options with lives under 1 year. Plots in Fig. 9.10 also suggest that small changes in interest rates, other things being equal, do not move the option value significantly.

9.4.2 Estimating the Volatility from Historical Data

The volatility parameter σ is the sole parameter not directly observable and has to be estimated. The Black–Scholes formula assumes that stock prices are lognormally distributed. In other words, the n continuously compounded rates of return per period,

$$u_i \equiv \ln\frac{S_i}{S_{i-1}}, \quad i = 1, 2, \ldots, n,$$

are independent samples from a normal distribution with mean $\mu\tau/n$ and variance $\sigma^2\tau/n$, where S_i denotes the stock price at time i. A good estimate of the standard deviation of the per-period rate of return is

$$s \equiv \sqrt{\frac{\sum_{i=1}^{n}(u_i - \bar{u})^2}{n - 1}},$$

where $\bar{u} \equiv (1/n)\sum_{i=1}^{n} u_i = (1/n)\ln(S_n/S_0)$. The preceding estimator may be biased in practice, however, notably because of the bid–ask spreads and the discreteness of stock prices [48, 201]. Estimators that utilize high and low prices can be superior theoretically in terms of lower variance [374]. We note that $\bar{u}$ and $s^2(n-1)/n$ are the ML estimators of μ and σ^2, respectively (see Section 20.1).

The **simple rate of return**, $(S_i - S_{i-1})/S_{i-1}$, is sometimes used in place of u_i to avoid logarithms. This is not entirely correct because $\ln x \approx x - 1$ only when x is small, and a small error here can mean huge differences in the option value [147].

If a period contains an ex-dividend date, its sample rate of return should be modified to

$$u_i = \ln \frac{S_i + D}{S_{i-1}},$$

where D is the amount of the dividend. If an n-for-m split occurs in a period, the sample rate of return should be modified to

$$u_i = \ln \frac{n S_i}{m S_{i-1}}.$$

Because the standard deviation of the rate of return equals $\sigma \sqrt{\tau/n}$, the estimate for σ is $s/\sqrt{\tau/n}$. This value is called **historical volatility**. Empirical evidence suggests that days when stocks were not traded should be excluded from the calculation. Some even count only trading days in the time to expiration τ [514].

Like interest rate, volatility is allowed to change over time as long as it is predictable. In the context of the binomial model, this means that u and d now depend on time. The variance of $\ln(S_\tau/S)$ is now $\int_0^\tau \sigma^2(t)\, dt$ rather than $\sigma^2 \tau$, and the volatility becomes $[\int_0^\tau \sigma^2(t)\, dt/\tau]^{1/2}$. A word of caution here: There is evidence suggesting that volatility is stochastic (see Section 15.5).

9.4.3 Implied Volatility

The Black–Scholes formula can be used to compute the market's opinion of the volatility. This is achieved by the solution of σ given the option price, S, X, τ, and r with the numerical methods in Subsection 3.4.3. The volatility thus obtained is called the **implied volatility** – the volatility implied by the market price of the option. Volatility numbers are often stored in a table indexed by maturities and strike prices [470, 482].

Implied volatility is often preferred to historical volatility in practice, but it is not perfect. Options written on the same underlying asset usually do not produce the same implied volatility. A typical pattern is a **"smile"** in relation to the strike price: The implied volatility is lowest for at-the-money options and becomes higher the further the option is in or out of the money [150]. This pattern is especially strong for short-term options [44] and cannot be accounted for by the early exercise feature of American options [97]. To address this issue, volatilities are often combined to produce a composite implied volatility. This practice is not sound theoretically. In fact, the existence of different implied volatilities for options on the same underlying asset shows that the Black–Scholes option pricing model cannot be literally true. Section 15.5 will survey approaches that try to explain the smile.

▶ **Exercise 9.4.1** Calculating the implied volatility from the option price can be facilitated if the option price is a monotonic function of volatility. Show that this is true of the Black–Scholes formula.

▶ **Exercise 9.4.2** Solving for the implied volatility of American options as if they were European overestimates the true volatility. Discuss.

▶ **Exercise 9.4.3 (Implied Binomial Tree).** Suppose that we are given m different European options prices, their identical maturity, their strike prices, their underlying asset's current price, the underlying asset's σ, and the riskless rate. (1) What should n be? (2) Assume that the path probabilities for all paths reaching the same node are equal. How do we compute the (implied) branching probabilities at each node of the binomial tree so that these options are all priced correctly?

▶ **Programming Assignment 9.4.4** Write a program to compute the implied volatility of American options.

9.5 American Puts on a Non-Dividend-Paying Stock

Early exercise has to be considered when pricing American puts. Because the person who exercises a put receives the strike price and earns the time value of money, there is incentive for early exercise. On the other hand, early exercise may render the put holder worse off if the stock subsequently increases in value.

The binomial tree algorithm starts with the terminal payoffs $\max(0, X - Su^j d^{n-j})$ and applies backward induction. At each intermediate node, it checks for early exercise by comparing the payoff if exercised with continuation. The complete quadratic-time algorithm appears in Fig. 9.13. Figure 9.14 compares an American put with its European counterpart.

Let us go through a numerical example. Assume that $S = 160$, $X = 130$, $n = 3$, $u = 1.5$, $d = 0.5$, and $R = e^{0.18232} = 1.2$. We can verify that $p = (R - d)/(u - d) = 0.7$, $h = (P_u - P_d)/S(u - d) = (P_u - P_d)/S$, and $P = [pP_u + (1 - p) P_d]/R = (0.7 \times P_u + 0.3 \times P_d)/1.2$. Consider node A in Fig. 9.15. The continuation value is

$$\frac{(0.7 \times 0) + (0.3 \times 70)}{1.2} = 17.5,$$

greater than the intrinsic value $130 - 120 = 10$. Hence the option should not be exercised even if it is in the money and the put value is 17.5. As for node B, the continuation value is

$$\frac{(0.7 \times 70) + (0.3 \times 110)}{1.2} = 68.33,$$

lower than the intrinsic value $130 - 40 = 90$. The option should be exercised, and the put value is 90.

Binomial tree algorithm for pricing American puts on a non-dividend-paying stock:

input: $S, u, d, X, n, \hat{r}$ $(u > e^{\hat{r}} > d$ and $\hat{r} > 0)$;
real $R, p, P[n+1]$;
integer i, j;
$R := e^{\hat{r}}$;
$p := (R - d)/(u - d)$;
for $(i = 0$ to $n)$ $\{ P[i] := \max(0, X - Su^{n-i}d^i); \}$
for $(j = n - 1$ down to $0)$
        for $(i = 0$ to $j)$
            $P[i] := \max((p \times P[i] + (1 - p) \times P[i+1])/R, X - Su^{j-i}d^i)$;
return $P[0]$;

Figure 9.13: Binomial tree algorithm for American puts on a non-dividend-paying stock.

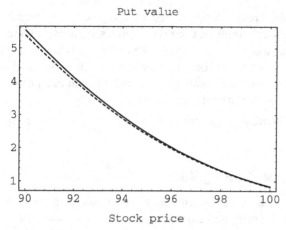

Figure 9.14: American put vs. European put. Plotted is the American put price at 1 month before expiration. The strike price is $95, and the riskless rate is 8%. The volatility of the stock is assumed to be 0.25. The corresponding European put is also plotted (dotted curve) for comparison.

➤ **Programming Assignment 9.5.1** Implement the algorithm in Fig. 9.13 for American puts.

9.6 Options on a Stock that Pays Dividends

9.6.1 European Options on a Stock that Pays a Known Dividend Yield

The BOPM remains valid if dividends are predictable. A known **dividend yield** means that the dividend income forms a constant percentage of the stock price. For a dividend yield of δ, the stock pays out $S\delta$ on each ex-dividend date. Therefore the

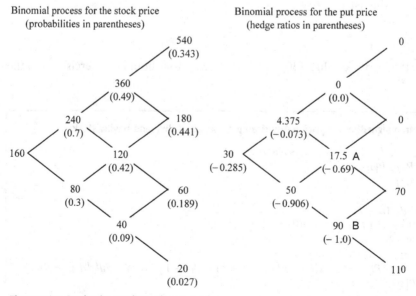

Figure 9.15: Stock prices and American put prices.

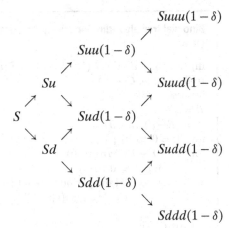

Figure 9.16: Binomial model for a stock that pays a known dividend yield. The ex-dividend date occurs in the second period.

stock price goes from S to $Su(1-\delta)$ or $Sd(1-\delta)$ in a period that includes an ex-dividend date. If a period does not contain an ex-dividend date, the binomial model is unchanged. See Fig. 9.16 for illustration.

For European options, only the number of ex-dividend dates matters, not their specific dates. This can be seen as follows. Let m denote the number of ex-dividend dates before expiration. The stock price at expiration is then of the form $(1-\delta)^m Su^j d^{n-j}$, independent of the timing of the dividends. Consequently we can use binomial tree algorithms for options on a non-dividend-paying stock but with the current stock price S replaced with $(1-\delta)^m S$. Pricing can thus be achieved in linear time and constant space.

▶ **Exercise 9.6.1** Argue that the value of a European option under the case of known dividend yields equals $(1-\delta)^m$ European option on a non-dividend-paying stock with the strike price $(1-\delta)^{-m} X$.

9.6.2 American Options on a Stock that Pays a Known Dividend Yield

The algorithm for American calls applies backward induction and pays attention to each ex-dividend date (see Fig. 9.17). It can be easily modified to value American puts. Early exercise might be optimal when the period contains an ex-dividend date. Suppose that $Sd(1-\delta) > X$. Then $C_u = Su(1-\delta) - X$ and $C_d = Sd(1-\delta) - X$. Therefore

$$\frac{pC_u + (1-p)C_d}{R} = (1-\delta)S - \frac{X}{R},$$

which is exceeded by $S - X$ for sufficiently large S. This proves that early exercise before expiration might be optimal.

▶ **Exercise 9.6.2** Start with an American call on a stock that pays d dividends. Consider a package of $d+1$ European calls with the same strike price as the American call such that there is a European call expiring just before each ex-dividend date and a European call expiring at the same date as the American call. In light of Theorem 8.4.2, is the American call equivalent to this package of European calls?

Binomial tree algorithm for pricing American calls on a stock that pays a known dividend yield:

input: $S, u, d, X, n, \delta\,(1 > \delta > 0), m, \hat{r}\,(u > e^{\hat{r}} > d$ and $\hat{r} > 0)$;
real $R, p, C[n+1]$;
integer i, j;
$R := e^{\hat{r}}$;
$p := (R - d)/(u - d)$;
for $(i = 0$ to $n)$ { $C[i] := \max(0, Su^{n-i}d^i(1-\delta)^m - X)$; }
for $(j = n - 1$ down to $0)$
          for $(i = 0$ to $j)$ {
                    if [the period $(j, j+1)$ contains an ex-dividend date] $m := m - 1$;
                    $C[i] := \max((p \times C[i] + (1-p) \times C[i+1])/R,\ Su^{j-i}d^i(1-\delta)^m - X)$;
          }
return $C[0]$;

Figure 9.17: Binomial tree algorithm for American calls on a stock paying a dividend yield. Recall that m initially stores the total number of ex-dividend dates at or before expiration.

➤ **Programming Assignment 9.6.3** Implement binomial tree algorithms for American options on a stock that pays a known dividend yield.

9.6.3 Options on a Stock that Pays Known Dividends

Although companies may try to maintain a constant dividend yield in the long run, a constant dividend is satisfactory in the short run. Unlike constant dividend yields, constant dividends introduce complications. Use D to denote the amount of the dividend. Suppose an ex-dividend date falls in the first period. At the end of that period, the possible stock prices are $Su - D$ and $Sd - D$. Follow the stock price one more period. It is clear that the number of possible stock prices is not three but four: $(Su - D)u$, $(Su - D)d$, $(Sd - D)u$, and $(Sd - D)d$. In other words, the binomial tree no longer combines (see Fig. 9.18). The fundamental reason is that timing of the dividends now becomes important; for example, $(Su - D)u$ is different from $Suu - D$. It is not hard to see that m ex-dividend dates will give rise to at least 2^m terminal nodes. The known dividends case thus consumes tremendous computation time and memory.

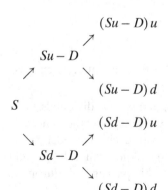

Figure 9.18: Binomial model for a stock that pays known dividends. The amount of the dividend is D, and the ex-dividend date occurs in the first period.

A Simplifying Assumption

One way to adjust for dividends is to use the Black–Scholes formula with the stock price reduced by the present value of the anticipated dividends. This procedure is valid if the stock price can be decomposed into a sum of two components, a riskless one paying known dividends during the life of the option and a risky one. The riskless component at any time is the PV of future dividends during the life of the option. The Black–Scholes formula is then applicable with S equal to the risky component of the stock price and σ equal to the volatility of the process followed by the risky component. The stock price, between two adjacent ex-dividend dates, follows the same lognormal distribution. This means that the Black–Scholes formula can be used provided the stock price is reduced by the PV of future dividends during the life of the option. We note that uncertainty about dividends is rarely important for options lasting less than 1 year.

With the above assumption, we can start with the current stock price minus the PV of future dividends before the expiration date and develop the binomial tree for the new stock price as if there were no dividends. Then we add to each stock price on the tree the PV of all future dividends before expiration. European option prices can be computed as before on this tree of stock prices. As for American options, the same procedure applies except for the need to test for early exercises at each node.

➤ **Programming Assignment 9.6.4** Implement the ideas described in this subsection.

9.6.4 Options on a Stock that Pays a Continuous Dividend Yield

In the **continuous-payout model**, dividends are paid continuously. Such a model approximates a broad-based stock market portfolio in which some company will pay a dividend nearly every day. The payment of a **continuous dividend yield** at rate q reduces the growth rate of the stock price by q. In other words, a stock that grows from S to S_τ with a continuous dividend yield of q would grow from S to $S_\tau e^{q\tau}$ without the dividends. Hence a European option on a stock with price S paying a continuous dividend yield of q has the same value as a European option on a stock with price $Se^{-q\tau}$ that pays no dividends. The Black–Scholes formulas thus hold, with S replaced with $Se^{-q\tau}$:

$$C = Se^{-q\tau} N(x) - Xe^{-r\tau} N(x - \sigma\sqrt{\tau}), \tag{9.20}$$

$$P = Xe^{-r\tau} N(-x + \sigma\sqrt{\tau}) - Se^{-q\tau} N(-x), \tag{9.20'}$$

where

$$x \equiv \frac{\ln(S/X) + \left(r - q + \sigma^2/2\right)\tau}{\sigma\sqrt{\tau}}.$$

Formulas (9.20) and (9.20'), which are due to Merton [660], remain valid even if the dividend yield is not a constant as long as it is predictable, in which case q is replaced with the average annualized dividend yield during the life of the option [470, 746].

To run binomial tree algorithms, pick the risk-neutral probability as

$$\frac{e^{(r-q)\Delta t} - d}{u - d}, \tag{9.21}$$

where $\Delta t \equiv \tau/n$. The quick reason is that the stock price grows at an expected rate of $r - q$ in a risk-neutral economy. Note that the u and d in Eqs. (9.15) now stand for stock price movements as if there were no dividends Other than the change in probability (9.21), binomial tree algorithms are identical to the no-dividend case.

▶ **Exercise 9.6.5** Prove that the put–call parity becomes $C = P + Se^{-q\tau} - \mathrm{PV}(X)$ under the continuous-payout model.

▶ **Exercise 9.6.6** Derive probability (9.21) rigorously by an arbitrage argument.

▶ **Exercise 9.6.7** (1) Someone argues that we should use $[(e^{r\Delta t} - d)/(u - d)]$ as the risk-neutral probability thus: Because the option value is independent of the stock's expected return $\mu - q$, it can be replaced with r. Show him the mistakes. (2) Suppose that we are asked to use the original risk-neutral probability $[(e^{r\Delta t} - d)/(u - d)]$. Describe the needed changes in the binomial tree algorithm.

▶ **Exercise 9.6.8** Give an example whereby the use of risk-neutral probability (9.21) makes early exercise for American calls optimal.

▶ **Programming Assignment 9.6.9** Implement the binomial tree algorithms for American options on a stock that pays a continuous dividend yield.

9.7 Traversing the Tree Diagonally

Can the standard quadratic-time backward-induction algorithm for American options be improved? Here an algorithm, which is due to Curran, is sketched that usually skips many nodes, saving time in the process [242]. Although only American puts are considered in what follows, the parity result in Exercise 9.7.1 can be used to price American calls as well.

Figure 9.19 mentions two properties in connection with the propagation of early-exercise nodes and non-early-exercise nodes during backward induction. The first property says that a node is an early-exercise node if both its successor nodes are exercised early. A terminal node that is in the money is considered an early-exercise node for convenience. The second property says if a node is a non-early-exercise node, then all the earlier nodes at the same horizontal level are also non-early-exercise nodes. An early-exercise node, once identified, is trivial to evaluate; it is just the difference of the strike price and the stock price. A non-early-exercise node, however, must be evaluated by backward induction.

Curran's algorithm adopts a nonconventional way of traversing the tree, as shown in Fig. 9.20. Evaluation at each node is the same as backward induction

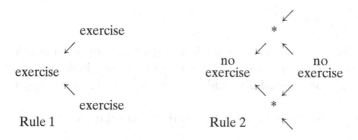

Figure 9.19: Two exercise rules. Rule 2 requires that $ud = 1$.

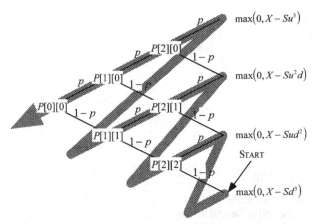

Figure 9.20: The diagonal method for binomial tree algorithms. In contrast to the vertical sweeps across time in the standard backward-induction algorithm such as Fig. 9.7, the new method traces the diagonals in the sequence shown.

before. Note that a node under evaluation always has both of its successors evaluated earlier.

Nothing would be gained if the whole tree needs to be explored. Enter the stopping rule. The process stops when a diagonal D consisting *entirely* of non-early-exercise nodes has been encountered. By Rule 2 of Fig. 9.19, all early-exercise nodes have been accounted for. When the algorithm finds an early-exercise node in traversing a diagonal, it can stop immediately and move on to the next diagonal. This is because, by Rule 1 and the sequence by which the nodes on the diagonals are traversed, the rest of the nodes on the current diagonal must all be early-exercise nodes, hence computable on the fly when needed. Also by Rule 1, the traversal can start from the zero-valued terminal node just above the strike price. Clearly $Su^a d^{n-a}$ is the stock price at that node, where a is defined in Eq. (9.9). The upper triangle above the strike price can be skipped because its nodes are all zero valued. See Fig. 9.21 for the overall strategy. The diagonal method typically skips many nodes (see Fig. 9.22).

The option value equals the sum of the discounted option values of the nodes on D, each multiplied by the probability that the stock price hits the diagonal for the *first* time at that node. It is impossible to go from the root to a node at which the option will be exercised without passing through D. For a node on D, which is the result of i up moves and j down moves from the root, the above-mentioned probability is

Figure 9.21: Search for a diagonal of non-early-exercise nodes. Only those nodes indicated will be visited. The **(optimal) exercise boundary** specifies the stock price at each time when it becomes optimal to exercise the option.

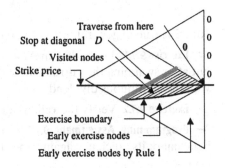

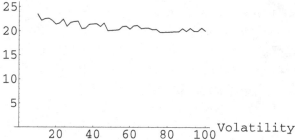

Percent of nodes visited
by the diagonal method

Figure 9.22: Performance of the diagonal method. The ratio of the number of nodes visited by the diagonal method to the total number of nodes is plotted. The parameters are $S = 50$, $X = 50$, $n = 100$, $r = 5\%$, $q = 1\%$, and $\tau = 1/3$. The volatility σ is in percentage terms.

$\binom{i+j-1}{i} p^i(1-p)^j$. This is because a valid path must pass through the node that is the result of i up moves and $j-1$ down moves. Call the option value on this node P_i. The desired option value then equals $\sum_{i=0}^{a-1} \binom{i+j-1}{i} p^i(1-p)^j P_i e^{-(i+j)r\Delta t}$. Because each node on D has been evaluated by that time, this part of the computation consumes $O(n)$ time. The space requirement is also linear in n because only the diagonal has to be allocated space. This idea can save computation time when D does not take long to find. See Fig. 9.23 for the algorithm.

It has been assumed up to now that the stock pays no dividends. Suppose now that the stock pays a continuous dividend yield $q \le r$ (or $r \le q$ for calls by parity). Therefore

$$p = \frac{e^{(r-q)\Delta t} - d}{u - d}.$$

Rule 1 of Fig. 9.19 continues to hold because, for a current stock price of $Su^i d^j$,

$$
\begin{aligned}
[\, p P_u + (1-p) P_d \,] e^{-r\Delta t} &= [\, p(X - Su^{i+1}d^j) + (1-p)(X - Su^i d^{j+1}) \,] e^{-r\Delta t} \\
&= Xe^{-r\Delta t} - Su^i d^j [\, pu + (1-p)d \,] e^{-r\Delta t} \\
&= Xe^{-r\Delta t} - Su^i d^j e^{-q\Delta t} \\
&\le Xe^{-r\Delta t} - Su^i d^j e^{-r\Delta t} \\
&\le X - Su^i d^j.
\end{aligned}
$$

Rule 2 is true in general, with or without dividends.

▶ **Exercise 9.7.1** Prove that an American call fetches the same price as an American put after swapping the current stock price with the strike price and the riskless rate with the continuous dividend yield.

▶ **Exercise 9.7.2** Verify the validity of Rule 2 under the binomial model.

▶ **Programming Assignment 9.7.3** Carefully implement the diagonal method and benchmark its efficiency against the standard backward-induction algorithm.

The diagonal method for American puts:

input: $S, u, d, \tau, X, n, r, q \ (r \geq q > 0)$;
real $p, P[\,n+1\,]$, cont, down;
integer i, j, a;
$p := (e^{(r-q)(\tau/n)} - d)/(u - d)$;
$a := \lceil \ln(X/Sd^n)/\ln(u/d)\rceil$;
$P[\,a\,] := 0$;   // Collect lower boundary of zero-valued nodes in one entry.
for $(i = 0 \text{ to } a - 1)$ $P[\,i\,] := -1$; //Upper boundary of early-exercise nodes.
for $(j = n - a \text{ down to } 0)$
        for $(i = a - 1 \text{ down to } 0)$ {
                down $:= P[\,i\,]$;   // Down move (computed in previous scan).
                if $[\,$down $< 0\,]$ down $:= X - Su^i d^{j+1}$;
                cont $:= (p \times P[\,i+1\,] + (1 - p) \times \text{down})/R$;
                if $[\,$cont $\geq X - Su^i d^j$ and $i > 0\,]$
                      $P[\,i\,] := $ cont;   // No early exercise.
                else if $[\,$cont $\geq X - Su^i d^j$ and $i = 0\,]$ { // Found D.
                      $P[\,i\,] := $ cont;
                      if $[\,j = 0\,]$ return $P[\,0\,]$;
                      else return $\sum_{k=0}^{a-1} \binom{k+j-1}{k} p^k (1 - p)^j \times P[\,k\,] \times e^{-(k+j)r(\tau/n)}$;
                }
                else break;   // Early-exercise node; exit the current loop.
        }
if $[\,P[\,0\,] < 0\,]$ $P[\,0\,] := X - S$;
return $P[\,0\,]$;

Figure 9.23: The diagonal method for American puts. $P[\,i\,]$ stores the put value when the stock price equals $Su^i d^j$, where j is the loop variable. As an early-exercide node's option value can be computed on the fly, -1 is used to state the fact that a node is an early-exercise node.

Additional Reading

The basic Black–Scholes model makes several assumptions. For example, margin requirements, taxes, and transactions costs are ignored, and only small changes in the stock price are allowed for a short period of time. See [86] for an early empirical work. Consult [236, 470] for various extensions to the basic model and [154, 423, 531, 683] for analytical results concerning American options. The Black–Scholes formula can be derived in at least four other ways [289]. A wealth of options formulas are available in [344, 423, 894]. Reference [613] considers predictable returns. Consult [48, 201, 346] for more information regarding estimating volatility from historical data. See [147, Subsection 9.3.5] and [514, Subsection 8.7.2] for more discussions on the "smile." To tackle multiple implied volatilities such as the smile, a generalized tree called the implied binomial tree may be used to price all options on the same underlying asset exactly (see Exercise 9.4.3) [215, 269, 299, 502, 503, 685]. Implied binomial trees are due to Rubinstein [770]. Wrong option pricing models and inaccurate volatility forecasts create great risk exposures for option writers [400]. See [56, 378, 420, 681, 753] for more information on the fundamental theorems of asset pricing.

The BOPM is generally attributed to Sharpe in 1975 [768] and appeared in his popular 1978 textbook, *Investments*. We followed the ideas put forth in [235, 738]. For American options, the BOPM offers a correct solution although its justification is delicate [18, 243, 576]. Several numerical methods for valuing American options are benchmarked in [127, 531, 834]. Convergences of binomial models for European and American options are investigated in [589, 590].

Many excellent textbooks cover options [236, 317, 346, 470, 878]. Read [494, 811] for intellectual developments that came before the breakthrough of Black and Scholes. To learn more about Black as a scientist, financial practitioner, and person, consult [345, 662].

NOTES

1. Their paper, "The Pricing of Options [and Corporate Liabilities]," was sent in 1970 to the *Journal of Political Economy* and was rejected immediately by the editors [64, 65].
2. Specifically, the **weak form of efficient markets hypothesis**, which says that current prices fully embody all information contained in historical prices [317]. This form of market efficiency implies that technical analysts cannot make above-average returns by reading charts of historical stock prices. It has stood up rather well [635].
3. A "clever" candidate once bought votes by issuing similar options, which paid off only when he was elected. Here is his reasoning: The option holders would not only vote for him but would also campaign hard for him, and in any case he kept the option premium if he lost the election, which he did.

CHAPTER TEN

Sensitivity Analysis of Options

> Cleopatra's nose, had it been shorter, the whole face of the world
> would have been changed.
> Blaise Pascal (1623–1662)

Understanding how the value of a security changes relative to changes in a given parameter is key to hedging. Duration, for instance, measures the rate of change of bond value with respect to interest rate changes. This chapter asks similar questions of options.

10.1 Sensitivity Measures ("The Greeks")

In the following, $x \equiv [\ln(S/X) + (r + \sigma^2/2)\,\tau\,]/(\sigma\sqrt{\tau})$, as in the Black–Scholes formula of Theorem 9.3.4, and $N'(y) = (1/\sqrt{2\pi})\,e^{-y^2/2} > 0$ is the density function of the standard normal distribution.

10.1.1 Delta

For a derivative such as option, **delta** is defined as $\Delta \equiv \partial f/\partial S$, where f is the price of the derivative and S is the price of the underlying asset. The delta of a portfolio of derivatives on the same underlying asset is the sum of the deltas of individual derivatives. The delta used in the BOPM to replicate options is the discrete analog of the delta here. The delta of a European call on a non-dividend-paying stock equals

$$\frac{\partial C}{\partial S} = N(x) > 0, \tag{10.1}$$

and the delta of a European put equals $\partial P/\partial S = N(x) - 1 < 0$. See Fig. 10.1 for an illustration. The delta of a long stock is of course one.

A position with a total delta equal to zero is said to be **delta-neutral**. Because a delta-neutral portfolio is immune to small price changes in the underlying asset, creating it can serve for hedging purposes. For example, a portfolio consisting of a call and $-\Delta$ shares of stock is delta-neutral. So one can short Δ shares of stock to hedge a long call. In general, one can hedge a long position in a derivative with a delta of Δ_1 by shorting Δ_1/Δ_2 units of another derivative with a delta of Δ_2.

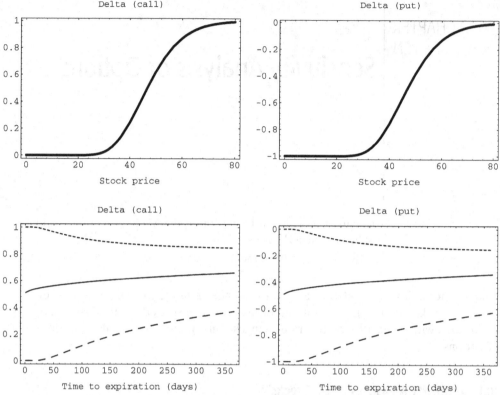

Figure 10.1: Option delta. The default parameters are $S = 50$, $X = 50$, $\tau = 201$ (days), $\sigma = 0.3$, and $r = 8\%$. The dotted curves use $S = 60$ (in-the-money call or out-of-the-money put), the solid curves use $S = 50$ (at-the-money option), and the dashed curves use $S = 40$ (out-of-the-money call or in-the-money put).

▶ **Exercise 10.1.1** Verify Eq. (10.1) and that the delta of a call on a stock paying a continuous dividend yield of q is $e^{-q\tau} N(x)$.

▶ **Exercise 10.1.2** Prove that $\partial P/\partial X = e^{-r\tau} N(-x + \sigma\sqrt{\tau})$.

▶ **Exercise 10.1.3** Show that at-the-money options have the maximum time value.

▶ **Exercise 10.1.4** What is the **charm**, defined as $\partial\Delta/\partial\tau$, of a European option?

10.1.2 Theta

Theta, or **time decay**, is defined as the rate of change of a security's value with respect to time, or $\Theta \equiv -\partial\Pi/\partial\tau$, where Π is the value of the security. For a European call on a non-dividend-paying stock,

$$\Theta = -\frac{SN'(x)\sigma}{2\sqrt{\tau}} - rXe^{-r\tau}N(x - \sigma\sqrt{\tau}) < 0.$$

The call hence loses value with the passage of time. For a European put,

$$\Theta = -\frac{SN'(x)\sigma}{2\sqrt{\tau}} + rXe^{-r\tau}N(-x + \sigma\sqrt{\tau}),$$

which may be negative or positive. See Fig. 10.2 for an illustration.

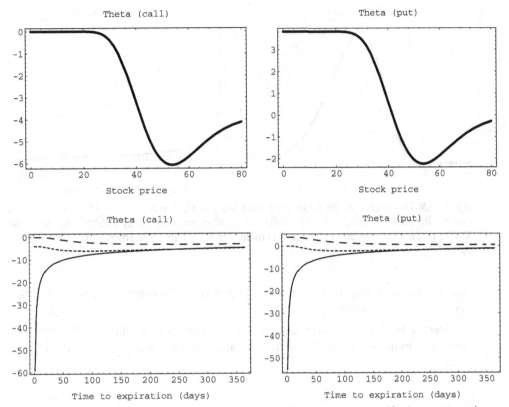

Figure 10.2: Option theta. The default parameters are $S = 50$, $X = 50$, $\tau = 201$ (days), $\sigma = 0.3$, and $r = 8\%$. The dotted curves uses $S = 60$ (in-the-money call or out-of-the-money put), the solid curves use $S = 50$ (at-the-money option), and the dashed curves use $S = 40$ (out-of-the-money call or in-the-money put).

▶ **Exercise 10.1.5** (1) At what stock price is the theta of a European call smallest? (2) Show that the theta of an American put is always negative.

10.1.3 Gamma

The **gamma** of a security is the rate of change of its delta with respect to the price of the underlying asset, or $\Gamma \equiv \partial^2 \Pi / \partial S^2$. The gamma measures how sensitive the delta is to changes in the price of the underlying asset. A portfolio with a high gamma needs in practice be rebalanced more often to maintain delta neutrality. The delta and the gamma have obvious counterparts in bonds: duration and convexity. The gamma of a European call or put on a non-dividend-paying stock is $N'(x)/(S\sigma\sqrt{\tau}) > 0$. See Fig. 10.3 for an illustration.

10.1.4 Vega

Volatility often changes over time. The **vega**[1] (sometimes called **lambda**, **kappa**, or **sigma**) of a derivative is the rate of change of its value with respect to the volatility of the underlying asset, or $\Lambda \equiv \partial \Pi / \partial \sigma$. A security with a high vega is very sensitive to small changes in volatility. The vega of a European call or put on a non-dividend-paying stock is $S\sqrt{\tau}\,N'(x) > 0$, which incidentally solves Exercise 9.4.1. A positive

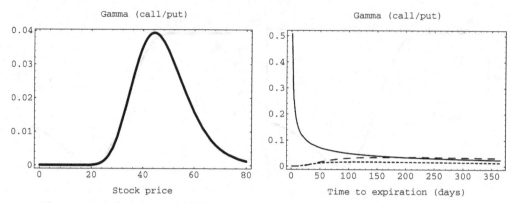

Figure 10.3: Option gamma. The default parameters are $S = 50$, $X = 50$, $\tau = 201$ (days), $\sigma = 0.3$, and $r = 8\%$. The dotted curve uses $S = 60$ (in-the-money call or out-of-the-money put), the solid curves use $S = 50$ (at-the-money option), and the dashed curve uses $S = 40$ (out-of-the-money call or in-the-money put).

vega is consistent with the intuition that higher volatility increases option value. See Fig. 10.4 for an illustration.

▶ **Exercise 10.1.6** Prove that the vega as a function of σ is unimodal for $\sigma > 0$. A function is **unimodal** if it is first increasing and then decreasing, thus having a single peak.

10.1.5 Rho

The **rho** of a derivative is the rate of change in its value with respect to interest rates, or $\rho \equiv \partial \Pi / \partial r$. The rhos of a European call and a European put on a non-dividend-paying stock are $X \tau e^{-r\tau} N(x - \sigma \sqrt{\tau}) > 0$ and $-X \tau e^{-r\tau} N(-x + \sigma \sqrt{\tau}) < 0$, respectively. See Fig. 10.5 for an illustration.

▶ **Exercise 10.1.7** (1) What is the **speed**, defined as $\partial \Gamma / \partial S$, of a European option? (2) What is the **color**, defined as $\partial \Gamma / \partial \tau$, of a European option?

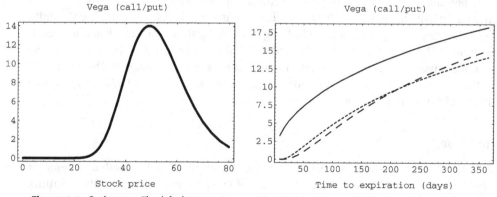

Figure 10.4: Option vega. The default parameters are $S = 50$, $X = 50$, $\tau = 201$ (days), $\sigma = 0.3$, and $r = 8\%$. The dotted curve uses $S = 60$ (in-the-money call or out-of-the-money put), the solid curves use $S = 50$ (at-the-money option), and the dashed curve uses $S = 40$ (out-of-the-money call or in-the-money put).

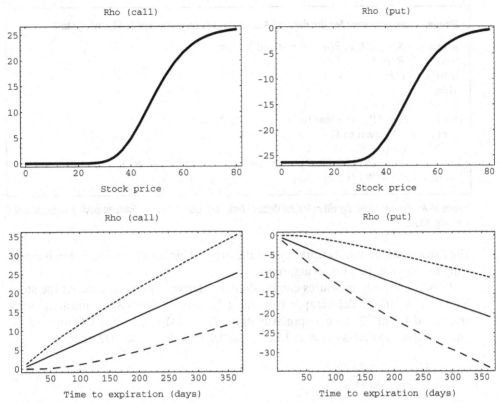

Figure 10.5: Option rho. The default parameters are $S = 50$, $X = 50$, $\tau = 201$ (days), $\sigma = 0.3$, and $r = 8\%$. The dotted curves use $S = 60$ (in-the-money call or out-of-the-money put), the solid curves use $S = 50$ (at-the-money option), and the dashed curves use $S = 40$ (out-of-the-money call or in-the-money put).

10.2 Numerical Techniques

Sensitivity measures of derivatives for which closed-form formulas do not exist have to be computed numerically. Take delta as an example. It is defined as $\Delta f/\Delta S$, where ΔS is a small change in the stock price and Δf is the resulting change in the derivative's price. A standard method computes $f(S - \Delta S)$ and $f(S + \Delta S)$ and settles for

$$\frac{f(S + \Delta S) - f(S - \Delta S)}{2\Delta S}.$$

The computation time for this numerical differentiation scheme roughly doubles that for evaluating the derivative security itself.

A preferred approach is to take advantage of the intermediate results of the binomial tree algorithm. When the algorithm reaches the end of the first period, f_u and f_d are computed. Recall that these values correspond to derivative values at stock prices Su and Sd, respectively. Delta is then approximated by

$$\frac{f_u - f_d}{Su - Sd}.$$

Binomial tree algorithm for the delta of American puts on a non-dividend-paying stock:

input: $S, u, d, X, n, \hat{r}$ ($u > e^{\hat{r}} > d$ and $\hat{r} > 0$);
real $R, p, P[n+1]$;
integer i, j;
$R := e^{\hat{r}}$;
$p := (R-d)/(u-d)$;
for ($i = 0$ to n) $\{P[i] := \max(0, X - Su^{n-i}d^i);\}$
for ($j = n-1$ down to 1)
          for ($i = 0$ to j)
                $P[i] := \max((p \times P[i] + (1-p) \times P[i+1])/R, X - Su^{j-i}d^i)$;
return $(P[0] - P[1])/(Su - Sd)$;

Figure 10.6: Binomial tree algorithm for the delta of American puts on a non-dividend-paying stock. Adapted from Fig. 9.13.

The extra computational effort beyond the original binomial tree algorithm is essentially nil. See Fig. 10.6 for an algorithm.

Other sensitivity measures can be similarly derived. Take gamma. At the stock price $(Suu + Sud)/2$, delta is approximately $(f_{uu} - f_{ud})/(Suu - Sud)$, and at the stock price $(Sud + Sdd)/2$, delta is approximately $(f_{ud} - f_{dd})/(Sud - Sdd)$. Gamma is the rate of change in deltas between $(Suu + Sud)/2$ and $(Sud + Sdd)/2$, that is,

$$\frac{\frac{f_{uu} - f_{ud}}{Suu - Sud} - \frac{f_{ud} - f_{dd}}{Sud - Sdd}}{(Suu - Sdd)/2}. \tag{10.2}$$

In contrast, numerical differentiation gives

$$\frac{f(S + \Delta S) - 2f(S) + f(S - \Delta S)}{(\Delta S)^2}.$$

As we shall see shortly, numerical differentiation may give inaccurate results.

Strictly speaking, the delta and the gamma thus computed are the delta at the end of the first period and the gamma at the end of the second period. In other words, they are not the sensitivity measures at the present time but at times τ/n and $2(\tau/n)$ from now, respectively, where n denotes the number of periods into which the time to expiration τ is partitioned. However, as n increases, such values should approximate delta and gamma well. The theta, similarly, can be computed as

$$\frac{f_{ud} - f}{2(\tau/n)}.$$

As for vega and rho, there is no alternative but to run the binomial tree algorithm twice. In Eq. (15.3), theta will be shown to be computable from delta and gamma.

10.2.1 Why Numerical Differentiation Fails

A careful inspection of Eq. (9.8) reveals why numerical differentiation fails for European options. First, the option value is a continuous piecewise linear function of the current stock price S. Kinks develop at prices $Xu^{-j}d^{-(n-j)}$, $j = 0, 1, \ldots, n$. As a result, if ΔS is suitably small, the delta computed by numerical differentiation will be a ladderlike function of S, hence not differentiable at the kinks. This bodes ill for

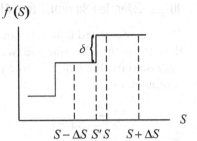

Figure 10.7: Numerical differentiation for delta and gamma.

numerical gamma. In fact, if ΔS is suitably small, gamma computed through numerical differentiation will be zero most of the time because $f'(S - \Delta S) = f'(S) = f'(S + \Delta S)$ unless S is near a kink. However, another problem arises when S is near a kink. Assume that S is to the right of the kink at S' and $S - \Delta S < S' < S$. Hence $f'(S) = f'(S + \Delta S)$ and $f'(S) - f'(S - \Delta S) = \delta$ for some constant $\delta > 0$ (see Fig. 10.2.1). Numerical gamma now equals

$$\frac{f(S + \Delta S) - 2f(S) + f(S - \Delta S)}{(\Delta S)^2} = \frac{\delta(S' - S + \Delta S)}{(\Delta S)^2}.$$

This number can become huge as ΔS decreases, and the common practice of reducing the step size ΔS will not help.

➤ **Exercise 10.2.1** Why does the numerical gamma in definition (10.2) not fail for the same reason?

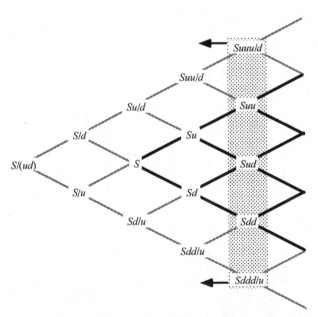

Figure 10.8: Extended binomial tree. The extended binomial tree is constructed from the original binomial tree (bold lines) but with time extended beyond the present by two periods.

10.2.2 Extended Binomial Tree Algorithms

It is recommended that delta and gamma be computed on the binomial tree, not at the current time, but one and two periods from now, respectively [718]. An improved method starts the binomial tree two periods *before* now, as in Fig. 10.8. Delta is then computed as

$$\frac{f_{u/d} - f_{d/u}}{(Su/d) - (Sd/u)},$$

and gamma is computed as

$$\frac{\frac{f_{u/d} - f}{(Su/d) - S} - \frac{f - f_{d/u}}{S - (Sd/u)}}{[(Su/d) - (Sd/u)]/2}.$$

➤ **Programming Assignment 10.2.2** Implement the extended binomial tree algorithm for numerical delta and gamma. Compare the results against numerical differentiation and closed-form solutions.

NOTE

1. Vega is not Greek.

CHAPTER ELEVEN

Extensions of Options Theory

As I never learnt mathematics, so I have had to think.
Joan Robinson (1903–1983)

This chapter samples various option instruments and presents important applications of the option pricing theory. Algorithms are described for a few nontrivial options.

11.1 Corporate Securities

With the underlying asset interpreted as the total value of the firm, Black and Scholes observed that the option pricing methodology can be applied to pricing corporate securities [87, 236, 658]. In the following analysis, it is assumed that (1) a firm can finance payouts by the sale of assets and (2) if a promised payment to an obligation other than stock is missed, the claim holders take ownership of the firm and the stockholders get nothing.

11.1.1 Risky Zero-Coupon Bonds and Stock

Consider a firm called XYZ.com. It has a simple capital structure: n shares of its own common stock S and zero-coupon bonds with an aggregate par value of X. The fundamental question is, what are the values of the bonds B and the stock?

On the bonds' maturity date, if the total value of the firm V^* is less than the bondholders' collective claim X, the firm declares bankruptcy and the stock becomes worthless. On the other hand, if $V^* > X$, then the bondholders obtain X and the stockholders $V^* - X$. The following table shows their respective payoffs:

	$V^* \leq X$	$V^* > X$
Bonds	V^*	X
Stock	0	$V^* - X$

The stock is therefore a call on the total value of the firm with a strike price of X and an expiration date equal to the maturity date of the bonds. It is this call that provides the limited liability for the stockholders. The bonds are a covered call on the total value of the firm.

Let C stand for this call and V stand for the total value of the firm. Then $nS = C$ and $B = V - C$. Knowing C thus amounts to knowing how the value of the firm is distributed between the stockholders and the bondholders. Whatever the value of C, the total value of the stock and bonds at maturity remains V^*. Hence the relative size of debt and equity is irrelevant to the firm's current value V, which is the expected PV of V^*.

From Theorem 9.3.4 and the put–call parity

$$nS = V N(x) - Xe^{-r\tau} N(x - \sigma \sqrt{\tau}),$$
$$B = V N(-x) + Xe^{-r\tau} N(x - \sigma \sqrt{\tau}),$$

where

$$x \equiv \frac{\ln(V/X) + (r + \sigma^2/2)\tau}{\sigma \sqrt{\tau}}.$$

The continuously compounded yield to maturity of the firm's bond is hence $(1/\tau) \ln(X/B)$. The **default premium** is defined as the yield difference between risky and riskless bonds:

$$(1/\tau) \ln(X/B) - r = -\frac{1}{\tau} \ln \left(N(-z) + \frac{1}{\omega} N(z - \sigma \sqrt{\tau}) \right), \tag{11.1}$$

where $\omega \equiv Xe^{-r\tau}/V$ and $z \equiv (\ln \omega)/(\sigma \sqrt{\tau}) + (1/2) \sigma \sqrt{\tau} = -x + \sigma \sqrt{\tau}$. Note that ω is the **debt-to-total-value ratio**. The volatility of the value of the firm, σ, can be looked on as a measure of **operating risk**. The default premium depends on only the firm's capital structure, operating risk, and debt maturity. The concept of default premium is a special case of static spread in Subsection 5.6.2.

▶ **Exercise 11.1.1** Argue that a loan guarantee that makes up any shortfalls in payments to the bondholders is a put with a strike price of B. The tacit assumption here is that the guarantor does not default.

▶ **Exercise 11.1.2** Prove Eq. (11.1).

▶ **Exercise 11.1.3** Verify the following claims and explain them in simple English: (1) $\partial B/\partial V > 0$, (2) $\partial B/\partial X > 0$, and (3) $\partial B/\partial \tau < 0$.

Numerical Illustrations

Suppose that XYZ.com's assets consist of 1,000 shares of Merck as of March 20, 1995, when Merck's market value per share is $44.5 (see Fig. 11.1). XYZ.com's securities consist of 1,000 shares of common stock and 30 zero-coupon bonds maturing on July 21, 1995. Each bond promises to pay $1,000 at maturity. Therefore $n = 1000$, $V = 44.5 \times n = 44500$, and $X = 30 \times n = 30000$. The Merck option relevant to our question is the July call with a strike price of $X/n = 30$ dollars. Such an option exists and is selling for $15.25. So XYZ.com's stock is worth $15.25 \times n = 15250$ dollars, and the entire bond issue is worth $B = 44500 - 15250 = 29250$ dollars, or $975 per bond.

The XYZ.com bonds are equivalent to a default-free zero-coupon bond with $\$X$ par value plus n written European puts on Merck at a strike price of $30 by the put–call parity. The difference between B and the price of the default-free bond is precisely the value of these puts. Figure 11.2 shows the total market values of

| Option | Strike | Exp. | -Call- | | -Put- | |
			Vol.	Last	Vol.	Last
Merck	30	Jul	328	15 1/4	...	...
44 1/2	35	Jul	150	9 1/2	10	1/16
44 1/2	40	Apr	887	4 3/4	136	1/16
44 1/2	40	Jul	220	5 1/2	297	1/4
44 1/2	40	Oct	58	6	10	1/2
44 1/2	45	Apr	3050	7/8	100	1 1/8
44 1/2	45	May	462	1 3/8	50	1 3/8
44 1/2	45	Jul	883	1 15/16	147	1 3/4
44 1/2	45	Oct	367	2 3/4	188	2 1/16

Figure 11.1: Merck option quotations. Source: Fig. 7.4.

the XYZ.com stock and bonds under various debt amounts X. For example, if the promised payment to bondholders is $45,000, the relevant option is the July call with a strike price of $45000/n = 45$ dollars. Because that option is selling for $1 15/16, the market value of the XYZ.com stock is $(1 + 15/16) \times n = 1937.5$ dollars. The market value of the stock decreases as the debt–equity ratio increases.

Conflicts between Stockholders and Bondholders

Options and corporate securities have one important difference: A firm can change its capital structure, but an option's terms cannot be changed after it is issued. This means that parameters such volatility, dividend, and strike price are under partial control of the stockholders.

Suppose XYZ.com issues 15 more bonds with the same terms in order to buy back stock. The total debt is now $X = 45,000$ dollars. Figure 11.2 says that the total market value of the bonds should be $42,562.5. The *new* bondholders therefore pay $42562.5 \times (15/45) = 14187.5$ dollars, which is used by XYZ.com to buy back shares. The remaining stock is worth $1,937.5. The stockholders therefore gain

$$(14187.5 + 1937.5) - 15250 = 875$$

dollars. The *original* bondholders lose an equal amount:

$$29250 - \left(\frac{30}{45} \times 42562.5\right) = 875. \tag{11.2}$$

This simple calculation illustrates the inherent conflicts of interest between stockholders and bondholders.

Promised payment to bondholders X	Current market value of bonds B	Current market value of stock nS	Current total value of firm V
30,000	29,250.0	15,250.0	44,500
35,000	35,000.0	9,500.0	44,500
40,000	39,000.0	5,500.0	44,500
45,000	42,562.5	1,937.5	44,500

Figure 11.2: Distribution of corporate value under alternative capital structures. Numbers are based on Fig. 11.1.

As another example, suppose the stockholders distribute $14,833.3 cash dividends by selling $(1/3) \times n$ Merck shares. They now have $14,833.3 in cash plus a call on $(2/3) \times n$ Merck shares. The strike price remains $X = 30000$. This is equivalent to owning two-thirds of a call on n Merck shares with a total strike price of $45,000. Because n such calls are worth $1,937.5 from Fig. 11.1, the total market value of the XYZ.com stock is $(2/3) \times 1937.5 = 1291.67$ dollars. The market value of the XYZ.com bonds is hence $(2/3) \times n \times 44.5 - 1291.67 = 28375$ dollars. As a result, the stockholders gain

$$(14833.3 + 1291.67) - 15250 \approx 875$$

dollars, and the bondholders watch their value drop from $29,250 to $28,375, a loss of $875.

Bondholders usually loathe the stockholders' taking unduly risky investments. The option theory explains it by pointing out that higher volatility increases the likelihood that the call will be exercised, to the financial detriment of the bondholders.

▶ **Exercise 11.1.4** If the bondholders can lose money in Eq. (11.2), why do they not demand lower bond prices?

▶ **Exercise 11.1.5** Repeat the steps leading to Eq. (11.2) except that, this time, the firm issues only five bonds instead of fifteen.

▶ **Exercise 11.1.6** Suppose that a holding company's securities consist of 1,000 shares of Microsoft common stock and 55 zero-coupon bonds maturing on the same date as the Microsoft April calls. Figure out the stockholders' gains (hence the original bondholders' losses) if the firm issues 5, 10, and 15 more bonds, respectively. Consult Fig. 7.4 for the market quotes.

▶ **Exercise 11.1.7** Why are dividends bad for the bondholders?

Subordinated Debts

Suppose that XYZ.com adds a **subordinated** (or **junior**) **debt** with a face value X_j to its capital structure. The original debt, with a face value of X_s, then becomes the **senior debt** and takes priority over the subordinated debt in case of default. Let both debts have the same maturity. The following table shows the payoffs of the various securities:

	$V^* \leq X_s$	$X_s < V^* \leq X_s + X_j$	$X_s + X_j < V^*$
Senior debt	V^*	X_s	X_s
Junior debt	0	$V^* - X_s$	X_j
Stock	0	0	$V^* - X_s - X_j$

The subordinated debt has the same payoff as a portfolio of a long X_s call and a short $X_s + X_j$ call – a bull call spread, in other words.

11.1.2 Warrants

Warrants represent the right to buy shares from the corporation. Unlike a call, a corporation issues warrants against its own stock, and *new* shares are issued when a warrant is exercised. Most warrants have terms between 5 and 10 years, although perpetual warrants exist. Warrants are typically protected against stock splits and cash dividends.

Consider a corporation with n shares of stock and m European warrants. Each warrant can be converted into one share on payment of the strike price X. The total value of the corporation is therefore $V = nS + mW$, where W denotes the current value of each warrant. At expiration, if it becomes profitable to exercise the warrants, the value of each warrant should be equal to

$$W^* = \frac{1}{n+m}(V^* + mX) - X = \frac{1}{n+m}(V^* - nX),$$

where V^* denotes the total value of the corporation just before the conversion. It will be optimal to exercise the warrants if and only if $V^* > nX$. A European warrant is therefore a European call on one $(n+m)$th of the total value of the corporation with a strike price of $Xn/(n+m)$ – equivalently, $n/(n+m)$ European call on one nth of the total value of the corporation (or $S + (m/n)W$) with a strike price of X. Hence

$$W = \frac{n}{n+m}C(W), \tag{11.3}$$

where $C(W)$ is the Black–Scholes formula for the European call but with the stock price replaced with $S + (m/n)W$. The value of W can be solved numerically given S.

➤ **Programming Assignment 11.1.8** (1) Write a program to solve Eq. (11.3). (2) Write a binomial tree algorithm to price American warrants.

11.1.3 Callable Bonds

Corporations issue callable bonds so that the debts can be refinanced under better terms if future interest rates fall or the corporation's financial situation improves. Consider a corporation with two classes of obligations: n shares of common stock and a single issue of callable bonds. The bonds have an aggregate face value of X, and the stockholders have the right to call the bonds at any time for a total price of X_c. Whenever the bonds are called before they mature, the payoff to the stockholders is $V - X_c$. The stock is therefore equivalent to an American call on the total value of the firm with a strike price of X_c before expiration and X at expiration.

11.1.4 Convertible Bonds

A **convertible bond (CB)** is like an ordinary bond except that it can be converted into new shares at the discretion of its owner. Consider a non-dividend-paying corporation with two classes of obligations: n shares of common stock and m zero-coupon CBs. Each CB can be converted into k newly issued shares at maturity (k is called the **conversion ratio**). If the bonds are not converted, their holders will receive X in aggregate at maturity.

The bondholders will own a fraction $\lambda \equiv (mk)/(n+mk)$ of the firm if conversion is chosen (λ is called the **dilution factor**). It makes sense to convert only if the part of the total market value of the corporation that is due the bondholders after conversion, or λV^*, exceeds X, i.e., $V^* > X/\lambda$. The payoff of the bond at maturity is (1) V^* if $V^* \leq X$, (2) X if $X < V^* \leq X/\lambda$ because it will not be converted, or (3) λV^* if $X/\lambda < V^*$.

It is in the interest of stockholders to pursue risky projects because higher volatility increases the stock price. The corporation may also withhold positive inside information from the bondholders; once the favorable information is released, the corporation calls the bonds. CBs solve both problems by giving the bondholders the option to take equity positions [837].

▶ **Exercise 11.1.9** Replicate the zero-coupon CB with the total value of the corporation and European calls on that value.

▶ **Exercise 11.1.10** (1) Replicate the zero-coupon CB with zero-coupon bonds and European calls on a fraction of the total value of the corporation. (2) Replicate the zero-coupon CB with zero-coupon bonds and warrants. (3) Show that early conversion is not optimal.

Convertible Bonds with Call Provisions

Many CBs contain call provisions. When the CBs are called, their holders can either convert the CB or redeem it at the call price. The call strategy is intended to minimize the value of the CBs. In the following analysis, assume that the CBs can be called any time before their maturity and that the corporation's value follows a continuous path without jumps.

Consider the same corporation again. In particular, the aggregate face value of the CBs is X, the aggregate call price is P, and $P \geq X$. The bondholders will own a fraction λ of the firm on conversion. We first argue that it is not optimal to call the CBs when $\lambda V < P$. As the following table shows, not calling the CBs leaves the bondholders at maturity with a value of V^*, X, or λV^* if the holders choose not to convert them earlier.

	$V^* \leq X$	$X < V^* \leq X/\lambda$	$X/\lambda < V^*$
Immediate call (PV)	P	P	P
No call throughout (FV at maturity)	V^*	X	λV^*
No call throughout (PV)	V	$\mathrm{PV}(X)$	λV

The PVs in all three cases are either less than or equal to P. Calling the CBs immediately is hence not optimal.

We now argue that it is not optimal to call the CBs after $\lambda V = P$ happens. Calling the CBs when $\lambda V = P$ leaves the bondholders with λV^* at maturity. The bondholders' terminal wealth if the CBs are not called is tabulated in the following table.

	$V^* < X$	$X \leq V^* < X/\lambda$	$X/\lambda \leq V^*$
No call throughout	V^*	X	λV^*
Call sometime *in the future*	λV^*	λV^*	λV^*

Not calling the CBs hence may result in a higher terminal value for the bondholders than calling them. In summary, the optimal call strategy is to call the CBs the first time $\lambda V = P$ happens. More general settings will be covered in Subsection 15.3.7.

▶ **Exercise 11.1.11** Complete the proof by showing that it is not optimal to call the CBs when $\lambda V > P$.

11.2 Barrier Options

Options whose payoff depends on whether the underlying asset's price reaches a certain price level H are called **barrier options**. For example, a **knock-out option** is like an ordinary European option except that it ceases to exist if the barrier H is reached by the price of its underlying asset. A call knock-out option is sometimes called a **down-and-out option** if $H < X$. A put knock-out option is sometimes called an **up-and-out option** when $H > X$. A **knock-in option**, in contrast, comes into existence if a certain barrier is reached. A **down-and-in option** is a call knock-in option that comes into existence only when the barrier is reached and $H < X$. An **up-and-in option** is a put knock-in option that comes into existence only when the barrier is reached and $H > X$. Barrier options have been traded in the United States since 1967 and are probably the most popular among the over-the-counter options [370, 740, 894].

The value of a European down-and-in call on a stock paying a dividend yield of q is

$$Se^{-q\tau}\left(\frac{H}{S}\right)^{2\lambda} N(x) - Xe^{-r\tau}\left(\frac{H}{S}\right)^{2\lambda-2} N(x - \sigma\sqrt{\tau}), \quad S \geq H, \qquad (11.4)$$

where

$$x \equiv \frac{\ln(H^2/(SX)) + (r - q + \sigma^2/2)\,\tau}{\sigma\sqrt{\tau}}$$

and $\lambda \equiv (r - q + \sigma^2/2)/\sigma^2$ (see Fig. 11.3). A European down-and-out call can be priced by means of the **in–out parity** (see Comment 11.2.1). The value of a European

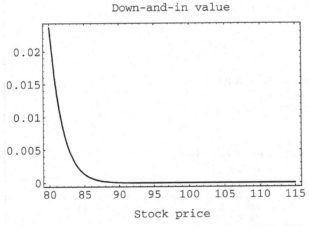

Down-and-in value

Figure 11.3: Value of down-and-in option. Plotted is the down-and-in option value as a function of the stock price with barrier $H = 80$. The other parameters are identical to those for the call in Fig. 7.3: $X = 95$, $\sigma = 0.25$, $\tau = 1/12$, and $r = 0.08$. Note the dramatic difference between the two plots.

up-and-in put is

$$Xe^{-r\tau}\left(\frac{H}{S}\right)^{2\lambda-2} N(-x+\sigma\sqrt{\tau}) - Se^{-q\tau}\left(\frac{H}{S}\right)^{2\lambda} N(-x),$$

where $S \le H$. A European up-and-out call can be priced by means of the in–out parity (see Exercise 11.2.1, part(2)). The formulas are due to Merton [660]. See [660] or Exercise 17.1.6 for proofs.

Backward induction can be used to price barrier options on a binomial tree. As the binomial tree algorithm works backward in time, it checks if the barrier price is reached by the underlying asset and, if so, replaces the option value with an appropriate value (see Fig. 11.4). In practice, the barrier is often monitored discretely, say at the end of the trading day, and the algorithm should reflect that.

▶ **Exercise 11.2.1** (1) Prove that a European call is equivalent to a portfolio of a European down-and-out option and a European down-and-in option with an identical barrier. (2) Prove that a European put is equivalent to a portfolio of a European up-and-out option and a European up-and-in option with an identical barrier.

Comment 11.2.1 The equivalence results in Exercise 11.2.1 are called the in–out parity [271]. Note that these results do not depend on the barrier's being a constant.

▶ **Exercise 11.2.2** Does the in–out parity apply to American-style options?

▶ **Exercise 11.2.3** Check that the formulas for the up-and-in and down-and-in options become the Black–Scholes formulas for standard European options when $S = H$.

Binomial tree algorithm for pricing down-and-out calls on a non-dividend-paying stock:

```
input:    S, u, d, X, H (H < X, H < S), n, r̂;
real      R, p, C[n+1];
integer   i, j, h;
R := e^r̂; p := (R − d)/(u − d);
h := ⌊ln(H/S)/ln u⌋; H := Su^h;
for (i = 0 to n) { C[i] := max(0, Su^{n−i}d^i − X); }
if [ n − h is even and 0 ≤ (n − h)/2 ≤ n ]
        C[(n−h)/2] := 0; // A hit.
for (j = n − 1 down to 0) {
        for (i = 0 to j)
            C[i] := (p × C[i] + (1 − p) × C[i + 1])/R;
        if [ j − h is even and 0 ≤ (j − h)/2 ≤ j ]
            C[(j−h)/2] := 0; // A hit.
}
return C[0];
```

Figure 11.4: Binomial tree algorithm for down-and-out calls on a non-dividend-paying stock. Because H may not correspond to a legal stock price, we lower it to Su^h, the highest stock price not exceeding H. The new barrier corresponds to $C[(j − h)/2]$ at times $j = n, n − 1, \ldots, h$. If the option provides a **rebate** K when the barrier is hit, simply change the assignment of zero to that of K.

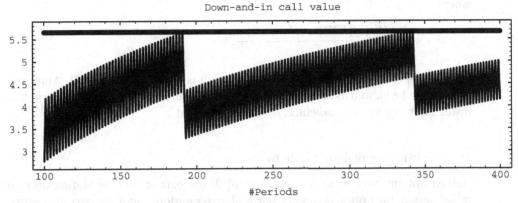

Down-and-in call value

Figure 11.5: Convergence of binomial model for down-and-in calls. Plotted are the option values against the number of time periods. The option's parameters are $S = 95$, $X = 100$, $H = 90$, $r = 10\%$ (continuously compounded), $\sigma = 0.25$, and $\tau = 1$ (year). The analytical value 5.6605 is also plotted for reference.

➤ **Exercise 11.2.4** A **reset option** is like an ordinary option except that the strike price is set to H when the stock price hits H. Assume that $H < X$. Create a synthetic reset option with a portfolio of barrier options.

➤ **Programming Assignment 11.2.5** (1) Implement binomial tree algorithms for European knock-in and knock-out options with rebates. Pay special attention to convergence (see Fig. 11.5). Here is a solved problem: A down-and-in European call without rebates has a value of $5.6605 given $S = 95$, $X = 100$, $H = 90$, $r = 10\%$ (continuously compounded), $\sigma = 0.25$, and $\tau = 1$ (year). (2) Extend the algorithms to handle American barrier options.

11.2.1 Bonds with Safety Covenants

Bonds with safety covenants can be evaluated with the help of knock-out options. Suppose that a firm is required to pass its ownership to the bondholders if its value falls below a specified barrier H, which may be a function of time. The bondholders therefore receive V the first time the firm's value falls below H. At maturity, the bondholders receive X if $V > X$ and V if $V < X$, where X is the aggregate par value. The value of the bonds therefore equals that of the firm minus a down-and-out option with a strike price X and barrier H.

11.2.2 Nonconstant Barrier

Consider the generalized barrier option with the barrier $H(t) = He^{-\rho\tau}$, where $\rho \geq 0$ and $H \leq X$. The standard barrier option corresponds to $\rho = 0$. The value of a European down-and-in call is

$$S\left[\frac{H(t)}{S}\right]^{2\lambda} N(x) - Xe^{-r\tau}\left[\frac{H(t)}{S}\right]^{2\lambda-2} N(x - \sigma\sqrt{\tau}),$$

where

$$x \equiv \frac{\ln(H(t)^2/(SX)) + (r + \sigma^2/2)\,\tau}{\sigma\sqrt{\tau}},$$

$\lambda \equiv (r - \rho + \sigma^2/2)/\sigma^2$, and $S \geq H(t)$. This result is due to Merton [660]. American options can be viewed as barrier options whose barrier is the exercise boundary that, instead of being given in advance, must be calculated.

11.2.3 Other Types of Barrier Options

Barrier options have many variations [158]. If the barrier is active during only an initial period, the option is called a **partial-barrier option**, and if the barrier is active during only the latter part of the option's life, it is called a **forward-starting-barrier option**. If the barrier must be breached for a particular length of time, we have a **Parisian option**. **Double-barrier options** have two barriers. In **rolling options**, a sequence of barriers is specified. For calls (puts), the strike price is lowered (raised, respectively) each time a barrier is hit, and the option is knocked out at the last barrier.

▶ **Exercise 11.2.6** A rolling call comes with barriers $H_1 > H_2 > \cdots > H_n$ (all below the initial stock price) and strike prices $X_0 > X_1 > \cdots > X_{n-1}$. This option starts as a European call with a strike price of X_0. When the first barrier H_1 is hit, the strike price is rolled down to X_1. In general, on hitting each barrier H_i, the strike is rolled down to X_i. The option knocks out when the last barrier H_n is hit. Replicate this option with a portfolio of down-and-out options.

11.3 Interest Rate Caps and Floors

In floating-rate debts, the borrower is concerned with rate rises and the lender is concerned with rate declines. They can seek protection in **interest rate caps** and **floors**, respectively. The writer of a cap pays the purchaser each time the contract's **reference rate** is above the contract's **cap rate** (or **ceiling rate**) on each settlement date. The writer of a floor pays the holder each time the contract's reference rate is below the contract's **floor rate** on each settlement date. The net effect is that a cap places a ceiling on the interest rate cost of a floating-rate debt, and a floor places a floor on the interest rate income of a floating-rate asset. The predetermined interest rate level such as the cap rate and the floor rate is called the **strike rate** [325]. One can also buy an interest rate cap and simultaneously sell an interest rate floor to create an **interest rate collar**. With a collar, the interest cost is bounded between the floor rate and the cap rate: When the reference rate rises above the cap rate, one is compensated by the cap seller, and when the reference rate dips below the floor rate, one pays the floor purchaser.

More formally, at each settlement date of a cap, the cap holder receives

$$\max(\text{reference rate} - \text{cap rate}, 0) \times \text{notional principal} \times t, \tag{11.5}$$

where t is the length of the payment period. Similarly, the payoff of the floor is

$$\max(\text{floor rate} - \text{reference rate}, 0) \times \text{notional principal} \times t.$$

	One month	Three months	Six months	One year
$ LIBOR FT London Interbank Fixing	61/16	61/8	61/8	63/16

Figure 11.6: Sample LIBOR rate quotations. Source: *Financial Times*, May 19, 1995.

Payoff (11.5) denotes a European call on the interest rate with a strike price equal to the cap rate. Hence a cap can be seen as a package of European calls (or **caplets**) on the underlying interest rate. Similarly, a floor can be seen as a package of European puts (or **floorlets**) on the underlying interest rate.

For example, if the reference rate is the 6-month LIBOR, t is typically either 181/360 or 184/360 as LIBOR uses the actual/360 day count convention. LIBOR refers to the lending rates on U.S. dollar deposits (**Eurodollars**) between large banks in London. Many short-term debts and floating-rate loans are priced off LIBOR in that the interest rate is quoted at a fixed margin above LIBOR. Differences between the LIBOR rate and the domestic rate are due to risk, government regulations, and taxes [767]. Non-U.S.-dollar LIBORs such as the German mark LIBOR are also quoted [646]. See Fig. 11.6 for sample quotations on LIBOR rates.

Unlike stock options, caps and floors are settled in cash. The premium is expressed as a percentage of the notional principal on which the cap or floor is written. For example, for a notional principal of $10 million, a premium of 20 basis points translates into $10 \times 20/10000 = 0.02$ million. The full premium is usually paid up front.

As a concrete example, suppose a firm issues a floating-rate note, paying the 6-month LIBOR plus 90 basis points. The firm's financial situation cannot allow paying an annual rate beyond 11%. It can purchase an interest rate cap with a cap rate of 10.1%. Thereafter, every time the rate moves above 10.1% and the firm pays more than 11%, the excess will be compensated for exactly by the dealer who sells the cap.

11.4 Stock Index Options

A stock index is a mathematical expression of the value of a portfolio of stocks. The New York Stock Exchange (NYSE) Composite Index (ticker symbol NYA), for example, is a weighted average of the prices of all the stocks traded on the NYSE; the weights are proportional to the total market values of their outstanding shares. Buying the index is thus equivalent to buying a portfolio of all the common stocks traded on the NYSE. This kind of average is called **capitalization weighted**, in which heavily capitalized companies carry more weights. The DJIA, S&P 100 Index (ticker symbol OEX), S&P 500 Index (ticker symbol SPX), and Major Market Index (ticker symbol XMI) are four more examples of stock indices. The SPX and OEX are capitalization-weighted averages, whereas the DJIA and XMI are **price-weighted** averages. A price-weighted index is calculated as $\sum_i P_i/\alpha$, where P_i is the price of stock i in the index and α is an adjustment factor that takes care of stock splits, stock dividends, bankruptcies, mergers, etc., so that the index is comparable over time. For example, $\alpha = 0.19740463$ for the DJIA as of October 26, 1999. A third weighting method is **geometric weighting**, in which every stock has the same influence on the index. The Value Line Index (ticker symbol VLE) for example is a geometrically

RANGES FOR UNDERLYING INDEXES
Monday, March 20, 1995

	High	Low	Close	Net Chg.	From Dec. 31	%Chg.
S&P 100 (OEX)	465.88	464.31	465.42	+0.57	+36.79	+8.6
S&P 500 -A.M.(SPX) .	496.61	495.27	496.14	+0.62	+36.87	+8.0
		. . .				
Nasdaq 100 (NDX) .	447.67	442.83	446.61	+2.67	+42.34	+10.5
		. . .				
Russell 2000 (RUT)	257.87	257.28	257.83	+0.51	+7.47	+3.0
		. . .				
Major Mkt (XMI) . .	435.14	432.82	434.90	+1.93	+34.95	+8.7
		. . .				
NYSE (NYA)	268.36	267.68	268.05	+0.21	+17.11	+6.8
Wilshire S-C (WSX)	332.50	331.69	332.11	+0.16	+11.05	+3.4
		. . .				
Value Line (VLE) . .	474.41	473.46	474.12	+0.58	+21.59	+4.8
		. . .				

Figure 11.7: Index quotations. The stock market's spectacular rise between 1995 and early 2000 is evident if we compare this table with the one in Fig. 8.2. Source: *Wall Street Journal*, March 21, 1995.

weighted index [646]. A geometrically weighted index is calculated as

$$I(t) = \prod_{i=1}^{n} \left[\frac{P_i(t)}{P_i(t-1)} \right]^{1/n} I(t-1),$$

where n is the number of stocks in the index, $I(t)$ is the index value on day t, and $P_i(t)$ is the price of stock i on day t. Stock indices are usually not adjusted for cash dividends [317]. See Fig. 11.7. Additional stock indices can be found in [95, 346, 470].

Stock indices differ also in their stock composition. For instance, the DJIA is an index of 30 blue-chip stocks,[1] whereas the S&P 500 is an index of 500 listed stocks from three exchanges. Nevertheless, their returns are usually highly correlated.

An index option is an option on an index value. Options on stock market portfolios were first offered by insurance companies in 1977. Exchange-traded index options started in March 1983 with the trading of the OEX option on the CBOE [346]. See Fig. 11.8 for sample quotations. Stock index options are settled in cash: Only the cash difference between the index's current market value and the strike price is exchanged when the option is exercised. Options on the SPX, XMI, and DJIA are European, whereas those on the OEX and NYA are American.

The cash settlement feature poses some risks to American option holders and writers. Because the exact amount to be paid when an option is exercised is determined by the closing price, there is an uncertainty called the **exercise risk**. Consider an investor who is short an OEX call and long the appropriate amounts of common stocks that comprise the index. If the call is exercised today, the writer will be notified the next business day and pay cash based on today's closing price. Because the index may open at a price different from today's closing price, the stocks may not fetch the same value as today's closing index value. Were the option settled in stocks, the writer would simply deliver the stocks the following business day without worrying about the change in the index value.

	Strike	Vol.	Last	Net Chg.	Open Int.		Strike	Vol.	Last	Net Chg.	Open Int.
		CHICAGO				Apr	490c	470	11	+3/8	23,903
		. . .				Apr	490p	1,142	3	−3/8	14,476
		S&P 100 INDEX(OEX)				May	490c	103	137/8	−5/8	3,086
May	380p	8	5/16	. . .	897	May	490p	136	53/8	. . .	13,869
Apr	385p	50	1/8	. . .	1,672	Jun	490c	2,560	167/8	+3/4	9,464
Apr	390p	335	1/8	. . .	4,406	Jun	490p	333	67/8	−3/8	8,288
		. . .						. . .			
Apr	410c	12	57	+12	37	Jun	550p	10	493/4	−1	575
		. . .				Call vol. 34,079			Open Int. 656,653		
Jun	490c	42	13/4	−1/4	719	Put vol. 59,582			Open Int. 806,961		
Call vol. 75,513			Open Int. 415,627					. . .			
Put vol. 104,773			Open Int. 447,094					**AMERICAN**			
		S&P 500 INDEX-AM(SPX)						MAJOR MARKET(XMI)			
Jun	350p	15	1/8	−1/8	2,027	Jun	325p	20	1/16	−3/16	45
Jun	375p	8	1/4	−1/8	5,307	Apr	350p	300	1/16	. . .	400
Apr	400p	50	1/16	. . .	5,949			. . .			
		. . .				Call vol. 1,913			Open Int. 6,011		
Apr	405c	2	913/4	. . .	. . .	Put vol. 4,071			Open Int. 27,791		
		. . .						. . .			

Figure 11.8: Index options quotations. Source: *Wall Street Journal*, March 21, 1995.

The **size** of a stock index option contract is the dollar amount equal to $100 times the index. A May OEX put with a strike price of $380 costs $100 \times (5/16) = 31.25$ dollars from the data in Fig. 11.8. This particular put is out of the money.

The valuation of stock index options usually relies on the Black–Scholes option pricing model with continuous dividend yields. Hence European stock index options can be priced by Eq. (9.20). This model actually approximates a broad-based stock market index better than it approximates individual stocks.[2]

One of the primary uses of stock index options is hedging large diversified portfolios. NYA puts, for example, can be used to protect a portfolio composed primarily of NYSE-listed securities against market declines. The alternative approach of buying puts for individual stocks would be cumbersome and more expensive by Theorem 8.6.1. Stock index options can also be used to create a long position in a portfolio of stocks by the put–call parity. It is easier to implement this synthetic security than it is to buy individual stocks.

➤ **Exercise 11.4.1** Verify the following claims: (1) A 1% price change in a lower-priced stock causes a smaller movement in the price-weighted index than that in a higher-priced stock. (2) A 1% price change in a lower capitalization issue has less of an impact on the capitalization-weighted index than that in a larger capitalization issue. (3) A 1% price change in a lower-priced stock has the same impact on the geometrically weighted index as that in a higher-priced stock.

11.5 Foreign Exchange Options

In the **spot** (or **cash**) **market** in which prices are for immediate payment and delivery, exchange rates between the U.S. dollar and foreign currencies are generally quoted

EXCHANGE RATES
Thursday, January 7, 1999

Country	U.S. $ equiv. Thu.	Wed.	Currency per U.S. $ Thu.	Wed.
	. . .			
Britain (Pound).....	1.6508	1.6548	.6058	.6043
1-month Forward.	1.6493	1.6533	.6063	.6049
3-months Forward	1.6473	1.6344	.6071	.6119
6-months Forward	1.6461	1.6489	.6075	.6065
	. . .			
Germany (Mark).....	.5989	.5944	1.6698	1.6823
1-month Forward.	.5998	.5962	1.6673	1.6771
3-months Forward	.6016	.5972	1.6623	1.6746
6-months Forward	.6044	.6000	1.6544	1.6666
	. . .			
Japan (Yen)........	.009007	.008853	111.03	112.95
1-month Forward	.009007	.008889	111.03	112.50
3-months Forward	.009008	.008958	111.02	111.63
6-months Forward	.009009	.009062	111.00	110.34
	. . .			
SDR.............	1.4106	1.4137	0.7089	0.7074
Euro	1.1713	1.1626	0.8538	0.8601

Figure 11.9: Exchange rate quotations. Source: *Wall Street Journal*, January 8, 1999.

with the **European terms**. This method measures the amount of foreign currency needed to buy one U.S. dollar, i.e., foreign currency units per dollar. The **reciprocal of European terms**, on the other hand, measures the U.S. dollar value of one foreign currency unit. For example, if the European-terms quote is .63 British pounds per $1 (£.63/$1), then the reciprocal-of-European-terms quote is $1.587 per British pound ($1/£.63 or $1.587/£1). The reciprocal of European terms is also called the **American terms**.

Figure 11.9 shows the spot exchange rates as of January 7, 1999. The **spot exchange rate** is the rate at which one currency can be exchanged for another, typically for settlement in 2 days. Note that the German mark is a **premium currency** because the 3-month forward exchange rate, $.6016/DEM1, exceeds the spot exchange rate, $.5989/DEM1; the mark is more valuable in the forward market than in the spot market. In contrast, the British pound is a **discount currency**. The **forward exchange rate** is the exchange rate for deferred delivery of a currency.

Foreign exchange (forex) options are settled by delivery of the underlying currency. A primary use of forex options is to hedge **currency risk**. Consider a U.S. company expecting to receive 100 million Japanese yen in March 2000. Because this company wants U.S. dollars, not Japanese yen, those 100 million Japanese yen will be exchanged for U.S. dollars. Although 100 million Japanese yen are worth 0.9007 million U.S. dollars as of January 7, 1999, they may be worth less or more in March 2000. The company decides to use options to hedge against the depreciation of the yen against the dollar. From Fig. 11.10, because the **contract size** for the Japanese yen option is JPY6,250,000, the company decides to purchase $100,000,000/6,250,000 = 16$ puts on the Japanese yen with a strike price of

	-Call-		-Put-			-Call-		-Put-	
	Vol.	Last	Vol.	Last		Vol.	Last	Vol.	Last
		...					...		
German Mark				59.31	Japanese Yen				88.46
62,500 German Marks-European Style.					6,250,000 J. Yen-100ths of a cent per unit.				
58 1/2	Jan	...	0.01	27	0.06	66 1/2	Mar	...	0.01	1	2.53
59	Jan	...	0.01	210	0.13	73	Mar	...	...	10	0.04
61	Jan	27	0.07	...	0.01	75	Mar	...	0.01	137	0.06
61 1/2	Jan	210	0.02	...	0.01	76	Mar	...	...	9	0.09
						77	Mar	...	...	17	0.09
						78	Mar	...	...	185	0.18
						79	Mar	...	...	10	0.16
						80	Mar	...	...	77	0.40
						81	Mar	...	...	60	0.36
						86	Jan	...	0.01	5	0.14
						88	Mar	...	...	10	2.14
						89	Mar	...	...	10	2.51
						90	Feb	...	0.01	12	2.30
						91	Feb	...	0.01	5	2.50
						100	Mar	2	0.86	...	0.01
							...				

Figure 11.10: Forex option quotations. Source: *Wall Street Journal*, January 8, 1999.

$.0088 and an exercise month in March 2000. This gives the company the right to sell 100,000,000 Japanese yen for $100,000,000 \times .0088 = 880,000$ U.S. dollars. The options command $100,000,000 \times 0.000214 = 21,400$ U.S. dollars in premium. The net proceeds per Japanese yen are hence $.88 - .0214 = 0.8586$ cent at the minimum.

A call to buy a currency is an insurance against the relative appreciation of that currency, whereas a put on a currency is an insurance against the relative depreciation of that currency. Put and call on a currency are therefore identical: A put to sell X_A units of currency A for X_B units of currency B is the same as a call to buy X_B units of currency B for X_A units of currency A. The above-mentioned option, for example, can be seen as the right to buy $6,250,000 \times 0.0088 = 55,000$ U.S. dollars for 6,250,000 Japanese yen, or equivalently a call on 55000 U.S. dollars with a strike price of $1/0.0088 \approx 113.6$ yen per dollar.

It is important to note that a DEM/$ call (an option to buy German marks for U.S. dollars) and a $/£ call do not a DEM/£ call make. This point is illustrated by an example. Consider a call on DEM1 for $.71, a call on $.71 for £.452, and a call on DEM1 for £.452. Suppose that the U.S. dollar falls to DEM1/$.72 and £1/$1.60. The first option nets a profit of $.01, but the second option expires worthless. Because $0.72/1.6 = 0.45$, we know that DEM1/£.45. The DEM/£ option therefore also expires worthless. Hence the portfolio of DEM/$ and $/£ calls is worth more than the DEM/£ call. This conclusion can be shown to hold in general (see Exercise 15.3.16). Options such as DEM/£ calls are called **cross-currency options** from the dollar's point of view.

Many forex options are over-the-counter options. One possible reason is that the homogeneity and liquidity of the underlying asset make it easy to structure custom-made deals. Exchange-traded forex options are available on the PHLX and the Chicago Mercantile Exchange (CME) as well as on many other exchanges [346]. Most exchange-traded forex options are denominated in the U.S. dollar.

> **Exercise 11.5.1** A **range forward contract** has the following payoff at expiration:

$$
\begin{array}{ll}
0 & \text{if the exchange rate } S \text{ lies within } [\,X_L, X_H\,], \\
S - X_H & \text{if } S > X_H, \\
S - X_L & \text{if } S < X_L.
\end{array}
$$

It guarantees that the effective future exchange rate will lie within X_L and X_H. Replicate this contract with standard options.

> **Exercise 11.5.2** In a **conditional forward contract**, the premium p is paid at expiration and only if the exchange rate is below a specified level X. The payoff at expiration is thus

$$
\begin{array}{ll}
S - X - p & \text{if the exchange rate } S \text{ exceeds } X, \\
-p & \text{if } S \le X.
\end{array}
$$

It guarantees that the effective future exchange rate will be, at most, $X + p$. Replicate this contract with standard options.

> **Exercise 11.5.3** A **participating forward contract** pays off at expiration

$$
\begin{array}{ll}
S - X & \text{if the exchange rate } S \text{ exceeds } X, \\
\alpha(S - X) & \text{if } S \le X.
\end{array}
$$

The purchaser is guaranteed an upper bound on the exchange rate at X and pays a proportion α of the decrease below X. Replicate this contract with standard options.

11.5.1 The Black–Scholes Model for Forex Options

Let S denote the spot exchange rate in domestic/foreign terms, σ the volatility of the exchange rate, r the domestic interest rate, and $\hat{r}$ the foreign interest rate. A foreign currency is analogous to a stock's paying a known dividend yield because the owner of foreign currencies receives a continuous dividend yield equal to $\hat{r}$ in the foreign currency. The formulas derived for stock index options in Eq. (9.20) hence apply with the dividend yield equal to $\hat{r}$:

$$
C = S e^{-\hat{r}\tau} N(x) - X e^{-r\tau} N(x - \sigma\sqrt{\tau}), \tag{11.6}
$$

$$
P = X e^{-r\tau} N(-x + \sigma\sqrt{\tau}) - S e^{-\hat{r}\tau} N(-x), \tag{11.6'}
$$

where

$$
x \equiv \frac{\ln(S/X) + (r - \hat{r} + \sigma^2/2)\,\tau}{\sigma\sqrt{\tau}}.
$$

The deltas of calls and puts are $e^{-\hat{r}\tau} N(x) \ge 0$ and $-e^{-\hat{r}\tau} N(-x) \le 0$, respectively. The Black–Scholes model produces acceptable results for major currencies such as the German mark, the Japanese yen, and the British pound [346].

> **Exercise 11.5.4** Assume the BOPM. (1) Verify that the risk-neutral probability for forex options is $[\,e^{(r-\hat{r})\Delta t} - d\,]/(u - d)$, where u and d denote the up and the down moves, respectively, of the domestic/foreign exchange rate in Δt time. (2) Let S be

the domestic/foreign exchange rate. Show that the delta of the forex call equals

$$h \equiv e^{-\hat{r}\Delta t} \frac{C_u - C_d}{Su - Sd}$$

if we use the foreign riskless asset (riskless in terms of foreign currency) and the domestic riskless asset to replicate the option. Above, h is the price of the foreign riskless asset held in terms of the foreign currency. (Hint: Review Eq. (9.1).)

▶ **Exercise 11.5.5** Prove that the European forex call and put are worth the same if $S = X$ and $r = \hat{r}$ under the Black–Scholes model.

11.5.2 Some Pricing Relations

Many of the relations in Section 8.2 continue to hold for forex options after the modifications required by the existence of the continuous dividend yield (equal to the foreign interest rate). To show that a European call satisfies

$$C \geq \max(Se^{-\hat{r}\tau} - Xe^{-r\tau}, 0), \tag{11.7}$$

consider the following strategies:

	Initial investment	Value at expiration	
		$S_\tau > X$	$S_\tau \leq X$
Buy a call	C	$S_\tau - X$	0
Buy domestic bonds (face value X)	$Xe^{-r\tau}$	X	X
Total	$C + Xe^{-r\tau}$	S_τ	X
Buy foreign bonds (face value 1 in foreign currency)	$Se^{-\hat{r}\tau}$	S_τ	S_τ

Hence the first portfolio is worth at least as much as the second in every scenario and therefore cannot cost less. Bound (11.7) incidentally generalizes Exercise 8.3.2.

A European call's price may approach the lower bound in bound (11.7) as closely as may be desired (see Exercise 11.5.6). As the intrinsic value $S - X$ of an American call can exceed the lower bound in bound (11.7), early exercise may be optimal. Thus we have the following theorem.

THEOREM 11.5.1 *American forex calls may be exercised before expiration.*

▶ **Exercise 11.5.6** Show how bound (11.7) may be approximated by the C in Eq. (11.6).

▶ **Exercise 11.5.7** (**Put–Call Parity**). Prove that (1) $C = P + Se^{-\hat{r}\tau} - Xe^{-r\tau}$ for European options and (2) $P \geq \max(Xe^{-r\tau} - Se^{-\hat{r}\tau}, 0)$ for European puts.

11.6 Compound Options

Compound options are options on options. There are four basic types of compound options: a call on a call, a call on a put, a put on a call, and a put on a put. Formulas for compound options can be found in [470].

If stock is considered a call on the total value of the firm as in Subsection 11.1.1, a stock option becomes a compound option. Renewable term life insurance offers another example: Paying a premium confers the right to renew the contract for the next term. Thus the decision to pay a premium is an option on an option [646]. A **split-fee option** provides a window on the market at the end of which the buyer can decide whether to extend it up to the notification date or to let it expire worthless [54, 346]. If the split-free option is extended up to the notification date; the second option can either expire or be exercised. The name comes from the fact that the user has to pay two fees to exercise the underlying asset [746].

Compound options are appropriate for situations in which a bid, denominated in foreign currency, is submitted for the sale of equipment. There are two levels of uncertainty at work here: the winning of the bid and the currency risk (even if the bid is won, a depreciated foreign currency may make the deal unattractive). What is needed is an arrangement whereby the company can secure a foreign currency option against foreign currency depreciation *if* the bid is won. This is an example of **contingent foreign exchange option**. A compound put option that grants the holder the right to purchase a put option in the future at prices that are agreed on today solves the problem. Forex options are not ideal because they turn the bidder into a speculator if the bid is lost.

▶ **Exercise 11.6.1** Recall our firm XYZ.com in Subsection 11.1.1. It had only two kinds of securities outstanding, shares of its own common stock and bonds. Argue that the stock becomes a compound option if the bonds pay interests before maturity.

▶ **Exercise 11.6.2** Why is a contingent forex put cheaper than a standard put?

▶ **Exercise 11.6.3** A **chooser option** (or an **as-you-like-it option**) gives its holder the right to buy for X_1 at time τ_1 either a call or a put with strike price X_2 at time τ_2. Describe the binomial tree algorithm for this option.

▶ **Programming Assignment 11.6.4** Implement binomial tree algorithms for the four compound options.

11.7 Path-Dependent Derivatives

Let $S_0, S_1, \ldots, S_n$ denote the prices of the underlying asset over the life of the option. S_0 is the known price at time zero and S_n is the price at expiration. The standard European call has a terminal value that depends on only the last price, $\max(S_n - X, 0)$. Its value thus depends on only the underlying asset's terminal price regardless of how it gets there; it is **path independent**. In contrast, some derivatives are **path dependent** in that their terminal payoffs depend critically on the paths. The (arithmetic) **average-rate call** has a terminal value given by

$$\max\left(\frac{1}{n+1}\sum_{i=0}^{n} S_i - X, 0\right),$$

and the **average-rate put's** terminal value is given by

$$\max\left(X - \frac{1}{n+1}\sum_{i=0}^{n} S_i, 0\right).$$

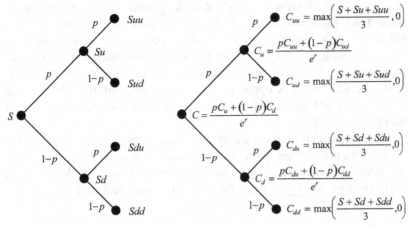

Figure 11.11: Binomial tree for average-rate call. The pricing tree on the right grows exponentially. Here a period spans one year.

Both are path dependent. At initiation, average-rate options cannot be more expensive than standard European options under the Black–Scholes option pricing model [379, 548]. Average-rate options also satisfy certain put–call parity identities [105, 598]. Average-rate options, also called **Asian options**, are useful hedging tools for firms that will make a stream of purchases over a time period because the costs are likely to be linked to the average price. They are mostly European for the same reason.

Average-rate options are notoriously hard to price. Take the terminal price $S_0 u^2 d$. Different paths to it such as $(S_0, S_0 u, S_0 u^2, S_0 u^2 d)$ and $(S_0, S_0 d, S_0 du, S_0 du^2)$ may lead to different averages and hence payoffs: $\max((S_0 + S_0 u + S_0 u^2 + S_0 u^2 d)/4 - X, 0)$ and $\max((S_0 + S_0 d + S_0 du + S_0 du^2)/4 - X, 0)$, respectively (see Fig. 11.11). The binomial tree for the averages therefore does not combine. A straightforward algorithm is to enumerate the 2^n price paths for an n-period binomial tree and then average the payoffs. However, the exponential complexity makes this naive algorithm impractical. As a result, the Monte Carlo method and **approximation algorithms** are the few alternatives left.

Not all path-dependent derivatives are hard to price, however. Barrier options, for example, are easy to price. When averaging is done *geometrically*, the option payoffs are

$$\max\left((S_0 S_1 \cdots S_n)^{1/(n+1)} - X, 0\right), \quad \max\left(X - (S_0 S_1 \cdots S_n)^{1/(n+1)}, 0\right).$$

For European geometric average-rate options, the limiting analytical solutions are the Black–Scholes formulas with the volatility set to $\sigma_a \equiv \sigma/\sqrt{3}$ and the dividend yield set to $q_a \equiv (r + q + \sigma^2/6)/2$, that is,

$$C = Se^{-q_a \tau} N(x) - Xe^{-r\tau} N(x - \sigma_a \sqrt{\tau}), \tag{11.8}$$

$$P = Xe^{-r\tau} N(-x + \sigma_a \sqrt{\tau}) - Se^{-q_a \tau} N(-x), \tag{11.8'}$$

where

$$x \equiv \frac{\ln(S/X) + (r - q_a + \sigma_a^2/2)\,\tau}{\sigma_a \sqrt{\tau}}.$$

In practice, average-rate options almost exclusively utilize arithmetic averages [894].

Another class of options, (**floating-strike**) **lookback options**, let the stock prices determine the **strike price** [225, 388, 833]. A lookback call option on the minimum has a terminal payoff of $S_n - \min_{0 \le i \le n} S_i$, and a lookback put option on the maximum has a terminal payoff of $\max_{0 \le i \le n} S_i - S_n$. The related **fixed-strike lookback option** provides a payoff of $\max(\max_{0 \le i \le n} S_i - X, 0)$ for the call and $\max(X - \min_{0 \le i \le n} S_i, 0)$ for the put. A perpetual American lookback option is called a **Russian option** [575]. One can also define lookback call and put options on the average. Such options are also called **average-strike options** [470].

An approximation algorithm for pricing arithmetic average-rate options, which is due to Hull and White [478], is described below. This algorithm is based on the binomial tree. Consider a node at time j with the underlying asset price equal to $S_0 u^{j-i} d^i$. Name such a node $N(j, i)$. The running sum $\sum_{m=0}^{j} S_m$ at this node has a maximum value of

$$\overbrace{S_0(1 + u + u^2 + \cdots + u^{j-i} + u^{j-i}d + \cdots + u^{j-i}d^i)}^{j} = S_0 \frac{1 - u^{j-i+1}}{1-u} + S_0 u^{j-i} d \frac{1 - d^i}{1-d}.$$

Divide this value by $j+1$ and call it $A_{\max}(j, i)$. Similarly, the running sum has a minimum value of

$$\overbrace{S_0(1 + d + d^2 + \cdots + d^i + d^i u + \cdots + d^i u^{j-i})}^{j} = S_0 \frac{1 - d^{i+1}}{1-d} + S_0 d^i u \frac{1 - u^{j-i}}{1-u}.$$

Divide this value by $j+1$ and call it $A_{\min}(j, i)$. Both $A_{\min}$ and $A_{\max}$ are running averages (see Fig. 11.12).

Although the possible running averages at $N(j, i)$ are far too many ($\binom{j}{i}$ of them), all lie between $A_{\min}(j, i)$ and $A_{\max}(j, i)$. Pick $k+1$ equally spaced values in this range and treat them as the true and only running averages, which are

$$A_m(j, i) \equiv \left(\frac{k-m}{k}\right) A_{\min}(j, i) + \left(\frac{m}{k}\right) A_{\max}(j, i), \quad m = 0, 1, \ldots, k.$$

Such "bucketing" introduces errors, but it works well in practice [366]. An alternative is to pick values whose logarithms are equally spaced [478, 555].

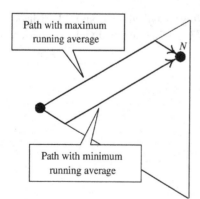

Path with maximum running average

N

Path with minimum running average

Figure 11.12: Paths with maximum and minimum running averages. Plotted are the paths with the maximum and minimum running averages from the root to node N.

During backward induction, we calculate the option values at each node for the $k+1$ running averages as follows. Suppose that the current node is $N(j, i)$ and the running average is a. Assume that the next node is $N(j+1, i)$, the result of an up move. Because the asset price there is $S_0 u^{j+1-i} d^i$, we seek the option value corresponding to the running average:

$$A_u \equiv \frac{(j+1)a + S_0 u^{j+1-i} d^i}{j+2}.$$

To be sure, A_u is not likely to be one of the $k+1$ running averages at $N(j+1, i)$. Hence we find the running averages that bracket it, that is,

$$A_\ell(j+1, i) \le A_u \le A_{\ell+1}(j+1, i).$$

Express A_u as a linearly interpolated value of the two running averages:

$$A_u = x A_\ell(j+1, i) + (1-x) A_{\ell+1}(j+1, i), \quad 0 \le x \le 1.$$

(An alternative is the quadratic interpolation [276, 555]; see Exercise 11.7.7.) Now, obtain the approximate option value given the running average A_u by means of

$$C_u \equiv x C_\ell(j+1, i) + (1-x) C_{\ell+1}(j+1, i),$$

where $C_\ell(t, s)$ denotes the option value at node $N(t, s)$ with running average $A_l(t, s)$. This interpolation introduces the second source of error. The same steps are repeated for the down node $N(j+1, i+1)$ to obtain another approximate option value C_d. We finally obtain the option value as $[pC_u + (1-p)C_d]e^{-r\Delta t}$. See Fig. 11.13 for the idea and Fig. 11.14 for the $O(kn^2)$-time algorithm.

▶ **Exercise 11.7.1** Achieve "selling at the high and buying at the low" with lookback options.

▶ **Exercise 11.7.2** Verify that the number of geometric averages at time n is $n(n+1)/2$.

▶ **Exercise 11.7.3** Explain why average-rate options are harder to manipulate. (They were first written on stocks traded on Asian exchanges, hence the name "Asian options," presumably because of their lighter trading volumes.)

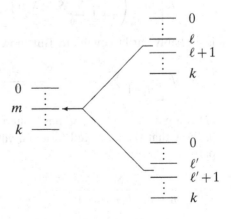

Figure 11.13: Backward induction for arithmetic average-rate options. The $k+1$ possible option values at $N(j, i)$ are stored in an array indexed by 0, 1,...,k, each corresponding to a specific arithmetic average. Each of the $k+1$ option values depends on two option values at node $N(j+1, i)$ and two option values at node $N(j+1, i+1)$.

Algorithm for pricing American arithmetic average-rate calls on a non-dividend-paying stock:

input: $S, u, d, X, n, \hat{r}\ (u > e^{\hat{r}} > d, \hat{r} > 0), k;$
real $p, a, A_u, x, A_d, C_u, C_d, C[n+1][k+1], D[k+1]; // C[\cdot][m]$ stores m th average.
integer $i, j, m, \ell;$
$p := (e^{\hat{r}} - d)/(u - d);$
for $(i = 0$ to $n) //$ Terminal price is $Su^{n-i}d^i$.
           for $(m = 0$ to $k)$
                    $C[i][m] := \max(0, A_m(n, i) - X);$
for $(j = n - 1$ down to $0) //$ Backward induction.
           for $(i = 0$ to $j) \{ //$ Price is $Su^{j-i}d^i$.
                    for $(m = 0$ to $k) \{$
                             $a := A_m(j, i); // $ "Average."
                             $A_u := ((j+1)a + Su^{j+1-i}d^i)/(j+2);$
                             Let ℓ be such that $A_\ell(j+1, i) \le A_u \le A_{\ell+1}(j+1, i);$
                             Let x be such that $A_u = x A_\ell(j+1, i) + (1-x) A_{\ell+1}(j+1, i);$
                             $C_u := x C[i][\ell] + (1-x) C[i][\ell+1]; //$ Linear interpolation.
                             $A_d := ((j+1)a + Su^{j-i}d^{i+1})/(j+2);$
                             Let ℓ be such that $A_\ell(j+1, i+1) \le A_d \le A_{\ell+1}(j+1, i+1);$
                             Let x be such that $A_d = x A_\ell(j+1, i+1) + (1-x) A_{\ell+1}(j+1, i+1);$
                             $C_d := x C[i+1][\ell] + (1-x) C[i+1][\ell+1]; //$ Linear interpolation.
                             $D[m] := \max(a - X, (p C_u + (1-p) C_d) e^{-\hat{r}});$
                    $\}$
                    Copy $D[0..k]$ to $C[i][0..k];$
           $\}$
return $C[0][0];$

Figure 11.14: Approximation algorithm for American arithmetic average-rate calls on a non-dividend-paying stock. The ideas in Exercise 11.7.5 can reduce computation time. For a fixed k, this algorithm may not converge as n increases [249, 251]; in fact, k may have to scale with n [555]. Note that $C[0][0], C[0][1], \ldots, C[0][k]$ are identical in value absent the rounding errors.

▶ **Exercise 11.7.4** Arithmetic average-rate options were assumed to be newly issued, and there was no historical average to deal with. Argue that no generality was lost in doing so.

▶ **Exercise 11.7.5** Let $\hat{r} \ne 0$ denote the continuously compounded riskless rate per period. (1) The future value of the average-rate call,

$$E\left[\max\left(\frac{1}{n+1} \sum_{i=0}^{n} S_i - X, 0 \right) \right],$$

is probably hard to evaluate. But show that

$$E\left[\frac{1}{n+1} \sum_{i=0}^{n} S_i \right] = \frac{S_0}{n+1} \frac{1 - e^{\hat{r}(n+1)}}{1 - e^{\hat{r}}}.$$

(2) If a path $S_0, S_1, \ldots, S_k$ has a running sum $\sum_{i=0}^{k-1} S_i$ equal to $(n+1)X + a, a \ge 0$, show that the expected terminal value of the average-rate call given this initial path is

$$\frac{a}{n+1} + \frac{S_k}{n+1} \frac{1 - e^{\hat{r}(n-k+1)}}{1 - e^{\hat{r}}}.$$

(So the tree growing out of node S_k can be pruned when pricing calls, leading to better efficiency.)

➤ **Exercise 11.7.6** Assume that $ud = 1$. Prove that the difference between the maximum running sum and the minimum running sum at a node with an asset price of $S_0 u^i d^{n-i}$ is an increasing function of i for $i < n/2$ and a decreasing function of i for $i > n/2$.

➤ **Exercise 11.7.7** Derive the quadratic polynomial $y = a + bx + cx^2$ that passes through three points (x_0, y_0), (x_1, y_1), and (x_2, y_2).

➤ **Exercise 11.7.8** Two sources of error were mentioned for the approximation algorithm. Argue that they disappear if all the asset prices on the tree are integers.

➤ **Exercise 11.7.9** Suppose that there is an algorithm that generates an upper bound on the true option value and an algorithm that generates a lower bound on the true option value. How do we design an approximate option pricing scheme with information on the pricing error?

➤ **Programming Assignment 11.7.10** Implement $O(n^4)$-time binomial tree algorithms for European and American geometric average-rate options.

➤ **Programming Assignment 11.7.11** Implement the algorithm in Fig. 11.14.

➤ **Programming Assignment 11.7.12** (1) Implement binomial tree algorithms for European and American lookback options. The time complexity should be, at most, cubic. (2) Improve the running time to quadratic time for newly issued floating-strike lookback options, which of course have no price history.

Additional Reading

Consult [565] for alternative models for warrants and [370] for an analytical approach to pricing American barrier options. See [346, Chap. 7] and [514, Chap. 11] for additional forex options and [886] for the term structure of exchange rate volatility. An $O(kn^2)$-time approximation algorithm with ideas similar to those in Fig. 11.14 can be found in [215, 470, 478]. More path-dependent derivatives are discussed in [377, 449, 470, 485, 746, 812]. There is a vast literature on average-rate option pricing [25, 105, 168, 169, 170, 241, 366, 379, 478, 530, 548, 555, 700, 755] as well as analytical approximations for such options [423, 596, 597, 664, 748, 851]. The $O(kn^2)$-time approximation algorithm for average-rate options in [10] is similar to, but slightly simpler than, the algorithm in Fig. 11.14. More important is its guarantee not to deviate from the naive $O(2^n)$-time binomial tree algorithm by more than $O(Xn/k)$ in the case of European average-rate options. The number k here is a parameter that can be varied for trade-off between running time and accuracy. The error bound can be further reduced to $O(X\sqrt{n}/k)$ [250]. An efficient convergent approximation algorithm that is due to Dai and Lyuu is based on the insight of Exercise 11.7.8 [251]. See [459] for a general treatment of approximation algorithms. The option pricing theory has applications in capital investment decisions [279, 672].

NOTES

1. As of February 2001, the 30 stocks were Allied Signal, Alcoa, American Express, AT&T, Boeing, Caterpillar, Citigroup, Coca-Cola, DuPont, Eastman Kodak, Exxon, General Electric, General Motors, Hewlett-Packard, Home Depot, Intel, IBM, International Paper, J.P. Morgan Chase,

Johnson & Johnson, McDonald's, Merck, Microsoft, Minnesota Mining & Manufacturing, Philip Morris, Procter & Gamble, SBC Communications, United Technologies, Wal-Mart Stores, and Walt Disney.

2. Strictly speaking, it is inconsistent to assume that both the stock index and its individual stock prices satisfy the Black–Scholes option pricing model because the sum of lognormal random variables is not lognormally distributed.

CHAPTER TWELVE

Forwards, Futures, Futures Options, Swaps

> It does not matter if I speak; the future has already been determined.
>
> Sophocles (496 B.C.–406 B.C.), *Oedipus Tyrannus*

This chapter continues the coverage of derivatives, financial contracts whose value depends on the value of some underlying assets or indices. Derivatives are essential to risk management, speculation, efficient portfolio adjustment, and arbitrage. Interest-rate-sensitive derivative securities require a separate chapter, Chap. 21.

12.1 Introduction

Many financial institutions take large positions in derivatives. For example, Chase Manhattan held U.S.$7.6 trillion (notional amount) in derivatives as of early 1998 [57]. Trading derivatives can be risky, however. The loss of U.S.$1.6 billion in the case of Orange County, California, led to its bankruptcy in December 1994 and the securities firms involved paying U.S.$739 million in subsequent settlements [527]; the value of an interest rate swap held by Sears in 1997 was a minus U.S.$382 million based on a notional principal of U.S.$996 million [884]; J.P. Morgan in 1997 had U.S.$659 million in nonperforming assets, 90% of which were defaults from Asian counterparties;[1] Union Bank of Switzerland (UBS) wrote off U.S.$699 million because of investment in the hedge fund, Long-Term Capital Management (LTCM).

Four types of derivatives stand out: futures contracts, forward contracts, options, and swaps. Futures contracts and forward contracts are contracts for future delivery of the underlying asset. The underlying asset can be a physical commodity (corn, oil, live cattle, pork bellies, precious metals, and so on) or financial instrument (bonds, currencies, stock indices, mortgage securities, other derivatives, and so on) [95, 470]. **Futures contracts** are essentially standardized forward contracts that trade on futures exchanges such as the Chicago Board of Trade (CBT) and the CME.[2] Futures and forward contracts can be used for speculation, hedging, or arbitrage between the spot and the deferred-delivery markets.

Futures and forward contracts are obligations on both the buyer and the seller. Options, we recall, are binding on only the seller. The buyer has the right, but not the obligation, to take a position in the underlying asset. Such a right naturally commands a premium. Options can be used to hedge downside risk, speculation, or arbitrage markets.

A **swap** is a contract between two parties to exchange cash flows in the future based on a formula. Typically, one party pays a fixed price to the other party in exchange for a market-determined floating price. Swaps can be used to reduce financing costs or to hedge. In reality, interest rate swaps and forex forward contracts make up banks' major derivatives holdings [60]. Like the forward contracts, swaps are traded by financial institutions and their corporate clients outside of organized exchanges.

The relevant riskless interest rate for many arbitragers operating in the futures market is the repo rate. A **repo** (**sale and repurchase agreement** or **RP**) is an agreement in which the owner of securities ("seller") agrees to sell them to a counterparty ("buyer") and buy them back at a slightly higher price later. The counterparty hence provides a loan. However, this loan has little risk as the lender keeps the securities if the seller fails to buy them back. (The lender essentially runs a pawnshop.) From the lender's perspective, this agreement is called a **reverse repo**. Overnight repo rates are lower than the federal funds rate. A loan of more than 1 day is called a **term repo**. The dollar interest is determined by

$$\text{dollar principal} \times \text{repo rate} \times \frac{\text{repo term}}{360}.$$

The Bank of England was the first central bank to introduce repos in 1830. The Federal Reserve uses the the repo market to influence short-term interest rates. When the Federal Reserve is doing repo, it is actually lending money, not borrowing it [827].

Throughout this chapter, r denotes the riskless interest rate. Other notations include, unless stated otherwise, the current time t, the maturity date of the derivative T, the remaining time to maturity $\tau \equiv T - t$ (all measured in years), the spot price S, the spot price at maturity S_T, the delivery price X, the forward or futures price F for a newly written contract, and the value of the contract f. A price with a subscript t usually refers to the price at time t. Continuous compounding will be assumed throughout this chapter.

12.2 Forward Contracts

Forward contracts are for delivery of the underlying asset for a certain **delivery price** on a specific time in the future. They are ideal for hedging purposes. Consider a corn farmer who enters into a forward contract with a food processor to deliver 100,000 bushels of corn for $2.5 per bushel on September 27, 1995. Assume that the cost of growing corn is $2.0 per bushel. Such a contract benefits both sides: the farmer, because he is assured of a buyer at an acceptable price, and the processor, because knowing the cost of corn in advance helps reduce uncertainty in planning. If the spot price of corn rises on the delivery date, the farmer will miss the opportunity of extra profits. On the other hand, if the price declines, the processor will be paying more than it would. A forward agreement hence limits both the risk and the potential rewards.

Problems may arise if one of the participants fails to perform. The food processor may go bankrupt, the farmer can go bust, the farmer might not be able to harvest 100,000 bushels of corn because of bad weather, or the cost of growing corn may skyrocket. More importantly, whichever way the corn price moves, either the food processor or the farmer has an incentive to default. Even corporate giants default on their forward contracts [767].

	EXCHANGE RATES Monday, March 20, 1995			
	U.S.$ equiv.		Currency per U.S.$	
Country	Mon.	Fri.	Mon.	Fri.
Germany (Mark)	.7126	.7215	1.4033	1.3860
30-Day Forward	.7133	.7226	1.4019	1.3839
90-Day Forward	.7147	.7242	1.3991	1.3808
180-Day Forward	.7171	.7265	1.3945	1.3765

Figure 12.1: German mark exchange rate quotations. The forward German marks are at a premium to the spot mark: The forward exchange rates in terms of $/DEM exceed the spot exchange rate, perhaps because of lower inflation in Germany. Source: *Wall Street Journal*, March 21, 1995.

12.2.1 Forward Exchange Rate

Along with forex options, forward contracts provide an avenue to hedging currency risk. Figure 12.1 shows the spot exchange rate and forward exchange rates for the German mark. Consider a U.S. company that is expecting to receive DEM10 million in 3 months' time. By using a forward sale at the 3-month forward exchange rate of $.7147/DEM1, the firm will receive exactly U.S.$7,147,000 in 3 months' time. Compared with hedging by use of forex options, the forward hedge insulates the firm from any movements in exchange rates whether they are favorable or not.

▶ **Exercise 12.2.1** Selling forward DEM10 million as in the text denies the hedger the profits if the German mark appreciates. Consider the "60:40" strategy, whereby only 60% of the German marks are sold forward with the remaining unhedged. Derive the payoff function in terms of the $/DEM exchange rate 3 months from now.

Spot and Forward Exchange Rates

Let S denote the spot exchange rate and F the forward exchange rate 1 year from now (both in domestic/foreign terms). Use r_f and r_l to denote the annual interest rates of the foreign currency and the local currency, respectively. First formulated by Keynes[3] in 1923 [303], arbitrage opportunities will arise unless these four numbers satisfy a definite relation known as the **interest rate parity**:

$$\frac{F}{S} = \frac{e^{r_l}}{e^{r_f}} = e^{r_l - r_f}. \tag{12.1}$$

Here is the argument. A holder of the local currency can either (1) lend the money in the domestic market to receive e^{r_l} one year from now or (2) convert the local currency in the spot market for the foreign currency, lend for 1 year in the foreign market, and convert the foreign currency into the local currency at the fixed forward exchange rate in the future, F, by selling forward the foreign currency now. As usual, no money changes hand in entering into a forward contract. One unit of local currency will hence become Fe^{r_f}/S 1 year from now in this case. If $Fe^{r_f}/S > e^{r_l}$, an arbitrage profit can result from borrowing money in the domestic market and lending it in the foreign market. Conversely, if $Fe^{r_f}/S < e^{r_l}$, an arbitrage profit can result from borrowing money in the foreign market and lending it in the domestic market. We conclude that $Fe^{r_{\mathrm{f}}}/S = e^{r_\ell}$. It is straightforward to check that

$$\frac{F}{S} = \frac{1+r_\ell}{1+r_{\mathrm{f}}} \tag{12.2}$$

under periodic compounding.

The interest rate parity says that if the domestic interest rate is higher than the foreign rate, the foreign country's currency will be selling at a premium in the forward market. Conversely, if the domestic interest rate is lower, the foreign currency will be selling at a discount in the forward market.

➤ **Exercise 12.2.2** (1) What does the table in Fig. 12.1 say about the relative interest rates between the United States and Germany? (2) Estimate German interest rates from Fig. 12.1 and Eq. (12.2) if the annualized U.S. rates for 30-day, 90-day, and 180-day T-bills are 5.66%, 5.9%, and 6.15%, respectively.

➤ **Exercise 12.2.3** Show that Eq. (11.6) can be simplified to

$$C = Fe^{-r\tau} N(x) - Xe^{-r\tau} N(x - \sigma\sqrt{\tau}),$$
$$P = Xe^{-r\tau} N(-x + \sigma\sqrt{\tau}) - Fe^{-r\tau} N(-x)$$

without the explicit appearance of the exchange rate, where

$$x \equiv \frac{\ln(F/X) + (\sigma^2/2)\,\tau}{\sigma\sqrt{\tau}}.$$

Note that S, F, and X above are in domestic/foreign terms.

12.2.2 Forward Price

The payoff from holding a forward contract at maturity is $S_T - X$ (see Fig. 12.2). Contrast this with the European call's $\max(S_T - X, 0)$. Forward contracts do not involve any initial cash flow. The **forward price** is the delivery price that makes the forward contract zero valued; in other words, $f = 0$ when $F = X$. The delivery price cannot change because it is written in the contract, but the forward price may change after the contract comes into existence. In other words, the value of a forward contract, f, is zero at the outset, and it will fluctuate with the spot price thereafter. Apparently this value is enhanced when the spot price climbs and depressed when the spot price declines. The forward price also varies with the maturity of the contract, which can be verified by the data in Fig. 12.1.

For example, a repo is a forward contract on a Treasury security. It has a zero value initially because the Treasury security is exchanged for its fair value in cash and the repurchase price is set to the forward price [510].

The Underlying Asset Pays No Income

LEMMA 12.2.1 *For a forward contract on an underlying asset providing no income,*

$$F = Se^{r\tau}. \tag{12.3}$$

Proof: If $F > Se^{r\tau}$, an investor can borrow S dollars for τ years, buy the underlying asset, and short the forward contract with delivery price F. At maturity, the asset

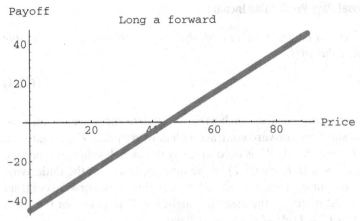

Figure 12.2: Payoff of forward contract. Shown is the payoff of a long forward contract with a delivery price of $45 at maturity.

is sold for F and $Se^{r\tau}$ is used to repay the loan, leaving an arbitrage profit of $F - Se^{r\tau} > 0$.

If $F < Se^{r\tau}$ instead, an investor can short the underlying asset, invest the proceeds for τ years, and take a long position in the forward contract with delivery price F. At maturity, the asset is bought for F to close out the short position, leaving a profit of $Se^{r\tau} - F$.

EXAMPLE 12.2.1 A new 3-month forward contract on a 6-month zero-coupon bond should command a delivery price of $Se^{r/4}$, where r is the annualized 3-month riskless interest rate and S is the spot price of the bond. For instance, if $r = 6\%$ and $S = 970.87$, then the delivery price is $970.87 \times e^{0.06/4} = 985.54$.

The forward price, as previously mentioned, may not maintain equality to the delivery price as time passes. In fact, the value of a forward contract providing no income at any time before maturity should be

$$f = S - Xe^{-r\tau}.$$

We can verify this by considering a portfolio of one long forward contract, cash amount $Xe^{-r\tau}$, and one short position in the underlying asset. The cash will grow to X at maturity, which can be used to take delivery of the forward contract. The delivered asset will then close out the short position. Because the value of the portfolio is zero at maturity, its PV must be zero. Lemma 12.2.1 can be proved alternatively by the preceding identity because X must equal $Se^{r\tau}$ in order for $f = 0$.

▶ **Exercise 12.2.4** (1) Prove that a newly written forward contract is equivalent to a portfolio of one long European call and one short European put on the same underlying asset and expiration date with a common strike price equal to the forward price. (An option is said to be **at the money forward** if $X = F$.) (2) Prove the alternative put–call parity that says $C = P$ for the call and put in (1). (3) Derive Lemma 12.2.1 from the put–call parity. (4) Verify that, by substituting the forward price in Eq. (12.3) for the strike price X in the Black–Scholes formula of Theorem 9.3.4, we obtain $C = P$.

The Underlying Asset Pays Predictable Income

LEMMA 12.2.3 *For a forward contract on an underlying asset providing a predictable income with a present value of I,*

$$F = (S - I)\, e^{r\tau}. \tag{12.4}$$

Proof: If $F > (S - I)\, e^{r\tau}$, an investor can borrow S dollars for τ years, buy the underlying asset, and short the forward contract with delivery price F. At maturity, the asset is sold for F, and $(S - I)\, e^{r\tau}$ is used to repay the loan, leaving an arbitrage profit of $F - (S - I)\, e^{r\tau} > 0$. If $F < (S - I)\, e^{r\tau}$, an investor can short the underlying asset, invest the proceeds for τ years, and take a long position in the forward contract with delivery price F. At maturity, the asset is bought for F to close out the short position, and a profit of $(S - I)\, e^{r\tau} - F > 0$ is realized.

EXAMPLE 12.2.4 The above results can be extended to nonflat yield curves. A 10-month forward contract on a $50 stock that pays a dividend of $1 every 3 months has

$$\left(50 - e^{-r_3/4} - e^{-r_6/2} - e^{-3 \times r_9/4}\right) e^{r_{10} \times (10/12)}$$

as the forward price, where r_i is the annualized i-month interest rate.

The value of a forward contract providing a predictable income with a PV of I is

$$f = (S - I) - Xe^{-r\tau}.$$

We can confirm this by considering a portfolio of one long forward contract, cash amount $Xe^{-r\tau} + I$, and one short position in the underlying asset. The cash will grow to X at maturity after paying the dividends to the original stockholder. There is a sufficient fund to take delivery of the forward contract, which then offsets the short position. Because the value of the portfolio is zero at maturity, its value must be zero at present, or $f - (S - I) + Xe^{-r\tau} = 0$.

The Underlying Asset Pays a Continuous Dividend Yield

A continuous dividend yield means that dividends are paid out continuously at an annual rate of q. The value of a forward contract at any time before maturity must equal

$$f = Se^{-q\tau} - Xe^{-r\tau}. \tag{12.5}$$

We can verify this by considering a portfolio of one long forward contract, cash amount $Xe^{-r\tau}$, and a short position in $e^{-q\tau}$ units of the underlying asset. All dividends are paid for by shorting additional units of the underlying asset. Hence the cash will grow to X at maturity, and the short position will grow to exactly one unit of the underlying asset. There is a sufficient fund to take delivery of the forward contract, which then offsets the short position. Because the value of the portfolio is zero at maturity, its PV must be zero. One consequence of Eq. (12.5) is that the forward price is

$$F = Se^{(r-q)\tau}. \tag{12.6}$$

▶ **Exercise 12.2.5** All the above cases satisfy the following relation:

$$f = (F - X)e^{-r\tau}. \tag{12.7}$$

Prove the preceding identity by an arbitrage argument.

12.3 Futures Contracts

Futures contracts are different from forward contracts in several ways. First, they are traded on a central exchange rather than on over-the-counter markets, leading to a more efficient and accurate price determination. Second, the establishment of a clearinghouse means that sellers and buyers do not face each other. As in the options market, the clearinghouse acts as a seller to all buyers and a buyer to all sellers. Credit risk inherent in forward contracts is hence minimized. Third, futures contracts are standardized instruments. They specify the delivery of a specific quantity of a specific commodity that meets quality standards at predetermined places and dates. This is in sharp contrast with forward contracts for which the only requirement is mutual agreement. Finally, gains and losses of futures contracts are **marked to market** daily. Hence the account is adjusted at the end of each trading day based on the **settlement price** to reflect the investor's gain or loss. The settlement price is the average of the prices at which the contract is traded immediately before the bell signaling the end of trading for the day.

EXAMPLE 12.3.1 The CBT July wheat contract specifies, among other things, that the wheat delivered be 5,000 bushels of no. 2 soft red wheat, no. 2 hard red winter wheat, no. 2 dark northern spring wheat, or no. 1 northern spring wheat on a date in the month of July chosen by the seller [95, 799].

The **contract size**, or simply the **size**, of a futures contract is the amount of the underlying asset to be delivered under the contract. For instance, it is 5,000 bushels for the corn futures contracts on the CBT and 1 million U.S. dollars for the Eurodollar futures contracts on the CME. A position can be **closed out** or **offset** by entering into a reversing trade to the original one. An investor who is long one November soybeans futures contract can close out the position by shorting one November contract, for example. A clearinghouse simply cancels offsetting positions from its book. Most futures contracts are closed out in this way rather than have the underlying asset delivered. In contrast, forward contracts are meant for delivery.

EXAMPLE 12.3.2 A farmer sold short corn futures, and the cost of growing corn rises now. He can offset his short futures position to reduce the losses. Consider another farmer who faces the problem that the crop is going to be different from the 100,000-bushel projection. Because corn futures trade in 5,000-bushel pieces, 20 contracts were sold to cover the anticipated 100,000-bushel crop. If the crop now appears to be only 80,000 bushels, the farmer can offset four of those contracts. Such flexibility is not available to forward contracts.

Because price changes in the futures contract are settled daily, the difference between them has been paid for in installments throughout the life of the contract. Hence the spot price rather than the initial futures price is paid on the delivery date. Marking to market nullifies any financial incentive for not making delivery. Suppose that a farmer enters into a forward contract to sell a food processor 100,000 bushels of corn at $2.00 per bushel in November. If the price of corn rises to $2.5 by November, the farmer has incentive to sell his harvest in the spot market at $2.5 rather than to the processor at $2.00. With marking to market, however, the farmer has transferred $0.5 per bushel from his futures account to that of the food processor by November. When the farmer makes delivery, he is paid the spot price, $2.5 per bushel. The farmer thus has little incentive to default. Note that the net price remains $2.00 per bushel, the original delivery price.

The prospect of delivery ties the futures price to the spot price (or **cash price**), which makes hedging possible. On the delivery date, the settlement price of the futures contract is determined by the spot price. Before the delivery date, however, the futures price could be above or below the spot price.

Financial futures are futures contracts based on a financial instrument or index. Since the first financial futures were launched in 1972, the trading of financial futures has surpassed that of agricultural futures. The most popular equity financial futures today, the S&P 500 Index futures, were created in 1982 [865]. Like options on stock index, some financial futures are settled in cash rather than by delivery of the underlying asset. The S&P 500 Index futures contract, for instance, is settled in cash rather than by delivering 500 stocks. Each S&P 500 contract is on $500 times the index (see Fig. 12.3). As another example, one futures contract on the Nikkei 225 Stock Average is on US$5 times the index. This amounts to fixing the dollar–yen exchange rate. The index is price weighted and on a portfolio of 225 of the largest stocks traded on the Tokyo Stock Exchange.

The difference between the spot price and the futures price, $S - F$, is called the **basis**. For example, if the soybeans cost $7.80 a bushel, and the November soybean futures contract is $7.90 on the same day, the basis would be 10 cents under (-10 cents) the November contract. As another example, if the cash market price of a T-bond is 88-16 and the June adjusted futures price is 90-22, then the basis is 2-06 under the June contract. Note that T-bond futures are quoted in 32nds, and prices like $88_{16/32}$ are written as 88-16. Basis can be positive or negative, but it should converge eventually to zero. If the basis moves toward zero, it is said to be **narrowing**, whereas it is said to be **widening** if it moves away from zero. Although basis cannot be predicted precisely, it is generally less volatile than either the futures price or the spot price.

EXAMPLE 12.3.3 Suppose that the spot price of wheat is $4.225 per bushel and the July futures price of wheat is $3.59 per bushel with a contract size of 5,000 bushels. The basis is $4.225 - 3.59 = 0.635$ per bushel. Imagine that the basis widens by $.1 to $.735 caused by, say, the futures price's falling to $3.54 and the spot price's rising to $4.275. A person with a short position in one futures contract and a long position in 5,000 bushels of wheat will make $5000 \times 0.1 = 500$ dollars in profit. If the basis narrows by $.1 to $0.535, the same investor will have a loss of an equal amount.

EXAMPLE 12.3.4 A firm to be paid £1 million in 60 days is worried that the pound will weaken. Suppose that a pound futures contract on the International Monetary Market (IMM) of the CME has a settlement date in 71 days. Because each contract controls 62,500 pounds, $1,000,000/62,500 = 16$ contracts are sold.

Monday, March 20, 1995
FUTURES PRICES

	Open	High	Low	Settle	Change	Lifetime High	Low	Open Interest
GRAINS AND OILSEEDS								
CORN (CBT) 5,000 bu.; cents per bu.								
Mar	240¼	241	239½	239¾	−1	282½	220½	2,432
May	246¾	248	246¼	246½	−¾	285	228	113,058
				. . .				
INTEREST RATE								
TREASURY BONDS (CBT)–$100,000; pts. 32nds of 100%								
Mar	104-31	105-05	104-18	104-20	−11	116-20	95-13	28,210
June	104-12	104-19	104-00	104-02	−11	113-15	94-27	328,566
				. . .				
LIBOR-1MO. (CME)–$3,000,000; points of 100%								
Apr	93.84	93.84	93.82	93.83	. . .	6.17	. . .	27,961
				. . .				
EURODOLLAR (CME)–$1 million; pts of 100%								
June	93.52	93.54	93.51	93.52	. . .	6.48	. . .	515,578
Sept	93.31	93.32	93.28	93.30	. . .	6.70	. . .	322,889
				. . .				
CURRENCY								
JAPAN YEN (CME)–12.5 million yen; $ per yen (.00)								
June	1.1363	1.1390	1.1240	1.1301	−.0030	1.1390	.9915	56,525
Sept	1.1490	1.1491	1.1385	1.1430	−.0029	1.1491	1.0175	2,282
				. . .				
DEUTSCHEMARK (CME)–125,000 marks; $ per mark								
June	.7249	.7255	.7124	.7147	−.0087	.7448	.5980	56,053
Sept	.7182	.7215	.7153	.7171	−.0087	.7415	.6290	1,776
				. . .				
BRITISH POUND (CME)–62,500 pds; $ per pound								
June	1.5910	1.5936	1.5680	1.5734	−.0102	1.6530	1.5330	21,050
Sept	1.5770	1.5830	1.5680	1.5704	−.0102	1.6480	1.5410	149
				. . .				
INDEX								
S&P 500 INDEX (CME)–$500 times index								
June	499.75	500.75	498.90	500.15	+.40	501.00	449.50	186,725
				. . .				

Figure 12.3: Futures quotations. Source: *Wall Street Journal*, March 21, 1995.

12.3.1 Daily Cash Flows

Consider a futures contract with n days to maturity. Let F_i denote the futures price at the end of day i for $0 \leq i \leq n$. The contract's cash flow on day i is $F_i - F_{i-1}$ because of daily settlement. Hence, the net cash flow over the life of the contract is

$$(F_1 - F_0) + (F_2 - F_1) + \cdots + (F_n - F_{n-1}) = F_n - F_0 = S_T - F_0. \qquad (12.8)$$

Recall that F_n equals the spot price at maturity, S_T. Although a futures contract has the same accumulated payoff $S_T - F_0$ as a forward contract, the actual payoff may differ because of the reinvestment of daily cash flows and how $S_T - F_0$ is distributed over the n-day period. In contrast, no cash flows occur until settlement for forward contracts.

▶ **Exercise 12.3.1** Suppose that the interest rate is such that \$1 grows to \$$R$ over a 1-day period and that this rate applies to both borrowing and lending. Derive the payoff for a futures contract if the cash flow is reinvested through lending when it is positive and financed through borrowing when it is negative.

12.3.2 Forward and Futures Prices

Somewhat surprisingly, Cox et al. proved that the futures price equals the forward price if interest rates are nonstochastic [233]. This result often justifies treating a futures contract as if it were a forward contract, ignoring its marking-to-market feature.

Consider forward and futures contracts on the same underlying asset with n days to maturity. Suppose that the interest rate for day i is r_i and that \$1 at the beginning of day i grows to $R_i \equiv e^{r_i}$ by day's end. Let F_i be the futures price at the end of day i. Note that \$1 invested in the n-day discount bond at the end of day zero will be worth $R \equiv \prod_{j=1}^{n} R_j$ by the end of day n.

Starting from day one, we maintain $\prod_{j=1}^{i} R_j$ long futures positions at the end of day $i-1$ and invest the cash flow at the end of day i in riskless bonds maturing on day n, the delivery date. The cash flow from the position on day i is $(F_i - F_{i-1}) \prod_{j=1}^{i} R_j$ because day i starts with $\prod_{j=1}^{i} R_j$ contracts. This amount will be compounded until the end of day n to become

$$(F_i - F_{i-1}) \prod_{j=1}^{i} R_j \prod_{j=i+1}^{n} R_j = (F_i - F_{i-1}) \prod_{j=1}^{n} R_j = (F_i - F_{i-1}) \, R.$$

The value at the end of day n is therefore

$$\sum_{i=1}^{n} (F_i - F_{i-1}) \, R = (F_n - F_0) \, R = (S_T - F_0) \, R.$$

Observe that no investment is required for the strategy.

Suppose that the forward price f_0 exceeds the futures price F_0. We can short R forward contracts, borrow $f_0 - F_0$, and carry out the above strategy. The initial cash flow is $f_0 - F_0 > 0$. At the end of day n, the debt grows to $(f_0 - F_0) \, R$, and the net value is

$$f_0 R - S_T R - (f_0 - F_0) \, R + (S_T - F_0) \, R = 0.$$

Therefore $f_0 - F_0 > 0$ is a pure arbitrage profit. The case of $f_0 < F_0$ is symmetrical. This completes the proof.

With stochastic interest rates, forward and futures prices are no longer theoretically identical. In fact, this is the major reason for the price differences in the forex forward and futures markets [176, 275]. For short-term contracts, however, the differences tend to be small. In fact, the differences are significant only for longer-term contracts on interest-rate-sensitive assets. We shall assume that forward and futures prices are identical.

▶ **Exercise 12.3.2** Complete the proof by considering the $f_0 < F_0$ case.

▶ **Exercise 12.3.3** Suppose that interest rates are uncertain and that futures prices move in the same direction as interest rates. Argue that futures prices will exceed forward prices. Similarly, argue that if futures prices move in a direction opposite from that of interest rates, then futures prices will be exceeded by forward prices.

12.3.3 Stock Index Futures

Stock index futures originated in 1982 when the Kansas City Board of Trade introduced the Value Line Stock Index futures. There are now stock index futures based on the S&P 500 Index, the Nikkei 225 Stock Average futures, the NYSE Composite Index futures (traded on the New York Futures Exchange for $500 times the index), the Major Market Index (traded on the CBT for $500 times the index), the DJIA Index (traded on the CBT for $10 times the index; ticker symbol DJ), and so on.

Indices can be viewed as dividend-paying securities, the security being the basket of stocks comprising the index and the dividends being those paid by the stocks. If the index is broadly based, dividends can be assumed to be paid continuously. With q denoting the average annualized dividend yield during the life of the contract, the futures price is then

$$F = Se^{(r-q)\tau}. \tag{12.9}$$

EXAMPLE 12.3.5 The S&P 500 Index futures contract is based on the S&P 500 Index. The minimum fluctuation (**tick size**) is 0.05 point. Because the value of a contract is $500 times the Index, a change of 0.05 represents a $500 \times 0.05 = $25 tick. Consider a 3-month futures contract on the S&P 500 Index. Suppose that the stocks underlying the index provide a dividend yield of 3% per annum, the current value of the index is 480, and the interest rate is 8%. The theoretical futures price is then $480 \times e^{0.05 \times 0.25} = 486.038$.

When Eq. (12.9) fails to hold, one can create arbitrage profits by trading the stocks underlying the index and the index futures. For example, when $F > Se^{(r-q)\tau}$, one can make profits by buying the stocks underlying the index and shorting futures contracts. One should do the reverse if $F < Se^{(r-q)\tau}$. These strategies are known as **index arbitrage** and are executed by computers, an activity known as **program trading**. Equation (12.9) is not applicable to the Nikkei 225 futures, however. Recall that one such contract is on the *dollar* amount equal to five times the index, which is measured in *yen*. But no securities whose value is $5 times the index exist; hence the arbitrage argument breaks down.

For indices that all stocks tend to pay dividends on the same date, we can estimate the dividends' dollar amount and timing. Then the index becomes a security providing known income, and Eq. (12.4) says the futures price is

$$F = (S - I)e^{r\tau}. \tag{12.10}$$

▶ **Exercise 12.3.4** Do Eqs. (12.9) and (12.10) assume that the stock index involved is not adjusted for cash dividends?

12.3.4 Forward and Futures Contracts on Currencies

Let S denote the domestic/foreign exchange rate and let X denote the delivery price of the forward contract. Use r_f to refer to the foreign riskless interest rate. A portfolio consisting of one long forward contract, cash amount $Xe^{-r\tau}$ in domestic currency, and one short position in the amount of $e^{-r_f\tau}$ in foreign currency is clearly worth zero at time T. Hence its current value must be zero, that is, $f + Xe^{-r\tau} - Se^{-r_f\tau} = 0$. The value X that makes $f = 0$ is the forward price (i.e., the forward exchange rate) $F = Se^{(r-r_f)\tau}$, which is exactly the interest rate parity.

12.3.5 Futures on Commodities and the Cost of Carry

Some commodities are held solely for investment (such as gold and silver), whereas others are held primarily for consumption. Arbitrage arguments can be used to obtain futures prices in the former case, but they give only upper bounds in the latter.

For a commodity held for investment purposes and with zero storage cost, the futures price is $F = Se^{r\tau}$ according to Eq. (12.3). In general, if U stands for the PV of the storage costs incurred during the life of a futures contract, then Eq. (12.4) implies that $F = (S + U)e^{r\tau}$ as storage costs are negative income. Alternatively, if u denotes the storage cost per annum as a proportion of the spot price, then Eq. (12.6) implies that $F = Se^{(r+u)\tau}$ as storage costs provide a negative dividend yield. For commodities held primarily for consumption, however, we can say only that

$$F \le (S + U)e^{r\tau}, \quad F \le Se^{(r+u)\tau},$$

respectively, because of the benefits of holding the physicals. These benefits are measured by the so-called **convenience yield** defined as the y such that

$$Fe^{y\tau} = (S + U)e^{r\tau}, \quad Fe^{y\tau} = Se^{(r+u)\tau}, \tag{12.11}$$

respectively.

We can frame the relation between the futures and spot prices in terms of the **cost of carry**, which is the storage cost plus the interest cost paid to carry the asset but less the income earned on the asset. For a stock paying no dividends, the cost of carry is r because it neither incurs storage costs nor generates any income; for a stock index, it is $r - q$ as income is earned at rate q; for a currency, it is $r - r_f$; for a commodity with storage costs, it is $r + u$. Suppose the cost of carry is c and the convenience yield is y. For an investment asset $F = Se^{c\tau}$, and for a consumption asset $F = Se^{(c-y)\tau}$.

The cost of carry is often cast in monetary terms, called the **carrying charge** or the **carrying cost**. It measures the dollar cost of carrying the asset over a period and consists of interest expense I, storage costs U, minus cash flows generated by the asset D:

$$C \equiv I + U - D. \tag{12.12}$$

The cost of carry will be in dollar terms from now on unless stated otherwise. Similarly, the convenience yield can also be expressed in dollar terms:

$$\text{convenience yield} \equiv S + C - F.$$

As a consequence, the basis $S - F$ is simply the convenience yield minus the cost of carry:

$$\text{basis} = \text{convenience yield} - C. \tag{12.13}$$

Recall that the convenience yield is negligible for financial instruments and commodities held primarily for investment purposes. For such assets, changes in basis are due entirely to changes in the cost of carry.

Look up the futures prices of corn in Fig. 12.3. Because the prices for the near months are lower than the distant months, this market is said to be **normal**. The premium that the distant months command over near months is due to the greater carrying costs. In an **inverted** or **discount** market, the distant months sell at lower prices than the near months. A strong demand for cash grains or the willingness of elevator owners to store grains at less than the full storage costs can create an inverted market.

When the forward price equals the sum of the spot price and the carrying charge (in other words, zero convenience yield), the forward price is said to be at **full carry**. Forward and futures prices should be set at full carry for assets that have zero storage cost and can be sold short or in ample supply (see Exercises 12.3.7 and 12.3.8). As previously mentioned, commodities held for investment purposes should also reflect full carry. If the total cost of storing corn is, say, four cents per bushel a month and if futures prices reflect the full carrying cost, the prices for the different delivery months might look like the following table:

December	March	May	July	September
$2.00	$2.12	$2.20	$2.28	$2.36

▶ **Exercise 12.3.5** For futures, the cost of carry may be called **cash and carry**, which is the strategy of buying the cash asset with borrowed funds. (1) Illustrate this point with futures contracts when the underlying asset pays no income. (2) Show that the cost of carry can be used to find the futures price in (1) set at full carry.

▶ **Exercise 12.3.6** A manufacturer needs to acquire gold in 3 months. The following options are open to her: (1) Buy the gold now or (2) go long one 3-month gold futures contract and take delivery in 3 months. If she buys the gold now, the money that has been tied up could be invested in money market instruments. This is the opportunity cost of buying physical gold. What is the cost of carry for owning 100 ounces of gold at $350 per ounce for a year if the T-bills are yielding an annually compounded rate of 6%?

▶ **Exercise 12.3.7** Prove that $F \leq S + C$, where C is the net carrying cost per unit of the commodity to the delivery date.

▶ **Exercise 12.3.8** For a commodity that can be sold short, such as a financial asset, prove that $F \geq S + C - U$, where U is the net storage cost for carrying one unit of the commodity to the delivery date.

12.4 **Futures Options and Forward Options**

The underlying asset of a futures option is a futures contract. On exercise, the option holder takes a position in the futures contract with a futures price equal to the option's strike price. In particular, a futures call (put) option holder acquires a long (short, respectively) futures position. The option writer does the opposite: A futures call (put, respectively) option writer acquires a short (long, respectively) futures position. The futures contract is then marked to market immediately, and the futures position of the two parties will be at the prevailing futures price. The option holder can withdraw in cash the difference between the prevailing futures price and the strike price. Of course, the option's expiration date should precede the futures contract's delivery date.

The whole process works as if the option writer delivered a futures contract to the option holder and paid the holder the prevailing futures price minus the strike price in the case of calls. In the case of puts, it works as if the option writer took delivery a futures contract from the option holder and paid the holder the strike price minus the prevailing futures price. Note that the amount of money that changes hands on exercise is only the difference between the strike price and the prevailing futures price. See Fig. 12.4 for sample quotations.

EXAMPLE 12.4.1 An investor holds a July futures call on 5,000 bushels of soybeans with a strike price of 600 cents per bushel. Suppose that the current futures price

Monday, March 20, 1995

· · ·

INTEREST RATE

T-BONDS (CBT)
$100,000; points and 64ths of 100%

Strike Price	Calls – Settle			Puts – Settle		
	Apr	May	Jun	Apr	May	Jun
102	2-06	2-26	2-47	0-03	0-23	0-43
103	1-12	1-44	. . .	0-08	0-40	. . .
104	0-30	1-05	1-29	0-26	1-01	1-25
105	0-07	0-39	. . .	1-03	1-34	. . .
106	0-01	0-21	0-40	1-61	. . .	2-35
107	0-01	0-10	. . .	. . .	. . .	. . .

· · ·

T-NOTES (CBT)
$100,000; points and 64ths of 100%

Strike Price	Calls – Settle			Puts – Settle		
	Apr	May	Jun	Apr	May	Jun
102	2-26	. . .	2-45	0-01	. . .	0-20
103	1-28	. . .	1-60	0-02	0-19	0-35
104	0-36	. . .	1-19	0-11	. . .	0-57
105	0-07	0-32	0-51	0-45	1-06	1-25
106	0-01	. . .	0-28	1-39	. . .	2-01
107	0-01	0-05	0-14	. . .	. . .	2-51

· · ·

LIBOR – 1 Mo. (CME)
$3 million; pts. of 100%

Strike Price	Calls – Settle			Puts – Settle		
	Apr	May	Jun	Apr	May	Jun
9325	0.58	0.51	0.45	0.00	0.01	0.03
9350	0.34	0.29	0.24	0.01	0.04	0.07
9375	0.11	0.10	0.09	0.03	. . .	0.17
9400	0.01	. . .	0.03	. . .	. . .	. . .
9425	. . .	. . .	. . .	. . .	. . .	. . .
9450	0.00	0.00	. . .	. . .	. . .	. . .

· · ·

INDEX

S&P 500 STOCK INDEX (CME)
$500 times premium

Strike Price	Calls – Settle			Puts – Settle		
	Apr	May	Jun	Apr	May	Jun
490	12.55	14.80	16.75	2.45	4.75	6.75
495	8.75	11.25	13.35	3.65	6.15	8.25
500	5.70	8.10	10.20	5.55	7.95	10.05
505	3.45	5.55	7.50	8.25	10.35	12.30
510	1.85	3.55	5.30	11.65	. . .	15.00
515	0.85	2.15	3.55	. . .	. . .	. . .

· · ·

Figure 12.4: Futures options quotations. Months refer to the expiration month of the underlying futures contract. Source: *Wall Street Journal*, March 21, 1995.

of soybeans for delivery in July is 610 cents. The investor can exercise the option to receive $500 (5,000 × 10 cents) plus a long position in a futures contract to buy 5,000 bushels of soybeans in July. Similarly, consider an investor with a July futures put on 5,000 bushels of soybeans with a strike price of 620 cents per bushel. Suppose the current futures price of soybeans for delivery in July is 610 cents. The investor can exercise the option to receive $500 (5,000 × 10 cents) plus a short position in a futures contract to buy 5,000 bushels of soybeans in July.

EXAMPLE 12.4.2 Suppose that the holder of a futures call with a strike price of $35 exercises it when the futures price is $45. The call holder is given a long position in the futures contract at $35, and the call writer is assigned the matching short position at $35. The futures positions of both are immediately marked to market by the exchange. Because the prevailing futures price is $45, the long futures position (the position of the call holder) realizes a gain of $10, and the short futures position (the position of the call writer) realizes a loss of $10. The call writer pays the exchange $10 and the call holder receives from the exchange $10. The call holder, who now has a long futures position at $45, can either liquidate the futures position at $45 without costs or maintain it.

Futures options were created in 1982 when the CBT began trading options on T-bond futures. Futures options are preferred to options on the cash instrument in some markets on the following grounds. In contrast to the cash markets, which are often fragmented and over the counter, futures trading takes place in competitive, centralized markets. Futures options have fewer liquidity problems associated with shortages of the cash assets – selling a commodity short may be significantly more difficult than selling a futures contract. Futures options are also useful in implementing certain strategies. Finally, futures options are popular because of their limited capital requirements [746].

Forward options are similar to futures options except that what is delivered is a forward contract with a delivery price equal to the option's strike price. In particular, exercising a call (put) forward option results in a long (short, respectively) position in a forward contract. Note that exercising a forward option incurs no immediate cash flows. Unlike futures options, forward options are traded not on organized exchanges but in over-the-counter markets.

EXAMPLE 12.4.3 Consider a call with strike $100 and an expiration date in September. The underlying asset is a forward contract with a delivery date in December. Suppose that the forward price in July is $110. On exercise, the call holder receives a forward contract with a delivery price of $100. If an offsetting position is then taken in the forward market, a $10 profit in September will be ensured. Were the contract a call on the futures, the $10 profit would be realized in July.

▶ **Exercise 12.4.1** With a conversion, the trader buys a put, sells a call, and buys a futures contract. The put and the call have the same strike price and expiration month. The futures contract has the same expiration month as the options, and its price is equal to the options' strike price. Argue that any initial positive cash flow of conversion is guaranteed profit.

12.4.1 Pricing Relations

Assume a constant, positive interest rate. This is acceptable for short-term contracts. Even under this assumption, which equates forward price with futures price, a forward option does not have the same value as a futures option. Let delivery take place at time T, the current time be zero, and the option on the futures or forward contract have expiration date t $(t \le T)$. Note that the futures contract can be marked to market when the option is exercised. Example 12.4.3 established the following identities for the futures options and forward options when they are exercised at time t:

$$\text{value of futures option} = \max(F_t - X, 0) \tag{12.14}$$
$$\text{value of forward option} = \max(F_t - X, 0)\, e^{-r(T-t)} \tag{12.15}$$

Furthermore, a European futures option is worth the same as the corresponding European option on the underlying asset if the futures contract has the same maturity as the option. The reason is that the futures price equals the spot price at maturity. This conclusion is independent of the model for the spot price.

The put–call parity is slightly different from the one in Eq. (8.1). Whereas the undiscounted stock price was used in the case of stock options, it is the discounted futures/forward prices that should be used here (see also Exercise 12.2.4).

THEOREM 12.4.4 (Put–Call Parity). *For European options on futures contracts, $C = P - (X - F)\,e^{-rt}$. For European options on forward contracts, $C = P - (X - F)\,e^{-rT}$.*

Proof: Consider a portfolio of one short call, one long put, one long futures contract, and a loan of $(X - F)\,e^{-rt}$. We have the following cash flow at time t.

	$F_t \le X$	$F_t > X$
A short call	0	$X - F_t$
A long put	$X - F_t$	0
A long futures	$F_t - F$	$F_t - F$
A loan of $(X - F)\,e^{-rt}$	$F - X$	$F - X$
Total	0	0

Because the net future cash flow is zero in both cases, the portfolio must have zero value today. This proves the theorem for futures option.

The proof for forward options is identical except that the loan amount is $(X - F)\,e^{-rT}$ instead. The reason is that the forward contract can be settled only at time T.

An American forward option should be worth the same as its European counterpart. In other words, the early exercise feature is not valuable.

THEOREM 12.4.5 *American forward options should not be exercised before expiration as long as the probability of their ending up out of the money is positive.*

Proof: Consider a portfolio of one long forward call, one short forward contract with delivery price F, and a loan of $(F - X)\,e^{-rT}$. If $F_t < X$ at t, the wealth at t is

$$0 + (F - F_t)\,e^{-r(T-t)} - (F - X)\,e^{-r(T-t)} = (X - F_t)\,e^{-r(T-t)} > 0.$$

If $F_t \ge X$ at t, the wealth at t is

$$(F_t - X)\,e^{-r(T-t)} + (F - F_t)\,e^{-r(T-t)} - (F - X)\,e^{-r(T-t)} = 0.$$

So the value of the forward call C satisfies $C - (F - X)e^{-rT} > 0$. On the other hand, if the call is exercised immediately, the PV at time zero is only $\max(F - X, 0)e^{-rT}$. The case of forward puts is proved in Exercise 12.4.3.

Early exercise may be optimal for American futures options. Hence an American futures option is worth more than the European counterpart even if the underlying asset generates no payouts [125].

THEOREM 12.4.6 *American futures options may be exercised optimally before expiration.*

▶ **Exercise 12.4.2** Prove Theorem 12.4.5 for American forward puts. (Hint: Show that $P - (X - F)e^{-rT} > 0$ first.)

▶ **Exercise 12.4.3** Prove that $Fe^{-rt} - X \le C - P \le F - Xe^{-rt}$ for American futures options.

12.4.2 The Black Model

Black developed the following formulas for European futures options in 1976:

$$C = Fe^{-rt}N(x) - Xe^{-rt}N(x - \sigma\sqrt{t}), \tag{12.16}$$
$$P = Xe^{-rt}N(-x + \sigma\sqrt{t}) - Fe^{-rt}N(-x),$$

where [81]

$$x \equiv \frac{\ln(F/X) + (\sigma^2/2)t}{\sigma\sqrt{t}}.$$

Formulas (12.16) are related to those for options on a stock paying a continuous dividend yield. In fact, they are exactly Eq. (9.20) with the dividend yield q set to the interest rate r and the stock price S replaced with the futures price F. This observation also proves Theorem 12.4.6 based on the discussions in Subsection 9.6.2. For European forward options, just multiply the above formulas by $e^{-r(T-t)}$ as forward options differ from futures options by a factor of $e^{-r(T-t)}$ based on Eqs. (12.14) and (12.15).

Black's formulas can be expressed in terms of S instead of F by means of the substitution $F = Se^{(r-q)T}$. (The original formulas have the advantage of not containing q or T. The delta for the call is then $\partial C/\partial F = e^{-rt}N(x)$ and that for the put is $\partial P/\partial F = e^{-rt}[N(x) - 1]$. The delta for the call can also be cast with respect to the spot price:

$$\frac{\partial C}{\partial S} = \frac{\partial C}{\partial F}\frac{\partial F}{\partial S} = e^{-rt}N(x)e^{(r-q)T} = e^{-r(t-T)-qT}N(x).$$

Other sensitivity measures can be easily derived [746, p. 345].

Besides index options and index futures, a third type of stock index derivative is the index futures option. European index futures options can be priced by Black's formulas. The S&P 500 Index and the DJIA span all three types of derivatives. The NYA has options and futures options. Although the SPX and the DJIA index options are European, their index futures options are American. The NYA index option and futures option are both American.

Binomial tree algorithm for pricing American futures calls:

input: F, σ, t, X, r, n;
real $R, p, u, d, C[n+1]$;
integer i, j;
$R := e^{r(t/n)}$;
$u := e^{\sigma\sqrt{t/n}}; d := e^{-\sigma\sqrt{t/n}}$;
$p := (1-d)/(u-d)$;   // Risk-neutral probability.
for $(i = 0$ to $n)$ $\{ C[i] := \max(0, Fu^{n-i}d^i - X); \}$
for $(j = n-1$ down to $0)$
        for $(i = 0$ to $j)$
                $C[i] := \max((p \times C[i] + (1-p) \times C[i+1])/R, Fu^{j-i}d^i - X)$;
return $C[0]$;

Figure 12.5: Binomial tree algorithm for American futures calls.

▶ **Exercise 12.4.4** (1) Verify that, under the Black–Scholes model, a European futures option is worth the same as the corresponding European option on the cash asset if the options and the futures contract have the same maturity. The cash asset may pay a continuous dividend yield. (2) Then argue that, in fact, (1) must hold under any model.

12.4.3 Binomial Model for Forward and Futures Options

The futures price behaves like a stock paying a continuous dividend yield of r. Under the binomial model, the risk-neutral probability for the futures price is $p_f \equiv (1-d)/(u-d)$ by formula (9.21). Here, the futures price moves from F to Fu with probability p_f and to Fd with probability $1-p_f$. Figure 12.5 contains a binomial tree algorithm for pricing futures options. The binomial tree algorithm for forward options is identical except that (12.15) is used for the payoff when the option is exercised. So we replace $Fu^{n-i}d^i$ with $Fu^{n-i}d^i e^{-r(T-t)}$ and $Fu^{j-i}d^i$ with $Fu^{j-i}d^i e^{-r(T-t(j/n))}$ in Fig. 12.5.

Recall that the futures price is related to the spot price by $F = Se^{rT}$ if the underlying asset does not pay dividends. The preceding binomial model for futures prices implies that the stock price moves from $S = Fe^{-rT}$ to $S_u = Fue^{-r(T-\Delta t)} = Sue^{r\Delta t}$ with probability p_f and to $S_d \equiv Sde^{r\Delta t}$ with probability $1 - p_f$ in a period of length Δt.

Options can be replicated by a portfolio of futures contracts and bonds. This avenue may be preferred to using stocks because the restrictions on shorting futures are looser than those on stocks. To set up an equivalent portfolio of h_f futures contracts and $\$B$ in riskless bonds to replicate a call that costs C_u if the stock price moves to S_u and C_d if the stock price moves to S_d, we set up

$$h_f(Fu - F) + e^{r\Delta t}B = C_u, \quad h_f(Fd - F) + e^{r\Delta t}B = C_d.$$

Solve the preceding equations to obtain

$$h_f = \frac{C_u - C_d}{(u-d)F} \geq 0,$$

$$B = \frac{(u-1)C_d - (d-1)C_u}{(u-d)e^{r\Delta t}}.$$

Compared with the delta in Eq. (9.1), repeated below,

$$h = \frac{C_u - C_d}{S_u - S_d} = \frac{C_u - C_d}{(Su - Sd)\, e^{r \Delta t}},$$

we conclude that

$$h_f = \frac{C_u - C_d}{(u-d)\, Se^{rT}} = he^{-r(T-\Delta t)} < h.$$

Hence the delta with futures never exceeds that with stocks.

As $0 < p_f < 1$, we have $0 < 1 - p_f < 1$ as well. This suggests the following method to solve the problem of negative risk-neutral probabilities mentioned in Subsection 9.3.1: Build the binomial tree for the futures price F of the futures contract expiring at the same time as the option; then calculate S from F at each node by means of $S = Fe^{-(r-q)(T-t)}$ if the stock pays a continuous dividend yield of q [470].

➤ **Exercise 12.4.6** Start with the standard tree for the underlying non-dividend-paying stock (i.e., a stock price S can move to Su or Sd with $(e^{r\Delta t} - d)/(u-d)$ as the probability for an up move). (1) Construct the binomial model for the futures prices based on that tree. (2) What if the stock pays a continuous dividend yield of q?

➤ **Programming Assignment 12.4.7** Write binomial tree programs to price futures options and forward options.

➤ **Programming Assignment 12.4.8** Write binomial tree programs to implement the idea of avoiding negative risk-neutral probabilities enunciated above.

12.5 Swaps

Swaps are agreements between two **counterparties** to exchange cash flows in the future according to a predetermined formula. There are two basic types of swaps: interest rate and currency. An **interest rate swap** occurs when two parties exchange interest payments periodically. **Currency swaps** are agreements to deliver one currency against another [767]. Currency swaps made their debut in 1979, and interest rate swaps followed suit in 1981. In the following decade the growth of their notional volume was so spectacular as to dwarf that of any other market. For instance, interest rate swaps alone stood at over U.S.$2 trillion in 1993. Swaps also spurred the growth of related instruments such as multiperiod options (interest rate caps and floors, etc.) and **forward-rate agreements**.

Currency and interest rate swaps are collectively called **rate swaps**. Swaps on commodities are also available. For example, a company that consumes 200,000 barrels of oil per annum may pay $2 million per year for the next 5 years and in return receive $200,000 \times S$, where S is the prevailing market price of oil per barrel. This transaction locks in the price for its oil at $10 per barrel. We concentrate on currency swaps here.

12.5.1 Currency Swaps

A currency swap involves two parties to exchange cash flows in different currencies. As an example, consider the following fixed rates available to party A and party B in U.S. dollars and Japanese yen:

	Dollars	Yen
A	$D_A\%$	$Y_A\%$
B	$D_B\%$	$Y_B\%$

Suppose A wants to take out a fixed-rate loan in yen, and B wants to take out a fixed-rate loan in dollars. A straightforward scenario is for A to borrow yen at $Y_A\%$ and for B to borrow dollars at $D_B\%$.

Assume further that A is *relatively* more competitive in the dollar market than the yen market, and vice versa for B – in other words, $Y_B - Y_A < D_B - D_A$. Now consider this alternative arrangement: A borrows dollars, B borrows yen, and they enter into a currency swap, perhaps with a bank as the financial intermediary. With a swap, the counterparties exchange principal at the beginning and the end of the life of the swap. This act transforms A's loan into a yen loan and B's yen loan into a dollar loan. The total gain to all parties is $[(D_B - D_A) - (Y_B - Y_A)]\%$ because the total interest rate is originally $(Y_A + D_B)\%$ and the new arrangement has a smaller total rate of $(D_A + Y_B)\%$. Of course, this arrangement will happen only if the total gain is distributed in such a way that the cost to each party is less than that of the original scenario.

EXAMPLE 12.5.1 Two parties, A and B, face the following borrowing rates:

	Dollars	Yen
A	9%	10%
B	12%	11%

Assume that A wants to borrow yen and B wants to borrow dollars. A can borrow yen directly at 10%, and B can borrow dollars directly at 12%. As the rate differential in dollars (3%) is different from that in yen (1%), a currency swap with a total saving of $3 - 1 = 2\%$ is possible. Note that A is relatively more competitive in the dollar market, and B in the yen market. Figure 12.6 shows an arrangement that is beneficial to all parties involved, in which A effectively borrows yen at 9.5% and B borrows dollars at 11.5%. The gain is 0.5% for A, 0.5% for B, and, if we treat dollars and yen identically, 1% for the bank.

With the arrangement in Fig. 12.6 and principal amounts of U.S.$1 million and 100 million yen, the bank makes an annual gain of $0.025 million and an annual loss of 1.5 million yen. The bank thus bears some currency risk, but neither A nor B bears any currency risk. Currency risk clearly can be redistributed but not eliminated.

▶ **Exercise 12.5.1** Use the numbers in Example 12.5.1 to construct the same effective borrowing rates without the bank as the financial intermediary.

Figure 12.6: Currency swap: It turns a dollar liability into a yen liability and vice versa.

▶ **Exercise 12.5.2** Redesign the swap with the rates in Example 12.5.1 so that the gains are 1% for A, 0.5% for B, and 0.5% for the bank.

12.5.2 Valuation of Currency Swaps

As a Package of Cash Market Instruments

In the absence of default risk, the valuation of currency swap is rather straightforward. Take B in Fig. 12.6 as an example. The swap is equivalent to a long position in a yen bond paying 11% annual interest and a short position in a dollar bond paying 11.5% annual interest. The general pricing formula is thus $SP_Y - P_D$, where P_D is the dollar bond's value in dollars, P_Y is the yen bond's value in yen, and S is the \$/yen spot exchange rate. The value of a currency swap therefore depends on the term structures of interest rates in the currencies involved and the spot exchange rate. The swap has zero value when $SP_Y = P_D$.

EXAMPLE 12.5.2 Take a 2-year swap in Fig. 12.6 with principal amounts of U.S.\$1 million and 100 million yen. The payments are made once a year. Assume that the spot exchange rate is 90 yen/\$ and the term structures are flat in both nations – 8% in the U.S. and 9% in Japan. The value of the swap is

$$\frac{1}{90} \times \left(11 \times e^{-0.09} + 11 \times e^{-0.09 \times 2} + 111 \times e^{-0.09 \times 3}\right)$$
$$- \left(0.115 \times e^{-0.08} + 0.115 \times e^{-0.08 \times 2} + 1.115 \times e^{-0.08 \times 3}\right) = 0.074$$

million dollars for B.

As a Package of Forward Contracts

Swaps can also be viewed as a package of forward contracts. From Eq. (12.5), the forward contract maturing i years from now has a dollar value of

$$f_i \equiv (SY_i) e^{-qi} - D_i e^{-ri}, \tag{12.17}$$

where Y_i is the yen inflow at year i, S is the \$/yen spot exchange rate, q is the yen interest rate, D_i is the dollar outflow at year i, and r is the dollar interest rate. This formulation may be preferred to the cash market approach in cases involving costs of carry and convenience yields because forward prices already incorporate them [514]. For simplicity, flat term structures are assumed, but generalization is straightforward.

Take the swap in Example 12.5.2. Every year, B receives 11 million yen and pays \$0.115 million. In addition, at the end of the third year, B receives 100 million yen and pays \$1 million. Each of these transactions represents a forward contract. In particular, $Y_1 = Y_2 = 11$, $Y_3 = 111$, $S = 1/90$, $D_1 = D_2 = 0.115$, $D_3 = 1.115$, $q = 0.09$, and $r = 0.08$. Plug in these numbers to get $f_1 + f_2 + f_3 = 0.074$ million dollars, as in Example 12.5.2.

Equation (12.17) can be equivalently cast in terms of forward exchange rates as

$$f_i \equiv (F_i Y_i - D_i) e^{-ri},$$

where F_i is the i-year forward exchange rate. Even though the swap may have zero value (equivalently, $\sum_i f_i = 0$), it does not imply that each of the forward contracts, f_i, has zero value.

▶ **Exercise 12.5.3** Derive Eq. (12.17) with a forward exchange rate argument.

Additional Reading

Consult [88, 95, 346, 369, 470, 514, 646, 698, 746, 878, 879] for more information on derivative securities. The introduction of derivatives makes the price of the underlying asset more informative [151]. Black's model is very popular [94]. See [423, 894] for more "exotic" options and [521] for option pricing with default risk. Pointers to empirical studies on the relation between futures and forward prices can be found in [514].

NOTES

1. J.P. Morgan was acquired by Chase Manhattan in 2000, which became J.P. Morgan Chase.
2. The CBT developed futures contracts in 1865. Futures contracts were traded on the Amsterdam exchange in the seventeenth century.
3. Keynes (1883–1946) was one of the greatest economists in history [805, 806, 807].

CHAPTER THIRTEEN

Stochastic Processes and Brownian Motion

> Of all the intellectual hurdles which the human mind has confronted
> and has overcome in the last fifteen hundred years, the one which
> seems to me to have been the most amazing in character and the
> most stupendous in the scope of its consequences is the one relating
> to the problem of motion.
> > Herbert Butterfield (1900–1979), *The Origins of Modern
> > Science*

This chapter introduces basic ideas in stochastic processes and Brownian motion. The
Brownian motion underlies the continuous-time models in this book.[1] We will often
return to earlier discrete-time binomial models to mark the transition to continuous
time.

13.1 Stochastic Processes

A **stochastic process** $X = \{ X(t) \}$ is a time series of random variables. In other words,
$X(t)$ is a random variable for each time t and is usually called the **state** of the process
at time t. For clarity, $X(t)$ is often written as X_t. A **realization** of X is called a **sample
path**. Note that a sample path defines an ordinary function of t. If the times t form
a countable set, X is called a discrete-time stochastic process or a **time series**. In
this case, subscripts rather than parentheses are usually used, as in $X = \{ X_n \}$. If the
times form a continuum, X is called a continuous-time stochastic process.

A continuous-time stochastic process $\{ X(t) \}$ is said to have **independent increments** if for all $t_0 < t_1 < \cdots < t_n$ the random variables

$$X(t_1) - X(t_0), \; X(t_2) - X(t_1), \ldots, \; X(t_n) - X(t_{n-1})$$

are independent. It is said to possess **stationary increments** if $X(t+s) - X(t)$ has the
same distribution for all t. That is, the distribution depends on only s.

The **covariance function** of a stochastic process $X = \{ X(t) \}$ is defined as

$$K_X(s, t) \equiv \mathrm{Cov}[\, X(s), X(t) \,].$$

Note that $K_X(s, t) = K_X(t, s)$. The **mean function** is defined as $m_X(t) \equiv E[\, X(t) \,]$.
A stochastic process $\{ X(t) \}$ is **strictly stationary** if for any n time points
$t_1 < t_2 < \cdots < t_n$ and h, the random-variable sets $\{ X(t_1), X(t_2), \ldots, X(t_n) \}$ and
$\{ X(t_1 + h), X(t_2 + h), \ldots, X(t_n + h) \}$ have the same joint probability distribution.

From this definition,

$$m_X(t) = E[\,X(t)\,] = E[\,X(t+h)\,] = m_X(t+h)$$

for any h; in other words, the mean function is a constant. Furthermore,

$$K_X(s, s+t) = E[\,\{\,X(s) - m_X\}\{\,X(s+t) - m_X\}\,]$$
$$= E[\,\{\,X(0) - m_X\}\{\,X(t) - m_X\}\,];$$

in other words, the covariance function $K_X(s,t)$ depends on only the **lag** $|s - t|$. A process $\{\,X(t)\,\}$ is said to be **stationary** if $E[\,X(t)^2\,] < \infty$, the mean function is a constant, and the covariance function depends on only the lag.

A **Markov process** is a stochastic process for which everything that we know about its future is summarized by its current value. Formally, a continuous-time stochastic process $X = \{\,X(t), t \geq 0\,\}$ is Markovian if

$$\mathrm{Prob}[\,X(t) \leq x \mid X(u), 0 \leq u \leq s\,] = \mathrm{Prob}[\,X(t) \leq x \mid X(s)\,]$$

for $s < t$.

Random walks of various kinds are the foundations of discrete-time probabilistic models of asset prices [334]. In fact, the binomial model of stock prices is a random walk in disguise. The following examples introduce two important random walks.

EXAMPLE 13.1.1 Consider a particle on the integer line, $0, \pm 1, \pm 2, \ldots$. In each time step, this particle can make one move to the right with probability p or one move to the left with probability $1 - p$ (see Fig. 13.1). Let $P_{i,j}$ represent the probability that the particle makes a transition at point j when currently in point i. Then $P_{i,i+1} = p = 1 - P_{i,i-1}$ for $i = 0, \pm 1, \pm 2, \ldots$. This random walk is **symmetric** when $p = 1/2$. The connection with the BOPM should be clear: The particle's position denotes the cumulative number of up moves minus that of down moves.

EXAMPLE 13.1.2 The **random walk with drift** is the following discrete-time process:

$$X_n = \mu + X_{n-1} + \xi_n, \tag{13.1}$$

where ξ_n are independent and identically distributed with zero mean. The drift μ is the expected change per period. This random walk is a Markov process. An

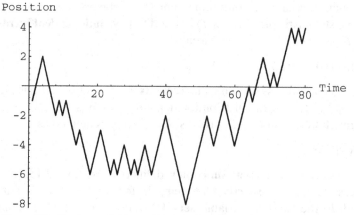

Figure 13.1: Random walk. The particle in each step can move up or down.

alternative characterization is $\{S_n \equiv \sum_{i=1}^n X_i, n \geq 1\}$, where X_i are independent, identically distributed random variables with $E[X_i] = \mu$.

▶ **Exercise 13.1.1** Prove that

$$E[X(t) - X(0)] = t \times E[X(1) - X(0)],$$

$$\text{Var}[X(t)] - \text{Var}[X(0)] = t \times \{\text{Var}[X(1)] - \text{Var}[X(0)]\}$$

when $\{X(t), t \geq 0\}$ is a process with stationary independent increments.

▶ **Exercise 13.1.2** Let $Y_1, Y_2, \ldots$, be mutually independent random variables and X_0 an arbitrary random variable. Define $X_n \equiv X_0 + \sum_{i=1}^n Y_i$ for $n > 0$. Show that $\{X_n, n \geq 0\}$ is a stochastic process with independent increments.

▶ **Exercise 13.1.3** Let $X_1, X_2, \ldots$, be a sequence of uncorrelated random variables with zero mean and unit variance. Prove that $\{X_n\}$ is stationary.

▶ **Exercise 13.1.4** (1) Use Eq. (13.1) to characterize the random walk in Example 13.1.1. (2) Show that the variance of the symmetric random walk's position after n moves is n.

▶ **Exercise 13.1.5** Construct two symmetric random walks with correlation ρ.

13.2 Martingales ("Fair Games")

A stochastic process $\{X(t), t \geq 0\}$ is a **martingale** if $E[|X(t)|] < \infty$ for $t \geq 0$ and

$$E[X(t) \mid X(u), 0 \leq u \leq s] = X(s). \tag{13.2}$$

In the discrete-time setting, a martingale means that

$$E[X_{n+1} \mid X_1, X_2, \ldots, X_n] = X_n. \tag{13.3}$$

If X_n is interpreted as a gambler's fortune after the nth gamble, identity (13.3) says the expected fortune after the $(n+1)$th gamble equals the fortune after the nth gamble regardless of what may have occurred before. A martingale is therefore a notion of fair games. Apply the law of iterated conditional expectations to both sides of Eq. (13.3) to yield

$$E[X_n] = E[X_1] \tag{13.4}$$

for all n. Similarly, $E[X(t)] = E[X(0)]$ in the continuous-time case.

EXAMPLE 13.2.1 Consider the stochastic process $\{Z_n \equiv \sum_{i=1}^n X_i, n \geq 1\}$, where X_i are independent random variables with zero mean. This process is a martingale because

$$E[Z_{n+1} \mid Z_1, Z_2, \ldots, Z_n] = E[Z_n + X_{n+1} \mid Z_1, Z_2, \ldots, Z_n]$$

$$= E[Z_n \mid Z_1, Z_2, \ldots, Z_n] + E[X_{n+1} \mid Z_1, Z_2, \ldots, Z_n]$$

$$= Z_n + E[X_{n+1}] = Z_n.$$

Note that $\{Z_n\}$ subsumes the random walk in Example 13.1.2.

A martingale is defined with respect to a probability measure under which the conditional expectation is taken. A **probability measure** assigns probabilities to states of the world. A martingale is also defined with respect to an information set [692].

In characterizations (13.2) and (13.3), the information set contains the current and the past values of X by default. However, it need not be so. In general, a stochastic process $\{X(t), t \geq 0\}$ is called a martingale with respect to information sets $\{I_t\}$ if, for all $t \geq 0$, $E[|X(t)|] < \infty$ and

$$E[X(u) \mid I_t] = X(t)$$

for all $u > t$. In the discrete-time setting, the definition becomes, for all $n > 0$,

$$E[X_{n+1} \mid I_n] = X_n,$$

given the information sets $\{I_n\}$. The preceding definition implies that $E[X_{n+m} \mid I_n] = X_n$ for any $m > 0$ by Eq. (6.6). A typical I_n in asset pricing is the price information up to time n. Then the definition says that the future values of X will not deviate systematically from today's value given the price history; in fact, today's value is their best predictor (see Exercise 13.2.2).

EXAMPLE 13.2.2 Consider the stochastic process $\{Z_n - n\mu, n \geq 1\}$, where $Z_n \equiv \sum_{i=1}^n X_i$ and $X_1, X_2, \ldots,$ are independent random variables with mean μ. As

$$E[Z_{n+1} - (n+1)\mu \mid X_1, X_2, \ldots, X_n] = E[Z_{n+1} \mid X_1, X_2, \ldots, X_n] - (n+1)\mu$$
$$= Z_n + \mu - (n+1)\mu$$
$$= Z_n - n\mu,$$

$\{Z_n - n\mu, n \geq 1\}$ is a martingale with respect to $\{I_n\}$, where $I_n \equiv \{X_1, X_2, \ldots, X_n\}$.

▶ **Exercise 13.2.1** Let $\{X(t), t \geq 0\}$ be a stochastic process with independent increments. Show that $\{X(t), t \geq 0\}$ is a martingale if $E[X(t) - X(s)] = 0$ for any $s, t \geq 0$ and $\text{Prob}[X(0) = 0] = 1$.

▶ **Exercise 13.2.2** If the asset return follows a martingale, then the best forecast of tomorrow's return is today's return as measured by the minimal mean-square error. Why? (Hint: see Exercise 6.4.3, part (2).)

▶ **Exercise 13.2.3** Define $Z_n \equiv \prod_{i=1}^n X_i, n \geq 1$, where $X_1, X_2, \ldots,$ are independent random variables with $E[X_i] = 1$. Prove that $\{Z_n\}$ is a martingale.

▶ **Exercise 13.2.4** Consider a martingale $\{Z_n, n \geq 1\}$ and let $X_i \equiv Z_i - Z_{i-1}$ with $Z_0 = 0$. Prove $\text{Var}[Z_n] = \sum_{i=1}^n \text{Var}[X_i]$.

▶ **Exercise 13.2.5** Let $\{S_n \equiv \sum_{i=1}^n X_i, n \geq 1\}$ be a random walk, where X_i are independent random variables with $E[X_i] = 0$ and $\text{Var}[X_i] = \sigma^2$. Show that $\{S_n^2 - n\sigma^2, n \geq 1\}$ is a martingale.

▶ **Exercise 13.2.6** Let $\{X_n\}$ be a martingale and let C_n denote the stake on the nth game. C_n may depend on $X_1, X_2, \ldots, X_{n-1}$ and is bounded. C_1 is a constant. Interpret $C_n(X_n - X_{n-1})$ as the gains on the nth game. The total gains up to game n are $Y_n \equiv \sum_{i=1}^n C_i(X_i - X_{i-1})$ with $Y_0 = 0$. Prove that $\{Y_n\}$ is a martingale with respect to $\{I_n\}$, where $I_n \equiv \{X_1, X_2, \ldots, X_n\}$.

13.2.1 Martingale Pricing and Risk-Neutral Valuation

We learned in Lemma 9.2.1 that the price of a European option is the expected discounted future payoff at expiration in a risk-neutral economy. This important principle can be generalized by use of the concept of martingale. Recall the recursive valuation of a European option by means of $C = [pC_u + (1 - p) C_d]/R$, where p is the risk-neutral probability and \$1 grows to \$R in a period. Let $C(i)$ denote the value of the option at time i and consider the **discount process** $\{ C(i)/R^i, i = 0, 1, \ldots, n \}$. Then,

$$E\left[\frac{C(i+1)}{R^{i+1}} \,\bigg|\, C(i) = C \right] = \frac{pC_u + (1 - p) C_d}{R^{i+1}} = \frac{C}{R^i}.$$

The above result can be easily generalized to

$$E\left[\frac{C(k)}{R^k} \,\bigg|\, C(i) = C \right] = \frac{C}{R^i}, \quad i \le k. \tag{13.5}$$

Hence the discount process is a martingale as

$$\frac{C(i)}{R^i} = E_i^\pi \left[\frac{C(k)}{R^k} \right], \quad i \le k, \tag{13.6}$$

where E_i^π means that the expectation is taken under the risk-neutral probability conditional on the price information up to time i.[2] This risk-neutral probability is also called the **equivalent martingale probability measure** [514]. Two probability measures are said to be **equivalent** if they assign nonzero probabilities to the same set of states.

Under general discrete-time models, Eq. (13.6) holds for all assets, not just options. In the general case in which interest rates are stochastic, the equation becomes [725]

$$\frac{C(i)}{M(i)} = E_i^\pi \left[\frac{C(k)}{M(k)} \right], \quad i \le k, \tag{13.7}$$

where $M(j)$ denotes the balance in the money market account at time j by use of the rollover strategy with an initial investment of \$1. For this reason, it is called the **bank account process**. If interest rates are stochastic, then $M(j)$ is a random variable. However, note that $M(0) = 1$ and $M(j)$ is known at time $j - 1$. Identity (13.7) is the general formulation of risk-neutral valuation, which says the discount process is a martingale under π. We thus have the following fundamental theorem of asset pricing.

THEOREM 13.2.3   *A discrete-time model is arbitrage free if and only if there exists a probability measure such that the discount process is a martingale. This probability measure is called the risk-neutral probability measure.*

▶ **Exercise 13.2.7** Verify Eq. (13.5).

▶ **Exercise 13.2.8** Assume that one unit of domestic (foreign) currency grows to R (R_f, respectively) units in a period, and u and d are the up and the down moves of the domestic/foreign exchange rate, respectively. Apply identity (13.7) to derive the risk-neutral probability $[(R/R_f) - d]/(u - d)$ for forex options under the BOPM in Exercise 11.5.4, part (1).

▶ **Exercise 13.2.9** Prove that the discounted stock price $S(i)/R^i$ follows a martingale under the risk-neutral probability; in particular, $S(0) = E^\pi[S(i)/R^i]$.

13.2.2 Futures Price under the Binomial Model

Futures prices form a martingale under the risk-neutral probability because the expected futures price in the next period is

$$
p_{\mathrm{f}} Fu + (1 - p_{\mathrm{f}})\, Fd = F \left(\frac{1-d}{u-d} u + \frac{u-1}{u-d} d \right) = F
$$

(review Subsection 12.4.3). The above claim can be generalized to

$$
F_i = E_i^{\pi}[\, F_k\,], \quad i \le k, \tag{13.8}
$$

where F_i is the futures price at time i. This identity holds under stochastic interest rates as well (see Exercise 13.2.11).

▶ **Exercise 13.2.10** Prove that $F = E^{\pi}[\, S_n\,]$, where S_n denotes the price of the underlying non-dividend-paying stock at the delivery date of the futures contract, time n. (The futures price is thus an unbiased estimator of the expected spot price *in a risk-neutral economy*.)

▶ **Exercise 13.2.11** Show that identity (13.8) holds under stochastic interest rates.

13.2.3 Martingale Pricing and the Choice of Numeraire

Martingale pricing formula (13.7) uses the money market account as **numeraire** in that it expresses the price of any asset relative to the money market account.[3] The money market account is not the only choice for numeraire, however. If asset S, whose value is positive at all times, is chosen as numeraire, martingale pricing says there exists a risk-neutral probability π under which the relative price of any asset C is a martingale:

$$
\frac{C(i)}{S(i)} = E_i^{\pi}\left[\frac{C(k)}{S(k)} \right], \quad i \le k, \tag{13.9}
$$

where $S(j)$ denotes the price of S at time j; the discount process remains a martingale.

Take the binomial model with two assets as an example. In a period, asset one's price can go from S to S_1 or S_2, whereas asset two's price can go from P to P_1 or P_2. Assume that $(S_1/P_1) < (S/P) < (S_2/P_2)$ for market completeness and to rule out arbitrage opportunities. For any derivative security, let C_1 be its price at time one if asset one's price moves to S_1 and let C_2 be its price at time one if asset one's price moves to S_2. Replicate the derivative by solving

$$
\alpha S_1 + \beta P_1 = C_1,
$$
$$
\alpha S_2 + \beta P_2 = C_2
$$

by using α units of asset one and β units of asset two. This yields

$$
\alpha = \frac{P_2 C_1 - P_1 C_2}{P_2 S_1 - P_1 S_2}, \quad \beta = \frac{S_2 C_1 - S_1 C_2}{S_2 P_1 - S_1 P_2}
$$

and the derivative costs

$$C = \alpha S + \beta P = \frac{P_2 S - P S_2}{P_2 S_1 - P_1 S_2} C_1 + \frac{P S_1 - P_1 S}{P_2 S_1 - P_1 S_2} C_2.$$

It is easy to verify that

$$\frac{C}{P} = p\frac{C_1}{P_1} + (1-p)\frac{C_2}{P_2}, \quad \text{where } p \equiv \frac{(S/P) - (S_2/P_2)}{(S_1/P_1) - (S_2/P_2)}. \tag{13.10}$$

The derivative's price with asset two as numeraire is thus a martingale under the risk-neutral probability p. Interestingly, the expected returns of the two assets are irrelevant.

EXAMPLE 13.2.4 For the BOPM in Section 9.2, the two assets are the money market account and the stock. Because the money market account is the numeraire, we substitute $P = 1$, $P_1 = P_2 = R$, $S_1 = Su$, and $S_2 = Sd$ into Eq. (13.10). The result,

$$p = \frac{S - (Sd/R)}{(Su/R) - (Sd/R)} = \frac{R - d}{u - d},$$

affirms the familiar risk-neutral probability. The risk-neutral probability changes if the stock is chosen as numeraire, however (see Exercise 13.2.12).

The risk-neutral probability measure therefore depends on the choice of numeraire, and switching numeraire changes the risk-neutral probability measure. Picking the "right" numeraire can simplify the task of pricing, especially for interest-rate-sensitive securities [25, 731, 783]. For the rest of the book, the money market account will continue to serve as numeraire unless stated otherwise.

▶ **Exercise 13.2.12** (1) Prove that $[(1/d) - (1/R)][(ud)/(u-d)]$ is the up-move probability for the stock price that makes the relative bond price a martingale under the binomial option pricing model in which the *stock* is chosen as numeraire. (2) Reinterpret Eq. (9.11).

▶ **Exercise 13.2.13** Show that for any $k > 0$ there exists a risk-neutral probability measure π under which the price of any asset C equals its discounted expected future price at time k, that is, $C(i) = d(k-i)\, E_i^\pi[\, C(k)\,]$, where $i \le k$. Recall that $d(\cdot)$ denotes the discount function at time i. This π is called the **forward-neutral probability measure**.

13.3 Brownian Motion

Brownian motion is a stochastic process $\{\, X(t), t \ge 0\,\}$ with the following properties:

(1) $X(0) = 0$, unless stated otherwise;
(2) for any $0 \le t_0 < t_1 < \cdots < t_n$, the random variables $X(t_k) - X(t_{k-1})$ for $1 \le k \le n$ are independent (so $X(t) - X(s)$ is independent of $X(r)$ for $r \le s < t$);
(3) for $0 \le s < t$, $X(t) - X(s)$ is normally distributed with mean $\mu(t - s)$ and variance $\sigma^2(t - s)$, where μ and $\sigma \ne 0$ are real numbers.

Such a process is called a (μ, σ) Brownian motion with **drift** μ and **variance** σ^2. Figure 13.2 plots a realization of a Brownian motion process. The existence and

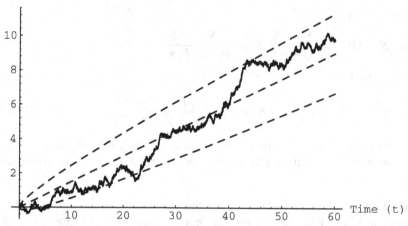

Figure 13.2: Drift and variance of Brownian motion. Shown is a sample path of a (0.15, 0.3) Brownian motion. The stochastic process has volatility, as evinced by the bumpiness of the path. The envelope is for one standard deviation, or $0.3\sqrt{t}$, around the mean function, which is the deterministic process with the randomness removed.

the uniqueness of such a process are guaranteed by **Wiener's theorem** [73]. Although Brownian motion is a continuous function of t with probability one, it is almost nowhere differentiable. The $(0, 1)$ Brownian motion is also called **normalized Brownian motion** or the **Wiener process**.

Any continuous-time process with stationary independent increments can be proved to be a Brownian motion process [419]. This fact explains the significance of Brownian motion in stochastic modeling. Brownian motion also demonstrates **statistical self-similarity** in that $X(rx)/\sqrt{r}$ remains a Wiener process if X is such. This means that if we sample the process 100 times faster and then shrink the result 10 times, the path will look statistically the same as the original one. This property naturally links Brownian motion to fractals [240, 784]. Finally, Brownian motion is Markovian.

Brownian motion, named after Robert Brown (1773–1858), was first discussed mathematically by Bachelier and received rigorous treatments by Wiener (1894–1964), who came up with the above concise definition. Therefore it is sometimes called the **generalized Wiener process** or the **Wiener–Bachelier process** [343, 543].

EXAMPLE 13.3.1 Suppose that the total value of a company, measured in millions of dollars, follows a $(20, 30)$ Brownian motion (i.e., with a drift of 20 per annum and a variance of 900 per annum). The starting total value is 50. At the end of 1 year, the total value will have a normal distribution with a mean of 70 and a standard deviation of 30. At the end of 6 months, as another example, it will have a normal distribution with a mean of 60 and a standard deviation of $\sqrt{450} \approx 21.21$.

From the definition, if $\{ X(t), t \geq 0 \}$ is the Wiener process, then $X(t) - X(s) \sim N(0, t - s)$. A (μ, σ) Brownian motion $Y = \{ Y(t), t \geq 0 \}$ can be expressed in terms of the Wiener process by

$$Y(t) = \mu t + \sigma X(t). \tag{13.11}$$

As $Y(t + s) - Y(t) \sim N(\mu s, \sigma^2 s)$, our uncertainty about the future value of Y as measured by the standard deviation grows as the square root of how far we look into the future.

► **Exercise 13.3.1** Prove that $\{(X(t) - \mu t)/\sigma, t \geq 0\}$ is a Wiener process if $\{X(t),$ $t \geq 0\}$ is a (μ, σ) Brownian motion.

► **Exercise 13.3.2** Verify that $K_X(t, s) = \sigma^2 \times \min(s, t)$ if $\{X(t), t \geq 0\}$ is a (μ, σ) Brownian motion.

► **Exercise 13.3.3** Let $\{X(t), t \geq 0\}$ represent the Wiener process. Show that the related process $\{X(t) - X(0), t \geq 0\}$ is a martingale. ($X(0)$ can be a random variable.)

► **Exercise 13.3.4** Let $p(x, y; t)$ denote the transition probability density function of a (μ, σ) Brownian motion starting at x; $p(x, y; t) = (1/\sqrt{2\pi t}\,\sigma) \exp[-(y - x - \mu t)^2/(2\sigma^2 t)]$. Show that p satisfies **Kolmogorov's backward equation** $\partial p/\partial t = (\sigma^2/2)(\partial^2 p/\partial x^2) + \mu(\partial p/\partial x)$, and **Kolmogorov's forward equation** (also called the **Fokker–Planck equation**) $\partial p/\partial t = (\sigma^2/2)(\partial^2 p/\partial y^2) - \mu(\partial p/\partial y)$.

► **Exercise 13.3.5** Let $\{X(t), t \geq 0\}$ be a $(0, \sigma)$ Brownian motion. Prove that the following three processes are martingales: (1) $X(t)$, (2) $X(t)^2 - \sigma^2 t$, and (3) $\exp[\alpha X(t) - \alpha^2\sigma^2 t/2]$ for $\alpha \in R$, called **Wald's martingale**.

13.3.1 Brownian Motion as the Limit of a Random Walk

A (μ, σ) Brownian motion is the limiting case of a random walk. Suppose that a particle moves Δx either to the left with probability $1 - p$ or to the right with probability p after Δt time. For simplicity assume that $n \equiv t/\Delta t$ is an integer. Its position at time t is

$$Y(t) \equiv \Delta x \,(X_1 + X_2 + \cdots + X_n),$$

where

$$X_i \equiv \begin{cases} +1, & \text{if the } i\text{th move is to the right} \\ -1, & \text{if the } i\text{th move is to the left} \end{cases},$$

and X_i are independent with $\text{Prob}[X_i = 1] = p = 1 - \text{Prob}[X_i = -1]$. Note that $E[X_i] = 2p - 1$ and that $\text{Var}[X_i] = 1 - (2p - 1)^2$. Therefore

$$E[Y(t)] = n(\Delta x)(2p - 1) \quad \text{and} \quad \text{Var}[Y(t)] = n(\Delta x)^2[1 - (2p - 1)^2].$$

Letting $\Delta x \equiv \sigma\sqrt{\Delta t}$ and $p \equiv (1 + (\mu/\sigma)\sqrt{\Delta t})/2$, we conclude that

$$E[Y(t)] = n\sigma\sqrt{\Delta t}\,(\mu/\sigma)\sqrt{\Delta t} = \mu t,$$

$$\text{Var}[Y(t)] = n\sigma^2\Delta t\,(1 - (\mu/\sigma)^2\Delta t) \to \sigma^2 t$$

as $\Delta t \to 0$. Thus $\{Y(t), t \geq 0\}$ converges to a (μ, σ) Brownian motion by the central limit theorem. In particular, Brownian motion with zero drift is the limiting case of symmetric random walk when $\mu = 0$ is chosen. Note also that

$$\text{Var}[Y(t + \Delta t) - Y(t)] = \text{Var}[\Delta x\, X_{n+1}] = (\Delta x)^2 \times \text{Var}[X_{n+1}] \to \sigma^2\Delta t.$$

The similarity to the BOPM is striking: The p above is identical to the probability in Eq. (9.15) and $\Delta x = \ln u$. This is no coincidence (see Subsection 14.4.3).

► **Exercise 13.3.6** Let dQ represent the probability that the random walk that converges to a $(\mu, 1)$ Brownian motion takes the moves $X_1, X_2, \ldots$. Let dP denote the probability that the symmetric random walk that converges to the Wiener process

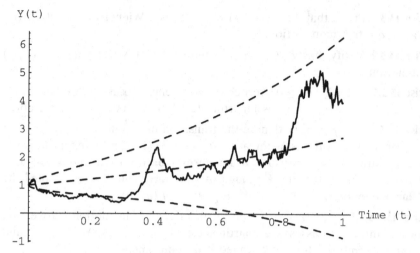

Figure 13.3: Sample path of geometric Brownian motion. The process is $Y(t) = e^{X(t)}$, where X is a $(0.5, 1)$ Brownian motion. The envelope is for one standard deviation, $\sqrt{(e^t - 1)e^{2t}}$, around the mean. Can you tell the qualitative difference between this plot and the stock price charts in Fig. 6.4?

makes identical moves. Derive dQ/dP. (The process dQ/dP in fact converges to Wald's martingale $\exp[\,\mu W(t) - \mu^2 t/2\,]$, where $W(t)$ is the Wiener process.)

13.3.2 Geometric Brownian Motion

If $X \equiv \{\, X(t), t \geq 0\,\}$ is a Brownian motion process, the process $\{\, Y(t) \equiv e^{X(t)}, t \geq 0\,\}$ is called **geometric Brownian motion**. Its other names are **exponential Brownian motion** and **lognormal diffusion**. See Fig. 13.3 for illustration. When X is a (μ, σ) Brownian motion, we have $X(t) \sim N(\mu t, \sigma^2 t)$ and the moment generating function

$$E\big[e^{s\,X(t)}\big] = E\big[\,Y(t)^s\,\big] = e^{\mu t s + (\sigma^2 t s^2/2)}$$

from Eq. (6.8). Thus

$$E[\,Y(t)\,] = e^{\mu t + (\sigma^2 t/2)}, \tag{13.12}$$

$$\text{Var}[\,Y(t)\,] = E[\,Y(t)^2\,] - E[\,Y(t)\,]^2 = e^{2\mu t + \sigma^2 t}\big(e^{\sigma^2 t} - 1\big). \tag{13.12$'$}$$

Geometric Brownian motion models situations in which percentage changes are independent and identically distributed. To see this point, let Y_n denote the stock price at time n and $Y_0 = 1$. Assume that relative returns $X_i \equiv Y_i/Y_{i-1}$ are independent and identically distributed. Then $\ln Y_n = \sum_{i=1}^{n} \ln X_i$ is a sum of independent, identically distributed random variables, and $\{\, \ln Y_n, n \geq 0\,\}$ is approximately Brownian motion. Thus $\{\, Y_n, n \geq 0\,\}$ is approximately geometric Brownian motion.

▶ **Exercise 13.3.7** Let $Y(t) \equiv e^{X(t)}$, where $\{\, X(t), t \geq 0\,\}$ is a (μ, σ) Brownian motion. Show that $E[\,Y(t)\,|\,Y(s)\,] = Y(s)\,e^{(t-s)(\mu + \sigma^2/2)}$ for $s < t$.

▶ **Exercise 13.3.8** Assume that the stock price follows the geometric Brownian motion process $S(t) \equiv e^{X(t)}$, where $\{\, X(t), t \geq 0\,\}$ is a (μ, σ) Brownian motion. (1) Show that the stock price is growing at a rate of $\mu + \sigma^2/2$ (not μ!) on the average if by this rate we mean $R_1 \equiv (t_2 - t_1)^{-1} \ln E[\, S(t_2)/S(t_1)\,]$ for the time period $[\, t_1, t_2\,]$. (2) Show that the alternative measure for the rate of return, $R_2 \equiv (t_2 - t_1)^{-1} E[\, \ln(S(t_2)/S(t_1))\,]$, gives rise to μ. (3) Argue that $R_2 < R_1$ independent of any assumptions about the process $\{\, X(t)\,\}$.

13.3.3 Stationarity

The Wiener process $\{\, X(t), t \geq 0\,\}$ is not stationary (see Exercise 13.3.2). However, it can be transformed into a stationary process by

$$Y(t) \equiv e^{-t} X(e^{2t}). \tag{13.13}$$

This claim can be verified as follows. Because $Y(t) \sim N(0, 1)$, the mean function is zero, a constant. Furthermore,

$$E[\, Y(t)^2\,] = E[\, e^{-2t} X(e^{2t})^2\,] = e^{-2t} e^{2t} = 1 < \infty.$$

Finally, the covariance function $K_Y(s, t), s < t,$ is

$$E[\, e^{-t} X(e^{2t}) e^{-s} X(e^{2s})\,] = e^{-s-t} E[\, X(e^{2t}) X(e^{2s})\,] = e^{-s-t} e^{2s} = e^{s-t},$$

where the next to last equality is due to Exercise 13.3.2. Therefore $\{\, Y(t), t \geq 0\,\}$ is stationary. The process Y is called the **Ornstein–Uhlenbeck process** [230, 261, 541].

13.3.4 Variations

Many formulas in standard calculus do not carry over to Brownian motion. Take the **quadratic variation** of any function $f : [\, 0, \infty) \to R$ defined by[4]

$$\sum_{k=0}^{2^n - 1} \left[f\left(\frac{(k+1)t}{2^n}\right) - f\left(\frac{kt}{2^n}\right) \right]^2.$$

It is not hard to see that the quadratic variation vanishes as $n \to \infty$ if f is differentiable. This conclusion no longer holds if f is Brownian motion. In fact,

$$\lim_{n \to \infty} \sum_{k=0}^{2^n - 1} \left[X\left(\frac{(k+1)t}{2^n}\right) - X\left(\frac{kt}{2^n}\right) \right]^2 = \sigma^2 t \tag{13.14}$$

with probability one, where $\{\, X(t), t \geq 0\,\}$ is a (μ, σ) Brownian motion [543]. This result informally says that $\int_0^t [\, dX(s)\,]^2 = \sigma^2 t$, which is frequently written as

$$(dX)^2 = \sigma^2 \, dt. \tag{13.15}$$

It can furthermore be shown that

$$(dX)^n = 0 \quad \text{for } n > 2 \tag{13.16}$$

and $dX \, dt = 0$.

From Eq. (13.14), the **total variation** of a Brownian path is infinite with probability one:

$$\lim_{n \to \infty} \sum_{k=0}^{2^n - 1} \left| X\left(\frac{(k+1)t}{2^n}\right) - X\left(\frac{kt}{2^n}\right) \right| = \infty. \tag{13.17}$$

Brownian motion is thus continuous but with highly irregular sample paths.

▶ **Exercise 13.3.9** To see the plausibility of Eq. (13.14), take the expectation of its left-hand side and drop $\lim_{n\to\infty}$ to obtain

$$\sum_{k=0}^{2^n-1} E\left[\, X\left(\frac{(k+1)\,t}{2^n}\right) - X\left(\frac{kt}{2^n}\right)\right]^2.$$

Show that the preceding expression has the value $\mu^2 t^2 2^{-n} + \sigma^2 t$, which approaches $\sigma^2 t$ as $n \to \infty$.

▶ **Exercise 13.3.10** We can prove Eq. (13.17) without using Eq. (13.14). Let

$$f_n(X) \equiv \sum_{k=0}^{2^n-1} \left| X\left(\frac{(k+1)\,t}{2^n}\right) - X\left(\frac{kt}{2^n}\right)\right|.$$

(1) Prove that $|X((k+1)\,t/2^n) - X(kt/2^n)|$ has mean $2^{-n/2}\sqrt{2/\pi}$ and variance $2^{-n}(1 - 2/\pi)$. ($f_n(X)$ thus has mean $2^{n/2}\sqrt{2/\pi}$ and variance $1 - 2/\pi$.) (2) Show that $f_n(X) \to \infty$ with probability one. (Hint: Prob$[\,|X - E[\,X\,]| \geq k\,] \leq \mathrm{Var}[\,X\,]/k^2$ by **Chebyshev's inequality**.)

13.4 Brownian Bridge

A **Brownian bridge** process $\{\, B(t), 0 \leq t \leq 1 \,\}$ is **tied-down** Brownian motion [544]. It is defined as the Wiener process plus the constraint $B(0) = B(1) = 0$. An alternative formulation is $\{\, W(t) - tW(1), 0 \leq t < 1 \,\}$, where $\{\, W(t), 0 \leq t \,\}$ is the Wiener process. For a general time period $[\, 0, T\,]$, a Brownian bridge process can be written as

$$B(t) \equiv W(t) - \frac{t}{T}\, W(T), \quad 0 \leq t \leq T,$$

where $W(0) = 0$ and $W(T)$ is known at time zero [193]. Observe that $B(t)$ is pinned to zero at both end points, times zero and T.

▶ **Exercise 13.4.1** Prove the following identities: (1) $E[\, B(t)\,] = 0$, (2) $E[\, B(t)^2\,] = t - (t^2/T)$, and (3) $E[\, B(s)\, B(t)\,] = \min(s, t) - (st/T)$.

▶ **Exercise 13.4.2** Write the Brownian bridge process with $B(0) = x$ and $B(T) = y$.

Additional Reading

The idea of martingale is due to Lévy (1886–1974) and received thorough development by Doob [205, 280, 541, 877]. See [817] for a complete treatment of random walks. Consult [631] for a history of Brownian motion from the physicist's point of view and [277] for adding Bachelier's contribution. Reference [104] collects results and formulas in connection with Brownian motion. Advanced materials can be found in [210, 230, 364, 543]. The backward and Fokker–Planck equations mentioned in Exercise 13.3.4 describe a large class of stochastic processes with continuous sample paths [373]. The heuristic arguments in Subsection 13.3.1 showing Brownian motion as the limiting case of random walk can be made rigorous by **Donsker's theorem** [73, 289, 541, 573]. The geometric Brownian motion model for stock prices is due to Osborne [709]. Models of stock returns are surveyed in [561].

NOTES

1. Merton pioneered the alternative **jump process** in pricing [80].
2. For standard European options, price information *at* time i suffices because they are path independent.
3. Regarded by Schumpeter as the greatest economist in his monumental *History of Economic Analysis* [786], Walras (1834–1910) introduced numeraire in his equilibrium analysis, recognizing that only *relative* prices matter [31].
4. For technical reasons, the partition of $[0, t]$ is **dyadic**, i.e., at points $k(t/2^n)$ for $0 < k < 2^n$.

Continuous-Time Financial Mathematics

> The pursuit of mathematics is a divine madness of the human spirit.
>
> Alfred North Whitehead (1861–1947), *Science and the Modern World*

This chapter introduces the mathematics behind continuous-time models. This approach to finance was initiated by Merton [290]. Formidable as the mathematics seems to be, it can be made accessible at some expense of rigor and generality. The theory will be applied to a few fundamental problems in finance.

14.1 Stochastic Integrals

From now on, we use $W \equiv \{ W(t), t \geq 0 \}$ to denote the Wiener process. The goal here is to develop stochastic integrals of X from a class of stochastic processes with respect to the Wiener process:

$$I_t(X) \equiv \int_0^t X \, dW, \quad t \geq 0.$$

We saw in Subsection 13.3.4 that classical calculus cannot be applied to Brownian motion. One reason is that its sample path, regarded as a function, has unbounded total variation. $I_t(X)$ is a random variable called the **stochastic integral** of X with respect to W. The stochastic process $\{ I_t(X), t \geq 0 \}$ is denoted here by $\int X \, dW$. Typical requirements for X in financial applications are (1) Prob$[\int_0^t X^2(s) \, ds < \infty] = 1$ for all $t \geq 0$ or the stronger $\int_0^t E[X^2(s)] \, ds < \infty$ and (2) that the information set at time t includes the history of X and W up to that point in time but nothing about the evolution of X or W after t (**nonanticipating**, so to speak). The future therefore cannot influence the present, and $\{ X(s), 0 \leq s \leq t \}$ is independent of $\{ W(t+u) - W(t), u > 0 \}$.

The **Ito integral** is a theory of stochastic integration. As with calculus, it starts with step functions. A stochastic process $\{ X(t) \}$ is **simple** if there exist $0 = t_0 < t_1 < \cdots$ such that

$$X(t) = X(t_{k-1}) \quad \text{for } t \in [t_{k-1}, t_k), \quad k = 1, 2, \ldots$$

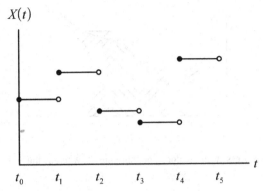

Figure 14.1: Simple stochastic process.

for any realization (see Fig. 14.1). The Ito integral of a simple process is defined as

$$I_t(X) \equiv \sum_{k=0}^{n-1} X(t_k)[\, W(t_{k+1}) - W(t_k)\,], \tag{14.1}$$

where $t_n = t$. Note that the integrand X is evaluated at t_k, not t_{k+1}.

The natural step to follow is to define the Ito integral of more general processes as a limiting random variable of the Ito integral of simple stochastic processes. Indeed, for a general stochastic process $X = \{\, X(t), t \geq 0\,\}$, there exists a random variable $I_t(X)$, unique almost certainly, such that $I_t(X_n)$ converges in probability to $I_t(X)$ for each sequence of simple stochastic processes $X_1, X_2, \ldots$, such that X_n converges in probability to X. In particular, if X is continuous with probability one, then $I_t(X_n)$ converges in probability to $I_t(X)$ as $\delta_n \equiv \max_{1 \leq k \leq n}(t_k - t_{k-1})$ goes to zero, written as

$$\int_0^t X \, dW = \text{st-lim}_{\delta_n \to 0} \sum_{k=0}^{n-1} X(t_k)[\, W(t_{k+1}) - W(t_k)\,]. \tag{14.2}$$

It is a fundamental fact that $\int X \, dW$ is continuous almost surely [419, 566]. The following theorem says the Ito integral is a martingale (see Exercise 13.2.6 for its discrete analog), and a corollary is the mean-value formula $E[\int_a^b X \, dW] = 0$.

THEOREM 14.1.1 *The Ito integral $\int X \, dW$ is a martingale.*

Let us inspect Eq. (14.2) more closely. It says the following simple stochastic process $\{\widehat{X}(t)\}$ can be used in place of X to approximate the stochastic integral $\int_0^t X \, dW$:

$$\widehat{X}(s) \equiv X(t_{k-1}) \quad \text{for } s \in [\, t_{k-1}, t_k), \ k = 1, 2, \ldots, n.$$

The key here is the nonanticipating feature of $\widehat{X}$; that is, the information up to time s,

$$\{\widehat{X}(t), W(t), 0 \leq t \leq s\},$$

cannot determine the future evolution of either X or W. Had we defined the stochastic integral as $\sum_{k=0}^{n-1} X(t_{k+1})[\, W(t_{k+1}) - W(t_k)\,]$, we would have been using the following different simple stochastic process in the approximation,

$$\widehat{Y}(s) \equiv X(t_k) \quad \text{for } s \in [\, t_{k-1}, t_k), \ k = 1, 2, \ldots, n,$$

which clearly anticipates the future evolution of X. See Fig. 14.2 for illustration.

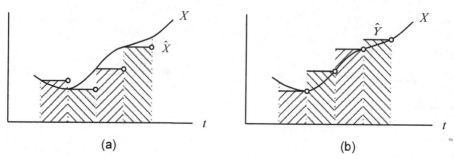

Figure 14.2: Stochastic integration. The simple process $\widehat{X}$ in (a) does not anticipate X, whereas the simple process $\widehat{X}$ in (b) does.

For example, $\int W\,dW$ can be approximated as follows:

$$\sum_{k=0}^{n-1} W(t_k)[\,W(t_{k+1}) - W(t_k)\,]$$

$$= \sum_{k=0}^{n-1} \frac{W(t_{k+1})^2 - W(t_k)^2}{2} - \sum_{k=0}^{n-1} \frac{[\,W(t_{k+1}) - W(t_k)\,]^2}{2}$$

$$= \frac{W(t)^2}{2} - \sum_{k=0}^{n-1} \frac{[\,W(t_{k+1}) - W(t_k)\,]^2}{2}.$$

Because the second term above converges to $t/2$ by Eq. (13.14),

$$\int_0^t W\,dW = \frac{W(t)^2}{2} - \frac{t}{2}. \tag{14.3}$$

One might have expected $\int_0^t W\,dW = W(t)^2/2$ from calculus. Hence the extra $t/2$ term may come as a surprise. It can be traced to the infinite total variation of Brownian motion. Another way to see the mistake of $\int_0^t W\,dW = W(t)^2/2$ is through Theorem 14.1.1: $W(t)^2/2$ is not a martingale (see Exercise 14.1.3), but $[\,W(t)^2 - t\,]/2$ is (see Exercise 13.3.5).

▶ **Exercise 14.1.1** Prove Theorem 14.1.1 for simple stochastic processes.

▶ **Exercise 14.1.2** Verify that using the following simple stochastic process,

$$Y(s) \equiv W(t_k) \quad \text{for } s \in [\,t_{k-1}, t_k), \quad k = 1, 2, \ldots, n,$$

to approximate W results in $\int_0^t W\,dW = (W(t)^2 + t)/2$.

Comment 14.1.2 The different results in Exercise 14.1.2 and Eq. (14.3) show the importance of picking the intermediate point for stochastic integrals (here, right end point vs. left end point). The simple stochastic process in Exercise 14.1.2 anticipates the future evolution of W. In general, the following simple stochastic process,

$$Z(s) \equiv W((1-a)\,t_{k-1} + at_k) \quad \text{for } s \in [\,t_{k-1}, t_k), \quad k = 1, 2, \ldots, n,$$

gives rise to $\int_0^t W\,dW = W(t)^2/2 + (a - 1/2)\,t$. The Ito integral corresponds to the choice $a = 0$. Standard calculus rules apply when $a = 1/2$, which gives rise to the **Stratonovich stochastic integral**.

➤ **Exercise 14.1.3** Verify that $W(t)^2/2$ is not a martingale.

➤ **Exercise 14.1.4** Prove that $E[\int_0^t W \, dW] = 0$.

➤ **Exercise 14.1.5** Prove that stochastic integration reduces to the usual Riemann–Stieltjes form for constant processes.

14.2 Ito Processes

The stochastic process $X = \{X_t, t \geq 0\}$ that solves

$$X_t = X_0 + \int_0^t a(X_s, s) \, ds + \int_0^t b(X_s, s) \, dW_s, \quad t \geq 0$$

is called an **Ito process**. Here, X_0 is a scalar starting point and $\{a(X_t, t) : t \geq 0\}$ and $\{b(X_t, t) : t \geq 0\}$ are stochastic processes satisfying certain regularity conditions. The terms $a(X_t, t)$ and $b(X_t, t)$ are the **drift** and the **diffusion**, respectively. A shorthand that is due to Langevin's work in 1904 is the following **stochastic differential equation** for the **Ito differential** dX_t,

$$dX_t = a(X_t, t) \, dt + b(X_t, t) \, dW_t, \tag{14.4}$$

or simply $dX_t = a_t \, dt + b_t \, dW_t$ [30, 386]. This is Brownian motion with an instantaneous drift of a_t and an instantaneous variance of b_t^2. In particular, X is a martingale if the drift a_t is zero by Theorem 14.1.1. Recall that dW is normally distributed with mean zero and variance dt. A form equivalent to Eq. (14.4) is the so-called **Langevin equation**:

$$dX_t = a_t \, dt + b_t \sqrt{dt} \, \xi, \tag{14.5}$$

where $\xi \sim N(0, 1)$. This formulation makes it easy to derive Monte Carlo simulation algorithms. Although $dt \ll \sqrt{dt}$, the deterministic term a_t still matters because the random variable ξ makes sure the fluctuation term b_t over successive intervals tends to cancel each other out.

There are regularity conditions that guarantee the existence and the uniqueness of solution for stochastic differential equations [30, 373, 566]. The solution to a stochastic differential equation is also called a **diffusion process**.

14.2.1 Discrete Approximations

The following finite-difference approximation follows naturally from Eq. (14.5):

$$\widehat{X}(t_{n+1}) = \widehat{X}(t_n) + a(\widehat{X}(t_n), t_n) \, \Delta t + b(\widehat{X}(t_n), t_n) \, \Delta W(t_n), \tag{14.6}$$

where $t_n \equiv n\Delta t$. This method is called the **Euler method** or the **Euler–Maruyama method** [556]. Under mild conditions, $\widehat{X}(t_n)$ indeed converges to $X(t_n)$ [572]. Note that $\Delta W(t_n)$ should be interpreted as $W(t_{n+1}) - W(t_n)$ instead of $W(t_n) - W(t_{n-1})$ because a and b are required to be nonanticipating. With the drift a and the diffusion b determined at time t_n, $\widehat{X}$ is expected to be $\widehat{X}(t_n) + a(\widehat{X}(t_n), t_n) \, \Delta t$ at time t_{n+1}. However, the new information $\Delta W(t_n)$, which is unpredictable given the information available at time t_n, dislodges $\widehat{X}$ from its expected position by adding $b(\widehat{X}(t_n), t_n) \, \Delta W(t_n)$. This procedure then repeats itself at $\widehat{X}(t_{n+1})$. See Fig. 14.3 for an illustration.

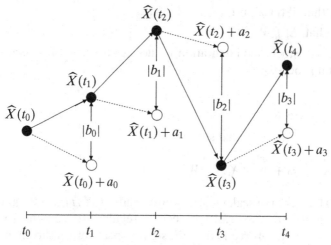

Figure 14.3: Dynamics of Ito process. The filled circles track the process, whereas the unfilled circles are the expected positions. In the plot, $a_i \equiv a(\widehat{X}(t_i), t_i) \Delta t$, the expected changes, and $b_i \equiv b(\widehat{X}(t_i), t_i) \Delta W(t_i)$, the random disturbances. Note that $\widehat{X}(t_{i+1}) = \widehat{X}(t_i) + a_i + b_i$.

The more advanced **Mil'shtein scheme** adds the $bb'[(\Delta W)^2 - \Delta t]/2$ term to Euler's method to provide better approximations [668, 669]. For geometric Brownian motion, for example, Euler's scheme yields

$$\widehat{X}(t_{n+1}) = \widehat{X}(t_n) + \mu\widehat{X}(t_n)\Delta t + \sigma\widehat{X}(t_n)\Delta W(t_n),$$

whereas Mil'shtein's scheme adds $\sigma^2\widehat{X}(t_n)\{[\Delta W(t_n)]^2 - \Delta t\}/2$ to the above.

Under fairly loose regularity conditions, approximation (14.6) can be replaced with

$$\widehat{X}(t_{n+1}) = \widehat{X}(t_n) + a(\widehat{X}(t_n), t_n)\Delta t + b(\widehat{X}(t_n), t_n)\sqrt{\Delta t}\, Y(t_n),$$

where $Y(t_0), Y(t_1), \ldots,$ are independent and identically distributed with zero mean and unit variance. This general result is guaranteed by Donsker's theorem [17]. The simpler discrete approximation scheme uses Bernoulli random variables instead:

$$\widehat{X}(t_{n+1}) = \widehat{X}(t_n) + a[\widehat{X}(t_n), t_n]\Delta t + b[\widehat{X}(t_n), t_n]\sqrt{\Delta t}\, \xi, \qquad (14.7)$$

where $\text{Prob}[\xi = 1] = \text{Prob}[\xi = -1] = 1/2$. Note that $E[\xi] = 0$ and $\text{Var}[\xi] = 1$. This clearly defines a binomial model. As Δt goes to zero, $\widehat{X}$ converges to X [294, 434].

14.2.2 Trading and the Ito Integral

Consider an Ito process $d\boldsymbol{S}_t = \mu_t\, dt + \sigma_t\, dW_t$, where $\boldsymbol{S}_t$ is the vector of security prices at time t. Let ϕ_t be a trading strategy denoting the quantity of each type of security held at time t. Clearly the stochastic process $\phi_t \boldsymbol{S}_t$ is the value of the portfolio ϕ_t at time t. Then $\phi_t\, d\boldsymbol{S}_t \equiv \phi_t(\mu_t\, dt + \sigma_t\, dW_t)$ represents the change in the value from security price changes occurring at time t. The equivalent Ito integral,

$$G_T(\phi) \equiv \int_0^T \phi_t\, d\boldsymbol{S}_t = \int_0^T \phi_t\mu_t\, dt + \int_0^T \phi_t\sigma_t\, dW_t,$$

measures the gains realized by the trading strategy over the period $[0, T]$. A strategy is self-financing if

$$\phi_t S_t = \phi_0 S_0 + G_t(\phi) \tag{14.8}$$

for all $0 \leq t < T$. In other words, the investment at any time equals the initial investment plus the total capital gains up to that time.

Discrete-time models can clarify the above concepts. Let $t_0 < t_1 < \cdots < t_n$ denote the trading points. As before, S_k is the price vector at time t_k, and the vector ϕ_k denotes the quantity of each security held during $[t_k, t_{k+1})$. Thus $\phi_k S_k$ stands for the value of portfolio ϕ_k right after its establishment at time t_k, and $\phi_k S_{k+1}$ stands for the value of ϕ_k at time t_{k+1} *before* any transactions are made to set up the next portfolio ϕ_{k+1}. The nonanticipation requirement of the Ito integral means that ϕ_k must be established before S_{k+1} is known. The quantity $\phi_k \Delta S_k \equiv \phi_k (S_{k+1} - S_k)$ represents the capital gains between times t_k and t_{k+1}, and the summation $G(n) \equiv \sum_{k=0}^{n-1} \phi_k \Delta S_k$ is the total capital gains through time t_n. Note the similarity of this summation to the Ito integral of simple processes (14.1). A trading strategy is self-financing if the investment at any time is financed completely by the investment in the previous period, i.e.,

$$\phi_k S_k = \phi_{k-1} S_k$$

for all $0 < k \leq n$. The preceding condition and condition (14.8) are equivalent (see Exercise 14.2.1).

When an Ito process $dX_t = a_t \, dt + b_t \, dW_t$ is Markovian, the future evolution of X depends solely on its current value. The nonanticipating requirement further says that a_t and b_t cannot embody future values of dW. The Ito process is hence ideal for modeling asset price dynamics under the weak form of efficient markets.

▶ **Exercise 14.2.1** Prove that the self-financing definition $\phi_k S_k = \phi_{k-1} S_k$ implies the alternative condition $\phi_k S_k = \phi_0 S_0 + G_k$ for $0 < k \leq n$, and vice versa.

14.2.3 Ito's Lemma

The central tool in stochastic differential equations is **Ito's lemma**, which basically says that a smooth function of an Ito process is itself an Ito process.

THEOREM 14.2.1 *Suppose that $f : R \to R$ is twice continuously differentiable[1] and that $dX = a_t \, dt + b_t \, dW$. Then $f(X)$ is the Ito process*

$$f(X_t) = f(X_0) + \int_0^t f'(X_s) \, a_s \, ds + \int_0^t f'(X_s) \, b_s \, dW + \frac{1}{2} \int_0^t f''(X_s) \, b_s^2 \, ds$$

for $t \geq 0$.

In differential form, Ito's lemma becomes

$$df(X) = f'(X) \, a \, dt + f'(X) \, b \, dW + \frac{1}{2} f''(X) \, b^2 \, dt. \tag{14.9}$$

Compared with calculus, the interesting part of Eq. (14.9) is the third term on the right-hand side. This can be traced to the positive quadratic variation of Brownian paths, making $(dW)^2$ nonnegligible. A convenient formulation of Ito's lemma suitable for generalization to higher dimensions is

$$df(X) = f'(X) \, dX + \frac{1}{2} f''(X) (dX)^2. \tag{14.10}$$

Here, we are supposed to multiply out $(dX)^2 = (a\,dt + b\,dW)^2$ symbolically according to the following multiplication table:

×	dW	dt
dW	dt	0
dt	0	0

Note that the $(dW)^2 = dt$ entry is justified by Eq. (13.15). This form is easy to remember because of its similarity to Taylor expansion.

THEOREM 14.2.2 (Higher-Dimensional Ito's Lemma) *Let* $W_1, W_2, \ldots, W_n$ *be independent Wiener processes and let* $X \equiv (X_1, X_2, \ldots, X_m)$ *be a vector process. Suppose that* $f : R^m \to R$ *is twice continuously differentiable and* X_i *is an Ito process with* $dX_i = a_i\,dt + \sum_{j=1}^{n} b_{ij}\,dW_j$. *Then* $df(X)$ *is an Ito process with the differential*

$$df(X) = \sum_{i=1}^{m} f_i(X)\,dX_i + \frac{1}{2}\sum_{i=1}^{m}\sum_{k=1}^{m} f_{ik}(X)\,dX_i\,dX_k,$$

where $f_i \equiv \partial f/\partial x_i$ *and* $f_{ik} \equiv \partial^2 f/\partial x_i \partial x_k$.

The multiplication table for Theorem 14.2.2 is

×	dW_i	dt
dW_k	$\delta_{ik}\,dt$	0
dt	0	0

in which

$$\delta_{ik} = \begin{cases} 1, & \text{if } i = k \\ 0, & \text{otherwise} \end{cases}.$$

In applying the higher-dimensional Ito's lemma, usually one of the variables, say X_1, is the time variable t and $dX_1 = dt$. In this case, $b_{1j} = 0$ for all j and $a_1 = 1$. An alternative formulation of Ito's lemma incorporates the interdependence of the variables $X_1, X_2, \ldots, X_m$ into that between the Wiener processes.

THEOREM 14.2.3 *Let* $W_1, W_2, \ldots, W_m$ *be Wiener processes and let* $X \equiv (X_1, X_2, \ldots, X_m)$ *be a vector process. Suppose that* $f : R^m \to R$ *is twice continuously differentiable and* X_i *is an Ito process with* $dX_i = a_i\,dt + b_i\,dW_i$. *Then* $df(X)$ *is the following Ito process,*

$$df(X) = \sum_{i=1}^{m} f_i(X)\,dX_i + \frac{1}{2}\sum_{i=1}^{m}\sum_{k=1}^{m} f_{ik}(X)\,dX_i\,dX_k,$$

with the following multiplication table:

×	dW_i	dt
dW_k	$\rho_{ik}\,dt$	0
dt	0	0

Here, ρ_{ik} *denotes the correlation between* dW_i *and* dW_k.

In Theorem 14.2.3 the correlation between $dW_i = \sqrt{dt}\,\xi_i$ and $dW_k = \sqrt{dt}\,\xi_k$ refers to that between the normally distributed random variables ξ_i and ξ_k.

▶ **Exercise 14.2.2** Prove Eq. (14.3) by using Ito's formula.

14.3 Applications

This section presents applications of the Ito process, some of which will be useful later.

EXAMPLE 14.3.1 A (μ, σ) Brownian motion is $\mu\,dt + \sigma\,dW$ by Ito's lemma and Eq. (13.11).

EXAMPLE 14.3.2 Consider the Ito process $dX = \mu(t)\,dt + \sigma(t)\,dW$. It is identical to Brownian motion except that the drift $\mu(t)$ and diffusion $\sigma(t)$ are no longer constants. Again,

$$X(t) \sim N\left(X(0) + \int_0^t \mu(s)\,ds, \int_0^t \sigma^2(s)\,ds \right)$$

is normally distributed.

EXAMPLE 14.3.3 Consider the geometric Brownian motion process $Y(t) \equiv e^{X(t)}$, where $X(t)$ is a (μ, σ) Brownian motion. Ito's formula (14.9) implies that

$$\frac{dY}{Y} = (\mu + \sigma^2/2)\,dt + \sigma\,dW.$$

The instantaneous rate of return is $\mu + \sigma^2/2$, not μ.

EXAMPLE 14.3.4 Consider the Ito process $U \equiv YZ$ with $dY = a\,dt + b\,dW$ and $dZ = f\,dt + g\,dW$. Processes Y and Z share the Wiener process W. Ito's lemma (Theorem 14.2.2) can be used to show that $dU = Z\,dY + Y\,dZ + dY\,dZ$, which equals

$$Z\,dY + Y\,dZ + (a\,dt + b\,dW)(f\,dt + g\,dW) = Z\,dY + Y\,dZ + bg\,dt.$$

If either $b \equiv 0$ or $g \equiv 0$, then integration by parts holds.

EXAMPLE 14.3.5 Consider the Ito process $U \equiv YZ$, where $dY/Y = a\,dt + b\,dW_y$ and $dZ/Z = f\,dt + g\,dW_z$. The correlation between W_y and W_z is ρ. Apply Ito's lemma (Theorem 14.2.3):

$$\begin{aligned}
dU &= Z\,dY + Y\,dZ + dY\,dZ \\
&= ZY(a\,dt + b\,dW_y) + YZ(f\,dt + g\,dW_z) \\
&\quad + YZ(a\,dt + b\,dW_y)(f\,dt + g\,dW_z) \\
&= U(a + f + bg\rho)\,dt + Ub\,dW_y + Ug\,dW_z.
\end{aligned}$$

Note that dU/U has volatility $\sqrt{b^2 + 2bg\rho + g^2}$ by formula (6.9). The product of two (or more) correlated geometric Brownian motion processes thus remains geometric Brownian motion. This result has applications in **correlation options**, whose value depends on multiple assets. As

$$Y = \exp[\,(a - b^2/2)\,dt + b\,dW_y],$$
$$Z = \exp[\,(f - g^2/2)\,dt + g\,dW_z],$$
$$U = \exp[\{a + f - (b^2 + g^2)/2\}\,dt + b\,dW_y + g\,dW_z\,],$$

ln U is Brownian motion with a mean equal to the sum of the means of ln Y and ln Z. This holds even if Y and Z are correlated. Finally, ln Y and ln Z have correlation ρ.

EXAMPLE 14.3.6 Suppose that S follows $dS/S = \mu \, dt + \sigma \, dW$. Then $F(S, t) \equiv Se^{y(T-t)}$ follows

$$\frac{dF}{F} = (\mu - y) \, dt + \sigma \, dW$$

by Ito's lemma. This result has applications in forward and futures contracts.

▶ **Exercise 14.3.1** Assume that $dX/X = \mu \, dt + \sigma \, dW$. (1) Prove that ln X follows $d(\ln X) = (\mu - \sigma^2/2) \, dt + \sigma \, dW$. (2) Derive the probability distribution of $\ln(X(t)/X(0))$.

▶ **Exercise 14.3.2** Let X follow the geometric Brownian motion process $dX/X = \mu \, dt + \sigma \, dW$. Show that $R \equiv \ln X + \sigma^2 t/2$ follows $dR = \mu \, dt + \sigma \, dW$.

▶ **Exercise 14.3.3** (1) What is the stochastic differential equation for the process W^n? (2) Show that

$$\int_s^t W^n \, dW = \frac{W(t)^{n+1} - W(s)^{n+1}}{n+1} - \frac{n}{2} \int_s^t W^{n-1} \, dt.$$

(Hint: Use Eqs. (13.15) and (13.16) or apply Ito's lemma.)

▶ **Exercise 14.3.4** Consider the Ito process $U \equiv (Y + Z)/2$, where $dY/Y = a \, dt + b \, dW$ and $dZ/Z = f \, dt + g \, dW$. Processes Y and Z share the Wiener process W. Derive the stochastic differential equation for dU/U.

▶ **Exercise 14.3.5** Redo Example 14.3.4 except that $dY = a \, dt + b \, dW_y$ and $dZ = f \, dt + g \, dW_z$, where dW_y and dW_z have correlation ρ.

▶ **Exercise 14.3.6** Verify that $U \equiv Y/Z$ follows $dU/U = (a - f - bg\rho) \, dt + b \, dW_y - g \, dW_z$, where Y and Z are drawn from Example 14.3.5.

▶ **Exercise 14.3.7** Given $dY/Y = \mu \, dt + \sigma \, dW$ and $dX/X = r \, dt$, derive the stochastic differential equation for $F \equiv X/Y$.

14.3.1 The Ornstein–Uhlenbeck Process

The **Ornstein–Uhlenbeck process** has the stochastic differential equation

$$dX = -\kappa X \, dt + \sigma \, dW, \tag{14.11}$$

where $\kappa, \sigma \geq 0$ (see Fig. 14.4). It is known that

$$E[X(t)] = e^{-\kappa(t-t_0)} E[x_0],$$

$$\text{Var}[X(t)] = \frac{\sigma^2}{2\kappa} \left[1 - e^{-2\kappa(t-t_0)}\right] + e^{-2\kappa(t-t_0)} \text{Var}[x_0],$$

$$\text{Cov}[X(s), X(t)] = \frac{\sigma^2}{2\kappa} e^{-\kappa(t-s)} \left[1 - e^{-2\kappa(s-t_0)}\right] + e^{-\kappa(t+s-2t_0)} \text{Var}[x_0]$$

for $t_0 \leq s \leq t$ and $X(t_0) = x_0$. In fact, $X(t)$ is normally distributed if x_0 is a constant or normally distributed [30]; X is said to be a **normal process**. Of course, $E[x_0] = x_0$ and $\text{Var}[x_0] = 0$ if x_0 is a constant. When $x_0 \sim N\left(0, \frac{\sigma^2}{2\kappa}\right)$, it is easy to see that X is

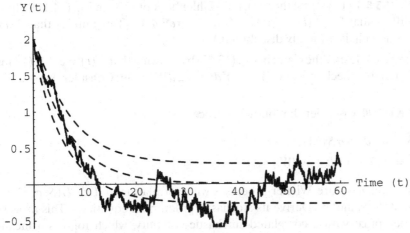

Figure 14.4: Sample path of Ornstein–Uhlenbeck process. Shown is a sample path of the Ornstein–Uhlenbeck process $dY = -0.15\, Y\, dt + 0.15\, dW$, starting at $Y(0) = 2$. The envelope is for one standard deviation $\sqrt{[\,(0.15)^2/0.3\,](1 - e^{-0.3t})}$ around the mean $2e^{-0.15t}$. In contrast to Brownian motion, which diverges to infinite values (see Fig. 13.2), the Ornstein–Uhlenbeck process converges to a stationary distribution.

stationary. The Ornstein–Uhlenbeck process describes the velocity of a tiny particle through a fluid in thermal equilibrium – in short, Brownian motion in nature [386].

The Ornstein–Uhlenbeck process has the following **mean-reversion** property. When $X > 0$, the dX term tends to be negative, pulling X toward zero, whereas if $X < 0$, the dX term tends to be positive, pulling X toward zero again.

EXAMPLE 14.3.7 Suppose that X is an Ornstein–Uhlenbeck process. Ito's lemma says that $V \equiv X^2$ has the differential

$$dV = 2X\, dX + (dX)^2 = 2\sqrt{V}\,(-\kappa\sqrt{V}\, dt + \sigma\, dW) + \sigma^2\, dt$$
$$= (-2\kappa V + \sigma^2)\, dt + 2\sigma\sqrt{V}\, dW,$$

a **square-root process**.

Consider the following process, also called the Ornstein–Uhlenbeck process:

$$dX = \kappa(\mu - X)\, dt + \sigma\, dW, \tag{14.12}$$

where $\sigma \geq 0$. Given $X(t_0) = x_0$, a constant, it is known that

$$E[\, X(t)\,] = \mu + (x_0 - \mu)\, e^{-\kappa(t-t_0)} \tag{14.13}$$

$$\mathrm{Var}[\, X(t)\,] = \frac{\sigma^2}{2\kappa}\left[1 - e^{-2\kappa(t-t_0)}\right] \tag{14.14}$$

for $t_0 \leq t$ [855]. Because the mean and the standard deviation are roughly μ and $\sigma/\sqrt{2\kappa}$, respectively, for large t, the probability of X's being negative is extremely unlikely in any finite time interval when $\mu > 0$ is large relative to $\sigma/\sqrt{2\kappa}$ (say $\mu > 4\sigma/\sqrt{2\kappa}$). Process (14.12) has the salient mean-reverting feature that X tends to move toward μ, making it useful for modeling term structure [855], stock price volatility [823], and stock price return [613].

▶ **Exercise 14.3.8** Let $X(t)$ be the Ornstein–Uhlenbeck process in Eq. (14.11). Show that the differential for $Y(t) \equiv X(t) e^{\kappa t}$ is $dY = \sigma e^{\kappa t} dW$. (This implies that $Y(t)$, hence $X(t)$ as well, is normally distributed.)

▶ **Exercise 14.3.9** Justify the claim in Eq. (13.13) by showing that $Y(t) \equiv e^{-t} W(e^{2t})$ is the Ornstein–Uhlenbeck process $dY = -Y dt + \sqrt{2} dW$. (Hint: Consider $Y(t + dt) - Y(t)$.)

▶ **Exercise 14.3.10** Consider the following processes:

$$dS = \mu S \, dt + \sigma S \, dW_1,$$
$$d\sigma = \beta(\overline{\sigma} - \sigma) \, dt + \gamma \, dW_2,$$

where dW_1 and dW_2 are Wiener processes with correlation ρ. Let $H(S, \sigma, \tau)$ be a function of S, σ, and τ. Derive its stochastic differential equation. (This process models stock price with a correlated stochastic volatility, which follows a mean-reverting process.)

▶ **Exercise 14.3.11** Show that the transition probability density function p of $dX = -(1/2) X \, dt + dW$ satisfies the backward equation

$$\frac{\partial p}{\partial s} = -\frac{1}{2} \frac{\partial^2 p}{\partial x^2} + \frac{1}{2} x \frac{\partial p}{\partial x}.$$

(Hint: $X(t) \sim N(x e^{-(t-s)/2}, 1 - e^{-(t-s)})$ when $X(s) = x$.)

14.3.2 The Square-Root Process

The square-root process has the stochastic differential equation

$$dX = \kappa(\mu - X) \, dt + \sigma \sqrt{X} \, dW,$$

where $\kappa, \sigma \geq 0$ and the initial value of X is a nonnegative constant. See Fig. 14.5 for an illustration. Like the Ornstein–Uhlenbeck process, the square-root process

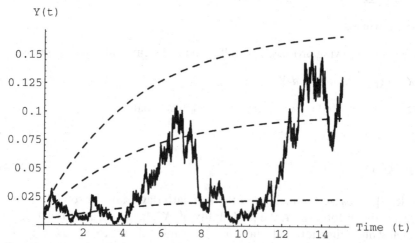

Figure 14.5: Sample path of square-root process. Shown is a sample path of the square-root process $dY = 0.2(0.1 - Y) \, dt + 0.15\sqrt{Y} \, dW$ with the initial condition $Y(0) = 0.01$. The envelope is for one standard deviation around the mean, which is $0.01 \, e^{-0.2t} + 0.1 \, (1 - e^{-0.2t})$.

possesses mean reversion in that X tends to move toward μ, but the volatility is proportional to $\sqrt{X}$ instead of a constant. When X hits zero and $\mu \geq 0$, the probability is one that it will not move below zero; in other words, zero is a **reflecting boundary**. Hence, the square-root process is a good candidate for modeling interest rate movements [234]. The Ornstein–Uhlenbeck process, in contrast, allows negative interest rates. The two processes are related (see Example 14.3.7).

Feller (1906–1970) showed that the random variable $2cX(t)$ follows the non-central chi-square distribution

$$\chi\left(\frac{4\kappa\mu}{\sigma^2}, 2cX(0)\,e^{-\kappa t}\right),$$

where $c \equiv (2\kappa/\sigma^2)(1 - e^{-\kappa t})^{-1}$ [234, 341]. Given $X(0) = x_0$, a constant, it can be proved that

$$E[\,X(t)\,] = x_0 e^{-\kappa t} + \mu(1 - e^{-\kappa t}),$$
$$\mathrm{Var}[\,X(t)\,] = x_0\,\frac{\sigma^2}{\kappa}\,(e^{-\kappa t} - e^{-2\kappa t}) + \mu\,\frac{\sigma^2}{2\kappa}(1 - e^{-\kappa t})^2$$

for $t \geq 0$.

14.4 Financial Applications

14.4.1 Transactions Costs

Transactions costs are a fact of life, never zero however negligible. Under the **proportional transactions cost model**, it is impossible to trade continuously. Intuitively, this is because the transactions cost per trade is proportional to $|dW|$, and $\int_0^T |dW| = \infty$ almost surely by Eq. (13.17). As a consequence, a continuous trader would be bankrupt with probability one [660]. Even stronger claims can be made. For instance, the cheapest trading strategy to dominate the value of European call at maturity is the covered call [814].

▶ **Exercise 14.4.1** Argue that an investor who has information about the entire future value of the Brownian motion's driving the stock price will have infinite wealth at any given horizon date. In other words, market fluctuations can be exploited.

14.4.2 Stochastic Interest Rate Models

Merton originated the following methodology to term structure modeling [493]. Suppose that the short rate r follows a Markov process $dr = \mu(r, t)\,dt + \sigma(r, t)\,dW$. Let $P(r, t, T)$ denote the price at time t of a zero-coupon bond that pays \$1 at time T. Its stochastic process must also be Markovian. Write its dynamics as

$$\frac{dP}{P} = \mu_p\,dt + \sigma_p\,dW$$

so that the expected instantaneous rate of return on a $(T - t)$-year zero-coupon bond is μ_p and the instantaneous variance is σ_p^2. Surely $P(r, T, T) = 1$ for any T. By Ito's lemma (Theorem 14.2.2),

$$
\begin{aligned}
dP &= \frac{\partial P}{\partial T}\, dT + \frac{\partial P}{\partial r}\, dr + \frac{1}{2}\frac{\partial^2 P}{\partial r^2}\,(dr)^2 \\
&= -\frac{\partial P}{\partial T}\, dt + \frac{\partial P}{\partial r}\,[\,\mu(r,t)\,dt + \sigma(r,t)\,dW\,] \\
&\quad + \frac{1}{2}\frac{\partial^2 P}{\partial r^2}\,[\,\mu(r,t)\,dt + \sigma(r,t)\,dW\,]^2 \\
&= \left[-\frac{\partial P}{\partial T} + \mu(r,t)\frac{\partial P}{\partial r} + \frac{\sigma(r,t)^2}{2}\frac{\partial^2 P}{\partial r^2} \right] dt + \sigma(r,t)\frac{\partial P}{\partial r}\, dW,
\end{aligned}
$$

where $dt = -dT$ in the second equality. Hence

$$
-\frac{\partial P}{\partial T} + \mu(r,t)\frac{\partial P}{\partial r} + \frac{\sigma(r,t)^2}{2}\frac{\partial^2 P}{\partial r^2} = P\mu_p, \quad \sigma(r,t)\frac{\partial P}{\partial r} = P\sigma_p. \tag{14.15}
$$

Models with the short rate as the only explanatory variable are called **short rate models**.

The Merton Model

Suppose we assume the local expectations theory, which means that μ_p equals the prevailing short rate $r(t)$ for all T, and we assume that μ and σ are constants. Then the partial differential equations (14.15) yield the following solution:

$$
P(r,t,T) = \exp\left[-r(T-t) - \frac{\mu(T-t)^2}{2} + \frac{\sigma^2(T-t)^3}{6} \right]. \tag{14.16}
$$

This model is due to Merton [660]. We make a few observations. First, $\sigma_p = -\sigma(T-t)$, which says sensibly that bonds with longer maturity are more volatile. The dynamics of P is $dP/P = r\,dt - \sigma(T-t)\,dW$. Now, P has no upper limits as T becomes large, which does not square with the reality. This happens because of negative rates in the model.

▶ **Exercise 14.4.2** Negative interest rates imply arbitrage profits for riskless bonds. Why?

Duration under Parallel Shifts

Consider duration with respect to parallel shifts in the spot rate curve. For convenience, assume that $t = 0$. Parallel shift means $S(r + \Delta r, T) = S(r, T) + \Delta r$ for any Δr; so $\partial S(r, T)/\partial r = 1$. This implies $S(r, T) = r + g(T)$ for some function g with $g(0) = 0$ because $S(r, 0) = r$. Consequently, $P(r, T) = e^{-[r+g(T)]\,T}$. Substitute this identity into the left-hand part of Eqs. (14.15) and assume the local expectations theory to obtain

$$
g'(T) + \frac{g(T)}{T} = \mu(r) - \frac{\sigma(r)^2}{2}\,T.
$$

As the left-hand side is independent of r, so must the right-hand side be. Because this must hold for all T, both $\mu(r)$ and $\sigma(r)$ must be constants, i.e., the Merton model. As mentioned before, this model is flawed, so must duration be, as such [496].

▶ **Exercise 14.4.3** Suppose that the current spot rate curve is flat. Under the assumption that only parallel shifts are allowed, what can we say about the parameters μ and σ governing the short rate process?

Immunization under Parallel Shifts Revisited

A duration-matched portfolio under parallel shifts in the spot rate curve begets arbitrage opportunities in that the portfolio value exceeds the liability for any instantaneous rate changes. This was shown in Subsection 5.8.2. However, this seeming inconsistency with equilibrium disappears if changes in portfolio value *through* time are taken into account. Indeed, for certain interest rate models, a liability immunized by a duration-matched portfolio exceeds the minimum portfolio value at any given time in the future. Thus the claimed arbitrage profit evaporates because the portfolio value does not always cover the liability.

We illustrate this point with the Merton model $dr = \mu\, dt + \sigma\, dW$, which results from parallel shifts in the spot rate curve and the local expectations theory. To immunize a \$1 liability that is due at time s, a two-bond portfolio is constructed now with maturity dates t_1 and t_2, where $t_1 < s < t_2$. Each bond is a zero-coupon bond with \$1 par value. The portfolio matches the present value of the liability today, and its value relative to the PV of the liability is minimum among all such portfolios (review Subsection 4.2.2). Consider any future time t such that $t < t_1$. With $A(t)$ denoting the portfolio value and $L(t)$ the liability value at time t, it can be shown that the asset/liability ratio $A(t)/L(t)$ is a convex function of the prevailing interest rate and $A(t) < L(t)$ (see Exercise 14.4.4). This conclusion holds for other interest rate models [52].

▶ **Exercise 14.4.4** (1) Prove that $A(t)/L(t)$ is a convex function of the prevailing interest rate. (2) Then verify $A(t) < L(t)$.

14.4.3 Modeling Stock Prices

The most popular stochastic model for stock prices has been geometric Brownian motion $dS/S = \mu\, dt + \sigma\, dW$. This model best describes an equilibrium in which expectations about future returns have settled down [660].

From the discrete-time analog $\Delta S/S = \mu\, \Delta t + \sigma\sqrt{\Delta t}\, \xi$, where $\xi \sim N(0,1)$, we know that $\Delta S/S \sim N(\mu\,\Delta t, \sigma^2 \Delta t)$. The percentage return for the next Δt time hence has mean $\mu\,\Delta t$ and variance $\sigma^2 \Delta t$. In other words, the percentage return per unit time has mean μ and variance σ^2. For this reason, μ is called the expected instantaneous rate of return and σ^2 the instantaneous variance of the rate of return. If there is no uncertainty about the stock price, i.e., $\sigma = 0$, then $S(t) = S(0)\, e^{\mu t}$.

Comment 14.4.1 It may seem strange that the rate of return is μ instead of $\mu - \sigma^2/2$. Example 14.3.3 says that $S(t)/S(0) = e^{X(t)}$, where $X(t)$ is a $(\mu - \sigma^2/2, \sigma)$ Brownian motion and the continuously compounded rate of return over the time period $[\,0, T\,]$ is

$$\frac{\ln[\,S(T)/S(0)\,]}{T} = \frac{X(T) - X(0)}{T} \sim N\left(\mu - \frac{\sigma^2}{2}, \sigma^2\right). \tag{14.17}$$

The expected continuously compounded rate of return is then $\mu - \sigma^2/2$! Well, they refer to alternative definitions of rates of return. Unless stated otherwise, it is the *former* (instantaneous rate of return μ) that we have in mind from now on. It should be pointed out that the μ used in the BOPM in Eq. (9.13) referred to the latter rate of return. In summary,

$$\frac{E[\{S(\Delta t) - S(0)\}/S(0)]}{\Delta t} \to \mu,$$

$$\frac{\ln E[S(T)/S(0)]}{T} = \mu,$$

$$\frac{E[\ln(S(T)/S(0))]}{T} = \mu - \frac{\sigma^2}{2}.$$

(See Comment 9.3.2, Lemma 9.3.1, Example 14.3.3, and Exercises 13.3.8 and 14.4.5.)

▶ **Exercise 14.4.5** Prove that $E[S(T)] = S(0)\, e^{\mu T}$.

▶ **Exercise 14.4.6** Suppose the stock price follows the geometric Brownian motion process $dS/S = \sigma\, dW$. Example 14.3.3 says that $S(t)/S(0) = e^{X(t)}$, where $X(t)$ is a $(-\sigma^2/2, \sigma)$ Brownian motion. In other words, the stock is expected to have a negative growth rate. Explain why the growth rate is not zero.

▶ **Exercise 14.4.7** Show that the simple rate of return $[S(t)/S(0)] - 1$ has mean $e^{\mu t} - 1$ and variance $e^{2\mu t}(e^{\sigma^2 t} - 1)$.

▶ **Exercise 14.4.8** Assume that the volatility σ is stochastic but driven by an independent Wiener process. Suppose that the average variance over the time period $[0, T]$ as defined by $\widehat{\sigma^2} \equiv \frac{1}{T}\int_0^T \sigma^2(t)\, dt$ is given. Argue that

$$\ln\frac{S(T)}{S(0)} \sim N(\mu T - (\widehat{\sigma^2}\, T/2),\, \widehat{\sigma^2}\, T).$$

(Thus $S(T)$ seen from time zero remains lognormally distributed.)

▶ **Exercise 14.4.9** Justify using $\Delta S/(S\sqrt{\Delta t})$ to estimate volatility.

▶ **Exercise 14.4.10** What are the shortcomings of modeling the stock price dynamics by $dS = \mu\, dt + \sigma\, dW$ with constant μ and σ?

Continuous-Time Limit of the Binomial Model

What is the Ito process for the stock's rate of return in a risk-neutral economy to which the binomial model in Section 9.2 converges? The continuously compounded rate of return of the stock price over a period of length τ is a sum of the following n independent identically distributed random variables:

$$X_i = \begin{cases} \ln u & \text{with probability } p \\ \ln d & \text{with probability } 1 - p \end{cases},$$

where $u \equiv e^{\sigma\sqrt{\tau/n}}$, $d \equiv e^{-\sigma\sqrt{\tau/n}}$, and $p \equiv (e^{r\tau/n} - d)/(u - d)$. The rate of return is hence the random walk $\sum_{i=1}^n X_i$. It is straightforward to verify that

$$E\left[\sum_{i=1}^n X_i\right] \to \left(r - \frac{\sigma^2}{2}\right)\tau, \qquad \text{Var}\left[\sum_{i=1}^n X_i\right] \to \sigma^2\tau. \tag{14.18}$$

The continuously compounded rate of return thus converges to a $(r - \sigma^2/2, \sigma)$ Brownian motion, and the stock price follows $dS/S = r\, dt + \sigma\, dW$ in a risk-neutral economy. Indeed, the discount process $\{Z(t) \equiv e^{-rt} S(t), t \geq 0\}$ is a martingale under this risk-neutral probability measure as $dZ/Z = \sigma\, dW$ is a driftless Ito process by Exercise 14.3.6.

▶ **Exercise 14.4.11** Verify Eqs. (14.18). (Hint: $e^x \approx 1 + x + x^2/2$.)

▶ **Exercise 14.4.12** From the above discussions, $E[\,X_i\,] \to (r - \sigma^2/2)\,\Delta t$ and $\sqrt{\mathrm{Var}[\,X_i\,]} \to \sigma\sqrt{\Delta t}$. Now $X_{i+1} = \ln(S_{i+1}/S_i)$, where $S_i \equiv S_0 e^{X_1 + X_2 + \cdots + X_i}$ is the stock price at time i. Hence

$$X_{i+1} = \ln\left(1 + \frac{S_{i+1} - S_i}{S_i}\right) \approx \frac{S_{i+1} - S_i}{S_i} \equiv \frac{\Delta S_i}{S_i}.$$

The above finding suggests that $dS/S = (r - \sigma^2/2)\,dt + \sigma\, dW$, contradicting the above! Find the hole in the argument and correct it.

▶ **Exercise 14.4.13** The continuously compounded rate of return $X \equiv \ln S$ follows $dX = (r - \sigma^2/2)\,dt + \sigma\, dW$ in a risk-neutral economy. Use this fact to show that

$$u = \exp[\,(r - \sigma^2/2)\,\Delta t + \sigma\sqrt{\Delta t}\,], \quad d = \exp[\,(r - \sigma^2/2)\,\Delta t - \sigma\sqrt{\Delta t}\,]$$

under the alternative binomial model in which an up move occurs with probability $1/2$.

Additional Reading

We followed [541, 543, 763, 764] in the exposition of stochastic processes and [30, 289, 419, 544] in the discussions of stochastic integrals. The Ito integral is due to Ito (1915–) [500]. Ito's lemma is also due to Ito under the strong influence of Bachelier's thesis [501, 512]. Rigorous proofs of Ito's lemma can be found in [30, 419], and informal ones can be found in [470, 492, 660]. *Mathematica* programs for carrying out some of the manipulations are listed in [854]. Consult [556, 557, 558, 774] for numerical solutions of stochastic differential equations. See [613] for the multivariate Ornstein–Uhlenbeck process. Other useful references include [822] (diffusion), [549] (Ito integral), [280, 761] (stochastic processes), [211, 364, 776] (stochastic differential equations), [261] (stochastic convergence), [542] (stochastic optimization in trading), [70] (nonprobabilistic treatment of continuous-time treading), [102, 181] (distribution-free competitive trading), and [112, 115, 262, 274, 681] (transactions costs).

NOTE

1. This means that all first- and second-order partial derivatives exist and are continuous.

> A proof is that which convinces a reasonable man; a rigorous proof is that which convinces an unreasonable man.
> Mark Kac (1914–1984)

CHAPTER FIFTEEN

Continuous-Time Derivatives Pricing

This problem of time in the art of music is of capital importance.
Igor Stravinsky (1882–1971), *Poetics of Music*

After a short introduction to partial differential equations, this chapter presents the partial differential equation that the option value should satisfy in continuous time. The general methodology is then applied to derivatives, including options on a stock that pays continuous dividends, futures, futures options, correlation options, exchange options, path-dependent options, currency-related options, barrier options, convertible bonds, and options under stochastic volatility. This chapter also discusses the correspondence between the partial differential equation and the martingale approach to pricing.

15.1 Partial Differential Equations

A two-dimensional second-order partial differential equation has the following form:

$$p\frac{\partial^2\theta}{\partial x^2} + q\frac{\partial^2\theta}{\partial x\,\partial y} + r\frac{\partial^2\theta}{\partial y^2} + s\frac{\partial\theta}{\partial x} + t\frac{\partial\theta}{\partial y} + u\theta + v = 0,$$

where p, q, r, s, t, u, and v may be functions of the two independent variables x and y as well as the dependent variable θ and its derivatives. It is called **elliptic**, **parabolic**, or **hyperbolic** according to whether $q^2 < 4pr$, $q^2 = 4pr$, or $q^2 > 4pr$, respectively, over the domain of interest. For this reason, $q^2 - 4pr$ is called the **discriminant**. Note that the solution to a partial differential equation is a function.

Partial differential equations can also be classified into **initial-value** and **boundary-value problems**. An initial-value problem propagates the solution forward in time from the values given at the initial time. In contrast, a boundary-value problem has known values that must be satisfied at both ends of the relevant intervals [391]. If the conditions for some independent variables are given in the form of initial values and those for others as boundary conditions, we have an **initial-value boundary problem**.

A standard elliptic equation is the two-dimensional **Poisson equation**:

$$\frac{\partial^2\theta}{\partial x^2} + \frac{\partial^2\theta}{\partial y^2} = -\rho(x, y).$$

The **wave equation**,

$$\frac{\partial^2 \theta}{\partial t^2} - \frac{1}{v^2} \frac{\partial^2 \theta}{\partial x^2} = 0,$$

is hyperbolic. The most important parabolic equation is the **diffusion (heat) equation**,

$$\frac{1}{2} D \frac{\partial^2 \theta}{\partial x^2} - \frac{\partial \theta}{\partial t} = 0,$$

which is a special case of the Fokker–Planck equation. Given the initial condition $\theta(x, 0) = f(x)$ for $-\infty < x < \infty$, its unique bounded solution for any $t > 0$ is

$$\frac{1}{\sqrt{2\pi Dt}} \int_{-\infty}^{\infty} f(z) e^{-(x-z)^2/(2Dt)} \, dz \qquad (15.1)$$

when $f(x)$ is bounded and piecewise continuous for all real x. The solution clearly depends on the entire initial condition. Any two-dimensional second-order partial differential equation can be reduced to generalized forms of the Poisson equation, the diffusion equation, or the wave equation according to whether it is elliptic, parabolic, or hyperbolic.

▶ **Exercise 15.1.1** Verify that the diffusion equation is indeed satisfied by integral (15.1).

15.2 **The Black-Scholes Differential Equation**

The price of any derivative on a non-dividend-paying stock must satisfy a partial differential equation. The key step is recognizing that the same random process drives both securities; it is **systematic**, in other words. Given that their prices are perfectly correlated, we can figure out the amount of stock such that the gain from it offsets exactly the loss from the derivative, and vice versa. This removes the uncertainty from the value of the portfolio of the stock and the derivative at the end of a short period of time and forces its return to be the riskless rate in order to avoid arbitrage opportunities.

Several assumptions are made: (1) the stock price follows the geometric Brownian motion $dS = \mu S \, dt + \sigma S \, dW$ with constant μ and σ, (2) there are no dividends during the life of the derivative, (3) trading is continuous , (4) short selling is allowed, (5) there are no transactions costs or taxes, (6) all securities are infinitely divisible, (7) there are no riskless arbitrage opportunities, (8) the term structure of riskless rates is flat at r, and (9) there is unlimited riskless borrowing and lending. Some of these assumptions can be relaxed. For instance, μ, σ, and r can be deterministic functions of time instead of constants. In what follows, t denotes the current time (in years), T denotes the expiration time, and $\tau \equiv T - t$.

15.2.1 Merton's Derivation

Let C be the price of a derivative on S. From Ito's lemma (Theorem 14.2.2),

$$dC = \left(\mu S \frac{\partial C}{\partial S} + \frac{\partial C}{\partial t} + \frac{1}{2} \sigma^2 S^2 \frac{\partial^2 C}{\partial S^2} \right) dt + \sigma S \frac{\partial C}{\partial S} \, dW.$$

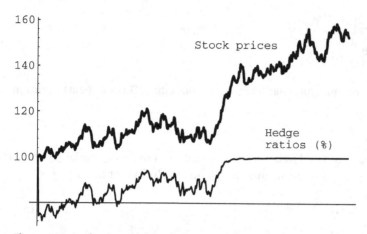

Figure 15.1: Stock price and delta (hedge ratio). Recall that delta is defined as $\partial C / \partial S$. The strike price here is $95.

Note that the same W drives both C and S. We now show that this random source can be eliminated by being short one derivative and long $\partial C / \partial S$ shares of stock (see Fig. 15.1). Define Π as the value of the portfolio. By construction, $\Pi = -C + S(\partial C / \partial S)$, and the change in the value of the portfolio at time dt is

$$d\Pi = -dC + \frac{\partial C}{\partial S}\, dS.$$

Substitute the formulas for dC and dS into the preceding equation to yield

$$d\Pi = \left(-\frac{\partial C}{\partial t} - \frac{1}{2}\sigma^2 S^2 \frac{\partial^2 C}{\partial S^2} \right) dt.$$

Because this equation does not involve dW, the portfolio is riskless during dt time and hence earns the instantaneous return rate r; that is, $d\Pi = r\Pi\, dt$, so

$$\left(\frac{\partial C}{\partial t} + \frac{1}{2}\sigma^2 S^2 \frac{\partial^2 C}{\partial S^2} \right) dt = r \left(C - S\frac{\partial C}{\partial S} \right) dt.$$

Equate the terms to obtain finally

$$\frac{\partial C}{\partial t} + rS\frac{\partial C}{\partial S} + \frac{1}{2}\sigma^2 S^2 \frac{\partial^2 C}{\partial S^2} = rC. \tag{15.2}$$

This is the celebrated **Black–Scholes differential equation** [87].

The Black–Scholes differential equation can be expressed in terms of sensitivity numbers:

$$\Theta + rS\Delta + \frac{1}{2}\sigma^2 S^2 \Gamma = rC. \tag{15.3}$$

(Review Section 10.1 for the definitions of sensitivity measures.) Identity (15.3) leads to an alternative way of computing Θ numerically from Δ and Γ. In particular, if a portfolio is delta-neutral, then the above equation becomes

$$\Theta + \frac{1}{2}\sigma^2 S^2 \Gamma = rC.$$

A definite relation thus exists between Γ and Θ.

▶ **Exercise 15.2.1** (1) Verify that the Black–Scholes formula for European calls in Theorem 9.3.4 indeed satisfies Black–Scholes differential equation (15.2). (2) Verify that the value of a forward contract on a non-dividend-paying stock satisfies Black–Scholes differential equation (15.2).

▶ **Exercise 15.2.2** It seems reasonable to expect the predictability of stock returns, as manifested in the drift of the Ito process, to have an impact on option prices. (One possibility is for the log price $R \equiv \ln S$ to follow a trendy Ornstein–Uhlenbeck process instead of Brownian motion.) The analysis in the text, however, implies that the drift of the process is irrelevant; the same Black–Scholes formula stands. Try to resolve the issue.

▶ **Exercise 15.2.3** Outline an argument for the claim that the Black–Scholes differential equation results from the BOPM by taking limits.

Continuous Adjustments

The portfolio Π is riskless for only an infinitesimally short period of time. If the delta $\partial C/\partial S$ changes with S and t, the portfolio must be continuously adjusted to ensure that it remains riskless.

Number of Random Sources

There is no stopping at the single-factor random source. In the presence of two random sources, three securities suffice to eliminate the uncertainty: Use two to eliminate the first source and the third to eliminate the second source. To make this work, the factors must be traded. A **traded security** is an asset that is held solely for investment by a significant number of individuals. Generally speaking, a market is complete only if the number of traded securities exceeds the number of random sources [76].

Risk-Neutral Valuation

Like the BOPM, the Black–Scholes differential equation does not depend directly on the risk preferences of investors. All the variables in the equation are independent of risk preferences, and the one that does depend on them, the expected return of the stock, does not appear in the equation. As a consequence, any risk preference can be used in pricing, including the risk-neutral one.

In a risk-neutral economy, the expected rate of return on all securities is the riskless rate r. Prices are then obtained by discounting the expected value at r. Lemma 9.2.1 says the same thing of the BOPM. The risk-neutral assumption greatly simplifies the analysis of derivatives. It is emphasized that it is the *instantaneous* return rate of the stock that is equal to r (see Comment 14.4.1 for the subtleties).

▶ **Exercise 15.2.4** Explain why the formula

$$e^{-r\tau} \int_X^\infty (y - X) \frac{1}{\sigma y \sqrt{2\pi\tau}} \exp\left[-\frac{\{\ln(y/S) - (r - \sigma^2/2)\tau\}^2}{2\sigma^2 \tau} \right] dy,$$

is equivalent to the Black–Scholes formula for European calls. (Hint: Review Eq. (6.10).)

15.2.2 Solving the Black–Scholes Equation for European Calls

The Black–Scholes differential equation can be solved directly for European options. After Eq. (15.2) is transformed with the change of variable $C(S, \tau) \equiv B(S, \tau) e^{-r\tau} X$, the partial differential equation becomes

$$-\frac{\partial B}{\partial \tau} + rS \frac{\partial B}{\partial S} + \frac{1}{2} \sigma^2 S^2 \frac{\partial^2 B}{\partial S^2} = 0,$$

where $B(0, \tau) = 0$ for $\tau > 0$ and $B(S, 0) = \max(S/X - 1, 0)$ for $S > 0$. With transformations $D(x, \tau) \equiv B(S, \tau)$ and $x \equiv (S/X) e^{r\tau}$, we end up with the diffusion equation:

$$-\frac{\partial D}{\partial \tau} + \frac{1}{2} (\sigma x)^2 \frac{\partial^2 D}{\partial x^2} = 0,$$

where $D(0, \tau) = 0$ for $\tau > 0$ and $D(x, 0) = \max(x - 1, 0)$ for $x > 0$. After one more transformation $u \equiv \sigma^2 \tau$, the function $H(x, u) \equiv D(x, \tau)$ satisfies

$$-\frac{\partial H}{\partial u} + \frac{1}{2} x^2 \frac{\partial^2 H}{\partial x^2} = 0,$$

where $H(0, u) = 0$ for $u > 0$ and $H(x, 0) = \max(x - 1, 0)$ for $x > 0$. The final transformation $\Theta(z, u) x \equiv H(x, u)$ where $z \equiv (u/2) + \ln x$, lands us at

$$-\frac{\partial \Theta}{\partial u} + \frac{1}{2} \frac{\partial^2 \Theta}{\partial z^2} = 0.$$

The boundary conditions are $|\Theta(z, u)| \leq 1$ for $u > 0$ and $\Theta(z, 0) = \max(1 - e^{-z}, 0)$. The above diffusion equation has the solution

$$\Theta(z, u) = \frac{1}{\sqrt{2\pi u}} \int_0^\infty (1 - e^{-y}) e^{-(z-y)^2/(2u)} \, dy$$

$$= \frac{1}{\sqrt{2\pi u}} \int_0^\infty e^{-(z-y)^2/(2u)} \, dy - \frac{1}{\sqrt{2\pi u}} \int_0^\infty e^{-y} e^{-(z-y)^2/(2u)} \, dy$$

$$= \frac{1}{\sqrt{2\pi}} \int_{-z/\sqrt{u}}^\infty e^{-\omega_1^2/2} \, d\omega_1 - \frac{1}{\sqrt{2\pi} x} \int_{-(z-u)/\sqrt{u}}^\infty e^{-\omega_2^2/2} \, d\omega_2$$

$$= N\left(\frac{z}{\sqrt{u}}\right) - \frac{1}{x} N\left(\frac{z-u}{\sqrt{u}}\right)$$

by formula (15.1) with the change of variables $\omega_1 \equiv (y - z)/\sqrt{u}$ and $\omega_2 \equiv (y - z + u)/\sqrt{u}$. Hence

$$H(x, u) = \Theta(\ln x + (u/2), u) x = xN\left(\frac{\ln x + (u/2)}{\sqrt{u}}\right) - N\left(\frac{\ln x - (u/2)}{\sqrt{u}}\right).$$

Retrace the steps to obtain

$$C(S, \tau) = H\left(\frac{S}{X} e^{r\tau}, \sigma^2 \tau\right) e^{-r\tau} X$$

$$= SN\left(\frac{\ln(S/X) + r\tau + \sigma^2\tau/2}{\sqrt{\sigma^2\tau}}\right) - e^{-r\tau} XN\left(\frac{\ln(S/X) + r\tau - \sigma^2\tau/2}{\sqrt{\sigma^2\tau}}\right),$$

which is precisely the Black–Scholes formula for the European call.

▶ **Exercise 15.2.5** Solve the Black–Scholes differential equation for European puts.

15.2.3 Initial and Boundary Conditions

Solving the Black–Scholes differential equation depends on the initial and the boundary conditions defining the particular derivative. These conditions spell out the values of the derivative at various values of S and t. For European calls (puts), the key terminal condition is $\max(S(T) - X, 0)$ ($\max(X - S(T), 0)$, respectively) at time T. There are also useful boundary conditions. The call value is zero when $S(t) = 0$, and the put is $Xe^{-r(T-t)}$ when $S(t) = 0$ (see Exercise 8.4.5). Furthermore, as S goes to infinity, the call value is S and the put value zero. The accuracy is even better if $S - Xe^{-r(T-t)}$ is used in place of S for the European call as $S \to \infty$. Although these boundary conditions are not mathematically necessary, they improve the accuracy of numerical methods [879].

The American put is more complicated because of early exercise whose boundary $\overline{S}(t)$ is unknown a priori. Recall that the exercise boundary specifies the stock price at each instant of time when it becomes optimal to exercise the option. The formulation that guarantees a unique solution is

$$\frac{\partial P}{\partial t} + rS\frac{\partial P}{\partial S} + \frac{1}{2}\sigma^2 S^2 \frac{\partial^2 P}{\partial S^2} = rP \quad \text{and} \quad P > X - S \quad \text{for } \overline{S} < S < \infty$$

$$P = X - S \quad \text{for } 0 \le S < \overline{S}$$

$$\frac{\partial P}{\partial S} = -1 \quad \text{and} \quad P = X - S \quad \text{for } S = \overline{S}$$

$$P = 0 \quad \text{for } S \to \infty$$

plus the obvious terminal condition [154]. The region $0 \le S < \overline{S}$ is where early exercise is optimal. The exercise boundary is a continuous decreasing function of τ for American puts and a continuous increasing function of τ for American calls [575].

▶ **Exercise 15.2.6** Verify that the Black–Scholes differential equation is violated where it is optimal to exercise the American put early; i.e., $X - S$ does not satisfy the equation.

15.3 Applications

15.3.1 Continuous Dividend Yields

The price for a stock that continuously pays out dividends at an annualized rate of q follows

$$\frac{dS}{S} = (\mu - q)\,dt + \sigma\,dW,$$

where μ is the stock's rate of return. This process was postulated in Subsection 9.6.4 for the stock index and the exchange rate. In a risk-neutral economy, $\mu = r$.

Consider a derivative security whose value f depends on a stock that pays a continuous dividend yield. From Ito's lemma (Theorem 14.2.2),

$$df = \left((\mu - q)\,S\frac{\partial f}{\partial S} + \frac{\partial f}{\partial t} + \frac{1}{2}\sigma^2 S^2 \frac{\partial^2 f}{\partial S^2} \right) dt + \sigma S\frac{\partial f}{\partial S}\,dW.$$

Set up a portfolio that is short one derivative security and long $\partial f/\partial S$ shares. Its value is $\Pi = -f + (\partial f/\partial S) S$, and the change in the value of the portfolio at time dt is given by $d\Pi = -df + (\partial f/\partial S) dS$. Substitute the formulas for df and dS to yield

$$d\Pi = \left(-\frac{\partial f}{\partial t} - \frac{1}{2} \sigma^2 S^2 \frac{\partial^2 f}{\partial S^2} \right) dt.$$

The total wealth change is simply the above amount plus the dividends, $d\Pi + qS(\partial f/\partial S) dt$. As this value is not stochastic, the portfolio must be instantaneously riskless:

$$\left(-\frac{\partial f}{\partial t} - \frac{1}{2} \sigma^2 S^2 \frac{\partial^2 f}{\partial S^2} \right) dt + qS \frac{\partial f}{\partial S} dt = r\Pi \, dt.$$

Simplify to obtain

$$\frac{\partial f}{\partial t} + (r-q) S \frac{\partial f}{\partial S} + \frac{1}{2} \sigma^2 S^2 \frac{\partial^2 f}{\partial S^2} = rf.$$

For European calls, the boundary conditions are identical to those of the standard option except that its value should be $Se^{-q(T-t)}$ as S goes to infinity. The solution appeared in Eq. (9.20). For American calls, the formulation that guarantees a unique solution is

$$\frac{\partial C}{\partial t} + (r-q) S \frac{\partial C}{\partial S} + \frac{1}{2} \sigma^2 S^2 \frac{\partial^2 C}{\partial S^2} = rC \quad \text{and} \quad C > S - X \quad \text{for } 0 \leq S < \overline{S},$$

$$C = S - X \quad \text{for } \overline{S} < S < \infty,$$

$$\frac{\partial C}{\partial S} = 1 \quad \text{and} \quad C = S - X \quad \text{for } S = \overline{S},$$

$$C = 0 \quad \text{for } S = 0,$$

plus the terminal condition $C = \max(S - X, 0)$, of course [879].

15.3.2 Futures and Futures Options

The futures price is related to the spot price by $F = Se^{(r-q)(T-t)}$. By Example 14.3.6, $dF/F = \sigma \, dW$. The futures price can therefore be treated as a stock paying a continuous dividend yield equal to r. This is the rationale behind the Black model.

▶ **Exercise 15.3.1** Derive the partial differential equation for futures options.

15.3.3 Average-Rate and Average-Strike Options

To simplify the notation, assume that the option is initiated at time zero. The arithmetic average-rate call and put have terminal values given by

$$\max \left(\frac{1}{T} \int_0^T S(u) \, du - X, 0 \right), \quad \max \left(X - \frac{1}{T} \int_0^T S(u) \, du, 0 \right),$$

respectively. Arithmetic average-rate options are notoriously hard to price. In practice, the prices are usually sampled at discrete points in time [598].

If the averaging is done geometrically, the payoffs become

$$
\max\left(\exp\left[\frac{\int_0^T \ln S(u)\, du}{T}\right] - X, 0\right), \quad \max\left(X - \exp\left[\frac{\int_0^T \ln S(u)\, du}{T}\right], 0\right),
$$

respectively. The geometric average $\exp\left[\frac{1}{T}\int_0^T \ln S(u)\, du\right]$ is lognormally distributed when the underlying asset's price is lognormally distributed (see Example 14.3.5). Lookback calls and puts on the average have terminal payoffs $\max(S(T) - \frac{1}{T}\int_0^T S(u)\, du, 0)$ and $\max(\frac{1}{T}\int_0^T S(u)\, du - S(T), 0)$, respectively.

The partial differential equation satisfied by the value V of a European arithmetic average-rate option can be derived as follows. Introduce the new variable $A(t) \equiv \int_0^t S(u)\, du$. It is not hard to verify that $dA = S\, dt$. Ito's lemma (Theorem 14.2.2) applied to V yields

$$
dV = \left(\mu S \frac{\partial V}{\partial S} + \frac{\partial V}{\partial t} + \frac{1}{2}\sigma^2 S^2 \frac{\partial^2 V}{\partial S^2} + S\frac{\partial V}{\partial A}\right) dt + \sigma S\frac{\partial V}{\partial S}\, dW.
$$

Consider the portfolio of short one derivative and long $\partial V/\partial S$ shares of stock. This portfolio must earn riskless returns because of lack of randomness. Therefore

$$
\frac{\partial V}{\partial t} + rS\frac{\partial V}{\partial S} + \frac{1}{2}\sigma^2 S^2 \frac{\partial^2 V}{\partial S^2} + S\frac{\partial V}{\partial A} = rV.
$$

▶ **Exercise 15.3.2** Show that geometric average-rate options satisfy

$$
\frac{\partial V}{\partial t} + rS\frac{\partial V}{\partial S} + \frac{1}{2}\sigma^2 S^2 \frac{\partial^2 V}{\partial S^2} + (\ln S)\frac{\partial V}{\partial A} = rV,
$$

where $A(t) \equiv \int_0^t \ln S(u)\, du$.

15.3.4 Options on More than One Asset: Correlation Options

For a correlation option whose value depends on the prices of two assets S_1 and S_2, both of which follow geometric Brownian motion, the partial differential equation is

$$
\frac{\partial C}{\partial t} + \sum_{i=1}^{2} rS_i\frac{\partial C}{\partial S_i} + \sum_{i=1}^{2} \frac{\sigma_i^2 S_i^2}{2}\frac{\partial^2 C}{\partial S_i^2} + \rho\sigma_1\sigma_2 S_1 S_2 \frac{\partial^2 C}{\partial S_1 \partial S_2} = rC. \tag{15.4}
$$

▶ **Exercise 15.3.3** (1) Justify Eq. (15.4). (2) Generalize it to n assets.

15.3.5 Exchange Options

An **exchange option** is a correlation option that gives the holder the right to exchange one asset for another. Its value at expiration is thus

$$
\max(S_2(T) - S_1(T), 0),
$$

where $S_1(T)$ and $S_2(T)$ denote the prices of the two assets at expiration. The payoff implies two ways of looking at the option: as a call on asset 2 with a strike price equal to the future price of asset 1 or as a put on asset 1 with a strike price equal to the future value of asset 2.

Assume that the two underlying assets do not pay dividends and that their prices follow

$$\frac{dS_1}{S_1} = \mu_1 \, dt + \sigma_1 \, dW_1, \qquad \frac{dS_2}{S_2} = \mu_2 \, dt + \sigma_2 \, dW_2,$$

where ρ is the correlation between dW_1 and dW_2. The option value at time t is

$$V(S_1, S_2, t) = S_2 N(x) - S_1 N(x - \sigma \sqrt{T - t}),$$

where

$$x \equiv \frac{\ln(S_2/S_1) + (\sigma^2/2)(T - t)}{\sigma \sqrt{T - t}},$$

$$\sigma^2 \equiv \sigma_1^2 - 2\rho\sigma_1\sigma_2 + \sigma_2^2. \tag{15.5}$$

This is called **Margrabe's formula** [640].

Margrabe's formula can be derived as follows. Observe first that $V(x, y, t)$ is **homogeneous of degree one** in x and y, meaning that $V(\lambda S_1, \lambda S_2, t) = \lambda V(S_1, S_2, t)$. An exchange option based on λ times the prices of the two assets is thus equal in value to λ original exchange options. Intuitively, this is true because of

$$\max(\lambda S_2(T) - \lambda S_1(T), 0) = \lambda \times \max(S_2(T) - S_1(T), 0)$$

and the perfect market assumption [660]. The price of asset 2 relative to asset 1 is $S \equiv S_2/S_1$. Hence the option sells for $V(S_1, S_2, t)/S_1 = V(1, S_2/S_1, t)$ with asset 1 as numeraire. The interest rate on a riskless loan denominated in asset 1 is zero in a perfect market because a lender of one unit of asset 1 demands one unit of asset 1 back as repayment of principal. Because the option to exchange asset 1 for asset 2 is a call on asset 2 with a strike price equal to unity and the interest rate equal to zero, the Black–Scholes formula applies:

$$\frac{V(S_1, S_2, t)}{S_1} = V(1, S, t) = SN(x) - 1 \times e^{-0 \times (T - t)} N(x - \sigma \sqrt{T - t}),$$

where

$$x \equiv \frac{\ln(S/1) + (0 + \sigma^2/2)(T - t)}{\sigma \sqrt{T - t}} = \frac{\ln(S_2/S_1) + (\sigma^2/2)(T - t)}{\sigma \sqrt{T - t}}.$$

Suppose the option holder sells $V_1 \equiv \partial V/\partial S_1$ units of asset 1 short and buys $-V_2 \equiv -\partial V/\partial S_2$ units of asset 2. Because $V(\cdot)$ is homogeneous of degree one in S_1 and S_2, the position has zero value because $V - V_1 S_1 - V_2 S_2 = 0$ by Euler's theorem (see Exercise 15.3.6). Hence $dV - V_1 \, dS_1 - V_2 \, dS_2 = 0$. From Ito's lemma (Theorem 14.2.2),

$$dV = V_1 \, dS_1 + V_2 \, dS_2 + \frac{\partial V}{\partial t} \, dt + \frac{V_{11}\sigma_1^2 S_1^2 + 2V_{12}\sigma_1\sigma_2\rho S_1 S_2 + V_{22}\sigma_2^2 S_2^2}{2} \, dt,$$

where $V_{ij} \equiv \partial^2 V/(\partial S_i \partial S_j)$. Hence

$$\frac{\partial V}{\partial t} + \frac{V_{11}\sigma_1^2 S_1^2 + 2V_{12}\sigma_1\sigma_2\rho S_1 S_2 + V_{22}\sigma_2^2 S_2^2}{2} = 0 \tag{15.6}$$

with the following initial and boundary conditions:

$$V(S_1, S_2, T) = \max(0, S_2 - S_1),$$

$$0 \leq V(S_1, S_2, t) \leq S_2 \quad \text{if } S_1, S_2 \geq 0.$$

Margrabe's formula is not much more complicated if S_i pays out a continuous dividend yield of q_i, $i = 1, 2$. We simply replace each occurrence of S_i with $S_i e^{-q_i(T-t)}$ to obtain

$$V(S_1, S_2, t) = S_2 e^{-q_2(T-t)} N(x) - S_1 e^{-q_1(T-t)} N(x - \sigma\sqrt{T-t}),$$

$$x \equiv \frac{\ln(S_2/S_1) + (q_1 - q_2 + \sigma^2/2)(T-t)}{\sigma\sqrt{T-t}},$$

$$\sigma^2 \equiv \sigma_1^2 - 2\rho\sigma_1\sigma_2 + \sigma_2^2. \tag{15.7}$$

▶ **Exercise 15.3.4** A call on the **maximum of two assets** pays $\max(S_1(T), S_2(T))$ at expiration. Replicate it by a position in one of the assets plus an exchange option.

▶ **Exercise 15.3.5** Consider a call on the **minimum of two assets with strike price** X. Its terminal value is $\max(\min(S_1(T), S_2(T)) - X, 0)$. Show that this option can be replicated by a long position in two ordinary calls and a short position in one call on the maximum of two assets at the same strike price X, which has a terminal payoff of $\max(\max(S_1(T), S_2(T)) - X, 0)$.

▶ **Exercise 15.3.6 (Euler's Theorem).** Prove that

$$\sum_{i=1}^{n} x_i \frac{\partial f(x_1, x_2, \ldots, x_n)}{\partial x_i} = f(x_1, x_2, \ldots, x_n)$$

if $f(x_1, x_2, \ldots, x_n)$ is homogeneous of degree one in $x_1, x_2, \ldots, x_n$.

▶ **Exercise 15.3.7** (1) Derive Margrabe's formula from the alternative view that a European exchange option is a put on asset 1 with a strike price equal to the future value of asset 2. (2) Derive the Black–Scholes formula from Margrabe's formula.

▶ **Exercise 15.3.8** Verify variance (15.5) for Margrabe's formula.

▶ **Exercise 15.3.9 (Put–Call Parity)** Prove that $V(S_2, S_1, t) - V(S_1, S_2, t) + S_2 = S_1$.

▶ **Exercise 15.3.10** Derive Eq. (15.6) from Eq. (15.4).

15.3.6 Options on Foreign Currencies and Assets

Correlation options involving foreign currencies and assets were first covered in Section 11.5. Analysis of such options can take place in either the domestic market or the foreign market before being converted back into the domestic currency [734].

In what follows, $S(t)$ denotes the spot exchange rate in terms of the domestic value of one unit of foreign currency. We know from Subsection 11.5.1 that foreign currency is analogous to a stock paying a continuous dividend yield equal to the foreign riskless interest rate r_f in foreign currency. Therefore $S(t)$ follows the geometric Brownian motion process,

$$\frac{dS}{S} = (r - r_f) \, dt + \sigma_s \, dW_s(t),$$

in a risk-neutral economy. The foreign asset is assumed to pay a continuous dividend yield of q_f, and its price follows

$$\frac{dG_f}{G_f} = (\mu_f - q_f) \, dt + \sigma_f \, dW_f(t)$$

in foreign currency. The correlation between the rate of return of the exchange rate and that of the foreign asset is ρ; in other words, ρ is the correlation between dW_s and dW_f.

Foreign Equity Options

From Eq. (9.20), European options on the foreign asset G_f with the terminal payoffs $S(T) \times \max(G_f(T) - X_f, 0)$ and $S(T) \times \max(X_f - G_f(T), 0)$ are worth

$$C_f = G_f e^{-q_f \tau} N(x) - X_f e^{-r_f \tau} N(x - \sigma_f \sqrt{\tau}),$$
$$P_f = X_f e^{-r_f \tau} N(-x + \sigma_f \sqrt{\tau}) - G_f e^{-q_f \tau} N(-x),$$

in foreign currency, where

$$x \equiv \frac{\ln(G_f / X_f) + (r_f - q_f + \sigma_f^2 / 2)\, \tau}{\sigma_f \sqrt{\tau}}$$

and X_f is the strike price in foreign currency. They will fetch SC_f and SP_f, respectively, in domestic currency. These options are called **foreign equity options** struck in foreign currency.

▶ **Exercise 15.3.11** The formulas of C_f and P_f suggest that a foreign equity option is equivalent to S domestic options on a stock paying a continuous dividend yield of q_f and a strike price of $X_f e^{(r - r_f)\tau}$. Verify that this observation is indeed valid.

▶ **Exercise 15.3.12** The dynamics of the foreign asset value in domestic currency, SG_f, depends on the correlation between the asset price and the exchange rate (see Example 14.3.5). (1) Why is ρ missing from the option formulas? (2) Justify the equivalence in Exercise 15.3.11 with (1).

Foreign Domestic Options

Foreign equity options fundamentally involve values in the foreign currency. However, consider this: Although a foreign equity call may allow the holder to participate in a foreign market rally, the profits can be wiped out if the foreign currency depreciates against the domestic currency. What is really needed is a call in *domestic* currency with a payoff of $\max(S(T) G_f(T) - X, 0)$. This is called a **foreign domestic option**.

To foreign investors, this call is an option to exchange X units of domestic currency (foreign currency to them) for one share of foreign asset (domestic asset to them) – an exchange option, in short. By formula (15.7), its price in foreign currency equals

$$G_f e^{-q_f \tau} N(x) - \frac{X}{S} e^{-r\tau} N(x - \sigma \sqrt{\tau}),$$

where

$$x \equiv \frac{\ln(G_f S / X) + (r - q_f + \sigma^2 / 2)\, \tau}{\sigma \sqrt{\tau}}$$

and $\sigma^2 \equiv \sigma_s^2 + 2\rho \sigma_s \sigma_f + \sigma_f^2$. The domestic price is therefore

$$C = SG_f e^{-q_f \tau} N(x) - X e^{-r\tau} N(x - \sigma \sqrt{\tau}).$$

Similarly, a put has a price of

$$P = Xe^{-r\tau} N(-x + \sigma \sqrt{\tau}) - SG_f e^{-q_f \tau} N(-x).$$

▶ **Exercise 15.3.13** Suppose that the domestic and the foreign bond prices in their respective currencies with a par value of one and expiring at T also follow geometric Brownian motion processes. Their current prices are B and B_f, respectively. Derive the price of a forex option to buy one unit of foreign currency with X units of domestic currency at time T. (This result generalizes Eq. (11.6), which assumes deterministic interest rates.)

Cross-Currency Options

A cross-currency option, we recall, is an option in which the currency of the strike price is different from the currency in which the underlying asset is denominated [775]. An option to buy 100 yen at a strike price of 1.18 Canadian dollars provides one example. Usually, a third currency, the U.S. dollar, is involved because of the lack of relevant exchange-traded options for the two currencies in question (yen and Canadian dollars in the above example) in order to calculate the needed volatility. For this reason, the notations below will be slightly different.

Let S_A denote the price of the foreign asset and S_C the price of currency C that the strike price X is based on. Both S_A and S_C are in U.S. dollars, say. If S is the price of the foreign asset as measured in currency C, then we have the **triangular arbitrage** $S = S_A/S_C$.[1] Assume that S_A and S_C follow the geometric Brownian motion processes $dS_A/S_A = \mu_A \, dt + \sigma_A \, dW_A$ and $dS_C/S_C = \mu_C \, dt + \sigma_C \, dW_C$, respectively. Parameters σ_A, σ_C, and ρ can be inferred from exchange-traded options. By Exercise 14.3.6,

$$\frac{dS}{S} = (\mu_A - \mu_C - \rho\sigma_A\sigma_C) \, dt + \sigma_A \, dW_A - \sigma_C \, dW_C,$$

where ρ is the correlation between dW_A and dW_C. The volatility of dS/S is hence $(\sigma_A^2 - 2\rho\sigma_A\sigma_C + \sigma_C^2)^{1/2}$.

▶ **Exercise 15.3.14** Verify that the triangular arbitrage must hold to prevent arbitrage opportunities among three currencies.

▶ **Exercise 15.3.15** Show that both forex options and foreign domestic options are special cases of cross-currency options.

▶ **Exercise 15.3.16** Consider a portfolio consisting of a long call on the foreign asset and X long puts on currency C. The strike prices in U.S. dollars of the call (X_A) and put (X_C) are such that $X = X_A/X_C$. Prove the portfolio is worth more than the cross-currency call when all options concerned are European. (A cross-currency call has a terminal payoff of $S_C \times \max(S - X, 0)$ in U.S. dollars.)

Quanto Options

Consider a call with a terminal payoff $\widehat{S} \times \max(G_f(T) - X_f, 0)$ in domestic currency, where $\widehat{S}$ is a constant. This amounts to fixing the exchange rate to $\widehat{S}$. For instance, a call on the Nikkei 225 futures, if it existed, fits this framework with $\widehat{S} = 5$ and G_f denoting the futures price. A guaranteed exchange rate option is called a **quanto option** or simply a **quanto**. The process $U \equiv \widehat{S} G_f$ in a risk-neutral economy follows

$$\frac{dU}{U} = (r_f - q_f - \rho\sigma_s\sigma_f)\, dt + \sigma_f\, dW \tag{15.8}$$

in domestic currency [470, 878]. Hence it can be treated as a stock paying a continuous dividend yield of $q \equiv r - r_f + q_f + \rho\sigma_s\sigma_f$. Apply Eq. (9.20) to obtain

$$C = \widehat{S}\,[\,G_f\, e^{-q\tau}\, N(x) - X_f\, e^{-r\tau}\, N(x - \sigma_f\sqrt{\tau})\,],$$
$$P = \widehat{S}\,[\,X_f\, e^{-r\tau}\, N(-x + \sigma_f\sqrt{\tau}) - G_f\, e^{-q\tau}\, N(-x)\,],$$

where

$$x \equiv \frac{\ln(G_f/X_f) + (r - q + \sigma_f^2/2)\,\tau}{\sigma_f\sqrt{\tau}}.$$

Note that the values do not depend on the exchange rate.

In general, a **quanto derivative** has nominal payments in the foreign currency that are converted into the domestic currency at a fixed exchange rate. A **cross-rate swap**, for example, is like a currency swap except that the foreign currency payments are converted into the domestic currency at a fixed exchange rate. Quanto derivatives form a rapidly growing segment of international financial markets [17].

▶ **Exercise 15.3.17** Justify Eq. (15.8).

15.3.7 Convertible Bonds with Call Provisions

When a CB with call provisions is called, its holder has the right either to convert the bond (**forced conversion**) or to redeem it at the call price. Assume that the firm and the investor pursue an optimal strategy whereby (1) the investor maximizes the value of the CB at each instant in time through conversion and (2) the firm minimizes the value of the CB at each instant in time through call.

Let the market value $V(t)$ of the firm's securities be determined exogenously and independent of the call and conversion strategies, which can be justified by the **Modigliani–Miller irrelevance theorem**. Minimizing the value of the CB therefore maximizes the stockholder value. The market value follows $dV/V = \mu\, dt + \sigma\, dW$. Assume that the firm in question has only two classes of obligations: n shares of common stock and m CBs with a conversion ratio of k. The stock may pay dividends, and the bond may pay coupon interests. The **conversion value** per bond is

$$C(V, t) = zV(t),$$

where $z \equiv k/(n + mk)$. Each bond has \$1,000 par value, and T stands for the maturity date.

Let $W(V, t)$ denote the market value at time t of one convertible bond. From assumption (1), the bond never sells below the conversion value as

$$W(V, t) \geq C(V, t). \tag{15.9}$$

In fact, the bond can never sell at the conversion value except immediately before a dividend date. This is because otherwise its rate of return up to the next dividend date would not fall below the stock's; actually, it would be higher because of the higher priority of bondholders. Therefore the bond sells above the conversion value, and the investor does not convert it. As a result, relation (15.9) holds with strict inequality between dividend dates, and conversion needs to be considered only at dividend or call dates.

We now consider the implications of the call strategy. When the bond is called, the investor has the option either to redeem at the call price $P(t)$ or convert it for $C(V, t)$. The value of the bond if called is hence given by

$$V_c(V, t) \equiv \max(P(t), C(V, t)).$$

There are two cases to consider.

1. $C(V, t) > P(t)$ **when the bond is callable:** The bond will be called immediately because, by a previous argument, the bond sells for at least the conversion value $C(V, t)$, which is the value if called. Hence,

$$W(V, t) = C(V, t). \tag{15.10}$$

2. $C(V, t) \le P(t)$ **when the bond is callable:** Note that the call price equals the value if called, V_c. The bond should be called when its value if not called equals its value if called. This holds because, in accordance with assumption (2), the firm will call the bond when the value if not called exceeds $V_c(V, t)$ and will not call it otherwise. Hence

$$W(V, t) \le V_c(V, t) = P(t), \tag{15.11}$$

and the bond will be called when its value if not called equals the call price.

Finally, the Black–Scholes differential equation implies that

$$\frac{\partial W}{\partial t} + rV \frac{\partial W}{\partial V} + \frac{1}{2} \sigma^2 V^2 \frac{\partial^2 W}{\partial V^2} = rW.$$

The boundary conditions for the above differential equation are summarized below.

- They include relation (15.9), Eq. (15.10), and relation (15.11) (the latter two when the bond is callable and under their respective conditions), and the maturity value

$$W(V, T) = \begin{cases} zV(T), & \text{if } zV(T) \ge 1000 \\ 1000, & 1000 \times m \le V(T) \le 1000/z. \\ V(T)/m, & V(T) \le 1000 \times m \end{cases}$$

These three conditions above correspond to the cases when the firm's total value (1) is greater than the total conversion value, (2) is greater than the total par value but less than the total conversion value, and (3) is less than the total par value.
- $0 \le mW(V, t) \le V(t)$ because the bond value cannot exceed the firm value.
- $W(0, t) = 0$.
- $W(V, t) \le B(V, t) + zV(t)$ because a CB is dominated by a portfolio of an otherwise identical fixed-rate bond $B(V, t)$ and stock with a total value equal to the conversion value. $B(V, t)$ is easy to calculate under constant interest rates.
- When the bond is not callable and $V(t)$ is high enough to make negligible the possibility of default, it behaves like an option to buy a fraction z of the firm. Hence $\lim_{V \to \infty} \partial W(V, t)/\partial V = z$.

- On a dividend date, $W(V, t^-) = \max(W(V - D, t^+), zV(t))$, where t^- denotes the instant before the event and t^+ the instant after. This condition takes into account conversion just before the dividend date.
- $W(V, t^-) = W(V - mc, t^+) + c$ on a coupon date and when the bond is not callable, where c is the amount of the coupon.
- $W(V, t^-) = \min(W(V - mc, t^+) + c, V_c(V, t))$ on a coupon date and when the bond is callable.

The partial differential equation has to be solved numerically by the techniques in Section 18.1.

➤ **Exercise 15.3.18** Suppose that the CB is continuously callable once it becomes callable, meaning that it is callable at any instant after a certain time t^*. Argue that the $C(V, t) > P(t)$ case needs to be considered only at $t = t^*$ and not thenceforth for the call provision. (Hence $W(V, t^-) = \min(W(V - mc, t^+) + c, P(t))$ on any coupon date $t > t^*$.)

15.4 General Derivatives Pricing

In general, the underlying asset S may not be traded. Interest rate, for instance, is not a traded security, whereas stocks and bonds are. Let S follow the Ito process $dS/S = \mu \, dt + \sigma \, dW$, where μ and σ may depend only on S and t. Let $f_1(S, t)$ and $f_2(S, t)$ be the prices of two derivatives with dynamics $df_i/f_i = \mu_i \, dt + \sigma_i \, dW$, $i = 1, 2$. Note that they share the same Wiener process as S.

A portfolio consisting of $\sigma_2 f_2$ units of the first derivative and $-\sigma_1 f_1$ units of the second derivative is instantaneously riskless because

$$\sigma_2 f_2 \, df_1 - \sigma_1 f_1 \, df_2 = \sigma_2 f_2 f_1(\mu_1 \, dt + \sigma_1 \, dW) - \sigma_1 f_1 f_2(\mu_2 \, dt + \sigma_2 \, dW)$$
$$= (\sigma_2 f_2 f_1 \mu_1 - \sigma_1 f_1 f_2 \mu_2) \, dt,$$

which is devoid of volatility. Therefore

$$(\sigma_2 f_2 f_1 \mu_1 - \sigma_1 f_1 f_2 \mu_2) \, dt = r(\sigma_2 f_2 f_1 - \sigma_1 f_1 f_2) \, dt,$$

or $\sigma_2 \mu_1 - \sigma_1 \mu_2 = r(\sigma_2 - \sigma_1)$. After rearranging the terms, we conclude that

$$\frac{\mu_1 - r}{\sigma_1} = \frac{\mu_2 - r}{\sigma_2} \equiv \lambda \quad \text{for some } \lambda.$$

Any derivative whose value depends on only S and t and that follows the Ito process $df/f = \mu \, dt + \sigma \, dW$ must thus satisfy

$$\frac{\mu - r}{\sigma} = \lambda \quad \text{or, alternatively, } \mu = r + \lambda \sigma. \tag{15.12}$$

We call λ the **market price of risk**, which is independent of the specifics of the derivative. Equation (15.12) links the excess expected return and risk. The term $\lambda \sigma$ measures the extent to which the required return is affected by the dependence on S.

Ito's lemma can be used to derive the formulas for μ and σ:

$$\mu = \frac{1}{f} \left(\frac{\partial f}{\partial t} + \mu S \frac{\partial f}{\partial S} + \frac{1}{2} \sigma^2 S^2 \frac{\partial^2 f}{\partial S^2} \right), \qquad \sigma = \frac{\sigma S}{f} \frac{\partial f}{\partial S}.$$

Substitute the preceding equations into Eq. (15.12) to obtain

$$\frac{\partial f}{\partial t} + (\mu - \lambda\sigma)\, S \,\frac{\partial f}{\partial S} + \frac{1}{2}\,\sigma^2 S^2 \,\frac{\partial^2 f}{\partial S^2} = rf. \tag{15.13}$$

The presence of μ shows that the investor's risk preference is relevant, and the derivative may be dependent on the underlying asset's growth rate and the market price of risk. Only when the underlying variable is the price of a traded security can we assume that $\mu = r$ in pricing.

Provided certain conditions are met, such as the underlying processes' being Markovian, the approach to derivatives pricing by solving partial differential equations is equivalent to the martingale approach of taking expectation under a risk-neutral probability measure [290, 692]. The fundamental theorem of asset pricing, Theorem 13.2.3, as well as other results in Subsection 13.2.1, also continues to hold in continuous time. This suggests the following risk-neutral valuation scheme for Eq. (15.13): Discount the expected payoff of f at the riskless interest rate under the revised process $dS/S = (\mu - \lambda\sigma)\, dt + \sigma\, dW$. Although the same symbol W is used, a convention adopted throughout the book for convenience, it is important to point out that W is no longer the original Wiener process. In fact, a change of probability measure has taken place, and W is a Wiener process with respect to the risk-neutral probability measure.

Assume a constant interest rate r. Then any European-style derivative security with payoff f_T at time T has value $e^{-r(T-t)} E_t^\pi[\, f_T\,]$, where E_t^π takes the expected value under a risk-neutral probability measure given the information up to time t. As a specific application, consider the futures price F. With a delivery price of X, a futures contract has value $f = e^{-r(T-t)} E_t^\pi[\, S_T - X\,]$. Because F is the X that makes f zero, it holds that $0 = E_t^\pi[\, S_T - F\,] = E_t^\pi[\, S_T\,] - F$, i.e., $F = E_t^\pi[\, S_T\,]$. This extends the result for the binomial model in Exercise 13.2.10 to the continuous-time case.

▶ **Exercise 15.4.1** Suppose that $S_1, S_2, \ldots, S_n$ pay no dividends and follow $dS_i/S_i = \mu_i\, dt + \sigma_i\, dW_i$. Let ρ_{jk} denote the correlation between dW_j and dW_k. Show that

$$\frac{\partial f}{\partial t} + \sum_i (\mu_i - \lambda_i\sigma_i)\, S_i \,\frac{\partial f}{\partial S_i} + \frac{1}{2} \sum_i \sum_k \rho_{ik}\sigma_i\sigma_k S_i S_k \,\frac{\partial^2 f}{\partial S_i\, \partial S_k} = rf \tag{15.14}$$

when the derivative f depends on more than one state variable $S_1, S_2, \ldots, S_n$.

▶ **Exercise 15.4.2** A **forward-start option** is like a standard option except that it becomes effective only at time τ^* from now and with the strike price set at the stock price then (the option thus starts at the money). Let $C(S)$ denote the value of an at-the-money European forward-start call, given the stock price S. (1) Show that $C(S)$ is a linear function in S under the Black–Scholes model. (2) Argue that the value of a forward-start option is $e^{-r\tau^*} C(E^\pi[\, S(\tau^*)\,]) = e^{-r\tau^*} C(Se^{(r-q)\tau^*})$, where q is the dividend yield.

15.5 Stochastic Volatility

The Black–Scholes formula displays bias in practice. Besides the smile pattern mentioned in Subsection 9.4.3, (1) volatility changes from month to month, (2) it is mean reverting in that extreme volatilities tend to return to the average over time, (3) it seems to fall as the price of the underlying asset rises [346, 823], and (4) out-of-the-money options and options on low-volatility assets are underpriced. These findings led to the study of stochastic volatility.

Stochastic volatility injects an extra source of randomness if this uncertainty is not perfectly correlated with the one driving the stock price. In this case, another traded security besides stock and bond is needed in the replicating portfolio. In fact, if volatility were the price of a traded security, there would exist a self-financing strategy that replicates the option by using stocks, bonds, and the volatility security.

Hull and White considered the following model:

$$\frac{dS}{S} = \mu \, dt + \sigma \, dW_1, \qquad \frac{dV}{V} = \mu_v \, dt + \sigma_v \, dW_2,$$

where $V \equiv \sigma^2$ is the instantaneous variance [471]. Assume that μ depends on S, σ, and t, that μ_v depends on σ and t (but not S), that dW_1 and dW_2 have correlation ρ, and that the riskless rate r is constant. From Eq. (15.14),

$$\frac{\partial f}{\partial t} + (\mu - \lambda \sigma) \, S \, \frac{\partial f}{\partial S} + (\mu_v - \lambda_v \sigma_v) \, V \, \frac{\partial f}{\partial V}$$
$$+ \frac{1}{2} \left(\sigma^2 S^2 \frac{\partial^2 f}{\partial S^2} + 2\rho \sigma \sigma_v SV \frac{\partial^2 f}{\partial S \partial V} + \sigma_v^2 V^2 \frac{\partial^2 f}{\partial V^2} \right) = rf.$$

Because stock is a traded security (but volatility is not), the preceding equation becomes

$$\frac{\partial f}{\partial t} + rS \frac{\partial f}{\partial S} + (\mu_v - \lambda_v \sigma_v) \, V \, \frac{\partial f}{\partial V}$$
$$+ \frac{1}{2} \left(\sigma^2 S^2 \frac{\partial^2 f}{\partial S^2} + 2\rho \sigma \sigma_v SV \frac{\partial^2 f}{\partial S \partial V} + \sigma_v^2 V^2 \frac{\partial^2 f}{\partial V^2} \right) = rf.$$

After two additional assumptions, $\rho = 0$ (volatility is uncorrelated with the stock price) and $\lambda_v \sigma_v = 0$ (volatility has zero systematic risk), the equation becomes

$$\frac{\partial f}{\partial t} + rS \frac{\partial f}{\partial S} + \mu_v V \frac{\partial f}{\partial V} + \frac{1}{2} \left(\sigma^2 S^2 \frac{\partial^2 f}{\partial S^2} + \sigma_v^2 V^2 \frac{\partial^2 f}{\partial V^2} \right) = rf.$$

A series solution is available for the model.

The volatility risk was assumed not to be priced [579]. To assume otherwise, we need to model risk premium on the variance process [440]. When the volatility follows an uncorrelated Ornstein–Uhlenbeck process, closed-form solutions exist [823].

Additional Reading

A rigorous derivation of the Black–Scholes differential equation can be found in [681]. See [212, 408, 446, 861, 883] for partial differential equations, [531] for the approximation of the early exercise boundary, [769] for the derivation of Margrabe's formula based on the binomial model, [575, 744, 746, 894] for currency-related options, [122, 221, 491, 697] for pricing CBs, and [424, 470, 615] for the bias of the Black–Scholes option pricing model. We followed [120] in Subsection 15.3.7. Martingale pricing in continuous time relies on changing the probability measure with the **Girsanov theorem** [289]. That using stochastic volatility models can result in some pricing improvement has been empirically documented [44]. In cases in which the

Black–Scholes model has been reasonably supported by empirical research, gains from complicated models may be limited, however [526]. Intriguingly, the Black–Scholes formula continues to hold as long as all traders believe that the stock prices are lognormally distributed, even if that belief is objectively wrong [194].

NOTE

1. Triangular arbitrage had been known for centuries. See Montesquieu's *The Spirit of Laws* [676, p. 179].

Hedging

Does an *instantaneous* cube exist?

H.G. Wells, *The Time Machine*

Hedging strategies appear throughout this book. This is to be expected because one of the principal uses of derivatives is in the management of risks. In this chapter, we focus on the use of non-interest-rate derivatives in hedging. Interest rate derivatives will be picked up in Chap. 21.

16.1 Introduction

One common thread throughout this book has been the management of risks. **Risk management** means selecting and maintaining portfolios with defined exposure to risks. Deciding which risks one is to be exposed to and which risks one is to be protected against is also an integral part of risk management. Evidence suggests that firms engaged in risk management not only are less risky but also perform better [813].

A **hedge** is a position that offsets the price risk of another position. A hedge reduces risk exposures or even eliminates them if it provides cash flows equal in magnitude but opposite in directions to those of the existing exposure. For hedging to be possible, the return of the derivative should be correlated with that of the hedged position. In fact, the more correlated their returns are, the more effective the hedge will be.

Three types of traders play in the markets. **Hedgers** set up positions to offset risky positions in the spot market. **Speculators** bet on price movements and hope to make a profit. **Arbitragers** lock in riskless profits by simultaneously entering into transactions in two or more markets, which is called arbitrage.

16.2 Hedging and Futures

The most straightforward way of hedging involves forward contracts. Because of daily settlements, futures contracts are harder to analyze than forward contracts. Luckily, the forward price and the futures price are generally close to each other; therefore results obtained for forwards will be assumed to be true of futures here.

16.2.1 Futures and Spot Prices

Two forces prevent the prevailing prices in the spot market and the futures market from diverging too much at any given time. One is the delivery mechanism, and the other is hedging. Hedging relates the futures price and the spot price through arbitrage. In fact, the futures price should equal the spot price by the carrying charges. Carrying charges, we recall, are the costs of holding physical inventories between now and the maturity of the futures contract. In practice, the futures price does not necessarily exceed the spot price by exactly the carrying charges for inventories that have what Kaldor (1908–1986) termed the convenience yield derived from their availability when buyers need them [468].

16.2.2 Hedgers, Speculators, and Arbitragers

A company that is due to sell an asset in the future can hedge by taking a short futures position. This is known as a **selling** or **short hedge**. The purpose is to lock in a selling price or, with fixed-income securities, a yield. If the price of the asset goes down, the company does not fare well on the sale of the asset, but it makes a gain on the short futures position. If the price of the asset goes up, the reverse is true. Clearly, a selling hedge is a substitute for a later cash market sale of the asset. A company that is due to buy an asset in the future can hedge by taking a long futures position. This is known as a **buying** or **long hedge**. Clearly, a buying hedge is used when one plans to buy the cash asset at a later date. The purpose is to establish a fixed purchase price. These strategies work because spot and futures prices are correlated.

A person who gains or loses from the difference between the spot and the futures prices is said to **speculate on the basis**. Simultaneous purchase and sale of futures contracts on two different yet related assets is referred to as a **spread**. A person who speculates by using spreads is called a **spreader** [95, 799]. A hedger is someone whose net position in the spot market is offset by positions in the futures market. A **short hedger** is long in the spot market and short in the futures market. A **long hedger** does the opposite. Those who are net long or net short are speculators. Speculators will buy (sell) futures contracts only if they expect prices to increase (decrease, respectively). Hedgers, in comparison, are willing to pay a premium to unload unwanted risk onto speculators. Speculators provide the market with liquidity, enabling hedgers to trade large numbers of contracts without adversely disrupting prices.

If hedgers in aggregate are short, speculators are net long and the futures price is set below the expected future spot price. On the other hand, if hedgers are net long, speculators are net short and the futures price is set above the expected future spot price. There seems to be evidence that short hedging exceeds long hedging in most of the markets most of the time. If hedgers are net short in futures, speculators must be net long. It has been theorized that speculators will be net long only if the futures price is expected to rise until it equals the spot price at maturity; speculators therefore extract a risk premium from hedgers. This is Keynes's theory of **normal backwardation**, which implies that the futures price underestimates the future spot price [295, 468, 470].

> **Exercise 16.2.1** If the futures price equals the expected future spot price, then hedging may in some sense be considered a free lunch. Give your reasons.

16.2.3 Perfect and Imperfect Hedging

Consider an investor who plans on selling an asset t years from now. To eliminate some of the price uncertainties, the investor sells futures contracts on the same asset with a delivery date in T years. After t years, the investor liquidates the futures position and sells the asset as planned. The cash flow at that time is

$$S_t - (F_t - F) = F + (S_t - F_t) = F + \text{basis},$$

where S_t is the spot price at time t, F_t is the futures price at time t, and F is the original futures price. The investor has replaced the price uncertainty with the smaller basis uncertainty; as a consequence, risk has been reduced. The hedge is **perfect** if $t = T$, that is, when there is a futures contract with a matching delivery date.

When the cost of carry and the convenience yield are known, the cash flow can be anticipated with complete confidence according to Eq. (12.13). This holds even when there is a maturity mismatch $t \neq T$ as long as (1) the interest rate r is known and (2) the cost of carry c and the convenience yield y are constants. In this case, $F_t = S_t e^{(r+c-y)(T-t)}$ by Eqs. (12.11). Let h be the number of futures contracts sold initially. The cash flow at time t, after the futures position is liquidated and the asset is sold, is

$$S_t - h(F_t - F) = S_t - h\left[S_t e^{(r+c-y)(T-t)} - F \right].$$

Pick $h = e^{-(r+c-y)(T-t)}$ to make the cash flow a constant hF and eliminate any uncertainty. The number h is the hedge ratio. Note that $h = 1$ may not be the best choice when $t \neq T$.

A number of factors make hedging with futures contracts less than perfect. The asset whose price is to be hedged may not be identical to the underlying asset of the futures contract; the date when the asset is to be transacted may be uncertain; the hedge may require that the futures contract be closed out before its expiration date. These problems give rise to basis risk. As shown above, basis risk does not exist in situations in which the spot price relative to the futures price moves in predictable manners.

Cross Hedge

A hedge that is established with a maturity mismatch, an asset mismatch, or both is referred to as a **cross hedge** [746]. Cross hedges are common practices. When firms want to hedge against price movements in a commodity for which there are no futures contracts, they can turn to futures contracts on related commodities whose price movements closely correlate with the price to be hedged.

EXAMPLE 16.2.1 We can hedge a future purchase price of 10,000,000 Dutch guilders as follows. Suppose the current exchange rate is U.S.$0.48 per guilder. At this rate, the dollar cost is U.S.$4,800,000. A regression analysis of the daily changes in the guilder rate and the nearby German mark futures reveals that the estimated slope is 0.95 with an R^2 of 0.92. Now that the guilder is highly correlated with the German mark, German mark futures are picked. The current exchange rate is U.S.$0.55/DEM1; hence the commitment of 10,000,000 guilders translates to $4,800,000/0.55 = 8,727,273$ German marks. Because each futures contract controls 125,000 marks, we trade $0.95 \times (8,727,273/125,000) \approx 66$ contracts.

EXAMPLE 16.2.2 A British firm expecting to pay DEM2,000,000 for purchases in 3 months would like to lock in the price in pounds. Besides the standard way of using mark futures contracts that trade in pounds, the firm can trade mark and pound futures contracts that trade in U.S. dollars as follows. Each mark futures contract controls 125,000 marks, and each pound futures contract controls 62,500 pounds. Suppose the payment date coincides with the last trading date of the currency futures contract at the CME. Currently the 3-month mark futures price is $0.7147 and the pound futures price is $1.5734. The firm buys $2,000,000/125,000 = 16$ mark futures contracts, locking in a purchase price of $1,429,400. To further lock in the price in pounds at the exchange rate of $1.5734/£1, the firm shorts $1,429,400/(1.5734 \times 62,500) \approx 15$ pound futures contracts. The end result is a purchase price of $2,000,000 \times (0.7147/1.5734) = 908,478$ pounds with a cross rate of £0.45424/DEM1.

Hedge Ratio (Delta)

In general, the futures contract may not track the cash asset perfectly. Let ρ denote the correlation between S_t and F_t, δ_S the standard deviation of S_t, and δ_F the standard deviation of F_t. For a short hedge, the cash flow at time t is $S_t - h(F_t - F)$, whereas for a long hedge it is $-S_t + h(F_t - F)$. The variance is $V \equiv \delta_S^2 + h^2 \delta_F^2 - 2h\rho\delta_S\delta_F$ in both cases. To minimize risk, the hedger seeks the hedge ratio h that minimizes the variance of the cash flow of the hedged position, V. Because $\partial V/\partial h = 2h\delta_F^2 - 2\rho\delta_S\delta_F$,

$$ h = \rho\, \frac{\delta_S}{\delta_F} = \frac{\mathrm{Cov}[\, S_t, F_t\,]}{\mathrm{Var}[\, F_t\,]}, \tag{16.1} $$

which was called beta in Exercise 6.4.1.

EXAMPLE 16.2.3 Suppose that the standard deviation of the change in the price per bushel of corn over a 3-month period is 0.4 and that the standard deviation of the change in the soybeans futures price over a 3-month period is 0.3. Assume further that the correlation between the 3-month change in the corn price and the 3-month change in the soybeans futures price is 0.9. The optimal hedge ratio is $0.9 \times (0.4/0.3) = 1.2$. Because the size of one soybeans futures contract is 5,000 bushels, a company expecting to buy 1,000,000 bushels of corn in 3 months can hedge by buying $1.2 \times (1,000,000/5,000) = 240$ futures contracts on soybeans.

The hedge ratio can be estimated as follows. Suppose that $S_1, S_2, \ldots, S_t$ and $F_1, F_2, \ldots, F_t$ are the daily closing spot and futures prices, respectively. Define $\Delta S_i \equiv S_{i+1} - S_i$ and $\Delta F_i \equiv F_{i+1} - F_i$. Now estimate $\rho, \delta_S,$ and δ_F with Eqs. (6.2) and (6.18).

▶ **Exercise 16.2.2** Show that if the linear regression of s on f based on the data

$$ (\Delta S_1, \Delta F_1), (\Delta S_2, \Delta F_1), \ldots, (\Delta S_{t-1}, \Delta F_{t-1}) $$

is $s = \beta_0 + \beta_1 f$, then β_1 is an estimator of the hedge ratio in Eq. (16.1).

16.2.4 Hedging with Stock Index Futures

Stock index futures can be used to hedge a well-diversified portfolio of stocks. According to the Capital Asset Pricing Model (CAPM), the relation between the return on a portfolio of stocks and the return on the market can be described by a parameter β, called beta. Approximately,

$$\Delta_1 = \alpha + \beta \times \Delta_2,$$

where Δ_1 (Δ_2) is the change in the value of \$1 during the holding period if it is invested in the portfolio (the market index, respectively) and α is some constant. The change in the portfolio value during the period is therefore $S \times \alpha + S \times \beta \times \Delta_2$, where S denotes the current value of the portfolio. The change in the value of one futures contract that expires at the end of the holding period is approximately $F \times \Delta_2$, where F is the current value of one futures contract. Recall that the value of one futures contract is equal to the futures price multiplied by the contract size. For example, if the futures price of the S&P 500 is 1,000, the value of one futures contract is $1,000 \times 500 = 500,000$.

The uncertain component of the change in the portfolio value, $S \times \beta \times \Delta_2$, is approximately $\beta S/F$ times the change in the value of one futures contract, $F \times \Delta_2$. The number of futures contracts to short in hedging the portfolio is thus $\beta S/F$. This strategy is called **portfolio immunization**. The same idea can be applied to change the beta of a portfolio. To change the beta from β_1 to β_2, we short

$$(\beta_1 - \beta_2) \frac{S}{F} \tag{16.2}$$

contracts. A perfectly hedged portfolio has zero beta and corresponds to choosing $\beta_2 = 0$.

EXAMPLE 16.2.4 Hedging a well-diversified stock portfolio with the S&P 500 Index futures works as follows. Suppose the portfolio in question is worth \$2,400,000 with a beta of 1.25 against the returns on the S&P 500 Index. So, for every 1% advance in the index, the expected advance in the portfolio is 1.25%. With a current futures price of 1200, $1.25 \times [2,400,000/(1,200 \times 500)] = 5$ futures contracts are sold short.

▶ **Exercise 16.2.3** Redo Example 16.2.4 if the goal is to change the beta to 2.0.

16.3 Hedging and Options

16.3.1 Delta Hedge

The delta (hedge ratio) of a derivative is defined as $\Delta \equiv \partial f/\partial S$. Thus $\Delta f \approx \Delta \times \Delta S$ for relatively small changes in the stock price, ΔS. A delta-neutral portfolio is hedged in the sense that it is immunized against small changes in the stock price. A trading strategy that dynamically maintains a delta-neutral portfolio is called **delta hedge**.

Because delta changes with the stock price, a delta hedge needs to be rebalanced periodically in order to maintain delta neutrality. In the limit in which the portfolio is adjusted continuously, perfect hedge is achieved and the strategy becomes self-financing [294]. This was the gist of the Black–Scholes–Merton argument in Subsection 15.2.1.

For a non-dividend-paying stock, the delta-neutral portfolio hedges N short derivatives with $N \times \Delta$ shares of the underlying stock plus B borrowed dollars such that

$$-N \times f + N \times \Delta \times S - B = 0.$$

This is called the **self-financing condition** because the combined value of derivatives, stock, and bonds is zero. At each rebalancing point when the delta is Δ', buy $N \times (\Delta' - \Delta)$ shares to maintain $N \times \Delta'$ shares with a total borrowing of $B' = N \times \Delta' \times S' - N \times f'$, where f' is the derivative's prevailing price. A delta hedge is a discrete-time analog of the continuous-time limit and will rarely be self-financing, if ever.

▶ **Exercise 16.3.1** (1) A delta hedge under the BOPM results in perfect replication (see Chap. 9). However, this is impossible in the current context. Why? (2) How should the value of the derivative behave with respect to that of the underlying asset for perfect replication to be possible?

A Numerical Example

Let us illustrate the procedure with a hedger who is short European calls. Because the delta is positive and increases as the stock price rises, the hedger keeps a long position in stock and buys (sells) stock if the stock price rises (falls, respectively) in order to maintain delta neutrality. The calls are replicated well if the cumulative cost of trading stock is close to the call premium's FV at expiration.

Consider a trader who is short 10,000 calls. This call's expiration is 4 weeks away, its strike price is $50, and each call has a current value of $f = 1.76791$. Because an option covers 100 shares of stock, $N = 1{,}000{,}000$. The underlying stock has 30% annual volatility, and the annual riskless rate is 6%. The trader adjusts the portfolio weekly. As $\Delta = 0.538560$, $N \times \Delta = 538{,}560$ shares are purchased for a total cost of $538{,}560 \times 50 = 26{,}928{,}000$ dollars to make the portfolio delta-neutral. The trader finances the purchase by borrowing

$$B = N \times \Delta \times S - N \times f = 26{,}928{,}000 - 1{,}767{,}910 = 25{,}160{,}090$$

dollars net. The portfolio has zero net value now.

At 3 weeks to expiration, the stock price rises to $51. Because the new call value is $f' = 2.10580$, the portfolio is worth

$$-N \times f' + 538{,}560 \times 51 - Be^{0.06/52} = 171{,}622 \tag{16.3}$$

before rebalancing. That this number is not zero confirms that a delta hedge does not replicate the calls perfectly; it is not self-financing as $171,622 can be withdrawn. The magnitude of the **tracking error** – the variation in the net portfolio value – can be mitigated if adjustments are made more frequently, say daily instead of weekly. In fact, the tracking error is positive ~68% of the time even though its expected value is essentially zero. It is furthermore proportional to vega [109, 537]. In practice tracking errors will cease to decrease to beyond a certain rebalancing frequency [45].

With a higher delta $\Delta' = 0.640355$, the trader buys $N \times (\Delta' - \Delta) = 101{,}795$ shares for $5,191,545, increasing the number of shares to $N \times \Delta' = 640{,}355$. The cumulative cost is

$$26{,}928{,}000 \times e^{0.06/52} + 5{,}191{,}545 = 32{,}150{,}634,$$

and the net borrowed amount is[1]

$$B' = 640{,}355 \times 51 - N \times f' = 30{,}552{,}305.$$

The portfolio is again delta-neutral with zero value. Figure 16.1 tabulates the numbers.

Weeks to expiration	Stock price S (1)	Option value f (2)	Delta Δ (3)	Gamma Γ (4)	Change in delta (3)−(3″) (5)	No. shares bought $N \times$ (5) (6)	Cost of shares (1)×(6) (7)	Cumulative cost FV(8″) +(7) (8)
4	50	1.76791	0.538560	0.0957074	–	538,560	26,928,000	26,928,000
3	51	2.10580	0.640355	0.1020470	0.101795	101,795	5,191,545	32,150,634
2	53	3.35087	0.855780	0.0730278	0.215425	215,425	11,417,525	43,605,277
1	52	2.24272	0.839825	0.1128601	−0.015955	−15,955	−829,660	42,825,960
0	54	4.00000	1.000000	0.0000000	0.160175	160,175	8,649,450	51,524,853

Figure 16.1: Delta hedge. The cumulative cost reflects the cost of trading stocks to maintain delta neutrality. The total number of shares is 1,000,000 at expiration (trading takes place at expiration, too). A doubly primed number refers to the entry from the previous row of the said column.

At expiration, the trader has 1,000,000 shares, which are exercised against by the in-the-money calls for $50,000,000. The trader is left with an obligation of

$$51,524,853 - 50,000,000 = 1,524,853,$$

which represents the replication cost. Compared with the FV of the call premium,

$$1,767,910 \times e^{0.06 \times 4/52} = 1,776,088,$$

the net gain is $1,776,088 - 1,524,853 = 251,235$. The amount of money to start with for perfect replication should converge to the call premium $1,767,910 as the position is rebalanced more frequently.

▶ **Exercise 16.3.2** (1) Repeat the calculations in Fig. 16.1 but this time record the weekly tracking errors instead of the cumulative costs. Verify that the following numbers result:

Weeks to expiration	Net borrowing (B)	Tracking error
4	25,160,090	–
3	30,552,305	171,622
2	42,005,470	367
1	41,428,180	203,874
0	50,000,000	−125,459

(This alternative view looks at how well the call is hedged.) (2) Verify that the FVs at expiration of the tracking errors sum to $251,235.

▶ **Exercise 16.3.3** A broker claimed the option premium is an arbitrage profit because he could write a call, pocket the premium, then set up a replicating portfolio to hedge the short call. What did he overlook?

▶ **Programming Assignment 16.3.4** Implement the delta hedge for options.

Weeks to expiration	Stock price S	Option value f_2	Delta Δ_2	Gamma Γ_2
4	50	1.99113	0.543095	0.085503
3	51	2.35342	0.631360	0.089114
2	53	3.57143	0.814526	0.070197
1	52	2.53605	0.769410	0.099665
0	54	4.08225	0.971505	0.029099

Figure 16.2: Hedging option used in delta–gamma hedge. This option is the same as the one in Fig. 16.1 except that the expiration date is 1 week later.

16.3.2 Delta–Gamma and Vega-Related Hedges

A delta hedge is based on the first-order approximation to changes in the derivative price Δf, which is due to small changes in the stock price ΔS. When ΔS is not small, the second-order term, gamma $\Gamma \equiv \partial^2 f / \partial S^2$, can help. A **delta–gamma hedge** is like delta hedge except that zero portfolio gamma, or **gamma neutrality**, is maintained. To meet this extra condition, in addition to self-financing and delta neutrality, one more security needs to be brought in.

The hedging procedure will be illustrated for the scenario in Fig. 16.1. A hedging call is brought in, and its properties along the same scenario are in Fig. 16.2. With the stock price at $50, each call has a value of $f = 1.76791$, delta $\Delta = 0.538560$, and gamma $\Gamma = 0.0957074$, whereas each hedging call has value $f_2 = 1.99113$, $\Delta_2 = 0.543095$, and $\Gamma_2 = 0.085503$. Note that the gamma of the stock is zero. To set up a delta–gamma hedge, we solve

$$-N \times f + n_1 \times 50 + n_2 \times f_2 - B = 0 \quad \text{(self-financing)},$$
$$-N \times \Delta + n_1 + n_2 \times \Delta_2 - 0 = 0 \quad \text{(delta neutrality)},$$
$$-N \times \Gamma + 0 + n_2 \times \Gamma_2 - 0 = 0 \quad \text{(gamma neutrality)}.$$

The solutions are $n_1 = -69{,}351$, $n_2 = 1{,}119{,}346$, and $B = -3{,}006{,}695$. We short 69,351 shares of stock, buy 1,119,346 hedging calls, and lend 3,006,695 dollars. The cost of shorting stock and buying calls is $n_1 \times 50 + n_2 \times f_2 = -1{,}238{,}787$ dollars.

One week later, the stock price climbs to $51. The new call values are $f' = 2.10580$ and $f_2' = 2.35342$ for the hedged and the hedging calls, respectively, and the portfolio is worth

$$-N \times f' + n_1 \times 51 + n_2 \times f_2' - Be^{0.06/52} = 1{,}757$$

before rebalancing. As this number is not zero, a delta–gamma hedge is not self-financing. Nevertheless, it is substantially smaller than delta hedge's 171,622 in Eq. (16.3). Now we solve

$$-N \times f' + n_1' \times 51 + n_2' \times f_2' - B' = 0,$$
$$-N \times \Delta' + n_1' + n_2' \times \Delta_2' - 0 = 0,$$
$$-N \times \Gamma' + 0 + n_2' \times \Gamma_2' - 0 = 0.$$

The solutions are $n_1' = -82{,}633$, $n_2' = 1{,}145{,}129$, and $B' = -3{,}625{,}138$. The trader therefore purchases $n_1' - n_1 = -82{,}633 + 69{,}351 = -13{,}282$ shares of stock and

Weeks to expiration	Stock price S	No. shares bought $n_1' - n_1$	Cost of shares $(1) \times (2)$	No. options bought $n_2' - n_2$	Cost of options $(4) \times f_2$	Net borrowing B	Cumulative cost $FV(7'') + (3) + (5)$
	(1)	(2)	(3)	(4)	(5)	(6)	(7)
4	50	−69,351	−3,467,550	1,119,346	2,228,763	−3,006,695	−1,238,787
3	51	−13,282	−677,382	25,783	60,678	−3,625,138	−1,856,921
2	53	91,040	4,825,120	−104,802	−374,293	810,155	2,591,762
1	52	−39,858	−2,072,616	92,068	233,489	−1,006,346	755,627
0	54	1,031,451	55,698,354	−1,132,395	−4,622,720	50,000,000	51,832,134

Figure 16.3: Delta–gamma hedge. The cumulative cost reflects the cost of trading stock and the hedging call to maintain delta–gamma neutrality. At expiration, the number of shares is 1,000,000, whereas the number of hedging calls is zero.

$n_2' - n_2 = 1,145,129 - 1,119,346 = 25,783$ hedging calls for $-13,282 \times 51 + 25,783 \times f_2' = -616,704$ dollars. The cumulative cost is

$$-1,238,787 \times e^{0.06/52} - 616,704 = -1,856,921.$$

The portfolio is again delta-neutral and gamma-neutral with zero value. The remaining steps are tabulated in Fig. 16.3.

At expiration, the trader owns 1,000,000 shares, which are exercised against by the in-the-money calls for $50,000,000. The trader is then left with an obligation of

$$51,832,134 - 50,000,000 = 1,832,134.$$

With the FV of the call premium at $1,776,088, the net loss is $56,046, which is smaller than the $251,235 with the delta hedge.

If volatility changes, a delta-gamma hedge may not work well. An enhancement is the **delta–gamma–vega hedge**, which maintains also **vega neutrality**, meaning zero portfolio vega. As before, to accomplish this, one more security has to be brought into the process. Because this strategy does not involve new insights, it is left to the reader. In practice, the **delta–vega hedge**, which may not maintain gamma neutrality, performs better than the delta hedge [44].

▶ **Exercise 16.3.5** Verify that any delta-neutral gamma-neutral self-financing portfolio is automatically theta-neutral.

▶ **Programming Assignment 16.3.6** Implement the delta–gamma hedge for options.

▶ **Programming Assignment 16.3.7** Implement the delta–gamma–vega hedge for options.

16.3.3 Static Hedging

Dynamic strategies incur huge transactions costs. A **static strategy** that trades only when certain rare events occur addresses this problem. This goal has been realized for hedging European barrier options and look back options with standard options [157, 158, 159, 270].

▶ **Exercise 16.3.8** Explain why shorting a bull call spread can in practice hedge a binary option statically.

Additional Reading

The literature on hedging is vast [470, 514, 569, 746]. Reference [809] adopts a broader view and considers instruments beyond derivatives. See [891] for mathematical programming techniques in risk management. They are essential in the presence of trading constraints or market imperfections. Consult [365, 369, 376, 646, 647] for more information on financial engineering and risk management.

NOTE

1. Alternatively, the number could be arrived at by $Be^{0.06/52} + 5,191,545 + 171,622 = 30,552,305$.

Trees

I love a tree more than a man.
Ludwig van Beethoven (1770–1827)

This chapter starts with combinatorial methods to speed up European option pricing. The influential and versatile trinomial model is also introduced. These tree are regarded as more "accurate" than binomial trees. Then an important maxim is brought up: The comparison of algorithms should be based on the actual running time. This chapter ends with multinomial trees for pricing multivariate derivatives.

17.1 Pricing Barrier Options with Combinatorial Methods

We first review the binomial approximation to the geometric Brownian motion $S = e^X$, where X is a $(\mu - \sigma^2/2, \sigma)$ Brownian motion. (Equivalently, $dS/S = \mu\,dt + \sigma\,dW$.) For economy of expression, we use S in place of $S(0)$ for the current time-zero price. Consider the stock price at time $\Delta t \equiv \tau/n$, where τ is the time to maturity. From Eq. (13.12),

$$E[\,S(\Delta t)\,] = Se^{\mu\Delta t}, \quad \mathrm{Var}[\,S(\Delta t)\,] = S^2 e^{2\mu\Delta t}\big(e^{\sigma^2\Delta t} - 1\big) \to S^2\sigma^2\Delta t.$$

Under the binomial model, the stock price increases to Su with probability q or decreases to Sd with probability $1 - q$ at time Δt. The expected stock price at time Δt is $qSu + (1 - q)\,Sd$. Our first requirement is that it converge to $Se^{\mu\Delta t}$. The variance of the stock price at time Δt is given by $q(Su)^2 + (1 - q)(Sd)^2 - (Se^{\mu\Delta t})^2$. Our second requirement is that it converge to $S^2\sigma^2\Delta t$. With $ud = 1$ imposed, the choice below works:

$$u = e^{\sigma\sqrt{\Delta t}}, \quad d = e^{-\sigma\sqrt{\Delta t}}, \quad q = \frac{e^{\mu\Delta t} - d}{u - d}. \tag{17.1}$$

In a risk-neutral economy, $\mu = r$ and q approaches

$$p \equiv \frac{1}{2} + \frac{1}{2}\frac{r - \sigma^2/2}{\sigma}\sqrt{\Delta t}$$

by convergence (9.17). We set $\mu' \equiv r - \sigma^2/2$ throughout.

The combinatorial method is as elementary as it is elegant. It can often cut the running time by an order of magnitude. The basic paradigm is to count the number of admissible paths that lead from the root to any terminal node. We first used this method in the linear-time algorithm for standard European option pricing in Fig. 9.9, and now we apply it to barrier option pricing. The reflection principle provides the necessary tool.

17.1.1 The Reflection Principle

Imagine a particle at position $(0, -a)$ on the integral lattice that is to reach $(n, -b)$. Without loss of generality, assume that $a, b \geq 0$. This particle is constrained to move to $(i + 1, j + 1)$ or $(i + 1, j - 1)$ from (i, j), as shown below:

$$(i, j) < \begin{matrix} (i+1, j+1) & \text{associated with the up move } S \to Su \\ (i+1, j-1) & \text{associated with the down move } S \to Sd \end{matrix}$$

How many paths touch the x axis?

For a path from $(0, -a)$ to $(n, -b)$ that touches the x axis, let J denote the first point at which this happens. When the portion of the path from $(0, -a)$ to J is reflected, a path from $(0, a)$ to $(n, -b)$ is constructed, which also hits the x axis at J for the first time (see Fig. 17.1). The one-to-one mapping shows that the number of paths from $(0, -a)$ to $(n, -b)$ that touch the x axis equals the number of paths from $(0, a)$ to $(n, -b)$. This is the celebrated **reflection principle** of André (1840–1917) published in 1887 [604, 686]. Because a path of this kind has $(n+b+a)/2$ down moves and $(n-b-a)/2$ up moves, there are

$$\binom{n}{\frac{n+a+b}{2}} \quad \text{for even } n+a+b \tag{17.2}$$

such paths. The convention here is $\binom{n}{k} = 0$ for $k < 0$ or $k > n$.

▶ **Exercise 17.1.1** What is the probability that the stock's maximum price is at least Su^k?

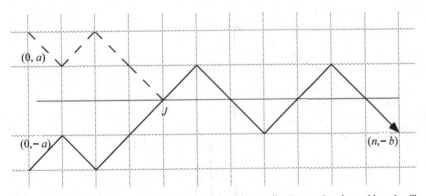

Figure 17.1: The reflection principle for binomial random walks. Two paths of equal length will be separated by a distance of $2k$ on the binomial tree if their respective accumulative numbers of up moves differ by k (see Eq. (17.4)).

17.1.2 Combinatorial Formulas for Barrier Options

We focus on the down-and-in call with barrier $H < X$. Assume that $H < S$ without loss of generality for, otherwise, the option is identical to a standard call. Define

$$a \equiv \left\lceil \frac{\ln(X/(Sd^n))}{\ln(u/d)} \right\rceil = \left\lceil \frac{\ln(X/S)}{2\sigma\sqrt{\Delta t}} + \frac{n}{2} \right\rceil,$$

$$h \equiv \left\lfloor \frac{\ln(H/(Sd^n))}{\ln(u/d)} \right\rfloor = \left\lfloor \frac{\ln(H/S)}{2\sigma\sqrt{\Delta t}} + \frac{n}{2} \right\rfloor.$$

(17.3)

Both a and h have straightforward interpretations. First, h is such that $\tilde{H} \equiv Su^h d^{n-h}$ is the terminal price that is closest to, but does not exceed, H. The true barrier is replaced with the **effective barrier** $\tilde{H}$ in the binomial model. Similarly, a is such that $\tilde{X} \equiv Su^a d^{n-a}$ is the terminal price that is closest to, but not exceeded by, X. A process with n moves hence ends up in the money if and only if the number of up moves is at least a.

The price $Su^k d^{n-k}$ is at a distance of $2k$ from the lowest possible price Sd^n on the binomial tree because

$$Su^k d^{n-k} = Sd^{-k} d^{n-k} = Sd^{n-2k}.$$

(17.4)

Given this observation, Fig. 17.2 plots the relative distances of various prices on the tree.

The number of paths from S to the terminal price $Su^j d^{n-j}$ is $\binom{n}{j}$, each with probability $p^j(1-p)^{n-j}$. With reference to Fig. 17.2, we can apply the reflection principle with $a = n - 2h$ and $b = 2j - 2h$ in formula (17.2) by treating the S line as the x axis. Therefore

$$\binom{n}{\frac{n+(n-2h)+(2j-2h)}{2}} = \binom{n}{n - 2h + j}$$

paths hit $\tilde{H}$ in the process for $h \le n/2$. We conclude that the terminal price $Su^j d^{n-j}$ is reached by a path that hits the effective barrier with probability

$$\binom{n}{n - 2h + j} p^j(1-p)^{n-j},$$

(17.5)

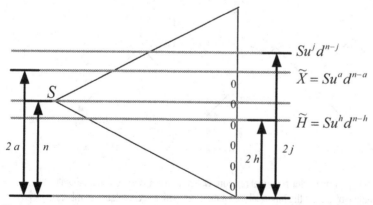

Figure 17.2: Down-and-in call and binomial tree. The effective barrier is the $\tilde{H}$ line, and the process starts on the S line.

Linear-time, constant-space algorithm for pricing down-and-in calls on a non-dividend-paying stock:

input: $S, \sigma, X, H(H < X, H < S), n, \tau, r$;
real p, u, d, b, D, C;
integer j, a, h;
$u := e^{\sigma\sqrt{\tau/n}}; d := e^{-\sigma\sqrt{\tau/n}}$;
$a := \lceil \ln(X/Sd^n)/\ln(u/d) \rceil; h := \lfloor \ln(H/Sd^n)/\ln(u/d) \rfloor$;
$p := (e^{r\tau/n} - d)/(u - d)$;   // Risk-neutral probability.
$b := p^{2h}(1 - p)^{n-2h}$;   // b_{2h} is computed.
$D := S \times u^{2h} d^{n-2h}; C := b \times (D - X)$;
for $(j = 2h - 1$ down to $a)$ {
$\quad b := b \times p \times (n - 2h + j + 1)/((1 - p) \times (2h - j))$;
$\quad D := D \times u/d$;
$\quad C := C + b \times (D - X)$;
}
return $C/e^{r\tau}$;

Figure 17.3: Optimal algorithm for European down-and-in calls on a stock that does not pay dividends. Variable b stores $b_j \equiv \binom{n}{n-2h+j} p^j(1 - p)^{n-j}$ for $j = 2h, 2h - 1, \ldots, a$, in that order, and variable C accumulates the summands in option value (17.6) for $j = 2h, 2h - 1, \ldots, a$. Note that $b_j = b_{j+1}[(1 - p)(n - 2h + j + 1)]/[p(2h - j)]$. The structure is similar to the one in Fig. 9.9.

and the option value equals

$$R^{-n} \sum_{j=a}^{2h} \binom{n}{n-2h+j} p^j(1 - p)^{n-j} \left(Su^j d^{n-j} - X\right), \tag{17.6}$$

where $R \equiv e^{r\tau/n}$ is the riskless return per period. Formula (17.6) is an alternative characterization of the binomial tree algorithm [624]. It also implies a linear-time algorithm (see Fig. 17.3). In fact, the running time is proportional to $2h - a$, which is close to $n/2$:

$$2h - a \approx \frac{n}{2} + \frac{\ln(H^2/(SX))}{2\sigma\sqrt{\tau/n}} = \frac{n}{2} + O(\sqrt{n}).$$

The preceding methodology has applications to exotic options whose terminal payoff is "nonstandard" and closed-form solutions are hard to come by. Discrete-time models may also be more realistic than continuous-time ones for contracts based on discrete sampling of the price process at regular time intervals [147, 597].

EXAMPLE 17.1.1 A binary call pays off $1 if the underlying asset finishes above the strike price and nothing otherwise. The price of a binary down-and-in call is formula (17.6) with $Su^j d^{n-j} - X$ replaced with 1.

EXAMPLE 17.1.2 A **power option** pays off $\max([S(\tau) - X]^p, 0)$ (sometimes $\max(S(\tau)^p - X, 0)$) at expiration [894]. To price a down-and-in power option, replace $Su^j d^{n-j} - X$ in formula (17.6) with $(Su^j d^{n-j} - X)^p$ $((Su^j d^{n-j})^p - X$, respectively).

➤ **Exercise 17.1.2** Derive pricing formulas similar to formula (17.6) for the other three barrier options: down-and-out, up-and-in, and up-and-out options.

➤ **Exercise 17.1.3** Use the reflection principle to derive a combinatorial pricing formula for the European lookback call on the minimum.

➤ **Exercise 17.1.4** Consider the **exploding call spread**, which has the same payoff as the bull call spread except that it is exercised promptly the moment the stock price touches the trigger price K [377]. (1) Write a combinatorial formula for the value of this path-dependent option. (2) Verify that the valuation of the option runs in linear time.

➤ **Exercise 17.1.5** Derive a combinatorial pricing formula for the reset call option.

➤ **Exercise 17.1.6** Prove that option value (17.6) converges to value (11.4) with $q = 0$.

➤ **Exercise 17.1.7** Derive a pricing formula for the European power option $\max(S(\tau)^2 - X, 0)$.

➤ **Programming Assignment 17.1.8** Design fast algorithms for European barrier options.

➤ **Programming Assignment 17.1.9** Implement $O(n^3)$-time algorithms for European geometric average-rate options with combinatorics, improving Programming Assignment 11.7.6.

➤ **Programming Assignment 17.1.10** (1) Implement $O(n^2)$-time algorithms for European lookback options, improving Programming Assignment 11.7.11, part (1). (2) Improve the running time to $O(n)$.

17.1.3 Convergence of Binomial Tree Algorithms

Option value (17.6) results in the sawtoothlike convergence shown in Fig. 11.5. Increasing n therefore does not necessarily lead to more accurate results. The reasons are not hard to see. The true barrier most likely does not equal the effective barrier. The same holds for the strike price and the effective strike price. Both introduce specification errors [271]. The issue of the strike price is less critical as evinced by the fast convergence of binomial tree algorithms for standard European options. The issue of the barrier is not negligible, however, because the barrier exerts its influence throughout the price dynamics.

Figure 17.4 suggests that convergence is actually good if we limit n to certain values – 191 in the figure, for example. These values make the true barrier coincide with or occur just above one of the stock price levels, that is, $H \approx Sd^j = Se^{-j\sigma\sqrt{\tau/n}}$ for some integer j [111, 196]. The preferred n's are thus

$$ n = \left\lfloor \frac{\tau}{\left[\ln(S/H)/(j\sigma)\right]^2} \right\rfloor, \quad j = 1, 2, 3, \ldots $$

There is only one minor technicality left. We picked the effective barrier to be one of the $n+1$ possible terminal stock prices. However, the effective barrier above, Sd^j, corresponds to a terminal stock price only when $n - j$ is even by Eq. (17.4).* To close this gap, we decrement n by one, if necessary, to make $n - j$ an even number.

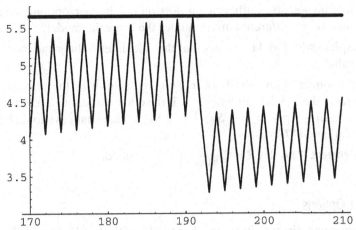

Figure 17.4: Convergence of binomial model for down-and-in calls. A detailed look of Fig. 11.5. Note that the approximation is quite close (5.63542 vs. the analytical value 5.6605) at $n = 191$.

The preferred n's are now

$$n = \begin{cases} \ell, & \text{if } \ell - j \text{ is even} \\ \ell - 1, & \text{otherwise} \end{cases}, \quad \ell \equiv \left\lfloor \frac{\tau}{\left[\ln(S/H)/(j\sigma) \right]^2} \right\rfloor \tag{17.7}$$

$j = 1, 2, 3, \ldots$. In summary, evaluate pricing formula (17.6) only with the n's above. The result is shown in Fig. 17.5.

Now that barrier options can be efficiently priced, we can afford to pick very large n's. This has profound consequences. For example, pricing seems prohibitively time consuming when $S \approx H$ because n, being proportional to $1/\ln^2(S/H)$, is large. This observation is indeed true of standard quadratic-time binomial tree algorithms like the one in Fig. 11.4. However, it no longer applies to the linear-time algorithm [624].

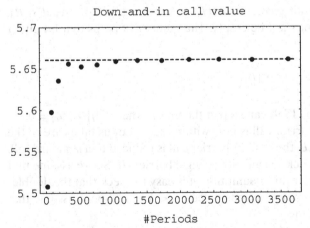

Figure 17.5: Convergence of binomial model for down-and-in calls at well-chosen n's. Formula (17.6) is evaluated at $n = 21$ (1), 84 (2), 191 (3), 342 (4), 533 (5), 768 (6), 1047 (7), 1368 (8), 1731 (9), 2138 (10), 2587 (11), 3078 (12), and 3613 (13), with the corresponding j in parentheses.

➤ **Exercise 17.1.11** How do we efficiently price a portfolio of barrier options with identical underlying assets but different barriers under the binomial model?

➤ **Exercise 17.1.12** Explain why Fig. 11.5 shows that the calculated values underestimate the analytical value.

➤ **Exercise 17.1.13** In formula (17.6), the barrier H is replaced with the effective barrier $\tilde{H}$, which is one of the $n+1$ terminal prices. If the effective barrier is allowed to be one of all possible $2n+1$ prices $Su^n, Su^{n-1}, \ldots, Su^{-n}$, what changes should be made to formulas (17.6) and (17.7)?

➤ **Programming Assignment 17.1.4** Try $\ell \equiv \lceil \frac{\tau}{[\ln(S/H)/(j\sigma)]^2} \rceil$ instead.

17.1.4 Double-Barrier Options

Double-barrier options contain two barriers, L and H, with $L < H$. Depending on how the barriers affect the security, various barrier options can be defined. We consider options that come into existence if and only if *either* barrier is hit.

A particle starts at position $(0, -a)$ on the integral lattice and is destined for $(n, -b)$. Without loss of generality, assume that $a, b \geq 0$. The number of paths in which a hit of the H line, $x = 0$, appears before a hit of the L line $x = -s$ is

$$\binom{n}{\frac{n+a-b+2s}{2}} \quad \text{for even } n+a-b. \tag{17.8}$$

In the preceding expression, we assume that $s > b$ and $a < s$ to make both barriers effective.

The preceding expression can be generalized. Let A_i denote the set of paths that hit the barriers with a hit sequence *containing* $\overbrace{H^+ L^+ H^+ \cdots}^{i}$, $i \geq 2$, where L^+ denotes a sequence of Ls and H^+ denotes a sequence of Hs. Similarly, let B_i denote the set of paths that hit the barriers with a sequence *containing* $\overbrace{L^+ H^+ L^+ \cdots}^{i}$, $i \geq 2$. For instance, a path with the hit pattern $LLHLLHH$ belongs to A_2, A_3, B_2, B_3, and B_4. Note that $A_i \cap B_i$ may not be empty. The number of paths that hit either barrier is

$$N(a, b, s) = \sum_{i=1}^{n} (-1)^{i-1}(|A_i| + |B_i|). \tag{17.9}$$

The calculation of summation (17.9) can stop at the first i when $|A_i| + |B_i| = 0$.

The value of the double-barrier call is now within reach. Let us take care of the degenerate cases first. If $S \leq L$, the double-barrier call is reduced to a standard call. If $S \geq H$, it is reduced to a knock-in call with a single barrier H. So we assume that $L < S < H$ from now on. Under this assumption, it is easy to check that the double-barrier option is reduced to simpler options unless $L < X < H$. So we assume that $L < X < H$ from now on. Define

$$h \equiv \left\lceil \frac{\ln(H/(Sd^n))}{\ln(u/d)} \right\rceil = \left\lceil \frac{\ln(H/S)}{2\sigma\sqrt{\Delta t}} + \frac{n}{2} \right\rceil,$$

$$l \equiv \left\lfloor \frac{\ln(L/(Sd^n))}{\ln(u/d)} \right\rfloor = \left\lfloor \frac{\ln(L/S)}{2\sigma\sqrt{\Delta t}} + \frac{n}{2} \right\rfloor.$$

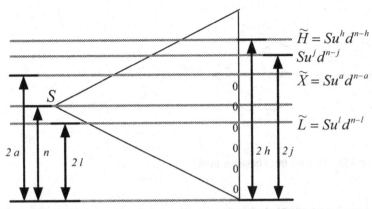

Figure 17.6: Double-barrier call under binomial model. The effective barriers are the $\tilde{H}$ line and the $\tilde{L}$ line, and the process starts on the S line.

The barriers will be replaced with the effective barriers $\tilde{H} \equiv S u^h d^{n-h}$ and $\tilde{L} \equiv S u^l d^{n-l}$. Note that in Eq. (17.9), only terminal nodes between $\tilde{L}$ and $\tilde{H}$ (inclusive) are considered. These terminal nodes together contribute

$$
A \equiv R^{-n} \sum_{j=a}^{h} N(2h - n, 2h - 2j, 2(h - l)) \, p^j (1 - p)^{n-j} \left(S u^j d^{n-j} - X \right)
$$

(17.10)

to the option value, where a is defined in Eqs. (17.3). See Fig. 17.6 for the relative positions of of the various parameters. As for the terminal nodes outside the above-mentioned range, they constitute a standard call with a strike price of $\tilde{H} u^2$. Let its value be D. The double-barrier call thus has value $A + D$. The convergence is sawtoothlike [179].

▶ **Exercise 17.1.15** Prove formula (17.8).

▶ **Exercise 17.1.16** Apply the reflection principle repetitively to verify that

$$
|A_i| = \begin{cases} \begin{pmatrix} n \\ \frac{n+a+b+(i-1)s}{2} \end{pmatrix} & \text{for odd } i \\ \begin{pmatrix} n \\ \frac{n+a-b+is}{2} \end{pmatrix} & \text{for even } i \end{cases} \, , \quad |B_i| = \begin{cases} \begin{pmatrix} n \\ \frac{n-a-b+(i+1)s}{2} \end{pmatrix} & \text{for odd } i \\ \begin{pmatrix} n \\ \frac{n-a+b+is}{2} \end{pmatrix} & \text{for even } i \end{cases} .
$$

Assume that $n + a - b$ is even for $|A_i|$ and $n - a + b$ is even for $|B_i|$.

▶ **Exercise 17.1.17** Prove Eq. (17.9).

▶ **Exercise 17.1.18** (1) Formulate the in–out parity for double-barrier options. (2) Replicate the double-barrier option by using knock-in options, knock-out options, and double-barrier options that come into existence if and only if both barriers are hit. (3) Modify Eq. (17.9) to price the double-barrier option in (2).

▶ **Exercise 17.1.19** Consider a generalized double-barrier option with the two barriers defined by functions f_ℓ and f_h, where $f_\ell(t) < f_h(t)$ for $t \geq 0$. Transform it into a double-barrier option with constant barriers by changing the underlying price process.

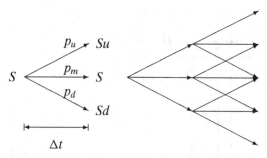

Figure 17.7: Trinomial model. There are three branches from each node.

> **Programming Assignment 17.1.20** Implement a linear-time algorithm for double-barrier options.

17.2 Trinomial Tree Algorithms

We now set up the trinomial approximation to the geometric Brownian motion $dS/S = r\,dt + \sigma\,dW$ [107]. The three stock prices at time Δt are S, Su, and Sd, where $ud = 1$ (see Fig. 17.7). Impose the matching of mean and that of variance to obtain

$$1 = p_u + p_m + p_d,$$
$$SM \equiv [\,p_u u + p_m + (p_d/u)\,]\,S,$$
$$S^2 V \equiv p_u(Su - SM)^2 + p_m(S - SM)^2 + p_d(Sd - SM)^2,$$

where $M \equiv e^{r\Delta t}$ and $V \equiv M^2(e^{\sigma^2 \Delta t} - 1)$ by Eqs. (6.11). It is easy to verify that

$$p_u = \frac{u(V + M^2 - M) - (M - 1)}{(u - 1)(u^2 - 1)},$$
$$p_d = \frac{u^2(V + M^2 - M) - u^3(M - 1)}{(u - 1)(u^2 - 1)}.$$

We need to make sure that the probabilities lie between zero and one. Use $u = e^{\lambda\sigma\sqrt{\Delta t}}$, where $\lambda \geq 1$ is a parameter that can be tuned. Then

$$p_u \to \frac{1}{2\lambda^2} + \frac{(r + \sigma^2)\sqrt{\Delta t}}{2\lambda\sigma}, \quad p_d \to \frac{1}{2\lambda^2} - \frac{(r - 2\sigma^2)\sqrt{\Delta t}}{2\lambda\sigma}.$$

A nice choice for λ is $\sqrt{\pi/2}$ [824].

> **Exercise 17.2.1** Verify the following: (1) $\ln(S(\Delta t)/S)$ has mean $\mu'\Delta t$, (2) the variance of $\ln(S(\Delta t)/S)$ converges to $\sigma^2\Delta t$, and (3) $S(\Delta t)$'s mean converges to $Se^{r\Delta t}$.

> **Exercise 17.2.2** The trinomial model no longer supports perfect replication of options with stocks and bonds as in the binomial model. Replicating an option with

h shares of stock and $\$B$ in bonds involves two unknowns h and B, but the three branches imply three conditions. Give an example for which resulting system of three equations in two knowns is inconsistent.

➤ **Programming Assignment 17.2.3** Recall the diagonal method in Section 9.7. Write a program to perform backward induction on the trinomial tree with the diagonal method.

17.2.1 Pricing Barrier Options

Binomial tree algorithms introduce a specification error by replacing the barrier with a nonidentical effective barrier. The trinomial tree algorithm that is due to Ritchken solves the problem by adjusting λ so that the barrier is hit exactly [745]. Here is the idea. It takes

$$h = \frac{\ln(S/H)}{\lambda \sigma \sqrt{\Delta t}}$$

consecutive down moves to go from S to H if h is an integer, which is easy to achieve by adjusting λ. Typically, we find the smallest $\lambda \geq 1$ such that h is an integer, that is,

$$\lambda = \min_{j=1,2,3,\ldots} \frac{\ln(S/H)}{j \sigma \sqrt{\Delta t}}.$$

This done, one of the layers of the trinomial tree coincides with the barrier. We note that such a λ may not exist for very small n's. The following probabilities may be used:

$$p_u = \frac{1}{2\lambda^2} + \frac{\mu' \sqrt{\Delta t}}{2\lambda\sigma}, \quad p_m = 1 - \frac{1}{\lambda^2}, \quad p_d = \frac{1}{2\lambda^2} - \frac{\mu' \sqrt{\Delta t}}{2\lambda\sigma}.$$

Note that this particular trinomial model reduces to the binomial model when $\lambda = 1$. See Fig. 17.8 for the algorithm. Figure 17.9 shows the trinomial model's convergence behavior. If the stock pays a continuous dividend yield of q, then we let $\mu' \equiv r - q - \sigma^2/2$.

➤ **Exercise 17.2.4** It was shown in Subsection 10.2.2 that binomial trees can be extended backward in time for two periods to compute delta and gamma. Argue that trinomial trees need to be extended backward in time for only one period to compute the same hedge parameters.

➤ **Exercise 17.2.5** Derive combinatorial formulas for European down-and-in, down-and-out, up-and-in, and up-and-out options.

➤ **Programming Assignment 17.2.6** Implement trinomial tree algorithms for barrier options. Add rebates for the knock-out type.

17.2.2 Remarks on Algorithm Comparison

Algorithms are often compared based on the n value at which they converge, and the one with the smallest n wins. This is a fallacy as it implies that giraffes are faster than cheetahs simply because they take fewer strides to travel the same distance,

---

Trinomial tree algorithm for down-and-out calls on a non-dividend-paying stock:

input: $S, \sigma, X, H (H < X, H < S), n, \tau, r;$
real $u, d, p_u, p_m, p_d, \lambda, \Delta t, C[2n+1];$
integer $i, j, h;$
$\Delta t := \tau/n;$
$h =: \lfloor \ln(S/H)/(\sigma\sqrt{\Delta t}) \rfloor;$
if $[h < 1 \text{ or } h > n]$ return failure;
$\lambda := \ln(S/H)/(h\sigma\sqrt{\Delta t});$
$p_u := 1/(2\lambda^2) + (r - \sigma^2/2)\sqrt{\Delta t}/(2\lambda\sigma);$
$p_d := 1/(2\lambda^2) - (r - \sigma^2/2)\sqrt{\Delta t}/(2\lambda\sigma);$
$p_m := 1 - p_u - p_d;$
$u := e^{\lambda\sigma\sqrt{\Delta t}};$
for $(i = 0 \text{ to } 2n)$ { $C[i] := \max(0, Su^{n-i} - X);$ }
$C[n+h] := 0;$ // A hit.
for $(j = n - 1 \text{ down to } 0)$ {
    for $(i = 0 \text{ to } 2j)$
        $C[i] := p_u C[i] + p_m C[i+1] + p_d C[i+2];$
        if $[j + h \le 2j]$ $C[j+h] := 0;$ // A hit.
}
return $C[0]/e^{r\tau};$

Figure 17.8: Trinomial tree algorithm for down-and-out calls on a non-dividend-paying stock. The barrier $H = Su^{-h}$ corresponds to $C[h+j]$ at times $j = n, n-1, \ldots, h$. It is not hard to show that h must be at least $\sigma^2\tau/\ln^2(S/H)$ to make $\lambda \ge 1$. This algorithm should be compared with the ones in Figs. 11.4 and 33.2.

forgetting that how fast the legs move is equally critical. like any race, an algorithm's performance must be based on its actual running time [717]. As a concrete example, Figs. 11.5 and 17.9 show that the trinomial model converges at a smaller n than the binomial model. It is in this sense when people say that trinomial models

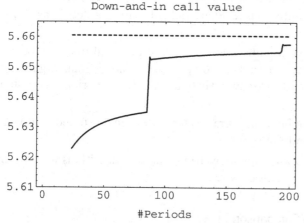

Down-and-in call value

Figure 17.9: Convergence of trinomial model for down-and-in calls. Plotted are the down-and-in call values as computed by the trinomial tree algorithm against the number of time periods. The parameters are identical to those used in Fig. 11.5. The analytical value 5.6605 is plotted for reference.

converge faster than binomial ones. However, the linear-time binomial tree algorithm for European barrier options actually performs better than the trinomial counterpart [610, 624].

17.3 Pricing Multivariate Contingent Claims

Multivariate derivatives such as correlation options are contingent claims that depend on two or more underlying assets. Consider the **basket option** on m assets. The basket call has the terminal payoff $\max(\sum_{i=1}^{m} \alpha_i S_i(\tau) - X, 0)$, whereas the basket put has the terminal payoff $\max(X - \sum_{i=1}^{m} \alpha_i S_i(\tau), 0)$, where α_i is the percentage of asset i [663]. Basket options are essentially options on a portfolio of stocks or index options on a capitalization- or a price-weighted index. Consider the **option on the best of two risky assets and cash** as another example. It has a terminal payoff of $\max(S_1(\tau), S_2(\tau), X)$, which guarantees a cash flow of X and the better of two assets, say a stock fund and a bond fund [833]. Because the terminal payoff can be written as $X + \max(\max(S_1(\tau), S_2(\tau)) - X, 0)$, the option is worth $Xe^{-r\tau} + C$, where C is the price of a call option on the maximum of two assets with strike price X. This section presents binomial and trinomial models for multiple underlying assets to price multivariate derivatives [107, 110]. The aim is to construct a multivariate discrete-time probability distribution with the desired means and variance–covariance values.

17.3.1 Construction of a Correlated Trinomial Model

Suppose that two risky assets S_1 and S_2 follow $dS_i/S_i = r\,dt + \sigma_i\,dW_i$ in a risk-neutral economy, $i = 1, 2$. Define $M_i \equiv e^{r\Delta t}$ and $V_i \equiv M_i^2(e^{\sigma_i^2 \Delta t} - 1)$, where $S_i M_i$ is the mean and $S_i^2 V_i$ is the variance of S_i at time Δt from now. The value of $S_1 S_2$ at time Δt has a joint lognormal distribution with mean $S_1 S_2 M_1 M_2 e^{\rho \sigma_1 \sigma_2 \Delta t}$, where ρ is the correlation between dW_1 and dW_2. We proceed to match the first and the second moments of the approximating discrete distribution to those of the continuous counterpart. At time Δt from now, there are five distinct outcomes. The five-point probability distribution of the asset prices is (as usual, we impose $u_i d_i = 1$)

Probability	Asset 1	Asset 2
p_1	$S_1 u_1$	$S_2 u_2$
p_2	$S_1 u_1$	$S_2 d_2$
p_3	$S_1 d_1$	$S_2 d_2$
p_4	$S_1 d_1$	$S_2 u_2$
p_5	S_1	S_2

The probabilities must sum to one, and the means must be matched, leading to

$$1 = p_1 + p_2 + p_3 + p_4 + p_5,$$
$$S_1 M_1 = (p_1 + p_2)\, S_1 u_1 + p_5 S_1 + (p_3 + p_4)\, S_1 d_1,$$
$$S_2 M_2 = (p_1 + p_4)\, S_2 u_2 + p_5 S_2 + (p_2 + p_3)\, S_2 d_2.$$

The following equations match the variances and the covariance:

$$S_1^2 V_1 = (p_1 + p_2)[\,(S_1 u_1)^2 - (S_1 M_1)^2\,] + p_5[\,S_1^2 - (S_1 M_1)^2\,]$$
$$+ (p_3 + p_4)[\,(S_1 d_1)^2 - (S_1 M_1)^2\,],$$
$$S_2^2 V_2 = (p_1 + p_4)[\,(S_2 u_2)^2 - (S_2 M_2)^2\,] + p_5[\,S_2^2 - (S_2 M_2)^2\,]$$
$$+ (p_2 + p_3)[\,(S_2 d_2)^2 - (S_2 M_2)^2\,],$$
$$S_1 S_2 R = (p_1 u_1 u_2 + p_2 u_1 d_2 + p_3 d_1 d_2 + p_4 d_1 u_2 + p_5)\, S_1 S_2,$$

where $R \equiv M_1 M_2 e^{\rho \sigma_1 \sigma_2 \Delta t}$. The solutions are

$$p_1 = \frac{u_1 u_2 (R-1) - f_1(u_1^2 - 1) - f_2(u_2^2 - 1) + (f_2 + g_2)(u_1 u_2 - 1)}{(u_1^2 - 1)\,(u_2^2 - 1)},$$

$$p_2 = \frac{-u_1 u_2 (R-1) + f_1(u_1^2 - 1)\,u_2^2 + f_2(u_2^2 - 1) - (f_2 + g_2)(u_1 u_2 - 1)}{(u_1^2 - 1)\,(u_2^2 - 1)},$$

$$p_3 = \frac{u_1 u_2 (R-1) - f_1(u_1^2 - 1)\,u_2^2 + g_2(u_2^2 - 1)\,u_1^2 + (f_2 + g_2)\,(u_1 u_2 - u_2^2)}{(u_1^2 - 1)\,(u_2^2 - 1)},$$

$$p_4 = \frac{-u_1 u_2 (R-1) + f_1(u_1^2 - 1) + f_2(u_2^2 - 1)\,u_1^2 - (f_2 + g_2)(u_1 u_2 - 1)}{(u_1^2 - 1)\,(u_2^2 - 1)},$$

where

$$f_1 = p_1 + p_2 = \frac{u_1(V_1 + M_1^2 - M_1) - (M_1 - 1)}{(u_1 - 1)\,(u_1^2 - 1)},$$

$$f_2 = p_1 + p_4 = \frac{u_2(V_2 + M_2^2 - M_2) - (M_2 - 1)}{(u_2 - 1)\,(u_2^2 - 1)},$$

$$g_1 = p_3 + p_4 = \frac{u_1^2(V_1 + M_1^2 - M_1) - u_1^3(M_1 - 1)}{(u_1 - 1)\,(u_1^2 - 1)},$$

$$g_2 = p_2 + p_3 = \frac{u_2^2(V_2 + M_2^2 - M_2) - u_2^3(M_2 - 1)}{(u_2 - 1)\,(u_2^2 - 1)}.$$

Because $f_1 + g_1 = f_2 + g_2$, we can solve for u_2 given $u_1 = e^{\lambda \sigma_1 \sqrt{\Delta t}}$ for an appropriate $\lambda > 0$.

Once the tree is in place, a multivariate derivative can be valued by backward induction. The expected terminal value should be discounted at the riskless rate.

▶ **Exercise 17.3.1** Show that there are $1 + 2n(n+1)$ pairs of possible asset prices after n periods.

17.3.2 The Binomial Alternative

In the binomial model for m assets, asset i's price S_i can in one period go up to $S_i u_i$ or down to $S_i d_i$. There are thus 2^m distinct states after one step. (This illustrates the **curse of dimensionality** because the complexity grows exponentially in the dimension m.) We fix $u_i = e^{\sigma_i \sqrt{\Delta t}}$ and $u_i d_i = 1$. As working with the log price $\ln S_i$ turns out to be easier, we let $R_i \equiv \ln S_i(\Delta t)/S_i$. From Subsection 14.4.3, we know that $R_i \sim N(\mu_i' \Delta t, \sigma_i^2 \Delta t)$, where $\mu_i' \equiv r - \sigma_i^2/2$.

We solve the $m = 2$ case first. Because (R_1, R_2) has a bivariate distribution, its moment generating function is

$$E\left[e^{t_1 R_1 + t_2 R_2}\right] = \exp\left[(t_1\mu_1' + t_2\mu_2')\,\Delta t + (t_1^2\sigma_1^2 + t_2^2\sigma_2^2 + 2t_1t_2\sigma_1\sigma_2\rho)\,\frac{\Delta t}{2}\right]$$

$$\approx 1 + (t_1\mu_1' + t_2\mu_2')\,\Delta t + (t_1^2\sigma_1^2 + t_2^2\sigma_2^2 + 2t_1t_2\sigma_1\sigma_2\rho)\,\frac{\Delta t}{2}.$$

Define the probabilities for up and down moves in the following table:

Probability	Asset 1	Asset 2
p_1	up	up
p_2	up	down
p_3	down	up
p_4	down	down

Under the binomial model, (R_1, R_2)'s moment generating function is

$$E\left[e^{t_1 R_1 + t_2 R_2}\right] = p_1 e^{(t_1\sigma_1 + t_2\sigma_2)\sqrt{\Delta t}} + p_2 e^{(t_1\sigma_1 - t_2\sigma_2)\sqrt{\Delta t}}$$

$$+ p_3 e^{(-t_1\sigma_1 + t_2\sigma_2)\sqrt{\Delta t}} + p_4 e^{(-t_1\sigma_1 - t_2\sigma_2)\sqrt{\Delta t}}$$

$$\approx (p_1 + p_2 + p_3 + p_4) + t_1\sigma_1(p_1 + p_2 - p_3 - p_4)\sqrt{\Delta t}$$

$$+ t_2\sigma_2(p_1 - p_2 + p_3 - p_4)\sqrt{\Delta t}$$

$$+ \left[t_1^2\sigma_1^2 + t_2^2\sigma_2^2 + 2t_1t_2\sigma_1\sigma_2(p_1 - p_2 - p_3 + p_4)\right]\frac{\Delta t}{2}$$

Match the preceding two equations to obtain

$$p_1 = \frac{1}{4}\left[1 + \rho + \sqrt{\Delta t}\left(\frac{\mu_1'}{\sigma_1} + \frac{\mu_2'}{\sigma_2}\right)\right], \quad p_2 = \frac{1}{4}\left[1 - \rho + \sqrt{\Delta t}\left(\frac{\mu_1'}{\sigma_1} - \frac{\mu_2'}{\sigma_2}\right)\right],$$

$$p_3 = \frac{1}{4}\left[1 - \rho + \sqrt{\Delta t}\left(-\frac{\mu_1'}{\sigma_1} + \frac{\mu_2'}{\sigma_2}\right)\right], \quad p_4 = \frac{1}{4}\left[1 + \rho + \sqrt{\Delta t}\left(-\frac{\mu_1'}{\sigma_1} - \frac{\mu_2'}{\sigma_2}\right)\right].$$

Note its similarity to univariate case (9.17).

For the general case, we simply present the result as the methodology is identical. Let $\delta_i(j) = 1$ if S_i in state j makes an up move and -1 otherwise. Let $\delta_{ik}(j) = 1$ if S_i and S_k in state j make a move in the same direction and -1 otherwise. Then the probability that state j is reached is

$$p_j = \frac{1}{2^m}\sum_{\substack{i,k=1 \\ i<k}}^{m} \delta_{ik}(j)\,\rho_{ik} + \frac{1}{2^m}\sqrt{\Delta t}\sum_{i=1}^{m}\delta_i(j)\,\frac{\mu_i'}{\sigma_i}, \quad j = 1, 2, \ldots, 2^m,$$

where ρ_{ik} denotes the correlation between dW_i and dW_k.

▶ **Exercise 17.3.2** It is easy to check that $\rho = 2(p_1 + p_4) - 1$. Show that this identity holds for any correlated binomial random walk defined by $R_1(i+1) - R_1(i) = \mu_1 \pm \sigma_1$ and $R_2(i+1) - R_2(i) = \mu_2 \pm \sigma_2$, where ρ denotes the correlation between R_1 and R_2 and "$\pm$" means "$+$" or "$-$" each with a probability of one half.

▶ **Exercise 17.3.3** With m assets, how many nodes does the tree have after n periods?

➤ **Exercise 17.3.4** (1) Write the combinatorial formula for a European call with terminal payoff $\max(S_1 S_2 - X, 0)$. (2) How fast can it be priced?

➤ **Exercise 17.3.5 (Optimal Hedge Ratio)** Derive the optimal number of futures to short in terms of minimum variance to hedge a long stock when the two assets are not perfectly correlated. Assume the horizon is Δt from now.

➤ **Programming Assignment 17.3.6** Implement the binomial model for the option on the best of two risky assets and cash.

Additional Reading

The combinatorial methods are emphasized in [342, 838]. Major ideas in lattice combinatorics can be found in [396, 675, 686]. See [434, 696, 835] for a more detailed analysis of the BOPM. An alternative approach to the convergence problem of binomial barrier option pricing is interpolation [271]. One more idea for tackling the difficulties in pricing various kinds of barrier options is to apply trinomial trees with varying densities [9]. More research on barrier options is discussed in [131, 158, 271, 370, 443, 444, 740, 752, 824, 841, 874]. See [68, 380, 423, 756] for numerical solutions of double-barrier options. The trinomial model is due to Parkinson [713]. The trinomial models presented here are improved in [202]. Consult [423, 519] and [114] for exact and approximation formulas for options on the maximum and the minimum of two assets, respectively. Further information on multivariate derivatives pricing can be found in [450, 451, 452, 539, 628, 876].

NOTE

1. We could have adopted the finer choice of the form Sd^j $(-n \leq j \leq n)$ for the effective barrier as the algorithm in Fig. 11.4 (see Exercise 17.1.13). This was not done in order to maintain similarity to binomial option pricing formula (9.10).

CHAPTER EIGHTEEN

Numerical Methods

All science is dominated by the idea of approximation.
Bertrand Russell

Stochastic differential equations are closely related to second-order elliptic and parabolic partial differential equations [572]. Besides the binomial tree algorithms already covered, two more prominent approaches are numerical methods for partial differential equations and Monte Carlo simulation. Both are investigated in this chapter. Techniques that reduce the variance of the Monte Carlo simulation are also explored. A deterministic version of the Monte Carlo simulation, called the **quasi-Monte Carlo method**, replaces probabilistic error bounds with deterministic bounds. It is covered in some detail.

18.1 Finite-Difference Methods

The **finite-difference method** places a grid of points on the space over which the desired function takes value and then approximates the function value at each of these points. See Fig. 18.1 for illustration. The method solves the equation numerically by introducing difference equations to approximate derivatives.

Take the Poisson equation $\partial^2\theta/\partial x^2 + \partial^2\theta/\partial y^2 = -\rho(x, y)$ as an example. By replacing the second derivatives with finite differences through central difference and introducing evenly spaced grid points with distance of Δx along the x axis and Δy along the y axis, the finite-difference form is

$$\frac{\theta(x_{i+1}, y_j) - 2\theta(x_i, y_j) + \theta(x_{i-1}, y_j)}{(\Delta x)^2} + \frac{\theta(x_i, y_{j+1}) - 2\theta(x_i, y_j) + \theta(x_i, y_{j-1})}{(\Delta y)^2}$$

$$= -\rho(x_i, y_j).$$

In the preceding equation, $\Delta x \equiv x_i - x_{i-1}$ and $\Delta y \equiv y_j - y_{j-1}$ for $i, j = 1, 2, \ldots$. When the grid points are evenly spaced in both axes so that $\Delta x = \Delta y = h$, the difference equation becomes

$$\theta(x_{i+1}, y_j) + \theta(x_{i-1}, y_j) + \theta(x_i, y_{j+1}) + \theta(x_i, y_{j-1}) - 4\theta(x_i, y_j)$$

$$= -h^2\rho(x_i, y_j).$$

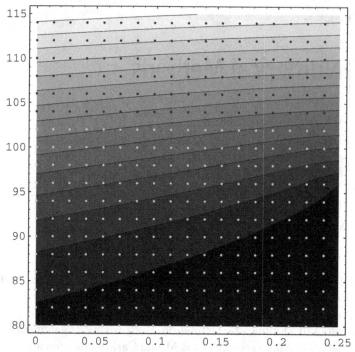

Figure 18.1: Finite-difference method. Grids are shown over the rectangle [0, 0.25] × [80, 115]. The analytical solution to the Black–Scholes differential equation for a European call is illustrated here with darker gray denoting smaller value. The strike price is $95, the expiration date is $x = 0.25$ (year), and the x axis denotes time. The finite-difference method finds values at the discrete grid points that match or approximate the analytical values.

Given the boundary values of $\theta(x, y)$ at $x = \pm L$ and $y = \pm L$, we can solve for the x_is and the y_js within the square $[\pm L, \pm L]$. From now on, $\theta_{i,j}$ denotes the finite-difference approximation to the exact $\theta(x_i, y_j)$ for clarity.

18.1.1 Explicit Methods

Consider another example, the diffusion equation $D(\partial^2\theta/\partial x^2) - (\partial\theta/\partial t) = 0$. Again, we use evenly spaced grid points (x_i, t_j) with distances Δx and Δt, where $\Delta x \equiv x_{i+1} - x_i$ and $\Delta t \equiv t_{j+1} - t_j$. We use the central difference for the second derivative and the forward difference for the time derivative to obtain

$$\left.\frac{\partial\theta(x, t)}{\partial t}\right|_{t=t_j} = \frac{\theta(x, t_{j+1}) - \theta(x, t_j)}{\Delta t} + O(\Delta t), \tag{18.1}$$

$$\left.\frac{\partial^2\theta(x, t)}{\partial x^2}\right|_{x=x_i} = \frac{\theta(x_{i+1}, t) - 2\theta(x_i, t) + \theta(x_{i-1}, t)}{(\Delta x)^2} + O[(\Delta x)^2]. \tag{18.2}$$

To assemble Eqs. (18.1) and (18.2) into a single equation at (x_i, t_j), we need to decide how to evaluate x in the first equation and t in the second. Because the central difference around x_i is used in Eq. (18.2), we might as well use x_i for x in Eq. (18.1). Two choices are possible for t in Eq. (18.2). The first choice uses $t = t_j$

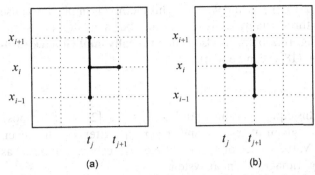

Figure 18.2: Explicit and implicit methods. Stencils of grid points for (a) explicit and (b) implicit finite-difference methods in solving a partial differential equation that is first order in t and second order in x.

to yield the following finite-difference equation:

$$\frac{\theta_{i,j+1} - \theta_{i,j}}{\Delta t} = D\frac{\theta_{i+1,j} - 2\theta_{i,j} + \theta_{i-1,j}}{(\Delta x)^2}. \tag{18.3}$$

The stencil of grid points involves four values, $\theta_{i,j+1}$, $\theta_{i,j}$, $\theta_{i+1,j}$, and $\theta_{i-1,j}$. We can therefore calculate $\theta_{i,j+1}$ from the other three, $\theta_{i,j}$, $\theta_{i+1,j}$, $\theta_{i-1,j}$, at the previous time t_j (see Fig. 18.2(a)). Starting from the initial conditions at t_0, that is, $\theta_{i,0} = \theta(x_i, t_0)$, $i = 1, 2, \ldots$, we calculate $\theta_{i,1}, i = 1, 2, \ldots$, and then $\theta_{i,2}, i = 1, 2, \ldots$, and so on. This approach is called the **explicit method** [883].

The explicit method is numerically unstable unless $\Delta t \le (\Delta x)^2/(2D)$. A numerical method is said to be **unstable** if the solution is highly sensitive to changes in initial conditions [391]. The stability condition may lead to high running times and memory requirements. For instance, doubling $(\Delta x)^{-1}$ would imply quadrupling $(\Delta t)^{-1}$, resulting in a running time eight times as much. This undesirable feature can be remedied by use of the **implicit method**, to be introduced shortly.

An interesting connection exists between the explicit method and the trinomial model. Rearrange Eq. (18.3) as

$$\theta_{i,j+1} = \frac{D\Delta t}{(\Delta x)^2}\theta_{i+1,j} + \left[1 - \frac{2D\Delta t}{(\Delta x)^2}\right]\theta_{i,j} + \frac{D\Delta t}{(\Delta x)^2}\theta_{i-1,j}.$$

When the stability condition is satisfied, the three coefficients for $\theta_{i+1,j}$, $\theta_{i,j}$, and $\theta_{i-1,j}$ all lie between zero and one and sum to one. They can therefore be interpreted as probabilities. Consequently the finite-difference equation becomes identical to backward induction on trinomial trees [473, 575].

▶ **Exercise 18.1.1** Sketch the finite-difference version of the Poisson equation in matrix form.

18.1.2 Implicit Methods

If we use $t = t_{j+1}$ in Eq. (18.2) instead, the finite-difference equation becomes

$$\frac{\theta_{i,j+1} - \theta_{i,j}}{\Delta t} = D\frac{\theta_{i+1,j+1} - 2\theta_{i,j+1} + \theta_{i-1,j+1}}{(\Delta x)^2}. \tag{18.4}$$

The stencil involves $\theta_{i,j}, \theta_{i,j+1}, \theta_{i+1,j+1}$, and $\theta_{i-1,j+1}$. This method is implicit because the value of any one of the three quantities at t_{j+1} cannot be calculated unless the other two are known (see Fig. 18.2(b)). It is also called the **fully implicit backward-difference scheme**. Equation (18.4) can be rearranged as

$$\theta_{i-1,j+1} - (2+\gamma)\,\theta_{i,j+1} + \theta_{i+1,j+1} = -\gamma\theta_{i,j},$$

where $\gamma \equiv (\Delta x)^2/(D\Delta t)$. This equation is unconditionally stable [28, 464]. Suppose the boundary conditions are given at $x = x_0$ and $x = x_{N+1}$. After $\theta_{i,j}$ has been calculated for $i = 1, 2, \ldots, N$, the values of $\theta_{i,j+1}$ at time t_{j+1} can be computed as the solution to the following tridiagonal linear system:

$$\begin{bmatrix} a & 1 & 0 & \cdots & \cdots & \cdots & 0 \\ 1 & a & 1 & 0 & \cdots & \cdots & 0 \\ 0 & 1 & a & 1 & 0 & \cdots & 0 \\ \vdots & \ddots & \ddots & \ddots & \ddots & \ddots & \vdots \\ \vdots & \ddots & \ddots & \ddots & \ddots & \ddots & \vdots \\ 0 & \cdots & \cdots & 0 & 1 & a & 1 \\ 0 & \cdots & \cdots & \cdots & 0 & 1 & a \end{bmatrix} \begin{bmatrix} \theta_{1,j+1} \\ \theta_{2,j+1} \\ \theta_{3,j+1} \\ \vdots \\ \vdots \\ \vdots \\ \theta_{N,j+1} \end{bmatrix} = \begin{bmatrix} -\gamma\theta_{1,j} - \theta_{0,j+1} \\ -\gamma\theta_{2,j} \\ -\gamma\theta_{3,j} \\ \vdots \\ \vdots \\ -\gamma\theta_{N-1,j} \\ -\gamma\theta_{N,j} - \theta_{N+1,j+1} \end{bmatrix},$$

where $a \equiv -2 - \gamma$. Tridiagonal systems can be solved in $O(N)$ time and $O(N)$ space [35]. The preceding matrix is nonsingular when $\gamma \geq 0$. Recall that a square matrix is **nonsingular** if its inverse exists.

Taking the average of explicit method (18.3) and implicit method (18.4) results in

$$\frac{\theta_{i,j+1} - \theta_{i,j}}{\Delta t} = \frac{1}{2}\left[D\frac{\theta_{i+1,j} - 2\theta_{i,j} + \theta_{i-1,j}}{(\Delta x)^2} + D\frac{\theta_{i+1,j+1} - 2\theta_{i,j+1} + \theta_{i-1,j+1}}{(\Delta x)^2} \right].$$

After rearrangement, the **Crank–Nicolson method** emerges:

$$\gamma\theta_{i,j+1} - \frac{\theta_{i+1,j+1} - 2\theta_{i,j+1} + \theta_{i-1,j+1}}{2} = \gamma\theta_{i,j} + \frac{\theta_{i+1,j} - 2\theta_{i,j} + \theta_{i-1,j}}{2}.$$

This is an unconditionally stable implicit method with excellent rates of convergence [810].

18.1.3 Numerically Solving the Black–Scholes Differential Equation

We focus on American puts; the technique, however, can be applied to any derivative satisfying the Black–Scholes differential equation as only the initial and the boundary conditions need to be changed.

The Black–Scholes differential equation for American puts is

$$\frac{1}{2}\sigma^2 S^2 \frac{\partial^2 P}{\partial S^2} + (r-q)\,S\,\frac{\partial P}{\partial S} - rP + \frac{\partial P}{\partial t} = 0$$

with $P(S, T) = \max(X - S, 0)$ and $P(S, t) = \max(\overline{P}(S, t), X - S)$ for $t < T$. $\overline{P}$ denotes the option value at time t if it is not exercised for the next instant of time.

After the change of variable $V \equiv \ln S$, the option value becomes $U(V, t) \equiv P(e^V, t)$ and

$$\frac{\partial P}{\partial t} = \frac{\partial U}{\partial t}, \quad \frac{\partial P}{\partial S} = \frac{1}{S} \frac{\partial U}{\partial V}, \quad \frac{\partial^2 P}{\partial^2 S} = \frac{1}{S^2} \frac{\partial^2 U}{\partial V^2} - \frac{1}{S^2} \frac{\partial U}{\partial V}.$$

The Black–Scholes differential equation is now transformed into

$$\frac{1}{2} \sigma^2 \frac{\partial^2 U}{\partial V^2} + \left(r - q - \frac{\sigma^2}{2} \right) \frac{\partial U}{\partial V} - rU + \frac{\partial U}{\partial t} = 0$$

subject to $U(V, T) = \max(X - e^V, 0)$ and $U(V, t) = \max(\overline{U}(V, t), X - e^V)$, $t < T$. Along the V axis, the grid will span from V_{min} to $V_{min} + N \times \Delta V$ at ΔV apart for some suitably small V_{min}; hence boundary conditions at the lower ($V = V_{min}$) and upper ($V = V_{min} + N \times \Delta V$) boundaries will have to be specified. Finally, S_0 as usual denotes the current stock price.

The Explicit Scheme

The explicit scheme for the Black–Scholes differential equation is

$$\frac{1}{2} \sigma^2 \frac{U_{i+1,j} - 2U_{i,j} + U_{i-1,j}}{(\Delta V)^2} + \left(r - q - \frac{\sigma^2}{2} \right) \frac{U_{i+1,j} - U_{i-1,j}}{2\Delta V}$$

$$- rU_{i,j} + \frac{U_{i,j} - U_{i,j-1}}{\Delta t} = 0$$

for $1 \leq i \leq N - 1$. Note that the computation moves backward in time. There are $N - 1$ difference equations. Regroup the terms to obtain

$$U_{i,j-1} = aU_{i-1,j} + bU_{i,j} + cU_{i+1,j},$$

where

$$a \equiv \left[\left(\frac{\sigma}{\Delta V} \right)^2 - \frac{r - q - \sigma^2/2}{\Delta V} \right] \frac{\Delta t}{2}, \quad b \equiv 1 - r\Delta t - \left(\frac{\sigma}{\Delta V} \right)^2 \Delta t,$$

$$c \equiv \left[\left(\frac{\sigma}{\Delta V} \right)^2 + \frac{r - q - \sigma^2/2}{\Delta V} \right] \frac{\Delta t}{2}.$$

These $N - 1$ equations express option values at time step $j - 1$ in terms of those at time step j. For American puts, we assume for U's lower boundary that the first derivative at grid point $(0, j)$ for every time step j equals $-e^{V_{min}}$. This essentially makes the put value $X - S = X - e^V$, so $U_{0,j-1} = U_{1,j-1} + (e^{V_{min}+\Delta V} - e^{V_{min}})$. For the upper boundary, we set $U_{N,j-1} = 0$. The put's value at any grid point at time step $j - 1$ is therefore an explicit function of its values at time step j. Finally $U_{i,j}$ is set to the greater of the value derived above and $X - e^{V_{min}+i \times \Delta V}$ for early-exercise considerations. Repeating this process as we move backward in time, we will eventually arrive at the solution at time zero, $U_{k,0}$, where k is the integer so that $V_{min} + k \times \Delta V$ is closest to $\ln S_0$. As implied by the stability condition, given ΔV, the value of Δt must be small enough for the method to converge (see Fig. 18.3). The formal conditions to satisfy are $a > 0$, $b > 0$, and $c > 0$.

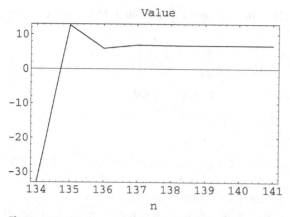

Figure 18.3: Convergence of the explicit method. Here $\Delta V =$ 0.031073 and $\Delta t = 5/(6n)$. With $n \geq 137$, the numerical solution converges to the analytical European put value 6.777986.

The explicit method evaluates all the grid points in a rectangle. In practice, we are interested in only the single grid point at time zero, $(0, k)$, that corresponds to the current stock price. The grid points that may influence the desired value form a triangular subset of the rectangle. This triangle could be truncated further by the two boundary conditions (see Fig. 18.4). Only those points within the truncated triangle need be evaluated.

➤ **Exercise 18.1.2** What are the terminal conditions?

➤ **Exercise 18.1.3** Repeat the steps for American calls.

➤ **Exercise 18.1.4** Derive the stability conditions for the explicit approach to solve the Black–Scholes differential equation. Assume $q = 0$ for simplicity.

➤ **Programming Assignment 18.1.5** Implement the explicit method for American puts.

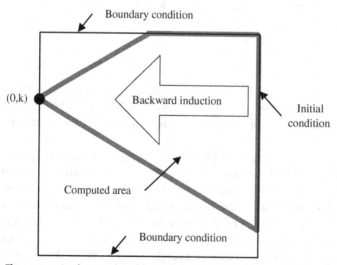

Figure 18.4: Implementation of the explicit method. Only the truncated triangle within the rectangle needs to have its grid points evaluated.

The Implicit Scheme

The partial differential equation now becomes the following $N-1$ difference equations,

$$\frac{1}{2}\sigma^2 \frac{U_{i+1,j}-2U_{i,j}+U_{i-1,j}}{(\Delta V)^2} + \left(r-q-\frac{\sigma^2}{2}\right)\frac{U_{i+1,j}-U_{i-1,j}}{2\Delta V}$$

$$-rU_{i,j} + \frac{U_{i,j+1}-U_{i,j}}{\Delta t} = 0$$

for $1 \le i \le N-1$. Regroup the terms to obtain

$$aU_{i-1,j} + bU_{i,j} + cU_{i+1,j} = U_{i,j+1},$$

where

$$a \equiv \left[-\left(\frac{\sigma}{\Delta V}\right)^2 + \frac{r-q-\sigma^2/2}{\Delta V}\right]\frac{\Delta t}{2}, \quad b \equiv 1+r\Delta t + \left(\frac{\sigma}{\Delta V}\right)^2 \Delta t,$$

$$c \equiv -\left[\left(\frac{\sigma}{\Delta V}\right)^2 + \frac{r-q-\sigma^2/2}{\Delta V}\right]\frac{\Delta t}{2}.$$

The system of equations can be written in matrix form:

$$\begin{bmatrix} b^* & c & 0 & \cdots & \cdots & \cdots & 0 \\ a & b & c & 0 & \cdots & \cdots & 0 \\ 0 & a & b & c & 0 & \cdots & 0 \\ \vdots & \ddots & \ddots & \ddots & \ddots & \ddots & \vdots \\ \vdots & \ddots & \ddots & \ddots & \ddots & \ddots & \vdots \\ 0 & \cdots & \cdots & 0 & a & b & c \\ 0 & \cdots & \cdots & \cdots & 0 & a & b \end{bmatrix} \begin{bmatrix} U_{1,j} \\ U_{2,j} \\ U_{3,j} \\ \vdots \\ \vdots \\ \vdots \\ U_{N-1,j} \end{bmatrix} = \begin{bmatrix} U_{1,j+1}-K \\ U_{2,j+1} \\ U_{3,j+1} \\ \vdots \\ \vdots \\ U_{N-2,j+1} \\ U_{N-1,j+1} \end{bmatrix},$$

where $b^* \equiv a+b$ and $K \equiv a(e^{V_{\min}+\Delta V} - e^{V_{\min}})$. We can obtain the values of $U_{1,j}, U_{2,j}, \ldots, U_{N-1,j}$ by inverting the tridiagonal matrix. As before, at every time step and before going to the next, we should set the option value obtained to the intrinsic value of the option if the latter is larger.

> **Programming Assignment 18.1.6** Implement the implicit method for American puts.

18.2 Monte Carlo Simulation

Monte Carlo simulation is a sampling scheme. In many important applications within finance and without, Monte Carlo simulation is one of the few feasible tools. It is also one of the most important elements of studying econometrics [550]. When the time evolution of a stochastic process is not easy to describe analytically, Monte Carlo simulation may very well be the only strategy that succeeds consistently [386].

Assume that $X_1, X_2, \ldots, X_n$ have a joint distribution and $\theta \equiv E[g(X_1, X_2, \ldots, X_n)]$ for some function g is desired. We generate

$$\left(x_1^{(i)}, x_2^{(i)}, \ldots, x_n^{(i)}\right), \quad 1 \le i \le N,$$

independently with the same joint distribution as $(X_1, X_2, \ldots, X_n)$ and set

$$Y_i \equiv g\left(x_1^{(i)}, x_2^{(i)}, \ldots, x_n^{(i)}\right).$$

Now $Y_1, Y_2, \ldots, Y_N$ are independent and identically distributed random variables, and each Y_i has the same distribution as that of $Y \equiv g(X_1, X_2, \ldots, X_n)$. Because the average of these N random variables, $\overline{Y}$, satisfies $E[\overline{Y}] = \theta$, it can be used to estimate θ. In fact, the **strong law of large numbers** says that this procedure converges almost surely [699]. The number of **replications** (or *independent* trials), N, is called the **sample size**.

EXAMPLE 18.2.1 To evaluate the definite integral $\int_a^b g(x)\,dx$ numerically, consider the random variable $Y \equiv (b - a)\,g(X)$, where X is uniformly distributed over $[a, b]$. Note that $\text{Prob}[\,X \leq x\,] = (x - a)/(b - a)$ for $a \leq x \leq b$. Because

$$E[\,Y\,] = (b - a)E[\,g(X)\,] = (b - a)\int_a^b \frac{g(x)}{b - a}\,dx = \int_a^b g(x)\,dx,$$

any unbiased estimator of $E[\,Y\,]$ can be used to evaluate the integral.

The Monte Carlo estimate and the true value may differ owing to two reasons: sampling variation and the discreteness of the sample paths. The former can be controlled by the number of replications, as we shall see in the following paragraph, and the latter can be controlled by the number of observations along the sample path [147].

The statistical error of the sample mean $\overline{Y}$ of the random variable Y grows as $1/\sqrt{N}$ because $\text{Var}[\,\overline{Y}\,] = \text{Var}[\,Y\,]/N$. In fact, this convergence rate is asymptotically optimal by the **Berry–Esseen theorem** [413]. As a result, the variance of the estimator $\overline{Y}$ can be reduced by a factor of $1/N$ by doing N times as much work [721]. This property is amazing because the same order of convergence holds independently of the dimension n. In contrast, classic numerical integration schemes have an error bound of $O(N^{-c/n})$ for some constant $c > 0$. The required number of evaluations thus grows exponentially in n to achieve a given level of accuracy. This is a case of the curse of dimensionality. The Monte Carlo method, for example, is more efficient than alternative procedures for securities depending on more than one asset, the multivariate derivatives [530].

The statistical efficiency of Monte Carlo simulation can be measured by the variance of its output. If this variance can be lowered without changing the expected value, fewer replications are needed. Methods that improve efficiency in this manner are called **variance-reduction techniques**. Such techniques, covered in Subsection 18.2.3, become practical when the added costs are outweighed by the reduction in sampling.

18.2.1 Monte Carlo Option Pricing

For the pricing of European options on a dividend-paying stock, we may proceed as follows. From Eq. (14.17), stock prices $S_1, S_2, S_3, \ldots$, at times $\Delta t, 2\Delta t, 3\Delta t, \ldots$, can be generated by

$$S_{i+1} = S_i e^{(\mu - \sigma^2/2)\,\Delta t + \sigma\sqrt{\Delta t}\,\xi}, \quad \xi \sim N(0, 1)$$

Monte Carlo method for pricing average-rate calls on a non-dividend-paying stock:

input: $S, X, n, r, \sigma, \tau, m;$
real $P, C, M;$
real $\xi(\,)$; $//\,\xi(\,) \sim N(0, 1).$
integer $i, j;$
$C := 0;$ // Accumulated terminal option value.
for $(i = 1$ to $m)$ { // Perform m replications.
      $P := S; M := S;$
      for $(j = 1$ to $n)$ {
            $P := P \times e^{(r - \sigma^2/2)(\tau/n) + \sigma\sqrt{\tau/n}\,\xi(\,)};$
            $M := M + P;$
      }
      $C := C + \max(M/(n + 1) - X, 0);$
}
return $Ce^{-r\tau}/m;$

Figure 18.5: Monte Carlo method for average-rate calls. m is the number of replications, and n is the number of periods.

when $dS/S = \mu\, dt + \sigma\, dW$. We can generate non-dividend-paying stock prices in a risk-neutral economy by setting $\mu = r$. Figure 18.5 contains a pricing algorithm for arithmetic average-rate calls.

The sample standard deviation of the estimation scheme in Fig. 18.5 is proportional to $1/\sqrt{m}$, where m is the number of replications. To narrow down the confidence interval by a factor of f, f^2 times as many replications need to be carried out. Although we do not know how small $\Delta t \equiv \tau/n$ should be to yield acceptable approximations, it is not hard to figure out m. Because the estimate is composed of a simple average across replications, the central limit theorem says that the error of the estimate is distributed as $N(0, s^2/m)$, with s^2 denoting the variance of each replication. Hence the confidence interval can be used to derive the desired m.

The discreteness of sample paths and the variance in prices do not necessarily make Monte Carlo results inferior to closed-form solutions. The judgment ultimately depends on the security being priced. In reality, for instance, a case may be made that, as prices do not move continuously, discrete-time models are more appropriate.

Monte Carlo simulation is a general methodology. It can be used to value virtually any European-style derivative security [147]. A standard Monte Carlo simulation, however, is inappropriate for American options because of early exercise: It is difficult to determine the early-exercise point based on one single path. Intriguingly, Tilley showed that Monte Carlo simulation can be modified to price American options [842]; the estimate is biased, however [108].

▶ **Exercise 18.2.1** How do we price European barrier options by Monte Carlo simulation?

▶ **Exercise 18.2.2** Consider the Monte Carlo method that estimates the price of the American call by taking the maximum discounted intrinsic value per simulated path and then averaging them: $E[\max_{i=0,1,\dots,n} e^{-ri\Delta t} \max(S_i - X, 0)]$. Show that it is biased high.

Monte Carlo simulation of Ito process:

```
input:   x_0, T, Δt;
real     X[0..⌈T/Δt⌉];
real     ξ();   // ξ() ~ N(0, 1).
integer  i;
X[0] := x_0;  // Initial state.
for (i = 1 to ⌈T/Δt⌉)
        X[i] := X[i − 1] + a(X[i − 1]) Δt + b(X[i − 1])√Δt ξ();
return X[];
```

Figure 18.6: Monte Carlo simulation of Ito process. The Ito process is $dX_t = a(X_t) \, dt + b(X_t) \, dW_t$. A run of the algorithm generates an approximate sample path for the process.

➤ **Programming Assignment 18.2.3** Implement the Monte Carlo method for arithmetic average-rate calls and puts.

18.2.2 Ito Processes

Consider the stochastic differential equation $dX_t = a(X_t) \, dt + b(X_t) \, dW_t$. Although it is often difficult to give an analytic solution to this equation, the simulation of the process on a computer is relatively easy [585]. Recall that Euler's method picks a small number Δt and then approximates the Ito process by

$$\widehat{X}(t_{n+1}) = \widehat{X}(t_n) + a(\widehat{X}(t_n))\Delta t + b(\widehat{X}(t_n))\sqrt{\Delta t}\, \xi,$$

where $\xi \sim N(0, 1)$. See Fig. 18.6 for the algorithm. This simulation is exact for any Δt if both the drift a and the diffusion b are constants as in Brownian motion because the sum of independent normal distributions remains normal.

➤ **Exercise 18.2.4** The Monte Carlo method for Ito processes in Fig. 18.6 may not be the most ideal theoretically. Consider the geometric Brownian motion $dX/X = \mu \, dt + \sigma \, dW$. Assume that you have access to a perfect random-number generator for normal distribution. Find a theoretically better algorithm to generate sample paths for X.

➤ **Programming Assignment 18.2.5** Simulate $dX_t = (0.06 − X_t) \, dt + 0.3 \, dW_t$ by using $\Delta t \equiv 0.01$. Explain its dynamics.

Discrete Approximations to Ito Processes with Brownian Bridge

Besides the Euler method and the related approximation methods in Subsection 14.2.1, a Brownian bridge is one more alternative. Let the time interval $[0, T]$ be partitioned at time points $t_0 = 0, t_1, t_2, \ldots$. Instead of using

$$W(t_n) = W(t_{n-1}) + \sqrt{t_n - t_{n-1}}\, \xi, \quad \xi \sim N(0, 1)$$

to generate the discrete-time Wiener process, the new method uses

$$W(t_n) = \frac{t_{n+1} - t_n}{t_{n+1} - t_{n-1}} W(t_{n-1}) + \frac{t_n - t_{n-1}}{t_{n+1} - t_{n-1}} W(t_{n+1}) + \sqrt{\frac{(t_{n+1} - t_n)(t_n - t_{n-1})}{t_{n+1} - t_{n-1}}}\, \xi,$$

given a past value $W(t_{n-1})$ and a future value $W(t_{n+1})$. In general, the method determines a sample path $W(i(T/2^m))$, $i = 0, 1, \ldots, 2^m$, over $[0, T]$ as follows. First, set $W(0) = 0$ and $W(T) = \sqrt{T}\, \xi$. Then set the midpoint $W(T/2)$ according to the preceding equation. From here, we find the midpoints for $[W(0), W(T/2)]$ and $[W(T/2), W(T)]$, that is, $W(T/4)$ and $W(3T/4)$, respectively. Iterate for $m - 2$ more times. This scheme increases the accuracy of quasi-Monte Carlo simulation to be introduced shortly by reducing its effective dimension [6, 142, 679, 876].

> **Programming Assignment 18.2.6** Implement the Brownian bridge approach to generate the sample path of geometric Brownian motion.

18.2.3 Variance-Reduction Techniques

The success of variance-reduction schemes depends critically on the particular problem of interest. Because it is usually impossible to know beforehand how great a reduction in variance may be realized, if at all, preliminary runs should be made to compare the results of a variance-reduction scheme with those from standard Monte Carlo simulation.

Antithetic Variates

Suppose we are interested in estimating $E[g(X_1, X_2, \ldots, X_n)]$, where $X_1, X_2, \ldots, X_n$ are independent random variables. Let Y_1 and Y_2 be random variables with the same distribution as $g(X_1, X_2, \ldots, X_n)$. Then

$$\mathrm{Var}\left[\frac{Y_1 + Y_2}{2}\right] = \frac{\mathrm{Var}[Y_1]}{2} + \frac{\mathrm{Cov}[Y_1, Y_2]}{2}.$$

Note that $\mathrm{Var}[Y_1]/2$ is the variance of the Monte Carlo method with two (independent) replications. The variance $\mathrm{Var}[(Y_1 + Y_2)/2]$ is smaller than $\mathrm{Var}[Y_1]/2$ when Y_1 and Y_2 are negatively correlated instead of being independent.

The **antithetic-variates** technique is based on the above observation. First, simulate $X_1, X_2, \ldots, X_n$ by means of the **inverse-transform technique**. That is, X_i is generated by $F_i^{-1}(U_i)$, where U_i is a random number uniformly distributed over $(0, 1)$ and F_i is the distribution function of X_i. Set

$$Y_1 \equiv g\left(F_1^{-1}(U_1), \ldots, F_n^{-1}(U_n)\right).$$

Because $1 - U$ is also uniform over $(0, 1)$ whenever U is, it follows that

$$Y_2 \equiv g\left(F_1^{-1}(1 - U_1), \ldots, F_n^{-1}(1 - U_n)\right)$$

has the same distribution as Y_1. When g is a monotone function, Y_1 and Y_2 are indeed negatively correlated and the antithetic-variates estimate,

$$\frac{g\left(F_1^{-1}(U_1), \ldots, F_n^{-1}(U_n)\right) + g\left(F_1^{-1}(1 - U_1), \ldots, F_n^{-1}(1 - U_n)\right)}{2},$$

has a lower variance than the Monte Carlo method with two replications [764]. Computation time is also saved because only n rather than $2n$ random numbers need to be generated, with each number used twice.

In general, for each simulated sample path X, we obtain a second one by reusing the random numbers on which the first path is based, yielding a second sample path Y. Two estimates are then obtained, one based on X and the other on Y. If a total of N independent sample paths are generated, the antithetic-variates estimator averages over $2N$ estimates.

EXAMPLE 18.2.2 Consider the Ito process $dX = a_t\, dt + b_t \sqrt{dt}\, \xi$. Let g be a function of n samples $X_1, X_2, \ldots, X_n$ on the sample path. We are interested in $E[\, g(X_1, X_2, \ldots, X_n)\,]$. Suppose that one simulation run has realizations $\xi_1, \xi_2, \ldots, \xi_n$ for the normally distributed fluctuation term ξ, generating samples $x_1, x_2, \ldots, x_n$. The estimate is then $g(x)$, where $x \equiv (x_1, x_2 \ldots, x_n)$. Instead of sampling n more numbers from ξ for the second estimate, the antithetic-variates method computes $g(x')$ from the sample path $x' \equiv (x'_1, x'_2 \ldots, x'_n)$ generated by $-\xi_1, -\xi_2, \ldots, -\xi_n$ and outputs $[\, g(x) + g(x')\,]/2$. Figure 18.7 implements the antithetic-variates method for average-rate options.

▶ **Exercise 18.2.7** Justify and extend the procedure in Example 18.2.2.

▶ **Programming Assignment 18.2.8** Implement the antithetic-variates method for arithmetic average-rate calls and puts. Compare it with the Monte Carlo method in Programming Assignment 18.2.3.

Antithetic variates for pricing average-rate calls on a non-dividend-paying stock:

```
input:   S, X, n, r, σ, τ, m;
real     P₁, P₂, C, M₁, M₂, a;
real     ξ( );    // ξ( ) ~ N(0, 1).
integer  i, j;
C := 0;
for (i = 1 to m) {
        P₁ := S; P₂ := S; M₁ := S; M₂ := S;
        for (j = 1 to n) {
                a := ξ( );
                P₁ := P₁e^((r−σ²/2)(τ/n)+σ√(τ/n) a);
                P₂ := P₂e^((r−σ²/2)(τ/n)−σ√(τ/n) a);
                M₁ := M₁ + P₁;
                M₂ := M₂ + P₂;
        }
        C := C + max(M₁/(n + 1) − X, 0);
        C := C + max(M₂/(n + 1) − X, 0);
}
return Ce^(−rτ)/(2m);
```

Figure 18.7: Antithetic-variates method for average-rate calls. P_1 keeps track of the first sample path, P_2 keeps track of the second sample path, m is the number of replications, and n is the number of periods.

Conditioning

Let X be a random variable whose expectation is to be estimated. There is another random variable Z such that the conditional expectation $E[X|Z=z]$ can be efficiently and precisely computed. We have $E[X] = E[E[X|Z]]$ by the law of iterated conditional expectations. Hence the random variable $E[X|Z]$ is also an unbiased estimator of μ. As $\text{Var}[E[X|Z]] \leq \text{Var}[X]$, $E[X|Z]$ indeed has a smaller variance than we obtain by observing X directly [764]. The computing procedure is to first obtain a random observation z on Z, then calculate $E[X|Z=z]$ as our estimate. There is no need to resort to simulation in computing $E[X|Z=z]$. The procedure can be repeated a few times to reduce the variance.

➢ **Programming Assignment 18.2.9** Apply conditioning to price European options when the stock price volatility is stochastic. The stock price and its volatility may be correlated.

Control Variates

The idea of **control variates** is to use the analytic solution of a similar yet simpler problem to improve the solution. Suppose we want to estimate $E[X]$ and there exists a random variable Y with a known mean $\mu \equiv E[Y]$. Then $W \equiv X + \beta(Y - \mu)$ can serve as a "controlled" estimator of $E[X]$ for any constant β that scales the deviation $Y - \mu$ to arrive at an adjustment for X. However β is chosen, W remains an unbiased estimator of $E[X]$. As

$$\text{Var}[W] = \text{Var}[X] + \beta^2 \text{Var}[Y] + 2\beta \text{Cov}[X, Y], \tag{18.5}$$

W is less variable than X if and only if

$$\beta^2 \text{Var}[Y] + 2\beta \text{Cov}[X, Y] < 0. \tag{18.6}$$

The success of the scheme clearly depends on both β and the choice of Y. For example, arithmetic average-rate options can be priced if Y is chosen to be the otherwise identical geometric average-rate option's price and $\beta = -1$ [548]. This approach is much more effective than the antithetic-variates method (see Fig. 18.8) [108].

Equation (18.5) is minimized when β equals $\beta^* \equiv -\text{Cov}[X, Y]/\text{Var}[Y]$, which was called beta earlier in Exercise 6.4.1. For this specific β,

$$\text{Var}[W] = \text{Var}[X] - \frac{\text{Cov}[X, Y]^2}{\text{Var}[Y]} = (1 - \rho_{X,Y}^2) \text{Var}[X],$$

where $\rho_{X,Y}$ is the correlation between X and Y. The stronger X and Y are correlated, the greater the reduction in variance. For example, if this correlation is nearly

Figure 18.8: Variance-reduction techniques for average-rate puts. An arithmetic average-rate put is priced with antithetic variates and control variates ($\beta = -1$). The parameters used for each set of data are $S = 50$, $\sigma = 0.2$, $r = 0.05$, $\tau = 1/3$, $X = 50$, $n = 50$, and $m = 10000$. Sample standard deviations of the computed values are in parentheses.

Antithetic variates		Control variates	
1.11039	1.11452	1.11815	1.11864
1.10952	1.10892	1.11788	1.11853
1.10476	1.10574	1.11856	1.11789
1.13225	1.10509	1.11852	1.11868
(0.009032505)		(0.000331789)	

perfect (± 1), we could control X almost exactly, eliminating practically all of its variance. Typically, neither Var[Y] nor Cov[X, Y] is known, unfortunately. Therefore we usually cannot obtain the maximum reduction in variance. One approach in practice is to guess at these values and hope that the resulting W does indeed have a smaller variance than X. A second possibility is to use the simulated data to estimate these quantities.

Observe that $-\beta^*$ has the same sign as the correlation between X and Y. Hence, if X and Y are positively correlated, $\beta^* < 0$; then X is adjusted downward whenever $Y > \mu$ and upward otherwise. The opposite is true when X and Y are negatively correlated, in which case $\beta^* > 0$.

➤ **Exercise 18.2.10** Pick $\beta = \pm 1$. The success of the scheme now depends solely on the choice of Y. Derive the conditions under which the variance is reduced.

➤ **Exercise 18.2.11** Why is it a mistake to use *independent* random numbers in generating X and Y?

➤ **Programming Assignment 18.2.12** Implement the control-variates method for arithmetic average-rate calls and puts.

Other Schemes

Two more schemes are briefly mentioned before this section closes. In **stratified sampling**, the support of the random variable being simulated is partitioned into a finite number of disjoint regions and a standard Monte Carlo simulation is performed in each region. When there is less variance within regions than across the regions, the sampling variance of the estimate will be reduced. **Importance sampling** samples more frequently in regions of the support where there is more variation.

➤ **Exercise 18.2.13** Suppose you are searching in set A for any element from set $B \subseteq A$. The Monte Carlo approach selects N elements randomly from A and checks if any one belongs to B. An alternative partitions the set A into m disjoint subsets $A_1, A_2, \ldots, A_m$ of equal size, picks N/m elements from each subset randomly, and checks if there is a hit. Prove that the second approach's probability of failure can never exceed that of the Monte Carlo approach.

18.3 Quasi-Monte Carlo Methods

The basic Monte Carlo method evaluates integration at randomly chosen points. There are several deficiencies with this paradigm. To start with, the error bound is probabilistic, not a concrete guarantee about the accuracy. The probabilistic error bound of $\sqrt{N}$ furthermore does not benefit from any additional regularity of the integrand function. Another fundamental difficulty stems from the requirement that the points be independent random samples. Random samples are wasteful because of clustering (see Fig. 18.9); indeed, Monte Carlo simulations with very small sample sizes cannot be trusted. Worse, truly random numbers do not exist on digital computers; in reality, **pseudorandom numbers** generated by completely *deterministic* means are used instead. Monte Carlo simulation exhibits a great sensitivity on the seed of the pseudorandom-number generator. The **low-discrepancy sequences**, also known as **quasi-random sequences**,[1] address the above-mentioned problems.

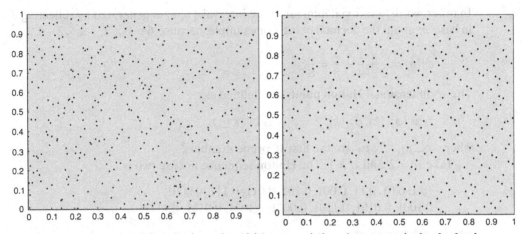

Figure 18.9: Random points (left) and Halton points (right) compared. The points are over the $[0,1] \times [0,1]$ rectangle [868]. Note the clustering of points on the left.

Proposed in the 1950s, the quasi-Monte Carlo method leaves nothing to chance [699]. In fact, it can be viewed as the deterministic version of the Monte Carlo method in that random samples are replaced with deterministic quasi-random points. If a smaller number of samples suffices as a result, efficiency has been gained. The main aim hence is to select deterministic points for which the deterministic error bound is smaller than Monte Carlo's probabilistic error bound.

The quasi-Monte Carlo method is not without limitations. Their theories are valid for integration problems, but may not be directly applicable to simulations because of the correlations between points in a quasi-random sequence. This problem can be overcome in many cases if the desired result is written as an integral. However, the resulting integral often has a very high dimension (e.g., 360 for a 30-year mortgage); in fact, the improved accuracy is generally lost for problems of high dimension or problems in which the integrand is not smooth. There is furthermore no theoretical basis for empirical estimates of their accuracy, a role played by the central limit theorem in the Monte Carlo method [142].

Although the results are somewhat mixed, the application of such methods in finance seems promising [108, 147]. A speed-up as high as 1,000 over the Monte Carlo method, for example, is reported in [711]. The success of the quasi-Monte Carlo method when compared with traditional variance-reduction techniques depends on the problem in question [715, 716]. For example, the antithetic-variates method outperforms the quasi-Monte Carlo method in bond pricing [679, 868].

18.3.1 The Halton Sequence

Every integer $k \geq 1$ can be represented uniquely as

$$k = a_0 + a_1 m + a_2 m^2 + \cdots + a_r m^r,$$

where $a_i \in [0, m-1]$ are integers and r is chosen such that $m^r \leq k < m^{r+1}$. The **radical inverse function in base** m is defined by

$$\phi_m(k) \equiv a_0 m^{-1} + a_1 m^{-2} + a_2 m^{-3} + \cdots + a_r m^{-(r+1)}.$$

In other words, a rational number in the interval $[0, 1)$ is generated by reflecting k in base m about the decimal point. For example, because $6 = 0 \times 2^0 + 1 \times 2 + 1 \times 2^2$,

$$\phi_2(6) = \frac{0}{2} + \frac{1}{4} + \frac{1}{8} = \frac{3}{8}.$$

The d-dimensional **Halton points** are defined as

$$z_k \equiv (\phi_{p_1}(k), \phi_{p_2}(k), \ldots, \phi_{p_d}(k)), \quad k \geq 1,$$

where $p_1, p_2, \ldots, p_d$ are the first d prime numbers.

Take $m = 2$ and $d = 1$. The Halton sequence in base two is

$$(0.1, 0.01, 0.11, 0.001, 0.101, 0.011, 0.111, 0.0001,$$
$$0.1001, 0.0101, 0.1101, 0.0011, 0.1011, 0.0111, 0.1111, \ldots),$$

which corresponds to

$$\left(\frac{1}{2}, \frac{1}{4}, \frac{3}{4}, \frac{1}{8}, \frac{5}{8}, \frac{3}{8}, \frac{7}{8}, \frac{1}{16}, \frac{9}{16}, \frac{5}{16}, \frac{13}{16}, \frac{3}{16}, \frac{11}{16}, \frac{7}{16}, \frac{15}{16}, \ldots \right).$$

Although of no concern to numerical integration, the sequence above does not look useful for certain simulations. Write the numbers in decimal as 0.5, 0.25, 0.75, 0.125, 0.625, 0.375, 0.875, ... Were we to use them in simulating a symmetric random walk, it would alternate between right and left moves because the sequence consists eventually of pairs of monotonically increasing numbers: 0.5, 0.25 < 0.75, 0.125 < 0.625, 0.375 < 0.875, ... This phenomenon holds in general as a Halton sequence consists eventually of size-m subsequences of monotonically increasing numbers. For these reasons, the **scrambled Halton sequence** that permutes the a_is in $\phi_m(k)$ may be needed. Many quasi-Monte Carlo simulations cannot work without such manipulation [868]. Typically, the first 10 to 200 Halton points are discarded [368].

▶ **Exercise 18.3.1** Compute the first 10 one-dimensional Halton points in base 3.

▶ **Programming Assignment 18.3.2** Implement the two-dimensional Halton sequence, apply it to numerically evaluating $\int_0^1 x^2 \, dx$, and compare it with Monte Carlo integration.

18.3.2 The Sobol' Sequence

Assume that $d = 1$ initially. A one-dimensional Sobol' sequence of length N is generated from $\omega \equiv \lceil \log_2 N \rceil$ **direction numbers** $0 < v_1, v_2, \ldots, v_\omega < 1$ through the bitwise **exclusive-OR** operation $\oplus$.[2] Each point has the form

$$b_1 v_1 \oplus b_2 v_2 \oplus \cdots \oplus b_\omega v_\omega,$$

where b_i are zero or one. Each point is therefore the result of exclusive-ORing a subset of the direction numbers, those v_i whose corresponding b_i is one. The sequence of $(b_1, b_2, \ldots, b_\omega)$ can either be $1, 2, \ldots, N$ in binary form or the **Gray code** [727]. The sequence of direction numbers is generated by a primitive polynomial with coefficients in the field Z_2 (whose elements are 0 and 1, and all operations are modulo 2). **Primitive polynomials** are, roughly speaking, polynomials that cannot be factored. Consider a primitive polynomial $P(x) \equiv x^n + a_1 x^{n-1} + \cdots + a_{n-1} x + 1$ of degree n. The direction numbers are obtained from this recurrence formula:

$$v_i = a_1 v_{i-1} \oplus a_2 v_{i-2} \oplus \cdots \oplus a_{n-1} v_{i-n+1} \oplus v_{i-n} \oplus (v_{i-n}/2^n), \quad i > n.$$

To jump-start the recurrence, we need to specify $v_1, v_2, \ldots, v_n$. We do this by setting $v_i = m_i/2^i$ for n arbitrary odd integers $0 < m_i < 2^i$ $(1 \le i \le n)$. Finally, the Sobol' sequence $z_0, z_2, \ldots, z_{N-1}$ is obtained recursively by

$$\begin{cases} z_0 = 0 \\ z_{k+1} = z_k \oplus v_c \end{cases},$$

where c is the position of the rightmost zero bit in the binary representation of k.

EXAMPLE 18.3.1 Consider the primitive polynomial $x^6 + x^4 + x^3 + x + 1$. For $i > 7$, the equation for the direction numbers is

$$v_i = 0 \cdot v_{i-1} \oplus 1 \cdot v_{i-2} \oplus 1 \cdot v_{i-3} \oplus 0 \cdot v_{i-4} \oplus 1 \cdot v_{i-5} \oplus v_{i-6} \oplus (v_{i-6}/2^6)$$
$$= v_{i-2} \oplus v_{i-3} \oplus v_{i-5} \oplus v_{i-6} \oplus (v_{i-6}/2^6).$$

Given the direction numbers, the Sobol' sequence can be easily generated. For instance, $z_{26} = z_{25} \oplus v_2$ because $25 = 11001$ (base two), whose rightmost 0 is at position two from the right.

The extension to a higher dimension is straightforward. Let $P_1, P_2, \ldots, P_d$ be primitive polynomials. Denote by z_k^i the sequence of one-dimensional Sobol' points generated by P_i. The sequence of d-dimensional Sobol' points is defined by

$$z_k \equiv \left(z_k^1, z_k^2, \ldots, z_k^d\right), \quad k = 0, 1, \ldots.$$

> **Programming Assignment 18.3.3** Implement the Sobol' sequence.

18.3.3 The Faure Sequence

The one-dimensional Faure sequence of quasi-random numbers coincides with the one-dimensional Halton sequence. To generate the d-dimensional Faure sequence, we proceed as follows. Let $p \ge 2$ be the smallest prime greater than or equal to d. The Faure sequence uses the same base p for each dimension. Denote the kth point by $z_k \equiv (c_1, c_2, \ldots, c_d)$. The first component c_1 is simply the one-dimensional Halton sequence $\phi_p(1), \phi_p(2), \ldots$. Inductively, if

$$c_{m-1} = a_0 p^{-1} + a_1 p^{-2} + \cdots + a_r p^{-(r+1)},$$

then

$$c_m = b_0 p^{-1} + b_1 p^{-2} + \cdots + b_r p^{-(r+1)},$$

where

$$b_j \equiv \sum_{i \ge j}^{r} \binom{i}{j} a_i \bmod p.$$

Note that $x \bmod p$ denotes the remainder of x divided by p. For instance, $24 \bmod 7 = 3$. The convention is $\binom{i}{0} = 1$. In practice, the sequence may start at $k = p^4$ instead of $k = 1$.

EXAMPLE 18.3.2 Take $d = 3$ and $p = 3$. Because the first component c_1 is merely the one-dimensional Halton sequence, it runs like

$$(0.1, 0.2, 0.01, 0.11, 0.21, 0.02, 0.12, 0.22, 0.001, \ldots)$$
$$= \left(\frac{1}{3}, \frac{2}{3}, \frac{1}{9}, \frac{4}{9}, \frac{7}{9}, \frac{2}{9}, \frac{5}{9}, \frac{8}{9}, \frac{1}{27}, \ldots \right).$$

The second component c_2 is

$$\left(\frac{1}{3}, \frac{2}{3}, \frac{4}{9}, \frac{7}{9}, \frac{1}{9}, \frac{8}{9}, \frac{2}{9}, \frac{5}{9}, \frac{16}{27}, \ldots \right).$$

Take the fourth entry, $7/9$, which is the c_2 in z_4, as an example. Because the corresponding number in the first component is 0.11, we have $a_0 = 1$, $a_1 = 1$, and $r = 2$. The number is thus calculated by

$$b_0 = \binom{0}{0} a_0 + \binom{1}{0} a_1 = 1 + 1 = 2 \bmod 3,$$

$$b_1 = \binom{1}{1} a_1 = 1 \bmod 3,$$

$$c_2 = b_0 3^{-1} + b_1 3^{-2} = \frac{2}{3} + \frac{1}{9} = \frac{7}{9}.$$

The sequence for the third component is $\left(\frac{1}{3}, \frac{2}{3}, \frac{7}{9}, \frac{1}{9}, \frac{4}{9}, \frac{5}{9}, \frac{8}{9}, \frac{2}{9}, \frac{13}{27}, \ldots \right)$. The combined three-dimensional sequence is therefore $\left(\frac{1}{3}, \frac{1}{3}, \frac{1}{3} \right), \left(\frac{2}{3}, \frac{2}{3}, \frac{2}{3} \right), \left(\frac{1}{9}, \frac{4}{9}, \frac{7}{9} \right), \ldots$.

➤ **Exercise 18.3.4** Verify the sequence for the third component in Example 18.3.2.

➤ **Programming Assignment 18.3.5** Implement the Faure sequence. Pay attention to evaluating $\binom{i}{j} \bmod p$ efficiently.

➤ **Programming Assignment 18.3.6** Price European options with quasi-random sequences. The computational framework is identical to the Monte Carlo method except that random numbers are replaced with quasi-random numbers with n-dimensional sequences for problems with n time periods.

Additional Reading

Subsection 18.1.3 followed [229]. Finite-difference methods for the Black–Scholes differential equation for multivariate derivatives can be found in [215]. Consult [15, 381, 391, 810, 883] for more information on solving partial differential equations, [706, 707] for algorithms on parallel computers, and [861] for *Mathematica* programs. The implicit method, the explicit method, and the standard binomial tree algorithm are compared in [229, 383, 472, 897].

The term "Monte Carlo simulation" was invented by von Neumann (1903–1957) and Ulam (1909–1984) when they worked on the Manhattan Project [570]. The first paper on the subject was by Metropolis and Ulam in 1949 [699]. Complete treatments of Monte Carlo simulations can be found in [353, 773]. Consult [560, 584, 621, 721, 727] for pseudorandom-number generators. On the topic of Monte Carlo option pricing, see [313] for fast Monte Carlo path-dependent derivatives pricing, [53, 108, 129, 130, 160, 730] for the Monte Carlo pricing of American options, [106, 472] for the control-variates approach, [214, 520, 579, 876] for handling stochastic volatility, [241] for pricing average-rate options by use of conditioning, [366] for comparing numerical integrations, analytical approximations, and control variates in average-rate option pricing, and [20, 108, 128] for general treatments of Monte Carlo simulations in computing sensitivities. Consult [6, 108] for surveys on option pricing by use of Monte Carlo simulation, quasi-Monte Carlo methods, and variance reductions. Subsection 18.2.3 drew on [147, 584, 727, 764]. One suggestion for speeding up Monte Carlo simulation is by limiting the sample space to a finite size and then conducting an exhaustive sampling [509]. In estimating sensitivities such as delta by $E[\,P(S+\epsilon)-P(S-\epsilon)\,]/(2\epsilon)$, $P(S+\epsilon)$ and $P(S-\epsilon)$ should use *common* random numbers to lower the variance.

See [142, 197, 530, 532, 711, 715, 716] for case studies on quasi-Monte Carlo methods, [368, 530, 711, 715, 716, 876] for evaluations of various quasi-random sequences, and [108, 337, 699] for their mathematical foundations. Biology-inspired approaches such as **artificial neural networks** [486, 612, 887] and **genetic algorithms** are other interesting approaches.

NOTES

1. This term is, strictly speaking, misleading as there is nothing random about such sequences.
2. The exclusive-OR operation takes two input bits. It returns 1 if the bits are different and 0 if they are identical. The operation when applied to two bit streams of the same length computes the exclusive-OR of bits at the same position. For example, $10110 \oplus 01100 = 11010$

CHAPTER
NINETEEN

Matrix Computation

> To set up a philosophy against physics is rash; philosophers who have
> done so have always ended in disaster.
> Bertrand Russell

Matrix computation pervades many discussions on numerical and statistical techniques. Two major concerns here are **multivariate statistical analysis** and **curve fitting** with **splines**. Both have extensive applications in statistical inference. **Factor analysis** is presented as an interesting application.

19.1 Fundamental Definitions and Results

Let $A \equiv [\, a_{ij} \,]_{1 \le i \le m, 1 \le j \le n}$, or simply $A \in \boldsymbol{R}^{m \times n}$, denote an $m \times n$ matrix. It can also be represented as $[\, a_1, a_2, \ldots, a_n \,]$, where $a_i \in \boldsymbol{R}^m$ are vectors. Vectors are column vectors unless stated otherwise. It is a **square matrix** when $m = n$. The **rank** of a matrix is the largest number of linearly independent columns. An $m \times n$ matrix is **rank deficient** if its rank is less than $\min(m, n)$; otherwise, it has **full rank**. It has full **column rank** if its rank equals n: All of its columns are linearly independent. A square matrix A is said to be **symmetric** if $A^{\mathrm{T}} = A$. A real $n \times n$ matrix $A \equiv [\, a_{ij} \,]_{i,j}$ is **diagonally dominant** if $|a_{ii}| > \sum_{j \neq i} |a_{ij}|$ for $1 \le i \le n$. Such matrices are nonsingular [224]. A **leading principal submatrix** of an $n \times n$ matrix is a submatrix consisting of the first k rows and the first k columns for some $1 \le k \le n$. The expression $\|x\| \equiv \sqrt{x_1^2 + x_2^2 + \cdots + x_n^2}$ denotes the length of vector $x \equiv [\, x_1, x_2, \ldots, x_n \,]^{\mathrm{T}}$. It is also called the **Euclidean norm**. A **diagonal** $m \times n$ **matrix** $D \equiv [\, d_{ij} \,]_{i,j}$ may be denoted by $\mathrm{diag}[\, D_1, D_2, \ldots, D_q \,]$, where $q \equiv \min(m, n)$ and $D_i = d_{ii}$ for $1 \le i \le q$ (see Fig. 19.1). The **identity matrix** is the square matrix $I \equiv \mathrm{diag}[\, 1, 1, \ldots, 1 \,]$.

A vector set $\{\, x_1, x_2, \ldots, x_p \,\}$ is **orthogonal** if all its vectors are nonzero and the inner products $x_i^{\mathrm{T}} x_j$ equal zero for $i \neq j$. It is **orthonormal** if, furthermore,

$$x_i^{\mathrm{T}} x_j = \begin{cases} 1, & \text{if } i = j \\ 0, & \text{otherwise} \end{cases}.$$

A real square matrix Q is said to be **orthogonal**[1] if $Q^{\mathrm{T}} Q = I$. For such matrices, $Q^{-1} = Q^{\mathrm{T}}$ and $QQ^{\mathrm{T}} = I$. A real symmetric matrix A is **positive definite** (**positive semidefinite**) if $x^{\mathrm{T}} A x = \sum_{i,j} a_{ij} x_i x_j > 0$ ($x^{\mathrm{T}} A x \ge 0$, respectively) for any nonzero vector x. It is known that a matrix A is positive semidefinite if and only if there

$$\begin{bmatrix} \times & 0 & 0 & 0 & 0 \\ 0 & \times & 0 & 0 & 0 \\ 0 & 0 & \times & 0 & 0 \end{bmatrix} \quad \begin{bmatrix} \times & 0 & 0 \\ 0 & \times & 0 \\ 0 & 0 & \times \end{bmatrix} \quad \begin{bmatrix} \times & 0 & 0 \\ 0 & \times & 0 \\ 0 & 0 & \times \\ 0 & 0 & 0 \\ 0 & 0 & 0 \end{bmatrix}$$

Figure 19.1: Diagonal matrices. Three basic forms of $m \times n$ diagonal matrices corresponding to (left to right) $m < n, m = n$, and $m > n$.

exists a matrix W such that $A = W^{\mathsf{T}} W$; A is positive definite if and only if this W has full column rank.

19.1.1 Gaussian Elimination

Gaussian elimination is a standard method for solving a linear system $Ax = b$, where $A \in R^{n \times n}$. It is due to Gauss in 1809 [825]. After $O(n^3)$ operations, the system is transformed into an equivalent system $A'x = b'$, where A' is an upper triangular $n \times n$ matrix. This is the **forward-reduction** phase. **Backward substitution** is then applied to compute x after $O(n^2)$ more operations. The total running time is therefore cubic. See Fig. 19.2 for the algorithm.

Efficiency can often be improved if A has special structures. A case in point is when A is **banded**, that is, if all the nonzero elements are placed near the diagonal of the matrix. We say that $A = [a_{ij}]_{i,j}$ has **upper bandwidth** u if $a_{ij} = 0$ for $j - i > u$ and **lower bandwidth** l if $a_{ij} = 0$ for $i - j > l$. A tridiagonal matrix, for instance, has upper bandwidth one and lower bandwidth one (see Fig. 19.3). For banded matrices, Gaussian elimination can be easily modified to run in $O(nul)$ time by skipping

Gaussian elimination algorithm:

```
input:   A[1..n][1..n], b[1..n];
real     x[1..n];
integer  i, j, k;
// Forward reduction.
for (k = 1 to n − 1)
        for (i = k + 1 to n) {
                c := A[i][k]/A[k][k];
                for (j = k + 1 to n)
                        A[i][j] := A[i][j] − c × A[k][j];
                b[i] := b[i] − c × b[k];
        }
// Backward substitution.
for (k = n down to 1)
        x[k] := (b[k] − ∑_{j=k+1}^{n} A[k][j] × x[j])/A[k][k];
return x[ ];
```

Figure 19.2: Gaussian elimination algorithm. This algorithm solves $Ax = b$ for x. It assumes that the diagonal elements $A[k][k]$ are nonzero throughout. This can be guaranteed if all of A's leading principal submatrices are nonsingular.

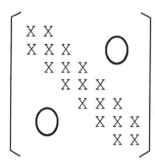

Figure 19.3: Banded matrix. Shown here is a tridiagonal matrix.

elements that are zero. This bound is substantially smaller than n^3 when u or l is small.

Gaussian elimination can be used to factor any square matrix whose leading principal submatrices are nonsingular into a product of a lower triangular matrix L and an upper triangular matrix U: $A = LU$. This is called the **LU decomposition**. The conditions are satisfied by positive definite matrices and diagonally dominant matrices. Positive definite matrices can in fact be factored as $A = LL^\mathsf{T}$, called the **Cholesky decomposition**. See Fig. 19.4 for a cubic-time algorithm.

19.1.2 Eigenvalues and Eigenvectors

An **eigenvalue** of a square matrix A is a complex number λ such that $Ax = \lambda x$ for some nonzero vector x, called an **eigenvector**. For example, because

$$\begin{bmatrix} 1 & 2 \\ 2 & 1 \end{bmatrix} = \begin{bmatrix} 0.707107 & -0.707107 \\ 0.707107 & 0.707107 \end{bmatrix} \begin{bmatrix} 3 & 0 \\ 0 & -1 \end{bmatrix} \begin{bmatrix} 0.707107 & 0.707107 \\ -0.707107 & 0.707107 \end{bmatrix},$$

the two eigenvalues are 3 and -1, and $[\,0.707107, -0.707107\,]^\mathsf{T}$ and $[\,0.707107, 0.707107\,]^\mathsf{T}$ are the corresponding eigenvectors. The eigenvalues for a real symmetric matrix are real numbers; in particular, the **Schur decomposition theorem** (also called the **principal-axes theorem** or the **spectral theorem**) says that there exists a real orthogonal matrix Q such that $Q^\mathsf{T}AQ = \mathrm{diag}[\,\lambda_1, \lambda_2, \ldots, \lambda_n\,]$. Note that Q's

---

Cholesky decomposition algorithm:

```
input:   A[1..n][1..n];
real     L[1..n][1..n];
integer  i, j, k;
for (j = 1 to n) {
        L[j][j] := (A[j][j] − ∑_{k=1}^{j−1} L[j][k]²)^{1/2};
        for (i = j + 1 to n) {
                L[i][j] := L[j][j]^{−1} × (A[i][j] − ∑_{k=1}^{j−1} L[i][k] × L[j][k]);
                L[j][i] := 0;
        }
}
return L[][];
```

Figure 19.4: Cholesky decomposition. The algorithm solves $A = L\,L^\mathsf{T}$ for positive definite A.

ith column is the eigenvector corresponding to λ_i, and the eigenvectors form an orthonormal set. The eigenvalues of positive definite matrices are furthermore positive.

Principal Components

Let $x \equiv [\, x_1, x_2, \ldots, x_n \,]^{\mathrm{T}}$ be a vector random variable with the covariance matrix

$$C \equiv [\, \mathrm{Cov}[\, x_i, x_j \,] \,]_{1 \leq i, j \leq n} = E[\, xx^{\mathrm{T}} \,] \neq 0,$$

where $E[\, x \,] = 0$ and 0 denotes the zero vector. Covariance matrices are positive definite provided that individual variances $\mathrm{Var}[\, x_i \,]$ are all positive, which will be assumed throughout [343]. C's eigenvalues $\lambda_1 \geq \lambda_2 \geq \cdots \geq \lambda_n > 0$ are therefore real.

We are interested in knowing which normalized linear combinations of the x_is give rise to the maximum variance, thus explaining most of the total variability of the system. Mathematically, we seek a vector $b \equiv [\, b_1, b_2, \ldots, b_n \,]^{\mathrm{T}}$ with $\| b \| = 1$ that maximizes

$$\mathrm{Var}[\, b^{\mathrm{T}}x \,] \equiv E[\, (b^{\mathrm{T}}x)^2 \,] = E[\, (b_1 x_1 + b_2 x_2 + \cdots + b_n x_n)^2 \,].$$

The answer turns out to be simple: The maximum variance equals C's largest eigenvalue λ_1, and its corresponding eigenvector u_1 is the b we are after. Among the normalized linear combinations of the x_i's uncorrelated with $u_1^{\mathrm{T}}x$, which gives rise to the maximum variance? The answer is similar: The maximum variance is λ_2, and its corresponding eigenvector u_2 produces the desired linear combination $u_2^{\mathrm{T}}x$. This process can be repeated for a total of n steps, with the jth time leading to a normalized linear combination uncorrelated with all the previous $j - 1$ combinations that has the maximum variance λ_j [23].

The proof goes like this. By the Schur decomposition theorem, the real orthogonal matrix B whose columns are the orthonormal eigenvectors satisfies

$$\Sigma \equiv \mathrm{diag}[\, \lambda_1, \lambda_2, \ldots, \lambda_n \,] = B^{\mathrm{T}}CB, \tag{19.1}$$

with $\lambda_1 \geq \lambda_2 \geq \cdots \geq \lambda_n$. The vector of **principal components** of x,

$$P \equiv [\, p_1, p_2, \ldots, p_n \,]^{\mathrm{T}} = B^{\mathrm{T}}x, \tag{19.2}$$

has the covariance matrix

$$E[\, PP^{\mathrm{T}} \,] = E[\, B^{\mathrm{T}}xx^{\mathrm{T}}B \,] = B^{\mathrm{T}}CB = \Sigma.$$

Each principal component p_j is a normalized linear combination of the x_is and corresponds to the jth combination mentioned previously (see Exercise 19.1.2). This identity also confirms that any two distinct components of P are uncorrelated and p_j has variance λ_j.

Principal components can be used to find linear combinations with large variances. Specifically, the p_js in Eq. (19.2) can be considered uncorrelated new variables, and those variables p_j whose corresponding eigenvalue λ_j is small are thrown away. The resulting variables can be much smaller in number than the original variables $x_1, x_2, \ldots, x_n$.

> **Exercise 19.1.1** Let $C \equiv [\, c_{ij} \,]$ denote the covariance matrix of $x \equiv [\, x_1, x_2, \ldots, x_n \,]^{\mathrm{T}}$. Show that C and x's correlation matrix P are related by $C = \Sigma P \Sigma$, where $\Sigma \equiv \mathrm{diag}[\, \sqrt{c_{11}}, \sqrt{c_{22}}, \ldots, \sqrt{c_{nn}} \,]$. (Hence, C is positive definite if and only if P is.)

> **Exercise 19.1.2** Prove that the jth principal component p_j has the maximum variance among all the normalized linear combinations of the x_is that are uncorrelated with $p_1, p_2, \ldots, p_{j-1}$.

Generation of the Multivariate Normal Distribution

Let $x \equiv [x_1, x_2, \ldots, x_n]^T$ be a vector random variable with a positive definite covariance matrix C. As usual, assume that $E[x] = 0$. This distribution can be generated by Py, where $C = PP^T$ is the Cholesky decomposition of C and $y \equiv [y_1, y_2, \ldots, y_n]^T$ is a vector random variable with a covariance matrix equal to the identity matrix. This holds because $\text{Cov}[Py] = P\,\text{Cov}[y]\,P^T = PP^T = C$ (see Exercise 6.1.3).

Suppose we want to generate the multivariate normal distribution with a covariance matrix $C = PP^T$. We start with independent standard normal distributions $y_1, y_2, \ldots, y_n$. Then $P[y_1, y_2, \ldots, y_n]^T$ has the desired distribution. Generating the multivariate normal distribution is essential for the Monte Carlo pricing of multivariate derivatives.

> **Exercise 19.1.3** Verify the correctness of the procedure for generating the bivariate normal distribution in Subsection 6.1.2.

> **Exercise 19.1.4** Prove that $WX \sim N(W\mu, WCW^T)$ if $X \sim N(\mu, C)$.

19.1.3 The Singular Value Decomposition

The following theorem is the basis for the **singular value decomposition (SVD)**.

THEOREM 19.1.1 (SVD). *For a real $m \times n$ matrix A,*

$$\Sigma \equiv U^T A V = \text{diag}[\sigma_1, \sigma_2, \ldots, \sigma_p] \in R^{m \times n}$$

*for some orthogonal matrices $U \equiv [u_1, u_2, \ldots, u_m] \in R^{m \times m}$ and $V \equiv [v_1, v_2, \ldots, v_n] \in R^{n \times n}$, where $p \equiv \min(m, n)$ and $\sigma_1 \geq \sigma_2 \geq \cdots \geq \sigma_p \geq 0$. In fact, $\sigma_1 \geq \cdots \geq \sigma_r > \sigma_{r+1} = \cdots = \sigma_p = 0$ with $\text{rank}(A) = r$. The σ_is are called the **singular values**.*

For example,

$$\begin{bmatrix} 1 & 2 \\ 1 & 4 \\ 1 & 6 \end{bmatrix} = \begin{bmatrix} -0.282970 & 0.867906 \\ -0.538373 & 0.208536 \\ -0.793777 & -0.450834 \end{bmatrix} \begin{bmatrix} 7.654435 & 0. \\ 0. & 0.640018 \end{bmatrix}$$

$$\times \begin{bmatrix} -0.211005 & -0.977485 \\ 0.977485 & -0.211005 \end{bmatrix}$$

$$= U\Sigma V^T. \tag{19.3}$$

Because U and V are orthogonal, the two singular values are 7.654435 and 0.640018. (The third column of U and the third row of Σ, which is a zero vector, were not shown.)

The SVD can be computed in cubic time and quadratic space in terms of m and n [392, 586]. It is implemented in every decent mathematical software library. Theorem 19.1.1 implies that

$$A = U\Sigma V^T = \sum_{i=1}^{p} \sigma_i u_i v_i^T. \tag{19.4}$$

▶ **Exercise 19.1.5** (1) Prove Eq. (19.4). (2) Show that A and A^T share the same singular values.

19.2 Least-Squares Problems

The **least-squares (LS) problem** is concerned with $\min_{x \in R^n} \|Ax - b\|$, where $A \in R^{m \times n}$, $b \in R^m$, $m \geq n$. The LS problem is called **regression analysis** in statistics and is equivalent to minimizing the mean-square error. Often stated as $Ax = b$, the LS problem is **overdetermined** when there are more equations than unknowns ($m > n$).

EXAMPLE 19.2.1 In polynomial regression, $\beta_0 + \beta_1 x + \cdots + \beta_n x^n$ is used to fit the data $\{(x_1, b_1), (x_2, b_2), \ldots, (x_m, b_m)\}$. This leads to the LS problem $Ax = b$, where

$$A \equiv \begin{bmatrix} 1 & x_1 & x_1^2 & \cdots & x_1^n \\ 1 & x_2 & x_2^2 & \cdots & x_2^n \\ \vdots & \vdots & \vdots & \ddots & \vdots \\ 1 & x_m & x_m^2 & \cdots & x_m^n \end{bmatrix}, \quad x \equiv \begin{bmatrix} \beta_0 \\ \beta_1 \\ \vdots \\ \beta_n \end{bmatrix}, \quad b \equiv \begin{bmatrix} b_1 \\ b_2 \\ \vdots \\ b_m \end{bmatrix}. \tag{19.5}$$

Linear regression corresponds to $n = 1$.

The LS problem can be solved by a geometric argument. Because Ax is a linear combination of A's columns with coefficients $x_1, x_2, \ldots, x_n$, the LS problem finds the minimum distance between b and A's column space. A solution x_{LS} must identify a point Ax_{LS} that is at least as close to b as any other point in the column space. Therefore the error vector $Ax_{LS} - b$ must be perpendicular to that space, that is,

$$(Ay)^T(Ax_{LS} - b) = y^T(A^T Ax_{LS} - A^T b) = \mathbf{0}$$

for all y. We conclude that any solution x must satisfy the **normal equations**

$$A^T Ax = A^T b. \tag{19.6}$$

▶ **Exercise 19.2.1** What are the normal equations for linear regression (6.13)?

▶ **Exercise 19.2.2** (1) Phrase multiple regression as an LS problem. (2) Write the normal equations.

▶ **Exercise 19.2.3** Let $\Phi(x) \equiv (1/2) \|Ax - b\|^2$. Prove that its **gradient vector**,

$$\nabla \Phi(x) \equiv \left[\frac{\partial \Phi(x)}{\partial x_1}, \frac{\partial \Phi(x)}{\partial x_2}, \ldots, \frac{\partial \Phi(x)}{\partial x_n} \right]^T,$$

equals $A^T(Ax - b)$, where $x \equiv [x_1, x_2, \ldots, x_n]^T$. (Normal equations are $\nabla \Phi(x) = \mathbf{0}$).

▶ **Exercise 19.2.4** Define the **bandwidth** of a banded matrix A as $l + u + 1$, where l is the lower bandwidth and u is the upper bandwidth. Prove that $A^T A$'s bandwidth is at most $\omega - 1$ if A has bandwidth ω.

Comment 19.2.2 The result in Exercise 19.2.4 holds under a more generous definition of banded matrices. In that definition, a matrix is banded with bandwidth ω if the nonzero elements of every row lie within a band with width ω [75]. This result can save some computational efforts.

19.2.1 The Full-Rank Case

The LS problem is called the **full-rank LS problem** when A has full column rank. Because $A^T A$ is then nonsingular, the unique solution for normal equations is $x_{LS} = (A^T A)^{-1} A^T b$, which is called the **ordinary least-squares (OLS) estimator**. As $A^T A$ is positive definite, the normal equations can be solved by the Cholesky decomposition. This approach is usually not recommended because its numerical stability is lower than the alternative SVD approach (see Theorem 19.2.3 below) [35, 586, 870].

Suppose the following linear regression model is postulated between x and b:

$$b = Ax + \epsilon, \tag{19.7}$$

where ϵ is a vector random variable with zero mean and finite variance. The Gauss–Markov theorem says that the OLS estimator x_{LS}, now a random variable, is unbiased and has the smallest variance among all unbiased *linear* estimators if ϵ's components have identical, known variance σ^2 and are uncorrelated (the covariance matrix of ϵ is hence $\sigma^2 I$) [802]. Under these assumptions,

$$\text{Cov}[x_{LS}] = \sigma^2 (A^T A)^{-1}. \tag{19.8}$$

Hence $(A^T A)^{-1}$, properly scaled, is an unbiased estimator of the covariance matrix of x_{LS}. If ϵ is moreover normally distributed, then x_{LS} has the smallest variance among all unbiased estimators, linear or otherwise [422].

▶ **Exercise 19.2.5** Show that the sample residuals of the OLS estimate, $Ax_{LS} - b$, are orthogonal to the columns of A.

▶ **Exercise 19.2.6** Suppose that $\sigma^2 C$ is the covariance matrix of ϵ for a positive definite C in linear regression model (19.7). How do we solve the LS problem $Ax = b$ by using the Gauss–Markov theorem?

▶ **Exercise 19.2.7** Verify Eq. (19.8).

19.2.2 The (Possibly) Rank-Deficient Case

When A is rank deficient, there are an infinite number of solutions to the LS problem. In this case, we are interested in the x with the minimum length such that $\|Ax - b\|$ is minimized. This x is unique because $\{ x \in R^m : \|Ax - b\| \text{ is minimized} \}$ is convex. The linkage between the SVD and the general LS problem when A may be rank deficient is established in the following theorem.

THEOREM 19.2.3 *Let* $U \equiv [u_1, u_2, \ldots, u_m] \in R^{m \times m}$ *and* $V \equiv [v_1, v_2, \ldots, v_n] \in R^{n \times n}$ *such that* $U^T A V$ *is the SVD of* $A \in R^{m \times n}$ *and* r *is* A's *rank. Then*

$$x_{LS} = \sum_{i=1}^{r} \left(\frac{u_i^T b}{\sigma_i} \right) v_i$$

is the solution to the LS problem and $\|Ax_{LS} - b\|^2 = \sum_{i=r+1}^{m} (u_i^T b)^2$.

Let $A = U \Sigma V^T$, where $\Sigma \equiv \text{diag}[\sigma_1, \sigma_2, \ldots, \sigma_n]$, be the SVD of $A \in R^{m \times n}$. Define

$$\Sigma^+ \equiv \text{diag}[\sigma_1^{-1}, \sigma_2^{-1}, \ldots, \sigma_r^{-1}, 0, 0, \ldots, 0] \in R^{n \times m}.$$

Then $x_{ls} = V\Sigma^+ U^\mathsf{T} b$. The matrix

$$A^+ \equiv V\Sigma^+ U^\mathsf{T} \tag{19.9}$$

is called the **pseudoinverse** of A. Note that $\|Ax_{ls} - b\| = \|(I - AA^+)b\|$ and Σ^+ is the pseudoinverse of Σ. For example, it is not hard to verify numerically that the pseudoinverse of matrix A in Eq. (19.3) is

$$\begin{bmatrix} 4/3 & 1/3 & -2/3 \\ -1/4 & 0 & 1/4 \end{bmatrix}.$$

▶ **Exercise 19.2.8** Prove that if the A in Theorem 19.2.3 has full column rank, then $A^+ = (A^\mathsf{T} A)^{-1} A^\mathsf{T}$. (This implies, in particular, that $A^+ = A^{-1}$ when A is square.)

▶ **Exercise 19.2.9** Prove that the pseudoinverse of the pseudoinverse is itself, i.e., $(A^+)^+ = A$.

▶ **Exercise 19.2.10 (Underdetermined Linear Equations)** Suppose as before that $A \in R^{m \times n}$ and $b \in R^m$, but $m \le n$. Assume further that $m = \text{rank}(A)$. Let $U\Sigma V^\mathsf{T}$ be the SVD of A. Argue that all solutions to $Ax = b$ are of the form

$$\hat{x} = V\Sigma^+ U^\mathsf{T} b + V\begin{bmatrix} \mathbf{0} \\ y \end{bmatrix} \begin{matrix} \}m \\ \}n-m \end{matrix} = A^+ b + V_2 y \tag{19.10}$$

for arbitrary $y \in R^{n-m}$, where $V \equiv [\ \underbrace{V_1}_{m},\ \underbrace{V_2}_{n-m}\ ]\}n$.

19.2.3 The Weighted Least-Squares Problem

The **weighted LS problem** is concerned with

$$\min_{x \in R^n} \|WAx - Wb\|, \tag{19.11}$$

where $W \in R^{m \times m}$ is nonsingular. Clearly all the above-mentioned results regarding the LS problem apply after we replace their A with WA and their b with Wb. In particular, the solution x satisfies the **weighted normal equations**:

$$(WA)^\mathsf{T} WAx = (WA)^\mathsf{T} Wb.$$

With $H \equiv W^\mathsf{T} W$ (hence positive definite), the above equations can be restated as

$$A^\mathsf{T} HAx = A^\mathsf{T} Hb. \tag{19.12}$$

In particular, if A has full column rank, then the unique solution for x is

$$(A^\mathsf{T} HA)^{-1} A^\mathsf{T} Hb. \tag{19.13}$$

Classic regression analysis assumes that the rows of $Ax - b$ have zero mean and a covariance matrix C. The regression model is hence $b = Ax + \epsilon$, where ϵ is a vector random variable with zero mean and $\text{Cov}[\epsilon] = C$. The optimal solution for x then satisfies

$$A^\mathsf{T} C^{-1} Ax = A^\mathsf{T} C^{-1} b. \tag{19.14}$$

(The H^{-1} in Eq. (19.12) thus plays the role of the covariance matrix.) When A has full column rank, the unique optimal solution $(A^{\mathsf{T}}C^{-1}A)^{-1}A^{\mathsf{T}}C^{-1}b$ is called the **generalized least-squares (GLS) estimator**, which is unbiased with covariance matrix $(A^{\mathsf{T}}C^{-1}A)^{-1}$.

▶ **Exercise 19.2.11** Prove Eq. (19.14). (Hint: Exercise 19.2.6.)

19.2.4 The Least-Squares Problem with Side Constraints

The **constrained LS problem** is an LS problem over a proper subset of R^n. We are mainly interested in problems with linear equality constraints:

$$\min_{x \in R^n,\, Bx=d} \|Ax - b\|, \quad B \in R^{p \times n},\, d \in R^p.$$

Assume that p is the rank of B. When the solution is not unique, we seek the unique x with minimum length. The solution is unique if and only if the $(m+p) \times n$ augmented matrix

$$\begin{bmatrix} B \\ A \end{bmatrix}$$

has rank n [586]. Algorithms for the problem generally adopt the paradigm of transforming it into an unconstrained LS problem from whose solution the desired x is constructed.

▶ **Exercise 19.2.12** Design an algorithm for the constrained LS problem by using the SVD.

▶ **Exercise 19.2.13 (Lawson–Hanson Algorithm)** It is known that any $A \in R^{m \times n}$ can be decomposed as $A = Q^{\mathsf{T}}R$ such that $Q \in R^{m \times m}$ is orthogonal and $R \in R^{m \times n}$ is zero below the main diagonal. This is called the **QR decomposition**. Solve the LS problem with linear constraints by using the QR decomposition instead of the SVD.

19.2.5 Factor Analysis

Factor analysis postulates that the multitude of influences that affect our concern, which will be interest rate changes here, can be summarized by a few variables called **factors** [91, 390, 804]. Factors are said to explain changes in interest rates, and the quantitative relations between interest rate changes and factors are called **factor loadings**.

Fundamentals

At each time t, where $t = 1, 2, \ldots, T$, a p-dimensional vector y_t is observed, such as the interest rates of various maturities. The orthogonal factor model assumes that these observations are linearly related to m underlying, unobserved factors f_t by means of

$$\begin{array}{cccccc} y_t - \mu & = & L & \times & f_t & + & \epsilon_t, \\ p \times 1 & & p \times m & & m \times 1 & & p \times 1 \end{array} \tag{19.15}$$

where $m < p$, $E[y_t] = \mu$, $E[f_t] = E[\epsilon_t] = \mathbf{0}$, $\mathrm{Cov}[f_t] = E[f_t f_t^{\mathrm{T}}] = I$, and

$$\Psi \equiv \mathrm{Cov}[\epsilon_t] = \mathrm{diag}[\Psi_1, \Psi_2, \ldots, \Psi_p] \in \mathbf{R}^{p \times p}.$$

Moreover, f_t and ϵ_t are independent. In Eq. (19.15) L contains the factor loadings, ϵ_t is the vector of **specific factors** or individual residual errors, and Ψ is the vector of **specific variances**. Each factor has zero mean and unit standard deviation and is uncorrelated with others. Note that L, f_t, and ϵ_t are unknown and the model is linear in the factors.

Equation (19.15) implies that

$$C \equiv E[(y_t - \mu)(y_t - \mu)^{\mathrm{T}}] = LL^{\mathrm{T}} + \Psi,$$

in which f_t and ϵ_t drop out. This makes it possible to compute L and Ψ as follows. Start with the estimated covariance matrix $C \in \mathbf{R}^{p \times p}$. By the Schur decomposition theorem, $C = B \times \mathrm{diag}[\lambda_1, \lambda_2, \ldots, \lambda_n] \times B^{\mathrm{T}}$ with $\lambda_1 \geq \lambda_2 \geq \cdots \geq \lambda_n \geq 0$. Let L be the first m columns of $B \times \mathrm{diag}[\sqrt{\lambda_1}, \sqrt{\lambda_2}, \ldots, \sqrt{\lambda_n}]$ and let Ψ be constructed by zeroing the off-diagonal elements of $C - LL^{\mathrm{T}}$. Of course, if some off-diagonal elements of the residual matrix $C - LL^{\mathrm{T}}$ are "large," there may be additional omitted factors, which calls for a larger m.[2]

The solution to L is not unique when $m > 1$: If A is an orthogonal matrix, then $\tilde{L} \equiv LA$ is also a solution as

$$\tilde{L}\tilde{L}^{\mathrm{T}} + \Psi = LAA^{\mathrm{T}}L^{\mathrm{T}} + \Psi = LL^{\mathrm{T}} + \Psi = C.$$

When L is replaced with $\tilde{L}$, the factors become $\tilde{f}_t \equiv A^{\mathrm{T}} f_t$ because

$$\tilde{L}\tilde{f}_t = LAA^{\mathrm{T}} f_t = Lf_t.$$

Orthogonal transformation of the factor loadings L, however, has no effects on Ψ and the diagonal elements of LL^{T}. This suggests we look for an A that gives the factor loadings $\tilde{L}$ intuitive economic interpretations. For instance, we might desire loadings that let rate changes at all maturities have approximately equal loading on the first factor, signifying a parallel shift. For this purpose, under the case of three factors ($m = 3$), each of the following orthogonal matrices called **Givens transformations** rotates through an angle θ, leaving one dimension unchanged:

$$A_1 \equiv \begin{bmatrix} \cos\theta_1 & \sin\theta_1 & 0 \\ -\sin\theta_1 & \cos\theta_1 & 0 \\ 0 & 0 & 1 \end{bmatrix}, \quad A_2 \equiv \begin{bmatrix} \cos\theta_2 & 0 & \sin\theta_2 \\ 0 & 1 & 0 \\ -\sin\theta_2 & 0 & \cos\theta_2 \end{bmatrix},$$

$$A_3 \equiv \begin{bmatrix} 1 & 0 & 0 \\ 0 & \cos\theta_3 & \sin\theta_3 \\ 0 & -\sin\theta_3 & \cos\theta_3 \end{bmatrix}.$$

Note that $A \equiv A_1 A_2 A_3$ is an orthogonal matrix with parameters θ_1, θ_2, and θ_3. Now find $\theta_1, \theta_2, \theta_3 \in [-\pi, \pi]$ that minimize the variance of the first column of $\tilde{L}$.

Not all positive definite matrices can be factored as $C = [C_{ij}] = LL^{\mathrm{T}} + \Psi$ for the given $m < p$. Even those that can be so factored may give statistically meaningless numbers such as a negative Ψ_i. Because $\mathrm{Cov}[y_t, f_t] = [L_{ij}]$ (see Exercise 19.2.15), $[L_{ij} C_{ij}^{-1/2}]_{i,j}$ is the correlation matrix of y_t and f_t, but there is no guarantee that $L_{ij} C_{ij}^{-1/2}$ will lie within the $[-1, 1]$ range. These are some of the problems associated with factor analysis.

▶ **Exercise 19.2.14** Once the appropriate estimated factor loadings have been obtained, verify that the factors themselves can be estimated by $f_t = (L^T \Psi^{-1} L)^{-1} L^T \Psi^{-1} (y_t - \mu)$.

▶ **Exercise 19.2.15** Prove $\text{Cov}[\, y_t, \, f_t \,] = L$.

Factors Affecting Interest Rate Movements

Because the U.S. Treasury spot rate curve varies more at shorter maturities than at longer maturities, more data are typically used for spot rates at shorter maturities. For instance, one may use the 11 rates at 3 months, 6 months, 1 year, 2 years, 3 years, 5 years, 7 years, 10 years, 15 years, 20 years, and 30 years (so $p = 11$). Research has shown that three factors can explain more than 90% of the variation in interest rate changes (so $m = 3$) [607]. These factors can be interpreted as **level**, **slope**, and **curvature**. The first factor has approximately equal effects on all maturities in that a change in it produces roughly parallel movements in interest rates of all maturities. The second factor affects the slope of the term structure but not the average level of interest rates. It produces movements in the long and the short ends of the term structure in opposite directions, twisting the curve, so to speak, with relatively smaller changes at intermediate maturities. For the third factor, the loadings are zero or negative at the shortest maturity, positive for intermediate maturities, and then they decline to become negative for the longest maturities. Thus a positive change in this factor tends to increase intermediate rates and decrease long rates or even short rates, altering the curvature of the term structure [91, 390, 804]. The factor analysis of interest rates is not without its problems, however [591].

▶ **Exercise 19.2.16** How would you define **slope duration** and **curvature duration**?

19.3 Curve Fitting with Splines

The purpose of curve fitting is to approximate the data with a "smooth" curve. Linear regression or its generalization, polynomial regression, is not ideal because it tries to fit a single polynomial over the entire data. The spline approach is different. It divides the domain interval into several subintervals and uses a different polynomial for each subinterval.

Suppose we want to fit a curve $f(x)$ over $n+1$ data points,

$$(x_0, y_0), (x_1, y_1), \ldots, (x_n, y_n),$$

with $x_0 < x_1 < \cdots < x_n$ such that the curve agrees with the data at each **breakpoint** (or **knot**) x_i. Divide $[x_0, x_n]$ into n subintervals: $[x_0, x_1], [x_1, x_2], \ldots, [x_{n-1}, x_n]$. If polynomials of degree zero are used for each subinterval, the curve is a step function. Its major disadvantage is the curve's discontinuity at the breakpoints. If polynomials of degree one or two are used instead, the discontinuity problem remains – it now applies, respectively, to the first and the second derivatives of the curve. The curves are therefore not "smooth" enough. If we require that the curve be a polynomial of degree three in each subinterval and that its first and second derivatives be continuous on $[x_0, x_n]$, then a (cubic) spline results. See Fig. 19.5 for illustration.

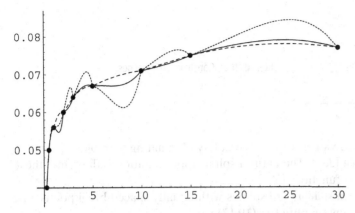

Figure 19.5: Splines of varying degrees. Plotted are splines of degrees two, three, and four. A higher degree leads to greater oscillation.

19.3.1 Cubic Splines

By considering the continuity of $f(x)$ and $f'(x)$, we find that the spline $f(x)$ over $x \in [x_{i-1}, x_i]$ is

$$f(x) = f'_{i-1} \frac{(x_i - x)^2(x - x_{i-1})}{h_i^2} - f'_i \frac{(x - x_{i-1})^2(x_i - x)}{h_i^2}$$

$$+ y_{i-1} \frac{(x_i - x)^2[2(x - x_{i-1}) + h_i]}{h_i^3} + y_i \frac{(x - x_{i-1})^2[2(x_i - x) + h_i]}{h_i^3},$$

where $h_i \equiv x_i - x_{i-1}$ with $n+1$ unknown values $f'_i \equiv f'(x_i)$, $i = 0, 1, \ldots, n$, to be determined. Now,

$$f''(x_i-) = \frac{2}{h_i}(f'_{i-1} + 2f'_i) - 6\frac{y_i - y_{i-1}}{h_i^2}, \tag{19.16}$$

$$f''(x_i+) = -\frac{2}{h_{i+1}}(2f'_i + f'_{i+1}) + 6\frac{y_{i+1} - y_i}{h_{i+1}^2}. \tag{19.16'}$$

If the second derivative of the spline is also continuous at each interior break-point, that is, $f''(x_i-) = f''(x_i+)$ for $i = 1, 2, \ldots, n-1$, then we have the following equations:

$$\frac{1}{h_i}f'_{i-1} + 2\left(\frac{1}{h_i} + \frac{1}{h_{i+1}}\right)f'_i + \frac{1}{h_{i+1}}f'_{i+1} = 3\frac{y_i - y_{i-1}}{h_i^2} + 3\frac{y_{i+1} - y_i}{h_{i+1}^2}, \tag{19.17}$$

$i = 1, 2, \ldots, n-1$. See Fig. 19.3.1 for illustration. The equations form a diagonally dominant, tridiagonal system, which admits linear-time and stable numerical procedures.

We need $n+1-(n-1) = 2$ more auxiliary conditions relating $f'_0, f'_1, \ldots, f'_n$ in order to solve them. Typical conditions specify the values of (1) $f''(x_0)$ and $f''(x_n)$ or (2) $f'(x_0)$ and $f'(x_n)$. For example, letting $f''(x_0) = f''(x_n) = 0$ leads to the so-called **natural splines**. With Eq. (19.16), these two conditions eliminate variables f'_0 and f'_n from Eq. (19.17) because f'_0 can be replaced with a linear function of f'_1 and f'_n can be replaced with a linear function of f'_{n-1}. Choice (2) forces the slope at each

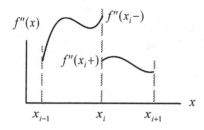

Figure 19.6: Construction of splines.

end to assume specified values, thus immediately eliminating variables f_0' and f_n'. Other alternatives exist [381]. The natural-spline approximation is the "smoothest" among all interpolating functions [391].

Splines with **uniform spacing** are splines with equally spaced breakpoints, $h_i \equiv x_i - x_{i-1} = h$. We can now simplify Eq. (19.17) to

$$f_{i-1}' + 4f_i' + f_{i+1}' = 3\frac{y_{i+1} - y_{i-1}}{h}, \quad i = 1, \ldots, n-1.$$

▶ **Exercise 19.3.1** Verify Eq. (19.16).

▶ **Exercise 19.3.2** Write the tridiagonal system for the natural spline.

An Alternative Formulation

The continuity of $f(x)$ and $f''(x)$ results in

$$f(x) = y_{i-1}\frac{x_i - x}{h_i} + y_i\frac{x - x_{i-1}}{h_i} - \frac{h_i^2}{6}f_{i-1}''\left[\frac{x_i - x}{h_i} - \left(\frac{x_i - x}{h_i}\right)^3\right]$$

$$- \frac{h_i^2}{6}f_i''\left[\frac{x - x_{i-1}}{h_i} - \left(\frac{x - x_{i-1}}{h_i}\right)^3\right], \quad x \in [x_{i-1}, x_i],$$

with $n+1$ values $f_i'' \equiv f''(x_i)$ to be determined. Imposing continuity of $f'(x)$ at each interior breakpoint results in

$$h_i f_{i-1}'' + 2(h_i + h_{i+1})f_i'' + h_{i+1}f_{i+1}'' = 6\left(\frac{y_{i+1} - y_i}{h_{i+1}} - \frac{y_i - y_{i-1}}{h_i}\right), \quad (19.18)$$

$i = 1, 2, \ldots, n-1$. The preceding linear equations form a diagonally dominant, tridiagonal system. As in Eq. (19.17), two more conditions are needed to solve it. This formulation is convenient for natural splines because, in this case, $f_0'' = f_n'' = 0$ [417].

EXAMPLE 19.3.1 For the data set $\{(0,0), (1,2), (2,1), (3,5), (4,4)\}$, we have $n = 4$, $h_1 = h_2 = h_3 = h_4 = 1$, $x_i = i$ ($1 \leq i \leq 4$), $y_0 = 0$, $y_1 = 2$, $y_2 = 1$, $y_3 = 5$, $y_4 = 4$. The spline is

$$p_i(x) \equiv y_{i-1}(x_i - x) + y_i(x - x_{i-1})$$

$$- \frac{1}{6}f_{i-1}''[(x_i - x) - (x_i - x)^3] - \frac{1}{6}f_i''[(x - x_{i-1}) - (x - x_{i-1})^3]$$

for $x \in [x_{i-1}, x_i]$. To fit the data with a natural spline, we solve

$$\begin{bmatrix} 4 & 1 & 0 \\ 1 & 4 & 1 \\ 0 & 1 & 4 \end{bmatrix}\begin{bmatrix} f_1'' \\ f_2'' \\ f_3'' \end{bmatrix} = \begin{bmatrix} -18 \\ 30 \\ -30 \end{bmatrix}.$$

The solutions are $f_1'' = -7.5$, $f_2'' = 12$, and $f_3'' = -10.5$. We finally obtain the spline by adding $f_0'' = f_4'' = 0$. Its four cubic polynomials are

$$p_1(x) = 2x + \frac{7.5}{6}(x - x^3),$$

$$p_2(x) = 2(2 - x) + (x - 1) + \frac{7.5}{6}[(2 - x) - (2 - x)^3]$$

$$- \frac{12}{6}[(x - 1) - (x - 1)^3],$$

$$p_3(x) = (3 - x) + 5(x - 2) - \frac{12}{6}[(3 - x) - (3 - x)^3]$$

$$+ \frac{10.5}{6}[(x - 2) - (x - 2)^3],$$

$$p_4(x) = 5(4 - x) + 4(x - 3) + \frac{10.5}{6}[(4 - x) - (4 - x)^3].$$

See the solid curve in Fig. 19.7. If y_2 is perturbed slightly to 1.3, the whole spline changes, to the dotted curve in Fig. 19.7.

▶ **Exercise 19.3.3** Verify Eq. (19.18).

▶ **Exercise 19.3.4** Construct the spline in Example 19.3.1 when $y_2 = 1.3$.

19.3.2 Cubic Splines and the Constrained Least-Squares Problem

In applications that have more data points than breakpoints, we find a spline that minimizes the distance to the data. Let the cubic polynomial for the interval $[x_{i-1}, x_i]$ be written as

$$p_i(x) \equiv a_i + b_i x + c_i x^2 + d_i x^3, \quad i = 1, 2, \ldots, n.$$

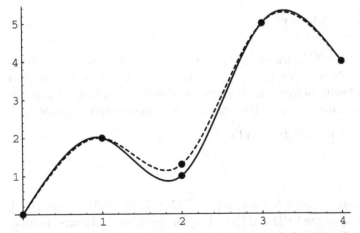

Figure 19.7: Curve fitting with cubic splines. The whole spline is changed, to the dotted curve, if just a data point is perturbed. See Example 19.3.1 for the details.

Because cubic splines are by definition twice continuously differentiable,

$$p_i(x_i) = p_{i+1}(x_i), \quad p_i'(x_i) = p_{i+1}'(x_i), \quad p_i''(x_i) = p_{i+1}''(x_i), \quad i = 1, 2, \dots, n-1,$$

where $x_0, x_1, \dots, x_n$ are the breakpoints. The preceding equations imply that

$$a_i + b_i x_i + c_i x_i^2 + d_i x_i^3 = a_{i+1} + b_{i+1} x_i + c_{i+1} x_i^2 + d_{i+1} x_i^3,$$
$$b_i + 2c_i x_i + 3d_i x_i^2 = b_{i+1} + 2c_{i+1} x_i + 3d_{i+1} x_i^2,$$
$$2c_i + 6d_i x_i = 2c_{i+1} + 6d_{i+1} x_i.$$

Let $x \equiv [a_1, b_1, c_1, d_1, \dots, a_n, b_n, c_n, d_n]^{\mathrm{T}} \in \mathbf{R}^{4n}$. Then the preceding equations can be written as $Bx = d$ for some $B \in \mathbf{R}^{3(n-1) \times 4n}$ and $d \in \mathbf{R}^{3(n-1)}$. Assume that $3(n-1)$ is the rank of B. Consider the data set $\{(\tilde{x}_1, \tilde{y}_1), (\tilde{x}_2, \tilde{y}_2), \dots, (\tilde{x}_m, \tilde{y}_m)\}$. Suppose $\tilde{x}_j$ falls within $[x_{i-1}, x_i]$. Then the LS formulation is

$$p_i(\tilde{x}_j) \equiv a_i + b_i \tilde{x}_j + c_i \tilde{x}_j^2 + d_i \tilde{x}_j^3 = \tilde{y}_j.$$

These m equations can be put into the form $Ax = b$ for some $A \in \mathbf{R}^{m \times 4n}$ and $b \in \mathbf{R}^m$. Thus we have the constrained LS problem $\min_{x \in \mathbf{R}^{4n}, \, Bx=d} \| Ax - b \|$. Note that A is banded (see Comment 19.2.2).

▶ **Exercise 19.3.5** Write the matrices for A, B, b, and d.

19.3.3 B-Splines and the Least-Squares Problem

A **B-spline** (**basic spline function**) B_i is a fixed cubic spline determined by five breakpoints $x_i, x_{i+1}, \dots, x_{i+4}$. It has zero value outside of $[x_i, x_{i+4}]$. Every cubic spline with breakpoints $x_0, x_1, \dots, x_n$ can be represented as

$$s(t) \equiv \sum_{i=-3}^{n-1} \alpha_i B_i(x), \quad x_0 \le x \le x_n$$

for unique α_is. (See [35] for the form of $B_i(x)$ and [391] for the simpler case of equally spaced breakpoints.) Let $\{(\tilde{x}_1, \tilde{y}_1), (\tilde{x}_2, \tilde{y}_2), \dots, (\tilde{x}_m, \tilde{y}_m)\}$ be the data. We seek $\alpha \equiv [\alpha_{-3}, \alpha_{-2}, \dots, \alpha_{n-1}]^{\mathrm{T}}$ that minimizes

$$\sum_{j=1}^{m} (s(\tilde{x}_j) - \tilde{y}_j)^2 = \| A\alpha - \tilde{y} \|^2 \tag{19.19}$$

for some matrix $A \in \mathbf{R}^{m \times (n+3)}$, where $\tilde{y} \equiv [\tilde{y}_1, \tilde{y}_2, \dots, \tilde{y}_m]^{\mathrm{T}}$. Because the only B-splines with nonzero values for $x \in [x_{i-1}, x_i]$ are $B_{i-3}, B_{i-2}, B_{i-1}$, and B_i, matrix A is banded with a bandwidth of four. The B-spline approach, in contrast to the cubic-spline approach to the same problem, does not lead to a constrained LS problem.

▶ **Exercise 19.3.6** Write the matrix A in Eq. (19.19).

Additional Reading

The literature on matrix computation is vast [391, 392, 417, 447, 465, 701, 830, 870]. See [825] for the history of the SVD and [392, 465] for applications. Comprehensive treatments of the LS problem can be found in [75, 586], and statistical properties of the LS estimator are covered in [422]. Reference [523] has a good coverage of factor analysis. Consult [417] for smoothing techniques. Splines in their present form are due to Schoenberg (1903–1990) in the 1940s [35, 447, 863].

NOTES

1. This term has become firmly entrenched even though "orthonormal" might be more consistent.
2. An alternative is to assume that the residuals have a multivariate normal distribution and then estimate the model parameters by using maximum likelihood [523, 632].

CHAPTER TWENTY

Time Series Analysis

The historian is a prophet in reverse.

Friedrich von Schlegel (1772–1829)

A sequence of observations indexed by the time of each observation is called a **time series**. Time series analysis is the art of specifying the most likely stochastic model that could have generated the observed series. The aim is to understand the relations among the variables in order to make better predictions and decisions. One particularly important application of time series analysis in financial econometrics is the study of how volatilities change over time because financial assets demand returns commensurate with their volatility levels.

20.1 Introduction

A time series of prices is called a financial time series: The prices can be those of stocks, bonds, currencies, futures, commodities, and countless others. Other time series of financial interest include those of prepayment speeds of MBSs and various economic indicators.

Models for the time series can be conjectured from studying the data or can be suggested by economic theory. Most models specify a stochastic process. Models should be consistent with past data and amenable to testing for specification error. They should also be simple, containing as few parameters as possible. Model parameters are to be estimated from the time series. However, because an observed series is merely a sample path of the proposed stochastic process, this is possible only if the process possesses a property called **ergodicity**. Ergodicity roughly means that sample moments converge to the population moments as the sample path lengthens.

The basic steps are illustrated with the maximum likelihood (ML) estimation of stock price volatility. Suppose that after the historical time series of prices $S_1, S_2, \ldots, S_{n+1}$, observed at Δt apart, is studied, the geometric Brownian motion process

$$\frac{dS}{S} = \mu \, dt + \sigma \, dW \tag{20.1}$$

is proposed. Transform the data to simplify the analysis. As the return process $r \equiv \ln S$ follows $dr = \alpha\, dt + \sigma\, dW$, where $\alpha \equiv \mu - \sigma^2/2$, we take the logarithmic transformation of the series and perform the difference transformation:

$$R_i \equiv \ln S_{i+1} - \ln S_i = \ln(S_{i+1}/S_i).$$

Clearly $R_1, R_2, \ldots, R_n$ are independent, identically distributed, normal random variables distributed according to $N(\alpha\,\Delta t, \sigma^2\Delta t)$. The log-likelihood function is

$$-\frac{n}{2}\ln(2\pi\sigma^2\Delta t) - \frac{1}{2\sigma^2\Delta t}\sum_{i=1}^{n}(R_i - \alpha\Delta t)^2.$$

Differentiate it with respect to α and σ^2 to obtain the ML estimators:

$$\widehat{\alpha} \equiv \frac{\sum_{i=1}^{n} R_i}{n\Delta t} = \frac{\ln(S_{n+1}/S_1)}{n\Delta t}, \tag{20.2}$$

$$\widehat{\sigma^2} \equiv \frac{\sum_{i=1}^{n}(R_i - \widehat{\alpha}\Delta t)^2}{n\Delta t}. \tag{20.3}$$

We note that the simple rate of return, $(S_{i+1} - S_i)/S_i$, and the continuously compounded rate of return, $\ln(S_{i+1}/S_i)$, should lead to similar conclusions because

$$\ln(S_{i+1}/S_i) = \ln(1 + (S_{i+1} - S_i)/S_i) \approx (S_{i+1} - S_i)/S_i.$$

EXAMPLE 20.1.1 Consider a time series generated by

$$S_{i+1} = S_i \times \exp[\,(\mu - \sigma^2/2)\Delta t + \sigma\sqrt{\Delta t}\,\xi\,], \quad \xi \sim N(0, 1),$$

with $S_1 = 1.0$, $\Delta t = 0.01$, $\mu = 0.15$, and $\sigma = 0.30$. Note that $\alpha = \mu - \sigma^2/2 = 0.105$. For the sample time series with $n = 5999$ in Fig. 20.1, the ML estimates of the parameters α and σ^2 are $\widehat{\alpha} = 0.118348$ and $\widehat{\sigma^2} = (0.299906)^2$. Because the variance of $\widehat{\sigma^2}$ is asymptotically $2\sigma^4/n$, increasing n definitely helps. However, because the variance for the estimator of α (hence μ) is asymptotically $\sigma^2/(n\Delta t)$, increasing n by sampling ever more frequently over the *same* time interval does not narrow

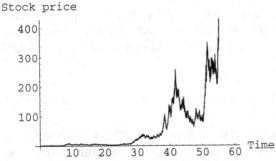

Figure 20.1: A simulated time series. The process is $dS = 0.15 \times S\,dt + 0.3 \times S\,dW$, starting at $S(0) = 1$.

the variance because the length of the sampling period, $n\Delta t$, remains unchanged [147, 611]. The variance can be reduced by sampling over a longer sampling period. For example, with $\Delta t = 2$ and $n = 999$ (the realized time series is not shown), $\widehat{\alpha}$ becomes 0.108039, a substantial improvement over the previous estimate even though the process is sampled *less* frequently (with a larger Δt).

As another illustration, consider the Ogden model for the short rate r,

$$dr = \beta(\mu - r)\,dt + \sigma r\,dW, \tag{20.4}$$

where $\beta > 0$, μ, and σ are the parameters [702]. Because approximately

$$\Delta r - \beta(\mu - r)\,\Delta t \sim N(0, \sigma^2 r^2 \Delta t),$$

conditional on r_1 the likelihood function for the n observations $\Delta r_1, \Delta r_2, \ldots,$ Δr_n is

$$\prod_{i=1}^{n} \left(2\pi\sigma^2 r_i^2 \Delta t\right)^{-1/2} \exp\left[-\frac{\{\Delta r_i - \beta(\mu - r_i)\,\Delta t\}^2}{2\sigma^2 r_i^2 \Delta t} \right],$$

where $\Delta r_i \equiv r_{i+1} - r_i$. The log-likelihood function after the removal of the constant terms and simplification is

$$-n \ln \sigma - \frac{1}{2\sigma^2 \Delta t} \sum_{i=1}^{n} [\,\Delta r_i - \beta(\mu - r_i)\,\Delta t\,]^2 r_i^{-2}.$$

Differentiating the log-likelihood function with respect to β, μ, and σ and equating them to zero gives rise to three equations in three unknowns:

$$0 = \sum_{i} [\,\Delta r_i - \beta(\mu - r_i)\,\Delta t\,](\mu - r_i)\,r_i^{-2}, \tag{20.5}$$

$$0 = \sum_{i} [\,\Delta r_i - \beta(\mu - r_i)\,\Delta t\,]\,r_i^{-2}, \tag{20.6}$$

$$\sigma^2 = \frac{1}{n\Delta t} \sum_{i} [\,\Delta r_i - \beta(\mu - r_i)\,\Delta t\,]^2 r_i^{-2}. \tag{20.7}$$

The ML estimators are not hard to obtain (see Exercise 20.1.3).

EXAMPLE 20.1.2 Consider the time series generated by $r_{i+1} = r_i + \beta(\mu - r_i)\,\Delta t + \sigma r_i \sqrt{\Delta t}\,\xi$, $\xi \sim N(0, 1)$, with $r_1 = 0.08$, $\Delta t = 0.1$, $\beta = 1.85$, $\sigma = 0.30$, $\mu = 0.08$, and $n = 999$. For the time series in Fig. 20.2, the ML estimates are $\widehat{\mu} = 0.0799202$, $\widehat{\beta} = 1.86253$, and $\widehat{\sigma} = 0.300989$.

▶ **Exercise 20.1.1** Derive the ML estimators for μ and σ^2 based on simple rates of returns, $\Delta S_i \equiv S_{i+1} - S_i$, $i = 1, 2, \ldots, n$.

▶ **Exercise 20.1.2** Assume that the stock price follows Eq. (20.1). The simple rate of return is defined as $[\,S(t) - S(0)\,]/S(0)$. Suppose that the volatility of the stock is that of simple rates of return, σ_s, instead of the instantaneous rates of return, σ. Express σ in terms of σ_s, the horizon t, and the expected simple rate of return at the horizon, μ_s.

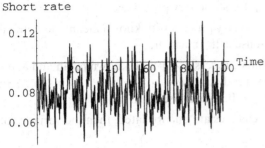

Short rate

Figure 20.2: A simulated time series of the Ogden model. The process is $dr = 1.85 \times (0.08 - r)\,dt + 0.3 \times r\,dW$, starting at $r(0) = 0.08$.

▶ **Exercise 20.1.3** Derive the formulas for β and μ.

▶ **Exercise 20.1.4** Use the process for $\ln r$ to obtain the ML estimators for the Ogden model.

▶ **Exercise 20.1.5** The **constant elasticity variance (CEV) process** follows $dS/S = \mu\,dt + \lambda S^\theta\,dW$, where $\lambda > 0$. Derive the ML estimators with $\mu = 0$.

▶ **Programming Assignment 20.1.6** Write the simulator in Fig. 18.5 to generate stock prices. Then experiment with the ML estimators for goodness of fit.

20.1.1 Basic Definitions and Models

Although processes in this chapter are discrete-time by default, most of the definitions have continuous-time counterparts in Chap. 13. To simplify the presentation, we rely on context instead of notation to distinguish between random variables and their realizations.

Consider a discrete-time stochastic process $X_1, X_2, \ldots, X_n$. The **autocovariance** of X_t at lag τ is defined as $\mathrm{Cov}[\, X_t, X_{t-\tau} \,]$, and the **autocorrelation** of X_t at lag τ is defined as $\mathrm{Cov}[\, X_t, X_{t-\tau} \,]/\sqrt{\mathrm{Var}[\, X_t \,] \times \mathrm{Var}[\, X_{t-\tau} \,]}$, which clearly lies between -1 and 1. In general, autocovariances and autocorrelations depend on the time t as well as on the lag τ. The process is stationary if it has an identical mean μ and the autocovariances depend on only the lag; in particular, the variance is a constant. Because stationary processes are easier to analyze, transformation should be applied to the series to ensure stationarity whenever possible. The process is said to be **strictly stationary** if $\{\, X_s, X_{s+1}, \ldots, X_{s+\tau-1} \,\}$ and $\{\, X_t, X_{t+1}, \ldots, X_{t+\tau-1} \,\}$ have the same distribution for any s, t, and $\tau > 0$. This implies that the X_ts have identical distributions. A strictly stationary process is automatically stationary. A general process is said to be (**serially**) **uncorrelated** if all the autocovariances with a nonzero lag are zero; otherwise, it is **correlated**.

A stationary, uncorrelated process is called **white noise**. A strictly stationary process $\{\, X_t \,\}$, where the X_t are independent, is called **strict white noise**. A stationary process is called **Gaussian** if the joint distribution of $X_{t+1}, X_{t+2}, \ldots, X_{t+k}$ is multivariate normal for every possible integer k. A Gaussian process is automatically strictly stationary; in particular, Gaussian white noise is strict white noise.

Denote the autocovariance at lag τ of a stationary process by λ_τ. The autocovariance λ_0 is then the same variance shared by all X_t. The autocorrelations at lag τ of a stationary process are denoted by $\rho_\tau \equiv \lambda_\tau/\lambda_0$.

➤ **Exercise 20.1.7** Prove that $\lambda_\tau = \lambda_{-\tau}$ for stationary processes.

➤ **Exercise 20.1.8** Show that, for a stationary process with known mean, the optimal predictor in the mean-square-error sense is the mean μ.

➤ **Exercise 20.1.9** Consider a stationary process $\{X_t\}$ with known mean and auto-covariances. Derive the optimal linear prediction $a_0 + a_1 X_t + a_2 X_{t-1} + \cdots + a_t X_1$ in the mean-square-error sense for X_{t+1}. (Hint: Exercise 6.4.1.)

➤ **Exercise 20.1.10** Show that if price changes are uncorrelated, then the variance of prices must increase with time.

➤ **Exercise 20.1.11** Verify that if $\{X_t\}$ is strict white noise, then so are $\{|X_t|\}$ and $\{X_t^2\}$.

20.1.2 The Efficient Markets Hypothesis

The **random-walk hypothesis** posits that price changes are random and prices therefore behave unpredictably. For example, Bachelier in his 1900 thesis assumed that price changes have independent and identical normal distributions. The more modern version asserts that the return process has constant mean and is uncorrelated. One consequence of the hypothesis is that expected returns cannot be improved by use of past prices.

The random walk hypothesis is a purely statistical statement. By the late 1960s, however, the hypothesis could no longer withstand the mounting evidence against it. It was the work of Fama and others that shifted the focus from the time series of returns to that of cost- and risk-adjusted returns: Excess returns should account for transactions costs, risks, and information available to the trading strategy. Thereafter, the efficient-markets debate became a matter of economics instead of a matter of pure statistics [666].

The economic theory to explain the randomness of security prices is called the **efficient markets hypothesis**. It holds that the market may be viewed as a great information processor and prices come to reflect available information immediately. A market is efficient with respect to a particular information set if it is impossible to make abnormal profits by using this set of information in making trading decisions [799]. If the set of information refers to past prices of securities, we have the weak form of market efficiency; if the set of information refers to all *publicly available* information, we have the **semistrong** form; finally, if the set of information refers to all public and *private* information, we have the **strong** form. These terms are due to Roberts in 1967 [147]. The evidence suggests that major U.S. markets are at least weak-form efficient [767].

➤ **Exercise 20.1.12** Why are near-zero autocorrelations important for returns to be unpredictable?

20.1.3 Three Classic Models

Let ϵ_t denote a zero-mean white-noise process with constant variance σ^2 throughout the section. The first model is the **autoregressive (AR) process**, whose simplest version is

$$X_t - b = a(X_{t-1} - b) + \epsilon_t.$$

This process is stationary if $|a| < 1$, in which case $E[X_t] = b$, $\lambda_0 = \sigma^2/(1 - a^2)$ and $\lambda_\tau = a^\tau \lambda_0$. The autocorrelations decay exponentially to zero because $\lambda_\tau/\lambda_0 = a^\tau$.[1] The general $AR(p)$ process follows

$$X_t - b = \epsilon_t + \sum_{i=1}^{p} a_i (X_{t-i} - b).$$

The next model is the **moving average (MA) process,**

$$X_t - b = \epsilon_t + c\epsilon_{t-1},$$

which has mean b and variance $\lambda_0 = (1 + c^2)\sigma^2$. The autocovariance λ_τ equals 0 if $\tau > 1$ and $c\sigma^2$ if $\tau = 1$. The process is clearly stationary. Note that the autocorrelation function drops to zero beyond $\tau = 1$. The general $MA(q)$ process is defined by

$$X_t - b = \epsilon_t + \sum_{j=1}^{q} c_j \epsilon_{t-j}.$$

It is not hard to show that observations more than q periods apart are uncorrelated. Repeated substitutions for the $MA(1)$ process with $|c| < 1$ yield

$$X_t - b = -\sum_{i=1}^{\infty} (-c)^i (X_{t-i} - b) + \epsilon_t.$$

This can be seen as an $AR(\infty)$ process in which the effect of past observations decrease with age. An MA process is said to be **invertible** if it can be represented as an $AR(\infty)$ process.

The third model, the **autoregressive moving average (ARMA) process,** combines the $AR(1)$ and $MA(1)$ processes; thus

$$X_t - b = a(X_{t-1} - b) + \epsilon_t + c\epsilon_{t-1}.$$

Assume that $|a| < 1$ so that the ARMA process is stationary. Then the mean is b and

$$\lambda_0 = \sigma^2 \frac{1 + 2ac + c^2}{1 - a^2},$$

$$\lambda_1 = \sigma^2 \frac{(1 + ac)(a + c)}{1 - a^2},$$

$$\lambda_\tau = a\lambda_{\tau-1} \quad \text{for } \tau \geq 2.$$

We define the more general $ARMA(p, q)$ process by combining $AR(p)$ and $MA(q)$:

$$X_t - b = \sum_{i=1}^{p} a_i (X_{t-i} - b) + \sum_{j=0}^{q} c_j \epsilon_{t-j},$$

where $c_0 = 1$, $a_p \neq 0$, and $c_q \neq 0$.

Repeated substitutions for the stationary AR(1) process, with $|a| < 1$, yield

$$X_t - b = \sum_{j=0}^{\infty} a^j \epsilon_{t-j},$$

an MA(∞) process. This is a special case of **Wold's decomposition**, which says that any stationary process $\{X_t\}$, after the linearly deterministic component has been removed, can be represented as an MA(∞) process:

$$X_t - b = \sum_{j=0}^{\infty} c_j \epsilon_{t-j}, \tag{20.8}$$

where $c_0 = 1$ and $\sum_{j=1}^{\infty} c_j^2 < \infty$. Both stationary AR and ARMA processes admit such a representation.

Stationary processes have nice asymptotic properties. Suppose that a stationary process $\{X_t\}$ is represented as Eq. (20.8), where $\sum_{j=1}^{\infty} |c_j| < \infty$ and ϵ_t are zero-mean, independent, identically distributed random variables with $E[\epsilon_t^2] < \infty$. A useful central limit theorem says that the sample mean is asymptotically normal in the sense that

$$\sqrt{n} \left(\frac{1}{n} \sum_{i=1}^{n} X_i - b \right) \to N \left(0, \sum_{j=-\infty}^{\infty} \lambda_j \right) \quad \text{as } n \to \infty.$$

We note that $\sum_{j=1}^{\infty} |c_j| < \infty$ implies that $\sum_{j=1}^{\infty} c_j^2 < \infty$ but not vice versa, and it guarantees ergodicity for MA(∞) processes [415].

▶ **Exercise 20.1.13** Let $\{X_t\}$ be a sequence of independent, identically distributed random variables with zero mean and unit variance. Prove that the process $\{Y_t \equiv \sum_{k=0}^{l} a_k X_{t-k}\}$ with constant a_k is stationary.

▶ **Exercise 20.1.14** Given Wold's decomposition (20.8), show that λ_τ, the autocovariance at lag τ, equals $\sigma^2 \sum_{j=0}^{\infty} c_j c_{j+\tau}$.

Conditional Estimation of Gaussian AR Processes

For Gaussian AR processes, the ML estimation reduces to OLS problems. Write the AR(p) process as

$$X_t = c + \sum_{i=1}^{p} a_i X_{t-i} + \epsilon_t,$$

where ϵ_t is a zero-mean Gaussian white noise with constant variance σ^2. The parameters to be estimated from the observations $X_1, X_2, \ldots, X_n$ are $a_1, a_2, \ldots, a_p, c$, and σ^2. Conditional on the first p observations, the log-likelihood function can be easily seen to be

$$-\frac{n-p}{2} \ln(2\pi) - \frac{n-p}{2} \ln(\sigma^2) - \frac{1}{2\sigma^2} \sum_{t=p+1}^{n} \left(X_t - c - \sum_{i=1}^{p} a_i X_{t-i} \right)^2.$$

The values of $a_1, a_2, \ldots, a_p$, and c that maximize the preceding function must minimize

$$\sum_{t=p+1}^{n} \left(X_t - c - \sum_{i=1}^{p} a_i X_{t-i} \right)^2, \tag{20.9}$$

an LS problem (see Exercise 20.1.15). This methodology is called the **conditional ML estimation**. The estimate will be consistent for any stationary ergodic AR process even if it is not Gaussian. The estimator of σ^2 is

$$\frac{1}{n-p} \sum_{t=p+1}^{n} \left(X_t - \widehat{c} - \sum_{i=1}^{p} \widehat{a}_i X_{t-i} \right)^2,$$

which can be found by differentiation of the log-likelihood function with respect to σ^2.

▶ **Exercise 20.1.15** Write the equivalent LS problem for function (20.9).

20.2 Conditional Variance Models for Price Volatility

Although a stationary model has constant variance, its *conditional* variance may vary. Take for example a stationary AR(1) process $X_t = a X_{t-1} + \epsilon_t$. Its conditional variance,

$$\text{Var}[\, X_t \mid X_{t-1}, X_{t-2}, \ldots \,],$$

equals σ^2, which is smaller than the *unconditional* variance $\text{Var}[\, X_t \,] = \sigma^2/(1 - a^2)$. Note that the conditional variance is independent of past information; this property holds for ARMA processes in general. Past information thus has no effect on the variance of prediction. To address this drawback, consider models for returns X_t consistent with a changing conditional variance in the form of

$$X_t - \mu = V_t U_t.$$

It is assumed that (1) U_t has zero mean and unit variance for all t, (2) $E[\, X_t \,] = \mu$ for all t, and (3) $\text{Var}[\, X_t \mid V_t = v_t \,] = v_t^2$. The process $\{\, V_t^2 \,\}$ thus models the conditional variance.

Suppose that $\{\, U_t \,\}$ and $\{\, V_t \,\}$ are independent of each other, which means that $\{\, U_1, U_2, \ldots, U_n \,\}$ and $\{\, V_1, V_2, \ldots, V_n \,\}$ are independent for all n. Then $\{\, X_t \,\}$ is uncorrelated because

$$\text{Cov}[\, X_t, X_{t+\tau} \,] = E[\, V_t U_t V_{t+\tau} U_{t+\tau} \,] = E[\, V_t U_t V_{t+\tau} \,] E[\, U_{t+\tau} \,] = 0 \tag{20.10}$$

for $\tau > 0$. Furthermore, if $\{\, V_t \,\}$ is stationary, then $\{\, X_t \,\}$ has constant variance because

$$E[\, (X_t - \mu)^2 \,] = E[\, V_t^2 U_t^2 \,] = E[\, V_t^2 \,] E[\, U_t^2 \,] = E[\, V_t^2 \,], \tag{20.11}$$

making $\{\, X_t \,\}$ stationary.

EXAMPLE 20.2.1 Here is a lognormal model. Let the processes $\{\, U_t \,\}$ and $\{\, V_t \,\}$ be independent of each other, $\{\, U_t \,\}$ is Gaussian white noise, and $\ln V_t \sim N(a, b^2)$. One simple way to achieve this as well as to make both $\{\, |X_t - \mu| \,\}$'s and $\{\, (X_t - \mu)^2 \,\}$'s autocorrelations positive is to posit the following AR(1) model for $\{\ln V_t\}$:

$$\ln(V_t) - \alpha = \theta(\ln(V_{t-1}) - \alpha) + \xi_t, \quad \theta > 0.$$

In the preceding equation, $\{\xi_t\}$ is zero-mean, Gaussian white noise independent of $\{U_t\}$. To ensure the above-mentioned variance for $\ln V_t$, let $\text{Var}[\xi_t] = b^2(1 - \theta^2)$. The four parameters in this model – $a, b, \theta,$ and α – can be estimated by the method of moments [839].

▶ **Exercise 20.2.1** Assume that the processes $\{V_t\}$ and $\{U_t\}$ are stationary and independent of each other. Show that the kurtosis of X_t exceeds that of U_t provided that both are finite.

▶ **Exercise 20.2.2** For the lognormal model, show that (1) the kurtosis of X_t is $3e^{4b^2}$, (2) $\text{Var}[V_t] = e^{2a+b^2}(e^{b^2} - 1)$, (3) $\text{Var}[|X_t - \mu|] = e^{2a+b^2}(e^{b^2} - 2/\pi)$, (4) $\text{Var}[V_t^2] = e^{4a+4b^2}(e^{4b^2} - 1)$, and (5) $\text{Var}[(X_t - \mu)^2] = e^{4a+4b^2}(3e^{4b^2} - 1)$.

20.2.1　ARCH and GARCH Models

One trouble with the lognormal model is that the conditional variance evolves independently of past returns. Suppose we assume that conditional variances are deterministic functions of past returns: $V_t = f(X_{t-1}, X_{t-2}, \ldots)$ for some function f. Then V_t can be computed given $I_{t-1} \equiv \{X_{t-1}, X_{t-2}, \ldots\}$, the information set of past returns. An influential model in this direction is the **autoregressive conditional heteroskedastic (ARCH) model**.

Assume that U_t is independent of $V_t, U_{t-1}, V_{t-1}, U_{t-2}, \ldots,$ for all t. Consequently $\{X_t\}$ is uncorrelated by Eq. (20.10). Assume furthermore that $\{U_t\}$ is a Gaussian white-noise process. Hence $X_t \mid I_{t-1} \sim N(\mu, V_t^2)$. The ARCH($p$) process is defined by

$$X_t - \mu = \left[a_0 + \sum_{i=1}^{p} a_i (X_{t-i} - \mu)^2\right]^{1/2} U_t,$$

where $a_0, a_1, \ldots, a_p$ are nonnegative and $a_0 > 0$ is usually assumed. The variance V_t^2 thus equals $a_0 + \sum_{i=1}^{p} a_i(X_{t-i} - \mu)^2$. This suggests that V_t depends on $U_{t-1}, U_{t-2}, \ldots, U_{t-p}$, which is indeed the case; $\{U_t\}$ and $\{V_t\}$ are hence not independent of each other. In practical terms, the model says that the volatility at time t as estimated at time $t-1$ depends on the p most recent observations on squared returns.[2]

The ARCH(1) process $X_t - \mu = [a_0 + a_1(X_{t-1} - \mu)^2]^{1/2}U_t$ is the simplest for which

$$\text{Var}[X_t \mid X_{t-1} = x_{t-1}] = a_0 + a_1(x_{t-1} - \mu)^2. \tag{20.12}$$

The process $\{X_t\}$ is stationary with finite variance if and only if $a_1 < 1$, in which case $\text{Var}[X_t] = a_0/(1 - a_1)$. The kurtosis is a finite $3(1 - a_1^2)/(1 - 3a_1^2)$ when $3a_1^2 < 1$ and exceeds three when $a_1 > 0$. Let $S_t \equiv (X_t - \mu)^2$. Then

$$E[S_t \mid S_{t-1}] = a_0 + a_1 S_{t-1}$$

by Eq. (20.12). This resembles an AR process. Indeed, $\{S_t\}$ has autocorrelations a_1^τ when the variance of S_t exists, i.e., $3a_1^2 < 1$. Because $X_t \mid I_{t-1} \sim N(\mu, a_0 + a_1 S_{t-1})$, the log-likelihood function equals

$$-\frac{n-1}{2}\ln(2\pi) - \frac{1}{2}\sum_{i=2}^{n}\ln(a_0 + a_1(X_{i-1} - \mu)^2) - \frac{1}{2}\sum_{i=2}^{n}\frac{(X_i - \mu)^2}{a_0 + a_1(X_{i-1} - \mu)^2}.$$

We can estimate the parameters by maximizing the above function. The results for the more general ARCH(p) model are similar.

A popular extension of the ARCH model is the **generalized autoregressive conditional heteroskedastic (GARCH) process**. The simplest GARCH$(1,1)$ process adds $a_2 V_{t-1}^2$ to the ARCH(1) model:

$$V_t^2 = a_0 + a_1(X_{t-1} - \mu)^2 + a_2 V_{t-1}^2.$$

The volatility at time t as estimated at time $t-1$ thus depends on the squared return and the estimated volatility at time $t-1$. By repeated substitutions, the estimate of volatility can be seen to average past squared returns by giving heavier weights to recent squared returns (see Exercise 20.2.3, part (1)). For technical reasons, it is usually assumed that $a_1 + a_2 < 1$ and $a_0 > 0$, in which case the unconditional long-run variance is given by $a_0/(1 - a_1 - a_2)$. The model also exhibits mean reversion (see Exercise 20.2.3, part (2)).

▶ **Exercise 20.2.3** Assume the GARCH$(1,1)$ model. Show that (1) $V_t^2 = a_0 \sum_{i=0}^{k-1} a_2^i + \sum_{i=1}^{k} a_1 a_2^{i-1}(X_{t-i} - \mu)^2 + a_2^k V_{t-k}^2$, where $k > 0$ and (2) $\text{Var}[X_{t+k} \mid V_t] = V + (a_1 + a_2)^k(V_t^2 - V)$, where $V \equiv a_0/(1 - a_1 - a_2)$.

Additional Reading

Many books cover time-series analysis and the important subject of model testing skipped here [22, 415, 422, 667, 839]. A version of the efficient markets hypothesis that is due to Samuelson in 1965 says asset returns are martingales (see Exercise 13.2.2) [594]. However, constant expected returns that a martingale entails have been rejected by the empirical evidence if markets are efficient [333]. See [148, 333, 424, 587, 594, 767] for more information on the efficient markets hypothesis. Wold's decomposition is due to Wold in 1938 [881]. The instability of the variance of returns has consequences for long-term investors [69]. Stochastic volatility has been extensively studied [49, 293, 440, 444, 450, 471, 520, 579, 790, 823, 839, 840]. Besides the Ito process approach to stochastic volatility in Section 15.5, jump processes have been proposed for the volatility process [177]. Consult [255] for problems with such approaches to volatility. One more approach to the smile problem is the implied binomial tree [502, 503]. The ARCH model is proposed by Engle [319]. The GARCH model is proposed by Bollerslev [99] and Taylor [839]. It has found widespread empirical support [464, 552, 578]. See [286, 287, 288, 442] for option pricing models based on the GARCH process and [749] for algorithms. Consult [517] for a survey on the estimation of Ito processes of the form $dX_t = \mu(X_t)\,dt + \sigma(X_t)\,dW_t$. As illustrated in Example 20.1.1, direct estimation of the drift μ is difficult in general from discretely observed data over a short time interval, however frequently sampled; the diffusion σ can be precisely estimated.

The **generalized method of moments (GMM)** is an estimation method that extends the method of moments in Subsection 6.4.3. Like the method of moments, the GMM formulates the moment conditions in which the parameters are implicitly defined. However, instead of solving equations, the GMM finds the parameters that jointly minimize the weighted "distance" between the sample and the population moments. The typical conditions for the GMM estimate to be consistent include stationarity, ergodicity, and the existence of relevance expectations. The GMM method of moments is due to Hansen [418] and is widely used in the analysis of time series [173, 248, 384, 526, 754, 819].

NOTES

1. The stationarity condition rules out the random walk with drift in Example 13.1.2 as a stationary AR process. Brownian motion, the limit of such a random walk, is also not stationary (review Subsection 13.3.3).
2. In practice μ is often assumed to be zero. This is reasonable when Δt is small, say 1 day, because the expected return is then insignificant compared with the standard deviation of returns.

> Someone who tried to use modern observations from London and Paris to judge mortality rates of the Fathers before the flood would enormously deviate from the truth.
> Gottfried Wilhelm von Leibniz (1646–1716)

CHAPTER TWENTY-ONE

Interest Rate
Derivative Securities

> I never gamble.
> J.P. Morgan, Sr. (1837–1913)

Interest-rate-sensitive securities are securities whose payoff depends on the levels and/or evolution of interest rates. The interest rate derivatives market is enormous. The global notional principal of over-the-counter derivative contracts was an estimated U.S.\$72 trillion as of the end of June 1998, of which 67% were interest rate instruments and 31% were forex instruments [51]. The use of such derivatives in portfolio risk management has made possible economical and efficient alteration of interest rate sensitivities [325]. Throughout this book, **interest rate derivative securities** exclude fixed-income securities with embedded options.

21.1 Interest Rate Futures and Forwards

An interest rate futures contract is a futures contract whose underlying asset depends solely on the level of interest rates. Figure 21.1 gives an idea of the diversity of interest rate futures.

21.1.1 Treasury Bill Futures

The first financial futures contract was based on a fixed-income instrument, the Government National Mortgage Association (GNMA or "Ginnie Mac") mortgage-backed certificates whose trading began in 1972 at the CBT. The IMM of the CME followed 3 months later with futures contracts based on the 13-week T-bill [95].

The T-bill futures contract traded on the IMM is based on the 13-week (3-month) T-bill with a face value of \$1 million. The seller of a T-bill futures contract agrees to deliver to the buyer at the delivery date a T-bill with 13 weeks remaining to maturity and a face value of \$1 million. The T-bill delivered can be newly issued or seasoned. The futures price is the price at which the T-bill will be sold by the short and purchased by the buyer. The contract allows for the delivery of 89-, 90-, or 91-day T-bills after price adjustments.

T-bills are quoted in the spot market in terms of the annualized discount rate of formula (3.9). In contrast, the futures contract is quoted, not directly in terms of

Monday, March 20, 1995
FUTURES PRICES

· · ·

	Open	High	Low	Settle	Change	Lifetime High	Lifetime Low	Open Interest

· · ·

INTEREST RATE

TREASURY BONDS (CBT) — $100,000; pts. 32nds of 100%

	Open	High	Low	Settle	Change	Lifetime High	Lifetime Low	Open Interest
Mar	104-31	105-05	104-18	104-20	−11	116-20	95-13	28,210
June	104-12	104-19	104-00	104-02	−11	113-15	94-27	328,566
Sept	104-06	104-06	103-19	103-21	−11	112-15	94-10	14,833
Dec	103-23	103-23	103-04	103-07	−12	111-23	93-27	1,372
Mr96	102-26	103-00	102-26	102-26	−13	103-17	93-13	232
June	102-15	102-15	102-13	102-13	−13	104-28	93-06	46

· · ·

TREASURY NOTES (CBT) — $100,000; pts. 32nds of 100%

	Open	High	Low	Settle	Change	Lifetime High	Lifetime Low	Open Interest
Mar	105-00	105-02	104-27	104-28	−3	111-07	98-11	24,870
June	104-16	104-20	104-11	104-13	−3	105-22	97-27	225,265

· · ·

5 YR TREAS NOTES (CBT) — $100,000; pts. 32nds of 100%

	Open	High	Low	Settle	Change	Lifetime High	Lifetime Low	Open Interest
Mar	103-31	03-315	103-27	03-275	−2.5	104-11	99-15	19,230
June	103-18	103-21	03-155	03-165	−2.0	104-01	99-06	179,928

· · ·

2 YR TREAS NOTES (CBT) — $200,000; pts. 32nds of 100%

	Open	High	Low	Settle	Change	Lifetime High	Lifetime Low	Open Interest
Mar	02-085	02-085	102-07	102-07	−1/4	02-105	99-252	5,776
June	103-30	01-302	01-282	101-29	+1/4	02-015	99-24	26,904

· · ·

30-DAY FEDERAL FUNDS (CBT) — $5 million; pts. of 100%

	Open	High	Low	Settle	Change	Lifetime High	Lifetime Low	Open Interest
Mar	94.05	94.05	94.05	94.05	−.01	94.44	93.28	2,905
Apr	93.97	93.97	93.96	93.97	· · ·	93.98	93.05	4,331

· · ·

TREASURY BILLS (CME) — $1 mil.; pts. of 100%

	Open	High	Low	Settle	Chg	Discount Settle	Discount Chg	Open Interest
June	94.04	94.06	94.04	94.05	+.02	5.95	−.02	16,964
Sept	93.83	93.83	93.80	93.82	+.01	6.18	−.01	10,146
Dec	93.64	93.66	93.64	93.66	−.01	6.34	+.01	9,082

· · ·

LIBOR-1 MO. (CME) — $3,000,000; points of 100%

	Open	High	Low	Settle	Chg	Discount Settle	Discount Chg	Open Interest
Apr	93.84	93.84	93.82	93.83	· · ·	6.17	· · ·	27,961

· · ·

EURODOLLAR (CME) — $1 million; pts of 100%

	Open	High	Low	Settle	Chg	Yield Settle	Yield Chg	Open Interest
June	93.52	93.54	93.51	93.52	· · ·	6.48	· · ·	515,578
Sept	93.31	93.32	93.28	93.30	· · ·	6.70	· · ·	322,889
				· · ·				
Mr04	91.65	91.65	91.65	91.65	−.01	8.35	+.01	1,910

· · ·

Figure 21.1: Interest rate futures quotations. Expanded from Fig. 12.3.

yield, but instead on an index basis that is related to the discount rate as

index price $= 100 - ($annualized discount yield $\times 100)$.

The price is therefore merely a different way of quoting the interest rate. For example, if the yield is 8%, the index price is $100 - (0.08 \times 100) = 92$. Alternatively, the discount yield for the futures contract can be derived from the price of the futures contract:

$$\text{annualized discount yield} = \frac{100 - \text{index price}}{100}.$$

The invoice price that a buyer of $1 million face-value of 13-week T-bills must pay at the delivery date is found by first computing the dollar discount:

dollar discount $=$ annualized discount yield $\times \$1,000,000 \times (T/360)$,

where T is the number of days to maturity. The invoice price is

invoice price $= \$1,000,000 -$ dollar discount.

Combining the two preceding equations, we arrive at

invoice price $= \$1,000,000 \times [\, 1 - $ annualized discount yield $\times (T/360)\,]$,

where $T = 89$, 90, or 91. For example, suppose that the index price for a T-bill futures contract is 92.52. The discount yield for this T-bill futures contract is $(100 - 92.52)/100 = 7.48\%$, and the dollar discount for the T-bill to be delivered with 91 days to maturity is

$$0.0748 \times \$1,000,000 \times \frac{91}{360} = \$18,907.78.$$

The invoice price is thus $1,000,000 - 18,907.78 = 981,092.22$ dollars.

The "tick" for the T-bill futures contract is 0.01. A change of 0.01 translates into a one-basis-point change in the discount yield. The dollar price change of a tick is therefore $0.0001 \times \$1,000,000 \times (90/360) = \$25$ for a 90-day contract. The contract is quoted and traded in half-tick increments.

Suppose that the futures contract matures in t years and its underlying T-bill matures in $t^* > t$ years, the difference between them being 90 days. Let $S(t)$ and $S(t^*)$ denote the continuously compounded riskless spot rates for terms t and t^*, respectively. Because no income is paid on the bill, the futures price equals $F = e^{S(t)t - S(t^*)t^*}$ by Lemma 12.2.1. This incidentally shows the duration of the T-bill futures to be $t^* - t$ if the yield curve is flat. Let $\bar{r} \equiv (t^* - t)^{-1} \ln(1/F) = [\, S(t^*)t^* - S(t)t\,]/(t^* - t)$ be the continuously compounded forward rate for the time period between t and t^*. The futures price is therefore the price the bill will have if the 90-day interest rate at the delivery date proves to be $\bar{r}$. Rearrange the terms to obtain

$$S(t) = \frac{S(t^*)t^* - \bar{r}(t^* - t)}{t}.$$

The rate derived by the right-hand-side formula above is called the **implied repo rate**. As argued above, arbitrage opportunities exist if the implied repo rate differs from the actual T-bill rate $S(t)$.

EXAMPLE 21.1.1 Suppose that the price of a T-bill maturing in 138 days (0.3781 year) is 95 and the futures price for a 90-day (0.2466-year) T-bill futures contract maturing in 48 days is 96.50. The implied repo rate is

$$\frac{138 \times (1/0.3781) \ln(1/0.95) - 90 \times (1/0.2466) \ln(1/0.965)}{48} = 0.1191,$$

or 11.91%.

➤ **Exercise 21.1.1** Derive the formula for the change in the T-bill futures' invoice price per tick with 91 days to maturity.

Duration-Based Hedging

Under continuous compounding, the duration of a T-bill is its term to maturity. Suppose that one anticipates a cash inflow of L dollars at time t and plans to invest it in 6-month T-bills. To address the concern that the interest rate may drop at time t by using Treasury bill futures, one should buy the following number of contracts:

$$\frac{L \times 0.5}{\text{T-bill futures contract price} \times 0.25}$$

by Eq. (4.13). The general formula is

$$\frac{L \times \text{duration of the liability}}{\text{interest rate futures contract price} \times \text{duration of the futures}}$$

under the assumption of parallel shifts.

EXAMPLE 21.1.2 A firm holds 6-month T-bills with a total par value of $10 million. The term structure is flat at 8%. The current T-bill value is hence $10,000,000 \times e^{-0.08/2} = \$9,607,894$, and the current T-bill futures price is $\$1,000,000 \times e^{-0.08/4} = \$980,199$. The firm hedges by selling $(9,607,894 \times 0.5)/(980,199 \times 0.25) = 19.6$ futures contracts.

21.1.2 The Eurodollar Market

Money deposited outside its nation of origin is called **Eurocurrency**. Eurocurrency trading involves the borrowing and the lending of time deposits. The most important Eurocurrency is the Eurodollar, and the interest rate banks pay on Eurodollar time deposits is LIBOR. The 1-month LIBOR is the rate offered on 1-month deposits, the 3-month LIBOR is the rate offered on 3-month deposits, and so on. LIBOR plays the role in international financial markets that prime rates do in domestic ones.

LIBOR is quoted actual over 360; therefore the interest rate is stated as if the year had 360 days even though interest is paid daily. Consider a $1 million loan with an annualized interest rate of the 6-month LIBOR plus 0.5%. The life of this loan is divided into 6-month periods. For each period, the rate of interest is set 0.5% above the 6-month LIBOR at the beginning of the period. For example, if the current LIBOR is 8% and the period has 182 days, the interest due at the end of the period is

$$\$1,000,000 \times 0.085 \times \frac{182}{360} = \$42,972.22.$$

The LIBOR rate underestimates the effective rate. For example, the effective annual rate corresponding to a 6-month LIBOR quote of 7% is the slightly higher 7.2231%

because

$$\left(1 + 0.07 \times \frac{182}{360}\right)\left(1 + 0.07 \times \frac{183}{360}\right) - 1 \approx 0.072231.$$

Because the LIBOR yields are quoted in terms of an add-on interest rate, they are directly comparable with those on domestic CDs.

21.1.3 Eurodollar Futures

Eurodollar futures started trading in 1981 and are now traded on both the IMM and the London International Financial Futures and Options Exchange (LIFFE). This contract is one of the most heavily traded futures contracts in the world. The 3-month Eurodollar is the underlying instrument for the Eurodollar futures contract, which has a delivery date ranging from 3 months to 10 years.

As with the T-bill futures, this contract is for $1 million of face value and is traded on an index price basis with a tick of 0.01. The quoted futures price is equal to 100 minus the annualized yield, which is also called the **implied** LIBOR **rate**. The contract month specifies the month and year in which the futures contract expires. For example, the June contract in Fig. 21.1 is quoted as 93.52. This implies a Eurodollar interest rate quote of 6.48% for the 3-month period beginning June and a contract price of

$$\$1,000,000 \times [\, 1 - (0.0648/4) \,] = \$983,800.$$

Unlike the T-bill futures, the Eurodollar futures contract is a cash-settlement contract. The parties settle in cash for the value of a Eurodollar time deposit based on the LIBOR at the delivery date. The final marking to market sets the contract price to

$$\$1,000,000 \times [\, 1 - (\text{LIBOR}/4) \,].$$

The LIBOR above is the actual 90-day rate on Eurodollar deposits with quarterly compounding. So the futures price converges to $100 \times (1 - \text{LIBOR})$ by design. Equivalently, the implied LIBOR rate converges to the spot LIBOR rate.

The T-bill futures is a contract on a price unlike that of the Eurodollar futures, which is a contract on an interest rate. Eurodollar futures prices move linearly with the bank discount yield; a 1% change in yield always causes a

$$\$1,000,000 \times \frac{1}{100} \times \frac{90}{360} = \$2,500$$

change in price, which implies a tick value of $25. This is in sharp contrast to T-bill futures, whose prices move linearly with T-bill prices.

EXAMPLE 21.1.3 Consider a floating-rate liability of $10 million with an interest rate at 1% above the prevailing 3-month LIBOR on the interest payment date in September. There happen to be Eurodollar futures expiring on the payment date with a futures price of 93.33. Selling 10 such contracts locks in a LIBOR rate of 6.67% for the 4-month period beginning September. Because the borrowing cost is 1% over LIBOR, the locked-in borrowing rate is 7.67%. The interest payment is $10 \times 0.0767/4 = 0.19175$ million dollars at this rate (see Fig. 21.2).

LIBOR	5%	6%	7%	8%	9%
Futures price	95	94	93	92	91
Interest expense	150,000	175,000	200,000	225,000	250,000
Loss on futures	41,750	16,750	−8,250	−33,250	−58,250
Net borrowing cost	191,750	191,750	191,750	191,750	191,750

Figure 21.2: Locking in the borrowing cost with Eurodollar futures. The interest expense is $10,000,000 × (1% + LIBOR)/4, and the loss on futures is 10 × $1,000,000 × (futures price − 93.33)/(4 × 100). See Example 21.1.3 for explanations.

The preceding single-period example readily extends to multiperiod situations. The way to transform a floating-rate liability into a fixed-rate liability is by selling a strip of futures whose expiration dates coincide with interest payment dates. The locked-in borrowing rates are the implied LIBOR rates plus the applicable spread. Here the interest generated by the marking-to-market feature is ignored.

A LIBOR term structure can be established from the implied 3-month LIBOR rates. Suppose now that June 1995 is the expiration date of the June 1995 Eurodollar futures contract. The Eurodollar futures prices are in Fig. 21.3. The actual 3-month LIBOR from June to September is 6.48%. Because the maturity is 92 days away, the return is $6.48\% \times (92/360) = 1.656\%$. The 6-month rate could be established by the June and the September Eurodollar futures contracts as follows. The implied rate from September to December is $6.70\% \times (91/360) = 1.694\%$. The implied rate over the 6-month period is thus $(1.01656 \times 1.01694) - 1 = 3.378\%$, which gives an annualized rate of $3.378\% \times (360/183) = 6.645\%$. This is also the borrowing rate that can be

	Futures Price	Implied LIBOR	Days Settle		Futures Price	Implied LIBOR	Days Settle
June	93.52	6.48	92	Dec	92.37	7.63	91
Sept	93.30	6.70	91	Mar00	92.38	7.62	92
Dec	93.11	6.89	91	June	92.31	7.69	92
Mar96	93.10	6.90	92	Sept	92.25	7.75	91
June	93.02	6.98	92	Dec	92.17	7.83	90
Sept	92.97	7.03	91	Mar01	92.18	7.82	92
Dec	92.87	7.13	90	June	92.11	7.89	92
Mar97	92.88	7.12	92	Sept	92.05	7.95	91
June	92.82	7.18	92	Dec	91.97	8.03	90
Sept	92.78	7.22	91	Mar02	92.01	7.99	92
Dec	92.70	7.30	90	June	91.96	8.04	92
Mar98	92.71	7.29	92	Sept	91.90	8.10	91
June	92.65	7.35	92	Dec	91.82	8.18	90
Sept	92.61	7.39	91	Mar03	91.83	8.17	92
Dec	92.53	7.47	90	June	91.76	8.24	92
Mar99	92.54	7.46	92	Sept	91.70	8.30	91
June	92.49	7.51	92	Dec	91.61	8.39	91
Sept	92.45	7.55	91	Mr04	91.65	8.35	92

Figure 21.3: Eurodollar futures prices. The yields are the hypothetical implied 3-month LIBOR rates as of June 1995.

locked in for the 6-month period. The procedure can be repeated to yield a LIBOR term structure.

EXAMPLE 21.1.4 What about the 12-month implied LIBOR rate beginning June 1996? It is determined by the June, September, December, and March Eurodollar futures contracts and is given by the value f satisfying

$$1 + f \times \frac{365}{360} = \left(1 + 6.98\% \times \frac{92}{360}\right)\left(1 + 7.03\% \times \frac{91}{360}\right)$$

$$\times \left(1 + 7.13\% \times \frac{90}{360}\right)\left(1 + 7.12\% \times \frac{92}{360}\right).$$

Solving this expression yields 7.257%.

▶ **Exercise 21.1.2** Calculate the annualized implied 9-month LIBOR rate between June 1995 and March 1996 from the data in Fig. 21.3.

21.1.4 Treasury Bond Futures

The CBT trades 2-year, 5-year, and 10-year T-note futures and T-bond futures. All require delivery of the underlying securities, but a number of eligible securities can be delivered against the contract.

Initiated in 1977, the T-bond futures contract calls for $100,000 face value in deliverable-grade U.S. T-bonds. The bond delivered must have at least 15 years to maturity or to the first call date if callable. The short position can choose any business day in the delivery month to deliver, although contracts are rarely settled by actual delivery. The underlying instrument for the T-bond futures contract is $100,000 par value of a hypothetical 20-year 8% coupon bond. For example, if the March contract settles at 104-20, the buyer is entitled to receive this coupon bond for a price of $104\frac{20}{32}\%$ of $100,000, or $104,625. The minimum price fluctuation for the T-bond futures contract is 1/32 of 1%. The dollar value of that is $100,000 \times (1/32)\% = \$31.25$; hence the minimum price fluctuation is thus $31.25.

Although prices and yields of the T-bond futures are quoted in terms of this hypothetical bond, the seller of the futures contract has the choice of several actual Treasury bonds that are acceptable for delivery. A well-defined mechanism computes from the quoted price the effective futures price for all the bonds that satisfy delivery requirements. The invoice amount is determined by

$$\text{invoice price} = (\text{settlement price} \times \text{contract size} \times \text{conversion factor})$$
$$+ \text{accrued interest.} \tag{21.1}$$

The **conversion factor** is the price of a $1 (face value) coupon bond with a maturity – measured from the first day in the delivery month – equal to the delivered bond if it were priced to yield 8%, compounded semiannually. Maturities are rounded down to the nearest quarter. For example, 24-years-and-5-months becomes 24-years-and-3-months. Let ω be the percentage of the coupon that due the holder as defined in Eq. (3.19). The bond is worth

$$\frac{c}{(1.04)^\omega} + \frac{c}{(1.04)^{\omega+1}} + \frac{c}{(1.04)^{\omega+2}} + \cdots + \frac{1+c}{(1.04)^{\omega+n-1}} - c(1-\omega),$$

where c is the semiannual coupon rate and n is the number of remaining coupon payments. The last term is the accrued interest. As coupon payments are made at 6-month intervals, ω is either $1/2$ or 1. For example, if the first coupon payment occurs 3 months hence, then $\omega = 1/2$.[1]

Consider a 13% coupon bond with 19 years and 2 months to maturity. In calculating the conversion factor, the bond is assumed to have exactly 19 years to maturity, and the nearest coupon payment will be made 6 months from now. On the assumption that the discount rate is 8% per annum with semiannual compounding, the bond has the value

$$\sum_{i=1}^{38} \frac{6.5}{(1.04)^i} + \frac{100}{(1.04)^{38}} = 148.42.$$

The conversion factor is therefore 1.4842. Consider an otherwise identical bond with 19 years and 4 months to maturity. For the purpose of calculating the conversion factor, the bond is assumed to have exactly 19 years and 3 months to maturity. There are 39 coupon payments, starting 3 months from now. The value of the bond is

$$\sum_{i=0}^{38} \frac{6.5}{(1.04)^{i+0.5}} + \frac{100}{(1.04)^{38.5}} - 3.25 = 148.66.$$

The conversion factor is therefore 1.4866.

EXAMPLE 21.1.5 Suppose the T-bond futures contract settles at 96 and the short elects to deliver a T-bond with a conversion factor of 1.15. The price is $\$100,000 \times 0.96 \times 1.15 = \$110,400$. The buyer of the contract must also pay the seller accrued interest on the bond delivered as dictated by Eq. (21.1).

The party with the short position can choose from among the deliverable bonds the "cheapest" one to deliver. Because the party with the short position receives the invoice price equal to

(quoted futures price × conversion factor) + accrued interest

and the cost of purchasing a bond is

quoted cash price + accrued interest,

the **cheapest-to-deliver bond** is the one for which

quoted cash price − (quoted futures price × conversion factor)

is least. One can find this by examining all deliverable bonds. The cheapest-to-deliver bond may change from day to day.

In addition to the option to deliver any acceptable Treasury issue (sometimes referred to as the **quality** or **swap option**), the short position has two more options. First, it decides when in the delivery month delivery actually takes place. This is called the **timing option**. (The futures contract stops trading 7 business days before the end of the delivery month.) The other option is the right to give notice of intent to deliver up to 8 P.M. Chicago time on the date when the futures settlement price has been fixed. This option is referred to as the **wild card option**. These three options, in sum referred to as the **delivery option**, mean that the long position can never be sure which T-bond will be delivered and when [325, 470]. Such complexity has helped provide liquidity.

▶ **Exercise 21.1.3** A deliverable 8% coupon bond must have a conversion factor of one regardless of its maturity as long as the accrued interest is zero. Why?

▶ **Exercise 21.1.4** Calculate the conversion factor for a 13% coupon bond with 19 years and 11 months to maturity at the delivery date.

Valuation

The delivery option creates value for the short position. It also makes pricing difficult. If both the cheapest-to-deliver bond and the delivery date are known, however, the T-bond futures contract becomes a futures contract on a security with known income. Lemma 12.2.3 says the futures price F is related to the cash bond price S by $F = (S - I)e^{r\tau}$, where I is the PV of the coupons during the life of the futures contract and r is the riskless interest rate for the period.[2]

EXAMPLE 21.1.6 Consider a cheapest-to-deliver bond with 9% coupon and a conversion factor of 1.0982. Its next coupon date is 120 days from now, and delivery will take place in 270 days' time followed by the next coupon date after 33 days. The term structure is flat at 8%, continuously compounded. Assume that the current quoted bond price is 108. We first figure out the accrued interest, say 1.533. The cash bond price is therefore 109.533. The futures price if the contract were written on the 9% bond would be

$$\left[109.533 - 4.5 \times e^{0.08 \times (120/365)} \right] \times e^{0.08 \times (270/365)} = 111.309.$$

At delivery, there are 150 days of accrued interest ($270 - 120 = 150$). The quoted futures price if the contract were written on the 9% bond would therefore be

$$111.309 - \left(4.5 \times \frac{150}{150 + 33} \right) = 107.620.$$

Because the contract is in fact written on a standard 8% bond and 1.0982 standard bonds are considered equivalent to each 9% bond, $107.620/1.0982 = 97.997$ should be the quoted futures price.

Hedging

Recall that a basis-point value (BPV) is the price change of a debt instrument given a one-basis-point (0.01%) change in its yield. For example, if the yield on a bond changes from 8% to 8.01% and the resulting price changes by $70, its BPV is $70. The greater the BPV, the greater the interest rate exposure. The BPV of a futures contract is the BPV of the cheapest-to-deliver instrument divided by its conversion factor. A conversion factor of α means that the price sensitivity of the bond is approximately α times that of the futures, and α futures contracts need to be sold to immunize the price change of every $100,000 face value of the underlying bond. The BPV of the T-bond futures contract increases as interest rates decline. This is in sharp contrast to Eurodollar futures for which each 0.01% change in rates changes the price by precisely $25.

In order to use futures to alter a portfolio's duration, it is necessary to calculate both the sensitivity of the bond portfolio to yield changes and the sensitivity of the futures prices to yield changes. For the futures contract to succeed as a hedge, its price movements should track those of the underlying bond closely. This is indeed the case [95, 746].

EXAMPLE 21.1.7 An investor holds a bond portfolio with a Macaulay duration (MD) of 5.0 and a market value of $100 million. Let the bond equivalent yield (BEY) be 10%. The number of futures contracts to buy in order to increase the MD to 9.0 can be determined as follows. Assume that the BPV of the futures is $85. The BPV of the portfolio equals

$$\frac{5.0}{1+(0.10/2)} \times \$100,000,000 \times 0.0001 = \$47,619.$$

The desired BPV can be derived from the targeted MD by

$$\frac{9.0}{1+(0.10/2)} \times \$100,000,000 \times 0.0001 = \$85,714.$$

As the BPV needs to be increased by $38,095, the number of futures contracts to buy is $38095/85 \approx 448$.

Bond futures and stock index futures can be combined to synthetically change the allocation of assets. A manager would like to achieve the equivalent of selling $100 million in bonds and buying $100 million in stock by using futures. First, follow the steps in Example 21.1.7 to figure out the BPV of $100 million worth of bonds and then the number of bond futures to sell with the same BPV. The number of stock index futures to buy is determined by $100,000,000 divided by the value of the contract, which equals $500 times the value of the S&P 500 Index for the S&P 500 Index futures contract, for example.

21.1.5 Treasury Note Futures

The T-note futures contract is modeled after the T-bond futures contract. For instance, the underlying instrument for the 10-year T-note futures contract is $100,000 par value of a hypothetical 10-year 8% T-note. Again, there are several acceptable Treasury issues that may be delivered. For the 10-year futures, for example, an issue is acceptable if the maturity is not less than 6.5 years and not more than 10 years from the first day of the delivery month. The delivery options granted to the short position and the minimum price fluctuation are the same as those of the T-bond futures [325]. An exception is the 2-year T-note futures, whose face value is $200,000 and whose prices are quoted in terms of one quarter of 1/32 of a dollar.

▶ **Exercise 21.1.5** A pension fund manager wants to take advantage of the high yield offered on T-notes, but unusually high policy payouts prohibit such investments. It is, however, expected that cash flow will return to normal in September. What can be done?

21.1.6 Forward Rate Agreements

First used in 1982, forward rate agreements (FRAs) are cash-settled forward contracts between two parties with a payoff linked to the future level of a reference rate [647]. The interest is based on a hypothetical deposit and paid at a predetermined future date. The buyer of an FRA pays the difference between interest on this hypothetical loan at a fixed rate and interest on the same loan at the prevailing rate. The market is primarily an interbank market and represents the over-the-counter equivalent of the exchange-traded futures contracts on short-term rates.

Formally, suppose that X represents the annualized fixed rate and y represents the actual annualized reference rate prevailing at the settlement date T. The net cash payment to the buyer of an FRA at the end of the contract period, $T + m$, is

$$(y - X) \times \mathcal{N} \times \frac{m}{360}. \tag{21.2}$$

Here m represents the "deposit" period in days, and $\mathcal{N}$ is the hypothetical loan amount (the **notional principal**). Generally, the net payment due is discounted back to the settlement date with the reference rate as the discount rate. The cash payment is thus

$$(y - X) \times \mathcal{N} \times \frac{m}{360} \times \frac{1}{1 + y(m/360)}$$

at the settlement date T.

The quote convention identifies the points in time when the contract begins and ends. Thus FRAs covering the period starting in 1 month and ending in 4 months are referred to as 1×4 (one-by-four) contracts. They are on the 3-month LIBOR (see the following diagram):

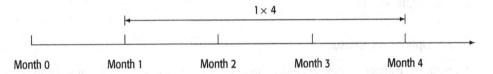

Similarly, FRAs on the 6-month LIBOR for settlement 1 month forward are 1×7 contracts. On any given day, forward rates are available for both 3- and 6-month LIBOR 1 month, 2 months, 3 months, 4 months, 5 months, and 6 months forward. On each subsequent day, new contracts are offered again.

EXAMPLE 21.1.8 A bank will, in 3 months, lend $1 million to a client for 6 months. To hedge the rate commitment the client demands, it uses FRAs to lock in the funding cost. The bank asks for a quote on 3×9 LIBOR and gets 5.5%. It then offers a fixed rate of 6% to the client. Suppose that the 6-month LIBOR becomes 6.2% 3 months from now. The loss 9 months from now on the actual lending is $(0.062 - 0.06) \times \$1,000,000 \times (182/360) = \$1,011.11$, whereas the gain from the FRA is $(0.062 - 0.055) \times \$1,000,000 \times (182/360) = \$3,538.89$.

Pricing FRAs amounts to deriving the fair fixed forward rate assuming there is no default risk on the part of the FRA writer. The forward rate for the time period $[T, T + m]$ equals

$$f_L(T, T + m) \equiv \left[\frac{d_L(T)}{d_L(T + m)} - 1 \right] \Big/ \Delta t.$$

Here $\Delta t \equiv m/360$, and $d_L(t)$ denotes the PV of a Eurodollar deposit that pays $1 t days from now (see Exercise 5.6.3). Hence $f_L(T, T + m)$ is the desired fixed contract rate X that makes the FRA zero valued now. In general, the PV of the FRA in

Eq. (21.2) equals

$$[f_L(T, T+m) - X] \times \Delta t \times d_L(T+m) = d_L(T) - (1 + X\Delta t) \, d_L(T+m)$$

$$(21.3)$$

per Eurodollar of notional principal.

▶ **Exercise 21.1.6** (1) Prove Exercise 5.6.3 with an arbitrage argument. (2) Verify Eq. (21.3).

21.2 Fixed-Income Options and Interest Rate Options

This section covers options on fixed-income securities and interest rates. We use "fixed-income options" for the former and "interest rate options" for the latter. The over-the-counter market for fixed-income options began in the mid-1970s with essentially put options on mortgages [724]. Almost all exchange-traded interest rate options are European.

With fixed-income options, one buys puts to hedge against rate rises and calls to hedge against rate falls. With yield-based interest rate options, the situation is reversed: A call buyer anticipates that interest rates will go up, whereas a put buyer anticipates that the rates will go down. Many of the trading strategies in Chap. 7 remain applicable here.

21.2.1 Options on Treasuries

Consider a European option that expires at date T and is written on a T-bill with a maturity of m days when the option expires. The strike price of the option, x (in percentage), is a discount rate whose corresponding number in percentage of par is

$$X = \left(1 - \frac{x}{100} \times \frac{m}{360} \right) \times 100$$

under the 360-day year. The payoff to a call at expiration is $\max(d(m) - X, 0)$, where $d(m)$ is the date-T value of a \$1 face-value zero-coupon bond maturing at date $T+m$. The payoff to a put at expiration is $\max(X - d(m), 0)$.

Treasury options are not very liquid. A thin market exists for exchange-traded options on specific T-bonds. Prices are quoted in points and $1/32$ of a point, with each point representing 1% of the principal value, or \$1,000. The amount paid on exercise is equal to the strike price times the underlying principal plus accrued interest. For example, the settlement price of an option with strike 90 is \$100,000 × (90/100) = \$90,000 plus accrued interest.

21.2.2 Interest Rate Options on Treasury Yields

The CBOE trades European-style, cash-settled interest rate options on the following Treasury yields: (1) the short-term rate based on the annualized discount rate on the most recently auctioned 13-week T-bill (ticker symbol IRX), (2) the 5-year rate based on the yield to maturity of the most recently auctioned 5-year T-note, (3) the 10-year rate based on the yield to maturity of the most recently auctioned 10-year T-note, and (4) the 30-year rate based on the yield to maturity of the most recently auctioned 30-year Treasury bond (ticker symbol TYX). See Fig. 21.4.

Friday, August 28, 1998

OPTIONS ON SHORT-TERM INTEREST RATES (IRX)

| Strike | Calls-Last | | | Puts-Last | | |
Price	Sep	Oct	Dec	Sep	Oct	Nov
50	...	...	...	1 7/16	...	...
55	...	...	5/16	...	...	...

...

5 YEAR TREASURY YIELD OPTION (FVX)

| Strike | Calls-Last | | | Puts-Last | | |
Price	Sep	Oct	Nov	Sep	Oct	Nov
52 1/2	...	...	7/8	...	...	...

...

10 YEAR TREASURY YIELD OPTION (TNX)

| Strike | Calls-Last | | | Puts-Last | | |
Price	Sep	Oct	Mar	Sep	Oct	Nov
55	...	...	7/8	...	...	...

...

30 YEAR TREASURY YIELD OPTION (TYX)

| Strike | Calls-Last | | | Puts-Last | | |
Price	Sep	Oct	Dec	Sep	Oct	Nov
50	...	...	4 1/8	3/16	...	...
52 1/2	...	...	...	5/8	...	...
55	7/16	...	...	1 15/16	...	...
57 1/2	...	5/16	...	...	...	...

...

Figure 21.4: Treasury yield quotations. Source: *Wall Street Journal*, August 31, 1998.

The underlying values for these options are 10 times the underlying rates. As a result, an annualized discount rate of 3.25% on the 13-week T-bill would place the underlying value of the IRX at 32.50, and a yield to maturity of 6.5% on the 30-year T-bond would place the underlying value of the TYX at 65.00. Clearly, every one-percentage-point change in interest rates makes the underlying value change by 10 points in the same direction. Like equity options, these options use the $100 multiplier for the contract size. The final settlement value is determined from quotes on the last trading day as reported by the Federal Reserve Bank of New York at 2:30 P.M. Central Time. The payoff, if exercised, is equal to $100 times the difference between the settlement value and the strike price. For example, an investor holding an expiring TYX July 66 call with a settlement value of 69 at expiration will exercise the call and receive $300.

▶ **Exercise 21.2.1** An investor owns T-bills and expects rising short-term rates and falling intermediate-term rates. To profit from this reshaping of the yield curve, he sells T-bills, deposits the cash in the bank, and purchases puts on the 10-year Treasury yield. Analyze this strategy.

21.2.3 Interest Rate Caps and Floors

A cap or a floor is a series of interest rate options. The underlying asset could be an interest rate or the price of a fixed-income instrument. Options on a cap are called **captions**, and options on a floor are called **flotions** [325, 346]. Captions and flotions are compound options.

Simple interest rates are often used in the specification of interest rate derivatives. The payoff to the cap at expiration is

$$\max\left(\frac{r-x}{1+r(m/360)}, 0\right) \times \frac{m}{360} \times \mathcal{N},$$

where x is the cap rate expressed in terms of simple interest rate and r is the m-day interest rate (both are annualized). Similarly, the payoff to a floor with a floor rate x written on the m-day interest rate is

$$\max\left(\frac{x-r}{1+r(m/360)}, 0\right) \times \frac{m}{360} \times \mathcal{N}.$$

The payoffs $\max(r - x, 0) \times (m/360) \times \mathcal{N}$ and $\max(x - r, 0) \times (m/360) \times \mathcal{N}$ are discounted because they are received at the maturity of the underlying, which is m days from the cap's or the floor's expiration date.

EXAMPLE 21.2.1 Take a 6-month European call on the 6-month LIBOR with a 7% strike rate. The face value is $10 million, and the expiration date is 6 months (183 days) from now. This call gives the buyer the right to receive $\max(r - 7\%, 0) \times (182/360) \times \$10,000,000$, where r is the 6-month LIBOR rate prevailing in 6 months' time. The payoff is received at the maturity date of the underlying interest rate, which is $183 + 182 = 365$ days from now. The PV of the above payoff at the expiration of the cap is therefore the above equation divided by $1 + r(182/360)$.

EXAMPLE 21.2.2 Consider the cap in Fig. 21.5. Although a total of 10 6-month periods are involved, the cap contains only 9 options whose payoffs are determined on reset dates. The underlying interest rate for the first period is the interest rate today, which is known; hence no option is involved here. See the following time line:

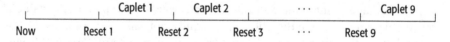

The cash flow is depicted in Fig. 21.6. The mechanics of floors are similar. Besides the 6-month LIBOR, other candidates for the underlying interest rate are the prime rate, the T-bill rate, the CD rate, and the commercial paper rate.

▶ **Exercise 21.2.2** Verify that an FRA to borrow at a rate of r can be replicated as a portfolio of one long caplet and one short floorlet with identical strike rate x and expiration date equal to the settlement date of the FRA.

Reference rate	six-month LIBOR
Strike rate	8%
Length of agreement	five years
Frequency of settlement	every six months
Notional principal amount	$100 million

Figure 21.5: An interest rate cap.

Year	LIBOR	Loan cash flow	Cap payoff	Net cash flow
0.0	8%	+100.0000		+100.0000
0.5	9%	−4.0556	0.0000	−4.0556
1.0	9%	−4.5625	0.5069	−4.0556
1.5	9%	−4.5625	0.5069	−4.0556
2.0	9%	−4.5625	0.5069	−4.0556
2.5	9%	−4.5625	0.5069	−4.0556
3.0	9%	−4.5625	0.5069	−4.0556
3.5	9%	−4.5625	0.5069	−4.0556
4.0	9%	−4.5625	0.5069	−4.0556
4.5	9%	−4.5625	0.5069	−4.0556
5.0		−104.5625	0.5069	−104.0556

Figure 21.6: Cash flow of a capped loan. Assume that the LIBOR starts at 8% and then moves to 9% half a year from now and stays there. The loan cash flow and the interest rate option's cash flow are calculated based on the prevailing LIBOR rate at the beginning of each half-year period. For instance, at year one the loan's interest is $100 \times 9\% \times (182.5/360) = 4.5625$ million dollars. A half-year is assumed to have exactly 182.5 days for simplicity.

21.2.4 Caps/Floors and Fixed-Income Options

Consider a caplet on the m-day LIBOR with strike rate x. Let the notional principal amount be \$1 for simplicity. At expiration, if the actual interest rate is r, the caplet pays

$$\max\left(\frac{r(m/360) - x(m/360)}{1 + r(m/360)}, 0\right).$$

Interestingly, this caplet is equivalent to $\alpha \equiv 1 + x(m/360)$ puts on the m-day zero-coupon bond with a strike price of $1/\alpha$ and the identical expiration date as the payoff of the puts is

$$\alpha \times \max\left(\frac{1}{1 + x(m/360)} - \frac{1}{1 + r(m/360)}, 0\right)$$

$$= \max\left(\frac{r(m/360) - x(m/360)}{1 + r(m/360)}, 0\right).$$

Similarly, a floorlet is equivalent to α calls with the same strike price. A cap is therefore a package of puts on zeros, and a floor is a package of calls on zeros. An interest rate collar, then, is equivalent to buying a package of puts and selling a package of calls.

The Black model is widely used in practice to price caps and floors. For caps, it applies Eqs. (12.16) with F denoting the implied forward rate for the period between the cap's expiration date T and date $T + m$. This amounts to assuming that the forward rate is lognormally distributed during the period $[T, T + m]$. The resulting formula is then multiplied by the notional principal and finally multiplied by $(m/360)/[1 + F(m/360)]$.

▶ **Exercise 21.2.3** Prove the equivalency for the floorlet.

21.2.5 Yield Curve Options

A **yield curve option** is a European option written on the difference between two reference rates. A call based on the difference between the yields on the 20-year T-bond and the 12-month T-bill, for example, has a payoff at expiration given by $\max((y_{20y} - y_{12m}) - X, 0)$, where X is the strike spread between the two reference yields [149, 616].

21.3 Options on Interest Rate Futures

The most popular exchange-traded interest rate options are those on T-bond futures, T-note futures (traded on the CBT), and Eurodollar futures (traded on the IMM). These futures options are all American style. The prices are quoted as a percentage of the principal amount of the underlying debt security. For options on Eurodollar futures, the price is quoted to two decimal places, and one contract is for the delivery of futures contracts with a face value of $1 million. For options on T-bond and T-note futures (except the 2-year note), the price is quoted to the nearest 1/64 of 1%, and one contract is for the delivery of futures contracts with a face value of $100,000. Figure 21.7 shows sample quotations. For example, an investor holding a December call with a strike price of 98, having paid 1-08 for it, will make a net profit of

$$\$100,000 \times \frac{101.00 - 98.00 - 1.125}{100} = \$1875$$

if the T-bond futures price rises to 101-00. Futures options on fixed-income securities have largely replaced options on the same securities as the vehicle of choice for institutional investors [325].

The option on T-bond futures is an option on the futures price itself, that is, the futures price of the fictitious 20-year 8% Treasury bond. The size of the contract is $100,000. For example, with futures prices at 95, a call struck at 94 has an intrinsic value of $1,000 and a put strike at 100 has an intrinsic value of $5,000. Prices are quoted in multiples of 1/64 of 1% of a $100,000 T-bond futures contract. Each 1/64 point (tick size) is worth $15.625 [375]. Options cease trading in the month before the delivery month of the underlying futures contract.

A 10-year T-note futures contract has a face value at maturity of $100,000. The tick size is 1/64 of a point ($15.625/contract). Options cease trading in the month before the delivery month of the underlying futures contract. The 5-year T-note futures option is identical. The 2-year T-note futures option is identical except that the face value is $200,000 and the tick size is 1/128 of a point ($15.625/contract).

Options on Eurodollar futures are based on the quoted Eurodollar futures price. Like the underlying futures, the size of the contract is $1 million, and each 0.01 change in price carries a value of $25. The option premium is quoted in terms of basis points. For example, a premium quoted as 20 implies an option price of $20 \times \$25 = \$500$. Take a 3-month put on the June Eurodollar futures contract at a strike price of 93. The expiration date of the put is in June, and the underlying asset is June Eurodollar futures. The terminal payoff is $[\max(93 - F, 0)/100] \times (90/360) \times \$1,000,000$, where F is the futures price in June.

Monday, March 20, 1995

...

INTEREST RATE

...

T-BONDS (CBT)

$100,000; points and 64ths of 100%

Strike Price	Calls – Settle Apr	May	Jun	Puts – Settle Apr	May	Jun
102	2-06	2-26	2-47	0-03	0-23	0-43
103	1-12	1-44	...	0-08	0-40	...
104	0-30	1-05	1-29	0-26	1-01	1-25
105	0-07	0-39	...	1-03	1-34	...
106	0-01	0-21	0-40	1-61	...	2-35
107	0-01	0-10	...	...	...	...

EURODOLLAR (CME)

$ million; pts. of 100%

Strike Price	Calls – Settle Jun	Sep	Dec	Puts – Settle Jun	Sep	Dec
9300	0.56	0.49	0.52	0.04	0.20	0.41
9325	0.34	0.34	0.39	0.07	0.29	0.52
9350	0.17	0.22	0.29	0.15	0.41	0.67
9375	0.05	0.11	0.19	0.28	0.55	...
9400	0.01	0.06	0.12	0.49	0.75	0.98
9425	0.00	0.03	0.07	0.73	0.96	1.18

...

T-NOTES (CBT)

$100,000; points and 64ths of 100%

Strike Price	Calls – Settle Apr	May	Jun	Puts – Settle Apr	May	Jun
102	2-26	...	2-45	0-01	...	0-20
103	1-28	...	1-60	0-02	0-19	0-35
104	0-36	...	1-19	0-11	...	0-57
105	0-07	0-32	0-51	0-45	1-06	1-25
106	0-01	...	0-28	1-39	...	2-01
107	0-01	0-05	0-14	...	...	2-51

LIBOR – 1 Mo. (CME)

$3 million; pts. of 100%

Strike Price	Calls – Settle Apr	May	Jun	Puts – Settle Apr	May	Jun
9325	0.58	0.51	0.45	0.00	0.01	0.03
9350	0.34	0.29	0.24	0.01	0.04	0.07
9375	0.11	0.10	0.09	0.03	...	0.17
9400	0.01	...	0.03	...	...	...
9425	...	...	...	...	...	...
9450	0.00	0.00	...	...	...	...

...

5 YR TREAS NOTES (CBT)

$100,000; points and 64ths of 100%

Strike Price	Calls – Settle Apr	May	Jun	Puts – Settle Apr	May	Jun
10200	1-33	...	1-47	0-01	0-07	0-15
10250	1-02	...	1-23	0-01	0-12	0-22
10300	0-36	...	1-01	0-03	0-21	0-32
10350	0-13	...	0-46	0-12	0-33	0-45
10400	0-03	0-20	0-31	0-34	...	0-62
10450	0-01	0-11	0-21	...	...	...

Figure 21.7: Interest rate futures option quotations. The months refer to the expiration month of the underlying futures contract. Source: *Wall Street Journal*, March 21, 1995.

21.3.1 Hedging Floating-Rate Liabilities

Exposure to fluctuations in short-term rates can be hedged with Eurodollar futures options. Let us redo the calculations behind the table in Fig. 21.2 (see Example 21.1.3), this time using Eurodollar futures options instead. The liability, we recall, is a floating-rate debt of $10 million with an interest rate at 1% above the prevailing 3-month LIBOR on the payment date. There happen to be Eurodollar futures options expiring on the payment date. Purchasing 10 futures put options at the strike price of 93 caps the borrowing cost at $200,000 when the liability is due (consult Fig. 21.8). Of course, there is a cost in the cap represented by the option premium.

LIBOR	5%	6%	7%	8%	9%
Futures price	95	94	93	92	91
Interest expense	150,000	175,000	200,000	225,000	250,000
Put payout	0	0	0	25,000	50,000
Net borrowing cost	150,000	175,000	200,000	200,000	200,000

Figure 21.8: Capping the borrowing cost with Eurodollar futures options. The interest expense is $10,000,000 \times (1\% + \text{LIBOR})/4$, and the put payout is $10 \times 1,000,000 \times (93 - \text{futures price})/(4 \times 100)$.

21.4 Interest Rate Swaps

Two parties enter into an interest rate swap to exchange periodic interest payments. The dollar amount each counterparty pays to the other is the agreed periodic interest rate times the notional principal. The benchmarks popular for the floating rate are those on various money market instruments [226].

21.4.1 "Plain Vanilla" Interest Rate Swaps

In a "plain vanilla" interest rate swap, one party periodically pays a cash flow determined by a fixed interest rate (the **fixed leg**) and receives a cash flow determined by a floating interest rate (the **floating leg**). The other party does the opposite. In other words, two parties swap floating-rate debt and fixed-rate debt. Unlike currency swaps, no principal is exchanged. The fixed rate that makes a swap's value zero is called the **swap rate**.

A swap has four major components: notional principal amount, interest rates for the parties, frequency of cash exchanges, and duration of the swap. A "$40 million, 2-year, pay fixed, receive variable, semi" swap, for example, means that the notional principal is $40 million, one party makes a fixed-rate payment every 6 months based on $40 million, and the counterparty makes a floating-rate payment every 6 months based on $40 million, for a period of 2 years.

The floating-rate payment is linked to some short-term interest rate such as the 6-month LIBOR. The fixed rate for a plain vanilla swap is usually quoted as some spread over benchmark U.S. Treasuries. For example, a quote of "30 over" for a 5-year swap says that the fixed rate will be set at the 5-year Treasury yield plus 30 basis points. Although the net cash flow is established at the beginning of the period, it is usually paid out at the end of the period (**in arrears**) instead of at the beginning of the period (**in advance**).

Consider a 2-year swap with a notional principal of $10 million and semiannual payments. The fixed rate is 20 basis points above the 2-year Treasury rate, and the floating rate is the 6-month LIBOR. Suppose the Treasury rate is currently 5%. The fixed rate is therefore 5.2%. The fixed-rate payer will be paid according to the 6-month LIBOR rates determined at dates 0, 0.5 year, 1 year, and 1.5 years. If the LIBOR rates at the above four dates are 5.5%, 6%, 5.9%, and 5%, the fixed-rate payer has the following cash flow:

Payment date	0.5 year	1 year	1.5 years	2 years
Received amount	275,000	300,000	295,000	250,000
Paid amount	260,000	260,000	260,000	260,000

Note that the applicable rates are divided by two because only one-half-year's interest is being paid. In reality, only the losing party pays the difference.

Treasury rates are quoted differently from LIBOR rates. Consider a fixed rate quoted as a BEY under the actual/365 day count convention. To compare this yield with a LIBOR rate, which differs in the number of days on which they are quoted, either the 6-month LIBOR rate must be multiplied by 365/360 or the BEY must be multiplied by 360/365. Hence, if a 0.4% spread over the LIBOR is on the basis of a 365-day year with semiannual compounding, the rate becomes

$$ \text{LIBOR} + 0.4 \times (360/365) = \text{LIBOR} + 0.3945\% $$

under the 360-day year. There are other complications. For instance, the timing of the cash flows for the fixed-rate payer and that of the floating-rate payer are rarely identical; the fixed-rate payer may make payments annually, whereas the floating-rate payer may make payments semiannually, say. The way in which interest accrues on each leg of the transaction may also differ because of different day count conventions.

Suppose A wants to take out a floating-rate loan linked to the 6-month LIBOR and B wants to take out a fixed-rate loan. They face the following borrowing rates:

	Fixed	Floating
A	$F_A\%$	6-month LIBOR + $S_A\%$
B	$F_B\%$	6-month LIBOR + $S_B\%$

Clearly A can borrow directly at LIBOR plus $S_A\%$ and B can borrow at $F_B\%$. The total interest rate is the LIBOR plus $(S_A + F_B)\%$. Suppose that $S_B - S_A < F_B - F_A$. In other words, A is *relatively* more competitive in the fixed-rate market than in the floating-rate market, and vice versa for B. Consider the alternative whereby A borrows in the fixed-rate market at $F_A\%$, B borrows in the floating-rate market at LIBOR plus $S_B\%$, and they enter into a swap, perhaps with a bank as the financial intermediary. These transactions transform A's loan into a floating-rate loan and B's loan into a fixed-rate loan, as desired. The new arrangement pays a total of the LIBOR plus $(S_B + F_A)\%$, a saving of $(S_A + F_B - S_B - F_A)\%$. Naturally these transactions will be executed only if the total gain is distributed in such a way that every party benefits; that is, A pays less than the LIBOR plus $S_A\%$, B pays less than $F_B\%$, and the bank enjoys a positive spread.

EXAMPLE 21.4.1 Consider the following borrowing rates that A and B face:

	Fixed	Floating
A	8%	6-month LIBOR + 1%
B	11%	6-month LIBOR + 2%

Party A desires a floating-rate loan, and B wants a fixed-rate loan. Clearly A can borrow directly at the LIBOR plus 1%, and B can borrow at 11%. As the rate differential in fixed-rate loans (3%) is different from that in floating-rate loans (1%), a swap with a total saving of $3 - 1 = 2\%$ is possible. Party A is relatively more competitive in the fixed-rate market, whereas B is relatively more competitive in the floating-rate market. So they borrow in the respective markets in which they are competitive. Then each enters into a swap with the bank. There are hence two separate swap agreements, as shown in Fig. 21.9. The outcome: A effectively borrows at the LIBOR

Figure 21.9: Plain vanilla interest rate swaps.

plus 0.5% and B borrows at 10%. The distribution of the gain is 0.5% for A, 1% for B, and 0.5% for the bank. From the bank's point of view, the swap with A is like paying 7.5% and receiving LIBOR **"flat"** (i.e., no spread to the LIBOR), and the swap with B is like receiving 8% and paying LIBOR flat. Suppose the swap's duration is 10 years and the 10-year Treasury yield is 7%. The bank would quote such a swap as 50–100, meaning it is willing to enter into a swap (1) to receive the LIBOR and pay a fixed rate equal to the 10-year Treasury rate plus 50 basis points and (2) to pay the LIBOR and receive a fixed rate equal to the 10-year Treasury rate plus 100 basis points. The difference between the Treasury rate paid and received (50 basis points here) is the bid–ask spread.

As in the preceding example, counterparties are seldom involved directly in swaps. Instead, swaps are usually executed between counterparties and market makers or between market makers. The market maker faces the risk of having to locate another counterparty or holding an unmatched position if one leg of the swap defaults. It does not, however, risk the loss of principal as there is no exchange of principal to begin with.

A **par swap curve** can be constructed from zero-valued swaps of various maturities. The bootstrapping algorithm in Fig. 5.4 may be applied to give the theoretical **zero-coupon swap curve**, which can be used to price any swap.

▶ **Exercise 21.4.1** Party A wants to take out a floating-rate loan, and B wants to take out a fixed-rate loan. They face the borrowing rates below:

	Fixed	Floating
A	F_A%	LIBOR $+S_A$%
B	F_B%	LIBOR $+S_B$%

Party A agrees to pay the bank a floating rate of (LIBOR $-S'_A$)% in exchange for a fixed rate of $(F_A + F'_A)$%, and B agrees to pay the bank a fixed rate of $(F_B + F'_B)$% in exchange for a floating rate of (LIBOR$-S'_B$)%. Prove that

$$0 < S_A + F'_A + S'_A < S_A + F_B - F_A + F'_B + S'_B < S_A + F_B - S_B - F_A$$

must hold for both A and B to enter into a swap with the bank in which A effectively takes out a floating-rate loan and B a fixed-rate loan.

Applications to Asset/Liability Management

By changing the cash flow characteristics of assets, a swap can provide a better match between assets and liabilities. Consider a commercial bank with short-term deposits that are repriced every 6 months at the 6-month LIBOR minus 20 basis points. It faces a portfolio mismatch problem because its customers borrow long term. To tackle this maturity mismatch, the bank enters into a swap agreement whereby it pays a fixed rate of 10% semiannually and receives the 6-month LIBOR. This arrangement

Figure 21.10: Interest rate swaps for asset/liability management.

transforms the floating-rate liability into a fixed-rate liability of 9.8%, as shown in Fig. 21.10.

As another application, suppose that a bank has a $100 million 10-year-term commercial loan paying a fixed rate of 10%. Interest is paid semiannually, and the principal is paid at the end of the 10-year period. The bank issues 6-month CDs to fund the loan. The interest rate that the bank plans to pay on such CDs is a 6-month LIBOR plus 50 basis points. Clearly, if the 6-month LIBOR rises above 9.5%, the bank loses money. A life insurance company faces a different problem. It has committed itself to paying 9% on a **guaranteed investment contract** (**GIC**) it issued for the next 10 years. The amount of the GIC is also $100 million. The insurance company invests this money on a floating-rate instrument that earns a rate equal to the 6-month LIBOR plus 150 basis points. The rate is reset semiannually. Clearly, if the 6-month LIBOR falls below 7.5%, the insurance company loses money.

Interest rate swaps may allow both parties to lock in a spread. Suppose there exists a 10-year interest rate swap with a notional principal of $100 million. The terms are for the bank to pay 8.5% annual rate and receive LIBOR every 6 months and for the insurance company to pay LIBOR and receive 8.4% every 6 months. The bank's and the insurance company's cash flows every 6 months now appear in Fig. 21.11. Hence, the bank locks in a spread of 100 basis points, however the 6-month LIBOR turns out. Similarly, the insurance company locks in a spread of 90 basis points.

Like currency swaps, an interest rate swap can be interpreted as either a package of cash flows from buying and selling cash market instruments or as a package of forward contracts. We conduct the following analysis in the absence of default risk.

Valuation of Swaps as a Package of Cash Market Instruments

Assume for the purpose of analysis that the counterparties exchange the notional principal of N dollars at the end of the swap's life. It is then easy to see that a fixed-rate payer is long a floating-rate bond and short a fixed-rate bond. The value of the swap is therefore $P_2 - P_1$ from the fixed-rate payer's perspective, where P_1 (P_2, respectively) is the value of the fixed-rate (floating-rate, respectively) bond underlying the swap. The value of the swap is $P_1 - P_2$ for the floating-rate payer. As shown in Subsection 4.2.3, the floating leg should be priced at par immediately after a payment date, i.e., $P_2 = N$, if the rate used for discounting the future cash flow is

	The bank			*The insurance company*		
	Loan	*Swap*	*Total*	*Investment/GIC*	*Swap*	*Total*
Inflow	10%	LIBOR	10% + LIBOR	LIBOR + 1.5%	8.4%	9.9% + LIBOR
Outflow	LIBOR + 0.5%	8.5%	9% + LIBOR	9%	LIBOR	9% + LIBOR
Spread			1%			0.9%

Figure 21.11: Locking in the spread with interest rate swaps.

the floating rate underlying the swap. For example, it was the LIBOR plus 1% for the swap in Fig. 21.9. Because the swap when first entered into has zero value, $P_1 = \mathcal{N}$.

Let the fixed-rate payments (C dollars each) and the floating-rate payments be made at times $t_1, t_2, \ldots, t_n$ from now. By the above analysis, $P_1 = \sum_{i=1}^{n} C e^{-r_i t_i} + \mathcal{N} e^{-r_n t_n}$, where r_i is the spot rate for time t_i. As for the floating-rate bond, $P_2 = (\mathcal{N} + C^*) e^{-r_1 t_1}$, where C^* is the known floating-rate payment to be made at the next payment time t_1.

EXAMPLE 21.4.2 A party agrees to pay the 6-month LIBOR plus 1% every 6 months and receive 9% annual interest rate paid every 6 months on a notional principal of $10 million. All rates are compounded semiannually. Assume that the 6-month LIBOR rate at the last payment date was 9%. There are two more payment dates, at 0.3 and 0.8 years from now, and the relevant continuously compounded rates for discounting them are 10.1% and 10.3%, respectively. From these data,

$$P_1 = 0.45 \times e^{-0.101 \times 0.3} + 10.45 \times e^{-0.103 \times 0.8} = 10.0600 \text{ (million)},$$

$$P_2 = (10 + 0.5) e^{-0.101 \times 0.3} = 10.1866 \text{ (million)}.$$

The swap's value is hence $P_2 - P_1 = 0.1266$ (million) for the fixed-rate payer.

The duration of an interest rate swap from the perspective of the fixed-rate payer is

duration of floating-rate bond − duration of fixed-rate bond.

Most of the interest rate sensitivity of a swap results from the duration of the fixed-rate bond because the duration of the floating-rate bond is less than the time to the next reset date (see Subsection 4.2.3).

A party who is long a floating-rate bond and short a fixed-rate bond loses the principal and must continue servicing its fixed-rate debt if the floating-rate note issuer defaults. In contrast, the fixed-rate payer in a swap does not need to continue payment if the counterparty defaults. Hence the observation that a swap is equivalent to a portfolio of floating-rate and fixed-rate bonds holds only in the absence of a default risk.

▶ **Exercise 21.4.2** A firm buys a $100 million par of a 3-year floating-rate bond that pays the 6-month LIBOR plus 0.5% every 6 months. It is financed by borrowing $100 million for 3 years on terms requiring 10% annual interest rate paid every 6 months. Show that these transactions create a synthetic interest rate swap.

Valuation of Swaps as a Package of Forward Rate Agreements

Consider a swap with a notional principal of $\mathcal{N}$ dollars. Let the fixed-rate payments (C dollars each) and the floating-rate payments (based on future annual rates $f_1, f_2, \ldots, f_n$) be made at times $t_1, t_2, \ldots, t_n$ from now. The rate f_i is determined at t_{i-1}. At time t_i, C dollars is exchanged for $(f_i / k) \times \mathcal{N}$, where $k \equiv 1/(t_i - t_{i-1})$ is the payment frequency per annum and f_i is compounded k times annually (e.g., $k = 2$ for semiannual payments). For the fixed-rate payer, this swap is essentially a forward contract on the floating rate, say the 6-month LIBOR, whereby it agrees to pay C dollars in exchange for delivery of the 6-month LIBOR. Similarly, the floating-rate payer is essentially short a forward contract on the 6-month LIBOR. Hence an interest rate swap is equivalent to a package of FRAs.

From Eq. (12.7), the value of the forward contract to take delivery of the floating rate equals $[(f_i/k)\mathcal{N} - C]e^{-r_i t_i}$, with r_i denoting the time-t_i spot rate and f_i the forward rate. The first exchange at time t_1 has PV of $(C^* - C)e^{-r_1 t_1}$, where C^* is based on a floating rate currently known. The value of the swap is hence

$$(C^* - C)e^{-r_1 t_1} + \sum_{i=2}^{n} \left(\frac{f_i}{k}\mathcal{N} - C\right)e^{-r_i t_i} \tag{21.4}$$

for the fixed-rate payer. For the floating-rate payer, simply reverse the sign.

EXAMPLE 21.4.3 Consider the swap between B and a bank in Fig. 21.9. Party B receives the 6-month LIBOR plus 1% and pays 9%. All rates are compounded semiannually. Payments occur every 6 months on a notional principal of $10 million. Assume the 6-month LIBOR rate at the last payment date was 9%. There are two more payment dates, at 0.3 and 0.8 years from now, and the relevant continuously compounded rates for discounting them – 6-month LIBOR plus 1% – are 10.1% and 10.3%. The annualized continuously compounded 6-month forward rate 0.3 year from now is $(0.8 \times 0.103 - 0.3 \times 0.101)/(0.8 - 0.3) = 0.1042$. It becomes $2 \times (e^{0.1042/2} - 1) = 0.106962$ under semiannual compounding. The swap value,

$$(0.5 - 0.45)e^{-0.101 \times 0.3} + \left(\frac{0.106962}{2} \times 10 - 0.45\right)e^{-0.103 \times 0.8} = 0.1266 \text{ (million)}$$

by formula (21.4), is in complete agreement with Example 21.4.2.

Because formula (21.4) defines the value of a swap, it can also serve as its **replacement value** for the fixed-rate payer. In other words, if the floating-rate payer terminates the swap, this is the amount the fixed-rate payer would request for compensation so that the floating side could be replaced without increasing the fixed rate. Let V stand for the replacement value. Clearly if $V < 0$, then the counterparty would choose not to default on this agreement because it can be sold at a profit. The risk exposure is hence $\max(0, V)$.

As with other synthetic securities, interest rate swaps are not redundant even though it can be replicated by forward contracts. Several reasons have been cited [325]. First, they offer longer maturities than forward contracts. Second, interest rate swaps incur less transactions costs than a package of forward contracts that involve multiple transactions. Third, interest rate swaps are more liquid than forward contracts, particularly long-term forward contracts.

▶ **Exercise 21.4.3** Use Example 21.4.3's data to calculate the fixed rate that makes the swap value zero.

▶ **Exercise 21.4.4** Verify the equivalence of the two views on interest rate swaps.

▶ **Exercise 21.4.5** Consider a swap with zero value. How much up-front premium should the fixed-rate payer pay in order to lower the fixed-rate payment from C to $\widehat{C}$?

▶ **Exercise 21.4.6** Replicate swaps with interest rate caps and floors.

21.4.2 More Interest Rate Swaps

The number of different types of swaps is almost limitless. A **callable swap** allows the fixed-rate payer to terminate the swap at no penalty and prevents losses during declining rates. The premium may be amortized over the term of the swap. A **putable swap** allows the floating-rate payer to terminate the swap early. An investor can sell swaps short (**reverse swap**) in order to pay floating rates and receive fixed rates. This strategy gains under declining-rate environments and loses under rising-rate environments. A **basis swap** is a swap in which the two legs of the swap are tied to two different floating rates. A basis swap becomes a **yield-curve swap** if the two legs are based on short- and long-term interest rates, say the 6-month LIBOR and the 10-year Treasury yield. Such a swap can be used to control the exposure to changes in the yield-curve shape. In **deferred swaps** (or **forward swaps**), parties do not begin to exchange interest payments until some future date. In an **extendible swap**, one party has the option to extend the life of the swap beyond the specified period. A swap can be an agreement to exchange a fixed interest rate in one currency for a floating interest rate in another currency, in other words, a combination of a plain vanilla interest rate swap and a currency swap. In an **amortizing swap**, the principal is reduced in a way that corresponds to, say, the amortization schedule on a loan. In an **accreting swap**, the principal increases according to a schedule.

▶ **Exercise 21.4.7** A firm holds long-duration corporate bonds. It uses swaps to create synthetic floating-rate assets at attractive spreads to LIBOR and to shorten the duration much as A does in Fig. 21.9. Why should the fact that most corporate bonds are callable trouble the firm? How may callable swaps help?

Swaptions

A **swaption** is an option to enter into an interest rate swap. It is almost always European. The swap rate and the duration of the interest rate swap as measured from the option expiration date are specified in the contract. The market generally quotes on the fixed-rate part of the swap. So swaptions can be either **receiver swaptions** (the right to receive fixed and pay floating rates, or **floating-for-fixed**) or **payer swaptions** (the right to pay fixed and receive floating rates, or **fixed-for-floating**). The buyer of a receiver swaption benefits as interest rates fall, and the buyer of a payer swaption benefits as interest rates rise.

EXAMPLE 21.4.4 A firm plans to issue a 5-year floating-rate loan in 1 year and then convert it into a fixed-rate loan using interest rate swaps. To establish a floor for the swap rate, it purchases a 1-year swaption for the right to swap a fixed rate, say 8% per year, for a floating rate for a period of 5 years starting 1 year from now. If the fixed rate on a 5-year swap in 1 year's time turns out to be less than 8%, the company will enter into a swap agreement in the usual way. However, if it turns out to be greater than 8%, the company will exercise the swaption.

A fixed-for-floating interest rate swap can be regarded as an agreement to exchange a fixed-rate bond for a floating-rate bond. At the start of a swap, the value of the floating-rate bond equals the principal of the swap. A payer swaption can therefore be regarded as an option to exchange a fixed-rate bond for the principal of the swap, in other words, a put on the fixed-rate bond with the principal as the strike price. Similarly, a receiver swaption is a call on the fixed-rate bond with the principal as the strike price. When the Black model is used in pricing swaptions, the underlying asset is the forward rate for the interest rate swap [844].

▶ **Exercise 21.4.8** Prove that a cap is more valuable than an otherwise identical swaption.

Index-Amortizing Swaps

Amortizing swaps whose principal declines (amortizes) when interest rates decline are called **index-amortizing swaps** (**IAS**s). The principal is reduced by an amortizing schedule based on the spot interest rate. The amortizing schedule may not apply until after a lockout period. As mortgage prepayments usually pick up with declining rates, these instruments can partially hedge the prepayment risk of MBSs [848].

Formally, let T be the maturity of the swap with initial principal $\mathcal{N}$. The contract receives a fixed rate c and pays a floating rate $r(t)$. Let the lockout period be T^* years during which the principal is fixed at $\mathcal{N}$. For $t > T^*$, the remaining principal at time t changes according to

$$\mathcal{N}_t = \mathcal{N}_{t-1}(1 - a_t),$$

where $\mathcal{N}_{T^*} = \mathcal{N}$ and a_t is the amortizing amount. Hence $\mathcal{N}_t$ is $\mathcal{N}_{t-1}$ reduced by the amortizing schedule amount a_t. An amortizing schedule may look like

$$a_t = \begin{cases} 0 & \text{if } r(t) > k_0 \\ b_0 & \text{if } k_0 \geq r(t) > k_1 \\ b_1 & \text{if } k_1 \geq r(t) > k_2 \\ b_2 & \text{if } k_2 \geq r(t) > k_3 \\ b_3 & \text{if } k_3 \geq r(t) > k_4 \\ b_4 & \text{if } k_4 \geq r(t) > k_5 \\ 1 & \text{if } k_5 \geq r(t) \end{cases},$$

where $k_0 > k_1 > \cdots > k_5$ and $b_0 < b_1 < \cdots < b_4 < 1$ are positive constants. The preceding amortizing schedule depends on the interest rate $r(t)$. If the rate is larger than k_0, no reduction in principal occurs and $a_t = 0$; if it lies between k_0 and k_1, a reduction of b_0 occurs; if it lies between k_1 and k_2, a reduction of b_1 occurs, and so on.

The time-t cash flow to the IAS can be written as $[c - r(t-1)]\mathcal{N}_{t-1}$, which is determined at time $t-1$. Clearly the principal $\mathcal{N}_j$ depends on the whole interest rate path before time j, making the IAS a path-dependent derivative. An efficient algorithm will be presented in Programming Assignment 29.1.3.

Differential Swaps

A **differential swap** is an interest rate swap in which the interest rates for the two legs are linked to different currencies and the actual interest payments are denominated in the same currency by fixed exchange rates. For example, consider a swap with a dollar-based interest rate x and a DEM-based interest rate y. Both payments are to be denominated in U.S. dollars. Let the \$/DEM exchange rate be fixed at $\widehat{s}$. Then the settlement amount is $(y\widehat{s} - x)\mathcal{N}$ dollars. Clearly, differential swaps are a type of quanto derivative, thus the alternative name **quanto swaps**.

Additional Reading

For more information on interest rate derivatives, consult [95, 155, 325, 470, 746, 827, 837] for interest rate futures, [397, 538] for interest rate options, [54, 369, 474, 514, 608, 746, 821] for interest rate swaps, [449, 510, 792] for IASs, and [873] for differential swaps. The duration of T-bond futures is discussed in [554, 737]. See [95, p. 267] for the origin of Eurodollars. Finally, see [175] for the pricing of **equity swaps**.

NOTES

1. The conversion factor is independent of the prevailing interest rates. The theoretically sounder conversion ratio should be the price of the delivered bond divided by the price of the 20-year 8% coupon bond, both discounted at the prevailing interest rates. View www.cbot.com/ourproducts/financial/convbond.html for the CBT's regularly updated table of conversion factors.
2. Bond futures prices are in general lower than bond forward prices (see Exercise 12.3.3).

CHAPTER
TWENTY-TWO

Term Structure Fitting

That's an old besetting sin; they think calculating is inventing.
Johann Wolfgang Goethe (1749–1832), *Der Pantheist*

Fixed-income analysis starts with the yield curve. This chapter reviews **term structure fitting**, which means generating a curve to represent the yield curve, the spot rate curve, the forward rate curve, or the discount function. The constructed curve should fit the data reasonably well and be sufficiently smooth. The data are either bond prices or yields, and may be raw or synthetic as prepared by reputable firms such as Salomon Brothers (now part of Citigroup).

22.1 Introduction

The yield curve consists of hundreds of dots. Because bonds may have distinct qualities in terms of tax treatment, callability, and so on, more than one yield can appear at the same maturity. Certain maturities may also lack data points. These two problems were referred to in Section 5.3 as the multiple cash flow problem and the incompleteness problem. As a result, both regression (for the first problem) and interpolation (for the second problem) are needed for constructing a continuous curve from the data.

A functional form is first postulated, and its parameters are then estimated based on bond data. Two examples are the exponential function for the discount function and polynomials for the spot rate curve [7, 317]. The resulting curve is further required to be continuous or even differentiable as the relation between yield and maturity is expected to be fairly smooth. Although functional forms with more parameters often describe the data better, they are also more likely to **overfit** the data. An economically sensible curve that fits the data relatively well should be preferred to an economically unreasonable curve that fits the data extremely well.

Whether we fit the discount function $d(t)$, the spot rate function $s(t)$, or the forward rate curve $f(t)$ makes no difference theoretically. They carry the same information because

$$d(t) = e^{-ts(t)} = e^{-\int_0^t f(s)\,ds},$$
$$f(t) = s(t) + ts'(t)$$

under continuous compounding. Knowing any one of the three therefore suffices to infer the other two. The reality is more complicated, however. Empirically speaking, after the fitting, the smoothest curve is the discount function, followed by the spot rate curve, followed by the forward rate curve [848].

Using stripped Treasury security yields for spot rate curve fitting, albeit sensible, can be misleading. The major potential problem is taxation. The accrued compound interest of a stripped security is taxed annually, rendering its yield after-tax spot rate unless tax-exempt investors dominate the market. The liquidity of these securities is also not as great as that of coupon Treasuries. Finally, stripped Treasuries of certain maturities may attract investors willing to trade yield for a desirable feature associated with that particular maturity sector [326, 653]. This may happen because of dedicated portfolios set up for immunization.

Tax is a general problem, not just for stripped securities. Suppose capital gains are taxed more favorably than coupon income. Then a bond selling at a discount, generating capital gains at maturity, should have a lower yield to maturity than a similar par bond in order to produce comparable after-tax returns [568]. As a result, these two bonds will have distinct yields to maturity. This is the coupon effect at play.

▶ **Exercise 22.1.1** Verify the relation between $s(t)$ and $d(t)$ given above.

22.2 Linear Interpolation

A simple fitting method to handle the incompleteness problem is **linear yield interpolation** [335]. This technique starts with a list of bonds, preferably those selling near par and whose prices are both available and accurate. Usually only the on-the-run issues satisfy the criteria.[1] The scheme constructs a yield curve by connecting the yields with straight lines (see Figs. 22.1 and 22.2). The yield curve alone does not contain enough information to derive spot rates or, for that matter, discount factors and forward rates. This problem disappears for the par yield curve as the yield of a par bond equals its coupon rate.

The spot rate curve implied by the linearly interpolated yield curve is usually unsatisfactory in terms of shape. It may contain convex segments, for instance. The forward rate curve also behaves badly: It is extremely bumpy, with each bump corresponding to a specific bond in the data set and may be convex where it should be concave [848]. Despite these reservations, this scheme enjoys better statistical properties than many others [90].

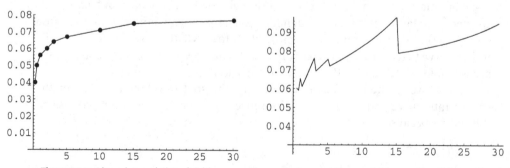

Figure 22.1: Linear interpolation of yield curve and forward rate curve. The par yield curve, linearly interpolated, is on the left, and the corresponding forward rate curve is on the right.

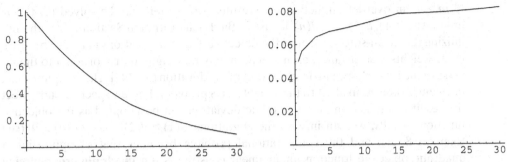

Figure 22.2: Discount function and spot rate curve. Plotted are the discount function and the spot rate curve as implied by the linearly interpolated par yield curve of Fig. 22.1.

A related scheme starts with the observation that the discount function is exponential in nature. It interpolates between known discount factors to obtain the discount function as follows. Let $t_1 < t < t_2$ and suppose that both $d(t_1)$ and $d(t_2)$ are available. The intermediate discount factor $d(t)$ is then interpolated by

$$d(t) = d(t_1)^{\frac{t(t_2-t)}{t_1(t_2-t_1)}} d(t_2)^{\frac{t(t-t_1)}{t_2(t_2-t_1)}}. \tag{22.1}$$

▶ **Exercise 22.2.1** Show that exponential interpolation scheme (22.1) for the discount function is equivalent to the linear interpolation scheme for the spot rate curve when spot rates are continuously compounded.

22.3 Ordinary Least Squares

Absent arbitrage opportunities, coupon bond prices and discount factors must satisfy

$$P_1 = (C_1 + 1) d(1),$$
$$P_2 = C_2 d(1) + (C_2 + 1) d(2),$$
$$P_3 = C_3 d(1) + C_3 d(2) + (C_3 + 1) d(3),$$
$$\vdots$$
$$P_n = C_n d(1) + C_n d(2) + \cdots + (C_n + 1) d(n).$$

In the preceding equations, the ith coupon bond has a coupon of C_i, a maturity of i periods, and a price of P_i. Once the discount factors $d(i)$ are solved for, the i-period spot rate $S(i)$ is simply $S(i) = d(i)^{-1/i} - 1$. This formulation makes clear what can derail in practice the bootstrapping procedure to extract spot rates from coupon bond prices in Section 5.2. A first case in point is when there are more data points P_i than variables $d(i)$. This scenario results in an overdetermined linear system – the multiple cash flow problem. A second case in point is when bonds of certain maturities are missing. Then we have an underdetermined linear system, which corresponds to the incompleteness problem.

The above formulation suggests that solutions be based on the principle of least squares. Suppose there are m bonds and the ith coupon bond has n_i periods to maturity, where $n_1 \le n_2 \le \cdots \le n_m = n$. Then we have the following system of m equations:

$$P_i = C_i d(1) + C_i d(2) + \cdots + (C_i + 1) d(n_i), \quad 1 \le i \le m. \tag{22.2}$$

If $m \geq n$, an overdetermined system results. This system can be solved for the n unknowns $d(1), d(2), \ldots, d(n)$ by use of the LS algorithm in Section 19.2 with minimizing the mean-square error as the objective. Certain equations may also be given more weights. For instance, each equation may be weighted proportional to the inverse of the bid–ask spread or the inverse of the duration [90, 147]. The computational problem is then equivalent to the weighted LS problem. If a few special bonds affect the result too much, the least absolute deviation can be adopted as the objective function. Finally, we can impose the conditions $d(1) \geq d(2) \geq \cdots \geq d(n) > 0$ (see Exercise 8.1.1), and the computational problem becomes that of optimizing a quadratic objective function under linear constraints – a **quadratic programming problem** [153].

We can handle the tax issue as follows. For each coupon bond, deduct the tax rate from the coupon rate. For each discount bond, reduce the principal repayment by the capital gains tax. For each premium bond, assume the loss will be amortized linearly over the life of the bond. For each zero-coupon bond, treat the income tax on imputed interest in each period as a negative cash flow. Finally, apply the methodology to the tax-adjusted cash flows to obtain the after-tax discount function [652].

Multiple-regression scheme (22.2) is called the McCulloch scheme. Other functional forms and target curves are clearly possible [653]. The Bradley–Crane scheme for example takes the form $\ln(1 + S(n)) = a + b_1 n + b_2 \ln n + \epsilon_n$ with three parameters, a, b_1, and b_2. The ϵ terms as usual represent errors. The Elliott–Echols scheme, as another example, adopts the form $\ln(1 + S(n_i)) = a + b_1/n_i + b_2 n_i + c_3 C_i + \epsilon_i$, where n_i is the term to maturity of the ith coupon bond. The explicit incorporation of the coupon rates is intended to take care of the coupon effect on yields due to tax considerations. Both schemes target the spot rate curve. See Fig. 22.3 for illustration.

Term structure fitting can also be model driven. Suppose we accept the Merton model for interest rate dynamics. It follows that the spot rate curve is a degree-two polynomial of the form $r + (\mu/2) t - (\sigma^2/6) t^2$ (see Fig. 22.4). This paradigm derives the model parameters – μ and σ in the Merton model – from regression [256, 511].

▶ **Exercise 22.3.1** Show how to fit a quadratic function $d(t) = a_0 + a_1 t + a_2 t^2$ to the discount factors by using multiple regression.

▶ **Exercise 22.3.2** Suppose we want to fit an exponential curve $y = ae^{bx}$ to the data, but we have only a linear-regression solver. How do we proceed?

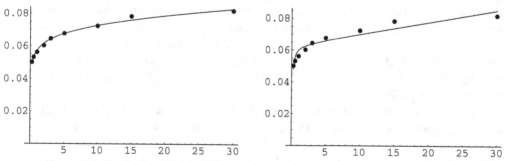

Figure 22.3: The Bradley–Crane and Elliott–Echols schemes. The Bradley–Crane scheme is on the left, and the Elliott–Echols scheme with the C_i set to zero is on the right.

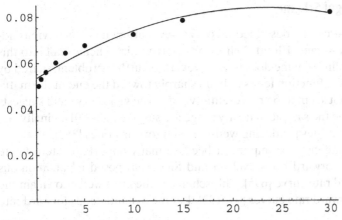

Figure 22.4: Polynomial regression.

22.4 Splines

The LS scheme of McCulloch, unlike those of Bradley–Crane and Elliott–Echols, cannot handle the incompleteness problem without imposing further restrictions. One proposal is to regard $d(i)$ as a linear combination of certain prespecified functions:

$$d(i) = 1 + \sum_{j=1}^{\ell} a_j f_j(i), \quad 1 \le i \le m, \tag{22.3}$$

where $f_j(i), 1 \le j \le \ell$, are known functions of maturity i and $a_1, a_2, \ldots, a_\ell$ are the parameters to be estimated [652]. Because $P(0) = 1$, $f_j(0)$ equals zero for every j. Substitute Eq. (22.3) into Eq. (22.2) to obtain the following overdetermined system:

$$P_i - 1 - C_i n_i = \sum_{k=1}^{\ell} a_k \left[f_k(n_i) + C_i \sum_{j=1}^{n_i} f_k(j) \right], \quad 1 \le i \le m. \tag{22.4}$$

There are now only ℓ unknown coefficients rather than n.

A lot of choices are open for the basis functions $f_j(\cdot)$. For instance, letting $f_j(x) = x^j$ makes the discount function a sum of polynomials. In this case, the problem is reduced to polynomial regression and is solved in the least-squares context (see Example 19.2.1). McCulloch suggested that the discount function be a cubic spline [652]. He also recommended picking the breakpoints that make each subinterval contain an equal number of maturity dates. This is probably the most well-known method.

▶ **Exercise 22.4.1** Verify Eq. (22.4).

▶ **Exercise 22.4.2** Assume continuous compounding. Justify the following claims. (1) If the forward rate curve should be a continuous function, a quadratic spline is the lowest-order spline that can fit the discount function. (2) If the forward rate curve should be continuously differentiable, a spline of at least cubic order is needed for the same purpose.

22.5 The Nelson–Siegel Scheme

A functional form that can be described by only a few parameters has obvious advantages. The Bradley–Crane, Elliott–Echols, and polynomial schemes fall into this category. The cubic-spline scheme does not, however. Its further problem, shared by others as well, is that the function tends to bend sharply toward the end of the maturity range. This does not seem to be representative of a true yield curve and suggests that predictions outside the sample maturity range are suspect [800, 801]. Finally, the spline scheme has difficulties producing well-behaved forward rates [352].

A good forward rate curve is important because many important interest rate models are based on forward rates. Nelson and Siegel proposed a parsimonious scheme for the forward rate curve [695]. This scheme is the most well known among families of smooth forward rate curves [77, 133]. Let the instantaneous forward rate curve be described by

$$f(\tau) = \beta_0 + \beta_1 e^{-\tau/\alpha} + \beta_2 \frac{\tau}{\alpha} e^{-\tau/\alpha},$$

where α is a constant (all rates are continuously compounded). The intent is to be able to measure the strengths of the short-, medium-, and long-term components of the forward rate curve. Specifically, the contribution of the long-term component is β_0, that of the short-term component is β_1, and that of the medium-term component is β_2. We can then find the quadruple $(\beta_0, \beta_1, \beta_2, \alpha)$ that minimizes the mean-square error between f and the data (see Fig. 22.5). The spot rate curve is

$$S(\tau) = \frac{\int_0^\tau f(s)\,ds}{\tau} = \beta_0 + (\beta_1 + \beta_2)\left(1 - e^{-\tau/\alpha}\right)\frac{\alpha}{\tau} - \beta_2 e^{-\tau/\alpha},$$

which is linear in the coefficients, given α. Both the forward rate curve and the spot rate curve converge to a constant, which has some appeal. All other functional forms seen so far have unbounded magnitude in the long end of the yield curve. The Nelson–Siegel scheme does not rule out negative forward rates.

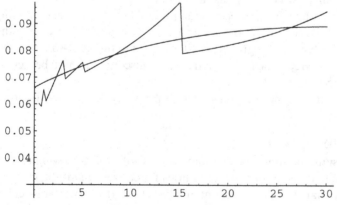

Figure 22.5: The Nelson–Siegel scheme. The forward rate curve in Fig. 22.1 (repeated here) is fitted by the Nelson–Siegel scheme.

Additional Reading

In reality we have bid and asked quotes. Sometimes the asked quotes are used to construct the curve, whereas the mean prices may be preferred in other times. Some schemes treat the fitting error as zero as long as the fitted value lies within the bid–ask spread [90]. See [551] for term structure modeling in Japan. Splines, being piecewise polynomials, seem ill-suited for the the discount function, which is exponential in nature. To tackle the mismatch, Fong and Vasicek proposed **exponential splines** [856], but the results seem little different from those of polynomial splines [801]. In [171] polynomials are proposed to fit each spot rate with the degrees of the polynomials chosen heuristically. Consult [39, 219, 352, 609, 801] for more fitting ideas and [135] for a linear programming approach to fitting the spot rate curve. Many fitting schemes are compared in [90].

NOTE

1. However, on-the-run bonds may be overvalued for a variety of reasons [890].

Introduction to Term Structure Modeling

> How much of the structure of our theories really tells us about things
> in nature, and how much do we contribute ourselves?
> Arthur Eddington (1882–1944)

The high interest rate volatility, especially since October 6, 1979 [401], calls for stochastic interest rate models. Models are also needed in managing interest rate risks of securities with interest-rate-sensitive cash flows. This chapter investigates stochastic term structure modeling with the **binomial interest rate tree** [779]. Simple as the model is, it illustrates most of the basic ideas underlying the models to come. The applications are also generic in that the pricing and hedging methodologies can be easily adapted to other models. Although the idea is similar to the one previously used in option pricing, the current task is complicated by two facts. First, the evolution of an entire term structure, not just a single stock price, is to be modeled. Second, interest rates of various maturities cannot evolve arbitrarily or arbitrage profits may result. The multitude of interest rate models is in sharp contrast to the single dominating model of Black and Scholes in option pricing.

23.1 Introduction ·

A stochastic interest rate model performs two tasks. First, it provides a stochastic process that defines future term structures. The ensuing dynamics must also disallow arbitrage profits. Second, the model should be "consistent" with the observed term structure [457]. Merton's work in 1970 marked the starting point of the continuous-time methodology to term structure modeling [493, 660]. This stochastic approach complements traditional term structure theories in that the unbiased expectations theory, the liquidity preference theory, and the market segmentation theory can all be made consistent with the model by the introduction of assumptions about the stochastic process [653].

Modern interest rate modeling is often traced to 1977 when Vasicek and Cox, Ingersoll, and Ross developed simultaneously their influential models [183, 234, 855]. Early models have fitting problems because the resulting processes may not price today's benchmark bonds correctly. An alternative approach pioneered by Ho and Lee in 1986 makes fitting the market yield curve mandatory [458]. Models based on such a paradigm are usually called **arbitrage-free** or **no-arbitrage models** [482]. The alternatives are **equilibrium models** and Black–Scholes models, which are covered in separate chapters.

▶ **Exercise 23.1.1** A riskless security with cash flow $C_1, C_2, \ldots, C_n$ has a market price of $\sum_{i=1}^{n} C_i d(i)$. The discount factor $d(i)$ denotes the PV of $1 at time i from now. Is the formula still valid if the cash flow depends on interest rates?

▶ **Exercise 23.1.2** Let i_t denote the period interest rate for the period from time $t - 1$ to t. Assume that $1 + i_t$ follows a lognormal distribution, $\ln(1 + i_t) \sim N(\mu, \sigma^2)$. (1) What is the value of $1 after n periods? (2) What are its distribution, mean, and variance?

23.2 The Binomial Interest Rate Tree

Our goal here is to construct a no-arbitrage interest rate tree consistent with the observed term structure, specifically the yields and/or yield volatilities of zero-coupon bonds of all maturities. This procedure is called **calibration**. We pick a binomial tree model in which the logarithm of the future short rate obeys the binomial distribution. The limiting distribution of the short rate at any future time is hence lognormal. In the binomial interest rate process, a binomial tree of future short rates is constructed. Every short rate is followed by two short rates for the following period. In Fig. 23.1, node A coincides with the start of period j during which the short rate r is in effect. At the conclusion of period j, a new short rate goes into effect for period $j + 1$. This may take one of two possible values: r_ℓ, the "low" short-rate outcome at node B, and r_h, the "high" short-rate outcome at node C. Each branch has a 50% chance of occurring in a risk-neutral economy.

We require that the paths combine as the binomial process unfolds. Suppose that the short rate r can go to r_h and r_ℓ with equal risk-neutral probability $1/2$ in a period of length Δt.[1] The volatility of $\ln r$ after Δt time is

$$\sigma = \frac{1}{2} \frac{1}{\sqrt{\Delta t}} \ln\left(\frac{r_h}{r_\ell}\right)$$

(see Exercise 23.2.3, part (1)). Above, σ is annualized, whereas r_ℓ and r_h are period based. As

$$\frac{r_h}{r_\ell} = e^{2\sigma\sqrt{\Delta t}}, \qquad (23.1)$$

greater volatility, hence uncertainty, leads to larger r_h/r_ℓ and wider ranges of possible short rates. The ratio r_h/r_ℓ may change across time if the volatility is a function of time. Note that r_h/r_ℓ has nothing to do with the current short rate r if σ is independent of r. The volatility of the short rate one-period forward is approximately $r\sigma$ (see Exercise 23.2.3, part (2)).

Figure 23.1: Binomial interest rate process. From node A there are two equally likely scenarios for the short rate: r_ℓ and r_h. Rate r is applicable to node A in period j. Rate r_ℓ is applicable to node B and rate r_h is applicable to node C, both in period $j + 1$.

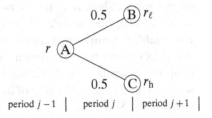

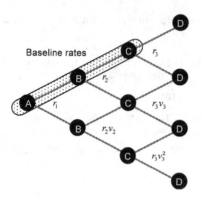

Baseline rates

Figure 23.2: Binomial interest rate tree. The sequence at each time point shows that the short rate will converge to the lognormal distribution.

In general there are j possible rates in period j:

$$r_j, \; r_j v_j, \; r_j v_j^2, \ldots, r_j v_j^{j-1},$$

where

$$v_j \equiv e^{2\sigma_j \sqrt{\Delta t}} \tag{23.2}$$

is the multiplicative ratio for the rates in period j (see Fig. 23.2). We call r_j the **baseline rates**. The subscript j in σ_j is meant to emphasize that the short rate volatility may be time dependent. In the limit, the short rate follows the following process:

$$r(t) = \mu(t) \, e^{\sigma(t) \, W(t)}, \tag{23.3}$$

in which the (percent) short-rate volatility $\sigma(t)$ is a deterministic function of time. As the expected value of $r(t)$ equals $\mu(t) \, e^{\sigma(t)^2 t/2}$, a declining short rate volatility is usually imposed to preclude the short rate from assuming implausibly high values. Incidentally, this is how the binomial interest rate tree achieves mean reversion.

One salient feature of the tree is path independence: The term structure at any node is independent of the path taken to reach it. A nice implication is that only the baseline rates r_i and the multiplicative ratios v_i need to be stored in computer memory in order to encode the whole tree. This takes up only $O(n)$ space. (Throughout this chapter, n denotes the depth of the tree, i.e., the number of discrete time periods.) The naive approach of storing the whole tree would take up $O(n^2)$ space. This can be prohibitive for large trees. For instance, modeling daily interest rate movements for 30 years amounts to keeping an array of roughly $(30 \times 365)^2/2 \approx 6 \times 10^7$ double-precision floating-point numbers. If each number takes up 8 bytes, the array would consume nearly half a gigabyte!

With the abstract process in place, the concrete numbers that set it in motion are the annualized rates of return associated with the various riskless bonds that make up the benchmark yield curve and their volatilities. In the United States, for example, the on-the-run yield curve obtained by the most recently issued Treasury securities may be used as the benchmark curve. The **term structure of (yield) volatilities** or simply the **volatility (term) structure** can be estimated from either historical data (historical volatility) or interest rate option prices such as cap prices (implied volatility) [149, 880]. The binomial tree should be consistent with both term structures. In this chapter we focus on the term structure of interest rates, deferring the handling of the volatility structure to Section 26.3.

For economy of expression, all numbers in algorithms are measured by the period instead of being annualized whenever applicable and unless otherwise stated. The relation is straightforward in the case of volatility: $\sigma(\text{period}) = \sigma(\text{annual}) \times \sqrt{\Delta t}$. As for the interest rates, consult Section 3.1.

An alternative process that also satisfies the path-independence property is the following arithmetic sequence of short rates for period j:

$$r_j, r_j + v_j, r_j + 2v_j, \ldots, r_j + (j-1)v_j.$$

Ho and Lee proposed this binomial interest rate model [458]. If the j possible rates for period j are postulated to be

$$ru^{j-1}, ru^{j-2}d, \ldots, rd^{j-1}$$

for some common u and d, the parameters are u and d and possibly the transition probability. This parsimonious model is due to Rendleman and Bartter [739].

▶ **Exercise 23.2.1** Verify that the variance of $\ln r$ in period k equals $\sigma_k^2(k-1)\Delta t$. (Consistent with Eq. (23.3), the variance of $\ln r(t)$ equals $\sigma(t)^2 t$ in the continuous-time limit.)

▶ **Exercise 23.2.2** Consider a short rate model such that the two equally probable short rates from the current rate r are $re^{\mu+\sigma\sqrt{\Delta t}}$ and $re^{\mu-\sigma\sqrt{\Delta t}}$, where μ may depend on time. Verify that this model can result from the binomial interest rate tree when the volatilities σ_j are all equal to some constant σ. (The μ is varied to fit the term structure.)

▶ **Exercise 23.2.3** Suppose the probability of moving from r to r_ℓ is $1-q$ and that of moving to r_h is q. Also assume that a period has length Δt. (1) Show that the variance of $\ln r$ after a period is $q(1-q)(\ln r_h - \ln r_\ell)^2$. (2) Hence, if we define σ^2 to be the above divided by Δt, then

$$\frac{r_h}{r_\ell} = \exp\left[\sigma\sqrt{\frac{\Delta t}{q(1-q)}}\right]$$

should replace Eq. (23.1). Now prove the variance of r after a time period of Δt is approximately $r^2\sigma^2\Delta t$.

23.2.1 Term Structure and Its Dynamics

With the binomial interest rate tree in place, the **model price** of a security can be computed by backward induction. Refer back to Fig. 23.1. Given that the values at nodes B and C are P_B and P_C, respectively, the value at node A is then

$$\frac{P_B + P_C}{2(1+r)} + \text{cash flow at node A.}$$

To save computer memory, we compute the values column by column without explicitly expanding the binomial interest rate tree (see Fig. 23.3 for illustration). Figure 23.4 contains the quadratic-time, linear-space algorithm for securities with a fixed cash flow. The same idea can be applied to any tree model.

We can compute an n-period zero-coupon bond's price by assigning \$1 to every node at period n and then applying backward induction. Repeating this step for $n = 1, 2, \ldots$, we obtain the market discount function implied by the tree. The tree

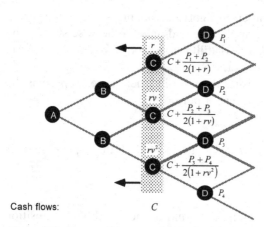

Cash flows: C

Figure 23.3: Sweep a line across time backward to compute model price.

therefore determines a term structure. Moreover, it encompasses **term structure evolution** or **dynamics** because taking any node in the tree as the current state induces a (smaller) binomial interest rate tree and, again, a term structure. The tree thus defines how the whole term structure evolves through time.

Comment 23.2.1 Suppose we want to know the m-period spot rate at time n in order to price a security whose payoff is linked to that spot rate. The tree has to be built all the way to time $n+m$ in order to obtain the said spot rate at time n, with dire performance implications. Later in Subsection 25.2.1 we will see cases in which the tree has to be built over the life of only the derivative (n periods) instead of over the life of the underlying asset ($n+m$ periods).

We shall construct interest rate trees consistent with the sample term structure in Fig. 23.5. For numerical demonstrations, we assume that the short rate volatility is such that $v \equiv r_{\mathrm{h}}/r_{\ell} = 1.5$, independent of time.

Algorithm for model price from binomial interest rate tree:

input: $m, n, r[1..n], C[0..n], v[1..n]$; // $m \le n$.
real $P[1..m+1]$;
integer i, j;
for ($i = 1$ to $m+1$) $P[i] := C[m]$; // Initialization.
for ($i = m$ down to 1)   // Backward induction.
        for ($j = 1$ to i)
                $P[j] := C[i-1] + 0.5 \times (P[j] + P[j+1])/(1 + r[i] \times v[i]^{j-1})$;
return $P[1]$;

Figure 23.4: Algorithm for model price. $C[i]$ is the cash flow occurring at time i (the end of the ith period), $r[i]$ is the baseline rate for period i, $v[i]$ is the multiplicative ratio for the rates in period i, and n denotes the number of periods. Array P stores the PV at each node. All numbers are measured by the period.

Period	1	2	3
Spot rate (%)	4	4.2	4.3
One-period forward rate (%)	4	4.4	4.5
Discount factor	0.96154	0.92101	0.88135

Figure 23.5: Sample term structure.

An Approximate Calibration Scheme

A scheme that easily comes to mind starts with the implied one-period forward rates and then equates the expected short rate with the forward rate. This certainly works in a deterministic economy (see Exercise 5.6.6). For the first period, the forward rate is today's one-period spot rate. In general, let f_j denote the forward rate in period j. This forward rate can be derived from the market discount function by $f_j = [d(j)/d(j+1)] - 1$ (see Exercise 5.6.3). Because the ith short rate, $1 \le i \le j$, occurs with probability $2^{-(j-1)} \binom{j-1}{i-1}$, this means that $\sum_{i=1}^{j} 2^{-(j-1)} \binom{j-1}{i-1} r_j v_j^{i-1} = f_j$, and thus

$$ r_j = \left(\frac{2}{1+v_j} \right)^{j-1} f_j. \tag{23.4} $$

The binomial interest rate tree is hence trivial to set up.

The ensuing tree for the sample term structure is shown in Fig. 23.6. For example, the price of the zero-coupon bond paying \$1 at the end of the third period is

$$ \frac{1}{4} \times \frac{1}{1.04} \times \left[\frac{1}{1.0352} \times \left(\frac{1}{1.0288} + \frac{1}{1.0432} \right) \right. $$
$$ \left. + \frac{1}{1.0528} \times \left(\frac{1}{1.0432} + \frac{1}{1.0648} \right) \right] = 0.88155, \tag{23.5} $$

which is very close to, but overestimates, the discount factor 0.88135. The tree is thus not calibrated. Indeed, this bias is inherent.

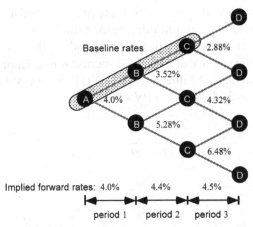

Figure 23.6: A binomial interest rate tree based on the unbiased expectations theory.

THEOREM 23.2.2 *The binomial interest rate tree constructed with Eq. (23.4) overestimates the prices of the benchmark securities in the presence of volatilities. This conclusion is independent of whether the volatility structure is matched.*

Theorem 23.2.2 implies that, under the binomial interest rate tree, the expected future spot rate exceeds the forward rate. As always, we took the money market account implicitly as numeraire. But it was argued in Subsection 13.2.1 that switching the numeraire changes the risk-neutral probability measure. Indeed, there exists a numeraire under which the forward rate equals the expected future spot rate (see Exercise 23.2.5, part (1)).

▶ **Exercise 23.2.4** (1) Prove Theorem 23.2.2 for two-period zero-coupon bonds. (2) Prove Theorem 23.2.2 in its full generality.

▶ **Exercise 23.2.5** Fix a period. (1) Show that the forward rate for that period equals the expected future spot rate under some risk-neutral probability measure. (2) Show further that the said forward rate is a martingale. (Hint: Exercise 13.2.13.)

23.2.2 Calibration of Binomial Interest Rate Trees

It is of paramount importance that the model prices generated by the binomial interest rate tree match the observed market prices. This may well be the most crucial aspect of model building. To achieve it, we can treat the backward-induction algorithm for the model price of the m-period zero-coupon bond in Fig. 23.4 as computing some function of the unknown baseline rate r_m called $f(r_m)$. A good root-finding method is then applied to solve $f(r_m) = P$ for r_m given the zero's market price P and $r_1, r_2, \ldots, r_{m-1}$. This procedure is carried out for $m = 1, 2, \ldots, n$. The overall algorithm runs in cubic time, thus hopelessly slow [508].

Calibration can be accomplished in quadratic time by the use of **forward induction** [508]. The scheme records how much $1 at a node contributes to the model price. This number is called the **state price** as it stands for the price of a state contingent claim that pays $1 at that particular node (state) and zero elsewhere. The column of state prices will be established by moving *forward* from time 1 to time n.

Let us be more precise. Suppose we are at time j and there are $j+1$ nodes. Let the baseline rate for period j be $r \equiv r_j$, the multiplicative ratio be $v \equiv v_j$, and $P_1, P_2, \ldots, P_j$ be the state prices a period prior, corresponding to rates $r, rv, \ldots, rv^{j-1}$. By definition, $\sum_{i=1}^{j} P_i$ is the price of the $(j-1)$-period zero-coupon bond. One dollar at time j has a known market value of $1/[1 + S(j)]^j$, where $S(j)$ is the j-period spot rate. Alternatively, this dollar has a PV of

$$g(r) \equiv \frac{P_1}{(1+r)} + \frac{P_2}{(1+rv)} + \frac{P_3}{(1+rv^2)} + \cdots + \frac{P_j}{(1+rv^{j-1})}.$$

So we solve

$$g(r) = \frac{1}{[1+S(j)]^j} \tag{23.6}$$

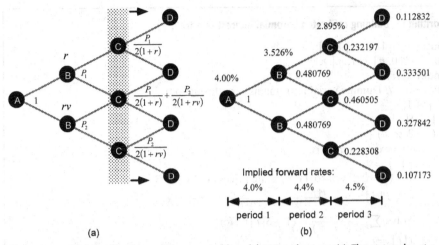

(a)   (b)

Figure 23.7: Sweep a line forward to compute binomial state price tree. (a) The state price at a node is a weighted sum of the state prices of its two predecessors. (b) The binomial state price tree calculated from the sample term structure and the resulting calibrated binomial interest rate tree. It prices the benchmark bonds correctly.

for r. Given a decreasing market discount function, a unique positive solution for r is guaranteed.[2] The state prices at time j can now be calculated as

$$\frac{P_1}{2(1+r)}, \ \frac{P_1}{2(1+r)} + \frac{P_2}{2(1+rv)}, \ \dots, \ \frac{P_{j-1}}{2(1+rv^{j-2})}$$

$$+ \frac{P_j}{2(1+rv^{j-1})}, \ \frac{P_j}{2(1+rv^{j-1})}$$

(see Fig. 23.7(a)). We call a tree with these state prices a **binomial state price tree**. Figure 23.7(b) shows one such tree. The calibrated tree is shown in Fig. 23.8.

The Newton–Raphson method can be used to solve for the r in Eq. (23.6) as $g'(r)$ is easy to evaluate. The monotonicity and the convexity of $g(r)$ also facilitate root finding. A good initial approximation to the root may be provided by Eq. (23.4), which

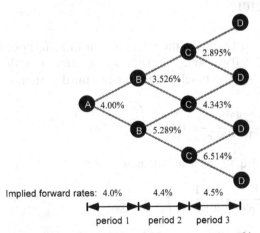

Figure 23.8: Calibrated binomial interest rate tree. This tree is from Fig. 23.7(b).

Algorithm for building calibrated binomial interest rate tree:

input: $n, S[1..n], \sigma[1..n]$;
real $P[0..n], r[1..n], r, v$;
integer i, j;
$P[0] := 0$;   // Dummy variable; remains zero throughout.
$P[1] := 1$;
$r[1] := S[1]$;
for ($i = 2$ to n) {
       $v := \exp[2 \times \sigma[i]]$;
       $P[i] := 0$;
       for ($j = i$ down to 1)   // State prices at time $i - 1$.
          $P[j] := \frac{P[j-1]}{2\times(1+r[i-1]v^{j-2})} + \frac{P[j]}{2\times(1+r[i-1]v^{j-1})}$;
       Solve $\sum_{j=1}^{i} \frac{P[j]}{(1+rv^{j-1})} = (1 + S[i])^{-i}$ for r;
       $r[i] := r$;
}
return $r[\ ]$;

Figure 23.9: Algorithm for building calibrated binomial interest rate tree. $S[i]$ is the i-period spot rate, $\sigma[i]$ is the percent volatility of the rates for period i, and n is the number of periods. All numbers are measured by the period. The period-i baseline rate is computed and stored in $r[i]$.

is guaranteed to underestimate the root (see Theorem 23.2.2). Using the previous baseline rate as the initial approximation to the current baseline rate also works well.

The preceding idea is straightforward to implement (see Fig. 23.9). The total running time is $O(Cn^2)$, where C is the maximum number of times the root-finding routine iterates, each consuming $O(n)$ work. With a good initial guess, the Newton–Raphson method converges in only a few steps [190, 625].

Let us follow up with some numerical calculations. One dollar at the end of the second period should have a PV of 0.92101 according to the sample term structure. The baseline rate for the second period, r_2, satisfies

$$\frac{0.480769}{1+r_2} + \frac{0.480769}{1+1.5 \times r_2} = 0.92101.$$

The result is $r_2 = 3.526\%$. This is used to derive the next column of state prices shown in Fig. 23.7(b) as 0.232197, 0.460505, and 0.228308, whose sum gives the correct market discount factor 0.92101. The baseline rate for the third period, r_3, satisfies

$$\frac{0.232197}{1+r_3} + \frac{0.460505}{1+1.5 \times r_3} + \frac{0.228308}{1+(1.5)^2 \times r_3} = 0.88135.$$

The result is $r_3 = 2.895\%$. Now, redo Eq. (23.5) using the new rates:

$$\frac{1}{4} \times \frac{1}{1.04} \times \left[\frac{1}{1.03526} \times \left(\frac{1}{1.02895} + \frac{1}{1.04343} \right) + \frac{1}{1.05289} \right.$$

$$\left. \times \left(\frac{1}{1.04343} + \frac{1}{1.06514} \right) \right] = 0.88135,$$

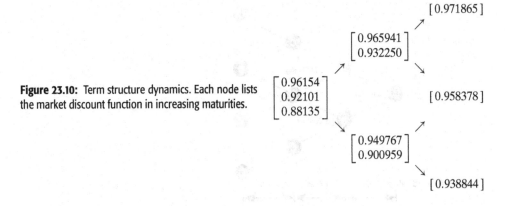

Figure 23.10: Term structure dynamics. Each node lists the market discount function in increasing maturities.

an exact match. The tree in Fig. 23.8 therefore prices without bias the benchmark securities. The term structure dynamics of the calibrated tree is shown in Fig. 23.10.

▶ **Exercise 23.2.6** (1) Based on the sample term structure and its associated binomial interest rate tree in Fig. 23.8, what is the next baseline rate if the four-period spot rate is 4.4%? (2) Confirm Theorem 23.2.2 by demonstrating that the baseline rate produced by Eq. (23.4) is smaller than the one derived in (1).

▶ **Exercise 23.2.7** (1) Suppose we are given a binomial state price tree and wish to price a security with the payoff function c at time j by using the risk-neutral pricing formula $d(j) E[c]$. What is the probability of each state's occurring at time j? (2) Take the binomial state price tree in Fig. 23.7(b). What are the probabilities of the C nodes in this risk-neutral economy?

▶ **Exercise 23.2.8** Compute the n discount factors implied by the tree in $O(n^2)$ time.

▶ **Exercise 23.2.9** Start with a binomial interest rate tree but *without* the branching probabilities, such as Fig. 23.2. (1) Suppose the state price tree is also given. (2) Suppose only the state prices at the terminal nodes are given and assume that the path probabilities for all paths reaching the same node are equal. How do we calculate the branching probabilities at each node in either case? (The result was called the implied binomial tree in Exercise 9.4.3.)

➤ **Programming Assignment 23.2.10** Program the algorithm in Fig. 23.9 with the Newton–Raphson method.

➤ **Programming Assignment 23.2.11** Calibrate the tree with the secant method.

23.3 Applications in Pricing and Hedging

23.3.1 Spread of Nonbenchmark Option-Free Bonds

Model prices calculated by the calibrated tree as a rule do not match market prices of nonbenchmark bonds. To gauge the incremental return, or **spread**, over the benchmark bonds, we look for the spread that, when added uniformly over the short rates in the tree, makes the model price equal the market price. Obviously the spread of a benchmark security is zero. We apply the spread concept to option-free bonds first

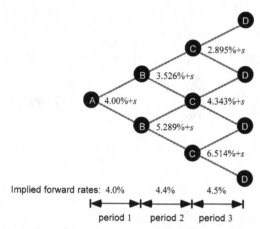

Figure 23.11: Spread over short rates of binomial interest rate tree. This tree is constructed from the calibrated binomial interest rate tree in Fig. 23.8 by the addition of a constant spread s to each short rate in the tree.

and return to bonds that incorporate embedded options in Subsection 27.4.3. The techniques are identical save for the possibility of early exercise.

It is best to illustrate the idea with an example. Start with the tree in Fig. 23.11. Consider a security with cash flow C_i at time i for $i = 1, 2, 3$. Its model price is

$$
\begin{aligned}
p(s) \equiv \frac{1}{1.04+s} \times \Bigg\{ & C_1 + \frac{1}{2} \times \frac{1}{1.03526+s} \\
& \times \left[C_2 + \frac{1}{2}\left(\frac{C_3}{1.02895+s} + \frac{C_3}{1.04343+s} \right) \right] \\
& + \frac{1}{2} \times \frac{1}{1.05289+s} \times \left[C_2 + \frac{1}{2}\left(\frac{C_3}{1.04343+s} + \frac{C_3}{1.06514+s} \right) \right] \Bigg\}.
\end{aligned}
$$

Given a market price of P, the spread is the s that solves $P = p(s)$.

In general, if we add a constant amount s to every rate in the binomial interest rate tree, the model price will be a monotonically decreasing, convex function of s. Call this function $p(s)$. For a market price P, we use the Newton–Raphson root-finding method to solve $p(s) - P = 0$ for s. However, a quick look at the preceding equation reveals that evaluating $p'(s)$ directly is infeasible. Fortunately the tree can be used to evaluate both $p(s)$ and $p'(s)$ during backward induction. Here is the idea. Consider an arbitrary node A in the tree associated with the short rate r. In the process of computing the model price $p(s)$, a price $p_A(s)$ is computed at A. Prices computed at A's two successor nodes, B and C, are discounted by $r + s$ to obtain $p_A(s)$:

$$
p_A(s) = c + \frac{p_B(s) + p_C(s)}{2(1+r+s)},
$$

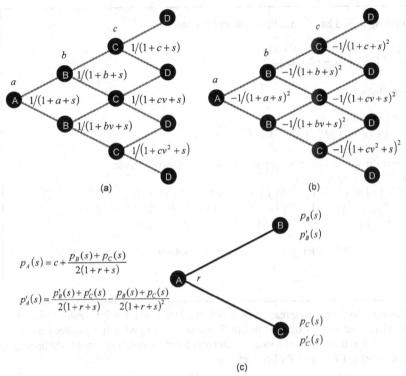

$$p_A(s) = c + \frac{p_B(s) + p_C(s)}{2(1+r+s)}$$

$$p'_A(s) = \frac{p'_B(s) + p'_C(s)}{2(1+r+s)} - \frac{p_B(s) + p_C(s)}{2(1+r+s)^2}$$

Figure 23.12: The differential tree method. (a) The original binomial interest rate tree with the short rates replaced with the discount factors, (b) the derivatives of the numbers on the tree, and (c) the simultaneous evaluation of a function and its derivative on the tree by use of the numbers from (a) and (b).

where c denotes the cash flow at A. To compute $p'_A(s)$ as well, node A calculates

$$p'_A(s) = \frac{p'_B(s) + p'_C(s)}{2(1+r+s)} - \frac{p_B(s) + p_C(s)}{2(1+r+s)^2}, \tag{23.7}$$

which is easy if $p'_B(s)$ and $p'_C(s)$ are also computed at nodes B and C. Applying the preceding procedure inductively will eventually lead to $p(s)$ and $p'(s)$ at the root. See Fig. 23.12 for illustration. This technique, which is due to Lyuu, is called the **differential tree method** and has wide applications [625]. It is also related to **automatic differentiation** in numerical analysis [602, 687].

Let us analyze the differential tree algorithm in Fig. 23.13. Given a spread, step 1 computes the PV, step 2 computes the derivative of the PV according to Eq. (23.7), and step 3 implements the Newton–Raphson method for the next approximation. If $\mathcal{C}$ represents the number of times the tree is traversed, which takes $O(n^2)$ time, the total running time is $O(\mathcal{C}n^2)$. In practice, $\mathcal{C}$ is a small constant. The memory requirement is $O(n)$.

Now we go through a numerical example. Consider a 3-year 5% bond with a market price of 100.569. For simplicity, assume that the bond pays annual interest. The spread can be shown to be 50 basis points over the tree (see Fig. 23.14). For comparison, let us compute the yield spread and the static spread of the nonbenchmark bond over an otherwise identical benchmark bond. Recall that the static spread is

Algorithm for computing spread based on differential tree method:

```
input:    n, P, r[1..n], C[0..n], v[1..n], ε;
real      P[1..n+1], P'[1..n+1], sold, snew;
integer   i, j;
snew := 0;  // Initial guess.
P[1] := ∞;
while [|P[1] − P| > ε] {
        for (i = 1 to n+1) { P[i] := C[n]; P'[i] := 0; }
        sold := snew;
        for (i = n down to 1)   // Sweep the column backward in time.
          for (j = 1 to i) {
            1. P[j] := C[i−1]+(P[j]+P[j+1])/(2×(1+r[i]×v[i]^{j−1}+sold));
            2. P'[j] := (P'[j]+P'[j+1])/(2×(1+r[i]×v[i]^{j−1}+sold))−
               (P[j]+P[j+1])/(2×(1+r[i]×v[i]^{j−1}+sold))^2;
          }
        3. snew := sold − (P[1]−P)/P'[1];   // Newton-Raphson.
}
return snew;
```

Figure 23.13: Algorithm for computing spread based on differential tree method. P is the market price, $r[i]$ is the baseline rate for period i, $C[i]$ contains the cash flow at time i, $v[i]$ is the multiplicative ratio for the rates in period j, and n is the number of periods. All numbers are measured by the period. The prices and their derivatives are stored in $P[\,]$ and $P'[\,]$, respectively.

the incremental return over the spot rate curve, whereas the spread based on the binomial interest rate tree is one over the future short rates. The yield to maturity of the nonbenchmark bond can be calculated to be 4.792%. The 3-year Treasury has a market price of

$$\frac{5}{1.04} + \frac{5}{(1.042)^2} + \frac{105}{(1.043)^3} = 101.954 \tag{23.8}$$

and a yield to maturity of 4.292%. The yield spread is thus 4.792% − 4.292% = 0.5%. The static spread can also be found to be 0.5%. So all three spreads turn out to be 0.5% up to round-off errors.

▶ **Exercise 23.3.1** Does the idea of spread assume parallel shifts in the term structure?

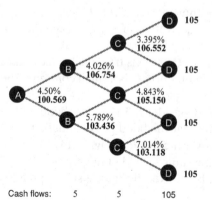

Cash flows: 5   5   105

Figure 23.14: Price tree with spread. Based on the tree in Fig. 23.11, the price tree is computed for a 3-year bond paying 5% annual interest. Each node of the tree signifies, besides the short rate, the discounted value of its future cash flows plus the cash flow at that node, if any. The model price is 100.569.

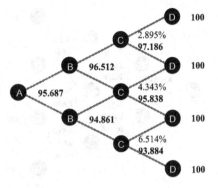

Figure 23.15: Futures price. The price tree is computed for a 2-year futures contract on a 1-year T-bill. C nodes store, besides the short rate, the discounted values of the 1-year T-bill under the model. A and B nodes calculate the expected values. The futures price is 95.687.

➤ **Programming Assignment 23.3.2** Implement the algorithm in Fig. 23.13.

➤ **Programming Assignment 23.3.3** Implement a differential tree method for the implied volatility of American options under the BOPM [172, 625].

23.3.2 Futures Price

The futures price is a martingale under the risk-neutral probability (see Exercise 13.2.11). To compute it, we first use the tree to calculate the underlying security's prices at the futures contract's delivery date to which the futures price converges. Then we find the expected value. In Fig. 23.15, for example, we are concerned with a 2-year futures contract on a 1 year T-bill. The futures price is found to be 95.687. The futures price can be computed in $O(n)$ time.

If the contract specification for a futures contract does not call for a quote that equals the result of our computation, steps have to be taken to convert the theoretical value into one consistent with the specification. The theoretical value above, for instance, corresponds to the invoice price of the T-bill futures traded on the CBT, but it is the index price that gets quoted.

➤ **Exercise 23.3.4** (1) How do we compute the forward price for a forward contract on a bond? (2) Calculate the forward price for a 2-year forward contract on a 1-year T-bill.

23.3.3 Fixed-Income Options

Determining the values of fixed-income options with a binomial interest rate tree follows the same logic as that of the binomial tree algorithm for stock options in Chap. 9. Hence only numerical examples are attempted here. Consider a 2-year 99 European call on the 3-year 5% Treasury. Assume that the Treasury pays annual interest. From Fig. 23.16 the 3-year Treasury's price minus the $5 interest could be $102.046, $100.630, or $98.579 2 years from now. Because these prices do not include the accrued interest, we should compare the strike price against them. The call is therefore in the money in the first two scenarios, with values of $3.046 and $1.630, and out of the money in the third scenario. The option value is calculated to be $1.458 in Fig. 23.16(a). European interest rate puts can be valued similarly. Consider a 2-year 99 European put on the same security. At expiration, the put is in the money only if the Treasury is worth $98.579 without the accrued interest. The option value is computed to be $0.096 in Fig. 23.16(b).

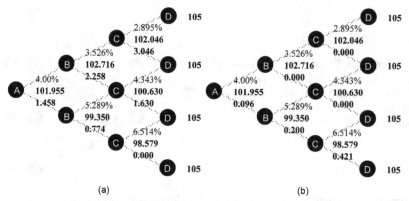

Figure 23.16: European options on Treasuries. The above price trees are computed for the 2-year 99 European (a) call and (b) put on the 3-year 5% Treasury. Each node of the tree signifies the short rate, the Treasury price without the $5 interest (except the D nodes), and the option value. The price $101.955 is slightly off compared with Eq. (23.8) because of round-off errors.

If the option is American and the underlying bond generates payments before the option's expiration date, early exercise needs to be considered. The criterion is to compare the intrinsic value against the option value at each node. The details are left to the reader.

The PV of the strike price is $PV(X) = 99 \times 0.92101 = 91.18$. The Treasury is worth $B = 101.955$. The PV of the interest payments during the life of the options is

$$PV(I) = 5 \times 0.96154 + 5 \times 0.92101 = 9.41275.$$

The call and the put are worth $C = 1.458$ and $P = 0.096$, respectively. Hence

$$C = P + B - PV(I) - PV(X).$$

The put–call parity is preserved.

▶ **Exercise 23.3.5** Prove that an American option on a zero-coupon bond will not be exercised early.

▶ **Exercise 23.3.6** Derive the put–call parity for options on coupon bonds.

▶ **Programming Assignment 23.3.7** Write a program to price European options on the Treasuries.

23.3.4 Delta (Hedge Ratio)

It is important to know how much the option price changes in response to changes in the price of the underlying bond. This relation is called the delta (or hedge ratio), defined as

$$\frac{O_h - O_\ell}{P_h - P_\ell}.$$

P_h and P_ℓ denote the bond prices if the short rate moves up and down, respectively. Similarly, O_h and O_ℓ denote the option values if the short rate moves up and down, respectively. Because delta measures the sensitivity of the option value to changes in

Short rate	Horizon price	Probability
2.895%	0.971865	0.25
4.343%	0.958378	0.50
6.514%	0.938844	0.25

Figure 23.17: HPRs. The horizon is two periods from now.

the underlying bond price, it shows how to hedge one with the other [84]. Take the call and put in Fig. 23.16 as examples. Their deltas are

$$\frac{0.774 - 2.258}{99.350 - 102.716} = 0.441, \qquad \frac{0.200 - 0.000}{99.350 - 102.716} = -0.059,$$

respectively.

23.3.5 Holding Period Returns

Analyzing the holding period return (HPR) with the binomial interest rate tree is straightforward. As an example, consider a two-period horizon for three-period zero-coupon bonds. Based on the price dynamics in Fig. 23.10, the HPRs are obtained in Fig. 23.17. If the bonds are coupon bearing, the interim cash flows should be reinvested at the prevailing short rate and added to the horizon price. The probability distribution of the scenario analysis is provided by the model, not exogenously.

23.4 Volatility Term Structures

The binomial interest rate tree can be used to calculate the yield volatility of zero-coupon bonds. Consider an n-period zero-coupon bond. First find its yield to maturity y_h (y_ℓ, respectively) at the end of the initial period if the rate rises (declines, respectively). The yield volatility for our model is defined as $(1/2) \ln(y_h/y_\ell)$. For example, based on the tree in Fig. 23.8, the 2-year zero's yield at the end of the first period is 5.289% if the rate rises and 3.526% if the rate declines. Its yield volatility is therefore

$$\frac{1}{2} \ln \left(\frac{0.05289}{0.03526} \right) = 20.273\%.$$

Now consider the 3-year zero-coupon bond. If the rate rises, the price of the zero 1 year from now will be

$$\frac{1}{2} \times \frac{1}{1.05289} \times \left(\frac{1}{1.04343} + \frac{1}{1.06514} \right) = 0.90096.$$

Thus its yield is

$$\sqrt{\frac{1}{0.90096}} - 1 = 0.053531.$$

If the rate declines, the price of the zero 1 year from now will be

$$\frac{1}{2} \times \frac{1}{1.03526} \times \left(\frac{1}{1.02895} + \frac{1}{1.04343} \right) = 0.93225.$$

Thus its yield is

$$\sqrt{\frac{1}{0.93225}} - 1 = 0.0357.$$

The yield volatility is hence

$$\frac{1}{2} \ln \left(\frac{0.053531}{0.0357} \right) = 20.256\%,$$

slightly less than the 1-year yield volatility. Interestingly, this is consistent with the reality that longer-term bonds typically have lower yield volatilities than shorter-term bonds. The procedure can be repeated for longer-term zeros to obtain their yield volatilities.

We started with v_i and then derived the volatility term structure. In practice, the steps are reversed. The volatility term structure is supplied by the user along with the term structure. The v_is – hence the short rate volatilities by Eq. (23.2) – and the r_is are then simultaneously determined. The result is the Black–Derman–Toy model, which is covered in Section 26.3.

Suppose the user supplies the volatility term structure that results in $(v_1, v_2, v_3, \dots)$ for the tree. The volatility term structure one period from now will be determined by $(v_2, v_3, v_4, \dots)$, not $(v_1, v_2, v_3, \dots)$. The volatility term structure supplied by the user is hence not maintained through time.

➤ **Exercise 23.4.1** Suppose we add a binomial process for the stock price to our binomial interest rate model. In other words, stock price S can in one period move to Su or Sd. What are the constraints on u and d?

➤ **Programming Assignment 23.4.2** Add the annualized term structure of yield volatilities to the output of the program of Programming Assignment 23.2.10.

NOTES

1. By designating the risk-neutral probabilities as $1/2$, we are obliged to adjust the state variable, the short rate, in order to match the desired distribution. This was done in Exercise 9.3.1, for example, in the case of the BOPM. An alternative is to prescribe the state variable's values on the tree and then to find the probabilities. This was the approach of the finite-difference method in Section 18.1.

2. This is because $g(r)$ is strictly decreasing with $g(0) = \sum_{i=1}^{j} P_i > 1/[1 + S(j)]^j$ and $g(\infty) = 0$.

Foundations of Term Structure Modeling

[The] foundations are the most controversial parts of many, if not all, sciences.

Leonard J. Savage (1917–1971), *The Foundations of Statistics*

This chapter introduces basic definitions and results in term structure modeling. It lays the theoretical foundations for interest rate models. A few simple models are presented at the end of the chapter.

24.1 Terminology

A period denotes a unit of elapsed time throughout this chapter. Hence, viewed at time t, the next time instant refers to time $t + dt$ in the continuous-time model and time $t + 1$ in the discrete-time case. If the discrete-time model results from dividing the time interval $[s, t]$ into n periods, then each period takes $(t - s)/n$ time. Here bonds are assumed to have a par value of one unless stated otherwise. We use the same notation for discrete-time and continuous-time models as the context is always clear. The time unit for continuous-time models is usually measured by the year. We standardize the following notation:

- t: a point in time.
- $r(t)$: the one-period riskless rate prevailing at time t for repayment one period later (the instantaneous spot rate, or **short rate**, at time t).
- $P(t, T)$: the PV at time t of \$1 at time T.
- $r(t, T)$: the $(T - t)$-period interest rate prevailing at time t stated on a per-period basis and compounded once per period – in other words, the $(T - t)$-period spot rate at time t. (This definition dictates that continuous-time models use continuous compounding and discrete-time models use periodic compounding.) The **long rate** is defined as $r(t, \infty)$, that is, the continuously compounded yield on a consol bond that pays out \$1 per unit time forever and never repays principal.
- $F(t, T, M)$: the forward price at time t of a forward contract that delivers at time T a zero-coupon bond maturing at time $M \geq T$.
- $f(t, T, L)$: the L-period forward rate at time T implied at time t stated on a per-period basis and compounded once per period.

$f(t, T)$: the one-period or instantaneous forward rate at time T as seen at time t stated on a per-period basis and compounded once per period. It is $f(t, T, 1)$ in the discrete-time model and $f(t, T, dt)$ in the continuous-time model. Note that $f(t, t)$ equals the short rate $r(t)$.

24.2 Basic Relations

The price of a zero-coupon bond is

$$P(t, T) = \begin{cases} [1 + r(t, T)]^{-(T-t)} & \text{in discrete time} \\ e^{-r(t, T)(T-t)} & \text{in continuous time} \end{cases}.$$

Recall that $r(t, T)$ as a function of T defines the spot rate curve at time t. By definition,

$$f(t, t) = \begin{cases} r(t, t+1) & \text{in discrete time} \\ r(t, t) & \text{in continuous time} \end{cases}.$$

Forward prices and zero-coupon bond prices are related by

$$F(t, T, M) = \frac{P(t, M)}{P(t, T)}, \quad T \leq M, \tag{24.1}$$

which says that the forward price equals the FV at time T of the underlying asset. Equation (24.1) can be verified with an arbitrage argument similar to the "locking in" of forward rates in Subsection 5.6.1 (see Exercise 24.2.1, part (1)). Equation (24.1) holds whether the model is discrete-time or continuous-time, and it implies that

$$F(t, T, M) = F(t, T, S) \, F(t, S, M), \quad T \leq S \leq M.$$

Forward rates and forward prices are related definitionally by

$$f(t, T, L) = \left[\frac{1}{F(t, T, T+L)} \right]^{1/L} - 1 = \left[\frac{P(t, T)}{P(t, T+L)} \right]^{1/L} - 1 \tag{24.2}$$

in discrete time (hence periodic compounding). In particular, $1 + f(t, T, 1) = 1/F(t, T, T+1) = P(t, T)/P(t, T+1)$. In continuous time (hence continuous compounding),

$$f(t, T, L) = -\frac{\ln F(t, T, T+L)}{L} = \frac{\ln(P(t, T)/P(t, T+L))}{L} \tag{24.3}$$

by Eq. (24.1). Furthermore, because

$$f(t, T, \Delta t) = \frac{\ln(P(t, T)/P(t, T+\Delta t))}{\Delta t} \rightarrow -\frac{\partial \ln P(t, T)}{\partial T} = -\frac{\partial P(t, T)/\partial T}{P(t, T)},$$

we conclude that

$$f(t, T) \equiv \lim_{\Delta t \to 0} f(t, T, \Delta t) = -\frac{\partial P(t, T)/\partial T}{P(t, T)}, \quad t \leq T. \tag{24.4}$$

Because Eq. (24.4) is equivalent to

$$P(t, T) = e^{-\int_t^T f(t,s)\,ds}, \tag{24.5}$$

the spot rate curve is $r(t, T) = [1/(T-t)] \int_t^T f(t, s) \, ds$. The discrete analog to Eq. (24.5) is

$$P(t, T) = \frac{1}{[1+r(t)][1+f(t, t+1)]\cdots[1+f(t, T-1)]}. \tag{24.6}$$

The liquidity premium is the difference between the forward rate and the expected spot rate, $f(t, T) - E_t[r(T) \mid r(t)]$. Finally, the short rate and the market discount function are related by

$$r(t) = -\left.\frac{\partial P(t, T)}{\partial T}\right|_{T=t}.$$

This can be verified with Eq. (24.4) and the observation that $P(t, t) = 1$ and $r(t) = f(t, t)$.

▶ **Exercise 24.2.1** (1) Supply the arbitrage argument for Eq. (24.1). (2) Generalize (1) by describing a strategy that replicates the forward contract on a coupon bond that may make payments before the delivery date.

▶ **Exercise 24.2.2** Suppose we sell one T-time zero-coupon bond and buy $P(t, T)/P(t, M)$ units of M-time zero-coupon bonds at time t. Proceed from here to justify Eq. (24.1).

▶ **Exercise 24.2.3** Prove Eq. (24.4) from Eq. (5.11).

▶ **Exercise 24.2.4** Prove Eq. (24.6) from Eq. (24.2).

▶ **Exercise 24.2.5** Show that the τ-period spot rate equals $(1/\tau) \sum_{i=0}^{\tau-1} f(t, t+i)$ (average of forward rates) if all the rates are continuously compounded.

▶ **Exercise 24.2.6** Verify that

$$f(t, T, L) = \frac{1}{L}\left[\frac{P(t, T)}{P(t, T+L)} - 1\right]$$

is the analog to Eq. (24.2) under simple compounding.

▶ **Exercise 24.2.7** Prove the following continuous-time analog to Eq. (5.9):

$$f(t, T, M-t) = \frac{(M-t)\,r(t, M) - (T-t)\,r(t, T)}{M-T}.$$

(Hint: Eq. (24.3).)

▶ **Exercise 24.2.8** Derive the liquidity premium and the forward rate for the Merton model. Verify that the forward rate goes to minus infinity as the maturity goes to infinity.

▶ **Exercise 24.2.9** Show that

$$\frac{P(t, T)}{M(t)} = \frac{1}{M(T)}$$

in a certain economy, where $M(t) \equiv e^{\int_0^t r(s) \, ds}$ is the money market account. (Hint: Exercise 5.6.6.)

24.2.1 Compounding Frequency

A rate can be expressed in different, yet equivalent, ways, depending on the desired compounding frequency (review Section 3.1). The convention in this chapter is to standardize on continuous compounding for continuous-time models and periodic compounding for discrete-time models unless stated otherwise.

The choice between continuous compounding and periodic compounding does have serious implications for interest rate models. Let r_e be the effective annual interest rate and let $r_c \equiv \ln(1 + r_e)$ be the equivalent continuously compounded rate. Both, we note, are instantaneous rates. When the continuously compounded interest rate is lognormally distributed, Eurodollar futures have negative infinite values [875]. However, this problem goes away if it is the effective rate that is lognormally distributed [781, 782].

▶ **Exercise 24.2.10** Suppose that the effective annual interest rate follows

$$\frac{dr_e}{r_e} = \mu(t)\,dt + \sigma(t)\,dW.$$

Prove that

$$\frac{dr_c(t)}{1 - e^{-r_c(t)}} = \left\{ \mu(t) - \frac{1}{2}\left[1 - e^{-r_c(t)}\right]\sigma(t)^2 \right\} dt + \sigma(t)\,dW.$$

(The continuously compounded rate is approximately lognormally distributed when $r_c(t) = o(dt)$ as $1 - e^{-r_c(t)} \approx r_c(t) + o(dt^2)$ and converges to a normal distribution when $r_c(t) \to \infty$.)

24.3 Risk-Neutral Pricing

The local expectations theory postulates that the expected rate of return of any riskless bond over a single period equals the prevailing one-period spot rate, i.e., for all $t + 1 < T$,

$$\frac{E_t[\,P(t+1,T)\,]}{P(t,T)} = 1 + r(t). \tag{24.7}$$

Relation (24.7) in fact follows from the risk-neutral valuation principle, Theorem 13.2.3, which is assumed to hold for continuous-time models. The local expectations theory is thus a consequence of the existence of a risk-neutral probability π, and we may use $E_t^\pi[\,\cdot\,]$ in place of $E_t[\,\cdot\,]$. Rewrite Eq. (24.7) as

$$\frac{E_t^\pi[\,P(t+1,T)\,]}{1 + r(t)} = P(t,T),$$

which says that the current spot rate curve equals the expected spot rate curve one period from now discounted by the short rate. Apply the preceding equality iteratively to obtain

$$P(t,T) = E_t^\pi\left[\frac{P(t+1,T)}{1 + r(t)}\right] = E_t^\pi\left[\frac{E_{t+1}^\pi[\,P(t+2,T)\,]}{\{1 + r(t)\}\{1 + r(t+1)\}}\right]$$

$$\cdots = E_t^\pi\left[\frac{1}{\{1 + r(t)\}\{1 + r(t+1)\}\cdots\{1 + r(T-1)\}}\right]. \tag{24.8}$$

Because Eq. (24.7) can also be expressed as

$$E_t[\, P(t+1, T)\,] = F(t, t+1, T),$$

the forward price for the next period is an unbiased estimator of the expected bond price.

In continuous time, the local expectations theory implies that

$$P(t, T) = E_t\left[\, e^{-\int_t^T r(s)\, ds}\,\right], \quad t < T. \tag{24.9}$$

In other words, the actual probability and the risk-neutral probability are identical. Note that $e^{\int_t^T r(s)\, ds}$ is the bank account process, which denotes the rolled-over money market account. We knew that bond prices relative to the money market account are constant in a certain economy (see Exercise 24.2.9). Equation (24.9) extends that proposition to stochastic economies. When the local expectations theory holds, riskless arbitrage opportunities are impossible [232]. The local expectations theory, however, is not the only version of expectations theory consistent with equilibrium [351].

The risk-neutral methodology can be used to price interest rate swaps. Consider an interest rate swap made at time t with payments to be exchanged at times $t_1, t_2, \ldots, t_n$. The fixed rate is c per annum. The floating-rate payments are based on the future annual rates $f_0, f_1, \ldots, f_{n-1}$ at times $t_0, t_1, \ldots, t_{n-1}$. For simplicity, assume that $t_{i+1} - t_i$ is a fixed constant Δt for all i, and that the notional principal is \$1. If $t < t_0$, we have a forward interest rate swap because the first payment is not based on the rate that exists when the agreement is reached. The ordinary swap corresponds to $t = t_0$.

The amount to be paid out at time t_{i+1} is $(f_i - c)\,\Delta t$ for the floating-rate payer. Note that simple rates are adopted here; hence f_i satisfies

$$P(t_i, t_{i+1}) = \frac{1}{1 + f_i \Delta t}.$$

The value of the swap at time t is thus

$$\sum_{i=1}^{n} E_t^{\pi}\left[\, e^{-\int_t^{t_i} r(s)\, ds}\,(f_{i-1} - c)\,\Delta t\,\right]$$

$$= \sum_{i=1}^{n} E_t^{\pi}\left[\, e^{-\int_t^{t_i} r(s)\, ds}\left\{\frac{1}{P(t_{i-1}, t_i)} - (1 + c\Delta t)\right\}\right]$$

$$= \sum_{i=1}^{n}[\, P(t, t_{i-1}) - (1 + c\Delta t) \times P(t, t_i)\,]$$

$$= P(t, t_0) - P(t, t_n) - c\Delta t \sum_{i=1}^{n} P(t, t_i).$$

So a swap can be replicated as a portfolio of bonds. In fact, it can be priced by simple PV calculations. The swap rate, which gives the swap zero value, equals

$$\frac{P(t, t_0) - P(t, t_n)}{\sum_{i=1}^{n} P(t, t_i)\,\Delta t}.$$

The swap rate is the fixed rate that equates the PVs of the fixed payments and the floating payments. For an ordinary swap, $P(t, t_0) = 1$.

▶ **Exercise 24.3.1** Assume that the local expectations theory holds. Prove that the $(T-t)$-time spot rate at time t is less than or equal to $E_t[\int_t^T r(s)\,ds]/(T-t)$, the expected average interest rate between t and T, with equality only if there is no uncertainty about $r(s)$.

▶ **Exercise 24.3.2** Under the local expectations theory, prove that the forward rate $f(t, T)$ is less than the expected spot rate $E_t[r(T)]$ provided that interest rates tend to move together in that $E_t[r(T) \mid \int_t^T r(s)\,ds]$ is increasing in $\int_t^T r(s)\,ds$. (The unbiased expectations theory is hence inconsistent with the local expectations theory. See also Exercise 5.7.3, part (2).)

▶ **Exercise 24.3.3** Show that the calibrated binomial interest rate tree generated by the ideas enumerated in Subsection 23.2.2 (hence the slightly more general tree of Exercise 23.2.3 as well) satisfies the local expectations theory. How about the uncalibrated tree in Fig. 23.6?

▶ **Exercise 24.3.4** Show that, under the unbiased expectations theory,

$$P(t, T) = \frac{1}{[1 + r(t)]\{1 + E_t[r(t+1)]\}\cdots\{1 + E_t[r(T-1)]\}}$$

in discrete time and $P(t, T) = e^{-\int_t^T E_t[r(s)]\,ds}$ in continuous time. (The preceding equation differs from Eq. (24.8), which holds under the local expectations theory.)

▶ **Exercise 24.3.5** The price of a consol that pays dividends continuously at the rate of $1 per annum satisfies the following expected discounted-value formula:

$$P(t) = E_t^\pi \left[\int_t^\infty e^{-\int_t^T r(s)\,ds}\,dT \right].$$

Compare this equation with Eq. (24.9) and explain the difference.

▶ **Exercise 24.3.6** Consider an amortizing swap in which the notional principal decreases by $1/n$ dollar at each of the n reset points. The initial principal is $1. Write a formula for the swap rate.

▶ **Exercise 24.3.7** Argue that a forward interest rate swap is equivalent to a portfolio of one long payer swaption and one short receiver swaption. (The situation is similar to Exercise 12.2.4, which said that a forward contract is equivalent to a portfolio of European options.)

▶ **Exercise 24.3.8** Use the risk-neutral methodology to price interest rate caps, caplets, floors, and floorlets as fixed-income options.

24.4 The Term Structure Equation

In arbitrage pricing, we start exogenously with the bank account process and a primary set of traded securities plus their prices and stochastic processes. We then price a security not in the set by constructing a replicating portfolio consisting of only the primary assets. For fixed-income securities, the primary set of traded securities comprises the zero-coupon bonds and the money market account.

Let the zero-coupon bond price $P(r, t, T)$ follow

$$\frac{dP}{P} = \mu_p\, dt + \sigma_p\, dW.$$

Suppose that an investor at time t shorts one unit of a bond maturing at time s_1 and at the same time buys α units of a bond maturing at time s_2. The net wealth change follows

$$-dP(r, t, s_1) + \alpha\, dP(r, t, s_2)$$
$$= [-P(r, t, s_1)\, \mu_p(r, t, s_1) + \alpha\, P(r, t, s_2)\, \mu_p(r, t, s_2)]\, dt$$
$$+ [-P(r, t, s_1)\, \sigma_p(r, t, s_1) + \alpha\, P(r, t, s_2)\, \sigma_p(r, t, s_2)]\, dW.$$

Hence, if we pick

$$\alpha \equiv \frac{P(r, t, s_1)\, \sigma_p(r, t, s_1)}{P(r, t, s_2)\, \sigma_p(r, t, s_2)},$$

then the net wealth has no volatility and must earn the riskless return, that is,

$$\frac{-P(r, t, s_1)\, \mu_p(r, t, s_1) + \alpha\, P(r, t, s_2)\, \mu_p(r, t, s_2)}{-P(r, t, s_1) + \alpha\, P(r, t, s_2)} = r.$$

Simplify this equation to obtain

$$\frac{\sigma_p(r, t, s_1)\, \mu_p(r, t, s_2) - \sigma_p(r, t, s_2)\, \mu_p(r, t, s_1)}{\sigma_p(r, t, s_1) - \sigma_p(r, t, s_2)} = r,$$

which becomes

$$\frac{\mu_p(r, t, s_2) - r}{\sigma_p(r, t, s_2)} = \frac{\mu_p(r, t, s_1) - r}{\sigma_p(r, t, s_1)}$$

after rearrangement. Because this equality holds for any s_1 and s_2, we conclude that

$$\frac{\mu_p(r, t, s) - r}{\sigma_p(r, t, s)} \equiv \lambda(r, t) \tag{24.10}$$

for some λ independent of the bond maturity s. As $\mu_p = r + \lambda \sigma_p$, all assets are expected to appreciate at a rate equal to the sum of the short rate and a constant times the asset's volatility.

The term $\lambda(r, t)$ is called the market price of risk because it is the increase in the expected instantaneous rate of return on a bond per unit of risk. The term $\mu_p(r, t, s) - r$ denotes the **risk premium**. Again it is emphasized that the market price of risk must be the same for all bonds to preclude arbitrage opportunities [76].

Assume a Markovian short rate model, $dr = \mu(r, t)\, dt + \sigma(r, t)\, dW$. Then the bond price process is also Markovian. By Eqs. (14.15),

$$\mu_p = \left[-\frac{\partial P}{\partial T} + \mu(r, t)\frac{\partial P}{\partial r} + \frac{\sigma(r, t)^2}{2}\frac{\partial^2 P}{\partial r^2} \right] \Big/ P, \quad \sigma_p = \left[\sigma(r, t)\frac{\partial P}{\partial r} \right] \Big/ P \tag{24.11}$$

subject to $P(\cdot, T, T) = 1$. Note that both μ_p and σ_p depend on P. Substitute μ_p and σ_p above into Eq. (24.10) to obtain the following parabolic partial differential equation:

$$-\frac{\partial P}{\partial T} + [\mu(r, t) - \lambda(r, t)\sigma(r, t)]\frac{\partial P}{\partial r} + \frac{1}{2}\sigma(r, t)^2\frac{\partial^2 P}{\partial r^2} = rP. \tag{24.12}$$

This is the **term structure equation** [660, 855]. Numerical procedures for solving partial differential equations were covered in Section 18.1. Once P is available, the spot rate curve emerges by means of

$$r(t, T) = -\frac{\ln P(t, T)}{T - t}.$$

The term structure equation actually applies to all interest rate derivatives, the difference being the terminal and the boundary conditions. The equation can also be expressed in terms of duration $D \equiv (\partial P/\partial r) P^{-1}$, convexity $C \equiv (\partial^2 P/\partial r^2) P^{-1}$, and **time value** $\Theta \equiv (\partial P/\partial t) P^{-1}$ as follows:

$$\Theta - [\mu(r, t) - \lambda(r, t)\sigma(r, t)]D + \frac{1}{2}\sigma(r, t)^2 C = r. \tag{24.13}$$

In sharp contrast to the Black–Scholes model, the specification of the short-rate process plus the assumption that the bond market is arbitrage free does *not* determine bond prices uniquely. The reasons are twofold: Interest rate is not a traded security and the market price of risk is not determined *within* the model.

The local expectations theory is usually imposed for convenience. In fact, a probability measure exists such that bonds can be priced as if the theory were true to preclude arbitrage opportunities [492, 493, 746]. In the world in which the local expectations theory holds, $\mu_p(r, t, s) = r$ and the market price of risk is zero (no risk adjustment is needed), and vice versa. The term structure equation becomes

$$-\frac{\partial P}{\partial T} + \mu(r, t)\frac{\partial P}{\partial r} + \frac{1}{2}\sigma(r, t)^2\frac{\partial^2 P}{\partial r^2} = rP, \tag{24.14}$$

and bond price dynamics (24.11) is simplified to

$$dP = rP\,dt + \sigma(r, t)\frac{\partial P}{\partial r}\,dW.$$

The market price of risk is usually assumed to be zero unless stated otherwise. We can also derive the bond pricing formula under local expectations theory (24.9) by assuming that the short rate follows the risk-neutral process:

$$dr = [\mu(r, t) - \lambda(r, t)\sigma(r, t)]\,dt + \sigma(r, t)\,dW.$$

▶ **Exercise 24.4.1** Suppose a liability has been duration matched by a portfolio. What can we say about the relations among their respective time values and convexities?

▶ **Exercise 24.4.2** Argue that European options on zero-coupon bonds satisfy the term structure equation subject to appropriate boundary conditions.

▶ **Exercise 24.4.3** Describe an implicit method for term structure equation (24.14). You may simplify the short rate process to $dr = \mu(r)\,dt + \sigma(r)\,dW$. Assume that $\mu(r) \geq 0$ and $\sigma(0) = 0$ to avoid negative short rates.

▶ **Exercise 24.4.4** Consider a futures contract on a zero-coupon bond with maturity date T_1. The futures contract expires at time T. Let $F(r, t)$ denote the futures price that follows $dF/F = \mu_f \, dt + \sigma_f \, dW$. Prove that F satisfies

$$-\frac{\partial F}{\partial T} + [\mu(r, t) - \lambda(r, t) \sigma(r, t)] \frac{\partial F}{\partial r} + \frac{1}{2} \sigma(r, t)^2 \frac{\partial^2 F}{\partial r^2} = 0$$

subject to $F(\cdot, T) = P(\cdot, T, T_1)$, where $\lambda \equiv \mu_f/\sigma_f$.

24.5 Forward-Rate Process

Assume that the zero-coupon bond price follows $dP(t, T)/P(t, T) = \mu_p(t, T) \, dt + \sigma_p(t, T) \, dW$ as before. Then the process followed by the instantaneous forward rate is [76, 477]

$$df(t, T) = \left[\sigma_p(t, T) \frac{\partial \sigma_p(t, T)}{\partial T} - \frac{\partial \mu_p(t, T)}{\partial T} \right] dt - \frac{\partial \sigma_p(t, T)}{\partial T} \, dW.$$

In a risk-neutral economy, the forward-rate process follows

$$df(t, T) = \left[\sigma(t, T) \int_t^T \sigma(t, s) \, ds \right] dt - \sigma(t, T) \, dW, \tag{24.15}$$

where

$$\sigma(t, T) \equiv \frac{\partial \sigma_p(t, T)}{\partial T},$$

because $\mu_p(t, T) = r(t)$ and

$$\sigma_p(t, T) = \int_t^T \frac{\partial \sigma_p(t, s)}{\partial s} \, ds.$$

▶ **Exercise 24.5.1** Justify Eq. (24.15) directly.

▶ **Exercise 24.5.2** What should $\sigma_p(t, T, P)$ be like for $df(t, T)$'s diffusion term to have the functional form $\psi(t) \, f(t, T)$? The dependence of σ_p on $P(t, T)$ is made explicit here.

24.6 The Binomial Model with Applications

The analytical framework can be nicely illustrated with the binomial model. Suppose the bond price P can move with probability q to Pu and probability $1 - q$ to Pd, where $u > d$:

$$P \begin{array}{l} \nearrow^{1-q} Pd \\[2ex] \searrow_{q} Pu \end{array}$$

Over the period, the bond's expected rate of return is

$$\hat{\mu} \equiv \frac{q \, Pu + (1 - q) \, Pd}{P} - 1 = qu + (1 - q) d - 1, \tag{24.16}$$

and the variance of that return rate is

$$\widehat{\sigma}^2 \equiv q(1-q)(u-d)^2. \tag{24.17}$$

Among the bonds, the one whose maturity is only one period away will move from a price of $1/(1+r)$ to its par value \$1. This is the money market account modeled by the short rate. The market price of risk is defined as $\lambda \equiv (\widehat{\mu} - r)/\widehat{\sigma}$, analogous to Eq. (24.10). The same arbitrage argument as in the continuous-time case can be used to show that λ is independent of the maturity of the bond (see Exercise 24.6.2).

Now change the probability from q to

$$p \equiv q - \lambda\sqrt{q(1-q)} = \frac{(1+r)-d}{u-d}, \tag{24.18}$$

which is independent of bond maturity and q. The bond's expected rate of return becomes

$$\frac{pPu+(1-p)\,Pd}{P} - 1 = pu+(1-p)\,d - 1 = r.$$

The local expectations theory hence holds under the new probability measure p.[1]

▶ **Exercise 24.6.1** Verify Eq. (24.18).

▶ **Exercise 24.6.2** Prove that the market price of risk is independent of bond maturity. (Hint: Assemble two bonds in such a way that the portfolio is instantaneously riskless.)

▶ **Exercise 24.6.3** Assume in a period that the bond price can go from \$1 to P_u or P_d and that the value of a derivative can go from \$1 to V_u or V_d. (1) Show that a portfolio of \$1 worth of bonds and $(P_d - P_u)/(V_u - V_d)$ units of the derivative is riskless. (2) Prove that these many derivatives are worth

$$\frac{(R-P_d)\,V_u + (P_u - R)\,V_d}{(P_u - P_d)\,R}$$

in total, where $R \equiv 1+r$ is the gross riskless return.

▶ **Exercise 24.6.4** Consider the symmetric random walk for modeling the short rate, $r(t+1) = \alpha + \rho r(t) \pm \sigma$. Let V denote the current value of an interest rate derivative, V_u its value at the next period if rates rise, and V_d its value at the next period if rates fall. Define $u \equiv e^{\alpha+\rho r(t)+\sigma}/P(t,t+2)$ and $d \equiv e^{\alpha+\rho r(t)-\sigma}/P(t,t+2)$, so they are the gross one-period returns on the two-period zero-coupon bond when rates go up and down, respectively. (1) Show that a portfolio consisting of \$B worth of one-period bonds and two-period zero-coupon bonds with face value \$Δ to match the value of the derivative requires

$$\Delta = \frac{V_u - V_d}{(u-d)\,P(t,t+2)}, \quad B = \frac{uV_d - dV_u}{(u-d)\,e^{r(t)}}.$$

(2) Prove that

$$V = \frac{pV_u + (1-p)\,V_d}{e^{r(t)}},$$

where $p \equiv (e^{r(t)} - d)/(u-d)$.

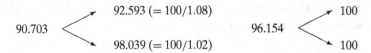

Figure 24.1: Bond price processes. The price process of the 2-year zero-coupon bond is on the left and that of the 1-year zero-coupon bond is on the right.

▶ **Exercise 24.6.5** To use the objective probability q in pricing, we should discount by the risk-adjusted discount factor, $1 + r + \lambda \widehat{\sigma} = 1 + \widehat{\mu}$. Prove this claim.

24.6.1 Numerical Examples

The following numerical examples involve the pricing of fixed-income options, MBSs, and derivative MBSs under this spot rate curve:

Year	1	2
spot rate	4%	5%

Assume that the 1-year rate (short rate) can move up to 8% or down to 2% after a year:

$$4\% \begin{cases} 8\% \\ 2\% \end{cases}$$

No real-world probabilities are specified. The prices of 1- and 2-year zero-coupon bonds are, respectively, $100/1.04 = 96.154$ and $100/(1.05)^2 = 90.703$. Furthermore, they follow the binomial processes in Fig. 24.1.

The pricing of derivatives can be simplified if we assume that investors are risk-neutral. If all securities have the same expected one-period rate of return, the riskless rate, then

$$(1 - p) \times \frac{92.593}{90.703} + p \times \frac{98.039}{90.703} - 1 = 4\%,$$

where p denotes the risk-neutral probability of an up move in rates. Solving this equation leads to $p = 0.319$. Interest rate contingent claims can be priced under this probability.

▶ **Exercise 24.6.6** We could not have obtained the unique risk-neutral probability had we not imposed a prevailing term structure that must be matched. Explain.

▶ **Exercise 24.6.7** Verify the risk-neutral probability $p = 0.319$ with Eq. (24.18) instead.

24.6.2 Fixed-Income Options

A 1-year European call on the 2-year zero with a $95 strike price has the payoffs

$$C \begin{cases} 0.000 \\ 3.039 \end{cases}$$

To solve for the option value C, we replicate the call by a portfolio of x 1-year and y 2-year zeros. This leads to the simultaneous equations,

$$x \times 100 + y \times 92.593 = 0.000,$$
$$x \times 100 + y \times 98.039 = 3.039,$$

which give $x = -0.5167$ and $y = 0.5580$. Consequently,

$$C = x \times 96.154 + y \times 90.703 \approx 0.93$$

to prevent arbitrage. Note that this price is derived without assuming any version of an expectations theory; instead, we derive the arbitrage-free price by replicating the claim with a money market instrument and the claim's underlying asset. The price of an interest rate contingent claim does not depend directly on the probabilities. In fact, the dependence holds only indirectly by means of the current bond prices (see Exercise 24.6.6).

An equivalent method is to utilize risk-neutral pricing. The preceding call option is worth

$$C = \frac{(1 - p) \times 0 + p \times 3.039}{1.04} \approx 0.93,$$

the same as before. This is not surprising, as arbitrage freedom and the existence of a risk-neutral economy are equivalent.

▶ **Exercise 24.6.8 (Dynamic Immunization)** Explain why the replication idea solves the problem of arbitrage opportunities in immunization against parallel shifts raised in Subsection 5.8.2.

24.6.3 Futures and Forward Prices

A 1-year futures contract on the 1-year rate has a payoff of $100 - r$, where r is the 1-year rate at maturity, as shown below:

$$F \left\langle \begin{array}{l} 92 \ (= 100 - 8) \\ \\ 98 \ (= 100 - 2) \end{array} \right.$$

As the futures price F is the expected future payoff (see Exercise 13.2.11), $F = (1 - p) \times 92 + p \times 98 = 93.914$. On the other hand, the forward price for a 1-year forward contract on a 1-year zero-coupon bond equals $90.703/96.154 = 94.331\%$. The forward price exceeds the futures price, as Exercise 12.3.3 predicted.

24.6.4 Mortgage-Backed Securities

Consider a 5%-coupon, 2-year MBS without amortization, prepayments, and default risk. Its cash flow and price process are illustrated in Fig. 24.2, and its fair price is

$$M = \frac{(1 - p) \times 102.222 + p \times 107.941}{1.04} = 100.045.$$

Identical results could have been obtained by no-arbitrage considerations.

In reality mortgages can be prepaid. Assume that the security in question can be prepaid at par and such decisions are rational in that it will be prepaid only when

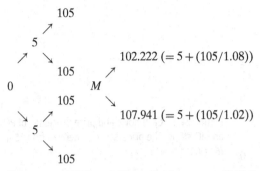

Figure 24.2: MBS's cash flow and price process. The left diagram depicts the cash flow, and the right diagram illustrates the price process.

its price is higher than par. Prepayment will hence occur only in the "down" state when the security is worth 102.941 (excluding coupon). The price therefore follows the process

$$
M \begin{cases} \nearrow & 102.222 \\ \searrow & 105 \end{cases} ,
$$

and the security is worth

$$
M = \frac{(1 - p) \times 102.222 + p \times 105}{1.04} = 99.142.
$$

We go on to price **stripped mortgage-backed securities (SMBSs)** derived from the above prepayable mortgage. The cash flow of the **principal-only (PO) strip** comes from the mortgage's principal cash flow, whereas that of the **interest-only (IO) strip** comes from the interest cash flow (see Fig. 24.3(a)). Their prices hence follow the processes in Fig. 24.3(b). The fair prices are

$$
\text{PO} = \frac{(1 - p) \times 92.593 + p \times 100}{1.04} = 91.304,
$$

$$
\text{IO} = \frac{(1 - p) \times 9.630 + p \times 5}{1.04} = 7.839.
$$

Of course, $\text{PO} + \text{IO} = M$.[2]

The above formulas reveal that IO and PO strips react to changes in p differently. The value of the PO strip rises with increasing p, whereas that of the IO strip declines with increasing p. Suppose the market price of risk is positive so that the real-world probability q exceeds the risk-neutral probability p. Then the market value of the PO strip, like that of the zero-coupon bond, is lower than its discounted expected value under q, which compensates the investors for its riskiness by earning more than the riskless return on average. The market value of the IO strip, however, is higher than its discounted expected value under q, making the security earn *less than* the riskless rate even though it is a risky security. The reason is that the IO's price correlates negatively with the zero-coupon bond's.

Suppose the mortgage is split into half **floater** and half **inverse floater**. Let the floater (FLT) receive the 1-year rate. Then the inverse floater (INV) must have a

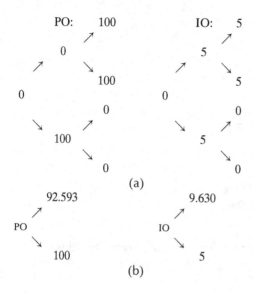

Figure 24.3: Cash flows and price processes of PO and IO strips. The price 9.630 is derived from $5 + (5/1.08)$.

coupon rate of $10\% - 1$-year rate to make the overall coupon rate 5%. Their cash flows as percentages of par and values are shown in Fig. 24.4. The current prices are

$$\text{FLT} = \frac{1}{2} \times \frac{104}{1.04} = 50,$$

$$\text{INV} = \frac{1}{2} \times \frac{(1-p) \times 100.444 + p \times 106}{1.04} = 49.142.$$

▶ **Exercise 24.6.9** Explain why all the securities covered up to now have the same 1-year return of 4% in a risk-neutral economy.

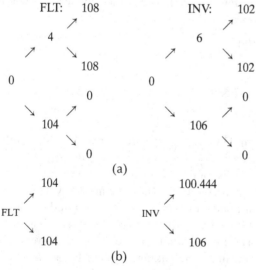

Figure 24.4: Cash flows and price processes of floater and inverse floater. The floater's price in the up node, 104, is derived from $4 + (108/1.08)$, and the inverse floater's price, 100.444, is derived from $6 + (102/1.08)$.

> **Exercise 24.6.10** Verify that the value of a European put, like that of the IO, declines with increasing p.

24.7 **Black–Scholes Models**

A few interest rate models are based on the Black–Scholes option pricing model or the related Black model. They differ mainly in whether it is the price or the yield that is being modeled. As simple as these models are and despite some difficulties, they usually provide adequate results for options with short maturities [305, 456].

24.7.1 Price Models

Suppose the long-term bond price follows geometric Brownian motion much like the stock price. As with stock options, options on the bond can be replicated by continuous trading of these bonds and borrowing at the prevailing short rate. Hence Black–Scholes formulas (12.16) apply.

This pricing model has several problems. It is inconsistent to assume that the short-term rate is a constant – as dictated by the Black–Scholes option pricing model – but the long bond price is uncertain. Another objection is about the volatility of bond prices. Although this volatility must first increase with the passage of time, it should eventually decrease toward zero because the bond converges to its par value at maturity. In other words, the price uncertainty is small in the immediate future and near bond maturity but large between these two extreme points (see Fig. 24.5). This unique property is not captured by the preceding model, which assumes that the variance of the bond price grows linearly in time.

The lognormal assumption for the zero-coupon bond price means that the continuously compounded interest rate is normally distributed. Three problems are associated with this distribution: the possibility of negative interest rates, the independence of interest rate volatility from the interest rate level, and the possibility of the bond price's rising above its sum of cash flows.

Bond price

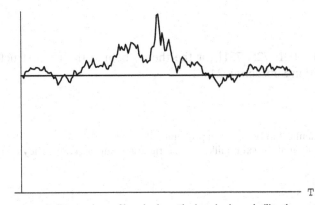

Figure 24.5: Pull toward par of bond prices. The bond price volatility changes over time.

▶ **Exercise 24.7.1** Assume that interest rates cannot be negative. (1) Why should a call on a zero-coupon bond with a strike price of $102 be worth zero given a par value of $100? (2) The Black–Scholes formula gives a positive call value. Why?

▶ **Exercise 24.7.2** Consider a call on a zero-coupon bond with an expiration date that coincides with the bond's maturity. Does the call premium depend on the interest rate movements between now and the expiration date?

24.7.2 Yield Models

Consider the alternative model that models the yield to maturity, not the bond price, as geometric Brownian motion. It solves a few problems that were poisoning the preceding model. To start with, because yield to maturity now has the lognormal distribution, negative interest rates are ruled out. Furthermore, as the bond price at any time is derived from the yield's probability distribution, it will reflect both the decrease in price volatility and the pull toward par as the bond matures.

This model has its own difficulties. In the binomial setting, it is known that the one-period riskless rate must be between the one-period bond returns of up and down yield shifts to avoid arbitrage (see Exercise 9.2.1). However, if the riskless rate is a constant, preventing such opportunities may be difficult, especially in light of the fact that both the up and the down returns must eventually approach one because of the pull toward par. Another problem is that the yield volatility is constant over the life of the bond; in reality, however, it decreases as the maturity increases.

24.7.3 Models Based on the Brownian Bridge

Because zero-coupon bonds move toward par at maturity, a Brownian bridge process seems ideal for modeling their price dynamics [50]. Recall that a Brownian bridge process $\{B(t), 0 \le t \le T\}$ can be defined as

$$B(t) = W(t) - \frac{t}{T} W(T).$$

Note that $B(0) = B(T) = 0$. The bond price model $P(t, T) = e^{r(t-T)+\sigma B(t)}$ clearly has the desirable pull-to-par property because $P(T, T) = 1$. However, certain models based on Brownian bridge are not arbitrage free, thus not sound [193].

Additional Reading

Consult [38, 76, 290, 510, 691, 725, 731] for the theory behind the term structure models. We followed [34] in Subsection 24.6.1.

NOTES

1. Note that Eq. (24.18) is identical to risk-neutral probability (9.5) of the BOPM.
2. You can order either whole milk or skim milk plus the right amount of cream. They cost the same!

CHAPTER TWENTY-FIVE

Equilibrium Term Structure Models

8. What's your problem? Any moron can understand bond pricing models.

Top Ten Lies Finance Professors Tell Their Students[1]

Many interest rate models have been proposed in the literature and used in practice. This chapter surveys equilibrium models, and the next chapter covers no-arbitrage models. Because the spot rates satisfy

$$r(t, T) = -\frac{\ln P(t, T)}{T - t},$$

the discount function $P(t, T)$ suffices to establish the spot rate curve. Most models to follow are short rate models, in which the short rate is the sole source of uncertainty. Unless stated otherwise, the market price of risk λ is assumed to be zero; the processes are hence risk-neutral to start with.

25.1 The Vasicek Model

Vasicek proposed the model in which the short rate follows [855]

$$dr = \beta(\mu - r) dt + \sigma dW.$$

The short rate is thus pulled to the long-term mean level μ at rate β. Superimposed on this "pull" is a normally distributed stochastic term σdW. The idea of mean reversion for interest rates dates back to Keynes [232]. This model seems relevant to interest rates in Germany and the United Kingdom [248].

Because the process is an Ornstein–Uhlenbeck process,

$$E[r(T) \mid r(t) = r] = \mu + (r - \mu) e^{-\beta(T-t)}$$

from Eq. (14.13). The term structure equation under the Vasicek model is

$$-\frac{\partial P}{\partial T} + \beta(\mu - r) \frac{\partial P}{\partial r} + \frac{1}{2} \sigma^2 \frac{\partial^2 P}{\partial r^2} = r P.$$

The price of a zero-coupon bond paying one dollar at maturity can be shown to be

$$P(t, T) = A(t, T) e^{-B(t,T) r(t)}, \tag{25.1}$$

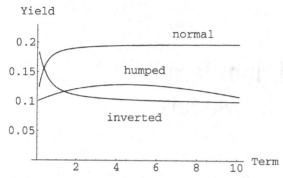

Figure 25.1: Term structure shapes. The parameters (β, μ, σ, r) are $(5.9, 0.2, 0.3, 0.1)$, $(3.9, 0.1, 0.3, 0.2)$, and $(0.1, 0.4, 0.11, 0.1)$ for normal, inverted, and humped term structures, respectively.

where

$$
A(t, T) = \begin{cases} \exp\left[\dfrac{\{ B(t, T) - T + t \}(\beta^2 \mu - \sigma^2/2)}{\beta^2} - \dfrac{\sigma^2 B(t, T)^2}{4\beta} \right], & \text{if } \beta \neq 0 \\[2mm] \exp\left[\dfrac{\sigma^2 (T-t)^3}{6} \right], & \text{if } \beta = 0 \end{cases},
$$

$$
B(t, T) = \begin{cases} \dfrac{1 - e^{-\beta(T-t)}}{\beta}, & \text{if } \beta \neq 0 \\[2mm] T - t, & \text{if } \beta = 0 \end{cases}.
$$

This model has some unpleasant properties; for example, if $\beta = 0$, then P goes to infinity as $T \to \infty$, like the Merton model. However, sensibly, P goes to zero as $T \to \infty$ if $\beta \neq 0$. Even so, P may exceed one for a finite T. See Fig. 25.1 for the shapes of the spot rate curve. The spot rate volatility structure is the curve $[\partial r(t, T)/\partial r] \sigma = \sigma B(t, T)/(T-t)$. When $\beta > 0$, the curve tends to decline with maturity. The speed of mean reversion, β, controls the shape of the curve; indeed, higher β leads to greater attenuation of volatility with maturity. It is not hard to verify that duration $-\dfrac{\partial P(t, T)/\partial r}{P(t, T)}$ equals $B(t, T)$. Duration decreases toward $1/\beta$ as the term lengthens if there is mean reversion ($\beta > 0$). On the other hand, duration equals the term to maturity $T - t$ if there is no mean reversion ($\beta = 0$), much like the static world. Interestingly, duration is independent of the interest rate volatility σ.

▶ **Exercise 25.1.1** Connect the Vasicek model with the AR(1) process.

▶ **Exercise 25.1.2** (1) Show that the long rate is $\mu - \sigma^2/(2\beta^2)$, independent of the current short rate. (2) Derive the liquidity premium for the $\beta \neq 0$ case.

▶ **Exercise 25.1.3** Show that Eq. (25.1) satisfies the term structure equation.

▶ **Exercise 25.1.4** Verify that $dP/P = r\, dt - B(t, T)\sigma\, dW$ is the bond price process for the Vasicek model.

▶ **Exercise 25.1.5** Show that the Ito process for the instantaneous forward rate $f(t, T)$ under the Vasicek model with $\beta \neq 0$ is

$$
df = \frac{\sigma^2}{\beta} e^{-\beta(T-t)} \left[1 - e^{-\beta(T-t)} \right] dt + \sigma e^{-\beta(T-t)}\, dW.
$$

(Hint: Section 24.5 and Exercise 25.1.4.)

25.1.1 Options on Zero-Coupon Bonds

Consider a European call with strike price X expiring at time T on a zero-coupon bond with par value \$1 and maturing at time $s > T$. Its price is given by the following Black–Scholes-like formula [506]:

$$P(t, s) N(x) - X P(t, T) N(x - \sigma_v),$$

where

$$x \equiv \frac{1}{\sigma_v} \ln\left(\frac{P(t, s)}{P(t, T) X}\right) + \frac{\sigma_v}{2},$$

$$\sigma_v \equiv v(t, T) B(T, s),$$

$$v(t, T)^2 \equiv \begin{cases} \frac{\sigma^2 \left[1 - e^{-2\beta(T-t)}\right]}{2\beta}, & \text{if } \beta \neq 0 \\ \sigma^2 (T - t), & \text{if } \beta = 0 \end{cases}.$$

Note that $v(t, T)^2$ is the variance of $r(t, T)$ by Eq. (14.14). The put–call parity says that

$$\text{call} = \text{put} + P(t, s) - P(t, T) X.$$

The price of a European put is thus $X P(t, T) N(-x + \sigma_v) - P(t, s) N(-x)$.

▶ **Exercise 25.1.6** Verify that the variance of $\ln P(t, T)$ is σ_v^2.

25.1.2 Binomial Approximation

We consider a binomial model for the short rate in the time interval $[0, T]$ divided into n identical pieces. Let $\Delta t \equiv T/n$ and

$$p(r) \equiv \frac{1}{2} + \frac{\beta(\mu - r)\sqrt{\Delta t}}{2\sigma}.$$

The following binomial model converges in distribution to the Vasicek model [696]:

$$r(k+1) = r(k) + \sigma\sqrt{\Delta t}\,\xi(k), \quad 0 \leq k < n,$$

where $\xi(k) = \pm 1$, with

$$\text{Prob}[\,\xi(k) = 1\,] = \begin{cases} p[\,r(k)\,], & \text{if } 0 \leq p(r(k)) \leq 1 \\ 0, & \text{if } p(r(k)) < 0 \\ 1, & \text{if } 1 < p(r(k)) \end{cases}.$$

Observe that the probability of an up move, p, is a decreasing function of the interest rate r. This is consistent with mean reversion.

The rate is the same whether it is the result of an up move followed by a down move or a down move followed by an up move; in other words, the binomial tree combines. The key feature of the model that makes it happen is its *constant* volatility, σ. For a general process Y with nonconstant volatility, the resulting binomial tree may not combine. Fortunately, if Y can be transformed into one with constant volatility, say X, then we can first construct a combining tree for X and then apply the inverse transformation on each node to obtain a combining tree for Y. This idea will be explored in Subsection 25.2.2.

> **Exercise 25.1.7** Prove that

$$\frac{E[r(k+1)-r(k)]}{\Delta t} = \begin{cases} \beta[\mu - r(k)], & \text{if } 0 \le p(r(k)) \le 1 \\ \sigma/\sqrt{\Delta t}, & \text{if } p(r(k)) < 0 \\ -\sigma/\sqrt{\Delta t}, & \text{if } 1 < p(r(k)) \end{cases}$$

and $\text{Var}[r(k+1) - r(k)] \to \sigma^2 \Delta t$.

> **Exercise 25.1.8** Show that discretizing the Vasicek model directly by (14.7) does not result in a combining binomial tree.

> **Programming Assignment 25.1.8** Write a program to implement the binomial tree for the Vasicek model. Price zero-coupon bonds and compare the results against Eq. (25.1).

25.2 The Cox-Ingersoll-Ross Model

Cox, Ingersoll, and Ross (CIR) proposed the following square-root short-rate model [234]:

$$dr = \beta(\mu - r)\,dt + \sigma\sqrt{r}\,dW. \tag{25.2}$$

Although the randomly moving interest rate is elastically pulled toward the long-term value μ, as in the Vasicek model, the diffusion differs by a multiplicative factor $\sqrt{r}$. The parameter β determines the speed of adjustment. The short rate can reach zero only if $2\beta\mu < \sigma^2$.

The price of a zero-coupon bond paying \$1 at maturity is [470]

$$P(t, T) = A(t, T)\,e^{-B(t,T)r(t)}, \tag{25.3}$$

where

$$A(t, T) = \left\{ \frac{2\gamma e^{(\beta+\gamma)(T-t)/2}}{(\beta+\gamma)\left[e^{\gamma(T-t)} - 1\right] + 2\gamma} \right\}^{2\beta\mu/\sigma^2},$$

$$B(t, T) = \frac{2\left[e^{\gamma(T-t)} - 1\right]}{(\beta+\gamma)\left[e^{\gamma(T-t)} - 1\right] + 2\gamma},$$

$$\gamma = \sqrt{\beta^2 + 2\sigma^2}.$$

A formula for consols is also available [266]. Figure 25.2 illustrates the shapes of the spot rate curve. In general, the curve is normal if the current short rate $r(t)$ is below the long rate $r(t, \infty)$, becomes inverted if $r(t) > \mu$, and is slightly humped if $r(t, \infty) < r(t) < \mu$ [493]. Figure 25.3 illustrates the long rate and duration of zero-coupon bonds. To incorporate the market price of risk into bond prices, replace each occurrence of β in $A(t, T)$ (except the exponent $2\beta\mu/\sigma^2$), $B(t, T)$, and γ with $\beta + \lambda$. Consult Subsection 14.3.2 for additional properties of the square-root process.

Two implications of the CIR model are at odds with empirical evidence: constant long rate and perfect correlation in yield changes along the term structure. The CIR model has been subject to many empirical studies (e.g., [11, 135, 138, 173, 257, 384, 835]). The model seemed to fit the term structure of *real* interest rates in the United Kingdom well until the end of 1992 [137]. (Real rates have been generally less volatile than nominal rates.) It also seems relevant to Denmark and Sweden [248].

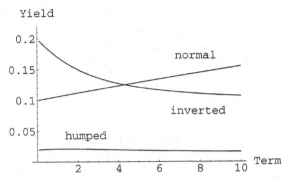

Figure 25.2: Term structure shapes. The values for the parameters (β, μ, σ, r) are $(0.02, 0.7, 0.02, 0.1)$, $(0.7, 0.1, 0.3, 0.2)$, and $(0.06, 0.09, 0.5, 0.02)$ for normal, inverted, and humped term structures, respectively. The long rates are 0.512436, 0.0921941, and 0.0140324, respectively.

> ▶ **Exercise 25.2.1** Show that the long rate is $2\beta\mu/(\beta+\gamma)$, independent of the short rate.

> ▶ **Exercise 25.2.2** Show that Eq. (25.3) satisfies the term structure equation.

> ▶ **Exercise 25.2.3** Verify that $dP/P = r\,dt - B(t, T)\,\sigma\sqrt{r}\,dW$ is the bond price process for the CIR model.

> ▶ **Exercise 25.2.4 (Affine Models)** For any short rate model $dr = \mu(r, t)\,dt + \sigma(r, t)\,dW$ that produces zero-coupon bond prices of the form $P(t, T) = A(t, T)\,e^{-B(t,T)\,r(t)}$, show that the spot rate volatility structure is the curve $\sigma(r, t)\,B(t, T)/(T-t)$.

> ▶ **Exercise 25.2.5** (1) Write the bond price formula in terms of $\phi_1 \equiv \gamma$, $\phi_2 \equiv (\beta + \gamma)/2$, and $\phi_3 \equiv 2\beta\mu/\sigma^2$. (2) How do we estimate σ, given estimates for ϕ_1, ϕ_2, and ϕ_3?

> ▶ **Exercise 25.2.6** Consider a yield curve option with payoff $\max(0, r(T, T_1) - r(T, T_2))$ at expiration T, where $T < T_1$ and $T < T_2$. The security is based on the yield spread of two different maturities, $T_1 - T$ and $T_2 - T$. Assume either the Vasicek or the CIR model. Show that this option is equivalent to a portfolio of caplets on the $(T_2 - T)$-year spot rate.

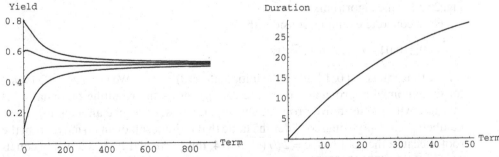

Figure 25.3: Long rates and duration under the CIR model. The parameters (β, μ, σ) are $(0.02, 0.7, 0.02)$. The long-rate plot uses $0.8, 0.6, 0.4$, and 0.1 as the initial rates. The duration of zero-coupon bonds uses $r = 0.1$,

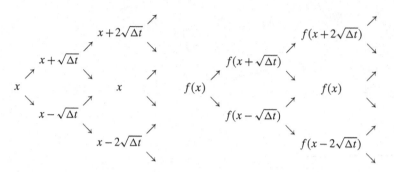

Figure 25.4: Binomial tree for the CIR model.

➤ **Programming Assignment 25.2.7** Implement the implicit method in Exercise 24.4.3 for zero-coupon bonds under the CIR model.

25.2.1 Binomial Approximation

Suppose we want to approximate the short rate process in the time interval $[0, T]$ divided into n periods of duration $\Delta t \equiv T/n$. Assume that $\mu, \beta \geq 0$. A direct discretization of the process is problematic because the resulting binomial tree will *not* combine (see Exercise 25.2.14, part (1)). Instead, consider the transformed process $x(r) \equiv 2\sqrt{r}/\sigma$. It follows

$$dx = m(x)\,dt + dW,$$

where $m(x) \equiv 2\beta\mu/(\sigma^2 x) - (\beta x/2) - 1/(2x)$. Because this new process has a constant volatility, its associated binomial tree combines.

The combining tree for r can be constructed as follows. First, construct a tree for x. Then transform each node of the tree into one for r by means of the inverse transformation $r = f(x) \equiv x^2\sigma^2/4$ (see Fig. 25.4). The probability of an up move at each node r is

$$p(r) \equiv \frac{\beta(\mu - r)\,\Delta t + r - r^-}{r^+ - r^-}, \tag{25.4}$$

where $r^+ \equiv f(x + \sqrt{\Delta t})$ denotes the result of an up move from r and $r^- \equiv f(x - \sqrt{\Delta t})$ the result of a down move [268, 696, 746]. Finally, set the probability $p(r)$ to one as r goes to zero to make the probability stay between zero and one. See Fig. 25.6 for the algorithm.

For a concrete example, consider the process

$$0.2\,(0.04 - r)\,dt + 0.1\sqrt{r}\,dW$$

for the time interval $[0, 1]$ given the initial rate $r(0) = 0.04$. We use $\Delta t = 0.2$ (year) for the binomial approximation. Figure 25.5(a) shows the resulting binomial short rate tree with the up-move probabilities in parentheses. To give an idea how these numbers come into being, consider the node that is the result of an up move from the root. Because the root has $x = 2\sqrt{r(0)}/\sigma = 4$, this particular node's x value equals $4 + \sqrt{\Delta t} = 4.4472135955$. Now use the inverse transformation to obtain the short rate $x^2 \times (0.1)^2/4 \approx 0.0494442719102$. Other short rates can be similarly obtained.

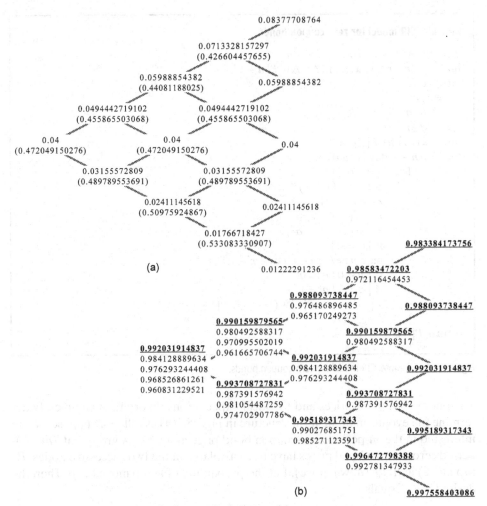

Figure 25.5: Short rate and bond price trees for the CIR model.

Once the short rates are in place, computing the probabilities is easy. Note that the up-move probability decreases as interest rates increase and decreases as interest rates decline. This phenomenon agrees with mean reversion.

➤ **Exercise 25.2.8** Derive $E[r(k+1) - r(k)]$ and $\mathrm{Var}[r(k+1) - r(k)]$.

➤ **Exercise 25.2.9** Show that $p(r) = (1/2) + (1/2) m(x(r)) \sqrt{\Delta t}$.

➤ **Programming Assignment 25.2.10** Write a program to implement the binomial short rate tree and the bond price tree for the CIR model. Compare the results against Eq. (25.3).

Term Structure Dynamics

The tree of short rates can be used to calculate the one-period bond prices (the underlined numbers in Fig. 25.5(b)). For example, the rate after two up moves, 0.05988854382, gives rise to

$$e^{-0.05988854382 \times 0.2} = 0.988093738447.$$

Binomial CIR model for zero-coupon bonds:

```
input:     τ, r, β, μ, σ, n;
real       r⁺, r⁻, r′, x, x′, p, Δt, Δx, P[n+1];
integer    i, j;
Δt := τ/n;
x := 2√r/σ;
Δx := √Δt;
for (i = 0 to n) P[i] := 1;
for (j = n − 1 down to 0)
        for (i = 0 to j) {
                x′ := x + (j − 2i) Δx;
                r′ := x′²σ²/4;
                r⁺ := (x′ + Δx)²σ²/4;
                r⁻ := (x′ − Δx)²σ²/4;
                if [r′ = 0] p := 1;
                else p := (β(μ − r′) Δt + r′ − r⁻)/(r⁺ − r⁻);
                if [p < 0] p := 0;
                if [p > 0] p := 1;
                P[i] := (p × P[i] + (1 − p) × P[i+1])/e^{r′Δt};
        }
return P[0];
```

Figure 25.6: Binomial CIR model for zero-coupon bonds.

The one-period bond prices and the local expectations theory then completely determine the evolution of the term structure in Fig. 25.5(b) as follows. Suppose we are interested in the m-period zero-coupon bond price at a node A given that $(m-1)$-period zero-coupon bond prices have been available in the two successive nodes P_u (up move) and P_d (down move). Let the probability of an up move be p. Then the desired price equals

$$P_A(p P_u + (1 - p) P_d),$$

where P_A is the one-period bond price at A. For instance, the five-period zero-coupon bond price at time zero, 0.960831229521, can be derived with $P_A = 0.992031914837$, $p = 0.472049150276$, $P_u = 0.961665706744$, and $P_d = 0.974702907786$. Once the discount factors are in place, they can be used to obtain the spot rates. For instance, the spot rates at time zero are $(0.04, 0.039996, 0.0399871, 0.0399738, 0.0399565)$ based on Fig. 25.5(b).

▶ **Exercise 25.2.11** Suppose we want to calculate the price of some interest-rate-sensitive security by using the binomial tree for the CIR or the Vasicek model. Assume further that we opt for the Monte Carlo method with antithetic variates. One difficulty with the standard paradigm covered in Subsection 18.2.3 is that, here, the probability at each node varies, and all paths are hence not equally probable. How do we handle this difficulty?

▶ **Exercise 25.2.12** (1) The binomial short rate tree as described requires $\Theta(n^2)$ memory space. How do we perform backward induction on the tree with only $O(n)$ space? (2) Describe a scheme that needs only $O(n^2)$ space for the bond price tree.

Convergence

The binomial approximation converges fast. For example, analytical formula (25.3) gives the following market discount factors at times zero and Δt after an up move.

	Discount factor	
Year	**(Now)**	**(Time Δt; up state)**
0.2	0.992032	0.990197
0.4	0.984131	0.980566
0.6	0.976299	0.971102
0.8	0.968536	0.961804
1.0	0.960845	

The numbers derived by the binomial model in Fig. 25.5(b) can be seen to stand up quite well even at $\Delta t = 0.2$. In fact, we could have started from the short rate tree and generated zero-coupon bond prices at the terminal nodes by Eq. (25.3). This nice feature, made possible by the availability of closed-form formulas, can speed up derivatives pricing. We need to build the tree up to only the maturity of the derivative instead of that of its underlying bond, which could be much longer (see Comment 23.2.1 for more on this point).

25.2.2 A General Method for Constructing Binomial Models

The binomial approximations for the Vasicek and the CIR models follow this general guideline. Given a continuous-time process $dy = \alpha(y, t) \, dt + \sigma(y, t) \, dW$, we first make sure the binomial model's drift and diffusion converge to those of the continuous-time process by setting the probability of an up move to

$$\frac{\alpha(y, t) \, \Delta t + y - y_u}{y_u - y_d},$$

where $y_u + \equiv y + \sigma(y, t)\sqrt{\Delta t}$ and $y_d \equiv y - \sigma(y, t)\sqrt{\Delta t}$ represent the two rates that follow the current rate y. Note that the displacements are identical at $\sigma(y, t)\sqrt{\Delta t}$.

As it stands, the binomial tree may not combine: An up move followed by a down move may not reach the same value as a down move followed by an up move as in general

$$\sigma(y, t)\sqrt{\Delta t} - \sigma(y_u, t)\sqrt{\Delta t} \neq -\sigma(y, t)\sqrt{\Delta t} + \sigma(y_d, t)\sqrt{\Delta t}.$$

When $\sigma(y, t)$ is a constant independent of y, equality holds and the tree combines. To achieve this, define the transformation $x(y, t) \equiv \int^y \sigma(z, t)^{-1} \, dz$. Then x follows $dx = m(y, t) \, dt + dW$ for some function $m(y, t)$ (see Exercise 25.2.13). The key is that the diffusion term is now a constant, and the binomial tree for x combines. The probability of an up move remains

$$\frac{\alpha(y(x, t), t) \, \Delta t + y(x, t) - y_d(x, t)}{y_u(x, t) - y_d(x, t)},$$

where $y(x, t)$ is the inverse transformation of $x(y, t)$ from x back to y. Note that $y_u(x, t) \equiv y(x + \sqrt{\Delta t}, t + \Delta t)$ and $y_d(x, t) \equiv y(x - \sqrt{\Delta t}, t + \Delta t)$ [696].

For example, the transformation is $\int^r (\sigma\sqrt{z})^{-1} \, dz = 2\sqrt{r}/\sigma$ for the CIR model. As another example, the transformation is $\int^S (\sigma z)^{-1} \, dz = (1/\sigma) \ln S$ for the Black–Scholes model. The familiar BOPM in fact discretizes $\ln S$, not S.

> **Exercise 25.2.13** Verify that the transformation $x(y, t) \equiv \int^y \sigma(z, t)^{-1} \, dz$ turns the process $dy = \alpha(y, t) \, dt + \sigma(y, t) \, dW$ into one for x whose diffusion term is one.

> **Exercise 25.2.14** (1) Show that the binomial tree for the untransformed CIR model does not combine. (2) Show that the binomial tree for the geometric Brownian motion $dr = r\mu \, dt + r\sigma \, dW$ does combine even though its volatility is *not* a constant.

25.2.3 Multifactor CIR Models

One-factor models such as the Vasicek and the CIR models are driven by a single source of uncertainty such as the short rate. To address the weaknesses of these models, reviewed in Section 25.5, multifactor models have been proposed. In one two-factor CIR model, the short rate is the sum of two factors r_1 and r_2: $r \equiv r_1 + r_2$ [474]. The risk-neutral processes for r_1 and r_2 are

$$dr_1 = \beta_1(\mu_1 - r_1) \, dt + \sigma_1\sqrt{r_1} \, dW_1,$$

$$dr_2 = \beta_2(\mu_2 - r_2) \, dt + \sigma_2\sqrt{r_2} \, dW_2,$$

and ρ is the correlation between dW_1 and dW_2. The partial differential equation for the zero-coupon bond is

$$-\frac{\partial P}{\partial T} + \beta_1(\mu_1 - r_1) \frac{\partial P}{\partial r_1} + \beta_2(\mu_2 - r_2) \frac{\partial P}{\partial r_2} + \frac{\sigma_1^2 r_1}{2} \frac{\partial^2 P}{\partial r_1^2} + \frac{\sigma_2^2 r_2}{2} \frac{\partial^2 P}{\partial r_2^2}$$

$$+ \rho\sigma_1\sigma_2\sqrt{r_1 r_2} \frac{\partial^2 P}{\partial r_1 \partial r_2} = r P.$$

Because both factors have an impact on yields from the very short end of the term structure, this model behaves like a one-factor model [149].

25.3 Miscellaneous Models

Ogden proposed the following short rate process:

$$dr = \beta(\mu - r) \, dt + \sigma r \, dW,$$

where $\beta \geq 0$ denotes the speed of adjustment and μ is the steady-state interest rate [702]. The predictable part of the change in rates, $\beta(\mu - r) \, dt$, incorporates the mean reversion toward the long-term mean. Clearly the size of the change in rates is greater the further the current rate deviates from its mean. The unpredictable part says that the interest rate is more volatile, in absolute terms, when it is high than when it is low. Dothan's model is lognormal [282]:

$$\frac{dr}{r} = \alpha \, dt + \sigma \, dW.$$

Because interest rates do not grow without bounds, the $dr = \sigma r \, dW$ version may be preferred.

Constantinides developed a family of models to address some of the shortcomings of the CIR model while maintaining positive interest rates and closed-form formulas for various prices of interest rate derivatives [223]. The simplest of the models is

$$dr = 2a \left(1 - \frac{\sigma^2}{a}\right) (\sigma^2 - 2axy) \, dt + 4a \left(1 - \frac{\sigma^2}{a}\right) y\sigma \, dW_1,$$

where $a, \sigma,$ and α are constants satisfying certain inequalities, $y \equiv x - \alpha + \frac{\alpha}{2(1-\sigma^2/a)}$, and x is the Ornstein–Uhlenbeck process $dx = -ax\,dt + \sigma\,dW_2$. The two Wiener processes W_1 and W_2 are uncorrelated. This model is able to produce inverted-humped yield curves, which are not possible with the CIR model.

Chan, Karolyi, Longstaff, and Sanders (CKLS) proposed the following model [173]:

$$dr = (\alpha + \beta r)\,dt + \sigma r^\gamma\,dW.$$

It subsumes the Vasicek model, the CIR model, the Ogden model, and the Dothan model, as well as many others. Using 1-month T-bills, they found that $\gamma \geq 1$ captures short rate dynamics better than $\gamma < 1$. They also reported positive relations between interest rate volatility and the level of interest rate. Their finding of weak evidence of mean reversion is not shared by the data from several European countries, however [248]. Other researchers suggest that $\gamma \geq 1$ overestimates the importance of rate levels on interest rate volatility [93, 126, 563].

Brennan and Schwartz proposed the following two-factor model:

$$d\ln r = \beta(\ln \ell - \ln r)\,dt + \sigma_1\,dW_1,$$

$$\frac{d\ell}{\ell} = a(r, \ell, b_2)\,dt + \sigma_2\,dW_2,$$

where ρ is the correlation between dW_1 and dW_2 [121, 123]. Unlike the two-factor CIR model, the two factors here are at the two ends of the yield curve, i.e., short and long rates. The short rate has mean reversion to the long rate and follows a lognormal process, whereas the long rate follows another lognormal process. This model seems popular [38, 653]. See [462] for problems with this model.

Fong and Vasicek proposed the following model whose two stochastic factors are the short rate and its instantaneous variance v:

$$dr = \beta(\mu_r - r)\,dt + \sqrt{v}\,dW_1,$$

$$dv = \gamma(\mu_v - v)\,dt + \xi\sqrt{v}\,dW_2,$$

where μ_r is the long-term mean of the short rate and μ_v is the long-term mean of the variance of the short rate [362]. See [387, 793] for additional information. A related three-factor model makes the long-term mean μ_r stochastic [46].

▶ **Exercise 25.3.1** To construct a combining binomial tree for the CKLS model, what function of r should be modeled?

25.4 Model Calibration

Two standard approaches to calibrating models are the time-series approach and the cross-sectional approach. In the **time-series** approach, the time series of short rates is used to estimate the parameters of the process. Although it may help in validating the proposed interest rate process, this approach alone cannot be used to estimate the risk premium parameter λ. The model prices based on the estimated parameters may also deviate a lot from those in the market.

The **cross-sectional** approach uses a cross section of observed bond prices. The parameters are to be such that the model prices closely match those in the market. After this procedure, the calibrated model can be used to price interest rate derivatives. Unlike the time-series approach, the cross-sectional approach is unable to separate out the interest rate risk premium from the model parameters. Furthermore, empirical evidence indicates that these estimates may not be stable over time [77, 746]. The common practice of repeated recalibration, albeit pragmatic, is not theoretically sound. A joint cross-section/time-series estimation is also possible [257].

If the model contains only a finite number of parameters, which is true of the Vasicek and the CIR models, a complete match with the market data must be the result of pure luck. This consideration calls for models that have an infinite parameter vector. One way to achieve this is to let some parameters in a finite-dimensional model be deterministic functions of time [234]. Many no-arbitrage models take this route. It must be emphasized that making parameters time dependent does *not* render a model multifactor. Each factor in a multifactor model must represent a distinct source of uncertainty, which a time-dependent parameter does not do, even though it does provide the model with greater flexibility [38].

Calibration cannot correct model specification error. The price of a derivative is the cost of carrying out a self-financing replicating strategy based on its delta, we recall. Delta hedges that fail to replicate the derivative will provide incorrect prices. Hence a misspecified model does not price or hedge correctly even if it has been calibrated [42, 149]. For instance, if the drift of the short rate is not linear, as some evidence suggests [11], then all models that postulate a linear drift err. Of course, it is possible for a wrong model to be useful as an interpolator of prices within a set of claims similar to the ones used in calibration. There is no support, however, for using such a model to price claims very different from the ones in the calibrating set.

▶ **Exercise 25.4.1** Two methods were mentioned for calibrating the Black–Scholes option pricing model: historical volatility and implied volatility. Which corresponds to the time-series approach and which to the cross-sectional approach?

25.5 One-Factor Short Rate Models

One-model short rate models have several shortcomings. To begin with, they throw away much information. By using only the short rate, they ignore other rates on the yield curve. Such models also restrict the volatility to be a function of interest rate *levels* only [126].

When changes in the term structure are driven by a single factor, the prices of all bonds move in the same direction at the same time even though their magnitudes may differ. The returns on all bonds thus become highly correlated. In reality, there seems to be a certain amount of independence between short- and long-term rates [38, 304].[2] One-factor models therefore cannot accommodate nondegenerate correlation structures across maturities. Not surprisingly, derivatives whose values depend on the correlation structure across distinct sectors of the yield curve, such as yield curve options, are mispriced by one-factor models [149].

In one-factor models, the shape of the term structure is typically limited to being monotonically increasing, monotonically decreasing, and slightly humped. The calibrated models also may not generate term structures as concave as the data suggest [41]. The term structure empirically changes in slope and curvature as well as makes parallel moves (review Subsection 19.2.5). This is inconsistent with the restriction that the movements of all segments of the term structure be perfectly correlated. One-factor models are therefore incomplete [607].

Generally speaking, one-factor models generate hedging errors for complex securities [91], and their hedging accuracy is poor [42, 149]. They may nevertheless be acceptable for applications such as managing portfolios of similar-maturity bonds or valuation of securities with cash flows determined predominantly by the overall level of interest rates [198].

Models in which bond prices depend on two or more sources of uncertainty lead to families of yield curves that can take a greater variety of shapes and can better represent reality [46, 793]. Multifactor models include the Brennan–Schwartz model, the Richard model [741], the Langetieg model, the Longstaff–Schwartz model, and the Chen–Scott model [183]. However, a multifactor model is much harder to think about and work with. It also takes much more computer time – the curse of dimensionality raises its head again. These practical concerns limit the use of multifactor models to two-factor ones [38]. Working with different one-factor models before moving on to multifactor ones may be a wise recommendation [84, 482].

The price of a European option on a coupon bond can be calculated from those on zero-coupon bonds as follows. Consider a European call expiring at time T on a bond with par value \$1. Let X denote the strike price. The bond has cash flows $c_1, c_2, \ldots, c_n$ at times $t_1, t_2, \ldots, t_n$, where $t_i > T$ for all i. The payoff for the option is clearly

$$\max\left(\sum_{i=1}^{n} c_i P(r(T), T, t_i) - X, 0\right).$$

At time T, there is a unique value r^* for $r(T)$ that renders the coupon bond's price equal to the strike price X. We can obtain this r^* by solving $X = \sum_i c_i P(r, T, t_i)$ numerically for r, which is straightforward if analytic formulas are known for zero-coupon bond prices. The solution is also unique for one-factor models as the bond price is a monotonically decreasing function of r. Let $X_i \equiv P(r^*, T, t_i)$, the value at time T of a zero-coupon bond with par value \$1 and maturing at time t_i if $r(T) = r^*$. Note that $P(r(T), T, t_i) >= X_i$ if and only if $r(T) <= r^*$. As $X = \sum_i X_i$, the option's payoff equals

$$\sum_{i=1}^{n} c_i \times \max(P(r(T), T, t_i) - X_i, 0).$$

Thus the call is a package of n options on the underlying zero-coupon bond [506].

▶ **Exercise 25.5.1** Suppose that the spot rate curve $r(r, a, b, t, T) \equiv r + a(T - t) + b(T - t)^2$ is implied by a three-factor model. Which of the factors, r, a, and b, affects slope, curvature, and parallel moves, respectively?

▶ **Exercise 25.5.2** Repeat the preceding argument for European puts on coupon bonds and show that the payoff equals $\sum_{i=1}^{n} c_i \times \max(X_i - P(r(T), T, t_i), 0)$.

Concluding Remarks and Additional Reading

When a financial series is described by a stochastic differential equation like

$$dX_t = \mu(X_t)\,dt + \sigma(X_t)\,dW_t,$$

the specific parametric forms chosen for μ and σ may be based more on analytic or computational tractability than economic considerations. This arbitrariness presents a potential problem for every **parametric model**: specification error from picking the wrong functional form. In fact, one study claims that none of the existing parametric interest rate models fit historical data well [11] (this finding is contested in [728]). **Nonparametric models** in contrast make no parametric assumptions about the functional forms of the drift μ and/or the diffusion σ [819]. Instead, one or both functions are to be estimated nonparametrically from the discretely observed data. The requirement is that approximations to the true drift and diffusion converge pointwise to μ and σ at a rate $(\Delta t)^k$, where Δt is the time between successive observations and $k > 0$. As a result, the approximation errors should be small as long as observations are made frequently enough. See [517, 611, 613] for the estimation of Ito processes.

This chapter surveyed equilibrium models and pointed out some of their weaknesses. One way to address them is the adoption of no-arbitrage models, to which we will turn in the next chapter. Another approach is the use of additional factors. Nonparametric models are yet another option. Unlike equity derivatives, no single dominant model emerges.

For the pricing of interest rate caps, consult [616] (the CIR case) and [617] (the Vasicek case). See [184, 185, 186, 257, 630, 645, 803] for more information on multifactor CIR models and parameter estimation techniques. Refer to [291, 477] for more discussions on one-factor models. Finally, see [301, 302] for discussions on long rates.

NOTES

1. www.cob.ohio-state.edu/~fin/journal/lies.htm.
2. Real rates seem to be more correlated [135].

CHAPTER TWENTY-SIX

No-Arbitrage Term
Structure Models

> The fox often ran to the hole by which they had come in, to find out
> if his body was still thin enough to slip through it.
> The Complete Grimm's Fairy Tales

This chapter samples no-arbitrage models pioneered by Ho and Lee. Some of the
salient features of such models were already covered, if implicit at that, in Chap. 23.

26.1 Introduction

Some of the difficulties facing equilibrium models were mentioned in Section 25.4.
For instance, they usually require the estimation of the market price of risk and cannot
fit the market term structure. However, consistency with the market is often mandatory in practice [457]. No-arbitrage models, in contrast, utilize the full information
of the term structure. They accept the observed term structure as consistent with an
unobserved and unspecified equilibrium. From there, arbitrage-free movements of
interest rates or bond prices over time are modeled. By definition, the market price
of risk must be reflected in the current term structure; hence the resulting interest
rate process is risk-neutral.

No-arbitrage models can specify the dynamics of zero-coupon bond prices, forward rates, or the short rate [477, 482]. Bond price and forward rate models are
usually non-Markovian (path dependent), whereas short rate models are generally
constructed to be explicitly Markovian (path independent). Markovian models are
easier to handle computationally than non-Markovian ones.

▶ **Exercise 26.1.1** Is the equilibrium or no-arbitrage model more appropriate in deciding which government bonds are overpriced?

26.2 The Ho–Lee Model

This path-breaking one-factor model enjoys popularity among practitioners [72].
Figure 26.1 captures the model's short rate process. The short rates at any given
time are evenly spaced. Let p denote the risk-neutral probability that the short rate
makes an up move. We shall adopt continuous compounding.

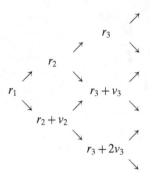

Figure 26.1: The Ho–Lee binomial interest rate tree. The distribution at any time converges to the normal distribution; the Ho–Lee model is a normal process in the limit. The v_is are related to short rate volatilities by Eq. (26.2).

The model starts with zero-coupon bond prices $P(t, t+1)$, $P(t, t+2)$, ..., at time t identified with the root of the tree. Let the discount factors in the next period be

$$P_d(t+1, t+2), P_d(t+1, t+3), \ldots, \quad \text{if the short rate makes a down move,}$$
$$P_u(t+1, t+2), P_u(t+1, t+3), \ldots, \quad \text{if the short rate makes an up move.}$$

By backward induction, it is not hard to see that, for $n \geq 2$,

$$P_u(t+1, t+n) = P_d(t+1, t+n) \, e^{-(v_2 + \cdots + v_n)} \tag{26.1}$$

(see Exercise 26.2.1) and the n-period zero-coupon bond has yields

$$y_d(n) \equiv -\frac{\ln P_d(t+1, t+n)}{n-1},$$

$$y_u(n) \equiv -\frac{\ln P_u(t+1, t+n)}{n-1} = y_d(n) + \frac{v_2 + \cdots + v_n}{n-1},$$

respectively. The volatility of the yield to maturity for this bond is therefore

$$\kappa_n \equiv \sqrt{p y_u(n)^2 + (1-p) \, y_d(n)^2 - [\, p y_u(n) + (1-p) \, y_d(n)\,]^2}$$

$$= \sqrt{p(1-p)} \, [\, y_u(n) - y_d(n)\,]$$

$$= \sqrt{p(1-p)} \, \frac{v_2 + \cdots + v_n}{n-1}.$$

In particular, we determine the short rate volatility by taking $n = 2$:

$$\sigma = \sqrt{p(1-p)} \, v_2. \tag{26.2}$$

The variance of the short rate therefore equals $p(1-p)(r_u - r_d)^2$, where r_u and r_d are the two successor rates.[1]

The volatility term structure is composed of $\kappa_2, \kappa_3, \ldots$, independent of the r_is. It is easy to compute the v_is from the volatility structure (see Exercise 26.2.2), and vice versa. The r_is can be computed by forward induction. The volatility structure in the original Ho–Lee model is flat because it assumes that v_i are all equal to some constant. For the general Ho–Lee model that incorporates a term structure of volatilities, the volatility structure is supplied by the market.

▶ **Exercise 26.2.1** Verify Eq. (26.1).

▶ **Exercise 26.2.2** Show that $v_i = [\,(i-1)\,\kappa_i - (i-2)\,\kappa_{i-1}\,]/\sqrt{p(1-p)}$.

26.2.1 Bond Price Process

In a risk-neutral economy, the initial discount factors satisfy

$$P(t, t+n) = [\, p P_u(t+1, t+n) + (1-p)\, P_d(t+1, t+n)\,]\, P(t, t+1).$$

Combine the preceding equation with Eq. (26.1) and assume that $p = 1/2$ to obtain[2]

$$P_d(t+1, t+n) = \frac{P(t, t+n)}{P(t, t+1)} \frac{2 \times \exp[\, v_2 + \cdots + v_n \,]}{1 + \exp[\, v_2 + \cdots + v_n \,]}, \tag{26.3}$$

$$P_u(t+1, t+n) = \frac{P(t, t+n)}{P(t, t+1)} \frac{2}{1 + \exp[\, v_2 + \cdots + v_n \,]}. \tag{26.3'}$$

This defines the bond price process. The above system of equations establishes the price relations that must hold to prevent riskless arbitrages [304, 504]. The bond price tree combines (see Exercise 26.2.3).

In the original Ho–Lee model, v_i all equal some constant v. Then

$$P_d(t+1, t+n) = \frac{P(t, t+n)}{P(t, t+1)} \frac{2 \delta^{n-1}}{1 + \delta^{n-1}},$$

$$P_u(t+1, t+n) = \frac{P(t, t+n)}{P(t, t+1)} \frac{2}{1 + \delta^{n-1}},$$

where $\delta \equiv e^v > 0$. The short rate volatility σ equals $v/2$ by Eq. (26.2). To annualize the numbers, simply apply $\sigma(\text{period}) = \sigma(\text{annual}) \times \sqrt{\Delta t}$ and $v(\text{period}) = v(\text{annual}) \times \Delta t$. As a consequence,

$$\delta(\text{annual}) = e^{2\sigma(\text{annual})(\Delta t)^{3/2}}. \tag{26.4}$$

The Ho–Lee model demonstrates clearly that no-arbitrage models price securities in a way consistent with the initial term structure. Furthermore, these models postulate dynamics that disallows intertemporal arbitrage opportunities. Derivatives are priced by taking expectations under the risk-neutral probability [359].

▶ **Exercise 26.2.3** Show that a rate rise followed by a rate decline produces the same term structure as that of a rate decline followed by a rate rise.

▶ **Exercise 26.2.4** Prove that Eqs. (26.3) and (26.3′) become

$$P_d(t+1, t+n) = \frac{P(t, t+n)}{P(t, t+1)} \frac{\exp[\, v_2 + \cdots + v_n \,]}{p + (1-p) \times \exp[\, v_2 + \cdots + v_n \,]},$$

$$P_u(t+1, t+n) = \frac{P(t, t+n)}{P(t, t+1)} \frac{1}{p + (1-p) \times \exp[\, v_2 + \cdots + v_n \,]}$$

for general risk-neutral probability p.

▶ **Exercise 26.2.5** Consider a portfolio of one zero-coupon bond with maturity T_1 and β zero-coupon bonds with maturity T_2. Find the β that makes the portfolio instantaneously riskless under the Ho–Lee model.

▶ **Programming Assignment 26.2.6** Write a linear-time program to calibrate the original Ho–Lee model. The inputs are Δt, the current market discount factors, and the short rate volatility σ, all annualized.

Yield Volatilities and Their Covariances

The one-period rate of return of an n-period zero-coupon bond is

$$r(t, t+n) \equiv \ln \left(\frac{P(t+1, t+n)}{P(t, t+n)} \right).$$

Because its value is either

$$\ln \frac{P_d(t+1, t+n)}{P(t, t+n)}$$

or

$$\ln \frac{P_u(t+1, t+n)}{P(t, t+n)},$$

the variance of the return is

$$\mathrm{Var}[\, r(t, t+n)\,] = p(1-p)[\,(n-1)\, v\,]^2 = (n-1)^2 \sigma^2.$$

The covariance between $r(t, t+n)$ and $r(t, t+m)$ is $(n-1)(m-1)\sigma^2$ (see Exercise 26.2.7). As a result, the correlation between any two one-period returns is unity. Strong correlation between rates is inherent in all one-factor Markovian models.

▶ **Exercise 26.2.7** Prove that under a general p, the variance of the one-period return of n-period zero-coupon bonds equals $(n-1)^2 \sigma^2$ and the covariance between the one-period returns of n- and m-period zero-coupon bonds equals $(n-1)(m-1)\sigma^2$.

26.2.2 Forward Rate Process

The forward rate at time t for money borrowed or lent from time $t+n$ to $t+n+1$ is

$$f(t, t+n) = -\ln \left(\frac{P(t, t+n+1)}{P(t, t+n)} \right)$$

from Eq. (24.3). The current state considered as the result of a downward rate move from time $t-1$ leads to

$$f(t, t+n) = -\ln \left(\frac{\frac{P(t-1,t+n+1)}{P(t-1,t)} \frac{2\delta^{n+1}}{1+\delta^{n+1}}}{\frac{P(t-1,t+n)}{P(t-1,t)} \frac{2\delta^n}{1+\delta^n}} \right)$$

$$= f(t-1, t+n) - \ln \left(\frac{1+\delta^n}{1+\delta^{n+1}} \right) - \ln \delta,$$

and the current state considered as the result of an upward rate move leads to

$$f(t, t+n) = -\ln \left(\frac{\frac{P(t-1,t+n+1)}{P(t-1,t)} \frac{2}{1+\delta^{n+1}}}{\frac{P(t-1,t+n)}{P(t-1,t)} \frac{2}{1+\delta^n}} \right) = f(t-1, t+n) - \ln \left(\frac{1+\delta^n}{1+\delta^{n+1}} \right).$$

The preceding two equations can be combined to yield this forward rate process:

$$f(t, t+n) = f(t-1, t+n) - \ln \left(\frac{1+\delta^n}{1+\delta^{n+1}} \right) - \frac{1}{2} \ln \delta + \xi_{t-1}, \qquad (26.5)$$

where ξ_s ($s \geq 0$) is the following zero-mean random variable:

$$\xi_s = \begin{cases} -(1/2)\ln\delta, & \text{if down move occurs at time } s \\ (1/2)\ln\delta, & \text{if up move occurs at time } s \end{cases}.$$

Because $\text{Var}[\,f(t, t+n) - f(t-1, t+n)\,] = \sigma^2$ (see Exercise 26.2.8), the volatility σ can be estimated from historical data without the need to estimate the risk-neutral probability [436]. Equation (26.5) can be applied iteratively to obtain

$$f(t, t+n) = f(0, t+n) - \ln\left(\frac{1+\delta^n}{1+\delta^{n+t}}\right) - \frac{t}{2}\ln\delta + \sum_{s=1}^{t}\xi_{s-1}.$$

▶ **Exercise 26.2.8** Verify that $\text{Var}[\,\xi_s\,] = \sigma^2$.

▶ **Exercise 26.2.9** Prove that

$$-\ln\left(\frac{1+\delta^n}{1+\delta^{n+1}}\right) - \frac{1}{2}\ln\delta \to \sigma^2(T-t)(\Delta t)^2$$

if we substitute $t/\Delta t$ for t and $(T-t)/\Delta t$ for n in Eq. (26.5) before applying Eq. (26.4). T, t, Δt, and σ above are annualized. (The forward rate process hence converges to $df(t, T) = \sigma^2(T-t)\,dt + \sigma\,dW$.)

26.2.3 Short Rate Process

Because the short rate $r(t)$ equals $f(t, t)$,

$$r(t) = f(0, t) - \ln\left(\frac{2}{1+\delta^t}\right) - \frac{t}{2}\ln\delta + \sum_{s=1}^{t}\xi_{s-1}.$$

This implies the following difference equation:

$$r(t) = r(t-1) + f(0, t) - f(0, t-1) - \ln\left(\frac{1+\delta^{t-1}}{1+\delta^t}\right) - \frac{1}{2}\ln\delta + \xi_{t-1}. \quad (26.6)$$

The continuous-time limit of the Ho–Lee model is $dr = \theta(t)\,dt + \sigma\,dW$. This is essentially Vasicek's model with the mean-reverting drift replaced with a deterministic, time-dependent drift. A nonflat term structure of volatilities can be achieved if the short rate volatility is also made time varying, i.e., $dr = \theta(t)\,dt + \sigma(t)\,dW$ [508]. This corresponds to the discrete-time model in which v_i are not all identical.

▶ **Exercise 26.2.10** Prove that

$$-\ln\left(\frac{1+\delta^{t-1}}{1+\delta^t}\right) - \frac{1}{2}\ln\delta \to \sigma^2 t\,(\Delta t)^2$$

if we substitute $t/\Delta t$ for t and apply Eq. (26.4). The t, Δt, and σ above are annualized. (Short rate process (26.6) thus converges to $dr = \{\,[\,\partial f(0, t)/\partial t\,] + \sigma^2 t\,\}\,dt + \sigma\,dW$.)

26.2.4 Problems with the Ho–Lee Model

Nominal interest rates must be nonnegative because we can hold cash. However, negative future interest rates are possible with the Ho–Lee model. This may not be a major concern for realistic volatilities and certain ranges of bond maturities [72]. More questionable is the fact that the short rate volatility is independent of the rate level [83, 173, 359, 645]. Given that the Ho–Lee model subsumes the Merton model and shares many of its unreasonable properties, how can it generate reasonable initial term structures? The answer lies in the model's unreasonable short rate dynamics.[3]

▶ **Exercise 26.2.11** Assess the claim that the problem of negative interest rates can be eliminated by making the short rate volatility time dependent.

Problems with No-Arbitrage Models in General

Interest rate movements should reflect shifts in the model's state variables (factors), not its parameters. This means that model parameters, such as the drift $\theta(t)$ in the continuous-time Ho–Lee model, should be stable over time. However, in practice, no-arbitrage models capture yield curve shifts through the recalibration of parameters. A new model is thus born everyday. This in effect says that the model estimated at some time does not describe the term structure of interest rates and their volatilities at other times. Consequently, a model's intertemporal behavior is suspect, and using it for hedging and risk management may be unreliable.

26.3 The Black–Derman–Toy Model

Black, Derman, and Toy (BDT) proposed their model in 1990 [84]. This model is extensively used by practitioners [72, 149, 215, 600, 731]. The BDT short rate process is the lognormal binomial interest rate process described in Chap. 23 and repeated in Fig. 26.2. The volatility structure is given by the market from which the short rate volatilities (thus v_i) are determined together with r_i. Our earlier binomial interest rate tree, in comparison, assumes that v_i are given a priori, and a related model of Salomon Brothers takes v_i to be constants [848]. Lognormal models preclude negative short rates.

The volatility structure defines the yield volatilities of zero-coupon bonds of various maturities. Let the yield volatility of the i-period zero-coupon bond be denoted by κ_i. Assume that P_u (P_d) is the price of the i-period zero-coupon bond one period from now if the short rate makes an up (down, respectively) move. Corresponding

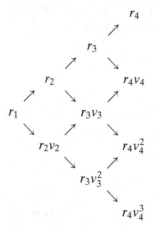

Figure 26.2: The BDT binomial interest rate tree. The distribution at any time converges to the lognormal distribution.

to these two prices are these yields to maturity:

$$y_u \equiv P_u^{-1/(i-1)} - 1, \quad y_d \equiv P_d^{-1/(i-1)} - 1.$$

The yield volatility is defined as $\kappa_i \equiv (1/2)\ln(y_u/y_d)$.

26.3.1 Calibration

The inputs to the BDT model are riskless zero-coupon bond yields and their volatilities. For economy of expression, all numbers are period based. Suppose inductively that we have calculated $r_1, v_1, r_2, v_2, \ldots, r_{i-1}, v_{i-1}$, which define the binomial tree up to period $i - 1$. We now proceed to calculate r_i and v_i to extend the tree to period i. Assume that the price of the i-period zero can move to P_u or P_d one period from now. Let y denote the current i-period spot rate, which is known. In a risk-neutral economy,

$$\frac{P_u + P_d}{2(1+r_1)} = \frac{1}{(1+y)^i}. \tag{26.7}$$

Obviously, P_u and P_d are functions of the unknown r_i and v_i. Viewed from now, the future $(i-1)$-period spot rate at time one is uncertain. Let y_u and y_d represent the spot rates at the up node and the down node, respectively, with κ^2 denoting the variance, or

$$\kappa_i = \frac{1}{2} \ln \left(\frac{P_u^{-1/(i-1)} - 1}{P_d^{-1/(i-1)} - 1} \right). \tag{26.8}$$

We use forward induction to derive a quadratic-time calibration algorithm [190, 625]. Recall that forward induction inductively figures out, by moving forward in time, how much $1 at a node contributes to the price (review Fig. 23.7(a)). This number is called the state price and is the price of the claim that pays $1 at that node and zero elsewhere.

Let the baseline rate for period i be $r_i = r$, let the multiplicative ratio be $v_i = v$, and let the state prices at time $i - 1$ be $P_1, P_2, \ldots, P_i$, corresponding to rates $r, rv, \ldots, rv^{i-1}$, respectively. One dollar at time i has a PV of

$$f(r, v) \equiv \frac{P_1}{1+r} + \frac{P_2}{1+rv} + \frac{P_3}{1+rv^2} + \cdots + \frac{P_i}{1+rv^{i-1}},$$

and the yield volatility is

$$g(r, v) \equiv \frac{1}{2} \ln \left(\frac{\left(\frac{P_{u,1}}{1+rv} + \frac{P_{u,2}}{1+rv^2} + \cdots + \frac{P_{u,i-1}}{1+rv^{i-1}} \right)^{-1/(i-1)} - 1}{\left(\frac{P_{d,1}}{1+r} + \frac{P_{d,2}}{1+rv} + \cdots + \frac{P_{d,i-1}}{1+rv^{i-2}} \right)^{-1/(i-1)} - 1} \right).$$

In the preceding equation, $P_{u,1}, P_{u,2}, \ldots$ denote the state prices at time $i - 1$ of the subtree rooted at the up node (like $r_2 v_2$ in Fig. 26.2), and $P_{d,1}, P_{d,2}, \ldots$ denote the state prices at time $i - 1$ of the subtree rooted at the down node (like r_2 in Fig. 26.2). Now solve

$$f(r, v) = \frac{1}{(1+y)^i}, \quad g(r, v) = \kappa_i$$

Algorithm for calibrating the BDT model:

input: $n, S[1..n], \kappa[1..n]$;
real $P[0..n], P_u[0..n-1], P_d[0..n-1], r[1..n], v[1..n], r, v$;
integer i, j;
$P[0] := 0; P_u[0] := 0; P_d[0] := 0$;   // Dummies; remain zero throughout.
$P[1] := 1; P_u[1] := 1; P_d[1] := 1$;
$r[1] := S[1]; v[1] := 0$;
for $(i = 2\,\text{to}\,n)$ {
  $P[i] := 0$;
  for $(j = i\,\text{down to}\,1)$ // State prices at time $i-1$.
   $P[j] := \frac{P[j-1]}{2\times(1+r[i-1]\times v[i-1]^{j-2})} + \frac{P[j]}{2\times(1+r[i-1]\times v[i-1]^{j-1})}$;
  Solve for r and v from
   $\sum_{j=1}^{i} \frac{P[j]}{(1+r\times v^{j-1})} = (1+S[i])^{-i}$ and
   $(\sum_{j=1}^{i-1} \frac{P_u[j]}{(1+r\times v^{j})})^{-1/(i-1)} - 1 = e^{2\times\kappa[i]} \times ((\sum_{j=1}^{i-1} \frac{P_d[j]}{(1+r\times v^{j-1})})^{-1/(i-1)} - 1)$;
  $r[i] := r; v[i] := v$;
  if $[i < n]$ {
   $P_u[i] := 0; P_d[i] := 0$;
   for $(j = i\,\text{down to}\,1)$ {   // State prices at time i.
    $P_u[j] := \frac{P_u[j-1]}{2\times(1+r[i]\times v[i]^{j-1})} + \frac{P_u[j]}{2\times(1+r[i]\times v[i]^{j})}$;
    $P_d[j] := \frac{P_d[j-1]}{2\times(1+r[i]\times v[i]^{j-2})} + \frac{P_d[j]}{2\times(1+r[i]\times v[i]^{j-1})}$;
   }
  }
}
return $r[\,]$ and $v[\,]$;

Figure 26.3: Algorithm for calibrating the BDT model. $S[i]$ is the i-period spot rate, $\kappa[i]$ is the yield volatility for period i, and n is the number of periods. All numbers are measured by the period. The period-i baseline rate and multiplicative ratio are stored in $r[i]$ and $v[i]$, respectively. The two-dimensional Newton–Raphson method of Eqs. (3.16) should be used to solve for r and v as the partial derivatives are straightforward to calculate.

for $r = r_i$ and $v = v_i$. This $O(n^2)$-time algorithm is given in Fig. 26.3. The continuous-time limit of the BDT model is $d\ln r = \{\theta(t) + [\sigma'(t)/\sigma(t)]\ln r\}\,dt + \sigma(t)\,dW$ [76, 149, 508]. Obviously the short rate volatility should be a declining function of time for the model to display mean reversion; in particular, constant volatility will not attain mean reversion.

▶ **Exercise 26.3.1** Describe the differential tree method with backward induction to calibrate the BDT model.

▶ **Programming Assignment 26.3.2** Implement the algorithm in Fig. 26.3.

▶ **Programming Assignment 26.3.3** Calibrate the BDT model with the secant method and evaluate its performance against the differential tree method.

26.3.2 The Black–Karasinski Model

The related Black–Karasinski model stipulates that the short rate follows

$$d\ln r = \kappa(t)[\theta(t) - \ln r]\,dt + \sigma(t)\,dW.$$

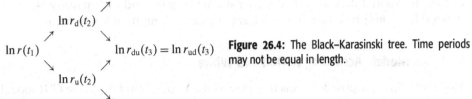

Figure 26.4: The Black–Karasinski tree. Time periods may not be equal in length.

This explicitly mean-reverting model depends on time through $\kappa(\cdot), \theta(\cdot)$, and $\sigma(\cdot)$. The Black–Karasinski model hence has one more degree of freedom than the BDT model. The speed of mean reversion $\kappa(t)$ and the short rate volatility $\sigma(t)$ are independent [85].

The discrete-time version of the Black–Karasinski model has the same representation as the BDT model. To maintain a combining binomial tree, however, requires some manipulations. These ideas are illustrated by Fig. 26.4 in which $t_2 \equiv t_1 + \Delta t_1$ and $t_3 \equiv t_2 + \Delta t_2$. Note that

$$\ln r_d(t_2) = \ln r(t_1) + \kappa(t_1)[\theta(t_1) - \ln r(t_1)]\Delta t_1 - \sigma(t_1)\sqrt{\Delta t_1},$$

$$\ln r_u(t_2) = \ln r(t_1) + \kappa(t_1)[\theta(t_1) - \ln r(t_1)]\Delta t_1 + \sigma(t_1)\sqrt{\Delta t_1}.$$

To ensure that an up move followed by a down move coincides with a down move followed by an up move, we impose

$$\ln r_d(t_2) + \kappa(t_2)[\theta(t_2) - \ln r_d(t_2)]\Delta t_2 + \sigma(t_2)\sqrt{\Delta t_2}$$
$$= \ln r_u(t_2) + \kappa(t_2)[\theta(t_2) - \ln r_u(t_2)]\Delta t_2 - \sigma(t_2)\sqrt{\Delta t_2},$$

which implies that

$$\kappa(t_2) = \frac{1 - [\sigma(t_2)/\sigma(t_1)]\sqrt{\Delta t_2/\Delta t_1}}{\Delta t_2}.$$

So from Δt_1, we can calculate the Δt_2 that satisfies the combining condition and then iterate.

▶ **Exercise 26.3.4** Show that the variance of r after Δt is approximately $[r(t)\sigma(t)]^2\Delta t$.

▶ **Programming Assignment 26.3.5** Implement a forward-induction algorithm to calibrate the Black–Karasinski model given a constant κ.

26.3.3 Problems with Lognormal Models

Lognormal models such as the BDT, Black–Karasinski, and Dothan models share the problem that $E^\pi[M(t)] = \infty$ for any finite t if it is the continuously compounded rate that is modeled (review Subsection 24.2.1) [76]. Hence periodic compounding should be used. Another issue is computational. Lognormal models usually do not give analytical solutions to even basic fixed-income securities. As a result, to price short-dated derivatives on longterm bonds, the tree has to be built over the life of the underlying asset – which can be, say, 30 years – instead of the life of the claim – possibly only 1–2 years (review Comment 23.2.1). This problem can be somewhat mitigated if different time steps are adopted: Use a fine time step up to the maturity of the short-dated derivative and a coarse time step beyond the maturity [477]. A down side of this procedure is that it has to be carried out for each derivative.

26.4 The Models According to Hull and White

Hull and White proposed models that extend the Vasicek model and the CIR model [474]. They are called the extended Vasicek model and the extended CIR model. The extended Vasicek model adds time dependence to the original Vasicek model:

$$dr = (\theta(t) - a(t)r) \, dt + \sigma(t) \, dW.$$

Like the Ho–Lee model, this is a normal model, and the inclusion of $\theta(\cdot)$ allows for an exact fit to the current spot rate curve. As for the other two functions, $\sigma(t)$ defines the short-rate volatility and $a(t)$ determines the shape of the volatility structure. Under this model, many European-style securities can be evaluated analytically, and efficient numerical procedures can be developed for American-style securities. The Hull–White model is the following special case:

$$dr = (\theta(t) - ar) \, dt + \sigma \, dW.$$

When the current term structure is matched,

$$\theta(t) = \frac{\partial f(0, t)}{\partial t} + af(0, t) + \frac{\sigma^2}{2a}(1 - e^{-2at})$$

[477]. In the extended CIR model the short rate follows

$$dr = [\theta(t) - a(t)r] \, dt + \sigma(t)\sqrt{r} \, dW.$$

The functions $\theta(\cdot)$, $a(\cdot)$, and $\sigma(\cdot)$ are implied from market observables. With constant parameters, there exist analytical solutions to a small set of interest-rate-sensitive securities such as coupon bonds and European options on bonds.

For the BDT and the Ho–Lee models, once the initial volatility structure is specified, the future short rate volatility is completely determined. Conversely, if the future short rate volatility is specified, then the initial volatility structure is fully determined. However, we may want to specify the volatility structure and the short rate volatility separately because the future short rate volatility may have little impact on the yield volatility, which is about the uncertainty over the spot rate's value in the next period [476]. The extended Vasicek model has enough degrees of freedom to accommodate this request.

▶ **Exercise 26.4.1** Between a normal and a lognormal model, which overprices out-of-the-money calls on bonds *and* underprices out-of-the-money puts on bonds?

26.4.1 Calibration of the Hull–White Model with Trinomial Trees

Now a trinomial forward-induction scheme is described to calibrate the Hull–White model given a and σ [477]. As with the Ho–Lee model, in this model the set of achievable short rates is evenly spaced. Let r_0 be the annualized, continuously compounded short rate at time zero. Every short rate on the tree takes on a value $r_0 + j\Delta r$ for some integer j. Time increments on the tree are also equally spaced at Δt apart. (Binomial trees should not be used to model mean-reverting interest rates when Δt is a constant [475].) Hence nodes are located at times $i\Delta t$ for $i = 0, 1, 2, \ldots$. We

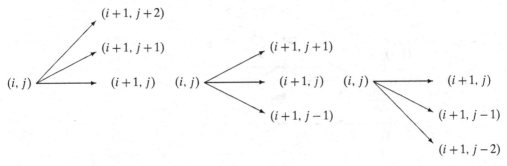

Figure 26.5: Three trinomial branching schemes in the Hull–White model. The choice is determined by the expected short rate at time t_{i+1} as seen from time t_i.

refer to the node on the tree with $t_i \equiv i\Delta t$ and $r_j \equiv r_0 + j\Delta r$ as the **(i, j) node**. The short rate at node (i, j), which equals r_j, is effective for the time period $[t_i, t_{i+1}]$. Use

$$\mu_{i,j} \equiv \theta(t_i) - ar_j \tag{26.9}$$

to denote the drift rate, or the expected change, of the short rate as seen from node (i, j). The three distinct possibilities for node (i, j) with three branches incident from it are shown in Fig. 26.5. The interest rate movement described by the middle branch may be an increase of Δr, no change, or a decrease of Δr. The upper and the lower branches bracket the middle branch. Define

$p_1(i, j) \equiv$ the probability of following the upper branch from node (i, j),

$p_2(i, j) \equiv$ the probability of following the middle branch from node (i, j),

$p_3(i, j) \equiv$ the probability of following the lower branch from node (i, j).

The root of the tree is set to the current short rate r_0. Inductively, the drift $\mu_{i,j}$ at node (i, j) is a function of $\theta(t_i)$. Once $\theta(t_i)$ is available, $\mu_{i,j}$ can be derived by means of Eq. (26.9). This in turn determines the branching scheme at every node (i, j) for each j, as we will see shortly. The value of $\theta(t_i)$ must thus be made consistent with the spot rate $r(0, t_{i+2})$.

The branches emanating from node (i, j) with their accompanying probabilities, $p_1(i, j)$, $p_2(i, j)$, and $p_3(i, j)$, must be chosen to be consistent with $\mu_{i,j}$ and σ. This is accomplished by letting the middle node be as close as possible to the current value of the short rate plus the drift. Let k be the number among $\{ j-1, j, j+1 \}$ that makes the short rate reached by the middle branch, r_k, closest to $r_j + \mu_{i,j}\Delta t$. Then the three nodes following node (i, j) are nodes $(i+1, k+1)$, $(i+1, k)$, and $(i+1, k-1)$. The resulting tree may have the geometry depicted in Fig. 26.6. The resulting tree combines.

The probabilities for moving along these branches are functions of $\mu_{i,j}, \sigma, j$, and k:

$$p_1(i, j) = \frac{\sigma^2\Delta t + \eta^2}{2(\Delta r)^2} + \frac{\eta}{2\Delta r}, \tag{26.10}$$

$$p_2(i, j) = 1 - \frac{\sigma^2\Delta t + \eta^2}{(\Delta r)^2}, \tag{26.10'}$$

$$p_3(i, j) = \frac{\sigma^2\Delta t + \eta^2}{2(\Delta r)^2} - \frac{\eta}{2\Delta r}, \tag{26.10''}$$

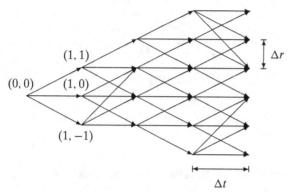

Figure 26.6: Trinomial tree for the Hull–White model. All the short rates at the nodes are known before *any* computation begins. They are simply $r_0 + j \Delta r$ for $j = 0, \pm 1, \pm 2, \ldots$. It is the branching schemes connecting the nodes that determine the term structure. Only the four nodes at times t_0 and t_1 are labeled here. The remaining nodes can be labeled similarly. The tree may not fully grow as some nodes are not reachable from the root $(0, 0)$ because of mean reversion.

where $\eta \equiv \mu_{i,j} \Delta t + (j - k) \Delta r$. As trinomial tree algorithms are but explicit methods in disguise (see Subsection 18.1.1), certain relations must hold for Δr and Δt to guarantee stability. It can be shown that their values must satisfy

$$\frac{\sigma \sqrt{3 \Delta t}}{2} \leq \Delta r \leq 2 \sigma \sqrt{\Delta t}$$

for the probabilities to lie between zero and one; for example, Δr can be set to $\sigma \sqrt{3 \Delta t}$ [473]. It remains only to determine $\theta(t_i)$, to which we now turn.

At this point at time $t_i, r(0, t_1), r(0, t_2), \ldots, r(0, t_{i+1})$ have already been matched. By construction, the state prices $Q(i, k)$ for all k are known by now, where $Q(i, k)$ denotes the value of the state contingent claim that pays \$1 at node (i, k) and zero otherwise. The value at time zero of a zero-coupon bond maturing at time t_{i+2} is then

$$e^{-r(0, t_{i+2})(i+2) \Delta t} = \sum_j Q(i, j) e^{-r_j \Delta t} E^{\pi} \left[e^{-\hat{r}(i+1) \Delta t} \mid \hat{r}(i) = r_j \right], \tag{26.11}$$

where $\hat{r}(i)$ refers to the short-rate value at time t_i. The right-hand side represents the value of \$1 obtained by holding a zero-coupon bond until time t_{i+1} and then reinvesting the proceeds at that time at the prevailing short rate $\hat{r}(i + 1)$, which is stochastic. The expectation above can be approximated by

$$E^{\pi} \left[e^{-\hat{r}(i+1) \Delta t} \mid \hat{r}(i) = r_j \right] \approx e^{-r_j \Delta t} \left[1 - \mu_{i,j} (\Delta t)^2 + \frac{\sigma^2 (\Delta t)^3}{2} \right]. \tag{26.12}$$

Substitute approximation (26.12) into Eq. (26.11) and replace $\mu_{i,j}$ with $\theta(t_i) - a r_j$ to obtain

$$\theta(t_i) \approx \frac{\sum_j Q(i, j) e^{-2 r_j \Delta t} [1 + a r_j (\Delta t)^2 + \sigma^2 (\Delta t)^3 / 2] - e^{-r(0, t_{i+2})(i+2) \Delta t}}{(\Delta t)^2 \sum_j Q(i, j) e^{-2 r_j \Delta t}}.$$

For the Hull–White model, the expectation in approximation (26.12) is actually known analytically:

$$E^\pi\left[e^{-\hat{r}(i+1)\,\Delta t}\mid \hat{r}(i)=r_j\right]=e^{-r_j\Delta t+[-\theta(t_i)+ar_j+\sigma^2\Delta t/2](\Delta t)^2}.$$

Therefore, alternatively,

$$\theta(t_i)=\frac{r(0,t_{i+2})(i+2)}{\Delta t}+\frac{\sigma^2\Delta t}{2}+\frac{\ln\sum_j Q(i,j)\,e^{-2r_j\Delta t+ar_j(\Delta t)^2}}{(\Delta t)^2}.$$

With $\theta(t_i)$ in hand, we can compute $\mu_{i,j}$, the probabilities, and finally the new state prices:

$$Q(i+1,j)=\sum_{(i,j^*)\text{ is connected to }(i+1,j)\text{ with probability }p_{j^*}}p_{j^*}e^{-r_{j^*}\Delta t}Q(i,j^*).$$

The total running time is quadratic. See Fig. 26.7 for an algorithm.

When using the Hull–White model, one can try different values of a and σ for each option or have an a value common to all options but use a different σ value for each option. Either approach can match all the option prices exactly. If the demand is for a single set of parameters that replicate all option prices, the Hull–White model can be calibrated to all the observed option prices by choosing a and σ that minimize the mean-square pricing error [482].

Algorithm for calibrating the Hull–White model:

```
input:    σ, a, Δt, n, S[1..n];
real      branch [n − 1][ −n..n], Q[ −n..n], q[ −n..n], θ, μ, r₀, Δr, p₁, p₂, p₃;
integer   i, j, k;
Q[0] := 1; r₀ := S[1];
Δr = σ√3Δt;
branch[i][j] := ∞ for 0 ≤ i < n − 1 and −n ≤ j ≤ n;   // Initial values ∉ {−1, 0, 1}.
for (i = 0 to n − 2) {
```
$$\theta:=\frac{S[i+2]\times(i+2)}{\Delta t}+\frac{\sigma^2\Delta t}{2}+\frac{\ln\left(\sum_{j=-i}^{i}Q[i][j]\times\exp\left[-2(r_0+j\Delta r)\,\Delta t+a(r_0+j\Delta r)(\Delta t)^2\right]\right)}{(\Delta t)^2};\quad//\,\theta(t_i).$$
```
       for (j = −i to i) { q[ j ] := 0; }
       for (j = −i to i)   { // Work on node (i, j)'s branching scheme.
          μ := θ − a(r₀ + jΔr);   // μ_{i,j}.
          Let k ∈ {−1, 0, 1} minimize | (r₀ + (j + k) Δr) − (r₀ + jΔr + μΔt) |;
          branch [i][ j ] := k;
          Use Eqs. (26.10) to calculate p₁, p₂, and p₃ with k = j+ branch [i][ j ];
          q[k+1] := p₁ × Q[ j ] × e^{−(r₀+jΔr)Δt} + q[k+1]; // Add contribution to Q.
          q[k] := p₂ × Q[ j ] × e^{−(r₀+jΔr)Δt} + q[k];
          q[k−1] := p₃ × Q[ j ] × e^{−(r₀+jΔr)Δt} + q[k−1];
       }
       for (j = −i to i) { Q[ j ] := q[ j ]; }   // Update Q.
}
return branch [ ][ ];
```

Figure 26.7: Algorithm for calibrating the Hull–White model. $S[i]$ is the annualized i-period spot rate, $Q[]$ stores the state prices with initial values of zero, branch[i][j] maintains the branching scheme for node (i, j), and n is the number of periods. Only the branching schemes are returned as they suffice to derive the short rates.

The algorithmic idea here is quite general and can be modified to apply to cases in which the diffusion term has the form σr^β [215]. A highly efficient algorithm exists that fully exploits the fact that the Hull–White model has a constant diffusion term [479].

➤ **Exercise 26.4.2** Verify approximation (26.12).

➤ **Programming Assignment 26.4.3** Implement the algorithm in Fig. 26.7.

➤ **Programming Assignment 26.4.4** Calibration takes the spot rate curve to reverse engineer the Hull–White model's parameters. However, just because the curve is matched by no means implies that the true model parameters and the estimated parameters are matched. Call a model **stable** if the model parameters can be approximated well by the estimated parameters. Verify that both the Hull–White model and the BDT model are stable [192, 885].

➤ **Programming Assignment 26.4.5** Implement a trinomial tree model for the Black–Karasinski model $d \ln r = (\theta(t) - a \ln r)\,dt + \sigma\,dW$.

26.4.2 Problems with the Models

When $\sigma(t)$ and $a(t)$ vary with time, the volatility structure will be nonstationary. Choosing $\sigma(t)$ and $a(t)$ to exactly fit the initial volatility structure then causes the volatility structure to evolve in unpredictable ways and makes option prices questionable. This observation holds for all Markovian models [76, 164]. Because it is in general dangerous to use time-varying parameters to match the initial volatility curve exactly, it has been argued that there should be no more than one time-dependent parameter in Markovian models and that it should be used to fit the initial spot rate curve only [482]. This line of reasoning favors the Hull–White model. Another way to maintain the volatility structure over time is to use the Heath–Jarrow–Morton (HJM) model.

26.5 The Heath–Jarrow–Morton Model

We have seen several Markovian short rate models. The Markovian approach, albeit computationally efficient, has the disadvantage that it is difficult to model the behavior of yields and bond prices of different maturities. The alternative **yield curve approach** regards the whole term structure as the state of a process and directly specifies how it evolves [725].

The influential model proposed by Heath, Jarrow, and Morton is a forward rate model [437, 511]. It is also a popular model [17, 198]. The HJM model specifies the initial forward rate curve and the forward rate volatility structure, which describes the volatility of each forward rate for a given maturity date. Like the Black–Scholes option pricing model, neither risk preference assumptions nor the drifts of forward rates are needed [515].

26.5.1 Forward-Rate Process

Within a finite time horizon $[0, U]$, we take as given the time-zero forward rate curve $f(0, t)$ for $t \in [0, U]$. Because this curve is used as the boundary value at

$t = 0$, perfect fit to the observed term structure is automatic. The forward rates are driven by n stochastic factors. Specifically the forward rate movements are governed by the stochastic process

$$df(t, T) = \mu(t, T)\, dt + \sum_{i=1}^{n} \sigma_i(t, T)\, dW_i,$$

where μ and σ_i may depend on the past history of the Wiener processes.

Take the one-factor model

$$df(t, T) = \mu(t, T)\, dt + \sigma(t, T)\, dW_t. \tag{26.13}$$

This is an infinite-dimensional system because there is an equation for each T. One-factor models seem to perform better than multifactor models empirically [19]. When is the bond market induced by forward rate model (26.13) arbitrage free in that there exists an equivalent martingale measure? For this to happen, there must exist a process $\lambda(t)$ such that for all $0 \leq t \leq T$, the drift equals

$$\mu(t, T) = \sigma(t, T) \int_t^T \sigma(t, s)\, ds + \sigma(t, T)\lambda(t). \tag{26.14}$$

The process $\lambda(t)$, which may depend on the past history of the Wiener process, is the market price of risk. Substitute this condition into Eq. (26.13) to yield the following arbitrage-free forward rate dynamics:

$$df(t, T) = \left[\sigma(t, T) \int_t^T \sigma(t, s)\, ds + \sigma(t, T)\lambda(t) \right] dt + \sigma(t, T)\, dW_t. \tag{26.15}$$

The market price of risk enters only into the drift. The short rate follows

$$dr(t) = \left[\sigma(t, t)\lambda(t) + \left. \frac{\partial f(t, T)}{\partial T} \right|_{T=t} \right] dt + \sigma(t, t)\, dW_t. \tag{26.16}$$

A unique equivalent martingale measure can be established under which the prices of interest rate derivatives do not depend on the market prices of risk. This fundamental result is summarized below.

THEOREM 26.5.1 *Assume that π is a martingale measure for the bond market and that the forward rate dynamics under π is given by $df(t, T) = \mu(t, T)\, dt + \sum_{i=1}^{n} \sigma_i(t, T)\, dW_i$. The volatility functions $\sigma_i(t, T)$ may depend on $f(t, T)$. (1) For all $0 < t \leq T$,*

$$\mu(t, T) = \sum_{i=1}^{n} \sigma_i(t, T) \int_t^T \sigma_i(t, u)\, du \tag{26.17}$$

holds under π almost surely. (2) The bond price dynamics under π is given by

$$\frac{dP(t, T)}{P(t, T)} = r(t)\, dt + \sum_{i=1}^{n} \sigma_{p,i}(t, T)\, dW_i,$$

where $\sigma_{p,i}(t, T) \equiv -\int_t^T \sigma_i(t, u)\, du$. (Choosing the volatility function $\sigma(t, T)$ of the forward rate dynamics under π uniquely determines the drift parameters under π and the prices of all claims.)

To use the HJM model, we first pick $\sigma(t, T)$. This is the modeling part. The drift parameters are then determined by Eq. (26.17). Now fetch today's forward rate curve $\{ f(0, T), T \geq 0 \}$ and integrate it to obtain the forward rates:

$$f(t, T) = f(0, T) + \int_0^t \mu(s, T) \, ds + \int_0^t \sigma(s, T) \, dW_s. \tag{26.18}$$

Compute the future bond prices by $P(t, T) = e^{-\int_t^T f(t,s) \, ds}$ if necessary. European-style derivatives can be priced by simulating many paths and taking the average.

From Eqs. (26.17) and (26.18),

$$r(t) = f(t, t) = f(0, t) + \int_0^t df(s, t)$$

$$= f(0, t) + \int_0^t \sigma_p(s, t) \sigma(s, t) \, ds + \int_0^t \sigma(s, t) \, dW_s,$$

where $\sigma_p(s, t) \equiv \int_s^t \sigma(s, u) \, du$. Differentiate with respect to t and note that $\sigma_p(t, t) = 0$ to obtain

$$dr(t) = \frac{\partial f(0, t)}{\partial t} \, dt + \left\{ \int_0^t \left[\sigma_p(s, t) \frac{\partial \sigma(s, t)}{\partial t} + \sigma(s, t)^2 \right] ds \right\} dt$$

$$+ \left[\int_0^t \frac{\partial \sigma(s, t)}{\partial t} \, dW_s \right] dt + \sigma(t, t) \, dW_t. \tag{26.19}$$

Because the second and the third terms on the right-hand side depend on the history of σ_p and/or dW, they can make r non-Markovian. In the special case in which $\sigma_p(t, T) = \sigma(T - t)$ for a constant σ, the short rate process r becomes Markovian and Eq. (26.19) reduces to

$$dr = \left[\frac{\partial f(0, t)}{\partial t} + \sigma^2 t \right] dt + \sigma \, dW.$$

Note that this is the continuous-time Ho–Lee model (review Exercise 26.2.10).

▶ **Exercise 26.5.1** What would $\mu(t, T)$ be if $\sigma(t, T) = \sigma e^{-\kappa(T-t)}$?

▶ **Exercise 26.5.2** Prove Eq. (26.16). (Hint: Use Eq. (26.15).)

▶ **Exercise 26.5.3** Consider the forward rate dynamics in Eq. (26.13) and define $r(t, \tau) \equiv f(t, t + \tau)$. Verify that

$$dr(r, \tau) = \left[\frac{\partial r(t, \tau)}{\partial \tau} + \mu(t, t + \tau) \right] dt + \sigma(t, t + \tau) \, dW_t.$$

Note that τ denotes the time to maturity.

Fixed-Income Option Pricing

For one-factor HJM models under the risk-neutral probability, the bond price process is

$$\frac{dP(t, T)}{P(t, T)} = r(t) \, dt + \sigma_p(t, T) \, dW_t. \tag{26.20}$$

For a European option to buy at time s and at strike price X a zero-coupon bond maturing at time $T \geq s$, its value at time t is

$$C_t = P(t, T) N(d_t) - XP(t, s) N(d_t - \sigma_t),\tag{26.21}$$

where

$$d_t \equiv \sigma_t^{-1} \ln\left(\frac{P(t, T)}{XP(t, s)}\right) + \frac{\sigma_t}{2}$$

and $\sigma_t^2 \equiv \int_t^s [\sigma \sigma_p(\tau, T) - \sigma_p(\tau, s)]^2 \, d\tau$ [164].

26.5.2 Markovian Short Rate Models

Markovian short rate models often simplify numerical procedures. First, the term structure at any time t is determined by t, the maturity, and the short rate at t. Second, the short rate dynamics can often be modeled by a combining tree. The HJM model's short rate process is usually non-Markovian. Under certain restrictions on volatility, however, the short rate contains all information relevant for pricing.

EXAMPLE 26.5.2 Suppose the volatility function is $\sigma(t, T) = \sigma$. Then the drift under π is $\mu(t, T) = \sigma \int_t^T \sigma \, ds = \sigma^2 (T - t)$ by Theorem 26.5.1. The forward rate process is $df(t, T) = \sigma^2 (T - t) \, dt + \sigma \, dW$, which is the Ho–Lee model (see Exercise 26.2.9). Integrate the above for each T to yield

$$f(t, T) = f(0, T) + \int_0^t \sigma^2 (T - s) \, ds + \int_0^t \sigma \, dW_s$$

$$= f(0, T) + \sigma^2 t \left(T - \frac{t}{2}\right) + \sigma W(t).$$

In particular, the short rate $r(t) \equiv f(t, t)$ is $r(t) = f(0, t) + (\sigma^2 t^2/2) + \sigma W(t)$. Because

$$\int_t^T f(t, s) \, ds = \int_t^T f(0, s) \, ds + \frac{\sigma^2}{2} t T(T - t) + \sigma(T - t) W(t),$$

the bond price is

$$P(t, T) = \frac{P(0, T)}{P(0, t)} e^{-(\sigma^2/2) t T(T-t) - \sigma(T-t) W(t)}.$$

Combining this with the short rate process above, we have

$$P(t, T) = \frac{P(0, T)}{P(0, t)} e^{f(0,t)(T-t) - (\sigma^2/2) t(T-t)^2 - r(t)(T-t)},$$

which expresses the bond price in terms of the short rate.

EXAMPLE 26.5.3 For the Hull–White model $dr = (\theta(t) - ar) \, dt + \sigma \, dW$, where $a, \sigma > 0$, the HJM formulation is $df(t, T) = \mu(t, T) \, dt + \sigma e^{-a(T-t)} \, dW$. The volatility structure of the forward rate, $\sigma(t, T) = \sigma e^{-a(T-t)}$, is exponentially decaying. The entire set of forward rates at any time t can be recovered from the short rate as follows:

$$f(t, T) = f(0, T) + e^{-a(T-t)}[r(t) - f(0, t) + \beta(t, T) \phi(t)],$$

where

$$\beta(t, T) \equiv \frac{1}{a}\big[1 - e^{-a(T-t)}\big], \qquad \phi(t) \equiv \frac{\sigma^2}{2a}(1 - e^{-2at}).$$

The bond prices at time t and the state variable $r(t)$ are also related:

$$P(t, T) = \frac{P(0, T)}{P(0, t)} e^{-\beta(t,T)[r(t)-f(0,t)]-[\beta(t,T)^2\phi(t)/2]}.$$

The duration of a zero-coupon bond at time t that matures at time T is $\beta(t, T)$. Finally,

$$dr(t) = \left[-a[r(t) - f(0, t)] + \frac{\partial f(0, t)}{\partial t} + \phi(t) \right] dt + \sigma\, dW_t$$

establishes the dynamics of the short rate.

EXAMPLE 26.5.4 When the σ_p in Eq. (26.20) is nonstochastic, $r(t)$ is Markovian if and only if $\sigma_p(t, T)$ has the functional form $x(t)[\, y(T) - y(t)\,]$. The process for r then has the general form of the extended Vasicek model $dr = [\, \theta(t) - a(t)r\,]\, dt + \sigma(t)\, dW$ [477].

▶ **Exercise 26.5.4** (1) Price calls on a zero-coupon bond under the Ho–Lee model. (2) Price calls on a zero-coupon bond under the Hull–White model.

▶ **Exercise 26.5.5** Derive the Hull–White model's volatility structure.

▶ **Exercise 26.5.6** Verify that the model in Example 26.5.3 converges to the Ho–Lee model as $a \to 0$.

26.5.3 Binomial Approximation

Let Δt denote the duration of a period. The initial forward rate curve is $\{\, f(0, T), T = 0, \Delta t, 2\Delta t, \dots\,\}$, and $f(t, T)$ is the forward rate implied at time t for the time period $[\, T, T + \Delta t\,]$. During the next time period Δt, new information may arrive and cause the term structure to move. Let $f(t + \Delta t, T)$ be the new forward rate. The change $f(t + \Delta t, T) - f(t, T)$ depends on the forward rate, its maturity date T, and a host of other factors. See Fig. 26.8 for illustration.

Consider the binomial process in the actual economy:

$$f(t + \Delta t, T) = \begin{cases} f(t, T) + \mu(t, T)\, \Delta t + \sigma(t, T)\sqrt{\Delta t} & \text{with probability } q \\ f(t, T) + \mu(t, T)\, \Delta t - \sigma(t, T)\sqrt{\Delta t} & \text{with probability } 1 - q \end{cases},$$

where $q = 1/2 + o(\Delta t)$. The mean and the variance of the change in forward rates are $\mu(t, T)\, \Delta t + o(\Delta t)$ and $\sigma(t, T)^2 \Delta t + o(\Delta t)$, respectively. Convergence is guaranteed under mild conditions [434]. To make the process arbitrage free, a probability measure p must exist under which all claims can be priced as if the local expectations theory holds. We may assume that $p = 1/2$ for ease of computation. It can be shown that

$$\mu(t, T)\, \Delta t = \tanh(x)\, \sigma(t, T)\sqrt{\Delta t}, \tag{26.22}$$

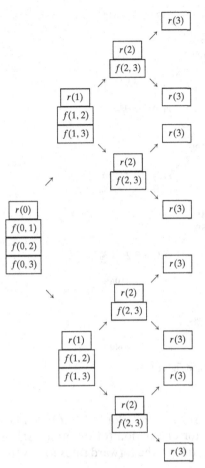

Figure 26.8: Binomial HJM model. For brevity, we use $f(i, j)$ to denote the forward rate for the period $[j\,\Delta t, (j+1)\,\Delta t]$ as seen from time $i\,\Delta t$. As always, $r(i) \equiv f(i,i)$ is the short rate. The binomial tree as a rule does not combine.

where

$$x \equiv \sqrt{\Delta t}\int_{t+\Delta t}^{T}\sigma(t, u)\,du, \quad \tanh(x) \equiv \frac{e^x - e^{-x}}{e^x + e^{-x}}.$$

This makes the price of any interest rate claim arbitrage free [746].

EXAMPLE 26.5.5 Consider $\sigma(t, T) = \sigma r(t)^\gamma e^{-\kappa(T-t)}$ for the forward rate volatility. Then

$$\mu(t, T)\,\Delta t = \tanh\left(\frac{\sigma r(t)^\gamma}{\kappa}\left[e^{-\kappa\Delta t} - e^{-\kappa(T-t)}\right]\sqrt{\Delta t}\right)\sigma r(t)^\gamma e^{-\kappa(T-t)}\sqrt{\Delta t}.$$

In particular, $\sigma(t, t)$ is the short rate volatility. This volatility is constant as in the Vasicek model when $\gamma = 0$. For $\gamma = 0.5$, the volatility is like that of the CIR model, whereas for $\gamma = 1$, the volatility is proportional to the short rate's level as in the BDT model. Take a three-period model with $\Delta t = 1$, $\gamma = 1$, $\kappa = 0.01$, and $\sigma = 0.3$. Then

$$\mu(t, T)\,\Delta t = \tanh\left(30 \times r(t)\left[e^{-0.01} - e^{-0.01\times(T-t)}\right]\right) \times 0.3 \times r(t)\,e^{-0.01\times(T-t)}.$$

Given a flat initial term structure at 5%, selected forward rates are shown in Fig. 26.9. The term structure at each node can be determined from the set of forward rates there.

▶ **Exercise 26.5.7** When pricing a derivative on bonds under the HJM model, does the tree have to be built over the life of the longer-term underlying bond or just over the life of the derivative?

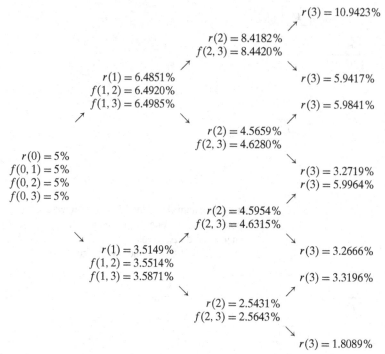

Figure 26.9: Sample binomial HJM tree. See Example 26.5.5 and Fig. 26.9.

A More Formal Setting

Assume that trades occur at times $\Delta t, 2\Delta t, 3\Delta t, \ldots$. As before, $f(t, T)$ denotes the forward interest rate contracted at time t for one period (of duration Δt) of borrowing or lending at time T. At each trading time t, the forward rates follow the stochastic difference equation:

$$f(t + \Delta t, T) - f(t, T) = \mu(t, T)\,\Delta t + \sigma(t, T)\,\xi(t + \Delta t)\sqrt{\Delta t}, \qquad (26.23)$$

where $\xi(\cdot)$ are independent random variables with zero mean and unit variance. The value of $\xi(t)$ is realized before the trading at time t but after time $t - \Delta t$. The drift μ and the diffusion σ can be functions of the current or past values of forward rates. There is a single source of uncertainty (or factor), represented by ξ, that influences forward rates of all maturities. To permit forward rates of different maturities to vary independently, simply add additional sources of uncertainty to Eq. (26.23).

The price at time t of a discount bond of maturity at T is given by

$$P(t, T) = \exp\left[-\sum_{i=t/\Delta t}^{(T/\Delta t)-1} f(t, i\Delta t)\,\Delta t \right], \quad t = 0, \Delta t, 2\Delta t, \ldots$$

Substitute Eq. (26.23) into the preceding equation to yield

$$P(t, T) = \exp\left[-\sum_{i=t/\Delta t}^{(T/\Delta t)-1} \left(f(0, i\Delta t) + \sum_{j=0}^{i-1} \{ \mu(j\Delta t, i\Delta t)\,\Delta t \right.\right.$$

$$\left.\left. + \sigma(j\Delta t, i\Delta t)\,\xi[(j+1)\,\Delta t]\sqrt{\Delta t} \} \right)\Delta t \right].$$

Now use the money market account $M(t)$ as numeraire and assume that it trades. Thus

$$M(0) = 1,$$

$$M(t) = \exp\left[\sum_{i=0}^{(t/\Delta t)-1} r(i\,\Delta t)\,\Delta t \right] = \exp\left[\sum_{i=0}^{(t/\Delta t)-1} f(i\,\Delta t, i\,\Delta t)\,\Delta t \right].$$

To avoid arbitrage, there must exist a probability measure π under which $P(t, T)/M(t)$ is a martingale; in particular,

$$P(t, T) = E_t^\pi[\, P(t + \Delta t, T)\, P(t, t + \Delta t)\,]. \tag{26.24}$$

The continuous-time limit under π yields $\mu(t, T) = \sigma(t, T) \int_t^T \sigma(t, u)\, du$.

We can take $\xi(\cdot) = \pm 1$, each with a probability of one-half, and then adjust the numbers on the nodes accordingly. This is what we did with the binomial interest rate tree, the Ho–Lee model, and the BDT model by fixing the risk-neutral probability to be 0.5. Because the martingale condition holds only as $\Delta t \to 0$, greater numerical accuracy may result by requiring it to hold under the binomial framework [19]. This leads to

$$\Delta t \int_{t+\Delta t}^T \mu(t, u)\, du = \ln \frac{e^x + e^{-x}}{2}, \tag{26.25}$$

where $x \equiv \sqrt{\Delta t} \int_{t+\Delta t}^T \sigma(t, u)\, du$. In Eq. (26.25) the convention is that $\mu(t, u)$ is constant for u between two adjacent trading times. The drifts $\mu(\cdot, \cdot)$ can be computed from Eq. (26.25) by solving them iteratively for $T = t + i\,\Delta t, i = 1, 2, \dots$.

▶ **Exercise 26.5.8** Verify Eq. (26.25).

▶ **Exercise 26.5.9** Show that Eqs. (26.22) and (26.25) are equivalent in the limit.

26.6 The Ritchken–Sankarasubramanian Model

For the Ritchken–Sankarasubramanian (RS) model proposed by Ritchken and Sankarasubramanian [747], the forward rate volatility $\sigma(t, T)$ is related to the short rate volatility $\sigma(t, t)$ through an exogenously provided deterministic function $\kappa(x)$ by

$$\sigma(t, T) = \sigma(t, t)\, e^{-\int_t^T \kappa(x)\, dx}.$$

No particular restrictions are imposed on the short rate volatility $\sigma(t, t)$. This model precludes certain volatility structures. For instance, it does not permit volatilities of different forward rates to fluctuate according to different spot rates [600].

The term structure dynamics can be made Markovian with respect to two state variables. Bond prices – hence forward rates as well – at time t can be expressed in terms of the price information at time zero, the short rate $r(t)$, and a path-dependent statistic that represents the accumulated variance for the forward rate up to time t:

$$\phi(t) \equiv \int_0^t \sigma(u, t)^2\, du.$$

Note that $\phi(t)$ depends on the path the rate takes from time zero to t. In fact,

$$P(t, T) = \frac{P(0, T)}{P(0, t)}\, e^{-\beta(t,T)[r(t) - f(0,t)] - \beta(t,T)^2 \phi(t)/2}, \tag{26.26}$$

where $\beta(t, T) \equiv \int_t^T e^{-\int_t^u \kappa(x)\,dx}\,du$. The risk-neutral process follows

$$dr(t) = \mu(r, \phi, t)\,dt + \sigma(t, t)\,dW,$$

$$d\phi(t) = [\sigma(t, t)^2 - 2\kappa(t)\,\phi(t)]\,dt, \tag{26.27}$$

where

$$\mu(r, \phi, t) \equiv \kappa(t)[f(0, t) - r(t)] + \frac{df(0, t)}{dt} + \phi(t). \tag{26.28}$$

Because the short rate volatility could depend on both state variables $r(t)$ and $\phi(t)$, it may be expressed by $\sigma(r(t), \phi(t), t)$. Calibration is achieved by the appropriate choice of $\kappa(\cdot)$.

26.6.1 Binomial Approximation

Throughout this subsection, the forward rate volatility structure is given by

$$\sigma(t, T) = \sigma r(t)^\gamma e^{-\kappa(T-t)}.$$

(Note that the short rate volatility $\sigma(t, t)$ equals $\sigma r(t)^\gamma$.) Distant forward rates are therefore less volatile than near forward rates. For instance, with $\gamma = 0$, the extended Vasicek model results, and bond pricing formula (26.26) reduces to that for the extended Vasicek model (see Exercise 26.6.1). The CIR model emerges with $\gamma = 0.5$.

The forward rate at time t is given by

$$f(t, T) = f(0, T) + e^{-\kappa(T-t)}[r(t) - f(0, t) + \beta(t, T)\,\phi(t)],$$

where $\beta(t, T) \equiv [1 - e^{-\kappa(T-t)}]/\kappa$. When $\gamma \neq 0$, the $\phi(t)$ variable is determined by the path of rate values over the period $[0, t]$. Although the knowledge of $r(t)$ alone is insufficient to characterize the term structure at time t, that of $r(t)$ *and* $\phi(t)$ suffices. At time t, the duration of a zero-coupon bond maturing at time T is $\beta(t, T)$ [746].

The binomial tree model can be developed much as we did with the CIR model. First, take the following transformation to a form that has constant volatility:

$$Y(t) = \int \frac{1}{\sigma(r(t), \phi(t), t)}\,dr(t).$$

Let $r(t) = h(Y(t))$ be the inverse function. Then

$$dY(t) = m(Y, \phi, t)\,dt + dW_t,$$

$$d\phi(t) = [\sigma(r(t), \phi(t), t)^2 - 2\kappa(t)\,\phi(t)]\,dt,$$

where

$$m(Y, \phi, t) \equiv \frac{\partial Y(t)}{\partial t} + \mu(r, \phi, t)\frac{\partial Y(t)}{\partial r(t)} + \frac{1}{2}\sigma(r(t), \phi(t), t)^2\frac{\partial^2 Y(t)}{\partial r(t)^2}.$$

For example, for the **proportional model** with $\gamma = 1$, $Y(t) = (1/\sigma)\ln r(t)$ and

$$\sigma(r(t), \phi(t), t)^2 = \sigma^2 e^{2\sigma Y(t)}$$

$$m(Y, \phi, t) = \frac{1}{\sigma}\left\{\left[\kappa\big(f(0, t) - e^{\sigma Y(t)}\big) + \phi(t) + \frac{df(0, t)}{dt}\right]e^{-\sigma Y(t)} - \frac{\sigma^2}{2}\right\}.$$

Now, partition the interval $[0, T]$ into n periods each of length $\Delta t \equiv T/n$ and set up a combining binomial tree for Y. Initially, $Y_0 = Y(0)$ and $\phi_0 = 0$. During each time increment, the approximating Y_i can move to one of two values,

$$
Y_i \underset{Y_{i+1}^- = Y_i - \sqrt{\Delta t}}{\overset{Y_{i+1}^+ = Y_i + \sqrt{\Delta t}}{<}}
$$

Let $r_i \equiv h(Y_i)$. Each node on the tree has two state variables. The tree evolves thus:

$$
r_i, \phi_i \underset{r_{i+1}^- = h(Y_{i+1}^-), \phi_{i+1}}{\overset{r_{i+1}^+ = h(Y_{i+1}^+), \phi_{i+1}}{<}}
$$

where $\phi_{i+1} = \phi_i + (\sigma^2 r_i^{2\gamma} - 2\kappa\phi_i)\,\Delta t$ by Eq. (26.27). The ϕ values are the same in both the up and the down nodes and have to be derived by forward induction. The probability of an up move to (r_{i+1}^+, ϕ_{i+1}) given (r_i, ϕ_i) is

$$
p(r_i, \phi_i, i) \equiv \frac{\mu(r_i, \phi_i, i\,\Delta t)\,\Delta t + r_i - r_{i+1}^-}{r_{i+1}^+ - r_{i+1}^-},
$$

where $\mu(\cdot, \cdot, \cdot)$ is the drift term in Eq. (26.28) with $\kappa(x) = \kappa$. Pricing can be carried out with backward induction, and the term structure at each node can be computed by Eq. (26.26).

EXAMPLE 26.6.1 For the proportional model, the transform is $r(t) = h(Y(t)) = e^{\sigma Y(t)}$. Let $r_i \equiv e^{\sigma Y_i}$. The binomial tree evolves according to

$$
r_i, \phi_i \underset{r_i e^{-\sigma\sqrt{\Delta t}}, \phi_{i+1}}{\overset{r_i e^{\sigma\sqrt{\Delta t}}, \phi_{i+1}}{<}}
$$

where $\phi_{i+1} = \phi_i + (\sigma^2 r_i^2 - 2\kappa\phi_i)\,\Delta t$.

The size of the tree as described grows exponentially in n. In fact, the number of ϕ values at a node equals the number of distinct paths leading to it from the root because the ϕ value is path dependent. This observation should be clear from the binomial process. One remedy is to keep only the maximum and the minimum ϕ values at each node, interpolating m intermediate values linearly on demand (see Subsection 11.7.1 for the same idea). The model is expected to converge for sufficiently large n and m.

▶ **Exercise 26.6.1** Show that $\phi(t) = \sigma^2(1 - e^{-2\kappa t})/(2\kappa)$ for the extended Vasicek model.

Additional Reading

See [858] for a survey of interest rate models, [754] for a comparison of models, [32, 78, 378, 670] for theoretical analysis, [47, 139, 677] for empirical studies of the short rate and the HJM models, and [276, 507, 779, 780, 850] for more information on derivatives pricing. References [40, 183, 470, 510, 681, 731] also cover interest rate models. Consult [359] for an empirical study of the Ho–Lee model. See [74, 625] for calibration of the BDT model and [479, 483, 735] for that of the Hull–White model. A two-factor extended Vasicek model is proposed in [480]. For the HJM model, see

[58, 116, 165, 166, 435, 437] for theoretical analysis, [163, 197, 477, 499, 515, 747] for the conditions that make the short rate Markovian, [682] for risk management issues, [510] for estimating volatilities by use of principal components, [436, 510] for discrete-time models, and [160] for the Monte Carlo approach to pricing American-style fixed-income options. For the RS model, see [600, 746, 747] for additional numerical ideas and [92] for an empirical study. Reference [77] asks whether an interest rate model generates term structures within the functionals (say, polynomials) used for fitting the term structure; in particular, it shows that the Ho–Lee model and the Hull–White model are inconsistent with the Nelson–Siegel scheme. Volatility structures are investigated in [136, 393, 722]. A universal trinomial tree algorithm for any Ito process is proposed in [187].

NOTES

1. Contrast this with the lognormal model in Exercise 23.2.3.
2. The value of p can be chosen rather arbitrarily because, in the limit, only the volatility matters [436].
3. Under the premise that the forward rate tends to a constant for large t, it follows from Eq. (26.6) that $r(t) - r(t-1)$ is either essentially zero or $|\ln \delta| > 0$ [300].

Fixed-Income Securities

> Neither a borrower nor a lender be.
>
> Shakespeare (1564–1616), *Hamlet*

Bonds are issued for the purpose of raising funds. This chapter concentrates on bonds, particularly those with embedded options. It ends with a discussion of key rate durations.

27.1 Introduction

A bond can be secured or unsecured. A **secured** issue is one for which the issuer pledges specific assets that may be used to pay bondholders if the firm defaults on its payments. Many bond issues are **unsecured**, however, with no specific assets acting as collateral. Long-term unsecured issues are called **debentures**, whereas short-term unsecured issues such as **commercial paper** are referred to as **notes**.

It is common for a bond issue to include in the indenture provisions that give either the bondholder and/or the issuer an option to take certain actions against the other party. The bond **indenture** is the master loan agreement between the issuer and the investor. A common type of embedded option in a bond is a call feature, which grants the issuer the right to retire the debt, fully or partially, before the maturity date. An issue with a put provision, as another example, grants the bondholder the right to sell the issue back to the issuer. Here the advantage to the investor is that if interest rates rise after the issue date, reducing the bond's price, the investor can force the issuer to redeem the bond at, say, par value. A convertible bond (CB) is an issue giving the bondholder the right to exchange the bond for a specified number of shares of common stock. Such a feature allows the bondholder to take advantage of favorable movements in the price of the issuer's stock. An **exchangeable bond** allows the bondholder to exchange the issue for a specified number of common stock shares of a corporation different from the issuer of the bond. Some bonds are issued with warrants attached as part of the offer. A warrant grants the holder the right to purchase a designated security at a specified price.

27.2 Treasury, Agency, and Municipal Bonds

Strictly speaking, only default-free bonds without options deserve the name "fixed-income security." In the U.S. market, almost all fixed-income securities in this narrow

Outstanding U.S. Treasury securities (U.S. $ billions)							
1980	616.4	1985	1,360.2	1990	2,195.8	1995	3,307.2
1981	683.2	1986	1,564.3	1991	2,471.6	1996	3,459.7
1982	824.4	1987	1,675.0	1992	2,754.1	1997	3,456.8
1983	1,024.4	1988	1,821.3	1993	2,989.5	1998	3,355.5
1984	1,176.6	1989	1,945.4	1994	3,126.0	1999	3,281.0

Figure 27.1: Outstanding U.S. Treasury securities 1980–1999. Prices are quoted in 1/32 of a percent. Source: *U.S. Treasury*.

sense are issued by the Treasury. Nevertheless, in reality the term fixed-income security is used rather loosely and has come to describe even bonds with uncertain payments. Figure 27.1 tabulates the U.S. Treasury securities in terms of outstanding volume, and Figure 27.2 provides a view of the immense U.S. Treasuries market.

In early 1996 the Treasury announced plans to issue bonds whose nominal payments are indexed to inflation so that their payments are fixed in real terms [147]. The index for measuring the inflation rate is the nonseasonally adjusted U.S. City Average All Items Consumer Price Index for All Urban Consumers (CPI-U) published monthly by the Bureau of Labor Statistics. When the bond matures, the principal will be adjusted to reflect all the inflation there is during the life of the bond. (The British government issued index-linked securities in 1981 [135].) On January 29, 1997, U.S.$7 billion of 3⅜% 10-year inflation-indexed Treasury notes were auctioned (see Fig. 27.3) [759].[1] The interest rate set at auction will remain fixed throughout the term of the security. Semiannual interest payments will be based on the inflation-adjusted principal at the time the interest is paid. At maturity, the securities will be redeemed at the greater of their inflation-adjusted principal or their par amount at original issue.

Federal agency debt can be issued by Federal agencies, which are direct arms of the U.S. government, or various **government-sponsored enterprises** (GSEs), which were created by Congress to fund loans to such borrowers as homeowners, farmers, and students [395]. GSEs are privately owned, publicly chartered entities that raise funds in the marketplace. Examples include Federal Home Loan Banks (FHLBanks), the Federal National Mortgage Association (FNMA or "Fannie Mae"), the Federal Home Loan Mortgage Corporation (FHLMC or "Freddie Mac"), and the Student Loan Marketing Association (SLMA or "Sallie Mae"). Although there are no Federal guarantees on the securities issued by the GSEs, the perception that the government would ultimately cover any defaults causes the yields on these securities to be below those on most corporate securities. This may change in the future, however [404].

Municipal bonds are fixed-income securities issued by state, state authorities, or local governments to finance capital improvements or support a government's general financing needs. There are two major categories of municipal bonds: revenue bonds and general obligation bonds. **Revenue bonds** are issued to raise funds for a particular project such as a toll road or a hospital that is projected to generate enough income to pay principal and interest to bondholders. **General obligation bonds** are backed by the taxing power of the issuer such as city, county, or state. They are often considered less risky than revenue bonds. Investors are attracted to municipal bonds because the interest is exempt from federal income taxes and, in some cases, state

TREASURY BONDS, NOTES & BILLS
Monday, March 20, 1995

GOVT. BONDS & NOTES

Rate	Maturity Mo/Yr	Bid	Asked	Chg.	Ask Yld.
37/8	Mar 95n	99:29	99:31	−1	5.05
83/8	Apr 95n	100:03	100:05	−1	5.79
37/8	Apr 95n	99:24	99:26	.....	5.54
57/8	May 95n	99:31	100:01	.....	5.55
81/2	May 95n	100:12	100:14	.....	5.38
103/8	May 95	100:20	100:22	.....	5.53
	. . .				
71/8	Feb 23	95:26	95:28	−11	7.48
61/4	Aug 23	85:21	85:23	−12	7.47
71/2	Nov 24	100:14	100:16	−15	7.46
75/8	Feb 25	102:23	102:25	−10	7.39

U.S. TREASURY STRIPS

Mat.	Type	Bid	Asked	Chg.	Ask Yld.
May 95	ci	99:04	99:04	+1	5.84
May 95	np	99:04	99:04	+1	5.95
Aug 95	ci	97:22	97:23	.....	5.82
Aug 95	np	97:19	97:19	.....	6.10
		. . .			
Nov 24	ci	11:17	11:20	−4	7.39
Nov 24	bp	11:21	11:24	−4	7.35
Feb 25	ci	11:25	11:29	−3	7.25
Feb 25	bp	12:04	12:08	−3	7.15

TREASURY BILLS

Maturity	Days to Mat.	Bid	Asked	Chg.	Ask Yld.
Mar 23 '95	1	5.43	5.33	+0.42	5.40
Mar 30 '95	8	4.96	4.86	−0.04	4.93
Apr 06 '95	15	5.61	5.51	+0.04	5.60
Apr 13 '95	22	5.50	5.40	+0.01	5.49
		. . .			
Dec 14 '95	267	5.93	5.91	−0.01	6.21
Jan 11 '96	295	5.95	5.93	−0.01	6.25
Feb 08 '96	323	5.96	5.94	−0.01	6.28
Mar 07 '96	351	5.98	5.96	.....	6.34

Figure 27.2: Treasuries quotations. Colons represent 32nds. The final column for T-bills shows the annualized BEYs as computed by Eq. (3.10). T-bill quotes are in hundredths and are on a discount basis. All yields are based on the asked quote. n, Treasury note; ci, stripped coupon interest; bp, T-bond, stripped principal; tt, T-note, stripped principal. Source: *Wall Street Journal*, March 21, 1995.

and local taxes as well. Municipal bonds are quoted in percent of par and 1/32 of a percent like T-bonds.

27.3 Corporate Bonds

Both stock and bond markets offer an efficient way for corporations to raise capital. Bonds have the advantage of not diluting the stockholder's equity. Compared with

TREASURY BONDS, NOTES & BILLS
Thursday, January 7, 1999

. . .

INFLATION-INDEXED TREASURY SECURITIES

Rate	Mat.	Bid/Asked	Chg.	*Yld.	Accr. Prin.
3.625	07/02	99-15/16	−01	3.768	1024
3.375	01/07	96-23/24	−02	3.850	1035
3.625	01/08	98-11/12	−05	3.831	1015
3.625	04/28	98-03/04	+17	3.734	1014

*-Yld. to maturity on accrued principal.

Figure 27.3: Inflation-indexed Treasuries quotations. Source: *Wall Street Journal*, January 8, 1999.

bank loans, bonds often allow corporations to borrow at a lower interest rate than the rates available from their banks. With bonds, a corporation also borrows money at a fixed rate for a longer term than it could at a bank because most banks do not make long-term fixed-rate loans.

Corporate bonds are quoted in points and eighths of a point. A bond with $10,000 par value quoted at 95$6/8$, for example, has a price of $9575. Because each bond must be customized to reflect the concerns of both the issuer and the investors, corporate bonds are not standardized. Bondholders, by making loans to the issuer, are legally the issuer's creditors, not owners like stockholders.

27.3.1 Callable and Putable Bonds

The holder of a callable bond sells the issuer an option to purchase the bond from the time it is first callable until the maturity date. The position of the bondholder is therefore

long a callable bond = long a noncallable bond + sold a call option.

In terms of price, we have

callable bond price = noncallable bond price − call option price.   (27.1)

The issuer may be entitled to call the bond at the first call date and any time thereafter (**continuously callable**) or at the first call date and all subsequent coupon payment dates (**discretely callable**). The call price may also vary over time. Typically, there is an initial call protection period, after which the bond is callable with a call price that declines to par over its remaining life [329]. Take for example a bond with 20 years to maturity and callable in 5 years at 106. The bondholder is essentially long a hypothetical 20-year noncallable bond and short a call option granting the issuer the right to call the bond 5 years from now for a price of 106.

The issuer will call a bond when the bond yield in the market for a new issue net of the underwriting fees and tax is lower than the current issue's coupon rate. Whether to exercise the option hinges on the future bond payments if the bond is not called. The price of the callable bond when it is callable will remain near its call

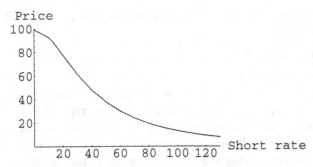

Figure 27.4: Price compression of callable bonds. The five-year 10% bond callable at par is priced by the CIR model.

price when interest rates are low, a phenomenon known as **price compression**. This is because the bond is likely to be called (see Fig. 27.4).

The holder of a **putable bond** has the right to sell the bond to the issuer at a designated price and time [238, 536]. The position of a putable bondholder can be described as

long a putable bond = long a nonputable bond + long a put option.

The price of a putable bond thus is

putable bond price = nonputable bond price + put option price.

A bond may carry both call and put options. If these options are exercisable at par on the same date, the bond is usually called an **extendible bond** [371].

27.3.2 Bonds with Sinking-Fund Provisions

The scheduled principal payments of a bond with sinking-fund provisions are spread out over many years. Two methods can be used to retire the required principal amount. Either the issuer can purchase the required amount in the open market and deliver the bonds to the trustee or it can call the required amount at par by random selection. Which method to execute depends on the prevailing interest rates. If rates are high, the issuer will choose to satisfy its sinking-fund requirement by market purchases. If rates are low, however, par calls will be chosen. When a sinking fund and a call option are both included in a bond issue, there are situations in which it is optimal to call only part of an issue. Almost all sinking-fund bond issues contain call options [848].

▶ **Exercise 27.3.1** Sketch a method to price a callable bond with sinking-fund provisions.

27.3.3 Convertible Bonds

CBs grant the bondholder the right to acquire the stock of the issuing corporation under specific conditions. The CB contract will state either a conversion ratio or a **conversion price**. A conversion ratio, we recall, specifies the number of shares to be obtained through conversion. The ratio is always adjusted proportionately for stock splits and stock dividends. Alternatively, the conversion ratio may be expressed in terms of a conversion price defined as

$$\text{conversion price} \equiv \frac{\text{par value of CB}}{\text{conversion ratio}}.$$

This price represents the cost per share through conversion. The conversion privilege may extend for all or only some portion of the bond's life. There are typically other embedded options in a CB, the most common being the right of the issuer to call or put the issue [328]. Conditions are usually imposed on the exercise of the options; for example, the stock price must be trading at a certain premium to the conversion price for the CB to be called. The contract may contain a **refix clause** in that the conversion price is set to the stock price if it is lower than the conversion price on the refix day [221].

The conversion value, we recall, is the value of the CB if it is converted immediately,

$$\text{conversion value} = \text{market price of stock} \times \text{conversion ratio}. \tag{27.2}$$

It is also called the **parity** in the market (but see Exercise 27.3.3). The market price of a CB must be at least its conversion value and **straight value** – the bond value without the conversion option. The price that an investor effectively pays for the stock if the CB is purchased and then converted is called the **market conversion price**:

$$\text{market conversion price} \equiv \frac{\text{market price of CB}}{\text{conversion ratio}}.$$

As the market conversion price cannot be lower than the market price of the stock, the bondholder pays a premium per share in the amount of

$$\text{market conversion price} - \text{market price of stock}.$$

The premium is usually expressed as a percentage of the market price of stock.

EXAMPLE 27.3.1 A CB with a par value of $10,000 and a conversion price of $80 would imply a conversion ratio of $10000/80 = 125$. Given the CB's quoted price, 103, the purchase price is $1.03 \times 10000 = 10300$ dollars. At the current stock price of $78, the premium per share is $10300/125 - 78 = 4.4$ dollars.

CBs exhibit the characteristics of both bond and stock. If the stock price is so low that the straight value is much higher than the conversion value, the CB behaves as a fixed-rate bond. On the other hand, if the stock price is so high that the conversion value is much higher than the straight value, the CB trades as an equity instrument. Between these two extremes, the CB trades as a hybrid security. Both the straight value and the parity act as a floor for the CB price, giving it a call-like behavior (see Fig. 27.5 for illustration).

It is simpler to price a CB on a *per-share basis*, that is, the hypothetical CB that can be converted for *one share*. The actual price, of course, equals the per-share price times the conversion ratio. As a reasonable first approximation, assume constant interest rates [847]. Then binomial tree algorithms for American options such as the one in Fig. 9.17 can be used to price CBs after the payoff function is modified. At maturity, the choice is between the conversion price plus coupon and the stock, whereas, before maturity, the choice is between the CB and the stock. See Fig. 27.6 for

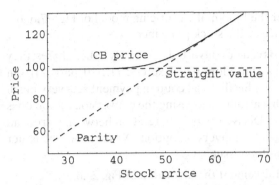

Figure 27.5: CB price vs. stock price. Both the straight value and the parity provide a floor. This particular CB has a conversion ratio of two and a conversion price of $50. The plot is similar to the one in Fig. 7.3 for calls on a stock.

an algorithm. Sensitivity measures can also be computed (review Chap. 10). Many of equity options' properties continue to hold for CBs. For example, higher stock price volatilities increase the CB's value.

▶ **Exercise 27.3.2** Use an arbitrage argument to show that a CB must trade for at least its conversion value.

▶ **Exercise 27.3.3** Can you find one fault with formula (27.2)?

▶ **Exercise 27.3.4** Prove that, much as with American calls, it never pays to convert the bond when the stock does not pay cash dividends and the interest rate remains constant.

Binomial tree algorithm for pricing convertible bonds on a stock that pays a known dividend yield:

input: $S, P, \sigma, t, n, \delta\,(1 > \delta > 0), m, r, c$;
real $R, p, u, d, C[n+1], v$;
integer i, j;
$R := e^{r(t/n)}$;
$u := e^{\sigma\sqrt{t/n}}; d := e^{-\sigma\sqrt{t/n}}$;
$p := (R - d)/(u - d) =:$;
for $(i = 0$ to $n)$ { $C[i] := \max\left(Su^{n-i}d^i(1-\delta)^m,\, P + c\right)$; }
for $(j = n - 1$ down to $0)$
    for $(i = 0$ to $j)$ {
        if [the period $(j, j+1]$ contains an ex-dividend date] $m := m - 1$;
        $v := (p \times C[i] + (1-p) \times C[i+1])/R$; // Backward induction.
        if [the period $[j, j+1)$ contains a coupon payment date] $v := v + c$;
        $C[i] := \max\left(Su^{j-i}d^i(1-\delta)^m,\, v\right)$;
    }
return $C[0]$; // On a per-share basis.

Figure 27.6: Binomial tree algorithm for CBs on a stock paying a dividend yield. S is the current stock price, m stores the total number of ex-dividend dates at or before expiration that occurs t years from now, δ is the dividend yield for each cash dividend, P is the conversion price, and c is the coupon payment per share. Note that the conversion price is already on a per-share basis. The partition should be fine enough that coupon payment dates and ex-dividend dates are separated by at least one period.

➤ **Exercise 27.3.5** Argue that under the binomial CB pricing model of Fig. 27.6, the CB price converges to the stock price as the stock price increases.

➤ **Exercise 27.3.6** Like warrants, CBs can be converted into newly issued shares; they are in fact equivalent under certain assumptions (see Exercise 11.1.10, part (2)). On a per-share basis, the conversion price plus the final coupon payment acts very much like the strike price. But unlike with warrants, exercising the conversion option does not require paying the "strike price." Derive the pricing relation between European warrants and CBs with a European-style conversion option. Assume that the issuer pays no dividends and ignore the dilution issue.

➤ **Programming Assignment 27.3.7** Implement the algorithm in Fig. 27.6.

27.3.4 Notes

A floating-rate note has coupon payments pegged to the yield of a particular interest rate such as LIBOR or the yield on a Treasury security. The interest rate that the borrower pays is reset periodically. For example, the rate might be reset every 6 months to the current T-bill rate plus 100 basis points. Variations on floating-rate notes include call features, issued by the firm, and conversion features that allow the investor to transfer to a fixed-rate note. In addition, the coupon rate may have a floor or a cap. Put features, whereby the holder can redeem the investment at par at particular coupon payment dates or after some predetermined date, may also be present. Of course, if there is no spread, a floating-rate note will sell at par on reset dates.

Structured notes are securities in which the issuer sells a note and simultaneously enters into a swap or derivative transaction to eliminate its exposure to the customized terms of the note structure [237]. Each structured note is customized with unique features that match the preferences of the investor. Consider an investor who believes that the yield curve will flatten. A security designed to reflect that view could have its coupon payments linked to the shape of the yield curve. Thus the coupon on the note might be reset semiannually to a rate that depends on the yield spread between the 30-year and the 2-year Treasury yields.

27.4 Valuation Methodologies

Several valuation methodologies were mentioned for corporate bonds before: yield to worst, yield to call, yield to par call, yield spread, and static spread. This section covers additional methodologies.

27.4.1 The Static Cash Flow Yield Methodology

In the **static cash flow yield methodology**, the yield to maturity of a bond is compared with that of the on-the-run Treasury security with a similar maturity. The difference is the yield spread. Because the yield must make certain assumptions about the future cash flow, it is also called the (**static**) **cash flow yield**. There are two obvious problems with this methodology. First, the yields fail to account for the term structure of interest rates. Second, interest rate volatility may alter the cash flow of bonds with embedded options.

▶ **Exercise 27.4.1** Assume any stochastic discrete-time short rate model. Consider a risky corporate zero that is not currently in default. When the firm defaults, it stays in the default state until the maturity when the investor receives zero dollar. Let p_i denote the risk-neutral probability that the bond defaults at time i given that it has not defaulted earlier; the p_is depend on only the time, not the short rate. Prove that

$$\frac{\text{price of } n\text{-period corporate zero}}{\text{price of } n\text{-period Treasury zero}} = \prod_{i=1}^{n} p_i.$$

(This result generalizes Eq. (5.6) in the static world. The algorithm in Fig. 5.9 can be used to retrieve the p_is.)

27.4.2 The Option Pricing Methodology

The value of an embedded option is the price difference between the bond with the option feature and an otherwise identical bond without the option. This insight leads the option pricing methodology to decompose fixed-income securities with option features into an option and an option-free component. The callable bond is a quintessential example, as demonstrated in Eq. (27.1). For instance, consider a coupon bond with a maturity date of June 2015 and callable in June 2012 at 102. Its bondholder as of April 1, 2000, effectively owns a noncallable bond with 15 years and 2 months to maturity and is short a call option, granting the issuer the right to call away 3 years of cash flows beginning June 1, 2012, for a strike price of 102.

The **option pricing methodology** applies the option pricing framework such as the Black–Scholes model in Section 24.7 to estimate the option price. The binomial tree algorithm for American options in Fig. 9.13 can be modified to price the embedded call option. After the option is priced, the **implied noncallable bond price** is then calculated as the sum of the callable bond price and the call option price. Finally, the **option-adjusted yield** is calculated as the yield that makes the PV of the cash flow of the hypothetical noncallable bond equal the implied price. The callable bond is said to be priced fairly if its option-adjusted yield equals the yield for an "equivalent" noncallable bond. It is said to be rich (overvalued) if the option-adjusted yield is lower, and cheap (undervalued) if the option-adjusted yield is higher [325, 330].

This methodology suffers from several difficulties. The Black–Scholes model is not satisfactory for pricing fixed-income securities as argued in Section 24.7. And there may not exist a benchmark with which to compare the option-adjusted yield to get the yield spread. Finally, this methodology does not incorporate the shape of the yield curve, which affects the value of all interest-rate-sensitive securities [329].

▶ **Exercise 27.4.2** Argue that when interest rates rise, the price of a callable bond will not fall as much as the price of its noncallable component.

▶ **Exercise 27.4.3** For bonds with embedded options, traditional duration measures such as modified duration lose relevance because of cash flow uncertainties. The duration of a callable bond after the call option is adjusted for is commonly referred to as the **option-adjusted duration** (**OAD**) and is defined as $\text{OAD}_c \equiv -\frac{\partial \text{price}_c}{\partial y} / \text{price}_c$. (The subscript "c" refers to callable measures.) The convexity of a callable bond after the call option is adjusted for is commonly referred to as the **option-adjusted convexity** (**OAC**) and is defined as $\text{OAC}_c \equiv -\frac{\partial^2 \text{price}_c}{\partial y^2} / \text{price}_c$. (1) Prove that

$OAD_c = (price_{nc}/price_c) \times duration_{nc} \times (1 - \Delta)$, where $\Delta \equiv \partial(\text{call price})/\partial price_{nc}$ is the delta of the embedded call option. (The subscript "nc" refers to noncallable measures.) (2) Show that

$$OAC_c = \frac{price_{nc}}{price_c} \times [\, \text{convexity}_{nc} \times (1 - \Delta) - price_{nc} \times \Gamma \times (duration_{nc})^2 \,],$$

where $\Gamma \equiv \partial^2(\text{call price})/(\partial price_{nc})^2$ is the gamma of the embedded call option.

27.4.3 The Option-Adjusted Spread Methodology

The final methodology is the **option-adjusted spread** (**OAS**) [34, 323]. This popular approach takes into account the embedded option features of fixed-income securities and tackles the difficulties faced by the previous methods. Unlike the previous methods, OAS analysis does not attempt to predict a bond's redemption date. The binomial interest rate tree of Chap. 23 is used to illustrate the main ideas. Generalization to other models is straightforward. Specifically, we use the calibrated binomial interest rate tree in Fig. 23.8 and callable bonds as the basis for our numerical calculations. Recall that a spread is defined as the incremental return applied to every short rate on the tree. It measures the extent to which the bond's rate of return exceeds riskless returns. The OAS generalizes the spread concept to bonds with uncertain cash flows.

Given an OAS, the procedure to calculate the model price is essentially the same as the one for option-free bonds except that, at each node, the exercise of the call must be considered. If the PV of *future* cash flows as determined by backward induction at any node exceeds the call price, the call will be exercised and the lower call price becomes the bond value. This usually happens when interest rates are low.

Consider a 3-year callable bond with a market price of 99.696 and an annual coupon rate of 5%. The call provision is a discrete par call that may be exercised on any coupon payment date. When the call provision is exercised, it eliminates all coupon payments scheduled to take place after the call date, and the par value plus the currently due coupon is paid at that time. We want to derive the OAS associated with the 99.696 observed price of the bond. In Fig. 27.7, the OAS is verified to be 50 basis points over the short rates. The same bond without the call option would have fetched 100.569 as shown in Fig. 23.14 under the same OAS. The call option depresses the bond value, as expected.

An algorithm for finding the OAS of callable bonds is given in Fig. 27.8. Besides callable bonds, OAS analysis can be extended to other corporate bond structures. The general procedure is summarized below.

1. Estimate the spot rate curve.
2. Calibrate the interest rate model.
3. Develop rules for exercising the embedded options.
4. Add the OAS to the short rates on the tree.
5. Compute the model price.
6. Iterate 4 and 5 by varying the OAS until the model price matches the market price.

There are many choices for the root-finding algorithms. The simple bisection method cannot fail. The Newton–Raphson method, albeit faster, may not apply at certain OASs because of nondifferentiability. See [727] and Programming Assignments 27.4.9 and 27.4.10 for additional choices.

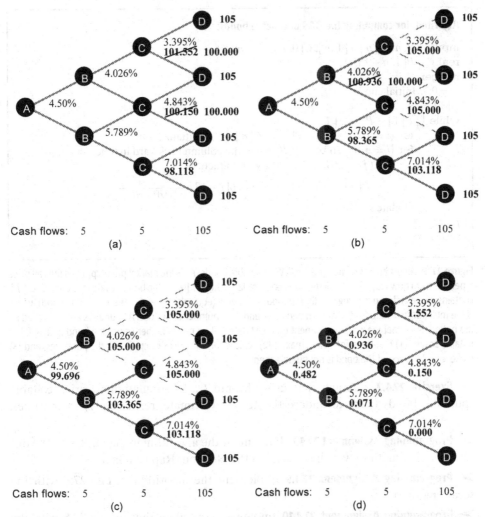

Figure 27.7: Calculation of the OAS of callable bonds. The price trees are based on a 3-year callable bond paying an annual coupon rate of 5%. The call provision is a discrete par (100) call that may be exercised on any coupon payment date. The underlying binomial interest rate tree is produced from Fig. 23.8 by the addition of a constant spread of 50 basis points to each short rate. The dotted lines signal early exercise. (a) PVs at C nodes are computed. Prices over 100 are replaced with 100 because of the exercise of the option. (b) PVs at B nodes are computed after the addition of 5 to prices at C nodes. Again, prices over 100 are replaced with 100. (c) Repeat the steps in (b) for node A. The price at each node has the PV of the *remaining* cash flows. Because the model price matches the market price, 99.696, 0.5% is the OAS. (d) Calculate the option value (this part is incorrect; see text).

▶ **Exercise 27.4.4** Argue that the OAS does not assume parallel shifts in the term structure.

▶ **Exercise 27.4.5** Explain why the OAS of a callable bond decreases as the interest rate volatility increases, other things being equal.

▶ **Exercise 27.4.6** For a putable bond, how does its OAS behave (1) when the market price decreases, other things being equal, and (2) when the coupon rate decreases, other things being equal and with the market price at par?

Algorithm for computing the OAS of callable bonds:

input: $P, n, \text{cp}[n], r[1..n], C[0..n], v[1..n], \epsilon$;
real $P[1..n+1], s$;
integer i, j;
$s := 0$; // Initial guess.
$P[1] := \infty$;
while $[\,|\,P[1] - P\,| > \epsilon\,]$ {
        for ($i = 1$ to $n+1$) { $P[i] := C[n]$; } // Initialization.
        for ($i = n$ down to 1) // Sweep the column backward in time.
          for ($j = 1$ to i) // Backward induction.
$$P[j] := C[i-1] + \min\left(\text{cp}[i-1], \frac{P[j]+P[j+1]}{2\times\left(1+r[i]\times v[i]^{j-1}+s\right)}\right);$$
        Update s;
}
return s;

Figure 27.8: Algorithm for computing the OAS of callable bonds. P is the market price, $\text{cp}[i]$ is the call price i periods from now ($\text{cp}[i] = \infty$ if the bond is not callable then), $r[i]$ is the baseline rate for period i, $C[i]$ contains the cash flow occurring exactly i periods from now (equivalently, end of the $(i-1)$th period), $v[i]$ is the multiplicative ratio for the rates in period j, and n is number of periods. All numbers are measured by the period. Note that the coupon payment on a call date will be paid whether or not the bond is called. For putable bonds, (1) replace min() with max(), (2) replace the call prices in cp[ ] with the put prices, and (3) make $\text{cp}[i] = -\infty$ if the bond is not putable then.

▶ **Exercise 27.4.7**   Argue that using Monte Carlo simulation to price callable (putable) bonds tends to underestimate (overestimate, respectively) their values.

▶ **Programming Assignment 27.4.8** Implement the algorithm in Fig. 27.8 with the differential tree method, which is based on the Newton–Raphson method.

▶ **Programming Assignment 27.4.9** Implement the algorithm in Fig. 27.8 with the secant method.

▶ **Programming Assignment 27.4.10** Implement the algorithm in Fig. 27.8 with the Ridders method.

Valuing the Embedded Option and Option-Adjusted Yield

In contrast with the explicit valuation of the embedded call option in the option pricing methodology, option price is a by-product of OAS analysis. It is calculated in the following way. First, the implied noncallable bond price is the PV of the bond afte the OAS is added to the short rates on the tree. Then the value of the embedded call option is determined by subtraction of the callable bond's market price from the implied noncallable bond price.

Refer to Fig. 27.7(d) for the calculation of the embedded option's value below. At C nodes the intrinsic value of the call option is the greater of zero and the price of the underlying bond minus the call price. For instance, the top C node's intrinsic value is

$$\max(0, 101.552 - 100) = 1.552.$$

Similarly, the third C node's intrinsic value is zero because the option is not exercised. At B nodes, we find the PV of the option values at C nodes and take as the option value the greater of the intrinsic value at B and the PV of the option value at the successor C nodes. At the top B node, for instance, the intrinsic value is 0.936 as it is exercised. The PV of the option values at the two successor C nodes is

$$\frac{1.552 + 0.150}{2 \times 1.04026} = 0.818.$$

The option value is therefore 0.936. Similarly, the option value at the bottom B node is

$$\frac{0.150}{2 \times 1.05789} = 0.071.$$

The option value is finally

$$\frac{0.936 + 0.071}{2 \times 1.045} = 0.482.$$

The implied price of the underlying fixed-rate bond is hence $99.696 + 0.482 = 100.178$. However, this cannot be right. Should not the price be 100.569, the price of the same bond without the call option as calculated in Fig. 23.14 under the same OAS of 50 basis points? And by Eq. (27.1), should not the embedded call's value be $100.569 - 99.696 = 0.873$ instead of 0.482? Indeed, the way 0.482 was calculated assumed an underlying bond that is callable. Hence 100.178 is the price of a bond that is *not* option free as desired.

Recall that the option-adjusted yield is the interest rate that makes the PV of the cash flows for the bond equal the implied price of the noncallable bond. Because the hypothetical noncallable bond with an implied price of 100.569 has 3 years to maturity and pays a 5% annual coupon, the option-adjusted yield is 4.792% compounded annually.

▶ **Exercise 27.4.11** Correct Fig. 27.7(d).

Option-Adjusted Spread Duration and Convexity

OAS analysis can be used to assess how prices move as interest rates change. This is done by first changing the short rate by a small amount Δr, say, plus 10 basis points. Next the interest rate tree is revised to reflect the new short rate (for example, Δr is added to every short rate on the Ho–Lee model's interest rate tree). With the OAS held constant, the new model price P_+ is computed. Finally, the market price P_0 and the new model price are used to estimate the effective duration $(P_0 - P_+)/(P_0 \Delta r)$ by Eqs. (4.6) [325, 848]. This duration is called the **OAS duration** [55, 330]. The effective convexity is more computation intensive as it requires the model price P_- after the short rate is decremented by a small amount Δr. It equals $(P_+ + P_- - 2 \times P_0)/(P_0 (\Delta r)^2)$ by Eq. (4.16). For multifactor interest rate models, the above procedures must be repeated for each factor.

A popular alternative is to add Δr to the whole spot rate curve and then calibrate the interest rate model with the same volatility. The rest of the computation is identical [36, 329, 429]. Specifying a term structure movement outside the interest rate model is certainly not theoretically sound. However, this method is easy to apply; besides, to do otherwise requires a lot of confidence in the model [731].

Holding Period Returns

The HPR assesses the bond over a holding period. The FV at the horizon consists of the projected principal and interest cash flows, the interest on the reinvestment thereof, and the projected horizon price. The monthly total return is then

$$\left(\frac{\text{total future amount}}{\text{price of the bond}} \right)^{1/\text{number of months}} - 1.$$

To calculate the above return, reinvestment rates and interest rate dynamics are all needed. The OAS can be easily combined with the HPR analysis. First we create a few static interest rate scenarios for the holding period. We then calculate the HPR for each scenario by assuming that the OAS remains unchanged at the horizon. This methodology is better than simply comparing the OASs of two securities.

➤ **Exercise 27.4.12** Assume a flat prevailing spot rate curve and continuous compounding. Prove that for an option-free nonbenchmark bond, any calibrated interest rate tree will compute a spread that equals the yield spread. (The yield spread is the difference between the yields to maturity of benchmark and nonbenchmark bonds. The spread is the incremental return over the short rate on the tree in the sense of Subsection 23.3.1.)

27.5 Key Rate Durations

Although duration is essential for identifying interest rate risk exposures, it has many limitations. For example, the assumed yield curve shift is usually not realistic. Also, when it becomes desirable to isolate a security's sensitivity to rate changes in various maturities, a vector of durations is needed instead of just a single number. Key rate durations were proposed by Ho to address these concerns [453] and have proved to be an effective and intuitive tool for risk management [260, 390, 455].

The idea is to break the effective duration into a vector of durations called key rate durations. This decomposition measures sensitivities to different segments of the yield curve so that when the key rate durations are added up, it gives approximately the original effective duration. Securities with identical effective duration can have very different key rate durations. By revealing what segments of the yield curve affect the security value most, key rate durations isolate the risks.

To define key rate durations, we start with a set of 11 key rates: 3 months, and 1, 2, 3, 5, 7, 10, 15, 20, 25, and 30 years. Other choices are also possible. Let $d(i)$ denote the amount of change at the ith key rate. A key rate's effect on other rates decline *linearly*, reaching zero at the adjacent key rates and beyond. The first and the last key rates need to be handled separately, of course. See Fig. 27.9, in which $t(i)$ denotes the ith key rate's term. Each key rate change induces a custom spot rate curve shift called the **basic key rate shift**. The corresponding key rate duration is defined as $(P_0 - P_+)/[\,P_0 d(i)\,]$. The price P_+ is calculated as follows. Use the custom shift defined by the basic key rate shift to derive the new spot rate curve, which is used to calibrate the interest rate model. The security is then priced, with the same OAS if necessary. The resulting 11 durations are the key rate durations.

EXAMPLE 27.5.1 Increase the 10-year key rate by 10 basis points. Because 5 years lie between the 10-year key rate and the next key rate to the right, the 15-year key rate, the effect of the change in the 10-year key rate falls off at a rate of $10/5 = 2$

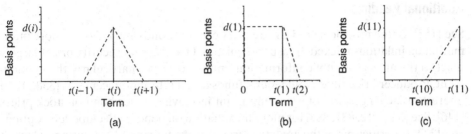

Figure 27.9: Term structure movements used by key rate durations. An increase of $d(i)$ basis points at (a) the ith key rate, $2 \leq i \leq 10$, (b) the first key rate, and (c) the last, eleventh key rate. Note that $t(1) = 0.25$ and $t(11) = 30$. Variations are possible. For example, we can break (b) into two segments such that the one between 0 and $t(1)$ is a mirror image of (c) and the one centered at $t(1)$ is a triangle like (a) [848].

basis points per year. Hence the 11-year rate increases by 8 basis points, the 12-year rate increases by 6 basis points, and so on. The 15-year rate and all the rates beyond 15 years are unchanged. The 10-year key rate move also affects rates of term less than 10 years. Because 3 years lie between the 7-year key rate and the 10-year key rate, the 10-basis-point increase in the 10-year rate falls off by $10/3 = 0.333$ basis points per year. Hence the 9-year rate increases by $10 - 0.333 = 9.667$ basis points, the 8.5-year rate increases by $10 - 1.5 \times 0.333 = 9.5$ basis points, and so on. The 7-year rate and all the rates below it are unchanged.

The sum of key rate durations can approximate the duration with respect to any custom shift to the spot rate curve as follows. Describe a custom shift by the function $s(t)$, $0 \leq t \leq 30$. Then $s(t)$ is linearly approximated by the sum of the basic key rate shifts defined by $d(i) \equiv s(t(i))$, $i = 1, 2, \ldots, 11$, at the 11 key rates (see Fig. 27.10 for illustration). Hence the duration with respect to $s(t)$ is expected to be roughly the sum of key rate durations.

One problem with key rate durations is that the basic key rate shifts are independent of the pricing model. Another is that the basic key rate shifts can introduce negative forward rates.

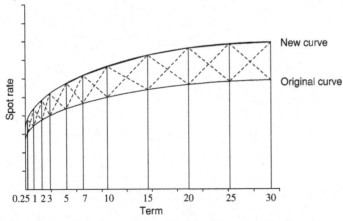

Figure 27.10: Linear interpolation of the term structure shift. The new spot rate curve is approximated when the 11 basic key rate shifts are added to the original spot rate curve. Although the new curve and the approximated curve are guaranteed to agree at only the key rates, they are quite close.

Additional Reading

See [152] for the mispricing of U.S. Treasury callable bonds and [777] for more information on inflation-indexed Treasury securities. U.S. Treasury security prices largely react to the arrival of public information on the economy, particularly the employment, Producer Price Index, and federal funds target rate announcements [358]. Real interest rate changes do not seem important in moving either bond or stock prices [146]. See [328, 712, 837, 889] for more information on bonds with embedded options and [372] for criticisms of the analytical frameworks for embedded options. Consult [117] for hedging with interest rate options and [317, 328, 799] for **bond swaps**. An empirical study of the call policy finds many inconsistencies with the theory [553]. Corporations issue CBs for various reasons [559]. CBs can serve as an alternative to venture capital [599]. See [21, 562, 651, 762, 880] for more information on the OAS, [218] for suggestions beyond the OAS, and [198] for various versions of OAS. The OAS is criticized in [349]. Consult [481, 513, 535, 583] for credit risk and [253, 254, 398] for **credit derivatives**, which protect investors from defaults or rating downgrades.

Reasons for the accumulation of public debts extend far beyond "wars and rebellions" [414]. Misgivings about national debts are common throughout history. Adam Smith (1723–1790) believed such debts would "in the long run probably ruin all the great nations of Europe." He was also concerned about the "pretended payment" that disguised public bankruptcy [808]. Montesquieu endorsed the use of a sinking fund to "procure the public confidence" [676]. J.S. Mill (1806–1873) asserted that the transfer of interest payment is a "serious evil" [665].

NOTE

1. Its CUSIP number is 9128272M3. All U.S. securities issued in book-entry or certificate form after 1970 are identified by a nine-digit **CUSIP number**. CUSIP stands for "Committee on Uniform Securities Identification Procedures." The CUSIP numbering system, developed by the American Bankers Association, was expanded in 1989 to include foreign securities, to be identified by the nine-digit CUSIP International Numbering System [355].

CHAPTER TWENTY-EIGHT

Introduction to Mortgage-Backed Securities

> Anyone stupid enough to promise to be responsible for a stranger's
> debts deserves to have his own property held to guarantee payment.
> —Proverbs 27:13

A **mortgage-backed security** (**MBS**) is a bond backed by an undivided interest in a pool of mortgages. MBSs traditionally enjoy high returns, wide ranges of products, high credit quality, and liquidity [432]. The mortgage market has witnessed tremendous innovations in product design [54]. The complexity of the products and the prepayment option mandate the deployment of advanced models and software techniques. In fact, the mortgage market probably could not have operated efficiently without them [659]. Although our focus will be mainly on residential mortgages, the underlying principles are applicable to other types of assets as well.

28.1 Introduction

A mortgage is a loan secured by the collateral of real estate property. The lender – the **mortgagee** – can foreclose the loan by seizing the property if the borrower – the **mortgagor** – defaults, that is, fails to make the contractual payments. An MBS is issued with pools of mortgage loans as the collateral. The cash flows of the mortgages making up the pool naturally reflect upon those of the MBS. There are three basic types of MBSs: **mortgage pass-through security** (**MPTS**), **collateralized mortgage obligation** (**CMO**), and **stripped mortgage-backed security** (**SMBS**).

The mortgage sector is by far the largest in the debt market (see Fig. 28.1). The mortgage market conceptually is divided between a primary market, also called the **origination market**, and a secondary market in which mortgages trade. The secondary market includes the market for loans that are not securitized, called **whole loans**, and the market for MBSs.

Individual mortgages are unattractive for many investors. To start with, often at hundreds of thousands of U.S. dollars or more, they demand too much investment. Most investors also lack the resources and knowledge to assess the credit risk involved. Furthermore, a traditional mortgage is fixed rate, level payment, and fully amortized with the percentage of **principal and interest** (**P&I**) varying from month to month, creating accounting headaches. Finally, prepayment levels fluctuate with a host of factors, making the size and the timing of the cash flows unpredictable.

Mortgage debt outstanding (U.S.$ millions)						
	1994	*1995*	*1996*	*1997*	*1998*	*1999*
Total outstanding	4,392,794	4,603,981	4,877,536	5,211,286	5,736,638	6,387,651
By holder:						
Commercial banks	1,012,711	1,090,189	1,145,389	1,245,315	1,337,217	1,495,717
Savings institutions	596,191	596,763	628,335	631,826	643,957	668,634
Life insurance cos	210,904	213,137	208,162	206,840	213,640	229,333
Federal/agency	315,580	308,757	295,192	286,167	292,636	320,105
Mortgage pools/trusts	1,730,004	1,863,210	2,040,848	2,239,350	2,589,764	2,954,836
Individuals/others	527,404	531,926	559,609	601,788	659,425	719,026

Figure 28.1: Mortgage debt outstanding 1994–1999. Source: Federal Reserve Bulletin.

A liquid market for individual mortgages did not appear until the mortgage institutions started securitizing their mortgage holdings in 1970. Individual, illiquid mortgages were then turned into marketable securities that were easier to analyze and trade. Today, financial intermediaries buy mortgages and place them in a pool. Interests in the pools are then sold to investors. These undivided ownership interests in the loans that collateralize the security are called **participation certificates** (**PCs**). The intermediary receives the mortgage payments from homeowners or servicing organizations and passes them to investors. The intermediary also guarantees that it will pay investors all the P&I that are due in case of default. Several of the above-mentioned problems are solved or alleviated by this arrangement. For instance, the minimum investment is reduced. The credit risk of the homeowners is virtually eliminated because of the intermediary's guarantees. As a result, the credit strength of the PC as seen by the investor is shifted from the homeowner and the property to the intermediary.

28.2 Mortgage Banking

The original lender is called the **mortgage originator**. It can be thrifts, commercial banks, mortgage bankers, life insurance companies, or pension funds. There are three revenues for the mortgage originator with regard to a new mortgage. It can hold the mortgage for investment or sell the mortgage to an investor or conduit. **Conduits** are either federally sponsored credit agencies or private companies that pool mortgages. Finally, it can use the mortgage as collateral for the issuance of a security. In this way, the mortgage becomes part of a pool of mortgages that are the collateral for a security – it is securitized.

Mortgage insurance is often required for guarding against default. Besides private mortgage insurers, three U.S. government agencies guarantee mortgages for qualified borrowers: the Federal Housing Administration (FHA), the Department of Veterans Affairs (VA), and the Rural Housing Service (RHS) [327]. Loans not guaranteed or insured by the FHA, VA, or RHS are called **conventional loans**. On the other hand, loans that comply with the underwriting standards for sale or conversion to MBSs issued and guaranteed by two federally sponsored credit agencies are called **conforming mortgages**. The two agencies are the Federal National Mortgage Association (FNMA or "Fannie Mae") and the Federal Home Loan Mortgage

```
Loan Information:
Balances:
Principal Balance on 10/03/97          $155,520.31
Escrow Balance on 10/03/97               $3,015.82
Payment Factors:
Interest Rate                            7.12500%
Principal & Interest                     $1,702.96
Escrow Payment                             $700.32
Total Payment:                           $2,403.28
Year-to-Date:
Interest                                 $8,514.63
Taxes                                    $5,665.60
Principal                                $6,817.07
```

Figure 28.2: Typical monthly mortgage statement.

Corporation (FHLMC or "Freddie Mac"). Both are now public companies. Mortgage bankers also originate FHA-insured and VA-guaranteed mortgage loans for sale in the form of Ginnie Mae pass-throughs. Ginnie Mae stands for Government National Mortgage Association (GNMA). MBSs issued by Fannie Mae or Freddie Mac are primarily sold by mortgage banking firms directly to securities dealers. FHA/VA/RHS mortgage loans are packaged for sale as pass-through securities guaranteed by Ginnie Mae and sold also primarily to securities dealers. Conventional loans exceeding the maximum amounts required for conformance are called **jumbo loans**.

A mortgage needs to be serviced. Principal, interest, and escrow funds for taxes and insurance are collected from the borrowers. Taxes and premiums are paid, and P&I are distributed to the investors of the loans. The issuer often has to advance P&I payments due if uncollected, which is referred to as **MBS servicing** [298]. Accounting and monthly reporting are also part of servicing. The **servicing fee** is a percentage of the remaining principal of the loan at the beginning of each month. It is part of the interest portion of the mortgage payment as far as the borrower is concerned. The monthly cash flow from the mortgage hence consists of three parts: servicing fee, interest payment net of the servicing fee, and the scheduled principal repayment. There is a secondary market for servicing rights. The cash flow of servicing right is uncertain because of the prepayment uncertainty. Figure 28.2 shows a typical monthly mortgage statement.

28.3 Agencies and Securitization

The existence of a secondary market is key to the liquidity of mortgages. Government agencies were created by Congress to foster the growth of this market. The means of providing such liquidity was the creation of securities backed by a pool of mortgages and guaranteed by these agencies. With the increase in liquidity and the reduction in credit risk comes the creation of products offering varieties of risk/return patterns. These products in turn attract investors to participate in the mortgage market (see Fig. 28.3).

Mortgage securitization commenced in February 1970 with the issuance of Ginnie Mae Pool #1, a mortgage pass-through. Explosive growth of the market came

Outstanding volume of agency MBSs (U.S.$ billions)								
	GNMA	*FNMA*	*FHLMC*	*Total*		*GNMA*	*FNMA*	*FHLMC*	*Total*
1980	93.9	–	17.0	110.9	**1990**	403.6	299.8	321.0	1,024.4
1981	105.8	0.7	19.9	126.4	**1991**	425.3	372.0	363.2	1,160.5
1982	118.9	14.4	43.0	176.3	**1992**	419.3	445.0	409.2	1,273.5
1983	159.8	25.1	59.4	244.3	**1993**	414.0	495.5	440.1	1,349.6
1984	180.0	36.2	73.2	289.4	**1994**	450.9	530.3	460.7	1,441.9
1985	212.1	55.0	105.0	372.1	**1995**	472.3	583.0	515.1	1,570.4
1986	262.7	97.2	174.5	534.4	**1996**	506.2	650.7	554.3	1,711.2
1987	315.8	140.0	216.3	672.1	**1997**	536.8	709.6	579.4	1,825.8
1988	340.5	178.3	231.1	749.9	**1998**	537.4	834.5	646.5	2,018.4
1989	369.9	228.2	278.2	876.3	**1999**	582.0	960.9	749.1	2,292.0

Issuance of agency MBSs (U.S.$ billions)								
	GNMA	*FNMA*	*FHLMC*	*Total*		*GNMA*	*FNMA*	*FHLMC*	*Total*
1980	20.6	–	2.5	23.1	**1990**	64.4	96.7	73.8	234.9
1981	14.3	0.7	3.5	18.5	**1991**	62.6	112.9	92.5	268.0
1982	16.0	14.0	24.2	54.2	**1992**	81.9	194.0	179.2	455.2
1983	50.7	13.3	21.4	85.4	**1993**	138.0	221.4	208.7	568.1
1984	28.1	13.5	20.5	62.1	**1994**	111.2	130.6	117.1	359.0
1985	46.0	23.6	41.5	111.1	**1995**	72.9	110.5	85.9	269.2
1986	101.4	60.6	102.4	264.4	**1996**	100.9	149.9	119.7	370.5
1987	94.9	63.2	75.0	233.1	**1997**	104.3	149.4	114.3	368.0
1988	55.2	54.9	39.8	149.9	**1998**	150.2	326.1	250.6	726.9
1989	57.1	69.8	73.5	200.4	**1999**	152.8	300.7	233.0	686.5

Figure 28.3: Agency MBSs 1980–1999. Source: Public Securities Association.

later in late 1981 when Fannie Mae and Freddie Mac started their mortgage swap programs. These developments allow mortgage holders – primarily thrifts – to sell their mortgages to agencies in return for agency-guaranteed pass-through securities backed by the same mortgages. Developments such as these have profound social implications. For example, they lower the cost of financing home ownership.

Among the three housing-related federal agencies, Ginnie Mae, Freddie Mac, and Fannie Mae, only Ginnie Mae is a government corporation within the Department of Housing and Urban Development (HUD).[1] Its guarantee hence carries the full faith and credit of the U.S. Treasury. MBSs with such a guarantee are perceived to have zero default risk. Ginnie Mae guarantees only government-insured or government-guaranteed loans in its programs, whereas Freddie Mac and Fannie Mae are government-sponsored enterprises that mainly use conventional mortgages in their programs. Securities offered by Ginnie Mae, Freddie Mac, and Fannie Mae are commonly referred to as "Ginnie Maes," "Freddie Macs," and "Fannie Maes," respectively.

Agency guarantees come in two forms. One type guarantees the *timely* payment of P&I. Under this guarantee, the P&I will be paid when due even if some of the mortgagors do not pay the monthly mortgage on time, if at all. Pass-throughs carrying this form of guarantee are called **fully modified pass-throughs**. For instance, Ginnie Mae either uses excess cash or borrows from the Treasury if the homeowner payments are late. All Ginnie Mae MBSs are fully modified pass-throughs. The second type guarantees the timely payment of interest and the ultimate payment of principal, say within a year. Pass-throughs carrying this form of guarantee are referred to as **modified pass-throughs**. Guarantees turn defaults into prepayments from the investor's point of view.

Although Fannie Mae and Freddie Mac buy only conforming mortgages, private conduits buy both conforming and nonconforming mortgages. Being **nonconforming** does not imply greater credit risk. Without explicit or implicit government guarantees on the underlying loans, the so-called **private-label** or **conventional pass-throughs**, which made their debut in 1977, receive high credit ratings through **credit enhancements**.

Traditional mortgages are fixed rate. Record-high fixed mortgage rates in the early 1980s led to the development of adjustable-rate mortgages (ARMs), which were first marketed in late 1983. ARMs are attractive for many reasons. First, the initial rate is typically several percentage points below that of fixed-rate mortgages. It is hence called the "**teaser**" **rate**. Because the home buyer qualifies for the mortgage at the initial loan rate, ARMs allow more people to qualify for a mortgage loan. By the same token, the home buyer can qualify for a larger loan with ARM financing. Second, the index used to adjust the rate is usually tied to a widely recognized and available index. This makes pricing and hedging practical. Third, the interest rate adjustments permitted by ARMs are capped, which insulates the mortgagor from loan payment shock during prolonged periods of rising interest rates. Fourth, ARMs represent an attractive investment for institutional investors such as thrifts and savings and loans because ARMs match their variable-rate liabilities better (review Subsection 4.2.3 for this point). Naturally, ARMs are less competitive against fixed-rate mortgages during the periods when the fixed mortgage rates are relatively low.

ARMs financing reduces the housing industry's sensitivity to interest rate fluctuations because borrowers can choose between fixed- and adjustable-rate mortgages based on the prevailing interest rate levels. MPTSs backed by ARMs were created by Fannie Mae in 1984.

28.4 Mortgage-Backed Securities

In the simplest kind of MBS, the MPTs, payments from the underlying mortgages are passed from the mortgage holders through the servicing agency, after a fee is subtracted, and distributed to the security holder on a pro rata basis (see Fig. 28.4). This means that the holder of a $25,000 certificate from a $1 million pool is entitled to 2 1/2% of the cash flow paid by the mortgagors. Because of higher marketability, a pass-through is easier to sell than its individual loans.

A pass-through still exposes the investor to the total prepayment risk associated with the underlying mortgages. Such risk is undesirable from an asset/liability perspective. To deal with prepayment uncertainty, CMOs were created in June 1983 by Freddie Mac with the help of the then First Boston. Unlike mortgage pass-throughs, which have a single maturity and are backed by individual mortgages, CMOs are *multiple*-maturity, *multi*class debt instruments collateralized by pass-throughs, SMBSs, and whole loans. The process of using pass-throughs and SMBSs to create CMOs is called **resecuritization**. The total prepayment risk is now divided among classes of bonds called **classes** or **tranches**.[2] The principal, scheduled and

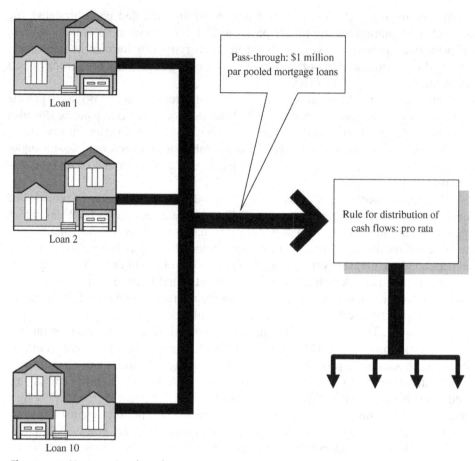

Figure 28.4: Mortgage pass-throughs.

prepaid, is allocated on a prioritized basis so as to redistribute the prepayment risk among the tranches in an unequal way.

In the **sequential tranche paydown structure**, for example, Class A receives principal paydown and prepayments before Class B, which in turn does it before Class C, and so on. Each tranche thus has a different effective maturity. Each tranche may even have a different coupon rate. CMOs were the first successful attempt to alter mortgage cash flows in a security form that attracts a wide range of investors (see Fig. 28.5).

EXAMPLE 28.4.1 Consider a two-tranche sequential pay CMO backed by $1,000,000 of mortgages with a 12% coupon and 6 months to maturity. The cash flow pattern for each tranche with zero prepayment and zero servicing fee is shown in Fig. 28.6. The calculation can be carried out first for the Total columns, which make up the amortization schedule, before the cash flow is allocated. Note that tranche A is retired after 4 months, and tranche B starts principal paydown at the end of month four.

EXAMPLE 28.4.2 (*Continued*) When prepayments are present the calculation is slightly more complex. Suppose the **single monthly mortality** (**SMM**) per month is 5%, which means that the prepayment amount is 5% of the remaining principal. The remaining principal at month i *after* prepayment then equals the scheduled remaining

Outstanding volume of agency collateralized mortgage obligations (U.S.$ billions)								
	GNMA	FNMA	FHLMC	Total		GNMA	FNMA	FHLMC	Total
1987	—	0.9	—	0.9	1994	—	315.0	263.7	578.7
1988	—	11.6	10.9	22.5	1995	—	294.0	247.0	540.9
1989	—	47.6	47.6	95.2	1996	—	283.4	237.6	521.0
1990	—	104.3	83.4	187.7	1997	17.5	328.6	233.6	579.7
1991	—	193.3	43.0	336.3	1998	29.0	311.4	260.3	600.8
1992	—	276.9	217.0	494.0	1999	52.5	293.6	316.1	662.1
1993	—	323.4	264.1	587.6					

Issuance of agency collateralized mortgage obligations (U.S.$ billions)								
	GNMA	FNMA	FHLMC	Total		GNMA	FNMA	FHLMC	Total
1987	—	0.9	—	0.9	1994	3.1	56.3	73.1	132.6
1988	—	11.2	13.0	24.2	1995	1.9	8.2	15.4	25.4
1989	—	37.6	39.8	77.3	1996	9.5	26.6	34.1	70.2
1990	—	60.9	40.5	101.4	1997	7.9	74.8	84.4	167.0
1991	—	101.8	72.0	173.8	1998	13.6	76.3	135.2	225.1
1992	—	154.8	131.3	286.1	1999	29.6	50.6	119.6	199.7
1993	—	168.0	143.3	311.3					

Figure 28.5: Agency CMOs 1987–1999. Source: Public Securities Association.

principal as computed by Eq. (3.8) times $(0.95)^i$. This done for all the months, the total interest payment at any month is the remaining principal of the previous month times 1%. And the prepayment amount equals the remaining principal times $0.05/0.95$ (the division by 0.95 yields the remaining principal *before* prepayment). Figure 28.7 tabulates the cash flows of the same two-tranche CMO under 5% SMM. For instance, the total principal payment at month one, $204,421, can be verified as follows. The scheduled remaining principal is $837,452 from Fig. 28.6. The remaining principal is hence $837452 \times 0.95 = 795579$, which makes the total principal payment $1000000 - 795579 = 204421$. Because tranche A's remaining principal is $500,000, all 204,421 dollars go to tranche A. Incidentally, the prepayment is $837452 \times 5\% = 41873$ (alternatively, $795579 \times 0.05/0.09$). Note that tranche A is retired after 3 months, and tranche B starts principal paydown at the end of month three.

	Interest			Principal			Remaining principal		
Month	A	B	Total	A	B	Total	A	B	Total
							500,000	500,000	1,000,000
1	5,000	5,000	10,000	162,548	0	162,548	337,452	500,000	837,452
2	3,375	5,000	8,375	164,173	0	164,173	173,279	500,000	673,279
3	1,733	5,000	6,733	165,815	0	165,815	7,464	500,000	507,464
4	75	5,000	5,075	7,464	160,009	167,473	0	339,991	339,991
5	0	3,400	3,400	0	169,148	169,148	0	170,843	170,843
6	0	1,708	1,708	0	170,843	170,843	0	0	0
Total	10,183	25,108	35,291	500,000	500,000	1,000,000			

Figure 28.6: CMO cash flows without prepayments. The total monthly payment is $172,548. Month-$i$ numbers reflect the ith monthly payment.

	Interest			Principal			Remaining principal		
Month	A	B	Total	A	B	Total	A	B	Total
							500,000	500,000	1,000,000
1	5,000	5,000	10,000	204,421	0	204,421	295,579	500,000	795,579
2	2,956	5,000	7,956	187,946	0	187,946	107,633	500,000	607,633
3	1,076	5,000	6,076	107,633	64,915	172,548	435,085	435,085	
4	0	4,351	4,351	0	158,163	158,163	0	276,922	276,922
5	0	2,769	2,769	0	144,730	144,730	0	132,192	132,192
6	0	1,322	1,322	0	132,192	132,192	0	0	0
Total	9,032	23,442	32,474	500,000	500,000	1,000,000			

Figure 28.7: CMO cash flows with prepayments. Month-i numbers reflect the ith monthly payment.

SMBSs were created in February 1987 when Fannie Mae issued its Trust 1 SMBS. For SMBSs, the P&I are divided between the PO strip and the IO strip. In the scenarios of Examples 28.4.1 and 28.4.2, the IO strip receives all the interest payments under the Interest/Total column, whereas the PO strip receives all the principal payments under the Principal/Total column. These new instruments allow investors to better exploit anticipated changes in interest rates. Because the collateral for an SMBS is a pass-through, this is yet another example of resecuritization. CMOs and SMBSs are usually called **derivative MBS**s

Exercise 28.4.1 Repeat the calculations in Example 28.4.2 under 3% SMM.

28.5 Federal Agency Mortgage-Backed Securities Programs

28.5.1 Government National Mortgage Association ("Ginnie Mae")

Security guaranteed by Ginnie Mae is called an MBS. Ginnie Mae issues its MBSs under one of two programs, GNMA I (established in 1970) and GNMA II (established in 1983). The two programs differ in terms of the collateral underlying the pass-throughs. For example, GNMA I MBSs require all loans in a pool to be approximately homogeneous [297]. A GNMA I MBS is issued with an annual coupon rate that is 0.50% lower than the coupon rate on the underlying mortgages because of guarantee and servicing fees. MBSs backed by **adjustable-payment mortgages** (**APM**s) are issued under the GNMA II program.

The issuer of a Ginnie Mae security passes through the scheduled P&I payments on the underlying mortgages to security holders each month even if the issuer does not collect payments from some mortgagors. It also passes through any additional principal prepayments because of foreclosure settlements. If the issuer defaults on the monthly payments, Ginnie Mae assumes responsibility for the timely payment of P&I.

Exercise 28.5.1 Even without prepayments, the scheduled monthly payment to MBS holders increases slightly over time. Why?

28.5.2 Federal Home Loan Mortgage Corporation ("Freddie Mac")

Freddie Mac was created on July 24, 1970, as a government-charted corporation. It became a public corporation like Fannie Mae in 1989. Freddie Mac seeks to increase liquidity and available credit for the conventional mortgage market by establishing and maintaining a secondary market for such mortgages. It started issuing pass-through securities in 1971, which was the first time conventional mortgages were securitized with a federal agency guarantee. Its mortgage pass-throughs are referred to as PCs. Unlike the Ginnie Mae pass-throughs, the Freddie Mac pass-throughs guarantee only eventual repayment of principal. In the fall of 1990, Freddie Mac introduced its Gold PC, which has stronger guarantees: All Gold PCs are fully modified pass-throughs. Freddie Mac securities are not backed by the full faith and credit of the U.S. government. The credit of its securities is perceived to be equivalent to that of securities issued by U.S. government agencies ("U.S. agency" status).

Freddie Mac issues CMOs and SMBSs besides PCs. All Freddie Mac CMOs have semiannual payments much like bonds. They also use only fixed-rate mortgages as collateral and a guaranteed sinking fund to establish minimum principal prepayments.

28.5.3 Federal National Mortgage Association ("Fannie Mae")

Established in 1938, Fannie Mae is the oldest of the three agencies and one of the largest corporations in the United States in terms of assets (U.S.$575 billion as of the end of 1999). It introduced the mortgage pass-through program in 1981. Pass-throughs issued by Fannie Mae are called MBSs. Fannie Mae guarantees the timely payment of both principal and interest on its MBS whether or not the payments have been collected from the borrower. The guarantee encompasses principal payments resulting from foreclosure or prepayment; the securities are fully modified pass-throughs, in other words. Although Fannie Mae obligations are not backed by the full faith and credit of the U.S. government, they carry "U.S. agency" status in the credit markets.

28.6 **Prepayments**

The prepayment option sets MBSs apart from other fixed-income securities. The exercise of options on most securities is expected to be "rational" in the sense that it will be executed only when it is profitable to do so. This kind of "rationality" is weakened when it comes to the homeowner's decision to prepay. For example, even when the prevailing mortgage rate, called the **current coupon**, exceeds the mortgage's loan rate, some loans remain prepaid.

Prepayment risk refers to the uncertainty in the amount and timing of the principal prepayments in the pool of mortgages that collateralize the security. This risk can be divided into **contraction risk** and **extension risk**. Contraction risk refers to the risk of having to reinvest the prepayments at a rate lower than the coupon rate when interest rates decline. Extension risk is due to the slowdown of prepayments when interest rates climb, making the investor earn the security's lower coupon rate rather than the market's higher rate. Prepayments can be in whole or in part; the former is called **liquidation**, and the latter **curtailment**. Prepayments, however, need not always result in losses (see Exercise 28.6.1). The holder of a pass-through security is exposed to the total prepayment risk associated with the underlying pool of mortgage loans, whereas the CMO is designed to alter the distribution of that risk among investors.

Besides prepayment risk, investors in mortgages are exposed to at least three other risks: interest rate risk, credit risk, and liquidity risk. Interest rate risk is inherent in any fixed-income security. Credit risk is the risk of loss from default. It is almost nonexisting for FHA-insured and VA-guaranteed mortgages. As for privately insured mortgage, the risk is related to the credit rating of the company that insures the mortgage. Liquidity risk is the risk of loss if the investment must be sold quickly.

> **Exercise 28.6.1** There are reasons prepayments arising from lower interest rates increase the return of a pass-through if it was purchased at a discount. What are they?

28.6.1 Causes and Characteristics

Prepayments have at least five components [4, 433].

Home sale ("housing turnover"). The sale of a home generally leads to the prepayment of mortgage because of the full payment of the remaining principal. This **due-on-sale** clause applies to most conventional loans. Exceptions are FHA/VA mortgages, which are **assumable**, meaning the buyer can assume the existing loan.

Refinancing. Mortgagors can refinance their home mortgage at a lower mortgage rate. This is the most volatile component of prepayment and constitutes the bulk of it when prepayments are extremely high.

Default. This type of prepayment is caused by foreclosure and subsequent liquidation of a mortgage. It is relatively minor in most cases.

Curtailment. As the extra payment above the scheduled payment, curtailment applies to the principal and shortens the maturity of fixed-rate loans. Its contribution to prepayments is minor.

Full payoff (liquidation). There is evidence that many mortgagors pay off their mortgage completely when it is very **seasoned** and the remaining balance is small. Full payoff can also be due to natural disasters. It is important for only very seasoned loans.

Prepayments exhibit certain characteristics [504]. They usually increase as the mortgage ages – first at an increasing rate and then at a decreasing rate. They are higher in the spring and summer and lower in the fall and winter. They vary by the geographic locations of the underlying properties. Prepayments increase when interest rates drop but with a time lag. If prepayments were higher for some time because of high refinancing rates, they tend to slow down. Perhaps homeowners who do not prepay when rates have been low for a prolonged time tend never to prepay.

Figure 28.8 illustrates the typical price/yield curves of the Treasury and passthrough. As yields fall and the pass-through's price moves above a certain price, it flattens and then follows a downward slope. This phenomenon is called the price compression of premium-priced MBSs. It demonstrates the negative convexity of such securities.

> **Exercise 28.6.2** Given that refinancing involves certain fixed costs, which will tend to prepay faster, mortgage securities backed by 15-year mortgages or 30-year mortgages?

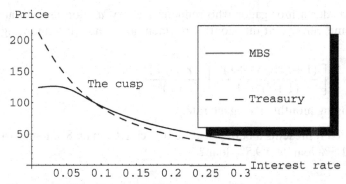

Figure 28.8: MBS vs. Treasury. Both are 15-year securities paying a 9% coupon rate in mortgage-equivalent yield. The segment above 100 means the security is premium-priced, whereas the segment below 100 signifies discount securities. Price compression occurs as yields fall through a threshold. The **cusp** represents that point.

28.6.2 An Analysis of the Incentive to Refinance

Consider a loan with a mortgage rate r_0 for a term of n months. Let the scheduled monthly payment of the original loan be C. At the time of refinancing, the mortgage rate for a new n-month loan is r_n, and a monthly payments have been remitted. Both r_0 and r_n are monthly rates.

From Eq. (3.8), the remaining principal at the time of refinancing is

$$C \frac{1 - (1 + r_0)^{-n+a}}{r_0}. \tag{28.1}$$

At the current rate r_n, the future cash flow of the original loan has a PV of

$$\sum_{i=1}^{n-a} C(1 + r_n)^{-i} = C \frac{1 - (1 + r_n)^{-n+a}}{r_n}.$$

Therefore the net monetary savings are

$$C \frac{1 - (1 + r_n)^{-n+a}}{r_n} - C \frac{1 - (1 + r_0)^{-n+a}}{r_0}. \tag{28.2}$$

Divide the preceding expression by expression (28.1) to obtain the savings per dollar of the remaining principal as

$$\frac{r_0}{r_n} \frac{1 - (1 + r_n)^{-n+a}}{1 - (1 + r_0)^{-n+a}} - 1.$$

For loans that have not seasoned sufficiently, the preceding expression is roughly

$$\frac{r_0}{r_n} - 1. \tag{28.3}$$

This heuristic argument points to using the *ratio* of loan rates rather than the *difference* to measure the incentive to refinancing [433].

▶ **Exercise 28.6.3** Does it make economic sense to refinance a mortgage if rates have not changed?

➤ **Exercise 28.6.4** Consider a mortgagor who refinances every a months with an n-month loan every time. Show that the monthly payment after the ith refinancing is

$$\text{original balance} \times \left[\frac{(1+r)^n - (1+r)^a}{(1+r)^n - 1} \right]^i \frac{r(1+r)^n}{(1+r)^n - 1},$$

where r is the unchanging monthly mortgage rate.

➤ **Exercise 28.6.5** Which represents a better deal, refinancing from an 8% loan to a 6% loan or from an 11.5% loan to a 9.5% loan?

Additional Reading

This chapter reviewed the mortgage markets, the institutions, the securitization of mortgages, and various mortgage products. Consult [54, 323, 325, 330, 331, 432, 469, 698, 799] for more background information and particularly [54] for a history of the MBS market. References [320, 324, 328] are also rich sources of information. See [54, Table 3.1] and [432, Exhibit 24-3] for other differences between Freddie Mac and Ginnie Mae pass-throughs. That securitization lowers the mortgage rates is not without its dissents [404].

NOTES

1. Fannie Mae used to be a government agency before being sold to the public in 1968.
2. *Tranche* is a French word for "slice."

CHAPTER TWENTY-NINE

Analysis of Mortgage-Backed Securities

Oh, well, if you cannot measure, measure anyhow.

Frank H. Knight (1885–1972)

Compared with other fixed-income securities, the MBS is unique in two respects. First, its cash flow consists of PRINCIPAL AND INTEREST (P&I). Second, the cash flow may vary because of prepayments in the underlying mortgages. This chapter covers the MBS's cash flow and valuation. We adopt the following time line when discussing cash flows:

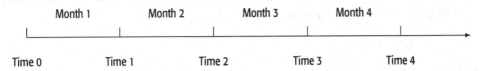

Because mortgage payments are paid in arrears, a payment for month i occurs at time i, that is, end of month i. The end of a month is identified with the beginning of the coming month.

29.1 Cash Flow Analysis

A traditional mortgage has a fixed term, a fixed interest rate, and a fixed monthly payment. Figure 29.1 illustrates the scheduled P&I for a 30-year, 6% mortgage with an initial balance of $100,000. Figure 29.2 shows how the remaining principal balance decreases over time. In the early years, the P&I consists mostly of interest. Then it gradually shifts toward principal payment with the passage of time. However, the total P&I payment remains the same each month, hence the term *level* pay. Identical characteristics hold for the pool's P&I payments in the absence of prepayments and servicing fees.

From the discussions in Section 3.3, we know that the remaining principal balance after the kth payment is

$$C \frac{1-(1+r/m)^{-n+k}}{r/m}, \tag{29.1}$$

where C is the scheduled P&I payment of an n-month mortgage making m payments per year and r is the annual mortgage rate. For mortgages, $m=12$. The remaining

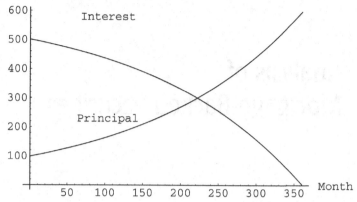

Figure 29.1: Scheduled P&I payments. The schedule is for a 30-year 6% mortgage with an original loan amount of $100,000.

principal balance after k payments can be expressed as a portion of the original principal balance; thus

$$\text{Bal}_k \equiv 1 - \frac{(1+r/m)^k - 1}{(1+r/m)^n - 1} = \frac{(1+r/m)^n - (1+r/m)^k}{(1+r/m)^n - 1}. \tag{29.2}$$

We can verify this equation by dividing balance (29.1) by Bal_0. The remaining principal balance after k payments is simply

$$\text{RB}_k \equiv \mathcal{O} \times \text{Bal}_k,$$

where $\mathcal{O}$ is the original principal balance.

The term **factor** denotes the portion of the remaining principal balance to its original principal balance expressed as a decimal [729]. So Bal_k is the monthly factor when there are no prepayments. It is also known as the **amortization factor**. When the idea of factor is applied to a mortgage pool, it is called the **paydown factor on the pool** or simply the **pool factor** [298].

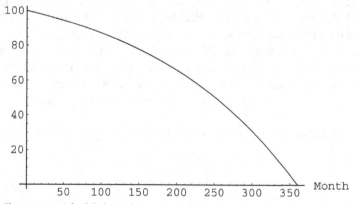

Figure 29.2: Scheduled remaining principal balances. Plotted are the remaining principal balances as percentages of par after each scheduled payment is made.

EXAMPLE 29.1.1 The remaining balance of a 15-year mortgage with a 9% mortgage rate after 54 months is

$$
\mathcal{O} \times \frac{[1+(0.09/12)]^{180}-[1+(0.09/12)]^{54}}{[1+(0.09/12)]^{180}-1} = \mathcal{O} \times 0.824866.
$$

In other words, roughly 82.49% of the original loan amount remains after 54 months.

By the amortization principle, the tth interest payment is

$$
I_t \equiv \mathrm{RB}_{t-1} \times \frac{r}{m} = \mathcal{O} \times \frac{r}{m} \times \frac{(1+r/m)^n-(1+r/m)^{t-1}}{(1+r/m)^n-1}.
$$

The principal part of the tth monthly payment is

$$
P_t \equiv \mathrm{RB}_{t-1} - \mathrm{RB}_t = \mathcal{O} \times \frac{(r/m)(1+r/m)^{t-1}}{(1+r/m)^n-1}. \tag{29.3}
$$

The scheduled P&I payment at month t, or $P_t + I_t$, is therefore

$$
(\mathrm{RB}_{t-1} - \mathrm{RB}_t) + \mathrm{RB}_{t-1} \times \frac{r}{m} = \mathcal{O} \times \left[\frac{(r/m)(1+r/m)^n}{(1+r/m)^n-1} \right], \tag{29.4}
$$

indeed a level pay independent of t. The term within the brackets, called the **payment factor** or **annuity factor**, represents the monthly payment for each dollar of mortgage.

EXAMPLE 29.1.2 The mortgage in Example 3.3.1 has a monthly payment of

$$
250,000 \times \frac{(0.08/12) \times [1+(0.08/12)]^{180}}{[1+(0.08/12)]^{180}-1} = 2,389.13
$$

by Eq. (29.4), in total agreement with the number derived there.

▶ **Exercise 29.1.1** Derive Eq. (29.4) from Eq. (3.6).

▶ **Exercise 29.1.2** Consider two mortgages with identical remaining principals but different mortgage rates. Show that their remaining principal balances after the next monthly payment will be different; in fact, the mortgage with a lower mortgage rate amortizes faster.

29.1.1 Pricing Adjustable-Rate Mortgages

We turn to ARM pricing as an interesting application of derivatives pricing and the analysis above. Consider a 3-year ARM with an interest rate that is 1% above the 1-year T-bill rate at the beginning of the year. This 1% is called the **margin**. For simplicity, assume that this ARM carries annual, not monthly, payments. The T-bill rates follow the binomial process, in boldface, in Fig. 29.3, and the risk-neutral probability is 0.5. How much is the ARM worth to the issuer?

Each new coupon rate at the reset date determines the level mortgage payment for the months until the next reset date as if the ARM were a fixed-rate loan with the new coupon rate and a maturity equal to that of the ARM. This implies, for example, that in the interest rate tree of Fig. 29.3 the scenario $A \rightarrow B \rightarrow E$ will leave our 3-year ARM with a remaining principal at the end of the second year different from

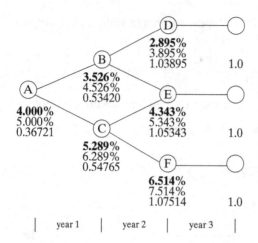

Figure 29.3: ARM's payment factors under stochastic interest rates. Stacked at each node are the T-bill rate, the mortgage rate (which is 1% above the T-bill rate), and the payment factor for a mortgage initiated at that node and ending at the end of year three (based on the mortgage rate at the same node, of course). The short rates are from Fig. 23.8.

that under the scenario $A \to C \to E$ (see Exercise 29.1.2). This path dependency calls for care in algorithmic design to avoid exponential complexity.

The idea is to attach to each node on the binomial tree the annual payment per $1 of principal for a mortgage initiated at that node and ending at the end of year three – in other words, the payment factor [546]. At node B, for example, the annual payment factor can be calculated by Eq. (29.4) with $r = 0.04526, m = 1$, and $n = 2$ as

$$\frac{0.04526 \times (1.04526)^2}{(1.04526)^2 - 1} = 0.53420.$$

The payment factors for other nodes in Fig. 29.3 are calculated in the same manner.

We now apply backward induction to price the ARM (see Fig. 29.4). At each node on the tree, the net value of an ARM of value $1 initiated at that node and ending at the end of the third year is calculated. For example, the value is zero at terminal nodes because the ARM is immediately repaid. At node D, the value is

$$\frac{1.03895}{1.02895} - 1 = 0.0097186,$$

which is simply the NPV of the payment 1.03895 next year (note that the issuer makes a loan of $1 at D). The values at nodes E and F can be computed similarly. At

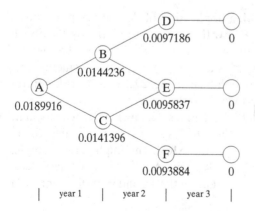

Figure 29.4: Backward induction for ARMs.

node B, we first figure out the remaining principal balance after the payment 1 year hence as

$$1 - (0.53420 - 0.04526) = 0.51106,$$

because $0.04526 of the payment of $0.53426 constitutes interest. The issuer will receive $0.01 above the T-bill rate next year, and the value of the ARM is either $0.0097186 or $0.0095837 per $1, each with probability 0.5. The ARM's value at node B thus is

$$\frac{0.51106 \times (0.0097186 + 0.0095837)/2 + 0.01}{1.03526} = 0.0144236.$$

The values at nodes C and A can be calculated similarly as

$$\frac{[1 - (0.54765 - 0.06289)] \times (0.0095837 + 0.0093884)/2 + 0.01}{1.05289} = 0.0141396,$$

$$\frac{[1 - (0.36721 - 0.05)] \times (0.0144236 + 0.0141396)/2 + 0.01}{1.04} = 0.0189916,$$

respectively. The value of the ARM to the issuer is hence $0.0189916 per $1 of loan amount. The complete algorithm appears in Fig. 29.5. The above idea of **scaling** has wide applicability for pricing certain classes of path-dependent securities [449, 546].

ARMs are indexed to publicly available indices such as LIBOR, the constant-maturity Treasury (CMT) rate, and the Cost of Funds Index (COFI). The CMT rates are based on the daily CMT yield curve constructed by the Federal Reserve Bank

---

Algorithm for pricing ARMs:

```
input:   n, r[n][n], s;
real     P[n], f, p;
integer  i, j;
for (j = 0 to n − 1) { // Nodes at time n − 1.
        f := 1 + r[n − 1][j] + s;   //(29.4) with n = 1.
        P[j] := f/(1 + r[n − 1][j]) − 1;
}
for (i = n − 2 down to 0) // Nodes at time i.
        for (j = 0 to i) {
                f := (r[i][j] + s)(1 + r[i][j] + s)^{n−i} ×
                        ((1 + r[i][j] + s)^{n−i} − 1)^{−1};  //See (29.4).
                p := 1 − (f − r[i][j] − s);
                P[j] := (p × (P[j] + P[j + 1]) × 0.5 + s) ×
                        (1 + r[i][j])^{−1};
        }
return P[0];
```

Figure 29.5: Algorithm for pricing ARMs. $r[i][j]$ is the $(j + 1)$th T-bill rate for period $i + 1$, the ARM has n periods to maturity, s is the margin, f stores the payment factors, and p stores the remaining principal amounts. All rates are measured by the period. In general, the floating rate may be based on the k-period Treasury spot rate plus a spread. Then Programming Assignment 29.1.3 can be used to generate the k-period spot rate at each node.

of New York and published weekly in the Federal Reserve's *Statistical Release* H.15 [525]. Cost of funds for thrifts indices are calculated based on the monthly weighted average interest cost for thrifts. The most popular cost of funds index is the 11th Federal Home Loan Bank Board District COFI [325, 330, 820].

If the ARM coupon reflects fully and instantaneously current market rates, then the ARM security will be priced close to par and refinancings rarely occur. In reality, adjustments are imperfect in many ways. At the reset date, a margin is added to the benchmark index to determine the new coupon. ARMs also often have **periodic rate caps** that limit the amount by which the coupon rate may increase or decrease at the reset date. They also have **lifetime caps** and **floors**. To attract borrowers, mortgage lenders usually offer a below-market initial rate (the "teaser" rate). The **reset interval**, the time period between adjustments in the ARM coupon rate, is often annual, which is not frequent enough. Note that these terms are easy to incorporate into the pricing algorithm in Fig. 29.5.

➢ **Programming Assignment 29.1.3**   Given an n-period binomial short rate tree, design an $O(kn^2)$-time algorithm for generating k-period spot rates on the nodes of the tree. This tree documents the dynamics of the k-period spot rate.

➢ **Programming Assignment 29.1.4** Implement the algorithm in Fig. 29.5. The binomial T-bill rate tree and the mortgage rate as a spread over the T-bill rate are parts of the input.

➢ **Programming Assignment 29.1.5** Consider an IAS with an amortizing schedule that depends solely on the prevailing k-period spot interest rate. This swap's cash flow depends on only the prevailing principal amount and the prevailing k-period spot interest rate. Design an efficient algorithm to price this swap on a binomial short rate tree.

29.1.2 Expressing Prepayment Speeds

The cash flow of a mortgage derivative is determined from that of the mortgage pool. The single most important factor complicating this endeavor is the unpredictability of prepayments. Recall that prepayment represents the principal payment made in excess of the scheduled principal amortization. We need only compare the amortization factor Bal_t of the pool with the reported factor to determine if prepayments have occurred. The amount by which the reported factor exceeds the amortization factor is the prepayment amount.

Single Monthly Mortality

An SMM of ω means that $\omega\%$ of the scheduled remaining balance at the end of the month will prepay. In other words, the SMM is the percentage of the remaining balance that prepays for the month. Suppose the remaining principal balance of an MBS at the beginning of a month is $50,000, the SMM is 0.5%, and the scheduled principal payment is $70. Then the prepayment for the month is $0.005 \times (50,000 - 70) \approx 250$ dollars. If the same monthly prepayment speed s is maintained since the issuance of the pool, the remaining principal balance at month i will be $\text{RB}_i \times (1 - s/100)^i$. It goes without saying that prepayment speeds must lie between 0% and 100%.

EXAMPLE 29.1.3 Take the mortgage in Example 29.1.1. Its amortization factor at the 54th month is 0.824866. If the actual factor is 0.8, then the SMM for the initial period of 54 months is

$$100 \times \left[1 - \left(\frac{0.8}{0.824866} \right)^{1/54} \right] = 0.0566677.$$

In other words, roughly 0.057% of the remaining principal is prepaid per month.

Conditional Prepayment Rate

The **conditional prepayment rate** (**CPR**) is the annualized equivalent of an SMM:

$$\text{CPR} = 100 \times \left[1 - \left(1 - \frac{\text{SMM}}{100} \right)^{12} \right].$$

Conversely,

$$\text{SMM} = 100 \times \left[1 - \left(1 - \frac{\text{CPR}}{100} \right)^{1/12} \right].$$

For example, the SMM of 0.0566677 in Example 29.1.3 is equivalent to a CPR of

$$100 \times \left\{ 1 - \left[1 - \left(\frac{0.0566677}{100} \right)^{12} \right] \right\} = 0.677897.$$

Roughly 0.68% of the remaining principal is prepaid annually. Figure 29.6 plots the P&I cash flows under various prepayment speeds. Observe that with accelerated prepayments, the principal cash flow is shifted forward in time.

PSA

In 1985 the Public Securities Association (PSA) standardized a prepayment model. The PSA standard is expressed as a monthly series of CPRs and reflects the increase in CPR that occurs as the pool seasons [619]. The PSA standard postulates the following prepayment speeds: The CPR is 0.2% for the first month, increases thereafter by 0.2% per month until it reaches 6% per year for the 30th month, and then stays at 6% for the remaining years. (At the time the PSA proposed its standard, a seasoned 30-year GNMA's typical prepayment speed was ~6% CPR [260].) The PSA benchmark is also referred to as **100 PSA**. Other speeds are expressed as some percentage of PSA. For example, 50 PSA means one-half the PSA CPRs, 150 PSA means one-and-a-half

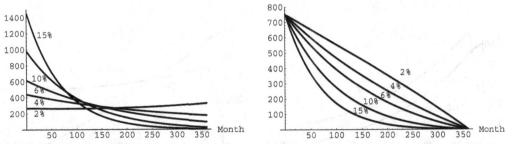

Figure 29.6: Principal (left) and interest (right) cash flows at various CPRs. The 6% mortgage has 30 years to maturity and an original loan amount of $100,000.

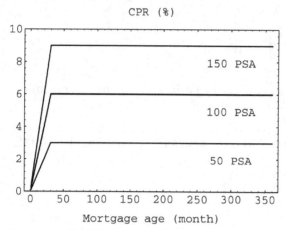

Figure 29.7: The PSA prepayment assumption.

the PSA CPRs, and so on. Mathematically,

$$\text{CPR} = \begin{cases} 6\% \times \frac{\text{PSA}}{100}, & \text{if the pool age exceeds 30 months} \\ 0.2\% \times m \times \frac{\text{PSA}}{100}, & \text{if the pool age } m \le 30 \text{ months} \end{cases} . \qquad (29.5)$$

See Fig. 29.7 for an illustration and Fig. 29.8 for the cash flows at 50 and 100 PSAs. Conversely,

$$\text{PSA} = \begin{cases} 100 \times \frac{\text{CPR}}{6}, & \text{if the pool age exceeds 30 months} \\ 100 \times \frac{\text{CPR}}{0.2 \times m}, & \text{if the pool age } m \le 30 \text{ months} \end{cases} .$$

See Fig. 29.9 for the conversion algorithm.

Conversion between PSA and CPR/SMM requires knowing the age of the pool. A prepayment speed of 150 PSA implies a CPR of $0.2\% \times 2 \times (150/100) = 0.6\%$ if the pool is 2 months old, but a CPR of $6\% \times 1.5 = 9\%$ if the pool age exceeds 30 months.

▶ **Exercise 29.1.6** Consider the following PSA numbers:

Month	6	12	18	24	30	36
PSA	100	130	154	230	135	125

Compute their equivalent CPRs.

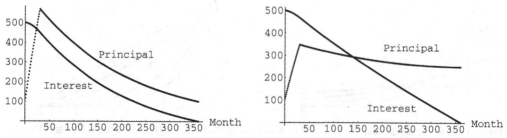

Figure 29.8: P&I payments at 100 PSA (left) and 50 PSA (right). The 6% mortgage has 30 years to maturity and an original loan amount of $100,000.

PSA-to-SMM algorithm:

```
input:    n, PSA, age;
real      SMM[ 1..n ], cpr;
integer   i;
PSA := PSA/100;
for (i = 1 to n) {
        if [ i + age ≤ 30 ]
                cpr := 0.2 × (i + age) × PSA/100 ;
        else  cpr := 6.0 × PSA/100;
        SMM[ i ] := 1 − (1 − PSA × cpr)^{1/12};
}
return SMM[ ];
```

Figure 29.9: PSA-to-SMM conversion. The pool has n more monthly cash flows, PSA is the prepayment speed, and age is the number of months since the pool's inception. SMM[ ] stores the prepayment vector in decimal, the ith of which denotes the SMM during month i as seen from now.

▶ **Exercise 29.1.7** Is the SMM assuming 200 PSA twice the SMM assuming 100 PSA?

29.1.3 Prepayment Vector and Cash Flow Analysis

Although it tries to capture, if crudely, how prepayments vary with age, the PSA should be viewed as a market convention rather than as a model. Instead of a single PSA number, a vector of PSAs generated by a prepayment model should be used to describe the monthly prepayment speed through time. The monthly cash flows can be derived thereof.

Similarly, the CPR should be seen purely as a measure of speed rather than a model. When we treat a single CPR number as the true prepayment speed, that number will be called the **constant prepayment rate** for obvious reasons. This simple model fails to address the empirical fact that pools with new production loans typically prepay at a slower rate than seasoned pools. As in the PSA case, a vector of CPRs should be preferred. In practice, a vector of CPRs or SMMs is easier to work with than a vector of PSAs because of the lack of dependence on the pool age. In any case, a CPR vector can always be converted into an equivalent PSA vector and vice versa.

To price an MBS, we start with its cash flow, that is, the periodic P&I under a static prepayment assumption as given by a prepayment vector. The invoice price is now $\sum_{i=1}^{n} C_i/(1+r)^{\omega-1+i}$, where C_i is the cash flow at time i, n is the **weighted average maturity** (**WAM**),[1] r is the discount rate, and ω is the fraction of period from settlement until the first P&I payment date. The WAM is the weighted average remaining term of the mortgages in the pool, where the weight for each mortgage is the remaining balance. The r that equates the above with the market price is called the (**static**) **cash flow yield**. The **implied PSA** is the single PSA speed producing the same cash flow yield.

MBSs are quoted in the same manner as U.S. Treasury notes and bonds. For example, a price of 94-05 means $94_{5/32}\%$ of par value. Sixty-fourth of a percent is expressed by appending "+" to the price. Hence, the price 94-05+ represents $94_{11/64}\%$ of par value.

Cash Flow

Each cash flow is composed of the principal payment, the interest payment, and the principal prepayment. Let B_k denote the actual remaining principal balance at month k. Given the pool's actual remaining principal balance at time $i-1$ (i.e., B_{i-1}), the P&I payments at time i are

$$\overline{P}_i \equiv B_{i-1} \left(\frac{\text{Bal}_{i-1} - \text{Bal}_i}{\text{Bal}_{i-1}} \right) = B_{i-1} \frac{r/m}{(1+r/m)^{n-i+1} - 1}, \tag{29.6}$$

$$\overline{I}_i \equiv B_{i-1} \frac{r-\alpha}{m}, \tag{29.7}$$

where α is the **servicing spread** (or servicing fee rate), which consists of the servicing fee for the servicer as well as the guarantee fee. The prepayment at time i is

$$\text{PP}_i = B_{i-1} \frac{\text{Bal}_i}{\text{Bal}_{i-1}} \times \text{SMM}_i,$$

where SMM_i is the prepayment speed for month i. If the total principal payment from the pool is $\overline{P}_i + \text{PP}_i$, the remaining principal balance is

$$
\begin{aligned}
B_i &= B_{i-1} - \overline{P}_i - \text{PP}_i \\
&= B_{i-1} \left[1 - \left(\frac{\text{Bal}_{i-1} - \text{Bal}_i}{\text{Bal}_{i-1}} \right) - \frac{\text{Bal}_i}{\text{Bal}_{i-1}} \times \text{SMM}_i \right] \\
&= \frac{B_{i-1} \times \text{Bal}_i \times (1 - \text{SMM}_i)}{\text{Bal}_{i-1}}.
\end{aligned} \tag{29.8}
$$

Equation (29.8) can be applied iteratively to obtain

$$B_i = \text{RB}_i \times \prod_{j=1}^{i} (1 - \text{SMM}_j). \tag{29.9}$$

Define $b_i \equiv \prod_{j=1}^{i}(1 - \text{SMM}_j)$. Then the scheduled P&I is

$$\overline{P}_i = b_{i-1} P_i, \quad \overline{I}_i = b_{i-1} I'_i \tag{29.10}$$

where $I'_i \equiv \text{RB}_{i-1} \times (r - \alpha)/m$ is the scheduled interest payment. The scheduled cash flow and the b_is determined from the prepayment vector are therefore all that are needed to calculate the projected actual cash flows. Note that if the servicing fees do not exist (that is, $\alpha = 0$), the projected monthly payment *before* prepayment at month i becomes

$$\overline{P}_i + \overline{I}_i = b_{i-1}(P_i + I_i) = b_{i-1} C, \tag{29.11}$$

where C is the scheduled monthly payment on the original principal. See Fig. 29.10 for a linear-time algorithm for generating the mortgage pool's cash flow.

Servicing and guarantee fees are deducted from the gross **weighted average coupon** (**WAC**) of the aggregate mortgage P&I to obtain the **pass-through rate**. The WAC is the weighted average of all the mortgage rates in the pool, in which the weight used for each mortgage is the remaining balance. The servicing spread

---

Mortgage pool cash flow under prepayments:

```
input:   n, r (r > 0), SMM[1..n];
real     B[n+1], P[1..n], Ī[1..n], PP[1..n], b;
integer  i;
b := 1;
B[0] := 1;
for (i = 1 to n) {
        b := b × (1 − SMM[i]);   //See (29.9).
        B[i] := b × (1+r)ⁿ−(1+r)ⁱ/(1+r)ⁿ−1;  // See (29.2).
        P[i] := B[i−1] − B[i];
        Ī[i] := B[i−1] × r;   //See (29.7).
        PP[i] := B[i] × SMM[i]/(1 − SMM[i]);
}
return B[ ], P[ ], Ī[ ], PP[ ];
```

Figure 29.10: Mortgage pool cash flow under prepayments. SMM is the prepayment vector, and the mortgage rate r is a monthly rate. The pool has n monthly cash flows, and its principal balance is $1. B stores the remaining principals, P are the principal payments (prepayments included), $\overline{I}$ are the interest payments, and PP are the prepayments. The prepayments are calculated based on Exercise 29.1.9, part (1).

for an MBS represents both the guarantee fee and the actual servicing fee itself. For example, a Ginnie Mae MBS with a 10.5% pass-through rate has a total servicing of 0.50%, of which 0.44% is retained by the servicer and 0.06% is remitted to Ginnie Mae. The figure most visible to the investor is the pass-through rate, but the amortization of P&I is a function of the gross mortgage rate of the individual loans making up the pool.

▶ **Exercise 29.1.8** Show that the scheduled monthly mortgage payment at month i is

$$B_{i-1} \frac{(r/m)(1+r/m)^{n-i+1}}{(1+r/m)^{n-i+1} - 1}.$$

▶ **Exercise 29.1.9** Verify that (1) $PP_i = B_i [SMM_i/(1 - SMM_i)]$ and (2) the actual principal payment $\overline{P}_i + PP_i$ is $b_{i-1}(P_i + RB_i \times SMM_i)$ (not $b_i P_i$).

▶ **Exercise 29.1.10** Verify Eqs. (29.9) and (29.10).

▶ **Exercise 29.1.11** Derive Eq. (29.11) by using Eqs. (29.2) and (29.4).

▶ **Exercise 29.1.12** Derive the PVs of the PO and IO strips based on current-coupon mortgages under constant SMM and zero servicing spread.

▶ **Exercise 29.1.13** Show that a pass-through backed by traditional mortgages with a mortgage rate equal to the market yield is priced at par regardless of prepayments. Assume either zero servicing spread or a pass-through rate equal to the market yield. (Prices of par-priced pass-throughs are hence little affected by variations in the prepayment speed.)

▶ **Programming Assignment 29.1.14** Implement the algorithm in Fig. 29.10.

29.1.4 Pricing Sequential-Pay CMOs

Consider a three-tranche sequential-pay CMO backed by $3,000,000 of mortgages with a 12% coupon and 6 months to maturity. The three tranches are called A, B, and Z. All three tranches carry the same coupon rate of 12%. The Z tranche consists of **Z bonds**. A Z bond receives no payments until all previous tranches are retired. Although a Z bond carries an explicit coupon rate, the owed interest is accrued and added to the principal balance of that tranche. For that reason, Z bonds are also called **accrual bonds** or **accretion bonds**. When a Z bond starts receiving cash payments, it becomes a pass-through instrument.

Assume that the ensuing monthly interest rates are 1%, 0.9%, 1.1%, 1.2%, 1.1%, and 1.0%. Assume further that the SMMs are 5%, 6%, 5%, 4%, 5%, and 6%. We want to calculate the cash flow and the fair price of each tranche.

We can compute the pool's cash flow by invoking the algorithm in Fig. 29.10 with $n = 6$, $r = 0.01$, and SMM $= [\,0.05, 0.06, 0.05, 0.04, 0.05, 0.06\,]$. We can derive individual tranches' cash flows and remaining principals thereof by allocating the pool's P&I cash flows based on the CMO structure. See Fig. 29.11 for the breakdown. Note that the Z tranche's principal is growing at 1% per month until all previous tranches are retired. Before that time, the interest due the Z tranche is used to retire A's and B's principals. For example, the $10,000 interest due tranche Z at month one is directed to tranche A instead, reducing A's remaining principal from $386,737 to $376,737 while increasing Z's from $1,000,000 to $1,010,000. At month four, the interest amount that goes into tranche Z, $10,303, is exactly what is required of Z's remaining principal of $1,030,301. The tranches can be priced

Month		1	2	3	4	5	6
Interest rate		1.0%	0.9%	1.1%	1.2%	1.1%	1.0%
SMM		5.0%	6.0%	5.0%	4.0%	5.0%	6.0%
Remaining principal (B_i)							
	3,000,000	2,386,737	1,803,711	1,291,516	830,675	396,533	0
A	1,000,000	376,737	0	0	0	0	0
B	1,000,000	1,000,000	783,611	261,215	0	0	0
Z	1,000,000	1,010,000	1,020,100	1,030,301	830,675	396,533	0
Interest ($\overline{I_i}$)		30,000	23,867	18,037	12,915	8,307	3,965
A		20,000	3,767	0	0	0	0
B		10,000	20,100	18,037	2,612	0	0
Z		0	0	0	10,303	8,307	3,965
Principal		613,263	583,026	512,195	460,841	434,142	396,534
A		613,263	376,737	0	0	0	0
B		0	206,289	512,195	261,215	0	0
Z		0	0	0	199,626	434,142	396,534

Figure 29.11: CMO cash flows. Month-i numbers reflect the ith monthly payment. "Interest" and "Principal" denote the pool's P&I and distributions to individual tranches. Interest payments may be used to make principal payments to tranches A, B, and C. The Z bond thus protects earlier tranches from extension risk.

as follows:

$$\text{tranche A} = \frac{20000 + 613263}{1.01} + \frac{3767 + 376737}{1.01 \times 1.009} = 1000369,$$

$$\text{tranche B} = \frac{10000 + 0}{1.01} + \frac{20100 + 206289}{1.01 \times 1.009} + \frac{18037 + 512195}{1.01 \times 1.009 \times 1.011}$$

$$+ \frac{2612 + 261215}{1.01 \times 1.009 \times 1.011 \times 1.012} = 999719,$$

$$\text{tranche Z} = \frac{10303 + 199626}{1.01 \times 1.009 \times 1.011 \times 1.012}$$

$$+ \frac{8307 + 434142}{1.01 \times 1.009 \times 1.011 \times 1.012 \times 1.011}$$

$$+ \frac{3965 + 396534}{1.01 \times 1.009 \times 1.011 \times 1.012 \times 1.011 \times 1.01} = 997238.$$

This CMO has a total theoretical value of \$2,997,326, slightly less than its par value of \$3,000,000. See the algorithm in Fig. 29.12.

We have seen that once the interest rate path and the prepayment vector for that interest rate path are available, a CMO's cash flow can be calculated and the CMO priced. Unfortunately, the remaining principal of a CMO under prepayments is, like an ARM, path dependent. For example, a period of high rates before dropping to the current level is not likely to result in the same remaining principal as a period of low rates before rising to the current level. This means that if we try to price a 30-year CMO on a binomial interest rate model, there will be $2^{360} \approx 2.35 \times 10^{108}$ paths to consider! As a result, Monte Carlo simulation is the computational method of choice. It works as follows. First, one interest rate path is generated. Based on that path, the prepayment model is applied to generate the pool's principal, prepayment, and interest cash flows. Now the cash flows of individual tranches can be generated and their PVs derived. The above procedure is repeated over many interest rate scenarios. Finally, the average of the PVs is taken.

▶ **Exercise 29.1.15** Calculate the monthly prepayment amounts for Fig. 29.11.

▶ **Programming Assignment 29.1.16** Implement the algorithm in Fig. 29.12 for the cash flows of a four-tranche sequential CMO with a Z tranche. Assume that each tranche carries the same coupon rate as the underlying pool's mortgage rate. Figures 29.13 and 29.14 plot the cash flows and remaining principal balances of one such CMO.

29.1.5 Weighted Average Life

The **weighted average life** (**WAL**) of an MBS is the average number of years that each dollar of unpaid *principal* due on the mortgages remains outstanding. It is computed by

$$\text{WAL} \equiv \frac{\sum_{i=1}^{m} i \, P_i}{12 \times P},$$

where m is the remaining term to maturity in months, P_i is the principal repayment i months from now, and P is the current remaining principal balance.[2] See Fig. 29.15 for an illustration. Usually, the greater the anticipated prepayment rate, the shorter

```
Sequential CMO cash flow generator:

input:    n, r (r > 0), SMM[1..n], O[1..4];
real      B[n+1], P[1..n], I[1..n]; // Pool cash flows.
real      B[1..4][n+1], P[1..4][1..n], I[1..4][1..n];
real      P, I;
integer   i, j;
Call the algorithm in Fig. 29.10 for B[n+1], P[1..n], I[1..n];
for (j = 1 to 4) { B[j][0] := O[j]; } // Original balances.
for (i = 1 to n) { // Month i.
        P := P[i]; I := I[i];  // Pool P&I for month i.
        for (j = 1 to 3)   { // Tranches A, B, C.
          I[j][i] := B[j][i−1] × r;   // Interest due tranche j.
          I := I − I[j][i];
          if [ B[j][i−1] ≤ P ]   { // Retire it.
            P := P − B[j][i−1]; P[j][i] := B[j][i−1];
            B[j][i] := 0;
          } else {
            B[j][i] := B[j][i−1] − P; P[j][i] := P; P := 0;
          }
        }
        for (j = 1 to 3)  { // Interest as prepayment for A, B, C.
          if [ B[j][i] ≤ I ]  { // Retire it.
            I[j][i] := I[j][i] + B[j][i]; I := I − B[j][i];
            B[j][i] := 0;
          } else {
            B[j][i] := B[j][i] − I; I[j][i] := I[j][i] + I;
            I := 0;
          }
        }
        // Tranche Z.
        I[4][i] := I; P[4][i] := P;
        B[4][i] := B[4][i−1] × (1+r) − P − I;
}
return B[][], P[][] I[][];
```

Figure 29.12: Sequential CMO cash flow generator. SMM is the prepayment vector, and the mortgage rate r is a monthly rate. The pool has n monthly cash flows, and its principal balance is assumed to be \$1. B stores the remaining principals, P are the principal payments (prepayments included), and I are the interest payments. Tranche 1 is the A tranche, tranche 2 is the B tranche, and so on. O stores the original balances of individual tranches as fractions of \$1.

the average life. Given a static prepayment vector, the WAL increases with coupon rates because a larger proportion of the payment in early years is then interest, delaying the repayment of principal. The implied PSA is sometimes defined as the single PSA speed that gives the same WAL as the static prepayment vector.

29.2 Collateral Prepayment Modeling

The interest rate level is the most important factor in influencing prepayment speeds. The MBS typically experiences accelerating prepayments after a lag when the prevailing mortgage rate becomes 200 basis points below the WAC. This event is known

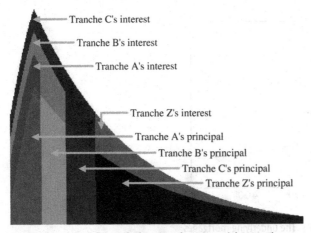

Figure 29.13: Cash flows of a four-tranche sequential CMO. The mortgage rate is 6%, the actual prepayment speed is 150 PSA, and each tranche has an identical original principal amount.

as the **threshold for refinancing**. The prepayment speed accelerates rapidly and then tends to "**burn out**" and settle at a lower speed. The subsequent times when rates fall through the refinancing threshold will not produce the same response. Over time, the pool is left mostly with mortgagors who do not refinance under any circumstances, and the pool's interest rate sensitivity falls. Next to refinancing incentive, loan size is also critical as the monetary savings are proportional to it [4].

The age of a pool has a general impact on prepayments. Refinancing rates are generally lower for new loans than seasoned ones. Interest rate changes and other human factors have little impact on prepayment speeds for the early years of the pool's life. Afterwards, the pool begins to experience such factors that can lead to higher prepayment speeds, such as the sale of the house. This increase in prepayment speeds will stabilize to a steady state with age. We must add that given sufficient refinancing incentives, prepayment speeds can rise sharply even for new loans.

Refinancing is not the only reason prepayments accelerate when interest rates decline. Lower interest rates make housing more affordable and may trigger the trade-up to a bigger house. However, by and large, very high prepayment speeds are primarily due to refinancings, not housing turnover.

In prepayment modeling, the WAC instead of the pass-through rate is the governing factor. To start with, MBSs with identical pass-through rate may have different WACs, which almost surely result in different prepayment characteristics. The

Figure 29.14: Remaining principal balances of a four-tranche sequential CMO. The CMO structure is identical to the one in Fig. 29.13. Tranche Zs principal balance grows until it becomes the current-pay tranche.

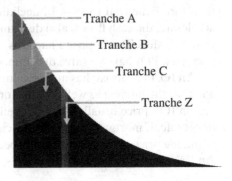

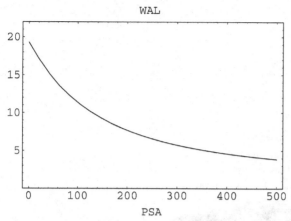

Figure 29.15: WAL under various PSAs. The underlying mortgages have 30 years to maturity and a 6% coupon rate.

WAC may also change over time because, absent prepayments, mortgages with lower coupons amortize faster than those with higher coupons (see Exercise 29.1.2). This makes the WAC increase over time. With prepayments, however, mortgages with higher coupons prepay faster, making the pool's WAC decline over time.

Each mortgage type (government-insured, conventional, and so forth) has a different prepayment behavior. For example, Freddie Mac and Fannie Mae pass-throughs seem to take longer to season than Ginnie Maes, and prepayment rates for 15-year mortgage pass-throughs usually exceed those of comparable-coupon 30-year pass-throughs [54, 325].

From the analysis above, a prepayment model needs at least the following factors: current and past interest rates, state of the economy (especially the housing market), WAC, current coupon rate, loan age, loan size, agency and pool type, month of the year, and burnout. Although we have discussed prepayment speeds at the pool level, a model may go into individual loans to generate the pool's cash flow if such information is available and the benefits outweigh the costs [4, 259]. A long-term average of the projected speeds is typically reported as the model's projected prepayment vector. This projection can be a weighted average of the projected speeds, the single speed that gives the same weighted average life as the vector, or the single speed that gives the same yield as the vector [433].

A PO is purchased at a discount. Because its cash flow is returned at par, a PO's dollar return is simply the difference between the par value and the purchase price. The faster that dollar return is realized, the higher the yield. Prepayments are therefore beneficial to POs. In declining mortgage rates, not only do prepayments accelerate, the cash flow is also discounted at a lower rate; consequently, POs appreciate in value. The opposite happens when mortgage rates rise (see Fig. 29.16). In summary, POs have positive duration and do well in bull markets.

An IO, in contrast, has no par value. Any prepayments reduce the pool principal and thus the interest as well. When mortgage rates decline and prepayments accelerate, an IO's price usually declines even though the cash flow will be discounted at a lower rate. If mortgage rates rise, the cash flow improves. However, beyond a certain point, the price of an IO will decline because of higher discount rates (see Fig. 29.17). An IO's price therefore moves in the same direction as the change in mortgage rates

PO price

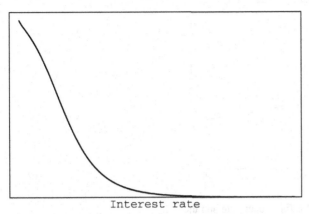

Interest rate

Figure 29.16: Price of PO.

over certain ranges (negative duration, in other words). Unlike most fixed-income securities, IOs do best in bear markets.

SMBSs are extremely sensitive to changes in prepayment speeds (see Exercise 29.2.2). These securities are often combined with other types of securities to alter the return characteristics. For example, because the PO thrives on the acceleration of prepayment speeds, it serves as an excellent hedge against MBSs whose price flattens or declines if prepayments accelerate, whereas IOs can hedge the interest rate risk of securities with positive duration.

▶ **Exercise 29.2.1** Divide the borrowers into slow and fast refinancers. (More refined classification is possible.) The slow refinancers are assumed to respond to refinancing incentive at a higher rate than fast refinancers. Describe how this setup models burnout.

▶ **Exercise 29.2.2** From Exercise 29.1.12, show that the prices of PO and IO strips are extremely sensitive to prepayment speeds.

IO price

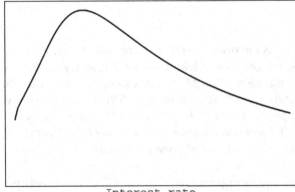

Interest rate

Figure 29.17: Price of IO. IOs and POs do not have symmetric exposures to rate changes.

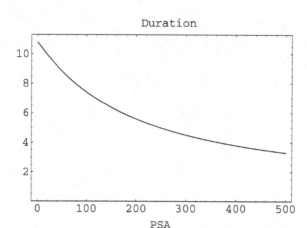

Figure 29.18: MD under various PSAs. The coupon rate and the market yield are assumed to be 6%. The underlying mortgages have 30 years to maturity.

▶ **Exercise 29.2.3** Firms that derive income from servicing mortgages can be viewed as taking a long position in IOs. Why?

29.3 Duration and Convexity

Duration is more important for the evaluation of pass-throughs than the WAL, which measures the time to the receipt of the principal cash flows [247, 619]. Figure 29.18 illustrates the Macaulay duration (MD) of a pass-through under various prepayment assumptions. The MD derived under a static prepayment vector, which does not change as yields change, is also called **static duration** or **cash flow duration**.

Duration is supposed to reveal how a change in yields affects the price, that is,

$$\text{percentage price change} \approx -\text{effective duration} \times \text{yield change.} \qquad (29.12)$$

Relation (29.12) has obvious applications in hedging. However, static duration is inadequate for that purpose because the cash flow of an MBS depends on the prevailing yield. The most relevant measure of price volatility is the effective duration,

$$\frac{\partial P}{\partial y} \approx \frac{P_- - P_+}{P_0(y_+ - y_-)},$$

where P_0 is the current price, P_- is the price if yield is decreased by Δy, P_+ is the price if yield is increased by Δy, y is the initial yield, $y_+ \equiv y + \Delta y$, and $y_- \equiv y - \Delta y$. Figure 29.19 plots the effective duration of an MBS. For example, it says that the effective duration is approximately six at 9%; a 1% change in yields will thus move the price by roughly 6%. The prices P_+ and P_- are often themselves expected values calculated by simulation. To save computation time, either $(P_- - P_0)/(P_0 \Delta t)$ or $(P_0 - P_+)/(P_0 \Delta t)$ may be used instead, as only one of P_- and P_+ needs to be calculated then.

Similarly, convexity $\partial^2 P/\partial y^2$ can be approximated by the effective convexity:

$$\frac{P_+ + P_- - 2 \times P_0}{P_0[\, 0.5 \times (y_+ - y_-)\,]^2}.$$

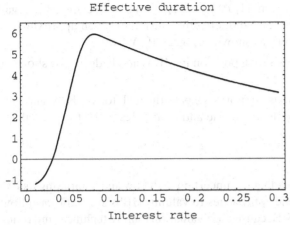

Figure 29.19: Effective duration. The MBS is from Fig. 28.8.

See Fig. 29.20 for an illustration. Convexity can improve first-order formula (29.12) by adding second-order terms,

$$\text{percentage price change} \approx -\text{effective duration} \times \text{yield change}$$
$$+ 0.5 \times \text{convexity} \times (\text{yield change})^2.$$

We saw in Fig. 28.8 that an MBS's price increases at a decreasing rate as the yield falls below the cusp because of accelerating prepayments, at which point it starts to decrease. This negative convexity is evident in Fig. 29.20. Therefore, even if the MD, which is always positive, is acceptable for current-coupon and moderately discount MBSs, it will not work for premium-priced MBSs.

▶ **Exercise 29.3.1** Suppose that MBSs are priced based on the premise that there are no prepayments until the 12th year, at which time the pool is repaid completely. This is called the **FHA 12-year prepaid-life concept**. Argue that premium-priced MBSs are overvalued and discount MBSs are undervalued if prepayments occur before the 12th year. (Studies have shown that the average life is much shorter than 12 years [577].)

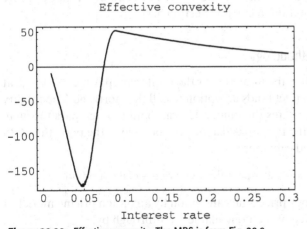

Figure 29.20: Effective convexity. The MBS is from Fig. 28.8.

➤ **Exercise 29.3.2** Modified duration $(1/P) \sum_{i=1}^{n} i C_i (1+y)^{-(i+1)}$ cannot be negative for pass-throughs. On the other hand, effective duration, which approximates modified duration, can be negative, as shown in Fig. 29.19. Why?

➤ **Exercise 29.3.3** A hedger takes a long position in MBSs and hedges it by shorting T-bonds. Assess this strategy.

➤ **Exercise 29.3.4** Consider options on mortgage pass-through forwards. Argue that Black's model tends to overstate the call value and to underestimate the put value.

29.4 Valuation Methodologies

Mortgage valuation involves modeling the uncertain cash flow and computing its PV. As in Section 27.4, the three basic approaches to valuing MBSs are static cash flow yield, option modeling, and OAS. Because their valuation is more technical and relies more on judgment than do other fixed-income securities, not to mention such issues as prepayment risk, credit quality, and liquidity, MBSs are priced to a considerable yield spread over the Treasuries and corporate bonds.

29.4.1 The Static Cash Flow Yield Methodology

When an internal rate of return is calculated with the static prepayment assumption over the life of the security, the result is the (static) cash flow yield, we recall. The static cash flow yield methodology compares the cash flow yield on an MBS with that on "comparable" bonds. For this purpose, it is inappropriate to use the stated maturity of the MBS because of prepayments. Instead, either the MD or the WAL under the same prepayment assumption can be used.

Although simple to use, this methodology sheds little light on the relative value of an MBS. Its problems, besides being static, are that (1) the projected cash flow may not be reinvested at the cash flow yield, (2) the MBS may not be held until the final payout date, and (3) the actual prepayment behavior is likely to deviate from the assumptions.

The static spread methodology goes beyond the cash flow yield by incorporating the Treasury yield curve. The static spread to the Treasuries is the spread that makes the PV of the projected cash flow from the MBS when discounted at the spot rate plus the spread equal its market price (review Section 5.4).

29.4.2 The Option Pricing Methodology

Virtually all mortgage loans give the homeowner the right to prepay the mortgage at any time. The homeowner in effect holds an option to call the mortgage. The totality of these rights to prepay constitutes the embedded call option of the pass-through. Because the homeowner has the right to call a pro rata portion of the pool, the MBS investor is short the embedded call; therefore,

pass-through price = noncallable pass-through price − call option price.

The option pricing methodology prices the call option by an option pricing model. It then estimates the market price of the noncallable pass-through by

noncallable pass-through price = pass-through price + call option price.

The preceding price is finally used to compute the yield on this theoretical bond that does not prepay. This yield is called the option-adjusted yield.

The option pricing methodology was criticized in Subsection 27.4.2. It has additional difficulties here. Prepayment options are often "irrationally" exercised. Furthermore, a partial exercise is possible as the homeowner can prepay a portion of the loan; there is not one option but many, one per homeowner. Finally, valuation of the call option becomes very complicated for CMO bonds.

29.4.3 The Option-Adjusted-Spread Methodology

The OAS methodology has four major parts [382]. The interest rate model is the first component. Then there is the prepayment model, which is the single most important component. Although the prepayment model may be deterministic or stochastic, there is evidence showing that deterministic models that are accurate on average are good enough for pass-throughs, IOs, and POs [428, 433]. The **cash flow generator** is the third component. It calculates the current coupon rates for the interest rate paths given by the interest rate model. It then generates the P&I cash flows for the pool as well as allocating them for individual securities based on the prepayment model and security information such as CMO rules. Note that the same pool cash flow drives many securities. Finally, the equation solver calculates the OAS. Because several paths of interest rates are used, many statistics are often computed as well. See Fig. 29.21 for the overall structure.

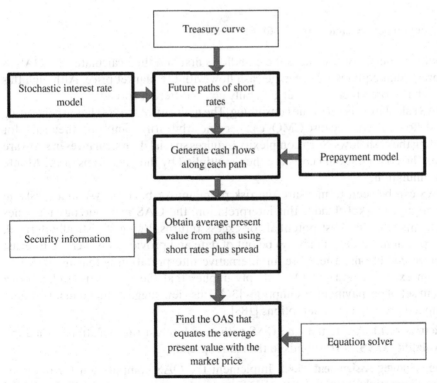

Figure 29.21: OAS computation framework for MBSs. Components boxed by thinner borders are supplied externally.

The general valuation formula for uncertain cash flows can be written as

$$PV = \lim_{N \to \infty} \frac{1}{N} \sum_{N \text{ paths } r^*} \frac{C_n^*}{(1+r_1^*)(1+r_2^*) \cdots (1+r_n^*)}, \tag{29.13}$$

where r^* denotes a risk-neutral interest rate path for which r_i^* is the ith one-period rate and C_n^* is the cash flow at time n under this scenario. The summation averages over a large number of scenarios whose distribution matches the interest rate dynamics. The average over scenarios must also match the current spot rate curve, i.e.,

$$\frac{1}{(1+f_1)(1+f_2) \cdots (1+f_n)} = \lim_{N \to \infty} \frac{1}{N} \sum_{N \text{ paths } r^*} \frac{1}{(1+r_1^*)(1+r_2^*) \cdots (1+r_n^*)},$$

$n = 1, 2, \ldots$, where f_i are the implied forward rates.

The Monte Carlo valuation of MBSs is closely related to Eq. (29.13). The interest rate model randomly produces a set of risk-neutral rate paths. The cash flow is then generated for each path. Finally, we solve for the spread s that makes the average discounted cash flow equal the market price:

$$P = \lim_{N \to \infty} \frac{1}{N} \sum_{N \text{ paths } r^*} \frac{C_n^*}{(1+r_1^*+s)(1+r_2^*+s) \cdots (1+r_n^*+s)}.$$

This spread s is the OAS. The implied cost of the embedded option is then calculated as

$$\text{option cost} = \text{static spread} - \text{OAS}.$$

A common alternative averages the cash flows first and then calculates the OAS as the spread that equates this average cash flow with the market price. Although this approach is more efficient, it will generally give a different spread.

OAS calculation is very time consuming. The majority of the cost lies in generating the cash flows. This is because CMOs can become arbitrarily complex in their rules for allocating the cash flows. Such complexity requires special **data structures** in software design. The computational costs are then multiplied by the many runs of the Monte Carlo simulation.

OAS can be seen to measure the risk premium for bearing systematic risks in the mortgage market. Under this interpretation, the OAS methodology identifies investments with the best potential for excess returns. Being statistically derived, the prepayment model will always be out of date and provide only a crude forecast for future conditions. Therefore an alternative interpretation is that no such risk premium exists: A nonzero OAS simply implies that the market is trading off a different set of prepayment assumptions [34]. This view suggests that one investigate the implied prepayment assumptions [188].

➤ **Exercise 29.4.1** Argue that the OAS with zero interest rate volatility, called the **zero-volatility OAS**, corresponds to the static spread.

➤ **Programming Assignment 29.4.2** Implement the OAS computation for the four-tranche sequential CMO under the BDT model. Assume a constant SMM.

Duration and Convexity

Effective duration and convexity can be computed if the OAS is held constant. The results are called the OAS duration and the **OAS convexity**, respectively [323, 325]. Key rate durations, introduced in Section 27.5 and calculated like the OAS duration, are most useful in identifying the segments of the yield curve that most affect the MBS value [260]. Note that the OAS duration is at least twice as expensive as the OAS in terms of computation time because at least one of P_+ and P_- has to be computed by simulation. The OAS convexity is three times as expensive because both P_+ and P_- have to be computed.

Prepayment risk can represent the risk that the market price reflects prepayment assumptions that are different from the model. An interesting measure of prepayment risk is the **prepayment duration**. It is the percentage change in price, with the OAS held constant, for a given percentage deviation in speeds from some base level projection (see Exercise 29.2.2) [198, 328, 433, 815].

Holding Period Returns

The HPR assesses the MBS over a holding period. The FV at the horizon consists of the projected P&I cash flows, the interest on the reinvestment thereof, and the projected horizon price. The monthly total return is

$$\left(\frac{\text{total future amount}}{\text{price of the MBS}} \right)^{1/\text{number of months}} - 1.$$

To calculate the preceding return, prepayment assumptions, reinvestment rates, and interest rate dynamics are all needed. These assumptions are not independent.

The OAS can be combined with the HPR analysis. First we create a few static interest rate and prepayment scenarios for the holding period. The prepayment assumptions are in the form of prepayment vectors. We then calculate the HPR for each scenario by assuming that the OAS remains unchanged at the horizon.

Additional Reading

See [54, 55, 260, 330, 829] for more information on MBSs, [54, 55, 124, 259, 260, 276, 297, 323, 325, 330, 595, 619, 649, 788, 789, 818, 896] for the valuation of MBSs, [134, 188, 197, 198, 438, 454] for OAS analysis, and [142, 715] for the Monte Carlo valuation of MBSs. Monte Carlo simulation typically provides an unbiased estimate [478]. Application of the variance-reduction techniques and quasi-Monte Carlo methods in Chap. 18 can result in less work [197, 354]. Parallel processing for much faster performance has been convincingly demonstrated [601, 794, 892, 893]. Additional information on duration measures can be found in [33, 258, 272, 394, 429, 504, 889]. Many yield concepts are discussed in [406]. See [118, 220, 268, 361, 411, 430, 431, 433, 540] for prepayment models, [260] for a historical account, and [296] for early models. Factors used in prepayment modeling are considered in [54, 330, 430, 433]. The FHA 12-year prepaid-life concept is discussed in [54, 363]. Valuation of MBSs may profit from two-factor models because prepayments tend to depend more on the long-term rate [456]. See [316] for the prepayments of multifamily MBSs and [203, 742, 864] for the empirical analysis of prepayments. Burnout modeling is discussed in [199]. The refinancing waves of 1991–1993 cast some doubts on the burnout concept, however

[433]. Consult [524] for hedging MBSs and [460] for options on MBSs. The 11th District COFI is analyzed in [347, 684], and the CMT rates are compared with the on-the-run yields in [525].

NOTES

1. Also known as the **weighted average remaining maturity (WARM)**.
2. **Payment delays** should be incorporated in the WAL calculation: 14 (actual) or 45 days (stated) for GNMA Is, 19 (actual) or 50 days (stated) for GNMA IIs, 24 (actual) or 55 (stated) for Fannie Mae MBSs, 44 (actual) or 75 (stated) for Freddie Mac non-Gold PCs, and 14 (actual) or 45 (stated) for Freddie Mac Gold PCs. The **stated payment delay** denotes the number of days between the first day of the month and the date the servicer actually remits the P&I to the investor [54, 330].

CHAPTER
THIRTY

Collateralized Mortgage Obligations

Capital can be understood only as motion, not as a thing at rest.
Karl Marx (1818–1883), *Das Kapital*

Mutual funds combine diverse financial assets into a portfolio and issue a single class of securities against it. CMOs reverse that process by issuing a diverse set of securities against a relatively homogeneous portfolio of assets [660]. This chapter surveys CMOs. The tax treatment of CMOs is generally covered under the provisions of the Real Estate Mortgage Investment Conduit (REMIC) rules of 1986. As a result, CMOs are often referred to as **REMICs** [162, 469].

30.1 Introduction

The complexity of a CMO arises from layering different types of payment rules on a prioritized basis. In the first-generation CMOs, the sequential-pay CMOs, each class of bond would be retired sequentially. A sequential-pay CMO with a large number of tranches will have very narrow cash flow windows for the tranches. To further reduce prepayment risk, tranches with a principal repayment schedule were introduced. They are called **scheduled bonds**. For example, bonds that guarantee the repayment schedule when the actual prepayment speed lies within a specified range are known as **planned amortization class** bonds (**PACs**). PACs were introduced in August 1986 [141]. Whereas PACs offer protection against both contraction and extension risks, some investors may desire protection from only one of these risks. For them, a bond class known as the **targeted amortization class** (**TAC**) was created.

Scheduled bonds expose certain CMO classes to less prepayment risk. However, this can occur only if the redirection in the prepayment risk is absorbed as much as possible by other classes referred to as the **support bonds** or **companion bonds**. Support bonds are a necessary by-product of the creation of scheduled tranches.

Pro rata bonds provide another means of layering. Principal cash flows to these bonds are divided proportionally, but the bonds can have different interest payment rules. Suppose the WAC of the collateral is 10%, tranche B1 receives 40% of the principal, and tranche B2 receives 60% of the principal. Given this pro rata structure, many choices of interest payment rules are possible for B1 and B2 as long as the interest payments are nonnegative and the WAC does not exceed 10%. The coupon rates can even be floating. One possibility is for B1 to have a coupon of 5% and B2 to have a coupon of 13.33%. Bonds with pass-through coupons that are higher and lower than the collateral coupon have thus been created. Bonds like B1 are called **synthetic discount securities** and bonds like B2 are called **synthetic premium securities**. An extreme case is for B1 to receive 99% of the principal and have a 5% coupon and B2 to receive only 1% of the principal and have a 505% coupon. In fact, first-generation IOs took the form of B2 in July 1986 [55].

IOs have either a **nominal principal** or a notional principal. A nominal principal represents actual principal that will be paid. It is called "nominal" because it is extremely small, resulting in an extremely high coupon rate. A case in point is the B2 class with a 505% coupon above. A notional principal, in contrast, is the amount on which interest is calculated. An IO holder owns none of the notional principal. Once the notional principal amount declines to zero, no further payments are made on the IO.

30.2 Floating-Rate Tranches

A form of pro rata bonds are floaters and inverse floaters whose combined coupon does not exceed the collateral coupon. A floater is a class whose coupon rate varies directly with the change in the reference rate, and an inverse floater is a class whose coupon rate changes in the direction opposite to the change in the reference rate. When the coupon on the inverse floater changes by x times the amount of the change in the reference rate, this multiple x is called its **slope**. Because the interest comes from fixed-rate mortgages, floaters must have a coupon cap. Similarly, inverse floaters must have a coupon floor. Floating-rate classes were created in September 1986.

Suppose the floater has a principal of P_f and the inverse floater has a principal of P_i. Define $\omega_f \equiv P_f/(P_f + P_i)$ and $\omega_i \equiv P_i/(P_f + P_i)$. To make the structure self-supporting, the coupon rates of the floater, c_f, and the inverse floater, c_i, must satisfy $\omega_f \times c_f + \omega_i \times c_i = \text{WAC}$, or

$$c_i = \frac{\text{WAC} - \omega_f \times c_f}{\omega_i}.$$

The slope is clearly ω_f/ω_i. To make sure that the inverse floater will not encounter a negative coupon, the cap on the floater must be less than WAC/ω_f. In fact, caps and floors are related by

$$\text{floor} = \frac{\text{WAC} - \omega_f \times \text{cap}}{\omega_i}.$$

EXAMPLE 30.2.1 Consider a CMO deal that includes a floater with a principal of $64 million and an inverse floater with a principal of $16 million. The coupon rate for the floating-rate class is LIBOR+ 0.65 and that for the inverse floater is $42.4 - 4 \times$ LIBOR The slope is thus four. The WAC of the two classes is

$$\frac{64}{80} \times \text{floater coupon rate} + \frac{16}{80} \times \text{inverse floater coupon rate} = 9\%,$$

regardless of the level of LIBOR. Consequently the coupon rate on the underlying collateral, 9%, can support the aggregate interest payments that must be made to these two classes. If we set a floor of 0% for the inverse floater, the cap on the floater is 11.25%.

A variant of the floating-rate CMO is the **superfloater** introduced in 1987. In a conventional floating-rate class, the coupon rate moves up or down on a one-to-one basis with the reference rate subject to caps and floors. A superfloater's coupon rate, in comparison, changes by some multiple of the change in the reference rate, thus magnifying any changes in the value of the reference rate. Superfloater tranches are bearish because their value generally appreciates with rising interest rates.

Suppose that the initial LIBOR is 7% and the coupon rate for a superfloater is based on this formula:

$$(\text{initial LIBOR} - 40 \text{ basis points}) + 2 \times (\text{change in LIBOR}).$$

The following table shows how the superfloater changes its coupon rate as LIBOR changes. The coupon rates for a conventional floater of LIBOR plus 50 basis points are also listed for comparison.

LIBOR Change (Basis Points)	−300	−200	−100	0	+100	+200	+300
Superfloater	0.6	2.6	4.6	6.6	8.6	10.6	12.6
Conventional floater	4.5	5.5	6.5	7.5	8.5	9.5	10.5

A superfloater provides a much higher yield than a conventional floater when interest rates rise and a much lower yield when interest rates fall or remain stable. We verify this by looking at the above table by means of spreads in basis points to LIBOR in the next table:

LIBOR Change (Basis Points)	−300	−200	−100	0	+100	+200	+300
Superfloater	−340	−240	−140	−40	60	160	260
Conventional floater	50	50	50	50	50	50	50

▶ **Exercise 30.2.1** Repeat the calculations in the text by using the following formula:

$$(\text{initial LIBOR} - 50 \text{ basis points}) + 1.5 \times (\text{change in LIBOR}).$$

▶ **Exercise 30.2.2** Argue that the maximum coupon rate that could be paid to a floater is higher than would be possible without the inclusion of an inverse floater.

30.3 PAC Bonds

PAC bonds may be the most important innovation in the CMO market [141]. They are created by calculation of the cash flows from the collateral by use of two prepayment speeds, a fast one and a slow one. Consider a **PAC band** of 100 PSA (the **lower collar**) to 300 PSA (the **upper collar**). Figure 30.1 shows the principal payments at the two collars. Note that the principal payments under the higher-speed scenario are higher in the earlier years but lower in later years. The shaded area represents the principal payment schedule that is guaranteed for every possible prepayment speed between 100% and 300% PSAs. It is calculated by taking the minimum of the principal paydowns at the lower collar and those at the upper collar. This schedule is called the **PAC schedule**. See Fig. 30.2 for a linear-time cash flow generator for a simple CMO containing a PAC bond and a support bond.

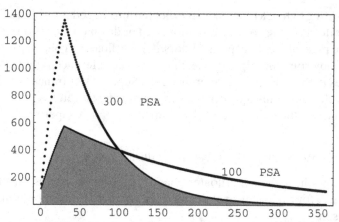

Figure 30.1: PAC schedule. The underlying mortgages are 30-year ones with a total original loan amount of $100,000,000 (the numbers on the y axis are in thousands) and a coupon rate of 6%. The PAC schedule is determined by the principal payments at 100 PSA and 300 PSA.

Adherence to the amortization schedule of the PAC takes priority over those of all other bonds. The cash flow of a PAC bond is therefore known as long as its support bonds are not fully paid off. Whether this happens depends to a large extent on the CMO structure, such as priority and the relative sizes of PAC and non-PAC classes. For example, a relatively small PAC is harder to break than a larger PAC, other things being equal.

If the actual prepayment speed is 150 PSA, the principal payment pattern of the PAC bond adheres to the PAC schedule. The cash flows of the support bond "flow around" the PAC bond (see Fig. 30.3). The cash flows are neither sequential nor pro rata; in fact, the support bond pays down *simultaneously* with the PAC bond. Because more than one class of bonds may be receiving principal payments at the same time, structures with PAC bonds are called **simultaneous-pay CMOs**. At the lower prepayment speed of 100 PSA, far less principal cash flow is available in the early years of the CMO. As all the principal cash flows go to the PAC bond in the early years, the principal payments on the support bond are deferred and the support bond extends. The support bond does, however, receive more interest payments.

If prepayments move outside the PAC band, the PAC schedule may not be met. At 400 PSA, for example, the cash flows to the support bond are accelerated. After the support bond is fully paid off, all remaining principal payments go to the PAC bond, shortening its life. See Fig. 30.4 for an illustration. The support bond thus absorbs part of the contraction risk. Similarly, should the actual prepayment speed fall below the lower collar, then in subsequent periods the PAC bond has priority on the principal payments. This reduces the extension risk, which is again absorbed by the support bond.

The PAC band guarantees that if prepayments occur at any single constant speed within the band *and* stay there, the PAC schedule will be met. However, the PAC schedule may not be met even if prepayments on the collateral always vary within the band over time. This is because the band that guarantees the original PAC schedule can expand and contract, depending on actual prepayments. This phenomenon is known as **PAC drift**.

PAC cash flow generator:

```
input:    n, r (r > 0), SMM[1..n], PSA_u, PSA_ℓ, 𝒪[2];
real      P[1..n], Ī[1..n]; // Pool cash flows.
real      smm[1..n], B[2][n+1], P[2][1..n], Ī[2][1..n], P, I;
integer   i;
Call the algorithm in Fig. 29.10 with SMM[1..n] for
          P[1..n] and Ī[1..n]; // Pool cash flows.
Call the algorithm in Fig. 29.9 for smm[1..n] based on PSA_u;
Call the algorithm in Fig. 29.10 with smm[1..n] and
          store the principal cash flow in P[0][1..n];
Call the algorithm in Fig. 29.9 for smm[1..n] based on PSA_ℓ;
Call the algorithm in Fig. 29.10 with smm[1..n] and
          store the principal cash flow in P[1][1..n];
for (i = 1 to n) {P[0][i] := min(P[0][i], P[1][i]); }
// PAC schedule per one dollar of original principal:
Normalize P[0][1..n] so that the n elements sum to one;
B[0][0] := 𝒪[0]; B[1][0] := 𝒪[1]; // Original balances.
for (i = 1 to n) { // Month i.
          P := P[i]; I := Ī[i]; // Pool P&I for month i.
          P[1][i] := min(0, P − 𝒪[0] × P[0][i], B[1][i−1]);
          B[1][i] := B[1][i−1] − P[1][i];
          Ī[1][i] := B[1][i−1] × r; // Support bond done.
          P := P − P[1][i];
          P[0][i] := P;
          B[0][i] := B[0][i−1] − P;
          Ī[0][i] := I − Ī[1][i]; // PAC bond done.
}
return B[ ][ ], P[ ][ ], Ī[ ][ ];
```

Figure 30.2: PAC cash flow generator. SMM[ ] stores the actual prepayment speeds, PSA$_u$ and PSA$_ℓ$ form the PAC band, and the mortgage rate r is a monthly rate. The pool has n monthly cash flows, and its principal balance is assumed to be \$1. $\mathcal{O}$ stores the original balances of individual bonds as fractions of \$1; in particular, bond 0 is the PAC bond, bond 1 is the support bond, and $\mathcal{O}[0] + \mathcal{O}[1] = 1$. B stores the remaining principals, P are the principal payments (prepayments included), and $\bar{I}$ are the interest payments. The CMO deal contains one PAC tranche and one support tranche.

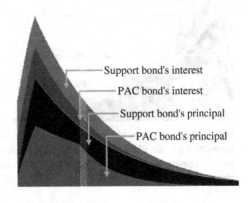

Figure 30.3: Cash flows of a PAC bond at 150 PSA. The mortgage rate is 6%, the PAC band is 100 PSA to 300 PSA, and the actual prepayment speed is 150 PSA.

Support bond's interest

PAC bond's interest

Support bond's principal

PAC bond's principal

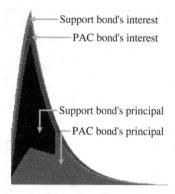

Figure 30.4: Cash flows of a PAC bond at 400 PSA. The mortgage rate is 6%, the PAC band is 100 PSA to 300 PSA, and the actual prepayment speed is 400 PSA.

PACs can be divided sequentially to provide narrower paydown structures. These **sequential PACs** narrow the range of years over which principal payments occur. See Fig. 30.5 for an illustration. Although these bonds are all structured with the same band, the actual range of speeds over which their schedules will be met may differ. We can take a CMO bond and further structure it. For example, the sequential PACs could be split by use of a pro rata structure to create high and low coupon PACs. We can also replace the second tranche in a four-tranche ABCZ sequential CMO with a PAC class that amortizes starting in year four, say. But note that tranche C may start to receive prepayments that are in excess of the schedule of the PAC bond. It may even be retired earlier than tranche B.

Support bonds themselves can have cash flows prioritized so as to reduce prepayment risk. Support bonds with schedules, also referred to as **PAC II bonds**, are supported by other support bonds without schedules. PACs in a structure in which there are PAC II level bonds are called **PAC I bonds**.

➤ **Programming Assignment 30.3.1** Implement the cash flow generator in Fig. 30.2.

➤ **Programming Assignment 30.3.2** Implement the cash flow generator for sequential PAC bonds.

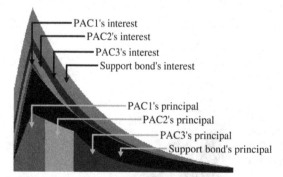

Figure 30.5: Cash flows of sequential PAC bonds. The mortgage rate is 6%, the PAC band is 100 PSA to 300 PSA, and the actual prepayment speed is 150 PSA. The three PAC bonds have identical original principal amounts.

30.4 TAC Bonds

Created in 1986, TAC bonds, just as PAC bonds, have priority over other bond classes that do not have a schedule for principal repayment. PACs have a higher priority over TACs, however. TAC bonds have a single PSA prepayment speed over which the principal repayment schedule is guaranteed. When prepayments exceed the speed, the excess principal is paid to the support bonds first. However, when prepayments fall short of the speed, TAC bonds will extend. TACs are therefore designed to provide protection against contraction risk but not extension risk.

30.5 CMO Strips

A class in a CMO structure can be a **CMO strip**. A CMO strip that is created when an IO is stripped from a CMO bond is called a **bond IO**. For example, this stripping mechanism creates an **inverse IO** from an inverse floater and a **PAC IO** from a PAC bond. Bond IOs lower the coupon of the CMO tranche. Some people call bond IOs **IOettes** to distinguish them from IO strips created off the entire collateral. A PO class that is neither a PAC nor a TAC is called a **super PO**. Like a PO strip, such bonds are purchased at a substantial discount from par and are returned at par. When prepayments accelerate as interest rates decline, "super" performance follows, hence the name.

30.6 Residuals

All CMOs contain a residual interest composed of the excess of collateral cash flows plus any reinvestment income over the payments for principal, interest, and expenses. This excess cash flow is called the **CMO residual**. The residual arises in part because credit rating agencies require CMOs to be overcollateralized in order to receive AAA credit rating: The cash flows must be sufficient to meet all the obligations under any prepayment scenario.

Another source of residual cash flow is reinvestment income. There is usually a delay between the time the payments from the collateral are received and the time they are remitted to the CMO bondholders. For example, whereas the mortgages in the collateral pay monthly, most CMOs pay quarterly or semiannually. The CMO trustee is therefore able to reinvest the pool cash flows before distribution dates. To be conservative in calculating the funds needed to meet future obligations, the rating agencies require that the trustee assume a relatively low reinvestment rate. CMO trustees have been able to reinvest at higher rates, and the excess is retained as a residual.

Additional Reading

See [14, 54, 161, 260, 325, 439, 758] for in-depth analyses of CMOs.

CHAPTER THIRTY-ONE

Modern Portfolio Theory

> Truly important and significant hypotheses will be found to have
> "assumptions" that are wildly inaccurate descriptive representations
> of reality.
>
> Milton Friedman,*"The Methodology of Positive Economics"*

This chapter starts with the **mean-variance theory** of portfolio selection. This theory
provides a tractable framework for quantifying the risk–return trade-off of assets. We
then investigate the equilibrium structure of asset prices. The result is the celebrated
Capital Asset Pricing Model (**CAPM**, pronounced cap-m). The CAPM is the foun-
dational quantitative model for measuring the risk of a security. Alternative asset
pricing models based on factor analysis are also presented. The practically important
concept of value at risk (VaR) for risk management concludes the chapter.

31.1 Mean–Variance Analysis of Risk and Return

Risk is the chance that expected returns will not be realized. We adopt standard
deviation of the rate of return as the measure of risk.[1] This choice, although not
without its critics, is standard in portfolio analysis and has nice statistical properties.
Investors are presumed to prefer higher expected returns and lower variances.

Assume that there are n assets with random rates of return, $r_1, r_2, \ldots, r_n$. The
expected values of these returns are $\bar{r}_i \equiv E[r_i]$. If we form a portfolio of these n
assets by using (capitalization) weights $\omega_1, \omega_2, \ldots, \omega_n$, the portfolio's rate of return
is

$$ r = \omega_1 r_1 + \omega_2 r_2 + \cdots + \omega_n r_n $$

with mean $\bar{r} = \sum_{i=1}^{n} \omega_i \bar{r}_i$ and variance

$$ \sigma^2 = \sum_{i=1}^{n} \sum_{j=1}^{n} \omega_i \omega_j \sigma_{ij} = \sum_{i \neq j} \omega_i \omega_j \sigma_{ij} + \sum_{i=1}^{n} \omega_i^2 \sigma_i^2, $$

where σ_i^2 represents the variance of r_i and σ_{ij} represents the covariance between
r_i and r_j. Note that $\sigma_{ii} = \sigma_i^2$.

The portfolio's total risk as measured by its variance consists of (1) $\sum_{i \neq j} \omega_i \omega_j \sigma_{ij}$,
the **systematic risk** associated with the correlations between the returns on the as-
sets in the portfolio, and (2) $\sum_{i=1}^{n} \omega_i^2 \sigma_i^2$, the **specific** or **unsystematic risk** associated

with the individual variances alone. Every possible weighting scheme $\omega_1, \omega_2, \ldots, \omega_n$ with $\sum_{i=1}^{n} \omega_i = 1$ corresponds to a portfolio, with negative weights meaning short sales. The constraints $\omega_i \geq 0$ can be added to exclude short sales. A portfolio $\omega \equiv [\omega_1, \omega_2, \ldots, \omega_n]^{\mathrm{T}}$ that satisfies all the specified constraints is said to be a **feasible portfolio**.

Interestingly, if the returns[2] of the assets are uncorrelated, i.e., $\sigma_{ij} = 0$ for $i \neq j$, the variance of the portfolio's return decreases toward zero as n increases, provided that the portfolio is well diversified. For example, with $\omega_i = 1/n$,

$$\sigma^2 = \sum_{i=1}^{n} \omega_i^2 \sigma_i^2 = \frac{\sum_{i=1}^{n} \sigma_i^2}{n^2} \leq \frac{\sigma_{\max}^2}{n},$$

where $\sigma_{\max} \equiv \max_i \sigma_i$. This shows the power of diversification. Diversification, however, has its limits when asset returns are correlated. To see this point, assume that (1) all the returns have the same variance s^2, (2) the return correlation is a constant z, hence $\sigma_{ij} = zs^2$ for $i \neq j$, and (3) $\omega_i = 1/n$. The variance of r then is

$$\sigma^2 = \sum_{i \neq j} \frac{zs^2}{n^2} + \sum_{i=1}^{n} \frac{s^2}{n^2} = n(n-1) \frac{zs^2}{n^2} + \frac{s^2}{n} = zs^2 + (1-z) \frac{s^2}{n},$$

which cannot be reduced below the average covariance zs^2.

These two examples demonstrate that specific risk and systematic risk behave very differently as the number of assets included in the portfolio grows. In general, as the portfolio gets larger and is well diversified, the specific risk tends to zero, whereas the systematic risk converges to the average of all the covariances for all pairs of assets in the portfolio. Markowitz called this phenomenon the **law of the average covariance** [644]. Systematic risk therefore does *not* disappear with diversification.

Consider a two-dimensional diagram with the horizontal axis denoting standard deviation and the vertical axis denoting mean. This is called the **mean–standard deviation diagram**. Every feasible portfolio with mean return rate $\bar{r}$ and standard deviation σ can be represented as a point at $(\sigma, \bar{r})$ on the diagram; it is an **obtainable mean–standard deviation combination**. The set of feasible points form the **feasible set**. In general, the feasible set is a solid two-dimensional region and convex to the left. Thus the straight line segment connecting any two points in the set does not cross the left boundary of the set. For a given expected rate of return, the feasible point with the smallest variance is the corresponding left boundary point. The left boundary of the feasible set is hence called the **minimum-variance set**, and the point on this set having the minimum variance is the **minimum-variance point** (MVP). Most investors will choose the portfolio with the smallest variance for a given mean. Such investors are risk averse because they seek to minimize risk as measured by the standard deviation. Similarly, most investors will choose the portfolio with the highest mean for a given level of standard deviation (i.e., the highest point on a given vertical line). Therefore only the subset of the minimum-variance set above the MVP will be of interest. An obtainable mean–standard deviation combination is **efficient** if no other obtainable combinations have either higher mean and no higher variance or less variance and no less mean. The set of efficient combinations is termed the **efficient frontier**, and the corresponding portfolios are termed the **efficient portfolios**. See Fig. 31.1.

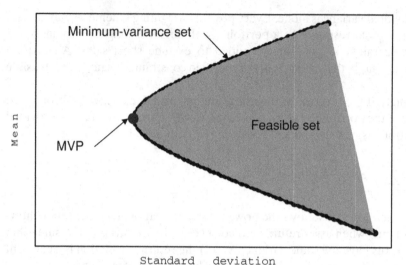

Figure 31.1: Feasible, minimum-variance, and efficient sets. The points in the minimum-variance set (that are above the MVP) form the efficient frontier, which is also called the **efficient set**. When short sales are not allowed, the feasible set is bounded because its mean lies within $[\min_i \bar{r}_i, \max_i \bar{r}_i]$ and its standard deviation lies within $[0, \max_i \sigma_i]$ (see Exercise 31.1.3).

Here is the mathematical formulation for the minimum-variance portfolio with a given mean value $\bar{r}$ that is due to Markowitz in 1952 [641]:

$$\text{minimize} \quad (1/2) \sum_{i=1}^{n} \sum_{j=1}^{n} \omega_i \omega_j \sigma_{ij},$$

$$\text{subject to} \quad \sum_{i=1}^{n} \omega_i \bar{r}_i = \bar{r},$$

$$\sum_{i=1}^{n} \omega_i = 1.$$

Short selling can be prohibited if $\omega_i \geq 0$ for $i = 1, 2, \ldots, n$, is added to the constraints. (The factor $1/2$ in front of the variance will simplify the analysis later.) The preceding **Markowitz problem** is a quadratic programming problem. It is a single-period investment theory that specifies the trade-off between the mean and the variance of a portfolio's rate of return.[3]

The Markowitz problem can be solved as follows. The weights ω_i and the two **Lagrange multipliers** λ and μ for an efficient portfolio satisfy

$$\sum_{j=1}^{n} \sigma_{ij} \omega_j - \lambda \bar{r}_i - \mu = 0 \quad \text{for } i = 1, 2, \ldots, n,$$

$$\sum_{i=1}^{n} \omega_i \bar{r}_i = \bar{r},$$

$$\sum_{i=1}^{n} \omega_i = 1.$$

There are $n+2$ equations with $n+2$ unknowns: $\omega_1, \omega_2, \ldots, \omega_n, \lambda, \mu$. Because the equations are linear, they can be easily solved (see Fig. 19.2). If the goal is to obtain the highest return for a given level of variance σ_p^2, then the problem becomes

$$\text{maximize } \sum_{i=1}^{n} \omega_i \bar{r}_i,$$

$$\text{subject to } \sum_{i=1}^{n} \sum_{j=1}^{n} \omega_i \omega_j \sigma_{ij} = \sigma_p^2,$$

$$\sum_{i=1}^{n} \omega_i = 1.$$

Sophisticated quadratic programming techniques are needed to solve it.

Striking conclusions can be drawn from the mean–variance framework. Suppose that two solutions are available: (1) $(\omega_1, \lambda_1, \mu_1)$ with expected return rate $\bar{r}_1$ and (2) $(\omega_2, \lambda_2, \mu_2)$ with expected return rate $\bar{r}_2$. Direct substitution shows that $(\alpha \omega_1 + (1-\alpha) \omega_2, \alpha \lambda_1 + (1-\alpha) \lambda_2, \alpha \mu_1 + (1-\alpha) \mu_2)$ is also a solution to the $n+2$ equations and corresponds to the expected return rate $\alpha \bar{r}_1 + (1-\alpha) \bar{r}_2$. Thus the combined portfolio $\alpha \omega_1 + (1-\alpha) \omega_2$ also represents a point in the minimum-variance set. To use this result, suppose that ω_1 and ω_2 are two different portfolios in the minimum-variance set. Then as α varies over $-\infty < \alpha < \infty$, the portfolios defined by $\alpha \omega_1 + (1-\alpha) \omega_2$ sweep out the entire minimum-variance set. In particular, if ω_1 and ω_2 are efficient, they will generate all other efficient points. This is the **two-fund theorem**. Hence all investors seeking efficient portfolios need consider investing in combinations of only these two funds instead of individual stocks. This conclusion rests on the assumptions, among others, that everyone cares about only means and variances, that everyone has the same assessment of the parameters (means, variances, and covariances), that short selling is allowed, and that a single-period framework is appropriate.

▶ **Exercise 31.1.1** Express the efficient portfolio in matrix form.

▶ **Exercise 31.1.2** Construct a portfolio with zero risk from two perfectly negatively correlated assets without short sales.

▶ **Exercise 31.1.3** Let $C \equiv [\sigma_{ij}]$ be a positive definite matrix. (1) Prove that $\max_i \sigma_{ii}$ is the maximum value of $\sum_i \sum_j \omega_i \omega_j \sigma_{ij}$ under the constraints $\sum_i \omega_i = 1$ and $\omega_i \geq 0$. (2) How about the minimum value under the same constraints? (You may assume that the row sums of C^{-1} are all nonnegative.)

▶ **Exercise 31.1.4** Let $P(t)$ denote the asset price at time t. Define $r(T) \equiv [P(T)/P(0)] - 1$ as the holding period rate of return for a period of length T and $r_c(T) \equiv \ln(P(T)/P(0))$ as the continuous holding period rate of return for the same period. Under the assumption that asset prices are lognormally distributed, derive the relations between the mean and the variance of $r(T)$ and those of $r_c(T)$.

▶ **Exercise 31.1.5** Consider a portfolio P of n assets each following an independent geometric Brownian motion process with identical mean and variance, $dS_i/S_i = \mu dt + \sigma dW_i$. Each asset has the same weight of $1/n$ in the portfolio. Show that this portfolio's expected rate of return, $E[\ln(P(t)/P(0))]/t$, exceeds each individual asset's expected rate of return, $E[\ln(S_i(t)/S_i(0))]/t$, by $(1-1/n)\sigma^2/2$. (Volatility is thus not synonymous with risk.)

31.1.1 Adding the Riskless Asset

The riskless asset by definition has a return that is certain; its return has zero volatility. The riskless return's covariance with any risky asset's return is thus zero. The presence of the riskless asset in a portfolio implies lending or borrowing cash at the riskless rate: Lending means a long position in the asset, whereas borrowing means a short position. Clearly the riskless asset has to be a zero-coupon bond whose maturity matches the investment horizon.

　　The shape of the feasible set changes dramatically when the riskless asset is available. Let r_f denote the riskless rate of return. Start with the feasible set defined by *risky* assets. Now for each portfolio in this set, say portfolio A, form combinations with the riskless asset. These new combinations trace out the infinite straight line originating at the riskless point, passing through the risky portfolio, and continuing indefinitely: the r_f–A ray in Fig. 31.2. There is a ray of this type for every portfolio in the feasible set. The totality of these rays forms a triangularly shaped feasible set. If borrowing of the riskless asset is not allowed, we can adjoin only the line segment between the riskless asset and points in the original feasible set but cannot extend the line further. The inclusion of these line segments leads to a feasible set with a straight-line front edge but a rounded top: the r_f–P–Q curve in Fig. 31.2. Note that investors who hold some riskless assets invest the remaining funds in portfolio P, as they are on the r_f–P segment.

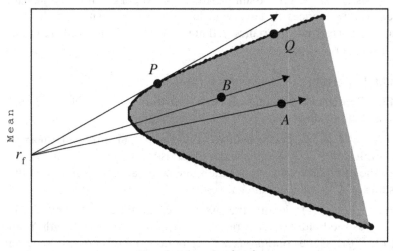

Figure 31.2: The efficient frontier with riskless lending and borrowing. The shaded area is the feasible set defined by risky assets. The line segment between r_f and A consists of combinations of portfolio A and lending, whereas the line segment beyond A consists of combinations of portfolio A and borrowing. The equation for the line is $y = r_f + x(\bar{r}_A - r_f)/\sigma_A$. The same observation can be made of any risky portfolio such as B, P, and Q. The ray through the tangent portfolio P defines the efficient frontier.

A special portfolio, denoted by P in Fig. 31.2, lies on the tangent point between the feasible set and a ray passing through r_f. When both borrowing and lending of the riskless asset are available, the efficient frontier is precisely this ray. Any efficient portfolio therefore can be expressed as a combination of P and the riskless asset. We have thus proved Tobin's **one-fund theorem**, which says there is a single fund of risky assets such that every efficient portfolio can be constructed as a combination of the fund and the riskless asset.

Identifying the tangent point P is computationally easy. For any point $(\sigma, \bar{r})$ in the feasible set defined by risky assets, we can draw a line between the riskless asset and that point as in Fig. 31.2. The slope is equal to $\theta \equiv (\bar{r} - r_f)/\sigma$, which has the interesting interpretation of the excess return per unit of risk. The tangent portfolio is the feasible point that maximizes θ. Assign weights $\omega_1, \omega_2, \ldots, \omega_n$ to the n risky assets such that $\sum_{i=1}^{n} \omega_i = 1$. The weight on the riskless asset in the tangent fund is zero. As a result, $\bar{r} - r_f = \sum_{i=1}^{n} \omega_i (\bar{r}_i - r_f)$, and

$$\theta = \frac{\sum_{i=1}^{n} \omega_i (\bar{r}_i - r_f)}{\sqrt{\sum_{i=1}^{n} \sum_{j=1}^{n} \sigma_{ij} \omega_i \omega_j}}.$$

Setting the derivative of θ with respect to each ω_j equal to zero leads to the equations

$$\lambda \sum_{i=1}^{n} \sigma_{ij} \omega_i = \bar{r}_j - r_f, \quad j = 1, 2, \ldots, n,$$

where $\lambda \equiv \sum_{i=1}^{n} \omega_i (\bar{r}_i - r_f)/(\sum_{i=1}^{n} \sum_{j=1}^{n} \sigma_{ij} \omega_i \omega_j) = (\bar{r} - r_f)/\sigma^2$. Making the substitution $v_i = \lambda \omega_i$ for each i simplifies the preceding equations to

$$\sum_{i=1}^{n} \sigma_{ij} v_i = \bar{r}_j - r_f, \quad j = 1, 2, \ldots, n. \tag{31.1}$$

We solve these linear equations for the v_is (see Fig. 19.2) and determine ω_i by setting $\omega_i = v_i/(\sum_{j=1}^{n} v_j)$. A negative ω_i means that asset i needs to be sold short.

If riskless lending and borrowing are disallowed, the whole efficient frontier can be traced out by solving Eq. (31.1) for all possible riskless rates, because an efficient portfolio is a tangent portfolio to a ray extending from *some* riskless rate (consult Fig. 31.2 again). However, there is a better way. Observe that v_i are linear in r_f; in other words, $v_i = c_i + d_i r_f$ for some constants c_i and d_i. We can find c_i and d_i by first solving Eq. (31.1) for v_i under two different r_fs, say r'_f and r''_f. The solutions v'_i and v''_i correspond to two efficient portfolios. Now we solve

$$v'_i = c_i + d_i r'_f,$$
$$v''_i = c_i + d_i r''_f$$

for the unknown c_i and d_i for each i. By treating r_f as a variable and varying it, we can trace out the entire frontier. Just as the two-fund theorem says, two efficient portfolios suffice to determine the frontier.

▶ **Exercise 31.1.6** What would the one-fund theorem imply about trading volumes?

31.1.2 Alternative Efficient Portfolio Selection Models

In the **Black model**, portfolios are chosen subject only to $\sum_{i=1}^{n} \omega_i = 1$. In the **standard portfolio selection model**, short sales are disallowed, and the constraints are

$$\sum_{i=1}^{n} \omega_i = 1,$$
$$\omega_i \geq 0, \quad i = 1, 2, \ldots, n.$$

By law or by policy, there may be restrictions on the amounts that can be invested in any one security. To handle them, one may augment the standard model with upper bounds:

$$\sum_{i=1}^{n} \omega_i = 1,$$
$$\omega_i \geq 0, \quad i = 1, 2, \ldots, n,$$
$$\omega_i \leq u_i, \quad i = 1, 2, \ldots, n.$$

In the **Tobin–Sharpe–Lintner model**, the portfolios are chosen subject to

$$\sum_{i=1}^{n+1} \omega_i = 1,$$
$$\omega_i \geq 0, \quad i = 1, 2, \ldots, n.$$

The variable ω_{n+1} represents the amount lent (or borrowed if ω_{n+1} is negative). The covariances $\sigma_{n+1,i}$ are of course zero for $i = 1, 2, \ldots, n + 1$. Limited borrowing can be modeled by the addition of the constraint $\omega_{n+1} \leq u_{n+1}$. In the **general portfolio selection model**, a portfolio is feasible if it satisfies

$$A\omega = b,$$
$$\omega \geq 0,$$

where A is any $m \times n$ matrix and b is an m-dimensional vector [642].

▶ **Exercise 31.1.7** Two portfolio selection models are **strictly equivalent** if they have the same set of obtainable mean–standard deviation combinations. Prove that any model that does not impose the nonnegative constraint on ω is strictly equivalent to some general portfolio selection model, which does.

31.2 The Capital Asset Pricing Model

Imagine a world in which all investors are mean–variance portfolio optimizers and they share the same expectation as to expected returns, variances, and covariances. Also assume zero transactions cost. By the one-fund theorem, every investor will hold some amounts of the riskless asset and the same portfolio of risky assets. As all risky assets must be held by somebody, an immediate implication is that every investor holds the **market portfolio** in equilibrium regardless of one's degree of risk aversion. The market portfolio, which consists of all *risky* assets, is furthermore efficient (see Exercise 31.2.1).[4]

Given that the single efficient fund of risky assets is the market portfolio, the efficient frontier consists of a single straight line emanating from the riskless point

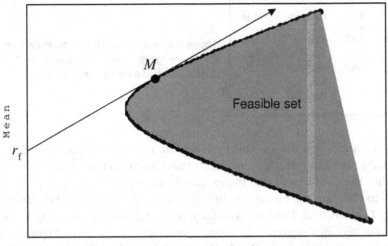

Figure 31.3: The capital market line. The point M stands for the market portfolio. The shaded set is the feasible set defined by risky assets. Investors can adjust the risk level by changing their holdings of riskless asset; for example, risk can be increased by holding negative amounts of the riskless asset.

and passing through the market portfolio. No complex computation is needed to determine the efficient frontier. This line is called the **capital market line**, which shows the relation between the expected rate of return and the risk for efficient portfolios (see Fig. 31.3). Prices should adjust so that efficient assets and portfolios fall on the line. Individual risky securities and inefficient portfolios, in contrast, will plot below the line. This is Sharpe's famous CAPM of 1964 [796],[5] which was independently arrived at by Lintner [605] and Mossin [680]. This model is fundamental to the equilibrium pricing of risky assets.

The capital market line states that

$$\bar{r} = r_{\mathrm{f}} + \frac{\bar{r}_M - r_{\mathrm{f}}}{\sigma_M}\, \sigma,$$

where $\bar{r}_M$ and σ_M are the expected value and the standard deviation of the market rate of return and $\bar{r}$ and σ are the expected value and the standard deviation of the rate of return of any efficient asset. Observe that as risk increases, the expected rate of return must also increase. The slope of the capital market line is $(\bar{r}_M - r_{\mathrm{f}})/\sigma_M$, which is called the **market price of risk**. It tells by how much the expected rate of return of an efficient portfolio must increase if the standard deviation of that rate increases by one unit. The market price of risk is also known as the **Sharpe ratio** [798].

The capital market line relates the expected rate of return of an efficient portfolio to its standard deviation, but it does not show how the expected rate of return of an individual asset relates to its individual risk. That relation is stated in the following theorem.

THEOREM 31.2.1 *If the market portfolio M is efficient, the expected return $\bar{r}_j$ of any asset j satisfies $\bar{r}_j - r_{\mathrm{f}} = \beta_j(\bar{r}_M - r_{\mathrm{f}})$, where $\beta_j \equiv \sigma_{j,M}/\sigma_M^2$ and $\sigma_{j,M} \equiv \mathrm{Cov}[r_j, r_M]$.*

The value β_i is referred to as the **beta** of an asset. An asset's beta is all that needs to be known about its risk characteristics. The value $\bar{r}_i - r_{\mathrm{f}}$ is the expected excess

Company	Beta	Company	Beta
America Online	2.43	Intel	1.03
AT&T	0.82	Merck	0.87
Citigroup	1.68	Microsoft	1.49
General Motors	1.01	Sun Micro.	1.19
IBM	1.07	Wal-Mart	1.20

Figure 31.4: Betas of some U.S. corporations. America Online merged with Time-Warner in 2001. Source: Standard & Poor's, May 8, 2000.

rate of return of asset i. It is the amount by which the rate of return is expected to exceed the riskless rate. Likewise, $\bar{r}_M - r_f$ is the expected excess rate of return of the market portfolio. The CAPM says that the expected excess rate of return of an asset is proportional to the expected excess rate of return of the market portfolio, and the proportionality factor is beta. Beta, not volatility, is the measure of a security's risk, and the method of beating the market is to assume greater risk, i.e., beta. Figure 31.4 shows the betas of some U.S. corporations. We can estimate beta by regressing the excess return on the asset against the excess return on the market.

The CAPM formula in Theorem 31.2.1 shows a linear relation between beta and the expected rate of return for all assets whether they are efficient or not. This relationship, when plotted on a beta expected-return diagram, is termed the **security market line**. All assets fall on the security market line; in particular, the market is the point at $\beta = 1$.

Essentially the same arguments go through even if there is no riskless asset (see Exercise 31.2.12). The role of the riskless rate of return is then played by the mean rate of return in which the line in Fig. 31.3 intercepts the axis of mean rate of return.

➤ **Exercise 31.2.1** Verify that the market portfolio is efficient.

➤ **Exercise 31.2.2** Prove the security market line formula in Theorem 31.2.1.

31.2.1 More on the CAPM

The portfolio beta is the weighted average of the betas of the individual assets in the portfolio. Specifically, suppose a portfolio contains n assets with the weights $\omega_1, \omega_2, \ldots, \omega_n$. The rate of return of the portfolio is $r \equiv \sum_i \omega_i r_i$. Hence $\text{Cov}[r, r_M] = \sum_i \omega_i \sigma_{i,M}$. It follows immediately that the portfolio beta equals $\sum_i \omega_i \beta_i$.

Write asset i's rate of return as

$$r_i = r_f + \beta_i(r_M - r_f) + \epsilon_i, \tag{31.2}$$

where $E[\epsilon_i] = 0$ by the CAPM. Now take the covariance of r_i with r_M in Eq. (31.2) to yield

$$\sigma_{i,M} = \beta_i^2 \sigma_M^2 + \text{Cov}[\epsilon_i, r_M] = \sigma_{i,M} + \text{Cov}[\epsilon_i, r_M].$$

Therefore $\text{Cov}[\epsilon_i, r_M] = 0$ and

$$\sigma_i^2 = \beta_i^2 \sigma_M^2 + \text{Var}[\epsilon_i]. \tag{31.3}$$

It is important to note that the total risk σ_i^2 is a sum of two parts. The first part, $\beta_i^2 \sigma_M^2$, is the systematic risk. This is the risk associated with the market as a whole, also called the **market risk**. It cannot be reduced by diversification because every asset with nonzero beta contains this risk. The second part $\text{Var}[\,\epsilon_i\,]$ is the specific risk. This risk is uncorrelated with the market and can be reduced by diversification. Only the systematic risk has any bearing on returns.

Consider an asset on the capital market line with a beta of β and an expected rate of return equal to $\bar{r} = r_f + \beta(\bar{r}_M - r_f)$. This asset, which is efficient, must be equivalent to a combination of the market portfolio and the riskless asset. Its standard deviation is therefore $\beta \sigma_M$, which implies it has only systematic risk but no specific risk by Eq. (31.3). Now consider another asset with the same beta β. According to the CAPM, its expected rate of return must be $\bar{r}$. However, if it carries specific risk, it will not fall on the capital market line. The specific risk is thus the distance by which the portfolio lies below the capital market line.

Although stated in terms of expected returns, the CAPM is also a pricing model. Suppose an asset is purchased at a known price P and later sold at price Q. The rate of return is $r \equiv (Q - P)/P$. By the CAPM,

$$\frac{\bar{Q} - P}{P} = r_f + \beta(\bar{r}_M - r_f),$$

where β is the beta of the asset. Solve for P to obtain

$$P = \frac{\bar{Q}}{1 + r_f + \beta(\bar{r}_M - r_f)}. \tag{31.4}$$

Hence the CAPM can be used to decide whether the price for a stock is "right." Note that the risk-adjusted interest rate is $r_f + \beta(\bar{r}_M - r_f)$, not r_f.

Equation (31.4) can take another convenient form. The value of beta is

$$\beta = \frac{\text{Cov}[r, r_M]}{\sigma_M^2} = \frac{\text{Cov}[\,(Q/P) - 1, r_M\,]}{\sigma_M^2} = \frac{\text{Cov}[\,Q, r_M\,]}{P\sigma_M^2}.$$

Substituting this into pricing formula (31.4) and dividing by P yields

$$1 = \frac{\bar{Q}}{P(1 + r_f) + \text{Cov}[\,Q, r_M\,](\bar{r}_M - r_f)/\sigma_M^2}.$$

Solve for P again to obtain

$$P = \frac{1}{1 + r_f}\left\{ \bar{Q} - \frac{\text{Cov}[\,Q, r_M\,](\bar{r}_M - r_f)}{\sigma_M^2} \right\}. \tag{31.5}$$

This demonstrates it is the asset's covariance with the market that is relevant for pricing.

▶ **Exercise 31.2.3** For an asset uncorrelated with the market (that is, with zero beta), the CAPM says its expected rate of return is the riskless rate even if this asset is very risky with a large standard deviation. Why?

▶ **Exercise 31.2.4** If an asset has a negative beta, the CAPM says its expected rate of return should be less than the riskless rate even if this asset is very risky with a large standard deviation. Why? (For example, we saw in Chap. 24 that IO strips earn less than the riskless rate despite their high riskiness.)

▶ **Exercise 31.2.5** Why must all portfolios with the same expected rate of return but different total risks fall on the same point on the security market line?

▶ **Exercise 31.2.6** (1) Verify that pricing formula (31.4) is linear (the price of the sum of two assets is the sum of their prices, and the price of a multiple of an asset is the same multiple of the price). (2) Derive the same results from the no-arbitrage principle.

31.2.2 Portfolio Insurance

Portfolio insurance is a trading strategy that protects a portfolio from market declines but without losing the opportunity to participate in market rallies – in a word, a protective put [772]. Using puts to protect a portfolio from falling below a specified floor is a simple example of *static* portfolio insurance. Alternatives to static schemes are dynamic strategies that create synthetic options with stocks and bonds. Dynamic strategies, however, generate high transactions costs. This problem was mitigated by the introduction of stock index futures. Compared with the underlying assets, futures can be traded at much lower transactions costs in achieving the desired mixture of risky and riskless assets.[6]

Let the value of the index be S and each put be on \$100 times the index. Consider a diversified portfolio with a beta of β. If for each $100 \times S$ dollars in the portfolio, one put contract is purchased with strike price X, the value of the portfolio is protected against the possibility of the index's falling below the floor of X. Our goal is to implement this protective put. Specifically, to protect each dollar of the portfolio against falling below W at time T, we buy β put contracts for each $100 \times S$ dollars in the portfolio. Note that the total number of puts bought is $\beta V/(S \times 100)$, where V is the current value of the portfolio. The strike price X is the index value when the portfolio value reaches W.

Let r be the interest rate and q the dividend yield. Suppose that the index reaches S_T at time T. The excess return of the index over the riskless interest rate is $(S_T - S)/S + q - r$, and the excess return of the portfolio over the riskless interest rate is $\beta((S_T - S)/S + q - r)$. The return from the portfolio is therefore $\beta[(S_T - S)/S + q - r] + r$, and the increase in the portfolio value net of the dividends is $\beta[(S_T - S)/S + q - r] + r - q$. Therefore the portfolio value per dollar of the original value is

$$1 + \beta \left(\frac{S_T - S}{S} + q - r \right) + r - q = \beta \frac{S_T}{S} + (\beta - 1)(q - r - 1). \tag{31.6}$$

Choose X to be the S_T that makes Eq. (31.6) equal W; in other words,

$$X = [W + (q - r - 1)(1 - \beta)] \frac{S}{\beta}.$$

From Eq. (31.6), the portfolio value is less than W by $\beta(\Delta S/S)$ if and only if the index value is less than X by $\beta(\Delta S/S)(S/\beta) = \Delta S$. Exercising the options therefore induces a matching cash inflow of

$$\beta \frac{\Delta S}{100 \times S} \times 100 = \beta \frac{\Delta S}{S}.$$

The strategy's cost is $P\beta V/(S \times 100)$, where P is the put premium with strike price X.

Clearly a higher strike price provides a higher floor of WV dollars at a greater cost. This trade-off between the cost of insurance and the level of protection is typical of any insurance. The total wealth of course has a floor of

$$WV - \frac{P\beta V}{S \times 100}.$$

EXAMPLE 31.2.2 Start with $S = 1000$, $\beta = 1.5$, $q = 0.02$, and $r = 0.07$ for a period of 1 year. We have the following relations between the index value and the portfolio value per dollar of the original value.

Index value in a year	1200	1100	1000	900	800
Portfolio value in a year	1.275	1.125	0.975	0.825	0.675

For example, if the portfolio starts at $1 million and the insured value is $0.825 million, then $(1.5 \times 1,000,000)/(100 \times 1,000) = 15$ put contracts with a strike price of 900 should be purchased.

▶ **Exercise 31.2.7** Redo Example 31.2.2 with $S = 1000$, $\beta = 2$, $q = 0.01$, and $r = 0.05$.

▶ **Exercise 31.2.8** Consider a portfolio worth $1,000 times the S&P 500 Index and with a beta of 1.0 against the index. Argue that buying 10 put index options with a strike price of 1,000 insures against the portfolio value's dropping below $1,000,000.

▶ **Exercise 31.2.9** A mutual fund manager believes that the market is going to be relatively calm in the near future and writes a covered index call. Analyze it by following the same logic as that of the protective put.

▶ **Exercise 31.2.10** A bank offers the following financial product to a mutual fund manager planning to buy a certain stock in the near future. If the stock price is over $50, the manager buys it at $50. If the stock price is below $40, the manager buys it at $40. If the stock price is between the two, the manager buys at the spot price. Analyze the underlying options.

31.2.3 Critical Remarks

> Fire those CAPM-peddling consultants.
> —Louis Lowenstein [620]

Although the CAPM is widely used by practitioners [592], many of its assumptions have been controversial. It assumes either normally distributed asset returns or quadratic utility functions. It furthermore assumes that investors care about only the mean and the variance of returns, which implies that they view upside and downside risks with equal distaste. In reality, portfolio returns are not, strictly speaking, normally distributed, and investors seem to distinguish between upside and downside risks. The theory posits, unrealistically, that everyone has identical information about the returns of all assets and their covariances. Even if this assumption were valid, it would not be easy to obtain accurate data. Usually, the variances and covariances can be accurately estimated, but not the expected returns (see Example 20.1.1). Unfortunately, errors in means are more critical than errors in variances, and errors in variances are more critical than errors in covariances [204]. The assumption that all investors share a common investment horizon is rarely the case in practice.

The CAPM assumes that all assets can be bought and sold on the market. The assets include not just securities, but also real estate, cash, and even human capital. Because the market portfolio is difficult to define, in reality proxies for the market portfolio are used [799]. The trouble is that different proxies result in different beta estimates for the same security (see Exercise 31.2.12). Finally, a single risk factor does not seem adequate for describing the cross section of expected returns [145, 336, 424, 635, 636, 666].

▶ **Exercise 31.2.11** Why are security analysts' 1-year forecasts worse than 5-year ones?

▶ **Exercise 31.2.12** Prove that using any efficient portfolio for the risky assets as the proxy for the market portfolio results in linear relations between the expected rates of return and the betas, just as in the CAPM.

31.3 Factor Models

The mean–variance theory requires that many parameters be estimated: n for the expected returns of the assets and $n(n+1)/2$ for their covariances. Luckily, asset returns can often be explained by a much smaller number of underlying sources of randomness called factors. A factor model represents the connection between factors and individual returns. In this section a factor model of the return process for asset pricing is presented, the **Arbitrage Pricing Theory** (**APT**).

31.3.1 Single-Factor Models

We start with single-factor models. Suppose there are n assets with rates of return, $r_1, r_2, \ldots, r_n$. There is a single factor f, which is a random quantity such as the return on a stock index for the holding period. The rates of return and the factor are related by

$$r_i = a_i + b_i f + \epsilon_i, \quad i = 1, 2, \ldots, n,$$

where a_i and b_i are constants. The b_is are the factor loadings or **factor betas**, and they measure the sensitivity of the return to the factor. Without loss of generality, let $E[\epsilon_i] = 0$. Assume that ϵ_i are uncorrelated with f: $\text{Cov}[f, \epsilon_i] = 0$. Furthermore, assume that they are uncorrelated with each other, i.e., $E[\epsilon_i \epsilon_j] = 0$ for $i \neq j$.[7] Any correlation between asset returns thus arises from a common response to the factor. The variance of ϵ_i is denoted by $\sigma_{\epsilon_i}^2$ and that of f by σ_f^2. There is a total of $3n+2$ parameters: $a_i, b_i, \sigma_{\epsilon_i}^2, \overline{f}$, and σ_f^2. The following results are straightforward:

$$\overline{r}_i = a_i + b_i \overline{f},$$
$$\sigma_i^2 = b_i^2 \sigma_f^2 + \sigma_{\epsilon_i}^2,$$
$$\text{Cov}[r_i, r_j] = b_i b_j \sigma_f^2, \quad i \neq j,$$
$$b_i = \text{Cov}[r_i, f]/\sigma_f^2.$$

The preceding simple covariance matrix leads to very efficient algorithms for the portfolio selection problems in Subsection 31.1.1 [317]. The single-factor model is due to Sharpe [795].

The return on a portfolio can be analyzed similarly. Consider a portfolio constructed with weights ω_i. Its rate of return is just

$$r = a + bf + \epsilon,$$

where $a \equiv \sum_{i=1}^{n} \omega_i a_i$, $b \equiv \sum_{i=1}^{n} \omega_i b_i$, and $\epsilon \equiv \sum_{i=1}^{n} \omega_i \epsilon_i$. The portfolio's beta b is hence the average of the underlying assets' betas b_i (recall the law of the average covariance). It is easy to verify that $E[\epsilon] = 0$, $\mathrm{Cov}[f, \epsilon] = 0$, and $\mathrm{Var}[\epsilon] = \sum_{i=1}^{n} \omega_i^2 \sigma_{\epsilon_i}^2$. Finally, the variance of r is

$$\sigma^2 = b^2 \sigma_f^2 + \mathrm{Var}[\epsilon],$$

similar to Eq. (31.3). Among the total risk above, the systematic part is $b^2 \sigma_f^2$. The systematic risk, which is due to the $b_i f$ terms, results from the factor f that influences every asset and is therefore present even in a diversified portfolio. The $\mathrm{Var}[\epsilon]$ term represents the specific risk. The specific risk, which is due to the ϵ_i terms, can be made to go to zero through diversification. It is also called the **diversifiable risk**.

▶ **Exercise 31.3.1** Assume that the single factor f is the market rate of return, r_M. Write the return processes as $r_i - r_f = \alpha_i + b_i(r_M - r_f) + \epsilon_i$. As usual, $E[\epsilon_i] = 0$ and ϵ_i is uncorrelated with the market return. Show that $b_i = \mathrm{Cov}[r_i, r_M]/\mathrm{Var}[r_M]$, as in the CAPM.

31.3.2 Multifactor Models

Now there are two factors f_1 and f_2, and the rate of return of asset i takes the form

$$r_i = a_i + b_{i1} f_1 + b_{i2} f_2 + \epsilon_i.$$

As in Subsection 31.3.1, assume that $E[\epsilon_i] = 0$ and that ϵ_i is uncorrelated with the factors and ϵ_j for $j \neq i$. The formulas for the expected rates of return and covariances are

$$\bar{r}_i = a_i + b_{i1} \bar{f}_1 + b_{i2} \bar{f}_2,$$
$$\sigma_i^2 = b_{i1}^2 \sigma_{f_1}^2 + b_{i2}^2 \sigma_{f_2}^2 + 2 b_{i1} b_{i2} \mathrm{Cov}[f_1, f_2] + \sigma_{\epsilon_i}^2,$$
$$\mathrm{Cov}[r_i, r_j] = b_{i1} b_{j1} \sigma_{f_1}^2 + b_{i2} b_{j2} \sigma_{f_2}^2 + (b_{i1} b_{i2} + b_{j1} b_{j2}) \mathrm{Cov}[f_1, f_2], \quad i \neq j.$$

From the preceding equations,

$$\mathrm{Cov}[r_i, f_1] = b_{i1} \sigma_{f_1}^2 + b_{i2} \mathrm{Cov}[f_1, f_2],$$
$$\mathrm{Cov}[r_i, f_2] = b_{i2} \sigma_{f_2}^2 + b_{i1} \mathrm{Cov}[f_1, f_2].$$

These give two equations that can be solved for b_{i1} and b_{i2}. Factor models with more than two factors are easy generalizations. For U.S. stocks, between 3 and 15 factors may be needed [623].

▶ **Exercise 31.3.2** Describe a procedure to convert a set of correlated factors into a set of uncorrelated factors, which are easier to handle.

31.3.3 The Arbitrage Pricing Theory (APT)

The factor-model framework leads to an alternative theory of asset pricing, Ross's APT, which is a theory about equilibrium under factor models [765]. The APT does not require that investors evaluate portfolios on the basis of means and variances. Neither is a quadratic utility function required. Instead, (1) the mean–variance framework is replaced with a factor model for returns, (2) investors are assumed to prefer a greater return to a lesser return when returns are certain, and (3) the universe of assets is assumed to be large.

Consider first a special case in which the rates of return observe the one-factor model:

$$r_i = a_i + b_i f.$$

This factor model has no residual errors, and the uncertainty associated with a return is due solely to the uncertainty in the factor f. Interestingly, the values of a_i and b_i must be related if arbitrage opportunities are to be excluded. Here is the argument. Consider two assets i and j with $b_i \neq b_j$. Now form a portfolio with weights $\omega_i \equiv \omega$ for asset i and $\omega_j \equiv 1 - \omega$ for asset j. Its rate of return is

$$r = \omega a_i + (1 - \omega) a_j + (\omega b_i + (1 - \omega) b_j) f.$$

If we select $\omega = b_j/(b_j - b_i)$ to make the coefficient of f zero, the rate of return r becomes

$$\lambda_0 \equiv \frac{a_i b_j - a_j b_i}{b_j - b_i}.$$

This portfolio is riskless because the equation for r contains no random elements. If there happens to be a riskless asset, then $\lambda_0 = r_f$. Even if riskless assets do not exist, all portfolios constructed without dependence on f must have the same rate of return, λ_0. Now $\lambda_0(b_j - b_i) = a_i b_j - a_j b_i$, which can be rearranged as

$$\frac{a_j - \lambda_0}{b_j} = \frac{a_i - \lambda_0}{b_i}.$$

As this relation holds for all i and j, there is a constant c such that

$$\frac{a_i - \lambda_0}{b_i} = c$$

for all i.[8] The values of a_i and b_i are thus related by $a_i = \lambda_0 + b_i c$. The expected rate of return of asset i is now

$$\bar{r}_i = a_i + b_i \bar{f} = \lambda_0 + b_i c + b_i \bar{f} = \lambda_0 + b_i \lambda_1, \tag{31.7}$$

where $\lambda_1 \equiv c + \bar{f}$. We see that once the constants λ_0 and λ_1 are known, the expected return of an asset is determined entirely by the factor betas b_i. The above analysis can be generalized (see Exercise 31.3.4).

THEOREM 31.3.1 *Let there be n assets whose rates of return are governed by $m < n$ factors according to the equations $r_i = a_i + \sum_{j=1}^{m} b_{ij} f_j$, $i = 1, 2, \ldots, n$. Then there exist constants $\lambda_0, \lambda_1, \ldots, \lambda_m$ such that $\bar{r}_i = \lambda_0 + \sum_{j=1}^{m} b_{ij} \lambda_j$, $i = 1, 2, \ldots, n$. The value λ_i is called the market price of risk associated with factor f_i or simply the* **factor price**.

Next we consider general multifactor models with residual errors,

$$r_i = a_i + \sum_{j=1}^{m} b_{ij} f_j + \epsilon_i,$$

where $E[\epsilon_i] = 0$ and $\sigma_{\epsilon_i}^2 \equiv E[\epsilon_i^2]$. As before, ϵ_i is assumed to be uncorrelated with the factors and with the residual errors of other assets. Let us form a portfolio by using the weights $\omega_1, \omega_2, \ldots, \omega_n$ with $\sum_{i=1}^{n} \omega_i = 1$. The rate of return of the portfolio is

$$r = a + \sum_{j=1}^{m} b_j f_j + \epsilon,$$

where $a \equiv \sum_{i=1}^{n} \omega_i a_i$, $b_j \equiv \sum_{i=1}^{n} \omega_i b_{ij}$, and $\epsilon \equiv \sum_{i=1}^{n} \omega_i \epsilon_i$. Let $\sigma_{\epsilon_i} \le S$ for some constant S for all i. Assume that the portfolio is *well diversified* in the sense that $\omega_i \le W/n$ for some constant W for all i – no single asset dominates the portfolio. Then

$$\text{Var}[\epsilon] = \sum_{i=1}^{n} \omega_i^2 \sigma_{\epsilon_i}^2 \le \frac{1}{n^2} \sum_{i=1}^{n} W^2 S^2 = \frac{W^2 S^2}{n} \to 0$$

as $n \to \infty$. Combined with the fact $E[\epsilon] = 0$, the residual error ϵ of a well-diversified portfolio selected from a very large number of assets is approximately zero.[9]

A riskless portfolio in terms of zero sensitivity to all factors was used in the proof of Theorem 31.3.1. We just showed that the portfolio remains riskless under the more general models as long as it is well diversified. The existence of a riskless well-diversified portfolio suffices to extend Theorem 31.3.1 to the more general models (see Exercise 31.3.5).

The APT and the CAPM are not directly comparable [289]. Neither of the two models' assumptions imply the other's. The CAPM makes strong assumptions about the probability distribution of assets' rates of return, agents' utility functions, or both. The APT on the other hand, makes strong assumptions about assets' equilibrium rates of return. However, because the APT does not identify the factors, the CAPM can be made consistent with the APT and vice versa. For example, consider a two-factor model $r_i = a_i + b_{i1} f_1 + b_{i2} f_2 + \epsilon_i$. Under the APT model, $\bar{r}_i = r_f + b_{i1} \lambda_1 + b_{i2} \lambda_2$. Let portfolio j ($j = 1, 2$) have an expected return rate of $\lambda_j + r_f$ and a beta value of β_{f_j}. Clearly portfolio j's only source of risk is factor f_j. Because the CAPM says that $\lambda_j = \beta_{f_j}(\bar{r}_M - r_f)$,

$$\bar{r}_i = r_f + b_{i1}\beta_{f_1}(\bar{r}_M - r_f) + b_{i2}\beta_{f_2}(\bar{r}_M - r_f) = r_f + (b_{i1}\beta_{f_1} + b_{i2}\beta_{f_2})(\bar{r}_M - r_f).$$

The beta is thus a weighted sum of the underlying factors' betas with the factor betas as the weights. Different factor betas are the reason different assets have different betas.

▶ **Exercise 31.3.3** For the one-factor APT, what will become of λ_1 if the CAPM holds?

▶ **Exercise 31.3.4** Prove Theorem 31.3.1.

▶ **Exercise 31.3.5** Complete the proof of the APT under the general factor models.

31.4 **Value at Risk**

> Anyone that relied on so-called
> value-at-risk models has been crucified.
> —The Economist, 1999 [309]

Introduced in 1983, the VaR is an attempt to provide a single number for senior management that summarizes the total risk in a portfolio of financial assets. The VaR calculation is aimed at making a statement of the form: "We are c percent certain not to lose more than V dollars in the next m days." The variable V is the VaR of the portfolio. The VaR is therefore an estimate, with a given degree of confidence, of how much one can lose from one's portfolio over a given time horizon, or

$$\text{Prob[change in portfolio value} \leq -\text{VaR]} = 1 - c,$$

where c is the confidence level (see Fig. 31.5). The VaR is usually calculated assuming "normal" market circumstances, meaning that extreme market conditions such as market crashes are not considered. It has become widely used by corporate treasurers and fund managers as well as financial institutions.

For the purposes of measuring the adequacy of bank capital, the Bank for International Settlements (BIS) sets the confidence level $c = 0.99$ and the time horizon $m = 10$ (days) [293]. Another interesting application is in investment evaluation. Here risk is viewed in terms of the impact of the prospective change on the overall value at risk, i.e., *incremental* VaR, and we go ahead with the investment if the incremental VaR is low enough relative to the expected return [283].

Suppose returns are normally distributed and independent on successive days.[10] We consider a single asset first. We assume that the stock price is S, whose daily volatility for the return rate $\Delta S / S$ is σ. Because the time horizon m is usually small, we assume that the expected price change is zero. The standard deviation of the stock price over this time horizon is $S\sigma\sqrt{m}$. The VaR of holding one unit of the stock is $2.326 \times S\sigma\sqrt{m}$ if the confidence level is 99% and $1.645 \times S\sigma\sqrt{m}$ if

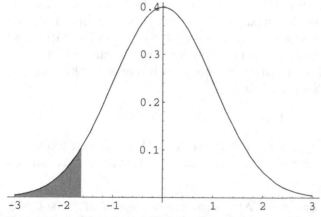

Figure 31.5: Confidence level and VaR. The diagram shows a confidence level of 95% under the standard normal distribution; the shaded area is 5% of the total area under the density function. It corresponds to 1.64485 standard deviations from the mean.

Confidence level (c)	Number of standard deviations
95%	1.64485
96%	1.75069
97%	1.88079
98%	2.05375
99%	2.32635

Figure 31.6: Confidence levels and standard deviations from the mean. The table samples confidence levels and their corresponding numbers of standard deviations from the mean when the random variable is normally distributed.

the confidence level is 95%. In general, the VaR is $-N^{-1}(1-c)$ times the standard deviation, or

$$-N^{-1}(1-c)\,S\sigma\sqrt{m},$$

where $N(\cdot)$ is the distribution function of the standard normal distribution (see Fig. 31.6). The preceding equation makes it easy to convert one horizon or confidence level to another. For example, the relation between 99% VaR and 95% VaR is

$$\text{VaR}\,(95\%) = \text{VaR}\,(99\%) \times (1.645/2.326).$$

Similarly, the variance of an m-day return should be m times the variance of a 1-day return. The m-day VaR thus equals $\sqrt{m}$ times the 1-day VaR, which is also called the **daily earnings at risk**. When m is not small, the expected annual rate of return μ needs to be considered. In that case, the drift $S\mu m/T$ is subtracted from the VaR when there are T trading days per annum.

Now consider a portfolio of assets. Assume that the changes in the values of asset prices have a multivariate normal distribution. Let the daily volatility of asset i be σ_i, let the correlation between the returns on assets i and j be ρ_{ij}, and let S_i be the market value of the positions in asset i. The VaR for the whole portfolio then is

$$-N^{-1}(1-c)\sqrt{m}\,\sqrt{\sum_i \sum_j S_i S_j \sigma_i \sigma_j \rho_{ij}}\,.$$

This way of computing the VaR is called the **variance–covariance approach** [518]. It was popularized by J.P. Morgan's RiskMetrics™ (1994).

The variance–covariance methodology may break down if there are derivatives in the portfolio because the returns of derivatives may not be normally distributed even if the underlying asset is. Nevertheless, if movements in the underlying asset are expected to be very small because, say, the time horizon is short, we may approximate the sensitivity of the derivative to changes in the underlying asset by the derivative's delta as follows. Consider a portfolio P of derivatives with a single underlying asset S. Recall that the delta of the portfolio, δ, measures the price sensitivity to S, or approximately $\Delta P/\Delta S$. The standard deviation of the distribution of the portfolio is $\delta S\sigma\sqrt{m}$, and its VaR is $-N^{-1}(1-c)\delta S\sigma\sqrt{m}$. In general when there are many underlying assets, the VaR of a portfolio containing options becomes

$$-N^{-1}(1-c)\sqrt{m}\,\sqrt{\sum_i \sum_j \delta_i \delta_j S_i S_j \sigma_i \sigma_j \rho_{ij}},$$

where δ_i denotes the delta of the portfolio with respect to asset i and S_i is the value of asset i. This is called the **delta approach** [527, 878]. The delta approach essentially treats a derivative as delta units of its underlying asset for the purpose of VaR calculation. This is not entirely unreasonable because such equivalence does hold instantaneously. It becomes questionable, however, as m increases.

Rather than using asset prices, VaR usually relies on a limited number of basic market variables that account for most of the changes in portfolio value [603]. As mentioned in Subsection 31.3.1, this greatly reduces the complexity related to the covariance matrix because only the covariances between the market variables are needed now. Typical market variables are yields or bond prices, exchange rates, and market returns. A basic instrument is then associated with each market variable. A security is now approximated by a portfolio of these basic instruments. Finally, its VaR is reduced to those of the basic instruments.

▶ **Exercise 31.4.1** What is the VaR of a futures contract on a stock?

▶ **Exercise 31.4.2** If the stock price follows $dS = S\mu\, dt + S\sigma\, dW$, what is its VaR τ years from now at c confidence?

31.4.1 Simulation

The Monte Carlo simulation is a general method to estimate the VaR, particularly for derivatives [571]. It works by computing the values of the portfolio over many sample paths, and the VaR is based on the distribution of the values. Figure 31.7 contains an algorithm for n asset prices following geometric Brownian motion:

$$\frac{dS_j}{S_j} = \mu_j\, dt + \sigma_j\, dW_j, \quad j = 1, 2, \ldots, n,$$

where the n factors, dW_j, are correlated. As always, *actual* returns, not risk-neutral returns, should be used. For short time horizons, this distinction is not critical for most cases. In practice, to save computation time, a stock with a beta of β is mapped to a position in β times the index. Of course, this approach ignores the stock's specific risk. A related simulation method, called **historical simulation**, utilizes historical data [518]. It is identical to the Monte Carlo simulation except that the sample paths are generated by sampling the historical data as if they are to be repeated in the future.

Brute-force Monte Carlo simulation is inefficient when the number of factors is large. Fortunately, factor analysis and principal components analysis can often reduce the number of factors needed in the simulation [2, 509]. Let C denote the covariance matrix of the n factors $dW_1, dW_2, \ldots, dW_n$. Let $u_i \equiv [\, u_{1i}, u_{2i}, \ldots, u_{ni}\,]^{\mathrm{T}}$ be the eigenvectors of C and $\lambda_1 \geq \lambda_2 \geq \cdots \geq \lambda_n$ be the corresponding positive eigenvalues. Hence $\lambda_i u_i = C u_i$ for $i = 1, 2, \ldots, n$. Recall that each eigenvalue indicates how much of the variation in the data its corresponding eigenvector explains. By the Schur decomposition theorem, the eigenvectors can be assumed to be orthogonal to each other. Normalize the eigenvectors such that $|u_i|^2 = \lambda_i$ and define

$$dZ_j \equiv \lambda_j^{-1} \sum_{k=1}^{n} u_{kj}\, dW_k.$$

It follows that

$$dW_i = \sum_{k=1}^{n} u_{ik}\, dZ_k,$$

VaR with Monte Carlo simulation:

input: $\overline{p}, c, n, C[n][n], S[n], \mu[n], \Delta t, m, N$;
real $S[m+1][n], y[n], dW[n], P[n][n], p[N]$;
real $\xi()$; $// \xi() \sim N(0,1)$.
integer i, j, k;
Let P be such that $C = PP^{\mathrm{T}}$;   // See p. 248.
for $(j = 0$ to $n-1)$ { $S[0][j] := S[j]$; }
for $(k = 0$ to $N-1)$ {
      for $(i = 1$ to $m)$ {
         for $(j = 0$ to $n-1)$
           $y[j] := \xi() \times \sqrt{\Delta t}$;
         $dW := Py$;
         for $(j = 0$ to $n-1)$ {
           $S[i][j] := S[i-1][j] \times ((1 + \mu[j]) \times \Delta t + \sqrt{C[j][j]} \times dW[j])$;
         }
      }
      Calculate the horizon portfolio value $p[k]$;
}
Sort $p[0], p[1], \dots, p[N-1]$ in non-decreasing order;
return $\overline{p} - p[\lfloor (1-c)N-1 \rfloor]$;

Figure 31.7: VaR with Monte Carlo simulation. The expected rates of return $\mu[\,]$ and the covariances are annualized. There are n assets, the portfolio's initial value is $\overline{p}$, c is the confidence level, C is the covariance matrix for the annualized asset returns, m is the number of days until the horizon, the number of replications is N, and $S[\,]$ stores the initial asset prices. Recall that $C = PP^{\mathrm{T}}$ is the Cholesky decomposition of C. The portfolio's values at the horizon date are calculated and stored in $p[\,]$. Here we need pricing models and assume that early exercise is not possible during the period. The appropriate percentile is returned after sorting.

where $dZ_k \, dZ_j = 0$ for $j \neq k$ and $dZ_k \, dZ_k = dt$ (see Exercise 31.4.3, part (3)). If the empirical analysis shows that all but the first m principal components are small, then

$$dW_i \approx \sum_{k=1}^{m} u_{ik} \, dZ_k.$$

As a result, the asset price processes can be approximated by

$$\frac{dS_j}{S_j} \approx \mu_j \, dt + \sigma_j \sum_{k=1}^{m} u_{ik} \, dZ_k, \quad j = 1, 2, \dots, n.$$

Only m orthogonal factors $dZ_1, dZ_2, \dots, dZ_m$ remain.

▶ **Exercise 31.4.3** Prove that (1) $C = PP^{\mathrm{T}}$, where P's ith column is the eigenvector u_i, (2) $P^{-1} = \mathrm{diag}[\lambda_1^{-1}, \lambda_2^{-1}, \dots, \lambda_n^{-1}] P^{\mathrm{T}}$, (3) $P[dZ_1, dZ_2, \dots, dZ_n]^{\mathrm{T}} = [dW_1, dW_2, \dots, dW_n]^{\mathrm{T}}$, and (4) $P^{\mathrm{T}} P = \mathrm{diag}[\lambda_1, \lambda_2, \dots, \lambda_n]$.

31.4.2 Critical Remarks

VaR relies on certain assumptions that are inconsistent with empirical evidence. Many implementations assume that asset returns are normally distributed. This simplifies the computation considerably but is inconsistent with the empirical evidence, which finds that many returns have fat tails, both left and right, at both daily and monthly time horizons. Extreme events are hence much more likely to occur in practice than would be predicted based on the assumption of normality [857]. A standard measure of tail fatness is kurtosis. Price jumps and stochastic volatility can be used to generate fat tails (see, e.g., Exercise 20.2.1) [293]. Although daily market returns are not normal [743], for longer periods, say 3 months, returns are quite close to being normally distributed [592].

The method of calculating the VaR depends on the horizon. A method yielding good results over a short horizon may not work well over longer horizons. The method of calculating the VaR also depends on asset types. If the portfolio contains derivatives, methods different from these used to analyze portfolios of stocks may be needed [464].

The ability to quantify risk exposure into a number represents the single most powerful advantage of the VaR. However, the VaR is extremely dependent on parameters, data, assumptions, and methodology. Although it should be part of an effective risk management program, the VaR is not sufficient to control risk [61, 464]. On occasion, it becomes necessary to quantify the magnitude of the losses that might accrue under events less likely than those analyzed in a standard VaR calculation. The procedures used to quantify potential loss exposures under such special circumstances are called **stress tests** [571]. A stress test measures the loss that could be experienced if a set of factors are exogenously specified.

31.4.3 VaR for Fixed-Income Securities

In contrast to stock prices, bond prices tend to move together because much of the movement is systematic, the common factor being the interest rate. For this reason, bond portfolio management does not require that the portfolios be well diversified. Instead, a few bonds of differing maturities can usually hedge the price fluctuations in any single bond or portfolio of bonds [91].

Duration (see Section 4.2) and key rate duration (see Section 27.5) were used to quantify the interest rate exposure of fixed-income portfolios and securities. A VaR methodology can also be based on duration. If S refers to the initial yield of a fixed-income instrument with duration D, the VaR for a long position in the instrument is $1.645 \times \sigma SD$ for a 95% confidence level. As before, VaR analysis requires parameters for the *actual* term structure dynamics. Simulation-based VaR usually conducts factor analysis before the actual simulation [804]. Three orthogonal factors seem to be sufficient (see Subsection 19.2.5).

The variance–covariance approach to the VaR is more complicated [470, 720]. First, an "equivalent" portfolio of standard zero-coupon bonds is obtained for each bond (this is called **cash flow mapping**). Then the historical volatility of spot rates and the correlations between them are used to construct a 95% confidence interval for the dollar return. It is difficult to apply this approach to securities with embedded options, however.

Additional Reading

In 1952, Markowitz and Roy independently published their papers that mark the era of modern portfolio theory [642, 644]. Our presentation of modern portfolio theory was drawn from [317, 623]. General treatment of mean–variance can be found in [643]. See [82, 174, 403, 760, 862] for additional information on beta and [81, 332, 399] for the issue of expected return and risk. The framework of modern portfolio theory can be applied to real estate [389, 407]. See [28] for modern portfolio theory's applicability in Japan. One of the reasons cited for the choice of standard deviation as the measure of risk is that it is easier to work with than the alternatives [799]. An interesting theory from experimental psychology, the **prospect theory**, says that an investor is much more sensitive to reductions in wealth than to increases, which is called **loss aversion** [533, 534]. Spreads can be used to profit from such behavioral "biases" [180]. Consult [592] for an approach beyond mean–variance analysis.

See [826] for development of the utility function. Optimization theory is discussed in [278, 687]. See [346, 470, 567, 646, 693] for additional information on portfolio insurance. Consult [317, 623, 673] for investment performance evaluation; there seems to be no consistent performance for mutual funds [634]. See [96, 360, 799] for security analysis, such as technical analysis and fundamental analysis, and [132] on market timing.

Consult [314] for the VaR of derivatives, [484, 691, 720, 771] for VaR when returns are not normally distributed, [8] for managing VaR using puts, [293, 390, 464, 484, 720, 804, 857] for the VaR of fixed-income securities. The **Cornish–Fisher expansion** is useful for correcting the skewness in distribution during VaR calculations [470, 522].

NOTES

1. Sometimes we use variance of return as the measure of risk for convenience.
2. "Rate of return" and "return" are often used interchangeably as only single-period analysis is involved.
3. Markowitz's Ph.D. dissertation was initially voted down by Friedman on the grounds that "It's not math, it's not economics, it's not even business administration." See [64, p. 60].
4. The S&P 500 Index often serves as the proxy for the market portfolio. **Index funds** are mutual funds that attempt to duplicate a stock market index. Offered in 1975 under the name of Vanguard Index Trust, the Vanguard 500 Index Fund is the first index mutual fund. It tracks the S&P 500 and became the largest mutual fund in April of 2000.
5. Sharpe sent the paper in 1962 to the *Journal of Finance*, but it was quickly rejected [64, p. 194].
6. Dynamic strategies rely on the market to supply the needed liquidity. The Crash of 1987 and the Russian and the LTCM crises of 1998 demonstrate that such liquidity may not be available at times of extreme market movements [308, 654]. As prices began to fall during the Crash of 1987, portfolio insurers sold stock index futures. This activity in the futures market led to more selling in the cash market as program traders attempted to arbitrage the spreads between the cash and futures markets. Further price declines led to more selling by portfolio insurers, and so on [567, 647].
7. The CAPM does not require that the residuals ϵ_i be uncorrelated; see Eq. (31.2).
8. See Eqs. (15.12) and (24.10) for similar arguments.
9. Chebyshev's inequality in Exercise 13.3.10, part (2), supplies the intuition.
10. "Return" means price change ΔS or simple rate of return $\Delta S/S$. This is consistent with the stochastic differential equation $\Delta S = S\mu\,\Delta t + S\sigma\sqrt{\Delta t}\,\xi$ when Δt is small.

Software

Test everything. Hold on to the good.
I Thessalonians 5:21

32.1 Web Programming

The software for the book is Web-centric in that a reasonably updated Web browser is all that is needed to run it. The software is transferred to the user when clicked; no installation is necessary. As the collection of software expands, *The Capitals* Web page will reflect that. This new medium of software distribution excels the traditional way of bundling software with each book in a floppy disk or a CD-ROM [705].

The Web promises to be a platform that is independent of the computer's operating system and hardware. That means a program or document written in HTML (Hypertext Markup Language [167]) can be run everywhere, and the author is relieved from worrying about the potentially infinite number of computer systems that may access the code. As of now, this promise is not yet fully realized. To start with, the same document or program often elicits different behaviors from browsers of various companies or even browsers of the same family but with different versions. Browsers may implement only a subset of the standard plus a few nonstandard features. Additional complications are the possible versions of Java shipped with the browser and window systems running on top of the operating system. Fortunately, in most cases these problems are either inessential or can be avoided by upgrading the browser and not using nonstandard features.

32.2 Use of *The Capitals* Software

Open *The Capitals* at

www.csie.ntu.edu.tw/~lyuu/capitals.html.

See Fig. 32.1 for a typical look. Now click on any program to run it. For example, click on "mortgage" to generate the calculator in Fig. 32.2. Many financial problems can be solved by using two or more programs simultaneously. For example, one can run "spot & forward rates from coupon bonds" to calculate the spot rates. Then copy these rates into "seq. CMO pricer (vector)" to price CMOs. As another example, consider pricing CMO tranches at 10 years before maturity. We can run

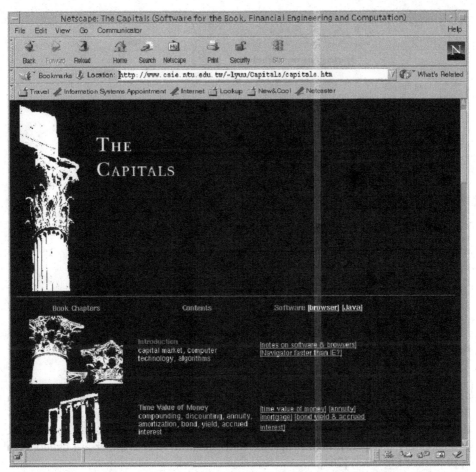

Figure 32.1: *The Capitals* page. This page is displayed by the Netscape browser in a Unix environment. The looks may differ from browser to browser and can be altered by changing the browser's settings.

"seq. CMO (vector)" to derive the tranches' remaining balances at that time. Then plug those numbers as original principals into "seq. CMO pricer (vector)" with 10 years remaining.

Some programs can be applied to situations not originally intended. For instance, CMO programs can be used for individual mortgages by allocating the entire principal to the first tranche; the cash flow of an SMBS can be tabulated by "pool P&I tabulator (vector)," and so on.

The following guidelines are recommended to run *The Capitals* software smoothly.

- Netscape Navigator 4.0 or higher, or Microsoft's Internet Explorer 4.0 or higher.
- Enable Java.
- Use Java 1.1.4 or higher.

Check "notes on software & browsers" for additional information.

The programs at *The Capitals* are written in JavaScript [357] and Java [356, 467].[1] Because it is the Java-enabled Web browser that interprets and executes the code, user interaction and processing are offloaded to the user's computer. This client/server

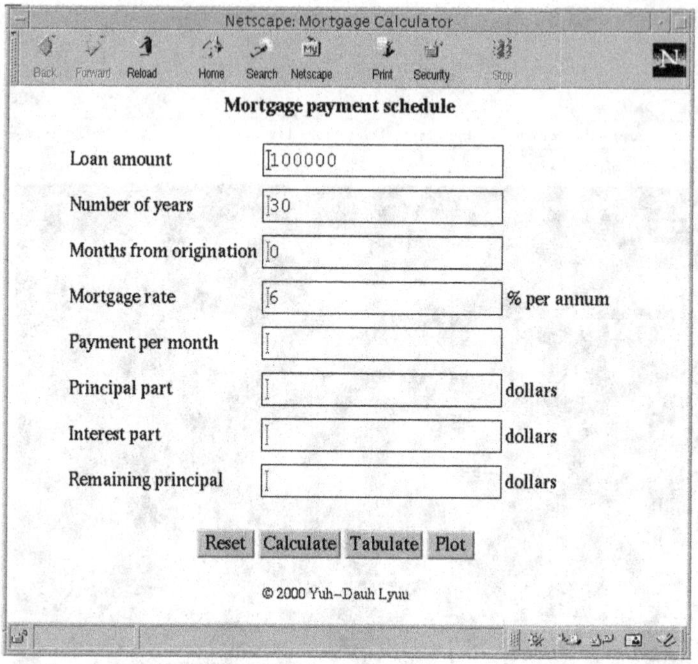

Figure 32.2: Mortgage calculator.

architecture is more efficient than having many clients' computing and interaction tasks running on the server, slowing it down for everybody. The unstated assumptions have been that the user's computer is reasonably powerful and the network is reasonably fast [719]. Java programs that run in a Web browser are called **Java applets** (see Fig. 32.3) [264].

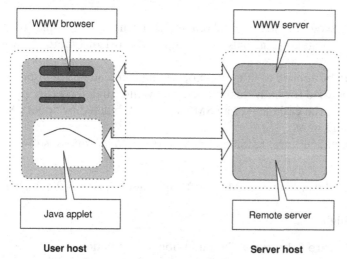

Figure 32.3: Java client/server architecture on the Web. The Java programming language was released by Sun in May of 1995. It promised platform-independent client/server software systems [356, 416].

32.3 **Further Topics**

Some computation-intensive tasks can take advantage of parallel processing for much faster performance. A good example is Monte Carlo MBS pricing. It starts by breaking the job into several tasks, each of which, on a different computer, simulates a fraction of the interest rate scenarios and calculates the average price. The averages are then collected to obtain the overall average price. Note that once the work has been divided, no communication among the tasks is needed before the collection stage. Good speed-ups have been obtained [528, 601, 794, 892, 893]. In contrast, a task that cannot be structured in such a way as to limit the amount of communication, hence dependency, among the tasks will not result in good performance [588]. Only computation-intensive problems are worthwhile to parallelize.

The Web technology is young and evolving quickly. Users and developers have been willing to tolerate many annoyances because they are witnessing something that promises to change the way society works. If the history of the auto industry is any guide, it will take decades for the technology to mature. Fortunately, thanks to the efforts and dedication of many corporations and computer professionals, the Web has become the most important and easy-to-use platform for software.

NOTE

1. JavaScript is not Java, but it has a similar syntax.

> There is nothing new to be discovered in physics now [1900].
> William Thomson (aka Lord Kelvin) (1824–1907)

Answers to Selected Exercises

> More questions may be easier to answer than just one question.
> Imre Lakatos (1922–1974), *Proofs and Refutations*

CHAPTER 2

Exercise 2.2.2: (1) Recall that $\sum_{i=1}^{n} i = n(n+1)/2$. (2) Use $\sum_{i=1}^{n} i^2 = (2n^3 + 3n^2 + n)/6$. (3), (4) Use $\sum_{i=0}^{k} 2^i = 2^{k+1} - 1$. (5) Use Euler's summation formula [461, p. 18],

$$\int_{a}^{b+1} g(x)\, dx \le \sum_{i=a}^{b} g(i) \le \int_{a-1}^{b} g(x)\, dx.$$

CHAPTER 3

Exercise 3.1.1: It is sufficient to show that $g(m) \equiv (1 + \frac{1}{m})^m$ is an increasing function of m. Note that

$$g'(m) = g(m) \left[\ln\left(1 + \frac{1}{m}\right) - \frac{1}{m+1} \right].$$

We can show the expression within the brackets to be positive by differentiating it with respect to m.

An alternative approach is to expand $g(m)$ and $g(m+1)$ as polynomials of $x \equiv 1/m$ using the binomial expansion. It is not hard to see that every term in $g(m+1)$, except the one of degree $m+1$ (which $g(m)$ does not have), is at least as large as the term of the same degree in $g(m)$. This approach does not require calculus.

Exercise 3.1.2: Monthly compounding, i.e., 12 times per annum. This can be verified by noting that $18.70/12 = 1.5583$.

Exercise 3.1.3: (1) The computing power's growth function is $(1.54)^n$, where n is the number of years since 1987. The equivalent continuous compounding rate is 43.18%. Because the memory capacity has quadrupled every 3 years since 1977, the function is $4^{n/3} = (1.5874)^n$, where n is the number of years since 1977. The equivalent continuous compounding rate is 46.21%. (2) It is $(500,000/300,000)^{1/4} - 1 \approx 13.6\%$. Data are from [574].

Exercise 3.2.1:

$$\mathrm{PV} = \sum_{i=0}^{nm-1} C\left(1 + \frac{r}{m}\right)^{-i} = C\, \frac{1 - (1 + \frac{r}{m})^{-nm}}{(r/m)} \left(1 + \frac{r}{m}\right).$$

Exercise 3.3.1: We derived Eq. (3.8) by looking forward into the future. We can also derive the same relation by looking back into the past: Right after the kth payment, the remaining principal is the value of the original principal minus the value of all the payments made to date, which is exactly what the formula says. Mathematically,

$$C \frac{1-\left(1+\frac{r}{m}\right)^{-nm}}{\frac{r}{m}}\left(1+\frac{r}{m}\right)^{k} - \sum_{i=1}^{k} C\left(1+\frac{r}{m}\right)^{i-1}$$

$$= C \frac{\left(1+\frac{r}{m}\right)^{k} - \left(1+\frac{r}{m}\right)^{-nm+k}}{\frac{r}{m}} - C \frac{1-\left(1+\frac{r}{m}\right)^{k}}{\frac{r}{m}}$$

$$= C \frac{1-\left(1+\frac{r}{m}\right)^{-nm+k}}{\frac{r}{m}}.$$

Exercise 3.3.2: Note that the PV of an ordinary annuity becomes $\sum_{i=1}^{n} Ce^{ir} = C \frac{1-e^{-nr}}{e^{r}-1}$ under continuous compounding. Without loss of generality, assume that the PV of the original mortgage is $1. The monthly payment is hence $\frac{e^{r}-1}{1-e^{-nr}}$. For an interest rate of $r-x$, the level payment is $D \equiv \frac{e^{r-x}-1}{1-e^{-n(r-x)}}$. The new instrument's cash flow is $De^{x}, De^{2x}, \ldots, De^{nx}$ by definition. The PV is therefore

$$\sum_{i=1}^{n} D \frac{e^{ix}}{e^{ir}} = D \frac{1-e^{-n(r-x)}}{e^{r-x}-1} = 1.$$

Consult [330, p. 120].

Exercise 3.4.1: Apply Eq. (3.11) with $y = 0.0755$, $n = 3$, $C_1 = 1000$, $C_2 = 1000$, $C_3 = 1500$, and $P = 3000$.

Exercise 3.4.2: The FV is $C \frac{(1+r)^{n}-1}{r}$ from Eq. (3.4). To guarantee a return of y, the PV should be $C \frac{(1+r)^{n}-1}{r} \frac{1}{(1+y)^{n}}$.

Exercise 3.4.3: Proposal A's NPV is now $2,010.014, and Proposal B's is $1,773.384. Proposal A wins out under this scenario.

Exercise 3.4.4: The iteration is

$$x_{k+1} = x_k - \frac{f(x_k)}{f'(x_k)} = x_k - \frac{x_k^3 - x_k^2}{3x_k^2 - 2x_k} = \frac{2x_k^2 - x_k}{3x_k - 2}$$

by Eq. (3.15), and the desired sequence is 1.5, 1,2, 1.05, 1.004348, and 1.00003732. The last approximation is very close to the root 1.0.

Exercise 3.4.5: $F'(x) = \frac{f(x) f''(x)}{f'(x)^2}$; hence $F'(\xi) = 0$.

Exercise 3.4.6: See Fig. 3.5 [656].

Exercise 3.4.8: Let $y_f(x)$ be the y that makes $f(x, y) = 0$ given x. Similarly, let $y_g(x)$ be the y that makes $g(x, y) = 0$ given x. Assume that y_f and y_g are continuous functions. Then the bisection method is applicable to the function $\phi(x) \equiv y_f(x) - y_g(x)$ if it starts with x_1 and x_2 such that $\phi(x_1)\phi(x_2) < 0$. In so doing, a two-dimensional problem is reduced to the standard one-dimensional problem. Note that $\phi(x) = 0$ if and only if $y_f(x) = y_g(x)$. Unfortunately, this method has no obvious generalizations to $n > 2$ dimensions. Note that $y_f(x)$ and $y_g(x)$ may have to be numerically solved. See [447, p. 585].

Exercise 3.5.1: For each Δt time, it pays out $c\Delta t$ dollars, which is discounted at the rate of r [462].

Exercise 3.5.2: The price that guarantees the return if the bond is called is, from Eq. (3.18),

$$5 \times \frac{1-[1+(0.12/2)]^{-2\times5}}{0.12/2} + \frac{100}{[1+(0.12/2)]^{2\times5}} = 92.6399.$$

Similarly, the price should be

$$5 \times \frac{1-[1+(0.12/2)]^{-2\times10}}{0.12/2} + \frac{100}{[1+(0.12/2)]^{2\times10}} = 88.5301$$

dollars if the bond is held to maturity. Hence the price to pay is $88.5301. (A more rigorous method yielding the same conclusion goes as follows. The formula for various call half-years $n = 10, \ldots, 20$ is $(250/3) + (50/3) \times (1.06)^{-n}$, which is minimized at $n = 20$.)

Exercise 3.5.3: It is

$$PV = \sum_{i=1}^{n} \frac{C(1-T)}{(1+r)^i} + \frac{F - \max((F-P)\,T_G, 0)}{(1+r)^n}$$

$$= C(1-T)\frac{1-(1+r)^{-n}}{r} + \frac{F - \max((F-P)\,T_G, 0)}{(1+r)^n}.$$

See [827, p. 123].

Exercise 3.5.4: (1) Because $P = F/(1+r)^n$,

$$\frac{\partial P}{\partial n} = -\frac{F\ln(1+r)}{(1+r)^n}, \qquad \frac{\partial P}{\partial r} = -\frac{Fn}{(1+r)^{n+1}}.$$

(2) For $r = 0.04$ and $n = 40$, we have

$$\Delta P \approx -0.00817 \times F \times \Delta n, \qquad \Delta P \approx -8.011 \times F \times \Delta r.$$

Exercise 3.5.5:

$$\sum_{i=1}^{n} \frac{Fr}{(1+r)^i} + \frac{F}{(1+r)^n} = F.$$

Exercise 3.5.6: Intuitively, this is true because the accrued interest is not discounted, whereas the next coupon to be received by the buyer is. Mathematically,

$$F + Fc(1-\omega) = \frac{Fc}{(1+r)^\omega} + \frac{Fc}{(1+r)^{\omega+1}} + \frac{Fc}{(1+r)^{\omega+2}} + \cdots + \frac{Fc+F}{(1+r)^{\omega+n-1}}.$$

If $r \geq c$, then $1 + c(1-\omega) \leq \frac{1+r}{(1+r)^\omega}$, which can be shown to be impossible.

Another approach is to observe that the buyer should have paid $Fc[\,(1+r)^{1-\omega} - 1\,]$ instead of $Fc(1-\omega)$ and to prove that the former is smaller than the latter.

Exercise 3.5.7: The number of days between the settlement date and the next coupon date is calculated as $30 + 31 + 1 = 62$. The number of days in the coupon period being $30 + 30 + 31 + 30 + 31 + 31 + 1 = 184$, the accrued interest is $100 \times (0.1/2) \times (184-62)/184 = 3.31522$. The yield to maturity can be calculated with the help of Eq. (3.20). Note that the PV should be $116 + 3.31522$.

Exercise 3.5.9: Let P_0 be the bond price now and P_1 be the bond price one period from now after the coupon is paid. We are asked to prove that $(c + P_1 - P_0)/P_0 = y$. Without loss of generality, we assume that the par value is $1. From Eq. (3.18),

$$P_0 = (c/y)[\,1 - (1+y)^{-n}\,] + (1+y)^{-n} = (c/y) + (1+y)^{-n}(1 - c/y),$$

$$P_1 = (c/y)\,[\,1 - (1+y)^{-(n-1)}\,] + (1+y)^{-(n-1)} = (c/y) + (1+y)^{-(n-1)}(1 - c/y).$$

So $P_1 - P_0 = (y-c)(1+y)^{-n}$. Also, $yP_0 = c + (y-c)(1+y)^{-n}$. Hence $c + P_1 - P_0 = yP_0$.

CHAPTER 4

Exercise 4.1.2: Equation (4.1) can be rearranged to become

$$-\frac{\partial P/P}{\partial y} = \frac{C\{-(n/y) + \frac{1}{y^2}[(1+y)^{n+1} - (1+y)]\} + nF}{C\{[(1+y)^{n+1} - (1+y)]/y\} + F(1+y)} = \frac{CA(y) + nF}{CB(y) + F(1+y)}.$$

To show that $-(\partial P/P)/\partial y$ increases monotonically if $y > 0$ as C decreases, it suffices to prove that

$$\frac{A(y)}{B(y)} < \frac{nF}{F(1+y)} = \frac{n}{(1+y)}.$$

We can verify that

$$\frac{A(y)}{B(y)} - \frac{n}{(1+y)} = \frac{(1+y)^2[(1+y)^n - 1 - ny(1+y)^{n-1}]}{B(y)(1+y)}.$$

The expression within the brackets is nonpositive because it is zero for $y = 0$ and its derivative is $-n(n-1)\,y(1+y)^{n-2} \leq 0$. (In fact, the expression within the brackets is strictly negative for $n > 1$.)

Exercise 4.1.3: (1) We observe that $-\frac{\partial P/P}{\partial y} = \frac{1-(1+y)^{-n}}{y}$, which is clearly a decreasing function of y. (2) We differentiate $-(\partial P/P)/\partial y$ with respect to yield y. After rearranging, we obtain

$$-(1+y)\left\{ \frac{\sum_i i^2(1+y)^{-i}C_i}{\sum_i(1+y)^{-i}C_i} - \left[\frac{\sum_i i(1+y)^{-i}C_i}{\sum_i(1+y)^{-i}C_i}\right]^2 \right\}.$$

The term within the braces can be interpreted in terms of probability theory: the variance of the random variable X defined by

$$\text{Prob}[\,X = j\,] = \frac{(1+y)^{-j}C_j}{\sum_i(1+y)^{-i}C_i}, \quad j \geq 1.$$

Hence it has to be positive. This shows that $-(\partial P/P)/\partial y$ is a decreasing function of yield. See [547, p. 318].

Exercise 4.2.1: 2.67 years.

Exercise 4.2.2: Let D be the modified duration as originally defined. We need to make sure that

$$100 \times D \times \Delta r = D_\% \times \Delta r_\%,$$

where $\Delta r_\%$ denotes the rate change in percentage. Because $\Delta r = \Delta r_\%/100$, the above identity implies that $D_\% = D$.

Exercise 4.2.3: The cash flow of the bond is $C, C, \ldots, C, C+F$, and that of the mortgage is $M, M, \ldots, M$. The conditions imply that $M > C$. The bond hence has a longer duration as its cash flow is more tilted to the rear. Mathematically, we want to prove that

$$C\sum_{i=1}^{n} i(1+y)^{-i} + nF(1+y)^{-n} > M\sum_{i=1}^{n} i(1+y)^{-i}$$

subject to

$$C\sum_{i=1}^{n}(1+y)^{-i} + F(1+y)^{-n} = M\sum_{i=1}^{n}(1+y)^{-i}.$$

The preceding equality implies that $C - M = -\frac{F(1+y)^{-n}}{\sum_{i=1}^{n}(1+y)^{-i}}$. Hence

$$C\sum_{i=1}^{n} i(1+y)^{-i} + nF(1+y)^{-n} - M\sum_{i=1}^{n} i(1+y)^{-i}$$

$$= (C - M)\sum_{i=1}^{n} i(1+y)^{-i} + nF(1+y)^{-n}$$

$$= -\frac{F(1+y)^{-n}}{\sum_{i=1}^{n}(1+y)^{-i}}\sum_{i=1}^{n} i(1+y)^{-i} + nF(1+y)^{-n}$$

$$= \frac{F}{(1+y)^n}\left[n - \frac{\sum_{i=1}^{n} i(1+y)^{-i}}{\sum_{i=1}^{n}(1+y)^{-i}}\right]$$

$$> 0.$$

We remark that the market prices of the two instruments are not necessarily equal. To show this, assume the market price of the bond is some $a > 0$ times that of the mortgage. Express M as a function now of $C, F, y, n,$ and a. Finally, substitute the expression for M into the formula of the MD for the mortgage.

Exercise 4.2.4: The MD is equal to $\sum_i \frac{iM}{(1+y)^i} / \sum_i \frac{M}{(1+y)^i}$, where M is the monthly payment. Note that M cancels out completely. The rest is just simple algebraic manipulation. See [348, p. 94].

Exercise 4.2.6: This is because convexity is additive. Check with Eq. (4.10) again and observe that it holds as long as each cash flow is nonnegative.

Exercise 4.2.7: Assume again that the liability is L at time m. The present value is therefore $L/(1+y)^m$. A coupon bond with an equal PV will grow to be exactly L at time m. In fact, it is not hard to show that at every point in time, the PV of the bond plus the cash incurred by reinvesting the coupon payments exactly matches the PV of the liability.

Exercise 4.2.8: Rearrange Eq. (4.8) to get

$$\frac{\partial FV}{\partial y} = (1+y)^{m-1} P \left[m + (1+y) \frac{\partial P/P}{\partial y} \right].$$

If m is less than the MD, the expression within the brackets is negative. This means that the FV decreases as y increases. The other cases can be handled similarly.

Exercise 4.2.9: Let Δy denote the rate change, which may be positive or negative. At time Δt, the PV of the liability will be $L/(1+y-\Delta y)^{m-\Delta t}$, whereas the PV of the bond plus any reinvestment of the interest will be $\sum_i C_i/(1+y-\Delta y)^{i-\Delta t}$. We are asked to prove that

$$\frac{L}{(1+y-\Delta y)^{m-\Delta t}} < \sum_i \frac{C_i}{(1+y-\Delta y)^{i-\Delta t}}.$$

After dividing both sides by $(1+y-\Delta t)^{\Delta t}$, we are left with the equivalent inequality

$$\frac{L}{(1+y-\Delta y)^m} < \sum_i \frac{C_i}{(1+y-\Delta y)^i}.$$

However, this must be true because we have shown in the text that any instantaneous change in the interest rate raises the bond's value over the liability because of convexity. Note that this conclusion holds for *any* $\Delta t < m$!

Exercise 4.2.10: Solve for ω_1 and ω_2 such that

$$1 = \omega_1 + \omega_2,$$
$$3 = \omega_1 + 4\omega_2.$$

The results: $\omega_1 = 1/3$ and $\omega_2 = 2/3$. So 1/3 of the portfolio's market value should be put in bond one, and the remaining 2/3 in bond two.

Exercise 4.2.11: (1) Observe that

$$
\begin{aligned}
P(y') &= A_1 e^{a_1 y'} + A_2 e^{-a_2 y'} - L_t \\
&= A_1 e^{a_1 y'} + A_2 e^{-a_2 y'} - (A_1 e^{a_1 y} + A_2 e^{-a_2 y}) \\
&= A_1 e^{a_1 y} \left[e^{a_1(y'-y)} + \frac{a_1}{a_2} e^{-a_2(y'-y)} - \left(1 + \frac{a_1}{a_2} \right) \right]
\end{aligned}
$$

after $A_1 a_1 e^{a_1 y} - A_2 a_2 e^{-a_2 y} = 0$ is applied. It is easy to show that

$$g(x) = e^{ax} + \frac{a}{b} e^{-bx} - \left(1 + \frac{a}{b} \right) > 0$$

when $x \neq 0$. The expression within the brackets is hence positive for $y' \neq y$. See [547, pp. 406–407]. A more concise proof is this. It is easy to see that

$$A_1 = \frac{a_2 e^{-a_1 y}}{a_1 + a_2} L_t > 0, \qquad A_2 = \frac{a_1 e^{a_2 y}}{a_1 + a_2} L_t > 0.$$

So $P''(y) = A_1 a_1^2 e^{a_1 y} + A_2 a_2^2 e^{a_2 y} > 0$.
  (2) It is not hard to see that a portfolio with cash inflows at t_1, t_2, t_3 can be constructed so that it is more valuable than one with cash inflows at $T < t_3$ after the shift [496, p. 635].

Exercise 4.2.12: To show that Eq. (4.12) is at most j, differentiate it with respect to y and prove that its first derivative is less than zero. Hence Eq. (4.12) is a decreasing function. Finally, show that it approaches j as $y \to 0$. An alternative approach uses the observation that $(1+y)^{-j} \geq 1 - jy$. Yet another alternative applies induction on j to show that Eq. (4.12) is at most j.

Exercise 4.3.1: Let C be the convexity in the original sense and $C_\%$ be the convexity in percentage terms. We need to make sure that

$$100 \times C \times (\Delta r)^2 = C_\% \times (\Delta r_\%)^2,$$

where $\Delta r_\%$ denotes the rate change in percentage. Because $\Delta r = \Delta r_\%/100$, it follows that $C_\% = C/100$. See [490, p. 29].

Exercise 4.3.2: This is just a simple application of the chain rule. Recall that duration $= -\partial P/\partial y$ and convexity $= (\partial^2 P/\partial y^2)(1/P)$. See [547, p. 333].

Exercise 4.3.4: From Eq. (4.15) we can verify that convexity is equal to $\frac{CA(y)+n(n+1)\,y^3}{CB(y)+y^3(1+y)^2}$, where

$$A(y) = 2(y+1)^{n+2} - 2(y+1)^2 - 2ny(y+1) - n(n+1)\,y^2,$$
$$B(y) = y^2(1+y)^2[\,(1+y)^n - 1\,].$$

To prove the claim, it suffices to show that $A(y)/B(y) < n(n+1)/(1+y)^2$. This is equivalent to proving that

$$G(y) = 2(1+y)^{n+1} - y^2 n(n+1)(1+y)^{n-1} - 2(1+y) - 2ny < 0$$

after simplification. It is easy to show that $G(y)$ is concave for $y > 0$ because $G''(y) < 0$ for $y > 0$. Hence

$$G'(y) = 2(n+1)(1+y)^n - 2yn(n+1)(1+y)^{n-1} - y^2 n(n+1)(n-1)(1+y)^{n-2} - 2 - 2n$$

is less than zero for $y > 0$ as it is a decreasing function of y with $G'(0) = 0$. This leads directly to our conclusion because $G(0) = 0$.

Exercise 4.3.5: A proof is presented with only elementary mathematics. Suppose the universe consists of three kinds of zero-coupon bonds ($n = 3$). Note that $C_i = D_i(D_i + 1)/(1 + y)^2$ by Exercise 4.3.3. The portfolio convexity, which is the objective function, is hence

$$\sum_{i=1}^{3} \omega_i C_i = (1+y)^{-2} \sum_{i=1}^{3} \omega_i\,(D_i^2 + D_i) = (1+y)^{-2}\left(D + \sum_{i=1}^{3} \omega_i D_i^2\right).$$

The original objective function can be replaced with the simpler

$$\sum_{i=1}^{3} \omega_i D_i^2. \tag{33.1}$$

It is not hard to show that, for distinct $i, j, k \in \{1, 2, 3\}$,

$$\omega_i = \frac{D - D_j + (D_j - D_k)\,\omega_k}{D_i - D_j},$$
$$\omega_j = \frac{D - D_i + (D_i - D_k)\,\omega_k}{D_j - D_i},$$

by manipulation of the two linear equality constraints. So only one variable ω_k remains. Objective function (33.1) becomes

$$\frac{D - D_j + (D_j - D_k)\,\omega_k}{D_i - D_j}\,D_i^2 + \frac{D - D_i + (D_i - D_k)\,\omega_k}{D_j - D_i}\,D_j^2 + \omega_k D_k^2$$

$$= \frac{(D - D_j)\,D_i^2 - (D - D_i)\,D_j^2}{D_i - D_j} + \omega_k\left[\frac{(D_j - D_k)\,D_i^2 - (D_i - D_k)\,D_j^2}{D_i - D_j} + D_k^2\right]$$

$$= D(D_i + D_j) - D_i D_j + \omega_k\left(D_i D_j - D_k(D_i + D_j) + D_k^2\right)$$

$$= D^2 - (D - D_i)(D - D_j) + \omega_k(D_k - D_i)(D_k - D_j). \tag{33.2}$$

Equation (33.2) equals $D^2 - (D - D_1)(D - D_3)$ by picking $i = 1, k = 2, j = 3$, and $\omega_k = 0$. We can confirm that this is a valid choice by checking $\omega_1 = (D - D_3)/(D_1 - D_3) > 0$ and $\omega_3 = (D - D_1)/(D_3 - D_1) > 0$ because $D_1 < D < D_3$. Note that it is a barbell portfolio.

To verify that no other valid portfolios have as high a convexity, we prove that Eq. (33.2) is indeed maximized with $\omega_2 = 0$ as follows. From Eq. (33.2) we must choose $\omega_k = 0$ for the $k = 2$ case because $(D_2 - D_i)(D_2 - D_j) < 0$. Now we consider the objective function with $k \neq 2$. First, we consider $k = 3$. Without loss of generality, we assume that $D_1 = D_i < D_j = D_2$. The formulas for ω_1 and ω_2 dictate that

$$\frac{D_2 - D}{D_2 - D_3} \leq \omega_3 \leq \frac{D_1 - D}{D_1 - D_3}.$$

We plug the preceding second inequality into objective function (33.2) to obtain an upper bound of

$$D^2 - (D - D_1)(D - D_2) + \frac{D_1 - D}{D_1 - D_3}(D_3 - D_1)(D_3 - D_2)$$

$$= D^2 - (D - D_1)(D - D_2) + (D - D_1)(D_3 - D_2)$$

$$= D^2 - (D - D_1)(D - D_3).$$

Now we consider $k = 1$. Without loss of generality, we assume that $D_2 = D_i < D_j = D_3$. The formulas for ω_2 and ω_3 dictate that

$$\frac{D_2 - D}{D_2 - D_1} \leq \omega_1 \leq \frac{D_3 - D}{D_3 - D_1}.$$

We plug the preceding second inequality into objective function (33.2) to obtain an upper bound of

$$D^2 - (D - D_2)(D - D_3) + \frac{D_3 - D}{D_3 - D_1}(D_1 - D_2)(D_1 - D_3)$$

$$= D^2 - (D - D_2)(D - D_3) + (D - D_3)(D_1 - D_2)$$

$$= D^2 - (D - D_1)(D - D_3).$$

Exercise 4.3.6: Let there be $n \geq 3$ kinds of zero-coupon bonds in the universe. Given a portfolio with more than two kinds of bonds, replace those with duration D, where $D_1 < D < D_n$, with bonds with durations D_1 and D_n with a *matching* duration. By Exercise 4.3.5, the new portfolio has a higher convexity. We repeat the steps for each such D until we end up with a barbell portfolio consisting solely of bonds with durations D_1 and D_n.

CHAPTER 5

Exercise 5.2.1: Because $P = \sum_i C_i[1 + S(i)]^{-i}$,

$$P \approx \sum_i \left\{ C_i(y) + \frac{\partial C_i(y)}{\partial y}[S(i) - y] \right\}$$

when $C_i[1 + S(i)]^{-i}$ is expanded in Taylor series at y. The preceding relation, when combined with $P = \sum_i C_i(y)$, leads to

$$\sum_i \frac{\partial C_i(y)}{\partial y}[S(i) - y] \approx 0.$$

Rearrange the above to obtain the result. See [38, pp. 23–24].

Exercise 5.3.1:

$$100 = \sum_{i=1}^{19} \frac{8/2}{(1 + 0.1)^i} + \frac{(8/2) + 100}{[1 + S(20)]^{20}}.$$

Thus

$$66.54 = \frac{104}{[1 + S(20)]^{20}},$$

and $S(20) = 2.258\%$.

Programming Assignment 5.4.1: See the algorithm in Fig. 33.1.

Exercise 5.5.1: (1) As always, assume $S(1) = y_1$ to start with; hence $S(1) \geq y_1$. From Eq. (5.1) and the definition of yield to maturity,

$$\sum_{i=1}^{k-1} \frac{C}{(1 + y_k)^i} + \frac{C + F}{(1 + y_k)^k} = \sum_{i=1}^{k-1} \frac{C}{[1 + S(i)]^i} + \frac{C + F}{[1 + S(k)]^k}$$

$$= \sum_{i=1}^{k-1} \frac{C}{(1 + y_{k-1})^i} + \left\{ \frac{F}{(1 + y_{k-1})^{k-1}} - \frac{F}{[1 + S(k-1)]^{k-1}} \right\} + \frac{C + F}{[1 + S(k)]^k}. \tag{33.3}$$

Static spread with the Newton-Raphson method:

input: $n, C, P, S[1..n]$;
real spread, price, priceD;
integer k;
spread := 0;
for $(k = 1$ to $10)$ {
    price := $\sum_{i=1}^{n} C/(1 + S[i] + \text{spread})^i + 100/(1 + S[n] + \text{spread})^n$;
    if [|price − P| < 0.000001] return spread;
    priceD := $- \sum_{i=1}^{n} iC/(1 + S[i] + \text{spread})^{i+1} - 100 \times n/(1 + S[n] + \text{spread})^{n+1}$;
    spread := spread − (price − P)/priceD;
}

Figure 33.1: Static spread with the Newton–Raphson method. P is the price (as a percentage of par) of the coupon bond maturing n periods from now, C is the coupon of the bond expressed as a percentage of par per period, and $S[i]$ denotes the i-period spot rate.

Because $S(k-1) \geq y_{k-1}$ by the induction hypothesis, the term inside the braces is nonnegative. The normality assumption about the yield curve further implies that

$$\sum_{i=1}^{k-1} \frac{C}{(1+y_k)^i} < \sum_{i=1}^{k-1} \frac{C}{(1+y_{k-1})^i}.$$

Therefore

$$\frac{C+F}{(1+y_k)^k} > \frac{C+F}{[1+S(k)]^k},$$

that is, $S(k) > y_k$, as claimed. (2) We can easily confirm the two claims by inspecting Eq. (33.3).

Exercise 5.5.2: Assume that the coupon rate is 100%, $y_1 = 0.05$, $y_2 = 0.5$, $y_3 = 0.6$, and $n = 3$. The spot rates are then $S(1) = 0.05$, $S(2) = 0.82093$, and $S(3) = 0.58753$.

Exercise 5.5.3: (1) The assumption means that we can use Eq. (33.3) with $F = 0$, which, after simplification, becomes

$$\sum_{i=1}^{k} \frac{1}{(1+y_k)^i} = \sum_{i=1}^{k} \frac{1}{[1+S(i)]^i}.$$

First, we assume that the spot rate curve is upward sloping but $y_{k-1} > y_k$. The preceding equation implies that

$$\sum_{i=1}^{k-1} \frac{1}{(1+y_{k-1})^i} + \frac{1}{(1+y_k)^k} = \sum_{i=1}^{k-1} \frac{1}{[1+S(i)]^i} + \frac{1}{(1+y_k)^k} < \sum_{i=1}^{k} \frac{1}{[1+S(i)]^i}.$$

Hence $y_k > S(k)$, a contradiction. (2) The given yield curve implies the following spot rates: $S(1) = 0.1$, $S(2) = 0.89242$, and $S(3) = 0.5036$. These spot rates are not increasing.

Exercise 5.6.1: $S(1) = 0.03$, $S(2) = 0.04020$, $S(3) = 0.04538$, $f(1,2) = 0.0505$, $f(1,3) = 0.0532$, and $f(2,3) = 0.0558$.

Exercise 5.6.2: One dollar invested for $b+c$ periods at the $(b+c)$-period forward rate starting from period a is the same as one dollar invested for b periods at the b-period forward rate starting from period a and reinvested for another c periods at the c-period forward rate starting from period $a+b$.

Exercise 5.6.3: The PV of \$1 at time T, or $d(T) = [1+S(T)]^{-T}$, is the same as that of $1 + f(T, T+1)$ dollars at time $T+1$. Mathematically,

$$[1 + f(T, T+1)] d(T+1) = \frac{[1+S(T+1)]^{T+1}}{[1+S(T)]^T} [1+S(T+1)]^{-(T+1)} = d(T).$$

A shorter proof:

$$f(T, T+1) = \frac{[1+S(T+1)]^{T+1}}{[1+S(T)]^T} - 1 = \frac{[1+S(T)]^{-T}}{[1+S(T+1)]^{-(T+1)}} - 1 = \frac{d(T)}{d(T+1)} - 1.$$

Exercise 5.6.4: The 10-year spot rate is 5.174%. Now we move the bond price to $60.6. The 10-year spot rate becomes $2 \times [(1/0.606)^{1/20} - 1] = 0.05072$. The percentage change is ~1.97%. By Exercise 5.6.3, the forward rate in question equals $f(19, 20) = 2 \times [\frac{0.62}{d(20)} - 1]$. The multiplicative factor 2 converts the forward rate into a semiannual yield. Note that each year has two periods. Simple calculations show that the forward rate moves from 6.667% to 4.620%. The percentage change is therefore 30.7%!

Exercise 5.6.5: From relation (5.3), $f(j, j+1) > S(j+1) > S(j)$, where the spot rate is upward sloping. Similarly from relation (5.4), $f(j, j+1) < S(j+1) < S(j)$, where the spot rate curve is downward sloping. And $f(j, j+1) = S(j+1) = S(j)$, where the spot rate curve is flat. See also [147, p. 400].

Exercise 5.6.6: (1) Otherwise, there are arbitrage opportunities. See [198], [234, p. 385], and [746, p. 526]. For an uncertain world, the arbitrage argument no longer holds: Although we are still able to lock in the forward rate, there is no a priori reason any future rate has to be known today for sure. (2) Because they are all realized by today's spot rate for that period according to (1) [731, p. 165].

Exercise 5.6.7: (1) From Eq. (3.6), we solve

$$1000 = \frac{1}{0.0255} \times \left[1 - \frac{1}{(1+0.0255)^{100}}\right] \times 27 + \frac{F}{(1+0.0255)^{100}}$$

for F. The answer is $F = 329.1686$. (2) They are equivalent because their cash flows match exactly. (3) Verify with Eq. (3.6) again. See [302].

Exercise 5.6.8: The probability of default is $1 - (0.92/0.94) = 0.0213$. The forward probability of default, f, satisfies $(1 - 0.0213)(1 - f) = (0.84/0.87)$. Hence $f = 0.0135$.

Exercise 5.6.9: Let $S(i)$ denote the probability that the corporation survives past time i and let $d_c(\cdot)$ denote the discount factors obtained by corporate zeros. (1) By definition, $d_c(i) = d(i) S(i)$. (2) As in Exercise 5.6.3, $1 + f_c(i-1, i) = d_c(i-1)/d_c(i)$. The forward probability of default for period i is

$$\frac{S(i-1) - S(i)}{S(i-1)} = 1 - \frac{S(i)}{S(i-1)} = 1 - \frac{d_c(i) \, d(i-1)}{d(i) \, d_c(i-1)}$$

$$= 1 - \frac{1 + f(i-1, i)}{1 + f_c(i-1, i)} \approx f_c(i-1, i) - f(i-1, i).$$

See [583, pp. 96–97]. The forward probability of default is also called the **hazard rate** [846].

Exercise 5.6.10: The forward rate is simply the f that equates $e^{i S(i)} e^{(j-i) f} = e^{j S(j)}$.

Exercise 5.6.11:

Period (n)	Spot Rate % Per Period	One-Period Forward Rate % Per Period
1	2.00	
2	2.50	3.00
3	3.00	4.00
4	3.50	5.00
5	4.00	6.00

Exercise 5.6.12: (1) Let $S(1) \rightarrow S(1) + \Delta y$ and, in general, $S(i) \rightarrow S(i) + \Delta y/i$. We confirm that this works by inspecting Eq. (5.9). (2) From Eq. (5.5), we know the n-period zero-coupon bond costs

$$\frac{1}{[1+S(1)][1+f(1, 2)] \cdots [1+f(n-1, n)]}$$

now. By the assumption, none of the forward rates in the preceding formula changes when $S(1)$ does. So $-\frac{\partial P/P}{\partial y} = \frac{1}{1+S(1)}$ for zero-coupon bonds. The same equation holds for coupon bonds as well by similar arguments. See [231, p. 53].

Exercise 5.6.13: An investor is assured that \$1 will grow to be $1 + jS(j)$ at time j. Suppose a person invests \$1 in riskless securities for i periods and, at time i, invests the proceeds in riskless securities for another $j - i$ periods ($j > i$). The implied $(j - i)$-period forward rate at time i is the f that satisfies

$$[1 + iS(i)](1 + (j - i)\, f) = 1 + jS(j).$$

See [731, p. 7].

Exercise 5.7.1: An n-period zero-coupon bond fetches $[1 + S(n)]^{-n}$ today and $[1 + S(k, n)]^{-(n-k)}$ at time k. The return is hence

$$\left\{ \frac{[1 + S(k, n)]^{-(n-k)}}{[1 + S(n)]^{-n}} \right\}^{1/k} - 1.$$

If the forward rate is realized, then $S(k, n) = f(k, n)$ and $[1 + S(k, n)]^{n-k} = \frac{[1+S(n)]^n}{[1+S(k)]^k}$. After substitution, we arrive at the desired result.

Exercise 5.7.2: By Eq. (5.14),

$$E[\{1 + S(1)\}]E[\{1 + S(1, 2)\}] \cdots E[\{1 + S(n - 1, n)\}]$$
$$= [1 + S(1)][1 + f(1, 2)] \cdots [1 + f(n - 1, n)],$$

which equals $[1 + S(n)]^n$ by Eq. (5.5).

Exercise 5.7.3: (1) Rearrange definition (5.16) to be

$$\frac{E[\{1 + S(1, n)\}^{-(n-1)}]}{1 + S(1)} = [1 + S(n)]^{-n}. \tag{33.4}$$

It implies that

$$E\left[\frac{1}{\{1 + S(1)\}\{1 + S(1, n)\}^{n-1}}\right] = [1 + S(n)]^{-n}.$$

Now recursively apply Eq. (33.4) to $[1 + S(1, n)]^{-(n-1)}$. (2) Take $n = 2$. The local expectations theory implies that $E[\{1 + S(1)\}^{-1}\{1 + S(1, 2)\}^{-1}] = [1 + S(2)]^{-2}$. However, this equals $[1 + S(1)]^{-1}[1 + f(1, 2)]^{-1} = [1 + S(1)]^{-1}E[1 + S(1, 2)]^{-1}$ under the unbiased expectations theory. We conclude that $E[\{1 + S(1, 2)\}^{-1}] = E[1 + S(1, 2)]^{-1}$, which is impossible unless there is no randomness by Jensen's inequality.

Exercise 5.7.4: If they were consistent, then

$$E[\{1 + S(1, 2)\} \cdots \{1 + S(n - 1, n)\}]\, E\left[\frac{1}{\{1 + S(1, 2)\} \cdots \{1 + S(n - 1, n)\}}\right] = 1$$

from Exercise 5.7.3(1). However, this is impossible from Jensen's inequality unless there is no randomness.

Exercise 5.7.5: From the assumption, $S(1) = S(2) = \cdots = s$. Hence

$$E[S(i, i + 1)] < f(i, i + 1) = s = S(i).$$

Exercise 5.8.1: (1) Suppose that the yields to maturity change by the same amount and that the spot rate curve shift is parallel. We prove that the spot rate curve is flat. Consider a security with cash flows $C_i \neq 0$ at times $t_i, i = 1, 2$. In the beginning, by definition,

$$\sum_{i=1}^{2} C_i e^{-S(t_i)\, t_i} = \sum_{i=1}^{2} C_i e^{-y t_i}. \tag{33.5}$$

Suppose that the spot rate curve witnesses a parallel shift by the amount of $\Delta \neq 0$. The yield to maturity must also shift by Δ because, in particular, a t_1-period zero-coupon bond's yield to maturity *is* the t_1-period spot rate (hence (2) holds, incidentally). Therefore,

$$\sum_{i=1}^{2} C_i e^{-(S(t_i)+\Delta)\,t_i} = \sum_{i=1}^{2} C_i e^{-(y+\Delta)\,t_i}.$$

The preceding two equations give rise to

$$C_1 e^{-\Delta \times t_1}[e^{-y t_1} - e^{-S(t_1)t_1}] = C_2 e^{-\Delta \times t_2}[e^{-S(t_2)t_2} - e^{-y t_2}]. \tag{33.6}$$

Because Eq. (33.6) reduces to Eq. (33.5) for $\Delta = 0$, it cannot hold for any other Δ because of the different growing rates between $e^{-\Delta \times t_1}$ and $e^{-\Delta \times t_2}$ when $t_1 \neq t_2$ unless $y = S(t_i)$ for all i. See [496, Theorem 1] for a different proof.

Exercise 5.8.2:

$$-\lim_{\Delta y \to 0} \frac{\sum_i \frac{C_i}{[1+S(i)]^i} - \sum_i \frac{C_i}{[[1+S(i)](1+\{\Delta y/[1+S(1)]\})]^i}}{\Delta y \sum_i \frac{C_i}{[1+S(i)]^i}} = \frac{\sum_i \frac{i C_i}{[1+S(i)]^i}}{[1+S(1)] \sum_i \frac{C_i}{[1+S(i)]^i}}.$$

Exercise 5.8.3: See [424, p. 561].

CHAPTER 6

Exercise 6.1.1: See [273, p. 196].

Exercise 6.1.2: $E[XY] = E[E[XY|Y]] = E[YE[X|Y]] = E[YE[X]] = E[X]E[Y]$. By Eq. (6.3), $\text{Cov}[X, Y] = E[XY] - E[X]E[Y]$; thus $\text{Cov}[X, Y] = 0$.

Exercise 6.1.3: See [75, p. 3].

Exercise 6.1.4: Simple manipulations will do [30, p. 10].

Exercise 6.1.5: Easy corollary from Eq. (6.11). See also [147, p. 16].

Exercise 6.1.6: See [492, pp. 14–15].

Exercise 6.2.1: See [273, p. 463].

Exercise 6.3.1: The estimated regression line is $0.75 + 0.32 x$. The coefficient of determination is $r^2 = 0.966038$. On the other hand, it is easy to show, by using Eq. (6.18), that $r = 0.982872$. Finally $0.982872 \times 0.982872 = 0.966038$.

Exercise 6.4.1: (1) Square both sides and take expectations to yield

$$E[Y^2] = \text{Var}[X_2] + \alpha^2 + \beta^2 \text{Var}[X_1] - 2\beta \text{Cov}[X_1, X_2].$$

We minimize the preceding equation by setting $\alpha = 0$ and $\beta = \frac{\text{Cov}[X_1, X_2]}{\text{Var}[X_1]}$. See [642, p. 19]. (2) It is because

$$\text{Cov}[X_1, Y] = E[(X_1 - E[X_1])(\{X_2 - E[X_2]\} - \beta\{X_1 - E[X_1]\})]$$
$$= \text{Cov}[X_1, X_2] - \beta \text{Var}[X_1] = 0.$$

Exercise 6.4.2: Differentiate Eq. (6.20) with respect to $\widehat{X}$ and set it to zero. The average emerges as an extremal value. To verify that it minimizes the function, check that the second derivative equals $2n > 0$. See [417, pp. 429–430].

Exercise 6.4.3: (1) This is because

$$(X - a)^2 = \{(X - E[X]) - (E[X] - a)\}^2$$
$$= (X - E[X])^2 + 2(E[X] - a)(X - E[X]) + (E[X] - a)^2.$$

Now take expectations of both sides to obtain

$$E[(X - a)^2] = \text{Var}[X] + (E[X] - a)^2,$$

which is clearly minimized at the said a value. See [195, p. 319]. (2) For any a,

$$E[\,(X_k - a)^2 \mid X_1, \ldots, X_{k-1}\,]$$

$$= E[\,X_k^2 \mid X_1, \ldots, X_{k-1}\,] - 2a\, E[\,X_k \mid X_1, \ldots, X_{k-1}\,] + a^2$$

$$= E[\,(X_k - E[\,X_k \mid X_1, \ldots, X_{k-1}\,])^2 \mid X_1, \ldots, X_{k-1}\,]$$

$$+ (a - E[\,X_k \mid X_1, \ldots, X_{k-1}\,])^2,$$

which we minimize by choosing $a = E[\,X_k \mid X_1, X_2, \ldots, X_{k-1}\,]$. See [413, p. 173].

CHAPTER 7

Exercise 7.4.1: It is bullish and defensive.

Exercise 7.4.2: The maximum profit is derived as follows. The initial cash outflow is $\mathrm{PV}(X) - P$. If the option is not exercised, then the cash grows to be X, the final cash inflow.

Exercise 7.4.3: Insurance works by diversification of risk: Fires do not burn down all insured houses at the same time. In contrast, when the market goes down, all diversified portfolios take a nose dive. See [64, p. 271].

Exercise 7.4.4: The payoff 1 year from now is clearly $100 + \alpha \times \max(S - X, 0)$, where α is the number of calls purchased. This is because the initial cost is $100, and the fund in the money market will grow to be $100, thus guaranteeing the preservation of capital. Because 90% of the money is to be put into the money market and the fund 1 year from now should be just sufficient to exercise the call, we have to make sure that $100 = 90 \times (1 + r)$. So $r = 11\%$.

Exercise 7.4.5: Consider the butterfly spread $C_{X - \Delta X} - 2C_X + C_{X + \Delta X}$ with the strike prices in subscript. Because the area under the spread's terminal payoff equals $(1/2)(2\Delta X)\,\Delta X = (\Delta X)^2$,

$$\frac{C_{X - \Delta X} - 2C_X + C_{X + \Delta X}}{(\Delta X)^2}$$

has area one at the expiration date. As $\Delta X \to 0$, the area is maintained at one, and the payoff function approaches the Dirac delta function. See [157].

CHAPTER 8

Exercise 8.1.1: Short the high-priced bonds and long the low-priced ones.

Exercise 8.1.2: Let $p \in R^n$ denote the prices of the n securities. One necessary condition for arbitrage freedom is that a portfolio of securities has a nonnegative market value if it has a nonnegative payoff in every state; in other words, $p^{\mathrm{T}}\gamma \geq 0$ if $D\gamma \geq 0$. This is called **weakly arbitrage-free** [289, p. 71]. The equivalent condition for this property to hold depends on **Farka's lemma** [67, 289, 581]. Another necessary condition for arbitrage freedom is $p^{\mathrm{T}}\gamma > 0$ if $D\gamma > 0$, which says that a portfolio of securities must have a positive market value if it has a positive payoff in every state. Interestingly, the two conditions combined are equivalent to the existence of a vector $\theta > 0$ satisfying $D^{\mathrm{T}}\theta = p$. This important result is called the **arbitrage theorem**, and the m elements of θ are called the state prices [692, p. 39]. In the preceding, $v > 0$ means every element of the vector v is positive, and $v \geq 0$ means that every element of the vector v is nonnegative.

Exercise 8.2.1: We want to show that $\mathrm{PV}(X_2 - X_1) < C_{X_1} - C_{X_2}$. Suppose it is not true. Then we can generate arbitrage profits by buying C_{X_2} and shorting C_{X_1}. This generates a positive cash flow. We deposit $\mathrm{PV}(X_2 - X_1)$ in a riskless bank account. At expiration, the funds will be sufficient to cover the calls because (1) $S < X_1$: the payoff is $X_2 - X_1 > 0$; (2) $X_1 \leq S < X_2$: the payoff is $X_2 - X_1 - (S - X_1) = X_2 - S > 0$; (3) $X_2 \leq S$: the payoff is $X_2 - X_1 - (X_2 - X_1) = 0$.

Another proof is to use the put–call parity:

$$C_{X_1} = P_{X_1} + S - \mathrm{PV}(X_1),$$

$$C_{X_2} = P_{X_2} + S - \mathrm{PV}(X_2).$$

We subtract to obtain

$$C_{X_1} - C_{X_2} = P_{X_1} - P_{X_2} - \mathrm{PV}(X_1) + \mathrm{PV}(X_2) > \mathrm{PV}(X_2 - X_1).$$

Exercise 8.2.2: It is $P \leq X$ if we consider a cash-secured put. An alternative is to observe that, because noone will put money in the bank, the interest rate can be treated as if it were zero and $\mathrm{PV}(X) = X$. Lemma 8.2.4 is thus applicable, and we have $P \leq X$.

Exercise 8.3.1: (1) The put–call parity shows the threshold interest to be $r = 12.905\%$ because it satisfies $95 \times e^{-r/3} = 3 + 94 - 6 = 91$. Because $r > 10\%$, $C - P > S - \mathrm{PV}(X)$. We can create an arbitrage profit by shorting the call, buying the put, buying the stock, and lending $\mathrm{PV}(X)$. (2) Apply the put–call parity $\mathrm{PV}(X) = P + S - C$. See [346].

Exercise 8.3.2: Consider the payoffs from the following two portfolios.

	Initial Investment	Value at Expiration Date	
		$S_1 > X$	$S_1 \leq X$
Buy a call	$-C$	$S_1 - X$	0
Buy bonds	$-\mathrm{PV}(X)$	X	X
Total	$-C - \mathrm{PV}(X)$	S_1	X
Buy stock	$-S_0$	S_1	S_1

(S_0 and S_1 denote the stock prices now and at the expiration date, respectively.) The table shows that whichever case actually happens, the first portfolio is worth at least as much as the second. It therefore cannot cost less, and $C + \mathrm{PV}(X) \geq S_0$. (A simpler alternative is to use the put–call parity.) American calls on a non-dividend-paying stock cannot be worth less than European ones. See [317, p. 577].

Exercise 8.3.3: The $C \geq S - \mathrm{PV}(X)$ inequality is derived under the no-arbitrage condition. Apparently, not every stock price series is arbitrage free. For instance, $S > \mathrm{PV}(X)$ is not for, otherwise, we can sell short the stock and invest the proceeds in riskless bonds, and at the option's expiration date, we close out the short position with X from the bonds. Margin requirements ignored, this is doable because we already assume that the stock price at expiration is less than the strike price.

Exercise 8.3.4: Consider the following portfolio: one short call, one long put, one share of stock, a loan of $\mathrm{PV}(X)$ maturing at time t, and a loan of $D^* d(t_1)$ maturing at time t_1. The initial cash flow is $C - P - S + \mathrm{PV}(X) + D^* d(t_1)$. The loan amount $D^* d(t_1)$ will be repaid by the dividend. The rest of the argument replicates that for the put–call parity at expiration. See [746, p. 148].

Exercise 8.3.5: It is equivalent to a long European call with strike price X and a short European call with exercise H, a vertical spread in short [111].

Exercise 8.3.6: Let the payoff function be

$$F(S) \equiv \begin{cases} 0, & \text{if } S < 0 \\ \alpha_i S + \beta_i, & \text{if } S_i \leq S < S_{i+1} \quad \text{for } 0 \leq i < n, \\ \alpha_n S + \beta_n, & \text{if } S_n \leq S \end{cases}$$

where $0 = S_0 < S_1 < \cdots < S_n$ are the breakpoints, $\alpha_{i-1} S_i + \beta_{i-1} = \alpha_i S_i + \beta_i$ for continuity, and $\beta_0 = 0$ for origin crossing. Clearly, $F(0) = 0$ and $F(S_i) = \sum_{j=1}^{i} \alpha_{j-1}(S_j - S_{j-1})$ for $i > 0$.

A generalized option can be replicated by a portfolio of α_0 European calls with strike price $S_0 = 0$, $\alpha_1 - \alpha_0$ European calls with strike price S_1, $\alpha_2 - \alpha_1$ European calls with strike price S_2, and so on, all with the same expiration date. When the stock price S finishes between S_i and S_{i+1}, the option has the payoff

$$F(S) = F(S_i) + \alpha_i(S - S_i) = \sum_{j=1}^{i} \alpha_{j-1}(S_j - S_{j-1}) + \alpha_i(S - S_i).$$

Among the options in the package, only those with the strike price not exceeding S_i finish in the money. The payoff is thus

$$\alpha_0(S - S_0) + \sum_{j=1}^{i}(\alpha_j - \alpha_{j-1})(S - S_j) = \alpha_i(S - S_i) + \sum_{j=1}^{i} \alpha_{j-1}(S_j - S_{j-1}),$$

the same as that of the generalized option.

For a payoff function that does not pass through the origin, say with an intercept of β, we add zero-coupon bonds with a total obligation in the amount of β at the expiration date to the portfolio. In general, when the payoff function could be any continuous function, we can use a piecewise linear function with enough breakpoints to approximate the payoff to the desired accuracy. See [236, pp. 371ff].

Exercise 8.4.1: From Exercise 8.3.2, the option value is at least $45 - 40 \times e^{-0.08/6} > 45 - 40 = 5$ dollars per share. Because the stock is considered overpriced and there is no dividend, the intrinsic value of the option will never exceed $5. Furthermore, other things being equal, the call becomes less valuable as its maturity approaches. Hence the investor should sell the option.

Exercise 8.4.2: In short, $X - \mathrm{PV}(X) > D$. Take a date just before an ex-dividend date. If a call holder exercises the option, the holdings just after that date will be worth $S - X + D'$ with D' denoting the dividend for holding the stock through the ex-dividend date. Note that $D' \le D$. If the holder chooses not to exercise the call, on the other hand, the holdings will then be worth by definition C after the dividend date. From Eq. (8.2) we conclude that

$$C \ge S - \mathrm{PV}(X) - (D - D') > S - X + D'.$$

Hence it is better to sell the call than to exercise it just before an ex-dividend date. Combine this conclusion with Theorem 8.4.2 to prove the result. See [236, p. 140].

Exercise 8.4.3: Because it is worse than exercising the option just after the ex-dividend date. More formally, let S be the stock price immediately before an ex-dividend date, let S' be the stock price immediately after an ex-dividend date, and let D be the amount of the dividend. S' should be $S - D$ within so short a time interval. So exercise immediately after an ex-dividend date fetches $X - S' = X - S + D$, and exercise immediately before an ex-dividend date fetches $X - S$. The interest gained from $X - S$ in such a short period of time can be ignored. The late-exercise strategy clearly dominates [236, p. 251].

Exercise 8.4.4: This inequality says that exercising this option now and investing the proceeds in riskless bonds fetches a terminal value exceeding X. However, Lemma 8.2.4 says that the American put can never be worth more than X. Exercising it is hence better.

Exercise 8.4.5: (1) The covered call strategy guarantees arbitrage profits otherwise. (2) Combine (1) with the put–call parity. (3) Lemma 8.2.4 says that $P \le X$. The case of $P < X$ can never happen because it would imply that the put is selling at less than its intrinsic value.

Exercise 8.4.6: Assume that $C - P < S - X$. Write the put, buy the call, sell the stock short (hence the need for the no-dividend assumption), and place X in a bank account. This generates a positive net cash flow. If the short put is exercised before expiration, withdraw the money from the bank account to pay for the stock, which is then used to close out the short position.

Exercise 8.6.1: Although the floor is the same, the portfolio of options offers a higher payoff when the terminal stock prices are such that some, but not all, put options finish in the money.

CHAPTER 9

Exercise 9.2.1: Suppose that $R > u$. Selling the stock short and investing the proceeds in riskless bonds for one period creates a pure arbitrage profit. A similar argument can be made for the $d > R$ case.

Exercise 9.2.2: Let P denote B's current price. Consider a portfolio consisting of one unit of A and h units of B, worth $100 + Ph$ now. It can fetch either $160 + (50 \times h)$ or $80 + (60 \times h)$ in a period. Pick $h = 8$ to make them both equal to 560. Note that h is simply the delta, because A's price in a period has a range of $160 - 80 = 80$ vs. B's $60 - 50 = 10$, a ratio of eight. Now that this particular portfolio's FV is no longer random, its PV must be $560/e^{0.1} = 506.71$. Hence $P = (506.71 - 100)/8 \approx 50.84$. The above methodology is apparently general.

Exercise 9.2.4: Suppose that $k > 0$ (the other case is symmetric). Sell short M/k options and use the proceeds to buy $(M/k) h$ shares of stock and $(M/k) B$ dollars in riskless bonds. This transaction nets a current value of

$$\frac{M}{k}(hS + B + k) - \frac{Mh}{k} S - \frac{MB}{k} = M.$$

The obligations after one period are nil because, modulo a multiplicative factor of M/k, the preceding levered position above is h shares of stock and M dollars in riskless bonds, which replicate the option. See [289, p. 7].

Exercise 9.2.5: The expected value of the call in a risk-neutral economy one period from now is $pC_u + (1 - p) C_d$. After being discounted by the riskless interest rate, the call value now is $\frac{pC_u + (1 - p) C_d}{R}$, which is equal to $C = hS + B$ by Eq. (9.3).

Exercise 9.2.6: If $Su \le X$, then $S - X < 0$ but $C = 0 > S - X$. If $X \le Sd$, then

$$C = \frac{p(Su - X) + (1 - p)(Sd - X)}{R} = S - \frac{X}{R} > S - X.$$

Finally, if $Sd < X < Su$, then $C = p(Su - X)/R$, which exceeds $S - X$ because

$$C = \frac{(R - d)(Su - X)}{R(u - d)} = \frac{Ru - ud}{Ru - Rd} S - \frac{R - d}{Ru - Rd} X > S - X.$$

See [236, p. 173]. A more compact derivation is to observe that, from Eq. (9.4),

$$hS + B = \frac{pC_u + (1 - p) C_d}{R} \ge \frac{p(Su - X) + (1 - p)(Sd - X)}{R} = S - \frac{X}{R} > S - X.$$

Exercise 9.2.7: This fact can be easily seen from Eq. (9.1) and the inductive steps in deriving the deltas.

Exercise 9.2.8: The stock price should grow at the riskless rate to SR^n at expiration in a risk-neutral economy. As $u > R > d$, we must have $R \to d$ as well. Now the formula says that $C = 0$ if $PV(X) = XR^{-n} > S$. This makes sense because $X > Sd^n$ and the call will finish out of the money. On the other hand, the formula says that $C = S - PV(X)$ if $PV(X) \le S$. This also makes sense as the call will have a terminal payoff of $SR^n - X$.

Exercise 9.2.10: (1) Consider the butterfly spread with strike prices X_L, X_M, and X_H such that

$$Su^{i-1} d^{n-i+1} < X_L < Su^i d^{n-i},$$
$$X_M = Su^i d^{n-i},$$
$$Su^i d^{n-i} < X_H < Su^{i+1} d^{n-i-1},$$

with $2X_M - X_H - X_L = 0$. This portfolio pays off $Su^i d^{n-i} - X_L$ dollars when the stock price reaches $Su^i d^{n-i}$. Furthermore, its payoff is zero if the stock price finishes at other prices. (2) The claim holds because calls can be replicated by continuous trading. Note that the continuous trading strategy described in Subsection 9.2.1 does *not* require that the strike price be one of the possible $n + 1$ stock prices.

Exercise 9.2.11: For p to be a risk-neutral probability, both securities must earn an expected return equal to R, or

$$p = \frac{R - d_1}{u_1 - d_1} = \frac{R - d_2}{u_2 - d_2}.$$

It is not hard to pick R, u_1, d_1, u_2, and d_2 such that the preceding identity does not hold. See [836, p. 91].

Programming Assignment 9.2.14: The limited precision of digital computers dictates that we compute and store $\ln b(j; n, p)$ instead of $b(j; n, p)$. The needed changes to the algorithm are

$$\cdots$$
$$b := \ln n! - \ln a! - \ln(n - a)! + a \times \ln p + (n - a) \times \ln(1 - p);$$
$$\cdots$$
1. $b := b + \ln p + \ln(n - j + 1) - \ln(1 - p) - \ln j;$
$$\cdots$$
3. $C := C + e^b \times (D - X)/R;$
$$\cdots$$

Exercise 9.3.1: (1) This approach, which is due to Jarrow and Rudd [690], differs from the one in the text in that we do not derive the risk-neutral probability [346, p. 546]. See [346, pp. 103–104, p. 544], [289, p. 146], and [290, p. 197]. (2) Theoretically, the risk-neutral probability formula $p \equiv (R - d)/(u - d)$ with the new choices of u and d should be used as is required by the replication argument. In fact, Lemma 9.3.3 will imply that the risk-neutral probability simplifies to

$$\frac{e^{\sigma^2(\tau/n)/2} - e^{-\sigma\sqrt{\tau/n}}}{e^{\sigma\sqrt{\tau/n}} - e^{-\sigma\sqrt{\tau/n}}} \to \frac{1}{2}.$$

However, if we treat the discrete-time economy as an approximation to its continuous-time limit and are concerned about its behavior only as $n \to \infty$, either p or $1/2$ should work in the limit.

Exercise 9.3.2: Use Eq. (9.15) with $n \to \infty$.

Exercise 9.3.4: Observe that

$$p = \frac{\left[1 + \frac{r\tau}{n} + \left(\frac{r\tau}{n}\right)^2/2! + \cdots\right] - \left[1 - \sigma\sqrt{\frac{\tau}{n}} + \left(\sigma\sqrt{\frac{\tau}{n}}\right)^2/2! - \cdots\right]}{\left[1 + \sigma\sqrt{\frac{\tau}{n}} + \left(\sigma\sqrt{\frac{\tau}{n}}\right)^2/2! + \cdots\right] - \left[1 - \sigma\sqrt{\frac{\tau}{n}} + \left(\sigma\sqrt{\frac{\tau}{n}}\right)^2/2! - \cdots\right]}.$$

Exercise 9.3.5: The probability for $S_\tau \geq X$ equals that for $\ln(S_\tau/S) \geq \ln(X/S)$. By Lemma 9.3.3, this event occurs with probability

$$1 - N\left(\frac{\ln(X/S) - (r - \sigma^2/2)\,\tau}{\sigma\sqrt{\tau}}\right) = N\left(-\frac{\ln(X/S) - (r - \sigma^2/2)\,\tau}{\sigma\sqrt{\tau}}\right)$$

$$= N\left(-\frac{\ln(S/X) + (r - \sigma^2/2)\,\tau}{\sigma\sqrt{\tau}}\right).$$

Multiply it by $e^{-r\tau}$ to get the desired result. See [495].

Exercise 9.3.6: The put–call parity says that $C - P = S - \mathrm{PV}(X)$. The exercise clearly holds because the right-hand side equals $S - Xe^{-rT}$. See also [853].

Exercise 9.3.7: Lemma 9.2.1 says that the call value equals $e^{-r\tau}\, E^\pi[\max(S_\tau - X)]$, where $\ln S_\tau$ is normally distributed with mean $\ln S + (r - \sigma^2/2)\,\tau$ and variance $\sigma^2\tau$ by Lemma 9.3.3. The expectation for European calls is

$$e^{-r\tau}\int_X^\infty (S - X)\,f(S)\,dS$$

$$= e^{-r\tau}\int_X^\infty S\,f(S)\,dS - e^{-r\tau}X\int_X^\infty f(S)\,dS$$

$$= e^{-r\tau}e^{\ln S + (r - \sigma^2/2)\,\tau + \sigma^2\tau/2}\,N\left(\frac{\ln S + (r - \sigma^2/2)\,\tau - \ln X}{\sigma\sqrt{\tau}} + \sigma\sqrt{\tau}\right)$$

$$- e^{-r\tau}XN\left(\frac{\ln S + (r - \sigma^2/2)\,\tau - \ln X}{\sigma\sqrt{\tau}}\right)$$

$$= SN\left(\frac{\ln(S/X) + (r + \sigma^2/2)\,\tau}{\sigma\sqrt{\tau}}\right) - e^{-r\tau}XN\left(\frac{\ln(S/X) + (r - \sigma^2/2)\,\tau}{\sigma\sqrt{\tau}}\right),$$

as desired. The second equality was due to Exercise 6.1.6 and Eq. (6.12).

Exercise 9.3.8: (1) This choice has the correct mean stock price $Se^{r\tau/n}$. The second moment also converges to that under the standard choice in the text. (2) When n is even, $Su^{n/2}d^{n/2} = X$, which places the strike price at the center of the tree at maturity. When n is odd, $Su^{(n+1)/2}d^{(n-1)/2} = Xe^{\sigma\sqrt{\tau/n}}$ and $Sd^{(n+1)/2}u^{(n-1)/2} = Xe^{-\sigma\sqrt{\tau/n}}$; the strike price is therefore between the two middle nodes of the tree at maturity. See [555, 589].

Exercise 9.4.3: (1) There are $m+2$ equations: the discounted expected payoffs of the options (m equations), the discounted expected stock price (one equation), and the summing of the terminal probabilities to one (one equation). So we can divide the time into $m+1$ periods, creating $m+2$ terminal nodes. (2) We solve for the terminal nodes' probabilities first by using the above-mentioned $m+2$ equations. Because each path leading up to the same node is equally likely, we divide each nodal probability by the number of paths leading up to that node for the path probability. Finally, we use backward induction to solve for the rest of the probabilities on the tree as follows. We assume that node A is followed by nodes B and C, which have path probabilities p_B and p_C, respectively. The path probability for node A is $p_B + p_C$, and the transition probability from node A to B is $p_B/(p_B + p_C)$. See [502, 503]. An algorithm for constructing the implied binomial tree under a more general setting is considered in [269].

Exercise 9.6.1: From the text we know the value can be computed by replacing the current stock price S with $(1 - \delta)^m S$. But this effectively implies a payoff function of

$$\max((1 - \delta)^m S - X) = (1 - \delta)^m \times \max(S - (1 - \delta)^{-m}X).$$

See [879, p. 97].

Exercise 9.6.2: No. When the American call is exercised, there is always a corresponding European call to exercise with the same payoff. Because there may be European calls remaining after the American call is exercised, the package is potentially more valuable than the American call.

Exercise 9.6.5: We know that $C = P + S - D - PV(X)$ from Eq. (8.2). Now, $S - D = Se^{-q\tau}$ because the stock price would be $S^{-q\tau}$ today to reach the same terminal price if there were no dividends. See [894, p. 72] for the formula.

Exercise 9.6.6: Just follow the same steps as before in setting up the replicating portfolio except that, now,

$$hSue^{\widehat{q}} + RB = C_u, \qquad hSde^{\widehat{q}} + RB = C_d.$$

The reason is that the stockholder is getting new shares at a rate of q per year. (Throughout this exercise, $\widehat{q} \equiv q\Delta t$ means the dividend yield per period.) Equations (9.1) and (9.2) are hence replaced with

$$h = \frac{C_u - C_d}{(Su - Sd)\, e^{\widehat{q}}}, \qquad B = \frac{uC_d - dC_u}{(u - d)\, R}.$$

After substitution and rearrangement,

$$hS + B = \left(\frac{Re^{-\widehat{q}} - d}{u - d}\, C_u + \frac{u - Re^{-\widehat{q}}}{u - d}\, C_d \right) \bigg/ R$$

in place of Eq. (9.3). Finally, Eq. (9.4) becomes $hS + B = [\, pC_u + (1 - p)\, C_d\,]/R$, where $p \equiv (Re^{-\widehat{q}} - d)/(u - d)$.

Another way to look at it is by observing that in a risk-neutral economy, the per-period return of holding the stock should be R. Now, because the stock is paying a dividend yield of $\widehat{q}$, the return of the stock price net of the dividends should be $Re^{-\widehat{q}}$.

Exercise 9.6.7: (1) The total *wealth* is growing at μ, not $\mu - q$! (2) It retains the binomial tree's backward-induction structure except that each S at j periods from now should be replaced with $Se^{-qj\Delta t}$. Of course, the result is not exactly the same as using Eq. (9.21) and pretending there were no dividends. However, it should converge to the same value. Interestingly, this is the same as retaining $(e^{r\Delta t} - d)/(u - d)$ and the original algorithm as if there were no dividends but with u and d multiplied by $e^{q\Delta t}$.

Exercise 9.6.8: Suppose there is one period to expiration and $Sue^{-\widehat{q}} < X$, where $\widehat{q}$ is the dividend yield per period. Clearly the option has zero value at present as it will not be exercised at expiration. However, if $S > X$, the option has positive intrinsic value, which means it should be exercised now. These two inequalities imply that $X < S < Xu^{-1}e^{\widehat{q}}$, which is possible when $u < e^{\widehat{q}}$.

Exercise 9.7.1: Consult [154, 156], [575, Subsection 4.1.4], or [783].

Exercise 9.7.2: Suppose the option is not exercised at price S but it is optimal to exercise it at the same price two periods earlier. It must hold that

$$e^{r\Delta t}(X - S) \le pP_u + (1 - p)\, P_d,$$
$$e^{r\Delta t}(X - S) > pP_u' + (1 - p)\, P_d'.$$

(Primed symbols are for values two periods earlier.) Hence,

$$pP_u' + (1 - p)\, P_d' < pP_u + (1 - p)\, P_d.$$

However, $P_d \le P_d'$ and $P_u \le P_u'$ by Lemma 8.2.1 because of the longer maturity. That lemma works whether there are dividends or not. We now have the contradiction $pP_u' + (1 - p)P_d' < pP_u' + (1 - p)P_d'$.

CHAPTER 10

Exercise 10.1.1: Note that $N'(y) = e^{-(y^2/2)}/\sqrt{2\pi}$ and $x' \equiv \partial x/\partial S$ in the following equation:

$$\partial C/\partial S = N(x) + SN'(x)\, x' - Xe^{-r\tau} N'(x - \sigma\sqrt{\tau})\, x'$$
$$= N(x) + SN'(x)\, x' - Xe^{-r\tau} N'(x)\, e^{x\sigma\sqrt{\tau} - \sigma^2\tau/2}\, x'$$
$$= N(x) + SN'(x)\, x' - Xe^{-r\tau} N'(x)\, e^{\ln(S/X) + r\tau}\, x' = N(x).$$

Exercise 10.1.2: Note that $N'(y) = e^{-(y^2/2)}/\sqrt{2\pi}$ and $x' \equiv \partial x/\partial X$ in the following equation:

$$\partial P/\partial X = e^{-r\tau} N(-x+\sigma\sqrt{\tau}) - Xe^{-r\tau} N'(-x+\sigma\sqrt{\tau})\,x' + SN'(-x)\,x'$$

$$= e^{-r\tau} N(-x+\sigma\sqrt{\tau}) - Xe^{-r\tau} N'(-x)\, e^{x\sigma\sqrt{\tau}-\sigma^2\tau/2}\,x' + SN'(-x)\,x'$$

$$= e^{-r\tau} N(-x+\sigma\sqrt{\tau}) - Xe^{-r\tau} N'(-x)\, e^{\ln(S/X)+r\tau}\,x' + SN'(-x)\,x'$$

$$= e^{-r\tau} N(-x+\sigma\sqrt{\tau}).$$

Exercise 10.1.3: We have to prove that the strike price X that maximizes the option's time value is the current stock price S. Recall that the time value is defined as $V \equiv C - \max(S - X, 0)$. It is not hard to verify that $\partial C/\partial X = -e^{-r\tau} N(x - \sigma\sqrt{\tau})$. So $\partial V/\partial X = \partial C/\partial X < 0$ if $X > S$, and $\partial V/\partial X = (\partial C/\partial X) + 1 > 0$ if $X \le S$. The time value is hence maximized at $X = S$. The case of puts is similar.

Exercise 10.1.4: It is $\frac{r\tau+\sigma^2\tau/2-\ln(S/X)}{2\sigma\tau\sqrt{\tau}}$ [894, p. 78].

Exercise 10.1.5: (1) It is $S = Xe^{(r+\sigma^2/2)\tau}$ (not $S = X$). To derive it, note that

$$\frac{\partial\Theta}{\partial S} = -\frac{N'(x)\sigma + SN''(x)\,x'\sigma}{2\sqrt{\tau}} - rXe^{-r\tau} N'(x - \sigma\sqrt{\tau})\,x'$$

$$= -\frac{N'(x)\sigma + SN''(x)\,x'\sigma}{2\sqrt{\tau}} - rSN'(x)\,x'.$$

The last equality above takes advantage of $Xe^{-r\tau} N'(x - \sigma\sqrt{\tau})\,x' = SN'(x)\,x'$, which can be verified with Eq. (10.1) and the Black–Scholes formula for the European call. With $N''(x) = -xN'(x)$ and $x' = 1/(S\sigma\sqrt{\tau})$, it is not hard to see that $\partial\Theta/\partial S = 0$ if and only if $-\sigma^2 + (x\sigma/\sqrt{\tau}) - 2r = 0$. From here, our claim follows easily. (2) This is by virtue of Lemma 8.2.1.

Exercise 10.1.6: Vega's derivative with respect to σ is $-xx'\Lambda$. Note that x can be expressed as $(A/\sigma) + B\sigma$. Hence $x'x = -2(A^2/\sigma^3) + 2B^2\sigma$. Clearly, $x'x = 0$ has a positive solution at $\sigma = \sigma^* \equiv \sqrt{|A/B|}$. It is not hard to see that Λ' begins at ∞ for $\sigma = 0$, penetrates the x axis at σ^* into the negative domain, and converges to zero at positive infinity. This confirms the unimodality of vega.

Exercise 10.1.7: (1) $-\frac{x+\sigma\sqrt{\tau}}{S\sigma\sqrt{\tau}}\,\Gamma$ [894, p. 78]. (2) $-\frac{\sigma\sqrt{\tau}+(\ln(X/S)+(r+\sigma^2/2)\tau)x}{2\sigma^2\tau^2 S}\,N(x)$. See [894, p. 78].

Exercise 10.2.7: The standard finite-difference scheme approximates the second derivative by using the function values at the three equally spaced stock prices $S - \Delta S$, S, and $S + \Delta S$. It purports to improve the accuracy by varying ΔS. Scheme (10.2) instead approximates the second derivative at the *fixed* prices Suu, S, and Sdd, and it improves the accuracy by varying n.

CHAPTER 11

Exercise 11.1.1: The guarantee generates a cash flow to the bondholders of $-\min(0, V^* - B)$, which is equal to $\max(0, B - V^*)$ [660, p. 630].

Exercise 11.1.3: Leibniz's rule might be useful [448]: If $F(x) \equiv \int_{a(x)}^{b(x)} f(x, z)\,dz$, then

$$F'(x) = \int_{a(x)}^{b(x)} \frac{\partial f(x, z)}{\partial x}\,dz + f(x, b(x))\,b'(x) - f(x, a(x))\,a'(x).$$

(1) The higher the firm value, the higher the bond price is. Intuitively, the higher the firm value, the less likely the firm is to default. (2) The more the firm borrows, the higher the bond price. However, note that the increase in the total bond value is less than the net increase in the face value. (3) The longer the time to maturity, the lower the bond price is. This is because the PV decreases and the default premium increases. See [746, p. 396].

Exercise 11.1.4: Any other bond price will lead to arbitrage profits by trading Merck stock, Merck calls, and XYZ.com's bonds.

Exercise 11.1.5: The stockholders gain $[\,35000 \times (5/35)\,] + 9500 - 15250 = -750$ dollars. The original bondholders lose the equal amount, $29250 - [\,(30/35) \times 35000\,] = -750$. So the bondholders gain.

Exercise 11.1.6: The results are tabulated below.

Promised Payment to Bondholders X	Current Market Value of Bonds B	Current Market Value of Stock nS	Current Total Value of Firm V	Stockholders' Gains from Issuing $\frac{X-30000}{1000}$ Bonds
55,000	54,375	16,750	71,125	0
60,000	59,375	11,750	71,125	−52.083
65,000	64,125	7,000	71,125	115.385
70,000	68,000	3,125.5	71,125	946.429

Exercise 11.1.7: Dividends reduce the total value by the equal amount. However, the stock, as an option on the total value, decreases by less than the dividend amount because of the convexity of option value by Lemma 8.5.1. See [424, p. 531].

Programming Assignment 11.1.8: (2) Guess a $V > nS$ and build the binomial tree for V/n. Price an American call with a strike price of X on that tree. Each warrant W is then $n/(n+m)$ of the call value. Stop if $V \approx nS + mW$. Otherwise, choose a new $V > nS$ and repeat the procedure.

Exercise 11.1.9: It is

$$V - C(X) + \lambda C(X/\lambda), \tag{33.7}$$

where $C(Y)$ represents a European call with a strike price of Y.

Exercise 11.1.10: (1) From expression (33.7) and that $V - C(X)$ is equivalent to a zero-coupon bond with a face value of X, a CB is basically a zero-coupon bond plus a call on λ times the total value of the corporation with a strike price of X/λ [746, p. 401]. (2) Warrants are calls (see Subsection 11.1.2). (3) By analyzation of the payoff at maturity, the terminal value is easily seen to be at least λV^* and sometimes more. On the other hand, conversion grants the owner a fraction $1/\lambda$ of the firm. The former strategy therefore dominates the latter. See [491, Theorem 1], [492, p. 430], and [697].

Exercise 11.1.11: In all three cases, the PVs are either less than or equal to P:

$$V < \mathrm{PV}(X) \le P,$$
$$\mathrm{PV}(X) \le X \le P,$$
$$\lambda V < P.$$

Exercise 11.2.1: (1) See [470, p. 463]. (2) See [470, p. 464].

Programming Assignment 11.2.5: Assume that the barrier option is a call in the following discussions. Our approach computes down-and-in and down-and-out options simultaneously, noting that they sum to the standard European call value at each node by the in–out parity.

Working from the terminal nodes toward the root, backward induction calculates the option value at each node *if the derivative is issued at that node*. Each node keeps two values: The in-value records the value of the down-and-in option and the out-value records the value of the down-and-out option. Each terminal node above the strike price starts with the payoff of the standard call option in the out-value and zero in the in-value, whereas each terminal node at or below the strike price starts with zeros in both values. Inductively, a node not on the barrier takes the expected PV of the in-values of its two successor nodes and that of the out-values of its two successor nodes and puts them into its respective value cells. For a node on the barrier, we do the same thing and then set the out-value to zero and the in-value to the sum of the in-value and out-value. The overall space requirement is linear, and the time complexity is quadratic. When there is a rebate for the down-and-out call, the rebate's cash flow has to be calculated *separately*. See Fig. 33.2 for the algorithm.

Exercise 11.2.6: It can be replicated as

$$D(X_0, H_1) + \sum_{i=1}^{n-1} [D(X_i, H_{i+1}) - D(X_i, H_i)].$$

Here $D(X, H)$ denotes a down-and-out option with strike price X and barrier H. This can be justified as follows. If the nearest barrier H_1 is never hit, then the first down-and-out option provides the necessary payoff and each term in the summation is zero because it contains two options with

Binomial tree algorithm for pricing down-and-out and down-and-in calls on a non-dividend-paying stock:

input: $S, u, d, X, H (H < X, H < S), n, \hat{r}, K$;
real $R, p, C_o[n+1], C_i[n+1], C_r[n+1]$;
integer i, j, h;
$R := e^{\hat{r}}; p := (R-d)/(u-d)$;
$h := \lfloor \ln(H/S)/\ln u \rfloor; H := Su^h$;
for ($i = 0$ to n) {
        $C_o[i] := \max(0, Su^{n-i}d^i - X)$;
        $C_i[i] := 0$;
        $C_i[r] := 0$;
}
if $[n - h$ is even and $0 \le (n-h)/2 \le n]$ {// A hit.
        $C_i[(n-h)/2] := C_o[(n-h)/2]$;
        $C_o[(n-h)/2] := 0$;
        $C_r[(n-h)/2] := K$;
}
for ($j = n-1$ down to 0) {
        for ($i = 0$ to j) {
                $C_o[i] := (p \times C_o[i] + (1-p) \times C_o[i+1])/R$;
                $C_i[i] := (p \times C_i[i] + (1-p) \times C_i[i+1])/R$;
                $C_r[i] := (p \times C_r[i] + (1-p) \times C_r[i+1])/R$;
        }
        if $[j - h$ is even and $0 \le (j-h)/2 \le j]$ {// A hit.
                $C_i[(j-h)/2] := C_i[(j-h)/2] + C_o[(j-h)/2]$;
                $C_o[(j-h)/2] := 0$;
                $C_r[(j-h)/2] := K$;
        }
}
return $C_o[0] + C_r[0], C_i[0]$;

Figure 33.2: Binomial tree algorithm for barrier calls on a non-dividend-paying stock. Because H may not correspond to a legal stock price, we lower it to Su^h, the highest stock price not exceeding H. The new barrier corresponds to $C_x[(j-h)/2]$ at times $j = n, n-1, \ldots, h$, where $x \in \{$"o", "i", "r"$\}$. The knock-out option provides a rebate K when the barrier is hit.

identical strike price. If H_1 is hit, then $D(X_0, H_1)$ and $D(X_1, H_1)$ are rendered worthless. The remaining portfolio is

$$D(X_1, H_2) + \sum_{i=2}^{n-1} [D(X_i, H_{i+1}) - D(X_i, H_i)].$$

The rest follows inductively. See [158, p. 16].

Exercise 11.4.1: See [95, p. 288].

Exercise 11.5.1: Buy a call with strike X_H and sell a put with strike X_L [346, p. 297].

Exercise 11.5.2: Buy a call with strike $X + p$, sell a put with strike $X + p$, and buy a put with strike X [346, p. 299].

Exercise 11.5.3: Buy a call with strike X and sell α puts with strike X [346, p. 302].

Exercise 11.5.4: (1) Use Eq. (9.21) [346, p. 165]. (2) Suppose we hold $\$h$ foreign riskless bond (denominated in foreign currency) and $\$B$ domestic riskless bond. The portfolio has the value $hS + B$ in domestic currency. Here S denotes the current domestic/foreign exchange rate. The same argument as that in Subsection 9.2.1 leads to the following equations:

$$he^{\widehat{r}\Delta t} Su + RB = C_u, \qquad he^{\widehat{r}\Delta t} Sd + RB = C_d.$$

The $e^{\widehat{r}\Delta t}$ term arises from foreign interest income. Solving them gives the result. See [514, p. 321].

Exercise 11.5.5: Use $N(x) = 1 - N(-x)$ and $r = \widehat{r}$.

Exercise 11.5.6: It holds when $S \gg X$ [746, p. 345].

Exercise 11.5.7: (1) See [746, p. 332] and Eq. (8.1). (2) See [746, p. 333] and Lemma 8.3.2.

Exercise 11.6.1: The stockholders, by paying the first interest payment, acquire the option to own the firm by making future interest and principal payments. See [424, p. 531] and [746, pp. 398–400].

Exercise 11.6.2: Because it can be exercised only when the contract is awarded [346, p. 305].

Exercise 11.6.3: See [879, p. 201].

Exercise 11.7.1: Construct a portfolio consisting of a lookback call on the minimum and a lookback put on the maximum [388].

Exercise 11.7.2: Let $r_k \equiv S_k / S_{k-1}$. Then $r_k \in \{u, d\}$, depending on whether the kth move is up or down. As $S_i = S_0 \prod_{j=1}^{i} r_j$, we have

$$\prod_{i=0}^{n} S_i = S_0^{n+1} \prod_{i=1}^{n} \prod_{j=1}^{i} r_j = S_0^{n+1} \prod_{i=1}^{n} r_i^{n-i+1}.$$

Our problem thus amounts to counting the distinct numbers (call it N) the expression $\prod_{i=1}^{n} r_i^{n-i+1}$ can take for $r_i \in \{u, d\}$; equivalently, we can ask the same question of $\sum_{i=1}^{n}(n-i+1)\ln r_i$. This implies that N equals the number of distinct sums of integers drawn from $\{1, 2, \ldots, n\}$. Because it is easy to see that every integer from 1 to $n(n+1)/2$ can be represented uniquely as the sum of distinct integers drawn from $\{1, 2, \ldots, n\}$, we conclude that $N = n(n+1)/2$. The preceding analysis did not assume that $ud = 1$.

Exercise 11.7.3: See [147, p. 382].

Exercise 11.7.4: Let the historical average from m prices be A as of time zero. The terminal payoff for a call is then

$$\max\left(\frac{mA + \sum_{i=0}^{n} S_i}{m+n+1} - X, 0\right)$$

$$= \max\left(\frac{\sum_{i=0}^{n} S_i}{m+n+1} - \left(X - \frac{mA}{m+n-1}\right), 0\right)$$

$$= \frac{n+1}{m+n+1} \times \max\left(\frac{\sum_{i=0}^{n} S_i}{n+1} - \frac{m+n+1}{n+1}\left(X - \frac{mA}{m+n-1}\right), 0\right).$$

So it becomes $\frac{n+1}{m+n+1}$ option with strike price $\frac{m+n+1}{n+1}\left(X - \frac{mA}{m+n-1}\right)$.

Exercise 11.7.5: (1) In a risk-neutral economy, $E[\,S_i\,] = S_0 e^{\widehat{r}i}$. So the expected average price is

$$\frac{S_0}{n+1} \sum_{i=0}^{n} e^{\widehat{r}i} = \begin{cases} \frac{S_0}{n+1} \frac{1 - e^{\widehat{r}(n+1)}}{1 - e^{\widehat{r}}} & \text{if } \widehat{r} \neq 0 \\ S_0 & \text{if } \widehat{r} = 0 \end{cases}.$$

(2) Assume that $\widehat{r} \neq 0$. The value of the running sum implies that the call will be in the money because any path extending that initial path will end up with an arithmetic average of at least $X + a/(n+1) \geq X$. The expected terminal value is thus $a/(n+1)$ plus $E[\frac{1}{n+1} \sum_{i=k}^{n} S_i]$, which is equal to

$$\frac{S_k}{n+1} \frac{1 - e^{\widehat{r}(n-k+1)}}{1 - e^{\widehat{r}}}$$

by (1).

Exercise 11.7.6: We prove the claim for the $i < n/2$ case. Assume $S_0 = 1$ for convenience. The said difference at $u^i d^{n-i} = u^{2i-n}$ equals

$$A \equiv (u + u^2 + \cdots + u^i + u^{i-1} + \cdots + u^{2i-n}) - (d + d^2 + \cdots + d^{n-i} + d^{n-i-1} + \cdots + d^{n-2i}).$$

Similarly, the said difference at $u^{i-1} d^{n-i+1} = u^{2i-n-2}$ equals

$$B \equiv (u + u^2 + \cdots + u^{i-1} + u^{i-2} + \cdots + u^{2i-n-2}) - (d + d^2 + \cdots + d^{n-i+1} + d^{n-i} + \cdots + d^{n-2i+2}).$$

Now,

$$A - B = u^i + u^{i-1} - u^{2i-n-1} - u^{2i-n-2} + d^{n-i+1} + d^{n-i} - d^{n-2i+1} - d^{n-2i}$$

$$= u^i + u^{i-1} - u^{2i-n-1} - u^{2i-n-2} - (u^{2i-n} + u^{2i-n-1} - u^{i-n} - u^{i-n-1})$$

$$= u^i + u^{i-1} - u^{2i-n-1} - u^{2i-n-2} - u^{i-n}(u^i + u^{i-1} - 1 - u^{-1})$$

$$> u^i + u^{i-1} - u^{2i-n-1} - u^{2i-n-2} - (u^i + u^{i-1} - 1 - u^{-1}) > 0,$$

as claimed.

Exercise 11.7.7: $y = y_0 + \frac{y_1 - y_0}{x_1 - x_0}(x - x_0) + [\frac{y_2 - y_0}{(x_2 - x_0)(x_2 - x_1)} - \frac{y_1 - y_0}{(x_1 - x_0)(x_2 - x_1)}](x - x_0)(x - x_1).$

Exercise 11.7.8: "Bucketing" is no longer necessary because all the running sums at node $N(j, i)$ are integers lying between the integers $(j + 1) A_{\min}(j, i)$ and $(j + 1) A_{\max}(j, i)$, hence finite in number. The second error, from interpolation, disappears as a consequence because the running sum – now an integer – that we seek in the next node exists. Specifically, let $V(j, i, k)$ denote the option value at node $N(j, i)$ given that the sum up to that node is k. Use $S(j, i)$ to represent the integral asset price at node $N(j, i)$. By backward induction,

$$V(j, i, k) = [pV(j + 1, i, k + S(j + 1, i)) + (1 - p) V(j + 1, i + 1, k + S(j + 1, i + 1))] e^{-r\Delta t},$$

where $0 \le j < n, 0 \le i \le j$, and $(j + 1) A_{\min}(j, i) \le k \le (j + 1) A_{\max}(j, i)$. See [251, 252].

Exercise 11.7.9: Run both algorithms and output their average. This average cannot deviate from the true value by more than the difference of their respective bounds.

Programming Assignment 11.7.11: We follow the terms in Subsection 11.7.1. Consider a lookback call on the minimum and let $S_{\min}$ denote the historical low as it stands now. (1) Figure 11.12 reveals that at node $N(j, i)$, the maximum minimum price between now and $N(j, i)$ is $S_{\max}(j, i) \equiv \min(S_0, S_0 u^{j-i} d^i)$. Similarly, the minimum minimum price is $S_{\min}(j, i) \equiv S_0 u^{-i}$, which is *independent* of j. The states at $N(j, i)$ obviously are $S_{\min}(j, i), u S_{\min}(j, i), u^2 S_{\min}(j, i), \ldots, S_{\max}(j, i)$. The number of states is either (a) $j - i$ when $j \ge 2i$ and $S_{\max}(j, i) = S_0 u^{j-i} d^i$ or (b) i when $j < 2i$ and $S_{\max}(j, i) = S_0$. It is then not hard to show that the total number of states over the whole tree is proportional to n^3, which is also the time bound. Use $N(j, i, k)$ to denote the state at node $N(j, i)$ when $u^k S_{\min}(j, i) = S_0 u^{k-i}$ is the minimum price between now and $N(j, i)$. Note that $k \le i$. Finally, let $C(j, i, k)$ denote the option value at state $N(j, i, k)$.

Backward induction starts with

$$C(n, i, k) = \max(S_0 u^{n-i} d^i - \min(u^k S_{\min}(n, i), S_{\min})),$$

at time n, where $0 \le i \le n$ and $0 \le k \le \ln_u \frac{S_{\max}(n, i)}{S_{\min}(n, i)}$. Inductively, each state at node $N(j, i)$ takes inputs from a state at the up node $N(j + 1, i)$ and a state at the down node $N(j + 1, i + 1)$. Specifically, for $j = n - 1, n - 2, \ldots, 0$, the algorithm carries out

$$C(j, i, k) = \begin{cases} [pC(j + 1, i, k) + (1 - p) C(j + 1, i + 1, k + 1)]/R, & \text{if } j > 2i \text{ (i),} \\ [pC(j + 1, i, k) + (1 - p) C(j + 1, i + 1, k + 1)]/R, & \text{if } j = 2i, \ k < i \text{ (ii),} \\ [pC(j + 1, i, k) + (1 - p) C(j + 1, i + 1, k)]/R, & \text{if } j = 2i, \ k = i \text{ (iii),} \\ [pC(j + 1, i, k) + (1 - p) C(j + 1, i + 1, k + 1)]/R, & \text{if } j < 2i, \ k < j - i \text{ (iv),} \\ [pC(j + 1, i, k) + (1 - p) C(j + 1, i + 1, k)]/R, & \text{if } j < 2i, \ k = j - i \text{ (v),} \end{cases}$$

where $0 \le i \le j$ and $0 \le k \le \ln_u \frac{S_{\max}(j, i)}{S_{\min}(j, i)}$. The observations to follow were utilized in deriving the preceding formula. First, the minimum price in the up-node state always equals $N(j, i, k)$'s minimum price. In case (i), $N(j, i, k)$'s stock price $S_0 u^{j-i} d^i = S_0 u^{j-2i}$ exceeds S_0; thus the minimum price in the down-node state equals $N(j, i, k)$'s minimum price. In both (ii) and (iii), $N(j, i, k)$'s stock price equals S_0. In case (ii), $N(j, i, k)$'s minimum price is less than S_0; thus the minimum price in the down-node state equals $N(j, i, k)$'s minimum price. In case (iii), $N(j, i, k)$'s minimum price equals S_0; thus the minimum price in the down-node state equals $S_0 d$. In both (iv) and (v), $N(j, i, k)$'s stock price $S_0 u^{j-i} d^i = S_0 u^{j-2i}$ is less than S_0. In case (iv), $N(j, i, k)$'s minimum price is less than $S_0 u^{j-2i}$; thus the minimum price in the down-node state equals $N(j, i, k)$'s minimum price. In case (v), $N(j, i, k)$'s minimum price equals $S_0 u^{j-2i}$; thus the minimum price in the down-node state equals $S_0 u^{j-2i-1}$. The returned value is $C(0, 0, 0)$. If the option is American, simply take the greater of $S_0 u^{j-i} d^i - \min(u^k S_{\min}(j, i), S_{\min})$ and the $C(j, i, k)$ above for the final $C(j, i, k)$.

(2) When the option is newly issued, the number of states at each node can be drastically reduced to one. The basic idea is to calculate at each node a newly issued lookback option when the current stock

price is one. See [470, pp. 475–477] (the algorithm in Fig. 29.5 uses a similar idea). Unfortunately, this idea may not work for lookback options with a price history.

CHAPTER 12

Exercise 12.2.1: $4,288,200 + 400,000 \times r$ in U.S. dollars, where r is the spot exchange rate in \$/DEM 3 months from now.

Exercise 12.2.2: (1) Germany has lower interest rates than the U.S.

Exercise 12.2.3: Apply the interest rate parity [514, pp. 328–329].

Exercise 12.2.4: (1) A newly written forward contract is equivalent to a portfolio of one long European call and one short European put on the same underlying asset with a common expiration date equal to the delivery date. This is true because $S_T - X = \max(S_T - X, 0) - \max(X - S_T, 0)$. Hence, if X equals the forward price, the payoff matches exactly that of a forward contract. (2) Because a new forward contract has zero value, the strike price must be such that the put premium is equal to the call premium. Alternatively, we may write put–call parity (8.1) as $C = P + \mathrm{PV}(F - X)$ with the help of Eq. (12.3). The conclusion is immediate as $F = X$. See [159]. (3) With (2), put–call parity (8.1) implies that

$$0 = C - P - S + Fe^{-r\tau} = -S + Fe^{-r\tau},$$

which is exactly Lemma 12.2.1. See [236, pp. 59–61] and [346, p. 244].

Exercise 12.2.5: Consider the case $f > (F - X)e^{-r\tau}$. We can create an arbitrage opportunity by buying one forward contract with delivery price F and shorting one forward contract with delivery price X, both maturing τ from now. This generates an initial cash inflow of f because the first contract has zero value by the definition of F. The cash flow at maturity is

$$(S_T - F) + (X - S_T) = -(F - X).$$

Hence a cash flow with a PV of $f - (F - X)e^{-r\tau} > 0$ has been ensured. To prove the identity under the remaining case $f < (F - X)e^{-r\tau}$, just reverse the above transactions.

Exercise 12.3.1: Repeat the argument leading to Eq. (12.8) but with lending when the cash flow is positive and borrowing when the cash flow is negative. The result is

$$(F_1 - F_0)R^{n-1} + (F_2 - F_1)R^{n-2} + \cdots + (F_n - F_{n-1}).$$

Clearly the result depends on how $F_n - F_0$ is distributed over the n-day period.

Exercise 12.3.3: Higher rates beget higher futures prices, generating positive cash flows that can be reinvested at higher rates. Similarly, lower rates beget lower futures prices, generating negative cash flows that can be financed at lower rates. Because both are advantageous to holders of futures contracts, their prices must be higher. See [402] or [514, p. 58] for more rigorous arguments.

Exercise 12.3.4: It assumes that the stock index is *not* adjusted for dividend payouts. First, note a crucial element in the proofs of Eqs. (12.4) and (12.6), that is, the dividends of the stock index – in fact, any underlying asset – are predictable. Therefore dividends must have predictable value. Now we come to the problem of adjustments for dividends. If the index were adjusted for dividends, the adjustment would have the effect of making the initial position in the underlying asset, a portfolio of stocks in the case of stock indices, not deliverable as is. This breaks the proofs, in which the only transactions that take place are related to the loan used to take a long position in the underlying asset or the cash outflow incurred when a short position is taken in the underlying asset.

Exercise 12.3.5: (1) The cost of carry is $Se^{r\tau}$ because it costs that much to carry the cash instrument. (2) $F = Se^{r\tau}$ by the condition of full carry. See [88, p. 170].

Exercise 12.3.6: If T-bills are yielding 6% per year and one owns gold outright for 1 year, then the "cost" of ownership is 6% of the cost, representing the interest one would have earned if one had bought a T-bill instead of the gold. At \$350 an ounce, the opportunity cost of owning 100 ounces for 1 year is $100 \times \$350 \times 0.06 = \$2,100$. Thus \$2,100 is the cost of carry. See [698, p. 192].

Exercise 12.3.7: Consider a portfolio consisting of borrowing S to buy a unit of the underlying commodity and a short position in a forward contract with the delivery price F (hence zero value). The initial net cash flow is zero. On the delivery date, the cash flow is $F - S - C$. Hence $F - S - C \leq 0$ must hold to prevent arbitrage. See [746, p. 40].

Exercise 12.3.8: Consider the strategy of shorting the commodity, investing the proceeds at the riskless rate, and buying the forward contract with the delivery price F. The initial net cash flow is zero. On the delivery date, the cash flow is $S + I - F - D$. This is because the investor has to pay F to get the commodity and close out the short position. The investor also has to pay any cash flow due the commodity holder. On the other hand, the investor receives $S + I$ for the investment. By Eq. (12.12),

$$S + I - F - D = S - F + C - U.$$

Hence $S - F + C - U \leq 0$ must hold to prevent arbitrage. See [746, p. 41].

Exercise 12.4.1: We have to show that there is no net cash outflow at expiration. Suppose the futures price decreases at expiration. We then exercise the put for a short position in futures, offsetting the long position in futures, and abandon the call. Similar actions can be taken for two other cases. See [95, p. 188] and [328].

Exercise 12.4.2: See [514, p. 204].

Exercise 12.4.3: Consider a portfolio of one long call, one short put, one short futures contract, and a loan of $Fe^{-rt} - X$. The initial portfolio value is $C - P - Fe^{-rt} + X$. At time t, the portfolio value is

$$0 - (X - F_t) - (F_t - F) - (F - Xe^{rt}) = X(e^{rt} - 1) \geq 0 \quad \text{if } F_t \leq X,$$
$$(F_t - X) - 0 - (F_t - F) - (F - Xe^{rt}) = X(e^{rt} - 1) \geq 0 \quad \text{if } F_t > X.$$

Suppose the put is exercised at time $s < t$. The value then is

$$C - (X - F_s) - (F_s - F) - [Fe^{-r(t-s)} - Xe^{rs}] = C + F[1 - e^{-r(t-s)}] + X(e^{rs} - 1) \geq 0.$$

We hence conclude that $C - P - Fe^{-rt} + X \geq 0$.

For the other bound, consider a portfolio of one long put, one futures contract, lending of $F - Xe^{-rt}$, and one short call. The initial portfolio value is $P + F - Xe^{-rt} - C$. At time t, the portfolio value is

$$(X - F_t) + (F_t - F) + (Fe^{rt} - X) - 0 = F(e^{rt} - 1) \geq 0 \quad \text{if } F_t \leq X,$$
$$0 + (F_t - F) + (Fe^{rt} - X) - (F_t - X) = F(e^{rt} - 1) \geq 0 \quad \text{if } F_t > X.$$

Suppose the call option is exercised at time $s < t$. The value then is

$$P + (F_s - F) + [Fe^{rs} - Xe^{-r(t-s)}] - (F_s - X) = P + F(e^{rs} - 1) + X[1 - e^{-r(t-s)}] \geq 0.$$

We hence conclude that $P + F - Xe^{-rt} - C \geq 0$. See [746, pp. 281–282].

Exercise 12.4.4: (1) We shall do it for calls only. Substitute $Se^{(r-q)t}$ for F into Eqs. (12.16) of Black's model to obtain Eq. (9.20). If the underlying asset does not pay dividends, then substitute Se^{rt} for F to obtain the original Black–Scholes formula for European calls. See [346, p. 201] and [470, p. 295]. (2) This holds because the futures price equals the cash asset's price at maturity.

Exercise 12.4.6: (1) From the binomial tree for the underlying stock, we replace the stock price S at each node of the tree with $Se^{r\tau}$, where τ is the time to the futures contract's maturity at that node. The binomial model for the futures price is then $Se^{r\tau} \to Sue^{r(\tau-\Delta t)}$ or $Sde^{r(\tau-\Delta t)}$; in other words, $F \to Fue^{-r\Delta t}$ or $Fde^{-r\Delta t}$. (2) If the underlying stock pays a continuous dividend yield of q, the binomial model for the futures price under the *same* risk-neutral probability is $Se^{(r-q)\tau} \to Sue^{-q\Delta t}e^{(r-q)(\tau-\Delta t)}$ or $Sde^{-q\Delta t}e^{(r-q)(\tau-\Delta t)}$; in other words, $F \to Fue^{-r\Delta t}$ or $Fde^{-r\Delta t}$, identical to (1).

Exercise 12.5.3: The forward dollar/yen exchange rate applicable to time $[0, i]$ is $F_i \equiv Se^{(r-q)i}$ by Eq. (12.1). Hence the PV of the forward exchange of cash flow i years from now is $(F_i Y_i - D_i) e^{-ri} = SY_i e^{-qi} - D_i e^{-ri}$, in total agreement with Eq. (12.17).

CHAPTER 13

Exercise 13.1.1:

$$E[X(t+s) - X(0)] = E[X(t+s) - X(s)] + E[X(s) - X(0)]$$
$$= E[X(t) - X(0)] + E[X(s) - X(0)].$$

Define $f(t) \equiv E[X(t) - X(0)]$. The above equality says that $f(t+s) = f(t) + f(s)$. Differentiate it to get $f'(t+s) = f'(s)$. In particular, $f'(t) = f'(t-1+1) = f'(1)$. Hence $f(t) = tf'(1) + a$. However, $a = 0$ because $f(0) = f(0+0) = 2f(0)$; hence $f(t) = tf'(1)$. This implies that $f(1) = f'(1)$ and $f(t) = tf(1)$.

We now proceed to the proof for the variance. Note that

$$\text{Var}[\, X(t+s) - X(0)\,] = \text{Var}[\, X(t+s) - X(s)\,] + \text{Var}[\, X(s) - X(0)\,]$$
$$= \text{Var}[\, X(t) - X(0)\,] + \text{Var}[\, X(s) - X(0)\,],$$

where the first equality is due to independent increments and the second to stationarity. Define $f(t) \equiv \text{Var}[\, X(t) - X(0)\,]$. The above equality says that $f(t+s) = f(t) + f(s)$. Differentiate it to get

$$f'(t+s) = f'(t), \qquad f'(t+s) = f'(s);$$

in particular, $f'(t) = f'(1)$. Hence $f(t) = t f'(1) + a$. However, $a = 0$ because $f(0) = 2f(0)$; hence $f(t) = t f'(1)$. This implies that $f(1) = f'(1)$ and $f(t) = t f(1)$. This proves the claim because

$$\text{Var}[\, X(t) - X(0)\,] = \text{Var}[\, X(t)\,] - \text{Var}[\, X(0)\,].$$

See [566, p. 61].

Exercise 13.1.2: See [280, p. 96].

Exercise 13.1.3: First, $m_X(n) = E[\, X_n\,] = 0$. Now, for $a \neq b$,

$$E[\, X_a X_b\,] = \text{Cov}[\, X_a, X_b\,] = 0$$

because X_a and X_b are uncorrelated. Finally,

$$K_X(m+n, m) = E[\, X_{m+n} X_m\,] = \begin{cases} 1 & \text{if } n = 0 \\ 0 & \text{otherwise} \end{cases}.$$

Exercise 13.1.4: (1) In Eq. (13.1), use $\mu = 0$ and

$$\xi_n = \begin{cases} 1 & \text{with probability } p \\ -1 & \text{with probability } q \equiv 1 - p \end{cases}.$$

(2) Because the mean is zero, the variance equals $n[\, (1/2) \times 1 + (1/2) \times (-1)^2\,] = n$.

Exercise 13.1.5: Consider two symmetric random walks whose joint displacement follows this distribution:

$$(X_1, X_2) = \begin{cases} (+1, +1) & \text{with probability } (1+\rho)/4 \\ (+1, -1) & \text{with probability } (1-\rho)/4 \\ (-1, +1) & \text{with probability } (1-\rho)/4 \\ (-1, -1) & \text{with probability } (1+\rho)/4 \end{cases}.$$

It is straightforward to verify that $\text{Var}[\, X_1\,] = \text{Var}[\, X_2\,] = 1$ and $E[\, X_1 X_2\,] = \rho$. See [254].

Exercise 13.2.1: In fact,

$$E[\, X(t_n) \mid X(t_{n-1}), X(t_{n-2}), \dots, X(t_1)\,]$$
$$= E[\, X(t_n) - X(t_{n-1}) \mid X(t_{n-1}), X(t_{n-2}), \dots, X(t_1)\,] + X(t_{n-1})$$
$$= X(t_{n-1}),$$

where the last equality is true because

$$E[\, X(t_n) - X(t_{n-1}) \mid X(t_{n-1}), X(t_{n-2}), \dots, X(t_1)\,]$$
$$= E[\, X(t_n) - X(t_{n-1}) \mid X(t_{n-1}) - X(t_{n-2}), \dots, X(t_2) - X(t_1), X(t_1) - X(0)\,]$$
$$= E[\, X(t_n) - X(t_{n-1})\,] = 0.$$

Exercise 13.2.2: Apply the result of Exercise 6.4.3(2), and the definition of a martingale, Eq. (13.3).

Exercise 13.2.3: See [763, p. 229].

Exercise 13.2.4: By the definition and Eq. (6.5),

$$\text{Var}[\, Z_n\,] = \text{Var}[\, Z_{n-1}\,] + \text{Var}[\, X_n\,] + \text{Cov}[\, Z_{n-1}, X_n\,].$$

We are done if we can prove that $\text{Cov}[\, Z_{n-1}, X_n\,] = 0$. Now,

$$\text{Cov}[\, Z_{n-1}, X_n\,] = E[\, Z_{n-1} X_n\,] - E[\, Z_{n-1}\,] E[\, X_n\,]$$
$$= E[\, Z_{n-1} X_n\,]$$
$$= E[\, Z_{n-1}(Z_n - Z_{n-1})\,]$$
$$= E[\, Z_{n-1} Z_n\,] - E[\, Z_{n-1}^2\,]$$
$$= E[\, E[\, Z_{n-1} Z_n \mid Z_{n-1}\,]\,] - E[\, Z_{n-1}^2\,]$$
$$= E[\, Z_{n-1} E[\, Z_n \mid Z_{n-1}\,]\,] - E[\, Z_{n-1}^2\,]$$
$$= E[\, Z_{n-1} Z_{n-1}\,] - E[\, Z_{n-1}^2\,] = 0,$$

where the first equality is due to Eq. (6.3), the second equality is due to Eq. (13.4) and $E[\,Z_0\,] = 0$, the fifth equality is due to the law of iterated conditional expectations, and the seventh equality is due to the definition of a martingale. See [763, p. 247].

Exercise 13.2.5: Observe that $\{S_n, n \geq 1\}$ is a martingale by Example 13.2.1. Let $Z_n \equiv S_n^2 - n\sigma^2$. Now,

$$
\begin{aligned}
E[\,Z_n \mid Z_1, \ldots, Z_{n-1}\,] \\
&= E[\,S_n^2 \mid Z_1, \ldots, Z_{n-1}\,] - n\sigma^2 \\
&= E[\,(S_{n-1} + X_n)^2 \mid S_1, \ldots, S_{n-1}\,] - n\sigma^2 \\
&= E[\,S_{n-1}^2 \mid S_1, \ldots, S_{n-1}\,] + 2\,E[\,S_{n-1}X_n \mid S_1, \ldots, S_{n-1}\,] \\
&\quad + E[\,X_n^2 \mid S_1, \ldots, S_{n-1}\,] - n\sigma^2 \\
&= S_{n-1}^2 + 2S_{n-1}\,E[\,X_n \mid X_1, \ldots, X_{n-1}\,] + E[\,X_n^2 \mid X_1, \ldots, X_{n-1}\,] - n\sigma^2 \\
&= S_{n-1}^2 + 2S_{n-1}E[\,X_n\,] + E[\,X_n^2\,] - n\sigma^2 \\
&= S_{n-1}^2 + E[\,X_n^2\,] - n\sigma^2 \\
&= S_{n-1}^2 - (n-1)\,\sigma^2 \\
&= Z_{n-1}.
\end{aligned}
$$

See [763, p. 247].

Exercise 13.2.6: $E[\,Y_n - Y_{n-1} \mid X_1, X_2, \ldots, X_{n-1}\,]$ equals

$$
E[\,C_n(X_n - X_{n-1}) \mid X_1, X_2, \ldots, X_{n-1}\,] = C_n E[\,X_n - X_{n-1} \mid X_1, X_2, \ldots, X_{n-1}\,] = 0.
$$

See [725, p. 94] or [877, p. 97].

Exercise 13.2.7: By induction. See also [419, p. 227].

Exercise 13.2.8: Let p denote the unknown risk-neutral probability. One unit of foreign currency will be R_f units in a period. Translated into domestic currency, the expected value is

$$
p R_f S u + (1 - p)\,R_f S d = R_f S[\,pu + (1 - p)\,d\,],
$$

where S is the current exchange rate. By Eq. (13.7), we must have $S = R_f S[\,pu + (1 - p)\,d\,]/R$; so

$$
\frac{R}{R_f} = pu + (1 - p)\,d = p(u - d) + d.
$$

See [514, p. 320].

Exercise 13.2.9: Because $S(i + 1) = S(i)\,up + S(i)\,d(1 - p)$, where $p = (R - d)/(u - d)$, we have $S(i + 1) = S(i)\,R$. The claim is then proved by induction.

Exercise 13.2.10: From relation (13.8), $F_i = E_i^\pi[\,F_n\,]$. However, $F_n = S_n$ because the futures price equals the spot price at maturity. See [514, p. 149].

Exercise 13.2.11: It suffices to show that $F_i = E_i^\pi[\,F_{i+1}\,]$ as the general case can be derived by recursion. Because futures contracts are marked to market daily, their value (not price) is zero. Hence $0 = E_i^\pi[\,(F_{i+1} - F_i)/M(i + 1)\,]$ from Eq. (13.7). Because $M(i + 1)$ is known at time i, this equality becomes $0 = \frac{E_i^\pi[\,F_{i+1}\,] - F_i}{M(i+1)}$, from which the claim easily follows. See [514, p. 171].

Exercise 13.2.12: (1) Assume that the bond's current price is $1/R$ and define $p \equiv \left(\frac{1}{d} - \frac{1}{R}\right)\frac{ud}{u-d}$. The bond's prices relative to the stock price one period from now equal $1/(Su)$ and $1/(Sd)$. Now

$$
p\,\frac{1}{Su} + (1 - p)\,\frac{1}{Sd} = \left(\frac{1}{d} - \frac{1}{R}\right)\frac{ud}{u-d}\frac{1}{Su} + \left(\frac{1}{R} - \frac{1}{u}\right)\frac{ud}{u-d}\frac{1}{Sd} = \frac{1}{SR},
$$

which is the bond price relative to the stock price today. An alternative is to evaluate Eq. (13.10) by using $S = 1$, $S_1 = S_2 = R$, $P = S$, $P_1 = Su$, and $P_2 = Sd$. See [681, p. 46]. (2) Prob_1 and Prob_2 are the risk-neutral probabilities that use the stock price and the bond price as numeraire, respectively. See [519] or [681, p. 48].

Exercise 13.2.13: Use the zero-coupon bond maturing at time k as numeraire in Eq. (13.9) to obtain

$$
\frac{C(i)}{P(i, k)} = E_i^\pi\left[\frac{C(k)}{P(k, k)}\right],
$$

where $P(j, k)$ denotes the zero-coupon bond's price at time $j \leq k$. We now prove the claim by observing that $P(i, k) = d(k - i)$ and $P(k, k) = 1$. See [731, p. 171].

Exercise 13.3.1: The first two conditions for Brownian motion remain satisfied. Now,

$$E\left[\frac{x(t) - \mu t}{\sigma} - \frac{x(s) - \mu s}{\sigma}\right] = \frac{E[x(t) - x(s)] - \mu(t - s)}{\sigma} = 0.$$

Furthermore, the variance of $Z \equiv \{[x(t) - \mu t] - [x(s) - \mu s]\}/\sigma$ equals

$$E[Z^2] = \sigma^{-2}(E[\{x(t) - x(s)\}^2] - 2E[\{x(t) - x(s)\}(\mu t - \mu s)] + \mu^2(t - s)^2)$$
$$= \sigma^{-2}(E[\{x(t) - x(s)\}^2] - E[x(t) - x(s)]^2)$$
$$= \sigma^{-2} \text{Var}[x(t) - x(s)]$$
$$= t - s.$$

(Note that Z's expected value is zero.)

Exercise 13.3.2: Without loss of generality, assume that $s < t$. Now,

$$K_X(t, s) \equiv E[\{X(t) - \mu t\}\{X(s) - \mu s\}]$$
$$= E[\{X(t) - X(s) - \mu(t - s) + X(s) - \mu s\}\{X(s) - \mu s\}]$$
$$= E[\{X(t) - X(s) - \mu(t - s)\}\{X(s) - \mu s\}] + E[\{X(s) - \mu s\}^2]$$
$$= E[\{X(t) - X(s) - \mu(t - s)\}X(s)] + s\sigma^2$$
$$= \{E[X(t) - X(s)] - \mu(t - s)\}E[X(s)] + s\sigma^2 \quad \text{from independence}$$
$$= s\sigma^2.$$

See [147, p. 345] or [364, p. 36]. Another method is to observe that [763, p. 187]

$$\text{Cov}[X(s), X(t)] = \text{Cov}[X(s), X(s) + X(t) - X(s)]$$
$$= \text{Cov}[X(s), X(s)] + \text{Cov}[X(s), X(t) - X(s)]$$
$$= \text{Var}[X(s)]$$
$$= s\sigma^2.$$

Exercise 13.3.3: Let $0 < t_1 < \cdots < t_n$. Then,

$$E[X(t_n) - X(0) \mid X(t_{n-1}) - X(0), \ldots, X(t_1) - X(0)]$$
$$= E[X(t_n) - X(t_{n-1}) + X(t_{n-1}) - X(0) \mid X(t_{n-1}) - X(0), \ldots, X(t_1) - X(0)]$$
$$= E[X(t_n) - X(t_{n-1}) \mid X(t_{n-1}) - X(0), \ldots, X(t_1) - X(0)] + X(t_{n-1}) - X(0)$$
$$= X(t_{n-1}) - X(0).$$

The last equality holds because $E[X(t_n) - X(t_{n-1})] = 0$ by the definition of the Wiener process and independent increments. This problem is in fact just a specialization of Exercise 13.2.1. See [280, p. 97] or [541, p. 233].

Exercise 13.3.5: For (2),

$$E[X(t)^2 - \sigma^2 t \mid X(u), 0 \leq u \leq s]$$
$$= E[X(t)^2 \mid X(s)] - \sigma^2 t$$
$$= E[\{X(t) - X(s)\}^2 \mid X(s)] + E[2X(s)\{X(t) - X(s)\} + X(s)^2 \mid X(s)] - \sigma^2 t$$
$$= \sigma^2(t - s) + X(s)^2 - \sigma^2 t$$
$$= X(s)^2 - \sigma^2 s.$$

For (3),

$$E[e^{\alpha X(t) - \alpha^2 \sigma^2 t/2} \mid X(u), 0 \leq u \leq s] = e^{\alpha X(s) - \alpha^2 \sigma^2 s/2} E[e^{\alpha\{X(t) - X(s)\} - \alpha^2(t-s)/2}]$$
$$= e^{\alpha X(s) - \alpha^2 \sigma^2 s/2},$$

where the last equality is due to Eq. (6.8). See [419, p. 7] and [543, p. 358].

Exercise 13.3.6: First, $dP = 2^{-n}$ because the random walk is symmetric. On the other hand, $dQ = \prod_{i=1}^{n} [2^{-1} + (p - 2^{-1}) X_i]$. Recall that $p \equiv (1 + \mu\sqrt{\Delta t})/2$. Hence $dQ/dP = \prod_{i=1}^{n} [1 + (2p-1) X_i] = \prod_{i=1}^{n} (1 + \mu\sqrt{\Delta t} X_i)$.

Exercise 13.3.7: Because $X(t) - X(s) \sim N(\mu(t-s), \sigma^2(t-s))$,

$$E[Y(t) \mid Y(s)] = E[e^{X(t)} \mid e^{X(s)}] = e^{X(s)} E[e^{X(t)-X(s)}] = Y(s) e^{(t-s)(\mu+\sigma^2/2)},$$

where the last equality is due to Eq. (13.12). See [543, p. 364] or [692, p. 349].

Exercise 13.3.8: (1) From Exercise 13.3.7, the rate of return is $\mu + \sigma^2/2$. Refer to Comment 14.4.1 for a discussion of why it is not μ. (2) It follows easily from the definition of $S(t)$. To establish (3), simply observe that $\ln(\cdot)$ is a concave function.

Exercise 13.3.9: Recall that $\mathrm{Var}[X] = E[X^2] - E[X]^2$. Hence, for $X \sim N(\mu, \sigma^2)$, we have $E[X^2] = \sigma^2 + \mu^2$. Returning to our exercise, simply observe that the expression within the brackets is normally distributed with mean $\mu t/2^n$ and variation $\sigma^2 t/2^n$.

Exercise 13.3.10: (1) Recall that $X((k+1)t/2^n) - X(kt/2^n) \sim N(0, 2^{-n})$. Note that if $X \sim N(0, \sigma^2)$, then $|X|$ has mean $\sigma\sqrt{2/\pi}$ and variance $\sigma^2(1 - 2/\pi)$. (2) Let $b \equiv 2^{n/2}\sqrt{2/\pi}$ and $c \equiv 1 - 2/\pi$. For any $\alpha > 0$, we can easily find an n_0 such that $\alpha < b - c^{1/2}2^{n/2}$ for $n > n_0$. Chebyshev's inequality implies that

$$
\begin{aligned}
\mathrm{Prob}[\, f_n(X) > \alpha \,] &\geq \mathrm{Prob}[\, f_n(X) \geq b - c^{1/2}2^{n/2} \,] \\
&\geq \mathrm{Prob}[\, |f_n(X) - b| \leq c^{1/2}2^{n/2} \,] \\
&\geq 1 - \frac{c}{(c^{1/2}2^{n/2})^2} \\
&= 1 - 2^{-n} \to 1.
\end{aligned}
$$

See [73, p. 64].

Exercise 13.4.1 For instance,

$$
\begin{aligned}
E[\, B(t)^2 \,] &= E\left[W(t)^2 - \frac{2t}{T} W(t) W(T) + \frac{t^2}{T^2} W(T)^2 \right] \\
&= t - \frac{2t}{T} E[\, W(t)\{ W(T) - W(t) \} \,] - \frac{2t}{T} E[\, W(t)^2 \,] + \frac{t^2}{T} \\
&= t - 0 - \frac{2t^2}{T} + \frac{t^2}{T} = t - \frac{t^2}{T}.
\end{aligned}
$$

See [193, p. 193] or [557, p. 59].

Exercise 13.4.2 It is $x + W(t) - (t/T)[\, W(T) - y + x \,], 0 \leq t \leq T$ [557, p. 59].

Chapter 14

Exercise 14.1.1: Consider s and t such that $t_k \leq s < t \leq t_{k+1}$ first. From Eq. (14.1),

$$
\begin{aligned}
&E[\, \mathrm{I}_t(X) \mid W(u), 0 \leq u \leq s \,] \\
&\quad = E[\, \mathrm{I}_s(X) + X(t_k)\{ W(t) - W(s) \} \mid W(u), 0 \leq u \leq s \,] \\
&\quad = E[\, \mathrm{I}_s(X) \mid W(u), 0 \leq u \leq s \,] + X(t_k) E[\, W(t) - W(s) \mid W(u), 0 \leq u \leq s \,] \\
&\quad = \mathrm{I}_s(X).
\end{aligned}
\tag{33.8}
$$

Simple induction can show that $E[\, \mathrm{I}_t(X) \mid W(u), 0 \leq u \leq t_i \,] = \mathrm{I}_{t_i}(X)$ for $t_i < t_k \leq t \leq t_{k+1}$. Hence, for $t_i \leq s < t_{i+1}, t_k \leq t < t_{k+1}$, and $i < k$,

$$
\begin{aligned}
E[\, \mathrm{I}_t(X) \mid W(u), 0 \leq u \leq s \,] &= E[\, E[\, \mathrm{I}_t(X) \mid W(u), 0 \leq u \leq t_{i+1} \,] \mid W(u), 0 \leq u \leq s \,] \\
&= E[\, \mathrm{I}_{t_{i+1}}(X) \mid W(u), 0 \leq u \leq s \,] \\
&= \mathrm{I}_s(X)
\end{aligned}
$$

by Eq. (33.8). See [566, p. 158] and [585, pp. 167–168].

Exercise 14.1.2: The approximating sum is now $\sum_{k=0}^{n-1} W(t_{k+1})[\, W(t_{k+1}) - W(t_k) \,]$ [30, p. 104].

Exercise 14.1.3:

$$E\left[\frac{W(t)^2}{2}\,\middle|\,W(u), 0 \le u \le s\right]$$

$$= E\left[\frac{W(s)^2}{2}\,\middle|\,W(u), 0 \le u \le s\right] + E\left[\frac{W(t)^2}{2} - \frac{W(s)^2}{2}\,\middle|\,W(u), 0 \le u \le s\right]$$

$$= \frac{W(s)^2}{2} + \frac{t-s}{2},$$

which is not a martingale [566, p. 160].

Exercise 14.1.4: Use Eq. (14.3) and the definition $E[\,W(t)^2\,] = \mathrm{Var}[\,W(t)\,] = t$.

Exercise 14.1.5: This is because $\int_{t_0}^{t} dW(s) = W(t) - W(t_0)$ [30, p. 59].

Exercise 14.2.1:

$$\phi_0 S_0 + G_k = \phi_0 S_0 + \sum_{i=0}^{k-1} \phi_i (S_{i+1} - S_i) = \phi_0 S_0 + \sum_{i=0}^{k-1} (\phi_{i+1} S_{i+1} - \phi_i S_i) = \phi_k S_k.$$

See [70] or [420, p. 226].

Exercise 14.2.2: Applying Ito's formula with $f(x) = x^2$, we have $W(t)^2 = \int_0^t ds + \int_0^t 2W(s)\,dW$. Hence $\int_0^t W(s)\,dW = [\,W(t)^2/2\,] - (t/2)$. See [585, p. 172].

Exercise 14.3.1: (1) See [746, p. 175]. (2) From Example 14.3.3, we know that $X(t) = e^{Y(t)}$, where Y is a $(\mu - \sigma^2/2, \sigma)$ Brownian motion. Thus $\ln(X(t)/X(0)) \sim N((\mu - \sigma^2/2)\,t, \sigma^2 t)$.

Exercise 14.3.2: Note that $dR = d(\ln X) + (\sigma^2/2)\,dt$. From the solution to Exercise 14.3.1, we arrive at

$$dR = \left(\mu - \frac{\sigma^2}{2}\right) dt + \sigma\,dW + \frac{\sigma^2}{2}\,dt = \mu\,dt + \sigma\,dW.$$

Exercise 14.3.3: (1) Apply Ito's formula (14.10) to the function $f(x) = x^n$ to obtain

$$dX^n = nX^{n-1}\,dX + \frac{1}{2}n(n-1)\,X^{n-2}\,(dX)^2.$$

Substitute W_t for X above to arrive at the stochastic differential equation,

$$dW_t^n = nW_t^{n-1}\,dW_t + \frac{n(n-1)}{2}\,W_t^{n-2}\,dt.$$

See [30, p. 94]. (2) Expand

$$dW(t)^n = [\,W(t) + dW(t)\,]^n - W(t)^n = nW(t)^{n-1}\,dW(t) + \frac{n(n-1)}{2}\,W(t)^{n-2}\,dt$$

with Eq. (13.15) and (13.16). See [373, p. 89].

Exercise 14.3.4: The multidimensional Ito's lemma (Theorem 14.2.2) can be used to show that $dU = (1/2)\,dY + (1/2)\,dZ$, which can be expanded into

$$\frac{dU}{U} = \left(\frac{Y}{Y+Z}a + \frac{Z}{Y+Z}f\right)dt + \left(\frac{Y}{Y+Z}b + \frac{Z}{Y+Z}g\right)dW.$$

Exercise 14.3.5: The Ito process $U = YZ$ is defined by Y and Z with differentials, respectively, $dY = a\,dt + b\,dW_y$ and $dZ = f\,dt + g\,dW_z$. Keep in mind that dW_y and dW_z have correlation ρ. The multidimensional Ito's lemma (Theorem 14.2.3) can be used to show that

$$dU = Z\,dY + Y\,dZ + (a\,dt + b\,dW_y)(f\,dt + g\,dW_z) = (Za + Yf + bg\rho)\,dt + Zb\,dW_y + Yg\,dW_z.$$

Exercise 14.3.7: It is $(1/F)\,dF = (1/X)\,dX + (1/Y)\,dY + \sigma^2\,dt$ [746, p. 176].

Exercise 14.3.8: View Y as $Y(X, t) \equiv Xe^{kt}$ and apply Ito's lemma [373, p. 106].

Exercise 14.3.9 Let $f(t) \ge 0$ be any continuous strictly increasing function. Then

$$W(f(t + \Delta t)) - W(f(t)) \sim N(0,\, f(t + \Delta t) - f(t)),$$

which approaches $N(0,\, f'(t)\Delta t) = \sqrt{f'(t)\Delta t}\,N(0, 1)$. Now,

$$dY(t) = \frac{\partial e^{-t}}{\partial t}\,W(e^{2t})\,dt + e^{-t}\,dW(f(t)) = -Y(t)\,dt + e^{-t}\sqrt{2e^{2t}\,dt}\,\xi = -Y(t)\,dt + \sqrt{2}\,dW,$$

where $\xi \sim N(0, 1)$. Hence $dY = -Y\,dt + \sqrt{2}\,dW$. The general formula for $a(t)\,W(f(t))$ appears in [230, p. 229].

Exercise 14.3.10: Ito's lemma (Theorem 14.2.3) says that H satisfies

$$
dH = \frac{\partial H}{\partial S} dS + \frac{\partial H}{\partial \sigma} d\sigma - \frac{\partial H}{\partial \tau} d\tau + \frac{1}{2} \sigma^2 S^2 \frac{\partial^2 H}{\partial S^2} dt + \sigma \gamma S \rho \frac{\partial^2 H}{\partial S \partial \tau} dt + \frac{1}{2} \gamma^2 \frac{\partial^2 H}{\partial \tau^2} dt
$$

$$
= \left[\mu S \frac{\partial H}{\partial S} + \beta(\overline{\sigma} - \sigma) \frac{\partial H}{\partial \sigma} - \frac{\partial H}{\partial \tau} + \frac{1}{2} \sigma^2 S^2 \frac{\partial^2 H}{\partial S^2} + \sigma \gamma S \rho \frac{\partial^2 H}{\partial S \partial \tau} + \frac{1}{2} \gamma^2 \frac{\partial^2 H}{\partial \tau^2} \right] dt
$$

$$
+ \sigma S \frac{\partial H}{\partial S} dW_1 + \gamma \frac{\partial H}{\partial \sigma} dW_2.
$$

Note that $d\tau = -dt$. See [790].

Exercise 14.3.11: Without loss of generality, we will verify, instead, that

$$
\frac{\partial p}{\partial \tau} = \frac{1}{2} \frac{\partial^2 p}{\partial x^2} - \frac{1}{2} x \frac{\partial p}{\partial x},
$$

where $\tau \equiv t - s$. This is admissible because the process is homogeneous, which we can see by observing that both the drift and the diffusion of the Ito process are independent of time t [30, Remark 2.6.8]. From the hint, we know that

$$
p(x, s; y, t) = \frac{1}{\sqrt{2\pi (1 - e^{-\tau})}} \exp\left[-\frac{(y - xe^{-\tau/2})^2}{2(1 - e^{-\tau})} \right] = B^{-1/2} A,
$$

where

$$
A \equiv \exp\left[-\frac{(y - xe^{-\tau/2})^2}{2(1 - e^{-\tau})} \right], \qquad B \equiv 2\pi(1 - e^{-\tau}).
$$

After we verify the following equations, we will be done:

$$
\frac{\partial p}{\partial \tau} = B^{-3/2} A \pi \left[-e^{-\tau} + (y - xe^{-\tau/2})^2 \frac{e^{-\tau}}{1 - e^{-\tau}} - (y - xe^{-\tau/2}) xe^{-\tau/2} \right],
$$

$$
\frac{1}{2} x \frac{\partial p}{\partial x} = B^{-3/2} A \pi (y - xe^{-\tau/2}) e^{-\tau/2} x,
$$

$$
\frac{1}{2} \frac{\partial^2 p}{\partial x^2} = B^{-3/2} A \pi \left[-e^{-\tau} + \frac{(y - xe^{-\tau/2})^2}{1 - e^{-\tau}} e^{-\tau} \right].
$$

Exercise 14.4.1: It follows from Eq. (13.17), the infinite total variation of Brownian motion (with probability one) [723].

Exercise 14.4.2: If rates are negative, then the price exceeds the total sum of future cash flows. To generate arbitrage profits, short the bond and reserve part of the proceeds to service future cash flow obligations.

Exercise 14.4.3: It must be that $g(T) = 0$; hence $\mu = \sigma = 0$. In particular, there are no random changes. See [207].

Exercise 14.4.4: Let the current time be zero and the portfolio contain $b_i > 0$ units of bond i for $i = 1, 2$. (1) Let $\theta(t) \equiv A(t)/L(t)$. To begin with,

$$
A(t) = \sum_{i=1}^{2} b_i P(r(t), t, t_i), \qquad L(t) = P(r(t), t, s),
$$

where $P(r(t), t, s)$ is from Eq. (14.16). Note that $r(t)$ is a random variable. Now,

$$
\theta(t) = \sum_{i=1}^{2} b_i \frac{P(r(t), t, t_i)}{P(r(t), t, s)}
$$

$$
= \sum_{i=1}^{2} b_i \times \exp\left[-r(t)(t_i - s) - \mu \frac{(t_i - t)^2 - (s - t)^2}{2} + \sigma^2 \frac{(t_i - t)^3 - (s - t)^3}{6} \right].
$$

Because $\partial^2 \theta(t)/\partial r^2$ equals

$$
\sum_{i=1}^{2} b_i (t_i - s)^2 \times \exp\left[-r(t)(t_i - s) - \mu \frac{(t_i - t)^2 - (s - t)^2}{2} + \sigma^2 \frac{(t_i - t)^3 - (s - t)^3}{6} \right] > 0,
$$

$\theta(t)$ is indeed a convex function of $r(t)$.

(2) Immunization requires that $\partial \theta / \partial r = 0$ at time zero, that is,

$$
\sum_{i=1}^{2} -(t_i - s) b_i \times \exp\left[-r(0)(t_i - s) - \mu \frac{t_i^2 - s^2}{2} + \sigma^2 \frac{t_i^3 - s^3}{6} \right] = 0. \tag{33.9}
$$

Incidentally, after some easy manipulations, Eq. (33.9) becomes the standard duration condition, $-[\partial A(t)/\partial r]/A(t) = s$. Another immunization condition is that the asset and the liability should match in value at time zero, or $\theta(0) = 1$. This leads to

$$\sum_{i=1}^{2} b_i \times \exp\left[-r(0)(t_i - s) - \mu \frac{t_i^2 - s^2}{2} + \sigma^2 \frac{t_i^3 - s^3}{6} \right] = 1. \tag{33.10}$$

With the definitions

$$w_i \equiv b_i \times \exp\left[-r(0)(t_i - s) - \mu \frac{t_i^2 - s^2}{2} + \sigma^2 \frac{t_i^3 - s^3}{6} \right],$$

we arrive at

$$\theta(t) = \sum_{i=1}^{2} w_i \times \exp\left[-\{r(t) - r(0)\}(t_i - s) - \mu \frac{(t_i - t)^2 - (s - t)^2 - t_i^2 + s^2}{2} \right.$$
$$\left. + \sigma^2 \frac{(t_i - t)^3 - (s - t)^3 - t_i^3 + s^3}{6} \right]$$
$$= \sum_{i=1}^{2} w_i \times \exp\left[-\{r(t) - r(0)\}(t_i - s) + \mu t(t_i - s) + \sigma^2 \frac{t - t_i - s}{2} t(t_i - s) \right] \tag{33.11}$$

The following lemma then implies that $\theta(t) < w_1 + w_2 = 1$, concluding the proof.

LEMMA

$$\exp\left[-\{r(t) - r(0)\}(t_i - s) + \mu t(t_i - s) + \sigma^2 \frac{t - t_i - s}{2} t(t_i - s) \right] < 1$$

for $i = 1, 2$.

Proof of Lemma: Equations (33.9) and (33.10) can be used to obtain $w_1 = [(s - t_2)/(t_1 - t_2)]$ and $w_2 = [(t_1 - s)/(t_1 - t_2)]$ because $\sum_{i=1}^{2} -(t_i - s)w_i = 0$ and $\sum_{i=1}^{2} w_i = 1$. From Eq. (33.11), the $[\partial\theta(t)/\partial r(t)] = 0$ condition becomes

$$\sum_{i=1}^{2} (s - t_i) w_i \times \exp\left[-\{r(t) - r(0)\}(t_i - s) + \mu t(t_i - s) + \sigma^2 \frac{t - t_i - s}{2} t(t_i - s) \right]$$
$$= 0.$$

Substitute the solution for (w_1, w_2) into the preceding equation to get

$$0 = (s - t_1) \frac{s - t_2}{t_1 - t_2} \exp\left[-\{r(t) - r(0)\}(t_1 - s) + \mu t(t_1 - s) + \sigma^2 \frac{t - t_1 - s}{2} t(t_1 - s) \right]$$
$$+ (s - t_2) \frac{t_1 - s}{t_1 - t_2} \exp\left[-\{r(t) - r(0)\}(t_2 - s) + \mu t(t_2 - s) + \sigma^2 \frac{t - t_2 - s}{2} t(t_2 - s) \right].$$

Therefore

$$-[r(t) - r(0)](t_1 - s) + \mu t(t_1 - s) + \sigma^2 \frac{t - t_1 - s}{2} t(t_1 - s)$$
$$= -[r(t) - r(0)](t_2 - s) + \mu t(t_2 - s) + \sigma^2 \frac{t - t_2 - s}{2} t(t_2 - s). \tag{33.12}$$

Identity (33.12) implies

$$r(t) = \left[r(0)(t_2 - t_1) + \mu t(t_2 - t_1) + \sigma^2 \frac{t - t_2 - s}{2} t(t_2 - s) - \sigma^2 \frac{t - t_1 - s}{2} t(t_1 - s) \right] \Big/ (t_2 - t_1)$$
$$= r(0) + \mu t + \frac{\sigma^2 t}{2}(t - t_2 - t_1).$$

Finally, substitute the preceding equation into the $i = 1$ term in Eq. (33.11) to get

$$\exp\left[-\left\{ \mu t + \frac{\sigma^2 t}{2}(t - t_2 - t_1) \right\}(t_1 - s) + \mu t(t_1 - s) + \sigma^2 \frac{t - t_1 - s}{2} t(t_1 - s) \right]$$
$$= \exp\left[\frac{\sigma^2 t}{2}(t_1 - s)(t_2 - s) \right] < 1.$$

The $i = 2$ case is redundant as its value is identical by identity (33.12).

Exercise 14.4.5: Direct from identity (13.12), it is $Se^{\mu T}$.

Exercise 14.4.6: Although $dS/S = \sigma \, dW$ looks "symmetric" around zero, it is not. We can see this most clearly by looking at its discrete version, which says that the percentage change has zero mean. However, a 2% decrease followed by a 2% increase results in $1.02 \times 0.98 = 0.9996 < 1$. In other words, a 2% decrease has to be followed by a more than 2% increase to be back to the original level.

Exercise 14.4.7: They follow from identity (13.12) and the fact that $\ln(S(t)/S(0))$ is a $(\mu - \sigma^2/2, \sigma)$ Brownian motion.

Exercise 14.4.8: This can be justified heuristically as follows. Partition $[0, T]$ into n equal periods. Let σ_{i-1} denote the (annualized) volatility during period i and S_i the stock price at time i. Consequently,

$$\ln \frac{S_i}{S_{i-1}} \sim N\left(\mu \Delta t - \frac{1}{2} \sigma_{i-1}^2 \Delta t, \sigma_{i-1}^2 \Delta t\right)$$

by relation (14.17). If S and σ are uncorrelated, preceding relation holds, given σ_{i-1}. The probability distribution of $\ln(S_n/S_0)$, given the path followed by σ (i.e., $\sigma_1, \sigma_2, \ldots, \sigma_n$), is thus normal with mean

$$\sum_i \left(\mu \Delta t - \frac{1}{2} \sigma_{i-1}^2 \Delta t\right) = \mu T - \frac{1}{2} \sum_i \sigma_{i-1}^2 \Delta t \rightarrow \mu T - \frac{1}{2} \widehat{\sigma^2} \, T$$

and variance $\sum_i (\sigma_{i-1}^2 \Delta t) \rightarrow \widehat{\sigma^2} \, T$.

Exercise 14.4.9: This is due to $(\Delta S)^2 \approx \sigma^2 S^2 \Delta t$ [514, p. 221].

Exercise 14.4.11: Note that

$$\ln u = -\ln 0, \qquad u \approx 1 + \sigma \sqrt{\frac{\tau}{n}} + \frac{\sigma^2 \tau}{2n}, \qquad d \approx 1 - \sigma \sqrt{\frac{\tau}{n}} + \frac{\sigma^2 \tau}{2n}.$$

Hence, $u + d \approx 2 + \sigma^2(\tau/n)$ and $u - d \approx 2\sigma \sqrt{\tau/n}$. Finally,

$$p \approx \frac{[1 + r(\tau/n)] - [1 - \sigma \sqrt{\tau/n} + (1/2) \sigma^2(\tau/n)]}{2\sigma \sqrt{\tau/n}} \approx \frac{1}{2} + \frac{r(\tau/n) - (1/2) \sigma^2(\tau/n)}{2\sigma \sqrt{\tau/n}}.$$

Now,

$$E[X_i] = p \ln u + (1-p) \ln d = (2p-1) \ln u \approx r\frac{\tau}{n} - \frac{1}{2} \sigma^2 \frac{\tau}{n}.$$

Hence $E[\sum_{i=1}^n X_i] = n E[X_i] \approx r\tau - \sigma^2 \tau/2$, as desired.

We proceed to calculate the variance:

$$\begin{aligned} \text{Var}[X_i] &= p(\ln u - E[X_i])^2 + (1-p)(\ln d - E[X_i])^2 \\ &= p[\ln u - (2p-1)\ln u]^2 + (1-p)[\ln d - (2p-1)\ln u]^2 \\ &= p[\ln u - (2p-1)\ln u]^2 + (1-p)[-\ln u - (2p-1)\ln u]^2 \\ &= 4p(1-p)\ln^2 u \\ &\approx 4\left[\frac{1}{2} + \frac{r(\tau/n) - (1/2)\sigma^2(\tau/n)}{2\sigma \sqrt{\tau/n}}\right]\left[\frac{1}{2} - \frac{r(\tau/n) - (1/2)\sigma^2(\tau/n)}{2\sigma \sqrt{\tau/n}}\right]\sigma^2 \frac{\tau}{n} \\ &\approx \sigma^2 \frac{\tau}{n}. \end{aligned}$$

Hence $\text{Var}[\sum_{i=1}^n X_i] = n \, \text{Var}[X_i] \approx \sigma^2 \tau$, as desired.

Exercise 14.4.12: The troubling step is

$$X_{i+1} = \ln\left(1 + \frac{S_{i+1} - S_i}{S_i}\right) \approx \frac{S_{i+1} - S_i}{S_i} \equiv \frac{\Delta S_i}{S_i}.$$

It should have been

$$X_{i+1} = \ln\left(1 + \frac{S_{i+1} - S_i}{S_i}\right) \approx \frac{\Delta S_i}{S_i} - \frac{1}{2}\left(\frac{\Delta S_i}{S_i}\right)^2.$$

Now, because $(dS)^2 = \sigma^2 S^2 \, dt$, the preceding approximation becomes $X_{i+1} \approx (\Delta S_i / S_i) - (1/2) \sigma^2 \Delta t$. This corrects the problem.

Exercise 14.4.13: Apply Eq. (14.7) to $dX = (r - \sigma^2/2) \, dt + \sigma \, dW$ [229].

CHAPTER 15

Exercise 15.2.2: Different specifications of the underlying process imply different ways of estimating the diffusion coefficient from the data, and they may not give the same value. Hence the σ that gets plugged into the Black–Scholes formula could be different for different models. Misspecification of the drift therefore may lead to misestimation of the diffusion of the Ito process [819]. See [613] for the complete argument.

Exercise 15.2.3: Write valuation formula (9.4) as

$$C(S,t) = \frac{pC(S^+, t+\Delta t) + (1-p)\,C(S^-, t+\Delta t)}{e^{r\Delta t}},$$

where $p \equiv (Se^{r\Delta t} - S^-)/(S^+ - S^-)$. Express $C(S^+, t+\Delta t)$ and $C(S^-, t+\Delta t)$ in Taylor expansion around (S,t) and substitute them into the preceding formula. See [696, p. 416].

Exercise 15.2.4: This is the risk-neutral valuation with the density function of the lognormal distribution [492, p. 312].

Exercise 15.2.5: Follow the same steps with slightly different boundary conditions: $B(0, \tau) = 1$ for $\tau > 0$ and $B(S, 0) = \max(1 - S/X, 0)$. In the end, we get $\Theta(z, 0) = \max(e^{-z} - 1, 0)$ instead. Now,

$$\Theta(z, u) = \frac{1}{\sqrt{2\pi u}} \int_{-\infty}^{0} (e^{-y} - 1)\, e^{-(z-y)^2/(2u)}\, dy = \cdots = -N\left(-\frac{z}{\sqrt{u}}\right) + \frac{1}{x} N\left(-\frac{z-u}{\sqrt{u}}\right).$$

The rest is straightforward.

Exercise 15.2.6:

$$\frac{\partial P}{\partial t} + rS \frac{\partial P}{\partial S} + \frac{1}{2}\sigma^2 S^2 \frac{\partial^2 P}{\partial S^2} - rP = rX < 0$$

when $X - S$ is substituted into the equation [879, p. 112].

Exercise 15.3.1: It is

$$\frac{\partial C}{\partial t} + \frac{1}{2}\sigma^2 F^2 \frac{\partial^2 C}{\partial F^2} = rC.$$

This can be derived by the hedging argument with the observation that it costs nothing to enter into a futures contract. An alternative replaces $\partial C/\partial t$ with $\partial C/\partial t - rF(\partial C/\partial F)$, $\partial C/\partial S$ with $e^{r(T-t)}(\partial C/\partial F)$, and $\partial^2 C/\partial S^2$ with $e^{2r(T-t)}(\partial^2 C/\partial F^2)$ in the original Black–Scholes differential equation. See [125], [575, Subsection 2.3.3], and [879, p. 100].

Exercise 15.3.3: (1) It is straightforward [492, p. 380]. (2)

$$\frac{\partial C}{\partial t} + \sum_{i=1}^{n} rS_i \frac{\partial C}{\partial S_i} + \sum_{i=1}^{n} \frac{\sigma_i^2 S_i^2}{2} \frac{\partial^2 C}{\partial S_i^2} + \frac{1}{2} \sum_{i=1}^{n} \sum_{j=1}^{n} \rho_{ij}\sigma_i\sigma_j S_i S_j \frac{\partial^2 C}{\partial S_i \partial S_j} = rC.$$

Exercise 15.3.4: $\max(S_1, S_2) = S_1 + \max(S_2 - S_1, 0)$.

Exercise 15.3.5: Its terminal value can be written as

$$\max(S_1(\tau) - X, 0) + \max(S_2(\tau) - X, 0) - \max(\max(S_1(\tau), S_2(\tau)) - X, 0).$$

The last term is the terminal value of a call on the maximum of two assets with strike price X. See [746, p. 376].

Exercise 15.3.6: See [492, p. 4].

Exercise 15.3.8: Ito's lemma applied to $f(S_1, S_2) = S_2/S_1$ gives

$$df = (\cdots)\, dt - \frac{S_2}{S_1}\sigma_1\, dW_1 + \frac{S_2}{S_1}\sigma_2\, dW_2.$$

From variance (6.9), we know that the variance for $df/f = d(S_2/S_1)/(S_2/S_1)$ is the σ^2 in Eq. (15.5).

Exercise 15.3.10: It is due to $dV - V_1\, dS_1 - V_2\, dS_2 = 0$.

Exercise 15.3.11: Just plug in the interpretation into formulas (9.20) for a stock paying a continuous dividend yield.

Exercise 15.3.12: (1) Let F denote the forward exchange rate τ years from now. Start with S units of domestic currency and convert it into one unit of foreign currency. Also sell forward the foreign currency. From then on, apply the standard hedging argument by using only foreign assets to construct replicating portfolios, earning the riskless rate r_f in foreign currency. To verify that the strategy is self-financing in terms of domestic value, observe that the forward price is $Se^{(r-r_f)\tau'}$ at τ' years remaining to expiration. The final wealth is $e^{r_f\tau}$, identical to a call option in a foreign risk-neutral economy. This can be converted for $e^{r_f\tau}F = Se^{r\tau}$ dollars in domestic currency because $F = Se^{(r-r_f)\tau}$. Correlation between the exchange rate and the foreign asset price is eliminated by the forward contract. (2) The results in (1) show that G_f grows at a rate of $r_f - q_f + r - r_f = r - q_f$ in a domestic risk-neutral economy, equivalent to a payout rate of q_f. Furthermore, in this economy, X_f becomes $X_f F/S = X_f e^{(r-r_f)\tau}$ at expiration.

Exercise 15.3.13: Let $S_f(t) \equiv S(t) B_f(t)$, which follows geometric Brownian motion. What we have is an exchange option that buys $S(T) B_f(T)$ with $XB(T)$ at T. Note that $B(T) = B_f(T) = 1$. Hence the desired formula is $SB_f N(x) - XBN(x - \sigma\sqrt{\tau})$, where $x \equiv \frac{\ln(SB_f/(XB)) + (\sigma^2/2)\tau}{\sigma\sqrt{\tau}}$, $\sigma^2 \equiv \sigma_{S_f}^2 - 2\rho\sigma_{S_f}\sigma_B + \sigma_B^2$, and ρ is the correlation between S_f and B. See [575, p. 110].

Exercise 15.3.15: Identity the U.S. dollar with the currency C in the case of foreign domestic options. The case of forex options is trivial.

Exercise 15.3.16: The portfolio has the following terminal payoff in U.S. dollars:

$$\max(S_A - X_A, 0) + X \times \max(X_C - S_C, 0).$$

Now consider all six possible relations between the spot prices (S_A, S_C, and S) and the strike prices. See [775].

Exercise 15.3.17: For simplicity, let $\widehat{S} = 1$. The exchange rate and the foreign asset's price follow

$$dS = \mu_s S \, dt + \sigma_s S \, dW_s, \qquad dG_f = \mu_f G_f \, dt + \sigma_f G_f \, dW_f,$$

respectively. The foreign asset pays a continuous dividend yield of q_f. Let C be the price of the quanto option. From Ito's lemma (Theorem 14.2.2),

$$dC = \left(\mu_s S \frac{\partial C}{\partial S} + \mu_f G_f \frac{\partial C}{\partial G_f} + \frac{\partial C}{\partial t} + \frac{1}{2} \sigma_s^2 S^2 \frac{\partial^2 C}{\partial S^2} + \frac{1}{2} \sigma_f^2 G_f^2 \frac{\partial^2 C}{\partial G_f^2} + \rho \sigma_s \sigma_f S G_f \frac{\partial^2 C}{\partial S \partial G_f} \right) dt$$

$$+ \sigma_s S \frac{\partial C}{\partial S} \, dW_s + \sigma_f G_f \frac{\partial C}{\partial G_f} \, dW_f.$$

Set up a portfolio that is long one quanto option, short δ_s units of foreign currency, and short δ_f units of the foreign asset. Its value is $\Pi = C - \delta_s S - \delta_f G_f S$. The total wealth change of the portfolio at time dt is given by

$$d\Pi = dC - \delta_s \, dS - \delta_f \, d(G_f S) - \delta_s Sr_f \, dt - \delta_f G_f Sq_f \, dt.$$

The last two terms are due to foreign interest and dividends. From Example 14.3.5,

$$d(G_f S) = G_f S(\mu_s + \mu_f + \rho\sigma_s\sigma_f) \, dt + \sigma_s G_f S \, dW_s + \sigma_f G_f S \, dW_f.$$

Substitute the formulas for dC, dS, and $d(G_f S)$ into $d\Pi$ to yield

$$d\Pi = \left[\mu_s S \frac{\partial C}{\partial S} + \mu_f G_f \frac{\partial C}{\partial G_f} + \frac{\partial C}{\partial t} + \frac{1}{2} \sigma_s^2 S^2 \frac{\partial^2 C}{\partial S^2} + \frac{1}{2} \sigma_f^2 G_f^2 \frac{\partial^2 C}{\partial G_f^2} \right.$$
$$\left. + \rho\sigma_s\sigma_f SG_f \frac{\partial^2 C}{\partial S \partial G_f} - \delta_s \mu_s S - \delta_f G_f S(\mu_s + \mu_f + \rho\sigma_s\sigma_f) - \delta_s Sr_f - \delta_f G_f Sq_f \right] dt.$$
$$+ \left(\sigma_s S \frac{\partial C}{\partial S} - \delta_s \sigma_s S - \delta_f \sigma_s G_f S \right) dW_s + \left(\sigma_f G_f \frac{\partial C}{\partial G_f} - \delta_f \sigma_f G_f S \right) dW_f.$$

Clearly, we have to pick $\delta_f = (\partial C/\partial G_f) S^{-1}$ and $\delta_s = (\partial C/\partial S) - (\partial C/\partial G_f)(G_f/S)$ to remove the randomness. Under these choices,

$$d\Pi = \left[\mu_s S \frac{\partial C}{\partial S} + \mu_f G_f \frac{\partial C}{\partial G_f} + \frac{\partial C}{\partial t} + \frac{1}{2}\sigma_s^2 S^2 \frac{\partial^2 C}{\partial S^2} + \frac{1}{2}\sigma_f^2 G_f^2 \frac{\partial^2 C}{\partial G_f^2} + \rho\sigma_s\sigma_f SG_f \frac{\partial^2 C}{\partial S \partial G_f} \right.$$
$$\left. - \left(\frac{\partial C}{\partial S} - \frac{\partial C}{\partial G_f}\frac{G_f}{S} \right) \mu_s S - \frac{\partial C}{\partial G_f} G_f(\mu_s + \mu_f + \rho\sigma_s\sigma_f) - \left(\frac{\partial C}{\partial S} - \frac{\partial C}{\partial G_f}\frac{G_f}{S} \right) Sr_f - \frac{\partial C}{\partial G_f} G_f q_f \right] dt$$

$$= r\Pi \, dt = r(C - \delta_s S - \delta_f G_f S) \, dt = r\left[C - \left(\frac{\partial C}{\partial S} - \frac{\partial C}{\partial G_f}\frac{G_f}{S} \right)S - \frac{\partial C}{\partial G_f} G_f \right] dt.$$

After simplification,

$$
rC = (r - r_f)\, S\, \frac{\partial C}{\partial S} + \frac{\partial C}{\partial t} + \frac{\sigma_s^2 S^2}{2}\, \frac{\partial^2 C}{\partial S^2} + \frac{\sigma_f^2 G_f^2}{2}\, \frac{\partial^2 C}{\partial G_f^2}
$$

$$
+ \rho \sigma_s \sigma_f S G_f\, \frac{\partial^2 C}{\partial S\, \partial G_f} + \frac{\partial C}{\partial G_f}\, G_f\, (r_f - \rho \sigma_s \sigma_f - q_f).
$$

Finally, C does not directly depend on the exchange rate S [734]. Hence the equation becomes

$$
rC = \frac{\partial C}{\partial t} + \frac{\sigma_f^2 G_f^2}{2}\, \frac{\partial^2 C}{\partial G_f^2} + \frac{\partial C}{\partial G_f}\, G_f\, (r_f - \rho \sigma_s \sigma_f - q_f).
$$

See [878, pp. 156–157].

Exercise 15.3.18: This is because the bond exists beyond time t^* only if $C(V, t^*) < P(t^*)$ *and* the market value if not called is exceeded by $P(t)$ at $t = t^*$ for, otherwise, it would be called at $t = t^*$. From that moment onward, the bond will be called the moment its market value if not called rises to the call price. Because the market value if not called cannot be exceeded by the conversion value, $C(V, t) \le P(t)$. As a consequence, on a coupon date and when the bond is callable (i.e., $t > t^*$),

$$
W(V, t^-) = \min(W(V - mc, t^+) + c,\ P(t))
$$

because, as argued in the text, the bond should be called only when its value if not called equals the call price, which equals the value if called under $C(V, t) \le P(t)$.

Exercise 15.4.1: Let $f_1, f_2, \dots, f_{n+1}$ denote the prices of securities whose value depends on $S_1, S_2, \dots, S_n$, and t. By Ito's lemma (Theorem 14.2.2),

$$
df_j = \left(\frac{\partial f_j}{\partial t} + \sum_i \mu_i S_i\, \frac{\partial f_j}{\partial S_i} + \sum_{i,k} \frac{1}{2} \rho_{ik} \sigma_i \sigma_k S_i S_k\, \frac{\partial^2 f_j}{\partial S_i\, \partial S_k} \right) dt + \sum_i \sigma_i S_i\, \frac{\partial f_j}{\partial S_i}\, dW_i
$$

$$
\equiv \mu_j f_j\, dt + \sum_i \sigma_{ij} f_j\, dW_i \tag{33.13}
$$

Maintain a portfolio of k_j units of f_j such that

$$
\sum_j k_j \sigma_{ij} f_j = 0 \quad \text{for } i = 1, 2, \dots, n. \tag{33.14}
$$

Because

$$
\sum_j k_j\, df_j = \sum_j k_j \mu_j f_j\, dt + \sum_j k_j \sum_i \sigma_{ij} f_j\, dW_i
$$

$$
= \sum_j k_j \mu_j f_j\, dt + \sum_i \sum_j k_j \sigma_{ij} f_j\, dW_i = \sum_j k_j \mu_j f_j\, dt,
$$

the portfolio is instantaneously riskless. Its return therefore equals the short rate, or $\sum_j k_j \mu_j f_j = r \sum_j k_j f_j$. After rearrangements,

$$
\sum_j k_j f_j (\mu_j - r) = 0. \tag{33.15}
$$

Equations (33.14) and (33.15) under the condition that not all k_js are zeros imply that $\mu_j - r = \sum_i \lambda_i \sigma_{ij}$ for some $\lambda_1, \lambda_2, \dots, \lambda_n$, which depend on only $S_1, S_2, \dots, S_n$, and t.[1] Therefore any derivative whose value depends only on $S_1, S_2, \dots, S_n$, and t and that follows

$$
\frac{df}{f} = \mu\, dt + \sum_i \sigma_i\, dW_i \tag{33.16}
$$

must satisfy

$$
\mu - r = \sum_i \lambda_i \sigma_i, \tag{33.17}
$$

where λ_i is the market price of risk for S_i. Equation (33.17) links the excess expected return and risk. The term $\lambda_i \sigma_i$ measures the extent to which the required return on a security is affected by its dependence on S_i. From Eq. (33.13), μ and σ_i in Eq. (33.16) are

$$
\mu = \frac{\partial f}{\partial t} + \sum_i \mu_i S_i\, \frac{\partial f}{\partial S_i} + \frac{1}{2} \sum_{i,k} \rho_{ik} \sigma_i \sigma_k S_i S_k\, \frac{\partial^2 f}{\partial S_i\, \partial S_k},
$$

$$
\sigma_i = \sum_i \sigma_i S_i\, \frac{\partial f}{\partial S_i}.
$$

Plugging them into Eq. (33.17) and rearranging the terms, we obtain Eq. (15.14). Risk-neutral valuation discounts the expected payoff of f at the riskless interest rate assuming that $dS_i/S_i = (\mu_i - \lambda_i \sigma_i)\, dt + \sigma_i\, dW_i$. The correlation between the dW_is is unchanged.

Exercise 15.4.2: (1) Just note that $S = X$. (2) Use the risk-neutral argument. See [894, p. 186].

Chapter 16

Exercise 16.2.1: It reduces risk (standard deviation of returns) without paying a risk premium [3, p. 38].

Exercise 16.2.2: Because the formula for $\widehat{\beta}_1$ in formula (6.14) is exactly the hedge ratio $h = \rho \delta_S / \delta_F$, resulting from plugging in Eq. (6.2) for δ_S and δ_F and plugging in Eq. (6.18) for ρ.

Exercise 16.2.3: By formula (16.2) we short $(1.25 - 2.0) \times [\,2,400,000/(600 \times 500)\,] = -6$ futures contracts, or, equivalently, long six futures contracts.

Exercise 16.3.1: (1) In the current setup, the stock price forms a continuum instead of only two states as in the binomial model. (2) This happens, for example, when their respective values are related linearly. See [119].

Exercise 16.3.3: Because of the convexity of option values, the value of the hedge, being a linear function of the stock price, loses money whichever way the price moves; one is buying stock when the stock price rises and selling it when its price falls. So the premium can be seen as the cash reserve to offset future hedging losses in order to maintain a self-financing strategy. See [593].

Alternatively, we can check that the equivalent portfolio at any time, Δ shares of stock plus B dollars of bonds, is *not* self-financing unless B exceeds the stock value by exactly the option premium. This means that we borrow money to buy the stock but have to initially invest our own money in the amount equal to the option premium.

Exercise 16.3.5: From Eq. (15.3), it is clear that $\Theta = 0$ when $\Delta = \Gamma = f = 0$.

Exercise 16.3.8: If the two strike prices used in the bull call spread are relatively close to each other, the payoff of the position will be approximately that of a binary option [514, p. 604].

CHAPTER 17

Exercise 17.1.1: Use the reflection principle [725, p. 106].

Exercise 17.1.3: Just note that for each terminal price, the number of paths that have the minimum price level M equals the number of paths that hit M minus the number of paths that hit $M-1$. Both numbers can be obtained by the reflection principle.

Exercise 17.1.4: Set

$$\kappa \equiv \left\lfloor \frac{\ln\left(K/(Sd^n)\right)}{\ln(u/d)} \right\rfloor = \left\lfloor \frac{\ln(K/S)}{2\sigma\sqrt{\Delta t}} + \frac{n}{2} \right\rfloor.$$

It is easy to see that $\check{K} \equiv Su^\kappa d^{n-\kappa}$ is the price among $Su^j d^{n-j}$ $(0 \le j \le n)$ closest to, but not exceeding, K. The role of the trigger price is played by the effective trigger price $\check{K}$ in the binomial model. Assume that the trigger price exceeds the current stock price, i.e., $2\kappa \ge n$.

Take any node A that is reachable from the root in l moves and with a price level equal to the trigger price, i.e., $Su^j d^{l-j} = \check{K}$ for some $0 \le j \le l$. The PV of the payoff is $R^{-l}(\check{K} - X)$. What we need to calculate is the probability that the price barrier $\check{K}$ has never been touched until now. Consider the node B that reaches A by way of two up moves. B's price level is $Su^{j-2} d^{l-j}$. Observe that the probability a path of length $l-2$ that reaches node B without touching the barrier, $\check{K} = Su^{j-1} d^{l-2-(j-1)}$, is precisely what we are after. However, this equivalent problem is the *complement* of the ballot problem! Write B's price level as $Su^{j-2} d^{(l-2)-(j-2)}$. The desired probability can be obtained from formula (17.5) as

$$\binom{l-2}{(l-2) - 2(j-1) + (j-2)} p^{j-2}(1-p)^{(l-2)-(j-2)} = \binom{l-2}{l-j-2} p^{j-2}(1-p)^{l-j}.$$

The nature of the process dictates that $n - l$ be even. By Eq. (17.4), we have $l \ge 2\kappa - n$. Furthermore, because $Su^j d^{l-j} = Su^\kappa d^{n-\kappa}$, we have $j = (l-n)/2 + \kappa$. The preceding probability therefore becomes

$$\binom{l-2}{(l+n)/2-\kappa-2} p^{(l-n)/2+\kappa-2}(1-p)^{(l+n)/2-\kappa}.$$

We conclude that the payoff from early exercise is

$$\sum_{\substack{2\kappa-n\le l\le n \\ n-j \text{ is even}}} R^{-l}\binom{l-2}{(l+n)/2-\kappa-2} p^{(l-n)/2+\kappa-2}(1-p)^{(l+n)/2-\kappa}(\check{K}-X).$$

The remaining part is to add the payoff from exercise at maturity, which is just the down-and-out option with the barrier set to the trigger price.

Exercise 17.1.6: First verify that

$$\left[\frac{up}{d(1-p)}\right]^{2h-n} \to \left(\frac{H}{S}\right)^{2\lambda}, \tag{33.18}$$

where $\lambda \equiv (r+\sigma^2/2)/\sigma^2$, by plugging in the formulas for u, d, and p from Eq. (17.1) and using $\ln(1+x)\approx x$. Now,

$$(17.6) = S\sum_{j=a}^{2h}\binom{n}{n-2h+j}\left(\frac{up}{R}\right)^j\left[\frac{d(1-p)}{R}\right]^{n-j} - XR^{-n}\sum_{j=a}^{2h}\binom{n}{n-2h+j}p^j(1-p)^{n-j}$$

$$= S\left[\frac{up}{d(1-p)}\right]^{2h-n}\sum_{j=a}^{2h}\binom{n}{n-2h+j}\left(\frac{up}{R}\right)^{n-2h+j}\left[\frac{d(1-p)}{R}\right]^{2h-j} - X\times(\cdots).$$

We analyze the first term; the second term can be handled analogously. Note that $h\le n/2$ because $H<S$. Thus,

$$\sum_{j=a}^{2h}\binom{n}{n-2h+j}\left(\frac{up}{R}\right)^{n-2h+j}\left[\frac{d(1-p)}{R}\right]^{2h-j} = \sum_{j=n-2h+a}^{n}\binom{n}{j}\left(\frac{up}{R}\right)^j\left[\frac{d(1-p)}{R}\right]^{n-j}$$

$$= \Phi(n-2h+a;n,\,pu/R).$$

It is straightforward to check that $n-2h+a\approx\ln(\frac{SX}{H^2d^n})/\ln(u/d)$. Finally, apply convergence (9.18) to obtain $\Phi(n-2h+a;n,\,pu/R)\to N(x)$, where $x\equiv\frac{\ln(H^2/(SX))+(r+\sigma^2/2)\tau}{\sigma\sqrt{\tau}}$. With convergence (33.18), we have proved the validity of Eq. (11.4) for the no-dividend-yield case.

Exercise 17.1.7: From Lemma 9.3.3, $\ln S_\tau \sim N(\ln S+(r-q-\sigma^2/2)\tau,\,\sigma^2\tau)$ in a risk-neutral economy, where q is the continuous dividend yield. Hence, $\ln S_\tau^2 \sim N(2\ln S+(2r-2q-\sigma^2)\tau,\,2\sigma^2\tau)$. The desired formula thus equals the Black–Scholes formula after the following substitutions: $S\to S^2$, $r\to 2r+\sigma^2$, $q\to 2q$, and $\sigma\to 2\sigma$. See [845].

Programming Assignment 17.1.8: See Programming Assignment 17.3 and [624].

Programming Assignment 17.1.9: We use the current stock price and the geometric average of past stock prices as the state information each node keeps. Based on Exercise 11.7.2, the number of states at time i is approximately $i^3/2$, and backward induction over n periods gives us a running time of $O(n^4)$. Consult [249] for the subtle issue of data structures.

To reduce the running time to $O(n^3)$, we prove that the number of paths of length n having the same geometric average is precisely the number of unordered partitions of some integer into unequal parts, none of which exceeds n. Let $q(m)$ denote the number of such partitions of integer $m>0$. Any legitimate partition of m, say $\lambda\equiv(x_1,x_2,\ldots,x_k)$, satisfies $\sum_i x_i=m$, where $n\ge x_1>x_2>\cdots>x_k>0$. Interpret λ as the path of length n that makes the first up move at step $n-x_1$ (i.e., during period $n-x_1+1$), the second up move at step $n-x_2$, and so on. Each up move at step $n-x_i$ contributes x_i to the sum m. This path has a terminal geometric average of $SM^{1/(n+1)}$, where $M\equiv u^m d^{n(n+1)/2-m}$, in which the ith up move contributes u^{x_i} to the u^m term. For completeness, let $q(0)=1$. (See [26] for more information on integer partitions.) The crucial step is to observe that

$$(1+x)(1+x^2)(1+x^3)\cdots(1+x^n) = \sum_{m=0}^{n(n+1)/2} q(m)\,x^m.$$

The above polynomial clearly can be expanded in time $O(n^3)$. Thus the $q(m)$s can be computed in cubic time as well. With the parameters from Exercise 9.3.1, the option value is

$$R^{-n} \sum_{m=0}^{n(n+1)/2} 2^{-n} q(m) \times \max\ (S[u^m d^{n(n+1)/2-m}]^{1/(n+1)} - X, 0).$$

Note that this particular parameterization makes calculating the probability of reaching any given terminal geometric average easy: $2^{-n} q(m)$. The standard parameters $u = e^{\sigma \sqrt{\tau/n}}$ and $d = 1/u$ would have led to complications because of the inequality between the probabilities of up and down moves.

Programming Assignment 17.1.10: (1) Let $H_i \equiv Sd^{n-i}$, $i = 0, 1, \ldots, n$, and $H_{-1} \equiv -\infty$. A lookback option on the minimum on an n-period binomial tree is equivalent to a portfolio of $n+1$ barrier-type options, the ith of which is an option that knocks in at H_i but knocks out at H_{i-1}. Because this option is basically long a knock-in option with barrier H_i and short a knock-out option with barrier H_{i-1}, it can be priced in $O(n)$ time. With $n+1$ such options the total running time is $O(n^2)$. See Exercise 17.1.3 and [285]. (2) For the barrier options mentioned in (1), successive pairs are related: Each can be priced from the prior one in $O(1)$ time.

Exercise 17.1.12: This holds because h is the result of taking the floor operation; the opposite would hold if h were the result of taking the ceiling. The influence of a is negligible in comparison.

Exercise 17.1.13: Consider the j that makes $H' = Su^j d^{n-1-j}$ the largest such number not exceeding H. Call this number h'. Use $2h + 1$ in place of $2h$ if $Su^{h'} d^{n-1-h'} > Su^h d^{n-h}$, as this effective barrier H' is "tighter" (closer to H) than $Su^h d^{n-h}$. An equivalent procedure is to test if $Su^h d^{n-1-h} \le H$ and, if true, replace $2h$ with $2h + 1$. The effect on the choice of n is that of simplification:

$$n = \left\lfloor \frac{\tau}{[\ln(S/H)/(j\sigma)]^2} \right\rfloor, \quad j = 1, 2, 3, \ldots.$$

Exercise 17.1.15: We prove formula (17.8) with reference to Fig. 33.3. Consider any legitimate path from $(0, -a)$ to $(n, -b)$ that hits H. Let J denote the first position at which this happens. (The path may have hit the L line earlier.) By reflecting the portion of the path from $(0, -a)$ to J, a path from $(0, a)$ to $(n, -b)$ is constructed. Note that this path hits H at J. The number of paths from $(0, -a)$ to $(n, -b)$ in which an L hit is preceded by an H hit is exactly the number of paths from $(0, a)$ to $(n, -b)$ that hits L. The desired number is as claimed by application of the reflection principle.

Exercise 17.1.16: See [675, p. 7]. Note that $|B_i|$ can be derived from $|A_i|$ if a is replaced with $s - a$ and b with $s - b$.

Exercise 17.1.17: Apply the inclusion–exclusion principle [604].

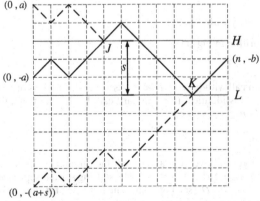

Figure 33.3: Repeated applications of the reflection principle. The random walk from $(0, -\boldsymbol{a})$ to $(n, -\boldsymbol{b})$ must hit either barrier, and there must exist an L hit preceded by an H hit. In counting the number of such walks, reflect the path first at J and then at K.

Trinomial backward induction with the diagonal method:

input: S, σ, τ, X, r, n;
real $p_u, p_m, p_d, u, d, C[n+1][2n+1]$;
integer i, j;
Calculate u and d;
Calculate the branching probabilities $p_u, p_m,$ and p_d;
for $(i = 2n$ down to $0)$
        for $(j = n$ down to $\lceil i/2 \rceil)$
            if $[j = n]$ // Expiration date.
                Calculate $C[j][i]$ based on Su^{j-i};
            else { // Backward induction.
                Calculate $C[j][i]$ based on $C[j+1][i]$,
                    $C[j+1][i+1], C[j+1][i+2]$,
                    $p_u, p_m, p_d,$ and Su^{j-i} (with
                    discounting factor $e^{-r\tau/n}$);
        }
return $C[0][0]$;

Figure 33.4: Trinomial backward induction with the diagonal method. $C[j][i]$ represents the derivative value at time j if the stock price makes $j - i$ more up moves than down moves; it corresponds to the node with the $(i + 1)$th largest underlying asset price. Recall that $ud = 1$. The space requirement can be further reduced by using a 1-dimensional array for C.

Exercise 17.1.18: (1) A standard European call is equivalent to a double-barrier option and a double-barrier option that knocks out if neither barrier is hit. (2) Construct a portfolio of long a down-and-in option, long an up-and-in option, and short the double-barrier option. (3) It equals $\sum_{i=2}^{n}(-1)^i$ $(|A_i| + |B_i|)$.

Exercise 17.1.19: Consider the process $Y(t) \equiv [X(t) - f_\ell(t)] / [f_h(t) - f_\ell(t)]$, where $X(t) \equiv \ln S(t)$. The payoff of the option at maturity becomes $\max(e^{[f_h(T) - f_\ell(T)] Y(T) + f_\ell(T)} - X, 0)$ with barriers at 0 and 1. See [757].

Programming Assignment 17.1.20: Calculate $\binom{n}{0}, \binom{n}{1}, \binom{n}{2}, \ldots, \binom{n}{n}$ and store them for random access when needed by the $|A_i|$s and the $|B_i|$s. Note that the calculation of $N(a, b, s)$ takes time $O(n/s)$ because $|A_i| = |B_i| = 0$ for $i > n/s$. The implication is that expression (17.10) only has $(h - a)\frac{n}{2(h-l)} < n/2$ nonzero terms because $l < a$.

Exerciset 17.2.2: Consider one period to maturity with $S = X$.

Programming Assignment 17.2.3: See the algorithm in Fig. 33.4.

Exercise 17.2.4: See [745, p. 21].

Exercise 17.2.5: We focus on the down-and-in call with barrier $H < X$. Assume without loss of generality that $H < S$ for, otherwise, the option is identical to a standard call. Under the trinomial model, there are $2n + 1$ stock prices Su^j $(-n \le j \le n)$. Let

$$a \equiv \left\lceil \frac{\ln(X/S)}{\lambda \sigma \sqrt{\Delta t}} \right\rceil, \qquad h \equiv \frac{\ln(S/H)}{\lambda \sigma \sqrt{\Delta t}} > 0.$$

A process with n moves ends up at a price at or above X if and only if the number of up moves exceeds that of down moves by at least a because $Su^{a-1} < X \le Su^a$. The starting price is separated from the barrier by h down moves because $Su^{-h} = H$ (see Fig. 33.5). The following formula is the pricing formula for European down-and-in calls:

$$R^{-n} \sum_{m=0}^{n-2h-a} \sum_{j \ge \max(a, m-n)}^{n-m-2h} \frac{n!}{[(n-m+j+2h)/2]! \, m! \, [(n-m-j-2h)/2]!}$$

$$p_u^{(n-m+j)/2} p_m^m p_d^{(n-m-j)/2} (Su^j - X). \tag{33.19}$$

This is an alternative characterization of the trinomial tree algorithm for down-and-in calls.

Figure 33.5: Down-and-in calls under trinomial tree. Note that the interpretations of a, j, and h differ from those in Fig. 17.2.

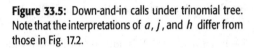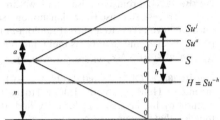

We use the reflection principle for trinomial random walks to prove the correctness of pricing formula (33.19). Consider paths that hit the barrier and end up at Su^j with m level moves. Any such path must contain, cumulatively, $(n-m+j)/2$ up moves, m level moves, and $(n-m-j)/2$ down moves with probability $p_u^{(n-m+j)/2} p_m^m p_d^{(n-m-j)/2}$. The reflection principle tells us that their number equals the number of paths from $(0, -h)$ to $(n, j+h)$, which is

$$\frac{n!}{[(n-m+j+2h)/2]!\, m!\, [(n-m-j-2h)/2]!}$$

because the number of up moves is $(n-m+j+2h)/2$, the number of down moves is $(n-m-j-2h)/2$, and the number of level moves is m.

The correctness of the bounds on m (the number of level moves) and j (the number of up moves minus the number of down moves) in pricing formula (33.19) can be verified as follows. The former cannot exceed $n-2h-a$ as it takes $2h+a$ nonlevel moves to hit the barrier and end up at a price of at least X. That $(n-m+j+2h)/2$, $(n-m-j-2h)/2$, $(n-m+j)/2$, and $(n-m-j)/2$ are integers means that $n-m+j$ must be even. That j must be at least a is obvious. The bound $j \geq m-n$ is necessitated by $(n-m+j)/2 \geq 0$. The bound $j \leq n-m-2h$ is needed because $(n-m-j-2h)/2 \geq 0$. Finally, we can easily check that all the terms are now nonnegative integers within their respective bounds.

Exercise 17.3.1: The number of nodes at time n is $1+4+8+12+16+\cdots+4n$.

Exercise 17.3.2: The covariance between R_1 and R_2, or $E[R_1 R_2] - \mu_1 \mu_2$, equals

$$\sigma_1 \sigma_2 (p_1 - p_2 - p_3 + p_4) = \sigma_1 \sigma_2 [p_1 - (1 - p_1 - p_4) + p_4] = \sigma_1 \sigma_2 (2p_1 + 2p_4 - 1).$$

Hence their correlation is $2p_1 + 2p_4 - 1$, as claimed.

Exercise 17.3.5: We use asset 1 to denote the stock S and asset 2 to denote the futures F. Let h denote the hedge ratio and the stock price has a correlation of ρ with the derivative's price. The variance at time Δt is $\delta_S^2 + h^2 \delta_F^2 - 2h\rho \delta_S \delta_F$, where δ_S denotes the standard deviation of the stock price and δ_F the standard deviation of the futures price at time Δt. The optimal hedge ratio is $h = \rho(\delta_S/\delta_F) = E[(S - S\mu_1)(F - F\mu_2)]/\delta_F^2$ as shown in Eq. (16.1), where μ_1 and μ_2 are the expected gross returns of the respective assets at Δt. The formula for h is

$$\frac{(Fu_2 - F\mu_2)[\, p_1(Su_1 - S\mu_1) + p_3(Sd_1 - S\mu_1)\,] + (Fd_2 - F\mu_2)[\, p_2(Su_1 - S\mu_1) + p_4(Sd_1 - S\mu_1)\,]}{(p_1 + p_3)(Fu_2 - F\mu_2)^2 + (p_2 + p_4)(Fd_2 - F\mu_2)^2}.$$

Of course, there is no stopping at stocks and futures. For instance, we can use index futures to hedge equity options dynamically, and the same formula applies!

Chapter 18

Exercise 18.1.1: See [706, p. 135].

Exercise 18.1.2: $U_{i,T} = \max(0, X - e^{V_{\min} + i \times \Delta V})$.

Exercise 18.1.4: After discretization, $U_{i,j-1} = a U_{i+1,j} + b U_{i,j} + c U_{i-1,j}$, where

$$a \equiv \left[\left(\frac{\sigma S_i}{\Delta S}\right)^2 + \frac{r S_i}{\Delta S}\right]\frac{1}{2\Delta t}, \qquad b \equiv \left[-r + \frac{1}{\Delta t} - \left(\frac{\sigma S_i}{\Delta S}\right)^2\right]\Delta t, \qquad c \equiv \left[\left(\frac{\sigma S_i}{\Delta S}\right)^2 - \frac{r S_i}{\Delta S}\right]\frac{1}{2\Delta t},$$

and S_i is the stock price for $U_{i,j}$. Two conditions must be met: (1) $1/\Delta t > (\sigma S_i/\Delta S)^2 + r$ to ensure that $b > 0$ and (2) $\sigma^2 S_i/r > \Delta S$ to ensure that $a > 0$ and $c > 0$. See [897].

Exercise 18.2.2: The node with the maximum discounted intrinsic value may not be exercised [129].

Exercise 18.2.4: From Exercise 14.3.1 we know that $d(\ln X) = (\mu - \sigma^2/2)\,dt + \sigma\,dW$, a Brownian motion process. Because Brownian motion's sample paths can be generated without loss of accuracy once we have access to a perfect random-number generator, a better algorithm generates sample paths for $\ln X$ and then turns them into ones for X by taking exponentiation. Specifically, $X_i = X_{i-1} e^{(\mu - \sigma^2/2)\,\Delta t + \sigma\sqrt{\Delta t}\,\xi}$, where $\xi \sim N(0, 1)$.

Exercise 18.2.7: Observe that $\text{Prob}[\, X < -y\,] = 1 - \text{Prob}[\, X < y\,]$ if X is normally distributed with mean zero. Hence the method alluded to does indeed fall into the category of the antithetic-variates method by the inverse-transform method. It also works for any random variable with a density function that is symmetric around zero.

Programming Assignment 18.2.9: Generalize the model for stochastic volatility in Section 15.5 as follows. In a risk-neutral economy, $dS = rS\,dt + \sigma S(\rho\,dW_v + \sqrt{1 - \rho^2}\,dW_s)$ and $d\sigma^2 = \mu\,dt + \sigma_v\,dW_v$, where $\mu(t)$ and $\sigma_v(t)$ may depend on the history of $\sigma^2(t)$, and $W_v(t)$ and $W_s(t)$ are uncorrelated. Now,

$$
\begin{aligned}
S_T &= S_0 \times \exp\left[rT - \frac{1}{2}\int_0^T \sigma^2\,du + \rho\int_0^T \sigma\,dW_v + \sqrt{1 - \rho^2}\int_0^T \sigma\,dW_s \right] \\
&= S_0 \times \exp\left[-\frac{\rho^2}{2}\int_0^T \sigma^2\,du + \rho\int_0^T \sigma\,dW_v \right] \\
&\quad \times \exp\left[rT - \frac{1 - \rho^2}{2}\int_0^T \sigma^2\,du + \sqrt{1 - \rho^2}\int_0^T \sigma\,dW_s \right].
\end{aligned}
$$

If the path of W_v is known, then $\xi \equiv \exp[\, -\frac{\rho^2}{2}\int_0^T \sigma^2\,du + \rho\int_0^T \sigma\,dW_v\,]$ is a constant and $\ln S_T$ is normally distributed with mean $\ln(S_0\xi) + rT - \frac{1 - \rho^2}{2}\int_0^T \sigma^2\,du$ and variance $(1 - \rho^2)\int_0^T \sigma^2\,du$, in which case the Black–Scholes formula applies. The algorithmic idea is now clear. Simulate W_v to obtain a path of σ values. Use the analytic solution to obtain the option value conditional on that path. Repeat a few times and average. There is no need to simulate W_s. See [876].

Exercise 18.2.10: They are $\text{Var}[\,Y\,] < 2\,\text{Cov}[\,X, Y\,]$ for $\beta = -1$ and $\text{Var}[\,Y\,] < -2\,\text{Cov}[\,X, Y\,]$ for $\beta = 1$. Both are more stringent than relation (18.6). See [584].

Exercise 18.2.11: This scheme will make $\text{Cov}[\,X, Y\,] = 0$ and thus $\text{Var}[\,W\,] > \text{Var}[\,X\,]$.

Programming Assignment 18.2.12: See the algorithm in Fig. 33.6.

Control variates for pricing average-rate calls on a non-dividend-paying stock:

```
input:    S, X, n, r, σ, τ, m;
real      P, C, M₁, M₂;
real      ξ();   //ξ() ~ N(0, 1).
integer   i, j;
C := 0;
for (i = 1 to m) {
        P := S; M₁ := S; M₂ := S^(1/(n+1));
        for (j = 1 to n) {
                P := Pe^((r−σ²/2)(τ/n)+σ√(τ/n) ξ());
                M₁ := M₁ + P;
                M₂ := M₂ × P^(1/(n+1));
        }
        C := C + e^(−rτ) × max(M₁/(n+1) − X, 0)−
                (e^(−rτ) × max(M₂ − X, 0) − analytic value);
}
return C/m;
```

Figure 33.6: Control-variates method for arithmetic average-rate calls. m is the number of replications, n is the number of periods, M_1 is the arithmetic average, and M_2 is the geometric average. The `analytic value` is computed by Eq. (11.8) for the geometric average-rate option. In practice, the option value under the binomial model may be preferred as the `analytic value` (but it takes more time, however, to calculate). Note carefully that the expected value of the control variate is not exactly M_2.

Exercise 18.2.13: Let $\epsilon \equiv |B|/|A|$. The Monte Carlo approach's probability of failure is clearly $(1 - \epsilon)^N$. The refined search scheme, on the other hand, has a probability of failure equal to $\prod_i (1 - \epsilon_i)^{N/m}$, where ϵ_i is the proportion of A_i that intersects B, in other words, $\epsilon_i \equiv |B \cap A_i|/|A_i|$. Clearly, $\sum_i \epsilon_i = m\epsilon$. Now, $[\prod_i (1 - \epsilon_i)]^{N/m} \le (\sum_i \frac{1-\epsilon_i}{m})^N = (1 - \epsilon)^N$, where the inequality is by the relation between the arithmetic average and the geometric average.

Exercise 18.3.1: They are $\frac{1}{3}, \frac{2}{3}, \frac{1}{9}, \frac{4}{9}, \frac{7}{9}, \frac{2}{9}, \frac{5}{9}, \frac{8}{9}, \frac{1}{27}, \frac{10}{27}$.

Chapter 19

Exercise 19.1.1: This follows from the definition and $c_{ii} = \sigma_i^2$, the variance of x_i.

Exercise 19.1.2: Consider any normalized linear combination $\sum_i b_i x_i = b^{\mathsf{T}} x$, where $\sum_i b_i^2 = 1$. Let $b = Be$ for some unit-length vector $e \equiv [e_1, e_2, \ldots, e_n]^{\mathsf{T}}$. Then $b^{\mathsf{T}} x = e^{\mathsf{T}} B^{\mathsf{T}} x = e^{\mathsf{T}} P$. Finally,

$$\text{Var}[e^{\mathsf{T}} P] = E[(e^{\mathsf{T}} P)^2] = \sum_{i=1}^n e_i^2 E[p_i^2] \quad \text{from the uncorrelatedness of the } p_i\text{s}$$

$$= \lambda_1 + \sum_{i=2}^n e_i^2 (\lambda_i - \lambda_1) \quad \text{because } E[p_i^2] = \lambda_i \text{ and } e_1^2 = 1 - \sum_{i=2}^n e_i^2.$$

The preceding equation achieves the maximum at λ_1 when $e_2 = e_3 = \cdots = e_n = 0$, that is, when $e^{\mathsf{T}} P = p_1$. In general, p_j is the normalized linear combination of the x_is, which is uncorrelated with $p_1, p_2, \ldots, p_{j-1}$ and has the maximum variance. This can be verified as follows. Uncorrelatedness implies that $e_1 = e_2 = \cdots = e_{j-1} = 0$. Hence

$$\text{Var}[e^{\mathsf{T}} P] = \sum_{i=j}^n e_i^2 E[p_i^2] = \lambda_j + \sum_{i=j+1}^n e_i^2 (\lambda_i - \lambda_j).$$

The preceding equation achieves the maximum at λ_j when $e_{j+1} = e_{j+2} = \cdots = e_n = 0$, that is, when $e^{\mathsf{T}} P = p_j$.

Exercise 19.1.3: Note that

$$\begin{bmatrix} 1 & \rho \\ \rho & 1 \end{bmatrix} = \begin{bmatrix} 1 & 0 \\ \rho & \sqrt{1-\rho^2} \end{bmatrix} \begin{bmatrix} 1 & \rho \\ 0 & \sqrt{1-\rho^2} \end{bmatrix}.$$

Exercise 19.1.4: See Exercise 6.1.3.

Exercise 19.1.5: (1) Let $B \equiv \sum_{i=1}^p \sigma_i u_i v_i^{\mathsf{T}}$. It is sufficient to prove that $Av_i = Bv_i$ for $i = 1, 2, \ldots, n$ because $\{v_1, v_2, \ldots, v_n\}$ forms a basis of R^n. Now $Av_i = \sigma_i u_i$ and $Bv_i = \sum_j \sigma_j u_j (v_j^{\mathsf{T}} v_i) = \sigma_i u_i$. See [870, p. 395]. (2) It is because $Av_i = \sigma_i u_i$ and $A^{\mathsf{T}} u_i = \sigma_i v_i$. See [392, p. 258], or [870, p. 392].

Exercise 19.2.1: From Eq. (19.5), we have the following LS problem for linear regression:

$$\begin{bmatrix} 1 & x_1 \\ 1 & x_2 \\ \vdots & \vdots \\ 1 & x_m \end{bmatrix} \begin{bmatrix} \beta_0 \\ \beta_1 \end{bmatrix} = \begin{bmatrix} y_1 \\ y_2 \\ \vdots \\ y_m \end{bmatrix}.$$

The normal equations are then

$$\begin{bmatrix} m & \sum_i x_i \\ \sum_i x_i & \sum_i x_i^2 \end{bmatrix} \begin{bmatrix} \beta_0 \\ \beta_1 \end{bmatrix} = \begin{bmatrix} \sum_i y_i \\ \sum_i x_i y_i \end{bmatrix}.$$

See [381, p. 263].

Exercise 19.2.2: (1) Let the data be

$$\{ (x_{1,1}, x_{1,2}, \ldots, x_{1,n}, y_1), (x_{2,1}, x_{2,2}, \ldots, x_{2,n}, y_2), \ldots, (x_{m,1}, x_{m,2}, \ldots, x_{m,n}, y_m) \}.$$

We desire to fit the data to the linear model $\beta_0 + \beta_1 x_1 + \beta_2 x_2 + \cdots + \beta_n x_n$. The LS formulation is

$$\begin{bmatrix} 1 & x_{1,1} & x_{1,2} & \cdots & x_{1,n} \\ 1 & x_{2,1} & x_{2,2} & \cdots & x_{2,n} \\ \vdots & \vdots & \vdots & \ddots & \vdots \\ 1 & x_{m,1} & x_{m,2} & \cdots & x_{m,n} \end{bmatrix} \begin{bmatrix} \beta_0 \\ \beta_1 \\ \vdots \\ \beta_n \end{bmatrix} = \begin{bmatrix} y_1 \\ y_2 \\ \vdots \\ y_m \end{bmatrix}.$$

(2) It is

$$
\begin{bmatrix}
m & \sum_{i=1}^{m} x_{i,1} & \sum_{i=1}^{m} x_{i,2} & \cdots & \sum_{i=1}^{m} x_{i,n} \\
\sum_{i=1}^{m} x_{i,1} & \sum_{i=1}^{m} x_{i,1}x_{i,1} & \sum_{i=1}^{m} x_{i,1}x_{i,2} & \cdots & \sum_{i=1}^{m} x_{i,1}x_{i,n} \\
\vdots & \vdots & \vdots & \ddots & \vdots \\
\sum_{i=1}^{m} x_{i,n} & \sum_{i=1}^{m} x_{i,n}x_{i,1} & \sum_{i=1}^{m} x_{i,n}x_{i,2} & \cdots & \sum_{i=1}^{m} x_{i,n}x_{i,n}
\end{bmatrix}
\begin{bmatrix} \beta_0 \\ \beta_1 \\ \vdots \\ \beta_n \end{bmatrix}
=
\begin{bmatrix} \sum_{i=1}^{m} y_i \\ \sum_{i=1}^{m} x_{i,1}y_i \\ \vdots \\ \sum_{i=1}^{m} x_{i,n}y_i \end{bmatrix}.
$$

Exercise 19.2.3: See [392, p. 223].

Exercise 19.2.4: Let $A \equiv [a_{ij}] \in R^{m \times n}$. Then $a_{ij}a_{ik} \neq 0$ only if $|j - k| < \omega$. Hence, the (j, k)th entry of $A^{\mathrm{T}} A$, which is $\sum_{i=1}^{m} a_{ij}a_{ik}$, equals 0 for $|j - k| \geq \omega$. See [75, p. 217].

Exercise 19.2.5: Because $A^{\mathrm{T}}(Ax_{\mathrm{LS}} - b) = 0$ [415, p. 201].

Exercise 19.2.6: Let $C = PP^{\mathrm{T}}$ be the Cholesky decomposition, where P is nonsingular. With $Py = b$ and $P\epsilon' = \epsilon$, we have transformed the problem into $Py = Ax + P\epsilon'$ or, equivalently, $y = P^{-1}Ax + \epsilon'$. We claim that $y = P^{-1}Ax$ is the LS problem to solve; in other words, we regress $P^{-1}b$ on $P^{-1}A$. This can be verified by

$$\mathrm{Cov}[\epsilon'] = \mathrm{Cov}[P^{-1}\epsilon] = P^{-1}\mathrm{Cov}[\epsilon](P^{\mathrm{T}})^{-1} = P^{-1}\sigma^2 C (P^{\mathrm{T}})^{-1} = \sigma^2 C.$$

See Exercise 6.1.3 and [802, p. 54].

Exercise 19.2.7: The covariance matrix of x_{LS} is

$$
\begin{aligned}
(A^{\mathrm{T}} A)^{-1} A^{\mathrm{T}} E[bb^{\mathrm{T}}] A[(A^{\mathrm{T}} A)^{-1}]^{\mathrm{T}} &= \sigma^2 (A^{\mathrm{T}} A)^{-1} A^{\mathrm{T}} A[(A^{\mathrm{T}} A)^{-1}]^{\mathrm{T}} \\
&= \sigma^2 (A^{\mathrm{T}} A)^{-1} [A^{\mathrm{T}} A (A^{\mathrm{T}} A)^{-1}]^{\mathrm{T}} \\
&= \sigma^2 (A^{\mathrm{T}} A)^{-1}.
\end{aligned}
$$

Exercise 19.2.8: Because both U and V are orthogonal, we have $U^{-1} = U^{\mathrm{T}}$ and $V^{-1} = V^{\mathrm{T}}$. Expand $(A^{\mathrm{T}} A)^{-1} A^{\mathrm{T}}$ and use the property $\Sigma^+ = \Sigma^{-1}$ to obtain the desired result.

Exercise 19.2.9: Just do the transformation $A = U\Sigma V^{\mathrm{T}} \to A^+ = V\Sigma^+ U^{\mathrm{T}}$ twice.

Exercise 19.2.10: This can be verified informally as follows. To begin with,

$$A\widehat{x} = AA^+ b + AV_2 y = U\Sigma V^{\mathrm{T}} V\Sigma^+ U^{\mathrm{T}} b + U\Sigma V^{\mathrm{T}} V_2 y = b + U\mathbf{0}y = b.$$

We complete the proof by noting that our solution set has dimension $n - m$. See [586].

Exercise 19.2.11: By Exercise 19.2.6 we should solve $y = P^{-1}Ax$ for x, where $C = PP^{\mathrm{T}}$ and $Py = b$. By normal equations (19.6), the solution must satisfy

$$(P^{-1}A)^{\mathrm{T}}(P^{-1}A)x = (P^{-1}A)^{\mathrm{T}} y = (P^{-1}A)^{\mathrm{T}} P^{-1}b.$$

After expansion, the preceding identity becomes $A^{\mathrm{T}}(P^{-1})^{\mathrm{T}} P^{-1} Ax = A^{\mathrm{T}}(P^{-1})^{\mathrm{T}} P^{-1}b$, which implies that $A^{\mathrm{T}} C^{-1} Ax = A^{\mathrm{T}} C^{-1}b$, as claimed.

Exercise 19.2.12: Let $B = U\Sigma V^{\mathrm{T}}$ be the SVD of B and $V \equiv [\underbrace{V_1}_{p}, \underbrace{V_2}_{n-p}]\}n$. The solution to the constrained problem is [586]

$$B^+ d + V_2(AV_2)^+ (b - AB^+ d). \tag{33.20}$$

Solution (33.20) clearly satisfies the constraint $Bx = d$ because all solutions to $Bx = d$ are of the form $B^+ d + V_2 y$ for arbitrary $y \in R^{n-p}$ by virtue of Eq. (19.10). An algorithm implementing solution (33.20) is given in Fig. 33.7. We verify its correctness as follows. After step 1, $U^{\mathrm{T}} BV = \Sigma$. Observe that $B^+ BV_1 = V_1$ even if $B^+ B \neq I$. This holds because, by pseudoinverse (19.9),

$$
\begin{aligned}
B^+ BV_1 &= V\Sigma^+ U^{\mathrm{T}} U\Sigma V^{\mathrm{T}} V_1 \\
&= V\Sigma^+ \Sigma V^{\mathrm{T}} V_1 \\
&= V\Sigma^+ \times \mathrm{diag}_{p \times p}[\sigma_1, \sigma_2, \ldots, \sigma_p] \\
&= V \times \mathrm{diag}_{n \times p}[\sigma_1^{-1}, \sigma_2^{-1}, \ldots, \sigma_p^{-1}] \times \mathrm{diag}_{p \times p}[\sigma_1, \sigma_2, \ldots, \sigma_p] \\
&= V \times \mathrm{diag}_{n \times p}[1, 1, \ldots, 1] \\
&= V_1.
\end{aligned}
$$

Therefore, after step 3, $B^+ d = B^+ BV_1 x_1 = V_1 x_1$. Step 4 makes $\widehat{b} = b - AV_1 x_1 = b - AB^+ d$. After step 5, $x_2 = (AV_2)^+ \widehat{b} = (AV_2)^+ (b - AB^+ d)$. Finally the returned value is $V_1 x_1 + V_2 x_2 = B^+ d + V_2(AV_2)^+ (b - AB^+ d)$, matching solution (33.20).

Exercise 19.2.13: See [586, p. 139].

Algorithm for the LS problem with linear equality constraints:

input: $A[m][n], b[m], B[p][n], d[p], m, n, p\,(p \le n)$;
real $x_1[p], x_2[n-p], \widehat{b}[m], U[p][p], V[n][n], \Sigma[p][n]$;
1.   Compute the SVD of B, $B = U \Sigma V^{\mathrm{T}}$;
2.   Partition $V \equiv [\ \underbrace{V_1}_{p}\ ,\ \underbrace{V_2}_{n-p}\ ]\}\,n$;
3.   Solve $BV_1 x_1 = d$ for x_1;
4. $\widehat{b} := b - AV_1 x_1$;
5.   Solve the LS problem $AV_2 x_2 = \widehat{b}$ for x_2;
6.   return $V_1 x_1 + V_2 x_2$;

Figure 33.7: Algorithm for the LS problem with linear equality constraints. See the text for the meanings of the input variables. Because only the diagonal elements of Σ are needed, a one-dimensional array suffices (the needed change to the algorithm is straightforward).

Exercise 19.2.14: Define $W \equiv \mathrm{diag}[\,1/\sqrt{\Psi_1}, 1/\sqrt{\Psi_2}, \ldots, 1/\sqrt{\Psi_p}\,]$. It plays the role of W in weighted LS problem (19.11). In other words, we claim that our problem is equivalent to

$$\min_{x \in R^n} \| WLf_t - W(y_t - \mu) \| . \tag{33.21}$$

This granted, because $\Psi^{-1} = W^{\mathrm{T}}W$, the solution for f_t is $(L^{\mathrm{T}}\Psi^{-1}L)^{-1}L^{\mathrm{T}}\Psi^{-1}(y_t - \mu)$ by formula (19.13), as desired. Hence we need to verify only equivalence (33.21). Because the covariance matrix of ϵ_t in Eq. (19.15) is Ψ, the discussion leading to Eq. (19.14) says that the inverse of $W^{\mathrm{T}}W$ should equal that matrix, which is true.

Exercise 19.2.15: $\mathrm{Cov}[\,y_t, f_t^{\mathrm{T}}\,] = E[\,(y_t - \mu)f_t^{\mathrm{T}}\,] = E[\,(Lf_t + \epsilon_t)f_t^{\mathrm{T}}\,] = E[\,Lf_t f_t^{\mathrm{T}}\,] = L$ [523, p. 399].

Exercise 19.2.16: See [239, 390].

Exercise 19.3.1: First we derive

$$f''(x) = f'_{i-1}\frac{6x - 4x_i - 2x_{i-1}}{h_i^2} - f'_i\frac{-6x + 2x_i + 4x_{i-1}}{h_i^2}$$

$$+ y_{i-1}\frac{12x - 6x_i - 6x_{i-1}}{h_i^3} + y_i\frac{-12x + 6x_i + 6x_{i-1}}{h_i^3} \tag{33.22}$$

for $x \in [x_{i-1}, x_i]$. For the $f''(x_i-)$ case, we simply evaluate $f''(x_i)$ as x_i is the right end point of Eq. (33.22) for $f''(x)$. As for the $f''(x_i+)$ case, we first derive an analogous formula for $f''(x)$, $x \in [x_i, x_{i+1}]$. It is obvious that the desired formula is merely Eq. (33.22) but with i replaced with $i+1$. Finally, we evaluate $f''(x_i)$ for the answer as x_i is the left end point of the formula in question. See [447, p. 479].

Exercise 19.3.2: With the help of Eq. (19.16), $f''(x_0) = f''(x_n) = 0$ implies that

$$f''(x_0) = -\frac{2}{h_1}(2f'_0 + f'_1) + 6\frac{y_1 - y_0}{h_1^2},$$

$$f''(x_n) = \frac{2}{h_n}(f'_{n-1} + 2f'_n) - 6\frac{y_n - y_{n-1}}{h_n^2}.$$

Hence

$$
\begin{bmatrix}
\frac{2}{h_1} & \frac{1}{h_1} & 0 & 0 & 0 & \cdots & 0 \\
\frac{1}{h_1} & \frac{2}{h_1}+\frac{2}{h_2} & \frac{1}{h_2} & 0 & 0 & \cdots & 0 \\
0 & \frac{1}{h_2} & \frac{2}{h_2}+\frac{2}{h_3} & \frac{1}{h_3} & 0 & \cdots & 0 \\
\vdots & \ddots & \ddots & \ddots & \ddots & \ddots & \vdots \\
\vdots & \ddots & \ddots & \ddots & \ddots & \ddots & \vdots \\
0 & \cdots & \cdots & 0 & \frac{1}{h_{n-1}} & \frac{2}{h_{n-1}}+\frac{2}{h_n} & \frac{1}{h_n} \\
0 & \cdots & \cdots & 0 & 0 & \frac{1}{h_n} & \frac{2}{h_n}
\end{bmatrix}
\begin{bmatrix}
f'_0 \\ f'_1 \\ \vdots \\ \\ \\ f'_n
\end{bmatrix}
=
\begin{bmatrix}
3\frac{y_1 - y_0}{h_1^2} \\
3\frac{y_1 - y_0}{h_1^2}+3\frac{y_2 - y_1}{h_2^2} \\
3\frac{y_2 - y_1}{h_2^2}+3\frac{y_3 - y_2}{h_3^2} \\
\vdots \\
3\frac{y_{n-1} - y_{n-2}}{h_{n-1}^2}+3\frac{y_n - y_{n-1}}{h_n^2} \\
3\frac{y_n - y_{n-1}}{h_n^2}
\end{bmatrix} .
$$

Exercise 19.3.4: 19.3.4: This time we solve

$$\begin{bmatrix} 4 & 1 & 0 \\ 1 & 4 & 1 \\ 0 & 1 & 4 \end{bmatrix} \begin{bmatrix} f_1'' \\ f_2'' \\ f_3'' \end{bmatrix} = \begin{bmatrix} -16.2 \\ 26.4 \\ -28.2 \end{bmatrix}.$$

The solutions are $f_1'' = -6.72857$, $f_2'' = 10.7143$, and $f_3'' = -9.72857$. Thus

$$p_1(x) = 2x + \frac{6.72857}{6}(x - x^3),$$

$$p_2(x) = 2(2 - x) + 1.3(x - 1) + \frac{6.72857}{6}[(2 - x) - (2 - x)^3] - \frac{10.7143}{6}[(x - 1) - (x - 1)^3],$$

$$p_3(x) = 1.3(3 - x) + 5(x - 2) - \frac{10.7143}{6}[(3 - x) - (3 - x)^3] + \frac{9.72857}{6}[(x - 2) - (x - 2)^3],$$

$$p_4(x) = 5(4 - x) + 4(x - 3) + \frac{9.72857}{6}[(4 - x) - (4 - x)^3].$$

Exercise 19.3.5: B is

$$\begin{bmatrix} 1 & x_1 & x_1^2 & x_1^3 & -1 & -x_1 & -x_1^2 & -x_1^3 & 0 & 0 & 0 & 0 & 0 & 0 & 0 & 0 & \cdots & \cdots \\ 0 & 0 & 0 & 0 & 0 & 0 & 0 & 0 & 1 & x_2 & x_2^2 & x_2^3 & -1 & -x_2 & -x_2^2 & -x_2^3 & 0 & \cdots \\ & \ddots & \ddots & \ddots & \ddots & \ddots & \ddots & \ddots & \ddots & \ddots & \ddots & \ddots & \ddots & \ddots & \ddots & \ddots & \ddots & \ddots \end{bmatrix},$$

d is the zero vector, the jth row of $A \in R^{m \times 4n}$ is

$$[\overbrace{0, \ldots, 0}^{4(i-1)}, 1, \tilde{x}_j, \tilde{x}_j^2, \tilde{x}_j^3, \overbrace{0, \ldots, 0}^{4(m-i)}]^\mathrm{T}$$

when $\tilde{x}_j$ falls within $[x_{i-1}, x_i]$, and $b \equiv [\tilde{y}_1, \tilde{y}_2, \ldots, \tilde{y}_m]^\mathrm{T}$.

Chapter 20

Exercise 20.1.1: As $\Delta S/S = \mu \Delta t + \sigma \sqrt{\Delta t}\, \xi$, we have $\Delta S_i - S_i \mu \Delta t \sim N(0, S^2 \sigma^2 \Delta t)$. The log-likelihood function is

$$\sum_{i=1}^{n} - \ln(2\pi S^2 \sigma^2 \Delta t) - \sum_{i=1}^{n} \frac{(\Delta S_i - S_i \mu \Delta t)^2}{2S^2 \sigma^2 \Delta t}.$$

After differentiation with respect to μ and σ^2, the estimators are found to be

$$\hat{\mu} = \frac{\sum_{i=1}^{n} (\Delta S_i / S_i)}{n \Delta t}, \qquad \hat{\sigma}^2 = \frac{\sum_{i=1}^{n} [(\Delta S_i / S_i) - \hat{\mu} \Delta t]^2}{n \Delta t}.$$

They are essentially the estimators of Eqs. (20.2) and (20.3) with R_i replaced with $(S_{i+1} - S_i)/S_i$ and α replaced with μ. Estimating μ remains difficult.

Exercise 20.1.2: Exercise 14.4.7 shows that the simple rate of return has mean $\mu_s = e^{t\mu} - 1$ and variance $\sigma_s^2 = (e^{t\sigma^2} - 1) e^{2t\mu}$. Hence

$$\sigma = \sqrt{\frac{1}{t} \ln\left(1 + \frac{\sigma_s^2}{(1 + \mu_s)^2}\right)}.$$

Published statistics for stock returns are mostly based on simple rates of return [147, p. 362]. Simple rates of return also should be used when calculating the VaR in Section 31.4.

Exercise 20.1.3: Equation (20.5) can be simplified to

$$0 = \sum_i (\Delta r_i - \beta(\mu - r_i) \Delta t) r_i^{-1} \tag{33.23}$$

with the help of Eq. (20.6) and the assumption that $\mu \neq 0$. Multiply out Eqs. (33.23) and (20.6) to obtain

$$0 = \sum_i \Delta r_i r_i^{-1} - \beta \mu \Delta t \sum_i r_i^{-1} + \beta n \Delta t,$$

$$0 = \sum_i \Delta r_i r_i^{-2} - \beta \mu \Delta t \sum_i r_i^{-2} + \beta \Delta t \sum_i r_i^{-1}.$$

Eliminate μ to gain a formula for β. Then plug the formula into one of the preceding equations to obtain the formula for μ. Finally, μ and β can be used in Eq. (20.7) to calculate σ^2.

Exercise 20.1.4: From Eq. (20.4) and Ito's lemma, $d \ln r = [\frac{\beta(\mu - r)}{r} - \frac{1}{2}\sigma^2] dt + \sigma \, dW$. The discrete approximation thus satisfies

$$\ln\left(1 + \frac{\Delta r}{r}\right) - \left(\frac{\beta(\mu - r)}{r} - \frac{1}{2}\sigma^2\right)\Delta t \sim N\left(0, \sigma^2 \Delta t\right).$$

The log-likelihood function of n observations $\Delta r_1, \Delta r_2, \ldots, \Delta r_n$ after the removal of the constant terms and simplification is

$$-n \ln \sigma - \left(2\sigma^2 \Delta t\right)^{-1} \sum_{i=1}^{n} \left\{ \ln\left(1 + \frac{\Delta r_i}{r_i}\right) - \left[\frac{\beta(\mu - r_i)}{r_i} - \frac{1}{2}\sigma^2\right]\Delta t \right\}^2.$$

We differentiate the log-likelihood function with respect to β, μ, and σ and equate them to zero. After simplification, we arrive at

$$0 = \sum_{i=1}^{n} \left\{ \ln\left(1 + \frac{\Delta r_i}{r_i}\right) - \left[\frac{\beta(\mu - r_i)}{r_i} - \frac{1}{2}\sigma^2\right]\Delta t \right\} r_i^{-1}, \tag{33.24}$$

$$0 = \sum_{i=1}^{n} \left\{ \ln\left(1 + \frac{\Delta r_i}{r_i}\right) - \left[\frac{\beta(\mu - r_i)}{r_i} - \frac{1}{2}\sigma^2\right]\Delta t \right\}, \tag{33.25}$$

$$\sigma^2 = \frac{1}{n\Delta t} \sum_{i=1}^{n} \left\{ \ln\left(1 + \frac{\Delta r_i}{r_i}\right) - \left[\frac{\beta(\mu - r_i)}{r_i} - \frac{1}{2}\sigma^2\right]\Delta t \right\} \ln\left(1 + \frac{\Delta r_i}{r_i}\right).$$

This set of three equations can be solved numerically. (The preceding formulas can be obtained as follows. We derive Eq. (33.24) first by differentiating the log-likelihood function with respect to μ and setting it to zero. Then we differentiate the function with respect to β and set it to zero. The resulting equation can be simplified with the help of Eq. (33.24) to obtain Eqs. (33.25). Finally, we differentiate the function with respect to σ and set it to zero. With the help of Eq. (33.24),

$$\frac{1}{n\Delta t} \sum_{i=1}^{n} \left\{ \ln\left(1 + \frac{\Delta r_i}{r_i}\right) - \left[\frac{\beta(\mu - r_i)}{r_i} - \frac{1}{2}\sigma^2\right]\Delta t \right\}^2 = \sigma^2.$$

Multiplying out the summand and using Eqs. (33.24) and (33.25) again leads to the last formula.)

Exercise 20.1.5: The log-likelihood function for a sample of size n is

$$-\frac{n}{2} \ln(2\pi) - \frac{n}{2} \ln(\lambda^2) - \theta \sum_{i=1}^{n} \ln S_i - \frac{1}{2\lambda^2} \sum_{i=1}^{n} \left(\frac{\Delta S_i}{S_i}\right)^2 S^{-2\theta}.$$

Straightforward maximization of the function with respect to λ^2 and θ results in

$$\lambda^2 = \frac{1}{n} \sum_i \left(\frac{\Delta S_i}{S_i}\right)^2 S_i^{-2\theta}, \qquad \lambda^2 = \sum_i \frac{\ln S_i}{\sum_i \ln S_i} \left(\frac{\Delta S_i}{S_i}\right)^2 S_i^{-2\theta}.$$

There are two nonlinear equations in two unknowns λ^2 and θ, that can be estimated by the Newton–Raphson method. See [208] and also [113] for a trinomial model for the CEV process.

Exercise 20.1.8: From Exercise 6.4.3, (2), the desired prediction of X_{t+1} given $X_1, X_2, \ldots, X_t$ should be $E[X_{t+1} \mid X_1, X_2, \ldots, X_t]$, which equals $E[X_{t+1}] = \mu$ by stationarity.

Exercise 20.1.9: Let $X \equiv [X_t, X_{t-1}, \ldots, X_1]^{\mathrm{T}}$ and $a \equiv [a_1, a_2, \ldots, a_t]^{\mathrm{T}}$. The linear prediction is then $a_0 + a^{\mathrm{T}}X$. The matrix version of Exercise 6.4.1 says that we should pick $a_0 = \mu$ and $a^{\mathrm{T}} = \mathrm{Cov}[X_{t+1}, X] \mathrm{Cov}[X]^{-1}$ [416, p. 75]. (See Exercise 19.2.2(2), for the finite-sample version and Exercise 6.4.1, (1), for the one-dimensional version called beta.) By definition $\mathrm{Cov}[X] = [\lambda_{|i-j|}]_{1 \le i, j \le t}$. As the covariance between X_{t+1} and X_{t-i} is λ_{i+1}, $\mathrm{Cov}[X_{t+1}, X] = [\lambda_1, \lambda_2, \ldots, \lambda_t]$. See [416, p. 86].

Exercise 20.1.10: Let S_t denote the price at time t and $S_t = S_{t-1} + \epsilon_t$. By the assumption, $\mathrm{Var}[S_t] = \mathrm{Var}[S_{t-1}] + \mathrm{Var}[\epsilon_t]$ based on Eq. (6.5).

Exercise 20.1.12: Take a positive autocorrelation for example. It implies that a higher-than-average (lower-than-average) return today is likely to be followed by higher-than-average (lower-than-average, respectively) returns in the future. In other words, today's returns can be used to predict future returns [424, 767].

Exercise 20.1.13: Note that $E[\,Y_t\,] = \sum_{k=0}^{l} E[\,a_k X_{t-k}\,] = 0$ and

$$\lambda_\tau = E\left[\left(\sum_{j=0}^{l} a_j X_{t+\tau-j}\right)\left(\sum_{k=0}^{l} a_k X_{t-k}\right)\right] = \begin{cases} a_l a_{l-\tau} + \cdots + a_\tau a_0, & \text{if } \tau \le l \\ 0, & \text{otherwise} \end{cases}.$$

See [541, p. 167].

Exercise 20.1.14: $\lambda_\tau \equiv E[\,(X_{t-\tau} - b)(X_t - b)\,]$ equals

$$E\left[\left(\sum_{j=0}^{\infty} c_j \epsilon_{t-j-\tau}\right)\left(\sum_{j=0}^{\infty} c_j \epsilon_{t-j}\right)\right] = \sigma^2(c_0 c_\tau + c_1 c_{\tau+1} + c_2 c_{\tau+2} + \cdots) = \sigma^2 \sum_{j=0}^{\infty} c_j c_{j+\tau}.$$

See [667, p. 68].

Exercise 20.1.15: Similar to Exercise 19.2.2(1), the LS problem is

$$\begin{bmatrix} 1 & X_p & X_{p-1} & \cdots & X_1 \\ 1 & X_{p+1} & X_p & \cdots & X_2 \\ \vdots & \vdots & \vdots & \ddots & \vdots \\ 1 & X_{n-1} & X_{n-2} & \cdots & X_{n-p} \end{bmatrix} \begin{bmatrix} c \\ a_1 \\ \vdots \\ a_p \end{bmatrix} = \begin{bmatrix} X_{p+1} \\ X_{p+2} \\ \vdots \\ X_n \end{bmatrix}.$$

Exercise 20.2.1: (1) Let k_X denote the kurtosis of X_t and k_U the kurtosis of U_t. First, $k_U = E[\,U_t^4\,]$ because $E[\,U_t\,] = 0$ and $E[\,U_t^2\,] = 1$. Now,

$$k_X \equiv \frac{E[\,(X_t - \mu)^4\,]}{E[\,(X_t - \mu)^2\,]^2} = \frac{E[\,V_t^4 U_t^4\,]}{E[\,V_t^2\,]^2 E[\,U_t^2\,]^2} = \frac{k_U\, E[\,V_t^4\,]}{E[\,V_t^2\,]^2} > k_U. \tag{33.26}$$

The inequality is due to Jensen's inequality. Note that we need to require only that V_t and U_t be independent. See [839, p. 72].

Exercise 20.2.2: Keep $E[\,V_t^n\,] = e^{na + n^2 b^2/2}$ (see p. 68) and $E[\,V_t^i U_t^i\,] = E[\,V_t^i\,]\,E[\,U_t^i\,]$ for any i in mind in the following. (1) By Eq. (33.26) and the fact that the kurtosis of the standard normal distribution is three. (3) $|X_t - \mu| = V_t\,|U_t|$ and the mean of $|U_t|$ is $\sqrt{2/\pi}$. (4) $(X_t - \mu)^2 = V_t^2 U_t^2$.

Exercise 20.2.3: (1) By repeated substitutions. (2) $V_{t+1}^2 = (1 - a_1 - a_2)\,V + a_1(X_t - \mu)^2 + a_2 V_t^2$ given V_t. Take expectations of both sides and rearrange to obtain $E[\,V_{t+1}^2 \mid V_t\,] = V + (a_1 + a_2)(V_t^2 - V)$. Repeat the above steps to arrive at $E[\,V_{t+k}^2 \mid V_t\,] = V + (a_1 + a_2)^k(V_t^2 - V)$. Finally, apply Eq. (20.11), which holds as long as U_t and V_t are uncorrelated. See [470, p. 379].

CHAPTER 21

Exercise 21.1.1: A change of 0.01 translates into a change in the yield on a bank discount basis of one basis point, or 0.0001, which will change the dollar discount, and therefore the invoice price, by

$$0.0001 \times \$1{,}000{,}000 \times \frac{t}{360} = \$0.2778 \times t,$$

where t is the number of days to maturity. As a result, the tick value is $\$0.2778 \times 91 = \$25.28$. See [88, p. 163] or [325, p. 393].

Exercise 21.1.2: The implied rate over the 9-month period is

$$1.01656 \times 1.01694 \times [\,1 + 0.0689 \times (91/360)\,] - 1 = 5.179\%.$$

The annualized rate is therefore $5.179\% \times (360/274) = 6.805\%$.

Exercise 21.1.4: In calculating the conversion factor, the bond is assumed to have exactly 19 years and 9 months to maturity. Therefore there are 40 coupon payments, starting 3 months from now. The value of the bond is

$$\sum_{i=0}^{39} \frac{6.5}{(1.04)^{i+.5}} + \frac{100}{(1.04)^{38.5}} - 3.25 = 149.19.$$

The conversion factor is therefore 1.4919.

Exercise 21.1.5: Buy the September T-note contracts and sell them in August [95, p. 107].

Exercise 21.1.6: (1) The forward rate for the Tth period is $f(T, T+1)$. Now sell one unit of the T-period zero-coupon bond and buy $d(T)/d(T+1)$ units of the $(T+1)$-period zero-coupon bond. The net cash flow today is zero. This portfolio generates a cash outflow of $1 at time T and a cash inflow of $d(T)/d(T+1)$ dollars at time $T+1$. As the forward rate can be locked in today, it must satisfy $1+f(T,T+1)=d(T)/d(T+1)$ to prevent arbitrage profits. (2) It is the cost to replicate the FRA now.

Exercise 21.2.2: See Exercise 12.2.4 or [346, p. 244].

Exercise 21.4.1: From A's point of view, by issuing a fixed-rate loan and entering into a swap with the bank, it effectively pays a floating rate of (LIBOR $-S'_A-F'_A$)%. It is better off if $S_A > -S'_A - F'_A$. Similarly, from B's point of view, by issuing a floating-rate loan and entering into a swap, it effectively pays a fixed rate of $(F_B + F'_B + S_B + S'_B)$%. It is better off if $F_B > F_B + F'_B + S_B + S'_B$. Finally, the bank is better off by entering into both swaps if $F_B + F'_B - S'_A - F_A - F'_A + S'_B > 0$. Put these inequalities together and rearrange terms to get the desired result. See [849].

Exercise 21.4.2: The cash flow is identical to that of the fixed-rate payer/floating-rate receiver. To start with, there is no initial cash flow. Furthermore, on all six payment dates, the net position is a cash inflow of LIBOR plus 0.5% and a cash outflow of $5 million.

Exercise 21.4.3: Because the swap value is

$$(0.5-x)e^{-0.101\times0.3} + \left(\frac{0.106962}{2}\times10-x\right)e^{-0.103\times0.8}$$

million, the x that makes it zero is 0.51695. The desired fixed rate is 10.339%.

Exercise 21.4.4: To start with,

$$P_2 - P_1 - (21.4) = (\mathcal{N}+C^*)e^{-r_1t_1} - \sum_{i=1}^{n}Ce^{-r_it_i} - \mathcal{N}e^{-r_nt_n}$$

$$-(C^*-C)e^{-r_1t_1} - \sum_{i=2}^{n}\left(\frac{f_i}{k}\mathcal{N}-C\right)e^{-r_it_i}$$

$$= \mathcal{N}e^{-r_1t_1} - \mathcal{N}e^{-r_nt_n} - \sum_{i=2}^{n}\frac{f_i}{k}\mathcal{N}e^{-r_it_i}.$$

So we need to prove only that

$$e^{-r_1t_1} - e^{-r_nt_n} - \sum_{i=2}^{n}\frac{f_i}{k}e^{-r_it_i} = 0. \tag{33.27}$$

The annualized, continuously compounded forward rate t_{i-1} years from now is

$$c_i \equiv \frac{r_it_i - r_{i-1}t_{i-1}}{t_i - t_{i-1}} = \frac{r_it_i - r_{i-1}t_{i-1}}{(1/k)}$$

from Eq. (5.9). Hence, $f_i = k(e^{c_i/k}-1) = k(e^{r_it_i-r_{i-1}t_{i-1}}-1)$ according to Eq. (3.3), as the desired rate needs to be one that is compounded k times per annum. Now, plug the formula for f_i into Eq. (33.27) to get $e^{-r_1t_1} - e^{-r_nt_n} - \sum_{i=2}^{n}(e^{r_it_i-r_{i-1}t_{i-1}}-1)e^{-r_it_i} = 0$, as claimed.

Exercise 21.4.5: It is $(C-\widehat{C})e^{-r_1t_1} + \sum_{i=2}^{n}(C-\widehat{C})e^{-r_it_i}$.

Exercise 21.4.7: Callable bonds are called away when rates decline. This leaves the institution with long interest rate swap positions that have a high fixed rate. Callable swaps can alleviate such a situation. See [821, p. 602].

Exercise 21.4.8: A cap gives the holder an option at each reset date to borrow at a capped rate. An option on a swap, in contrast, offers the holder the one-time option to borrow at a fixed rate over the remaining lifetime of the swap. Clearly, swaptions are less flexible. More rigorously, a swaption can be viewed as an option on a portfolio and a cap as a portfolio of options. Theorem 8.6.1 says that a portfolio of options is more valuable than an option on a portfolio. See [346, p. 248] or [746, p. 513].

Chapter 22

Exercise 22.1.1: See Exercise 5.6.3 and Eq. (5.11).

Exercise 22.2.1: As $d(t)=e^{-ts(t)}$, Eq. (22.1) becomes $e^{-ts(t)} = e^{-ts(t_1)\frac{t_2-t}{t_2-t_1}}e^{-ts(t_2)\frac{t-t_1}{t_2-t_1}}$, which can be simplified to $s(t)=s(t_1)\frac{t_2-t}{t_2-t_1} + s(t_2)\frac{t-t_1}{t_2-t_1}$.

Exercise 22.3.1: Consistent with the numbers in Eq. (22.2), the LS problem is

$$P_i = \sum_{j=1}^{n_i} C_i(a_0 + a_1 j + a_2 j^2) + (a_0 + a_1 n_i + a_2 n_i^2)$$

$$= a_0(n_i C_i + 1) + a_1 \left(n_i + C_i \sum_{j=1}^{n_i} j\right) + a_2 \left(n_i^2 + C_i \sum_{j=1}^{n_i} j^2\right), \quad 1 \le i \le m,$$

which can be solved by multiple regression.

Exercise 22.3.2: Suppose that $y \approx ae^{bx}$ for each piece of data. Then $\ln y \approx bx + \ln a$. Transform the data into pairs $(x_i, \ln y_i)$ and perform linear regression to obtain a line $z = a' + b'x$. Finally, set $a = e^{a'}$ and $b = b'$ so that $y = e^z = e^{a'+b'x} = ae^{bx}$. See [846, p. 545].

Exercise 22.4.2: Note that

$$f(T) = \frac{\partial[-\ln d(T)]}{\partial T} = -\frac{1}{d(T)} \frac{\partial d(T)}{\partial T}$$

by Eq. (5.7). For the forward rate curve to be continuous, the discount function must have at least one continuous derivative, hence claim (1). The preceding identities also show that if the forward rate curve should be continuously differentiable, then a cubic spline is needed. See [147, p. 411].

Chapter 23

Exercise 23.1.1: No [492, p. 388].

Exercise 23.1.2: The accumulated value of a $1 investment is $a(n) \equiv \prod_{t=1}^n (1+i_t)$. As $\ln(1+i_t) \sim N(\mu, \sigma^2)$, the mean and the variance of this lognormal variable are given by $e^{\mu+\sigma^2/2}$ and $e^{2\mu+\sigma^2}(e^{\sigma^2} - 1)$, respectively, according to Eq. (6.11). Now $\ln a(n) = \sum_{t=1}^n \ln(1+i_t)$ is normal with mean $n\mu$ and variance $n\sigma^2$. Equivalently, $a(n)$ is lognormal with mean $e^{n\mu+n\sigma^2/2}$ and variance $e^{2n\mu+n\sigma^2}(e^{n\sigma^2} - 1)$. See [547, p. 368].

Exercise 23.2.1: The logarithms of the short rates are $\ln r_k$, $\ln r_k + \ln v_k$, $\ln r_k + 2\ln v_k$, ..., $r_k + (k-1)\ln v_k$. As far as variance is concerned, the term $\ln r_k$ is irrelevant; hence it is deleted from the list. The mean of the logarithms of the short rates is

$$\sum_{i=0}^{k-1} (i \ln v_k) \frac{\binom{k-1}{i}}{2^{k-1}} = \frac{k-1}{2} \ln v_k.$$

Hence the variance equals

$$\sum_{i=0}^{k-1} (i \ln v_k)^2 \frac{\binom{k-1}{i}}{2^{k-1}} - \left(\frac{k-1}{2} \ln v_k\right)^2 = \frac{(\ln v_k)^2}{2^{k-1}} \sum_{i=0}^{k-1} i^2 \binom{k-1}{i} - \left(\frac{k-1}{2} \ln v_k\right)^2$$

$$= \frac{(\ln v_k)^2}{2^{k-1}} (k-1)[2^{k-2} + (k-2) 2^{k-3}] - \left(\frac{k-1}{2} \ln v_k\right)^2$$

$$= (k-1) \left(\frac{\ln v_k}{2}\right)^2,$$

which equals $\sigma_k^2 (k-1)\Delta t$ based on Eq. (23.2). See [731, p. 440].

Exercise 23.2.2: Consider rate $r \equiv r_j v_j^i$ for period j and its two subsequent rates $r_{j+1} v_{j+1}^i$ and $r_{j+1} v_{j+1}^{i+1}$. By Eq. (23.2) and the constant-volatility assumption, we have $v_j = v_{j+1} = e^{2\sigma \sqrt{\Delta t}}$. Hence the dynamics becomes $r \to r(r_{j+1}/r_j) e^{2\sigma \sqrt{\Delta t} \, \xi}$, where $\xi = 0, 1$, each with a probability of one-half. We establish the claim by defining μ such that $r_{j+1}/r_j = e^{\mu - \sigma \sqrt{\Delta t}}$.

Exercise 23.2.3: (1) It is equal to

$$q(\ln r_\ell)^2 + (1-q)(\ln r_h)^2 - [q \ln r_\ell + (1-q) \ln r_h]^2 = q(1-q)(\ln r_h - \ln r_\ell)^2.$$

(2) Use $r(\Delta t)$ to denote the r after a time period of Δt. From (1),

$$\sigma^2 \Delta t = \text{Var}[\ln r(\Delta t)] = \text{Var}[\ln(r + \Delta r)] = \text{Var}\left[\ln\left(1 + \frac{\Delta r}{r}\right)\right] \approx \text{Var}\left[\frac{\Delta r}{r}\right] = \frac{\text{Var}[\Delta r]}{r^2}.$$

So $\text{Var}[r + \Delta r] = \text{Var}[\Delta r] = r^2 \sigma^2 \Delta t$.

Exercise 23.2.4: (1) Let the rate for the first period be r and the forward rate one period from now be f. Suppose the binomial interest rate tree gives r_ℓ and r_h for the forward rates applicable in the second period. By construction, $r_h/r_\ell = v$ and $(r_h + r_\ell)/2 = f$. The price of a zero-coupon bond two periods from now is priced by the tree as

$$\frac{1}{1+r}\frac{1}{2}\left(\frac{1}{1+r_h} + \frac{1}{1+r_\ell}\right).$$

It is not difficult to see that the above number exceeds $1/[(1+r)(1+f)]$ unless $v = 1$.

(2) By (1) we know that the claim holds for trees with two periods. Assume that the claim holds for trees with $n-1$ periods and proceed to prove its validity for trees with n periods. Suppose the tree has baseline rates $r_1, r_2, \ldots, r_n$, where r_i is the baseline rate for the ith period. Denote this tree by $T(r_1, \ldots, r_n)$. Split the tree into two $(n-1)$-period subtrees by taking out the root. The first tree can be denoted by $T(r_2, \ldots, r_n)$ and the second by $T(r_2 v_2, \ldots, r_n v_n)$. If $V(T)$ is used to signify the value of a security as evaluated by the binomial interest rate tree T, clearly

$$V(T(r_1, \ldots, r_n)) = \frac{V(T(r_2, \ldots, r_n)) + V(T(r_2 v_2, \ldots, r_n v_n))}{2(1+r_1)}. \tag{33.28}$$

It suffices to prove the claim for zero-coupon bonds by the additivity of the valuation process. Hence the problem is reduced to proving

$$V(T(r_1, \ldots, r_n)) > [(1+f_1)(1+f_2)\cdots(1+f_n)]^{-1},$$

where f_i is the one-period forward rate for period i.

By Eq. (23.4), $f_j = r_j(1+v_j/2)^{j-1}$. So $T(r_2, \ldots, r_n)$ implies that its ith-period forward rate (counting from its root) is

$$f_i' = r_{i+1}\left(\frac{1+v_{i+1}}{2}\right)^{i-1} = f_{i+1}\frac{2}{1+v_{i+1}}.$$

Similarly, $T(r_2 v_2, \ldots, r_n v_n)$ implies that its ith-period forward rate is

$$f_i'' = r_{i+1}v_{i+1}\left(\frac{1+v_{i+1}}{2}\right)^{i-1} = f_{i+1}\frac{2v_{i+1}}{1+v_{i+1}}.$$

Apply the induction hypothesis to each of the subtrees to obtain

$$V(T(r_2, \ldots, r_n)) > [(1+f_1')(1+f_2')\cdots(1+f_{n-1}')]^{-1}$$
$$V(T(r_2 v_2, \ldots, r_n v_n)) > [(1+f_1'')(1+f_2'')\cdots(1+f_{n-1}'')]^{-1}$$

Add them up:

$$V(T(r_2, \ldots, r_n)) + V(T(r_2 v_2, \ldots, r_n v_n))$$
$$> [(1+f_1')(1+f_2')\cdots(1+f_{n-1}')]^{-1} + [(1+f_1'')(1+f_2'')\cdots(1+f_{n-1}'')]^{-1}.$$

By Eq. (33.28) we are done if we can show that

$$[(1+f_1')(1+f_2')\cdots(1+f_{n-1}')]^{-1} + [(1+f_1'')(1+f_2'')\cdots(1+f_{n-1}'')]^{-1} > 2[(1+f_2)\cdots(1+f_n)]^{-1}.$$

Recall that $r_1 = f_1$ by definition. Now,

$$[(1+f_1')(1+f_2')\cdots(1+f_{n-1}')]^{-1} + [(1+f_1'')(1+f_2'')\cdots(1+f_{n-1}'')]^{-1}$$
$$\geq 2\sqrt{[(1+f_1')(1+f_2')\cdots(1+f_{n-1}')]^{-1}[(1+f_1'')(1+f_2'')\cdots(1+f_{n-1}'')]^{-1}}$$

because of the relation between arithmetic and geometric means. It is sufficient to prove that

$$(1+f_1')(1+f_2')\cdots(1+f_{n-1}')(1+f_1'')(1+f_2'')\cdots(1+f_{n-1}'') < [(1+f_2)\cdots(1+f_n)]^2.$$

We now prove the validity of the preceding inequality by showing that $(1+f_i')(1+f_i'') < (1+f_{i+1})^2$ for each $1 \leq i \leq n-1$.

Because

$$(1+f_i')(1+f_i'') = \left(1+f_{i+1}\frac{2}{1+v_{i+1}}\right)\left(1+f_{i+1}\frac{2v_{i+1}}{1+v_{i+1}}\right) = \frac{4v_{i+1}}{(1+v_{i+1})^2}f_{i+1}^2 + 2f_{i+1} + 1,$$

we have

$$(1+f_i')(1+f_i'') - (1+f_{i+1})^2 = -\left(\frac{1-v_{i+1}}{1+v_{i+1}}\right)^2 f_{i+1}^2 < 0,$$

as desired.

Exercise 23.2.5: 23.2.5: Fix a period $[k-1, k]$. (1) Let f be the forward rate for that period. Initiate an FRA at time zero for that period. Its payoff at time k equals $s - f$, where s is the future spot rate for the period. The forward rate f is the fixed contract rate that makes the FRA zero valued. Adopt the forward-neutral probability measure π_k with the zero-coupon bond maturing at time k as numeraire. By Exercise 13.2.13, $0 = d(k) E_0^{\pi_k}[s - f]$, which implies that $f = E_0^{\pi_k}[s]$. (2) Let f' denote the same forward rate one period from now. Again, $f' = E_1^{\pi_k}[s]$. By applying $E_0^{\pi_k}$ to both sides of the equation and then Eq. (6.6), the law of iterated conditional expectations, we obtain $E_0^{\pi_k}[f'] = E_0^{\pi_k}[E_1^{\pi_k}[s]] = E_0^{\pi_k}[s]$, which equals f. See [691, p. 398] and [731, p. 177].

Exercise 23.2.6: Solve for

$$\frac{0.112832}{1+r} + \frac{0.333501}{1+1.5 \times r} + \frac{0.327842}{1+(1.5)^2 \times r} + \frac{0.107173}{1+(1.5)^3 \times r} = \frac{1}{(1+0.044)^4}.$$

The result is $r = 0.024329$. The forward rate for the fourth period is $(1+0.044)^4/(1+0.043)^3 - 1 = 1.047006$. So the baseline rate would have been $2^3 \times 0.047006/(2.5)^3 = 0.024067$ had we used Eq. (23.4). It is lower than 0.024329.

Exercise 23.2.7: (1) Because $d(j) = \sum_i P_i$, where P_i is the state price in state i at time j, simply define the risk-neutral probabilities as $q_i \equiv P_i/d(j)$. Clearly q_i sum to one. These probabilities are called forward-neutral probabilities in Exercise 13.2.13. (2) They are $0.232197/0.92101$, $0.460505/0.92101$, and $0.228308/0.92101$. See [40].

Exercise 23.2.8: Calculate all the state prices in $O(n^2)$ time by using forward induction. Then sum the state prices of each column.

Exercise 23.2.9: (1) We do it inductively. Suppose we are at time j and there are $j+1$ nodes with the state prices $P_1', P_2', \ldots, P_{j+1}'$. Let the baseline rate for period j be $r \equiv r_j$, the multiplicative ratio be $v \equiv v_j$, and $P_1, P_2, \ldots, P_j$ be the state prices a period prior, corresponding to rates $r, rv, \ldots, rv^{j-1}$. Each P_i has branching probabilities p_i for the up move and $1 - p_i$ for the down move. Add $P_0 \equiv 0$ for convenience. Clearly, $P_i' = (1 - p_{i-1}) P_{i-1}/(1+rv^{i-2}) + p_i P_i/(1+rv^{i-1})$, $i = 1, 2, \ldots, j+1$ (see Fig. 23.7). We are done because there are $j+1$ equations (one of which is redundant) and j unknowns $p_1, p_2, \ldots, p_j$. (2) Although the j state prices $P_1, P_2, \ldots, P_j$ are now unknown, $p_1 = p_2 = \cdots = p_j$ from the assumption. Hence $j+1$ unknowns remain. We use the $j+1$ equations, none of which is redundant, to solve for the unknowns.

Exercise 23.3.1: No [427].

Exercise 23.3.4: (1) It is identical to the calculation for the futures price except that (a) discounting should be used during backward induction and (b) the final price, the PV of a bond delivered at the delivery date T, should be divided by the discount factor $d(T)$ to obtain its future value [623, p. 390]. (2) The forward price is

$$\frac{97.186 \times 0.232197 + 95.838 \times 0.460505 + 93.884 \times 0.228308}{0.92101} = 95.693.$$

Note that the forward price exceeds the futures price 95.687 as Exercise 12.3.3 predicts.

Exercise 23.3.6: Consider the strategy of buying a coupon bond, selling a call on this bond struck at X, and buying a put on this bond also struck at X. The options have the same expiration date t. The cash flow is $C, C, \ldots, C, X$, where C is the periodic coupon payment of the bond. Note that there is no cash flow beyond time t. Hence $\text{PV}(I) + \text{PV}(X) = B + P - C$, where $\text{PV}(I)$ denotes the present value of the bond's cash flow on and before time t. See [731, p. 269].

Exercise 23.4.1: At any node on the correlated binomial tree with the stock price–short rate pair (S, r), it must hold that $d < 1 + r < u$ by Exercise 9.2.1. Unless there are prior restrictions on r, this implies that u and d must be state, time, and path dependent. See [779].

Chapter 24

Exercise 24.2.1: (1) If the forward price satisfies it, the following two strategies yield the same dollar amount after M time for the same cost: (1) Buying an M-time zero-coupon bond for $P(t, M)$ dollars and (2) spending $P(t, M)$ dollars to buy $P(t, M)/P(t, T)$ units of the T-time zero-coupon bond and entering into a T-time forward contract for the $(M - T)$-time zero-coupon bond. Observe that we have $F(t, T, M)$ dollars at the end of time T in strategy (2), exactly what is needed to take delivery of one $(M - T)$-time zero-coupon bond. If $P(t, M) > P(t, T) F(t, T, M)$, arbitrage opportunities exist by taking short positions in the M-time zero-coupon bonds and long positions in both the T-time zero-coupon bonds and the T-time forward contracts for $(M - T)$-time zero-coupon bonds. Reverse the positions if $P(t, M) < P(t, T) F(t, T, M)$. Note the affinity between Eq. (24.1) and Eq. (5.2).

(2) Suppose the cash flows before the forward contract's delivery date are $C_1, C_2, \ldots$, at times $T_1, T_2, \ldots$. Let F denote the forward price. Our replicating strategy is this: Borrow $FP(t, T)$ dollars for T years, borrow $C_i P(t, T_i)$ dollars for T_i years for $i = 1, 2, \ldots$, and buy one unit of the underlying bond for $B(t, M)$ dollars. Each of the loans' obligations before T can be paid off by the bond's coupon. On the delivery date T, the position is worth the bond's FV minus the forward price, $B(T, M) - F$, replicating the forward contract's payoff. Because the forward contract has zero value at time t, we must have $FP(t, T) + \sum_i C_i P(t, T_i) - B(t, M) = 0$. Thus $F = \frac{B(t,M) - \sum_i C_i P(t,T_i)}{P(t,T)}$. In words, the forward price equals the FV of the underlying bond's current invoice price minus the PV of the bond's cash flow until the delivery date. The forward price includes the accrued interest of the underlying bond payable at time T. See [848, p. 169].

Exercise 24.2.2: There is no net investment at time t, and we pay \$1 at time T and receive $P(t, T)/P(t, M)$ dollars at time M. In other words, the investment at time T results in a certain gross return of $P(t, T)/P(t, M)$ at time M. So the said forward price must be $P(t, M)/P(t, T)$. See [76, p. 22].

Exercise 24.2.3: Note that Eq. (5.11) is equivalent to $f(T) = \partial(TS(T))/\partial T$ and $TS(T) = -\ln d(T)$ by Eq. (5.7). The rest is simple translation between the two systems of notations.

Exercise 24.2.4: Equation (24.2) is

$$1 + f(t, s, 1) = 1 + f(t, s) = \frac{P(t, s)}{P(t, s+1)}.$$

With the preceding identity applied recursively, we have

$$P(t, T) = \frac{P(t, t+1)}{[1 + f(t, t+1)] \cdots [f(t, T-1)]}.$$

Now,

$$P(t, t+1) = [1 + r(t, t+1)]^{-1} = [1 + r(t)]^{-1}.$$

Combine the preceding two equations above to obtain the desired result. See [492, p. 388].

Exercise 24.2.5: We know that the spot rate s satisfies $P = e^{-s\tau}$. On the other hand, Eq. (24.6)'s continuous compounding analog, Eq. (5.8), says that $P = e^{-\sum_{i=0}^{\tau-1} f(t,t+i)}$. See [746, p. 422].

Exercise 24.2.6: Straightforward from Eqs. (5.12) and (5.13). An alternative starts from Eq. (24.1) with $F(t, T, T + L) = \frac{P(t,T+L)}{P(t,T)}$. Then observe that

$$f(t, T, L) = \frac{1}{L} \left[\frac{1}{F(t, T, T+L)} - 1 \right] = \frac{1}{L} \left[\frac{P(t, T)}{P(t, T+L)} - 1 \right].$$

Exercise 24.2.8: Apply Eqs. (14.16) and (24.4) to obtain $f(t, T) = r(t) + \mu(T - t) - \sigma^2(T - t)^2/2$. From the definition, $E_t[r(T) | r(t)] = r(t) + \mu(T - t)$. So the premium is

$$f(t, T) - E_t[r(T) | r(t)] = -\frac{\sigma^2(T - t)^2}{2} < 0.$$

As for the forward rate, use Eqs. (14.16) and (24.4) to obtain $r + \mu(T - t) - \sigma^2(T - t)^2/2$.

Exercise 24.2.9: Exercise 5.6.6 and Eq. (24.5) imply that $P(t, T) = e^{-\int_t^T r(s)\,ds}$ for a certain economy [746, p. 527].

Exercise 24.2.10: From $r_c \equiv \ln(1 + r_e)$, we have $r_e/(1 + r_e) = 1 - e^{-r_c}$. Hence,

$$dr_c = (1 + r_e)^{-1} dr_e - \frac{1}{2}(1 + r_e)^{-2} (dr_e)^2$$

$$= (1 + r_e)^{-1} r_e(\mu\,dt + \sigma\,dW) - \frac{1}{2}(1 + r_e)^{-2} r_e^2 \sigma^2\,dt$$

$$= (1 - e^{-r_c})(\mu\,dt + \sigma\,dW) - \frac{1}{2}(1 - e^{-r_c})^2 \sigma^2\,dt,$$

from which the equation follows. See [781, Theorem 3].

Exercise 24.3.1: From $P(t, T) = e^{-r(t, T)(T-t)}$ and Eq. (24.9), $r(t, T) = -\frac{\ln E_t[e^{-\int_t^T r(s)\, ds}]}{T-t}$. Jensen's inequality says it is less than $\frac{E_t[\int_t^T r(s)\, ds]}{T-t}$ unless there is no uncertainty. See [302, Theorem 1].

Exercise 24.3.2: From Eqs. (24.4) and (24.9) and Leibniz's rule,

$$f(t, T) = -\frac{\partial P(t, T)/\partial T}{P(t, T)} = \frac{E_t[r(T)e^{-\int_t^T r(s)\, ds}]}{E_t[e^{-\int_t^T r(s)\, ds}]}.$$

So the forward rate is a weighted average of future spot rates. Finally, the assumption says that the above average is less than the simple average $E_t[r(T)]$. See [302, Theorem 2].

Exercise 24.3.3: Yes for the calibrated tree and independent of the term structure of volatilities. Consider an n-period bond with $n > 1$. One period from now (time t), the bond will have two prices, P_u and P_d, such that $(P_u + P_d)/\{2[1 + r(t)]\} = P(t, n)$, the market price of n-period zero-coupon bonds. The expected one-period return for this bond is, by construction,

$$\frac{(P_u + P_d)/2}{P(t, n)} = 1 + r(t),$$

exactly as demanded by the local expectations theory. As for the uncalibrated tree, it does not satisfy the theory. The reason is simple:

$$\frac{P_u + P_d}{2(1 + r(t))} > P(t, n),$$

as the tree is known to overestimate the discount factor by Theorem 23.2.2.

Exercise 24.3.4: From Eqs. (24.5) and (24.6) and the theory, which says that $f(s, t) = E_s[r(t)]$.

Exercise 24.3.5: Every cash flow has to be discounted by the appropriate discount function [292].

Exercise 24.3.7: See [725, p. 231].

Exercise 24.3.8: An interest rate cap, we recall, is a contract in which the seller promises to pay a certain amount of cash to the holder if the interest rate exceeds a certain predetermined level (the cap rate) at certain future dates. In the same way, the seller of a floor contract promises to pay cash when future interest rates fall below a certain level. Technically, a cap contract is a sum of caplets. We now give a precise description of the caplet.

Let t stand for the time at which the contract is written and $[t_0, t_1]$ be the period for which the caplet is in effect with $\Delta t \equiv t_1 - t_0$. Denote the cap rate by x. For simplicity, assume that the notional principal is \$1.

The interest rate that in real life determines the payments of the cap is some market rate such as LIBOR. The rate is quoted as a simple rate over the period $[t_0, t_1]$. This simple rate, which we denote by f, is determined at t_0 and defined by the relation $P(t_0, t_1) = 1/(1 + f\Delta t)$. Finally a caplet is a contingent t_1 claim that at time t_1 will pay $\max(f - x, 0) \times \Delta t$ to the holder of the contract. The payment is in arrears. Specifically, the payoff at time t_1 is

$$\max\left(\frac{1 - P(t_0, t_1)}{P(t_0, t_1)\, \Delta t} - x, 0\right) \times \Delta t = \max\left(\frac{1}{P(t_0, t_1)} - (1 + x\Delta t), 0\right)$$

$$= \max\left(\frac{1}{P(t_0, t_1)} - \alpha, 0\right),$$

where $\alpha \equiv 1 + x\Delta t$. The price of the caplet at time t can be easily proved to be

$$E_t^\pi\left[e^{-\int_t^{t_1} r(s)\, ds} \max\left(\frac{1}{P(t_0, t_1)} - \alpha, 0\right)\right] = \alpha\, E_t^\pi\left[e^{-\int_t^{t_0} r(s)\, ds} \max\left(\frac{1}{\alpha} - P(t_0, t_1), 0\right)\right].$$

Thus a caplet is equivalent to α put options on a t_1 bond with delivery date t_0 and strike price $1/\alpha$ (see also Subsection 21.2.4).

Exercise 24.4.1: From Eq. (24.13) we have $\Theta_1 - \Theta_2 = \sigma(r, t)^2 (C_2 - C_1)/2$ [207].

Exercise 24.4.3: Rearrange the equation as

$$\frac{1}{2}\sigma(r)^2 \frac{\partial^2 P}{\partial r^2} + \mu(r)\frac{\partial P}{\partial r} - rP - \frac{\partial P}{\partial T} = 0$$

and assume that μ and σ are independent of time for simplicity. The partial differential equation becomes the following $N-1$ difference equations:

$$\frac{1}{2}\sigma_i^2 \frac{P_{i+1,j} - 2P_{i,j} + P_{i-1,j}}{(\Delta r)^2} + \mu_i \frac{P_{i+1,j} - P_{i,j}}{\Delta r} - rP_{i,j} - \frac{P_{i,j} - P_{i,j-1}}{\Delta t} = 0$$

for $1 \le i \le N-1$, where $\mu_i \equiv \mu(i\Delta r)$ and $\sigma_i \equiv \sigma(i\Delta r)$. (In the preceding equation, we could have used $(P_{i+1,j} - P_{i-1,j})/(2\Delta r)$ for $\partial P/\partial r$.) Regroup the terms by $P_{i,j-1}$, $P_{i-1,j}$, $P_{i,j}$, and $P_{i+1,j}$ to obtain the following system of equations at every time step:

$$a_i P_{i-1,j} + b_i P_{i,j} + c_i P_{i+1,j} = P_{i,j-1},$$

where

$$a_i \equiv -\left(\frac{\sigma_i}{\Delta r}\right)^2 \frac{\Delta t}{2}, \qquad b_i \equiv 1 + i\Delta r\,\Delta t + \left(\frac{\sigma_i}{\Delta r}\right)^2 \Delta t + \frac{\mu_i \Delta t}{\Delta r}, \qquad c_i \equiv -\left(\frac{\sigma_i}{\Delta r}\right)^2 \frac{\Delta t}{2} - \frac{\mu_i \Delta t}{\Delta r}.$$

Initially, the terminal conditions $P_{i,0}$ are given. The condition $\sigma(0) = 0$ implies that $a_0 = 0$. The other condition, $\lim_{r \to \infty} P(r, T) = 0$, leads to $P_{N,j} = 0$. The system of equations can be written as

$$
\begin{bmatrix}
b_0 & c_0 & 0 & \cdots & & \cdots & 0 \\
a_1 & b_1 & c_1 & 0 & \cdots & & 0 \\
0 & a_2 & b_2 & c_2 & 0 & \cdots & 0 \\
\vdots & \ddots & \ddots & \ddots & \ddots & & \vdots \\
\vdots & & \ddots & \ddots & \ddots & \ddots & \vdots \\
0 & \cdots & \cdots & 0 & a_{N-2} & b_{N-2} & c_{N-2} \\
0 & \cdots & \cdots & & 0 & a_{N-1} & b_{N-1}
\end{bmatrix}
\begin{bmatrix}
P_{0,j} \\
P_{1,j} \\
P_{2,j} \\
\vdots \\
\vdots \\
P_{N-1,j}
\end{bmatrix}
=
\begin{bmatrix}
P_{0,j-1} \\
P_{1,j-1} \\
P_{2,j-1} \\
\vdots \\
\vdots \\
P_{N-2,j-1} \\
P_{N-1,j-1}
\end{bmatrix}.
$$

We can obtain the $P_{i,j}$s by inverting the tridiagonal matrix given the $P_{i,j-1}$s. See Section 18.1 or [38, pp. 81–84]. Note that the time partial derivative of the Black–Scholes differential equation in Section 18.1 is with respect to t, not T here. This accounts for the slight notational differences from the implicit equations in Subsection 18.1.3.

In practice, a finite interval in the r axis such as $[0\%, 30\%]$ suffices to obtain good approximations. A better approach is to consider a transform like $x \equiv 1/(1+r)$ and solve the differential equation in terms of x, whose domain $(0, 1]$ is finite [38]. See also [859].

Exercise 24.4.4: The argument is the same as that leading to the term structure equation except that the futures contract's return is zero, not r [746, p. 565].

Exercise 24.5.1: In a risk-neutral probability measure, $\mu_p(t, T) = r(t)$ and

$$d\ln P(t, T) = \left[r(t) - \frac{1}{2}\sigma_p(t, T, P(t, T))^2\right] dt + \sigma_p(t, T, P(t, T))\, dW_t.$$

Recall from Eq. (24.3) that

$$f(t, T, \Delta t) = \frac{\ln P(t, T) - \ln P(t, T + \Delta t)}{\Delta t}.$$

Hence $df(t, T, \Delta t)$ equals

$$\frac{\sigma_p(t, T+\Delta t, P(t, T+\Delta t))^2 - \sigma_p(t, T, P(t, T))^2}{2\Delta t}\, dt$$

$$+ \frac{\sigma_p(t, T, P(t, T)) - \sigma_p(t, T+\Delta t, P(t, T+\Delta t))}{\Delta t}\, dW_t.$$

As $\Delta t \to 0$ the preceding equation becomes

$$d(f, T) = \sigma_p(t, T, P(t, T))\frac{\partial \sigma_p(t, T, P(t, T))}{\partial T}\, dt - \frac{\partial \sigma_p(t, T, P(t, T))}{\partial T}\, dW_t.$$

See [692, Subsection 19.3.2].

Exercise 24.5.2: Let $\sigma_p(t, T, P) = \psi(t) \ln P(t, T)$. The desired diffusion term, $-\partial \sigma_p(t, T, P)/\partial T$, equals $-\psi(t)(\partial \ln P(t, T)/\partial T) = \psi(t) f(t, T)$ because $f(t, T) = -\partial \ln P(t, T)/\partial T$. See [731, p. 378].

Exercise 24.6.2: In a period, bond one's price can go to $P_1 u_1$ or $P_1 d_1$, and bond two's price can go to $P_2 u_2$ or $P_2 d_2$. A portfolio consisting of one short unit of bond one and $\frac{P_1 u_1 - P_1 d_1}{P_2 u_2 - P_2 d_2}$ long units of bond two is riskless because its value equals $\frac{P_1(u_1 d_2 - u_2 d_1)}{u_2 - d_2}$ in either state. Hence

$$\left[-B_1 + \frac{P_1(u_1 - d_1)}{P_2(u_2 - d_2)} B_2 \right](1 + r) = \frac{P_1(u_1 d_2 - u_2 d_1)}{u_2 - d_2},$$

in which B_i is bond i's current price. Use the preceding equation and Eqs. (24.16) and (24.17) in the formula for λ to make sure that the final result is indeed independent of the bond. See [38, §3.3] and [781, Footnote 3].

Exercise 24.6.3: (1) In a period the portfolio's value is either $P_u + V_u(P_d - P_u)/(V_u - V_d)$ or $P_d + V_d(P_d - P_u)/(V_u - V_d)$. Because both equal $\frac{-P_u V_d + P_d V_u}{V_u - V_d}$, the return must be riskless. (2) Solve $(1 + x \frac{P_d - P_u}{V_u - V_d})R = \frac{-P_u V_d + P_d V_u}{V_u - V_d}$ for x. See [739].

Exercise 24.6.4: (1) To match the payoff one period from now, we need

$$u P(t, t+2) \Delta + e^r B = V_u,$$
$$d P(t, t+2) \Delta + e^r B = V_d.$$

It is easy to deduce from the preceding equations that the values are correct as listed. (2) Plug the values in (1) into $V = \Delta \times P(t, t+2) + B$. See [441].

Exercise 24.6.5: Simple algebraic manipulations [38, p. 44].

Exercise 24.6.6: Without the current term structure, the step $100/(1.05)^2 = 90.703$ would not have been taken, and everything that followed would have broken down. See [725, Example 6.1] for more information.

Exercise 24.6.7: $\frac{(90.703 \times 1.04) - 92.593}{98.039 - 92.593}$.

Exercise 24.6.8: When a security's value is matched in every state by the bond portfolio which is set up dynamically, it will be immunized. This notion admits of no arbitrage profits.

Exercise 24.7.1: (1) Because the bond can never be worth more than $100 if interest rates are non-negative. (2) This is due to the lognormal assumption for bond prices and the result that there is some probability that the bond will reach *any* given positive price. However, in fact, bond prices must lie between zero and the sum of the remaining cash flows if interest rates are nonnegative. See [304].

Exercise 24.7.2: No. The value of this option equals the maximum of zero and the discounted value of par minus the strike price, i.e., $\max(0, P(t, T)(100 - X))$, where T is the expiration date. See [304].

Chapter 25

Exercise 25.1.1: Rearrange the AR(1) process $X_t - b = a(X_{t-1} - b) + \epsilon_t$ to yield $X_t - X_{t-1} = (1 - a)(b - X_{t-1}) + \epsilon_t$, which is the discrete-time analog of the Vasicek model [41, p. 94]. Recall that the autocorrelation of X at lag one is a, which translates into $1 - \beta \, dt$ in the Vasicek model.

Exercise 25.1.2: (2) Observe that $\partial B(t, T)/\partial T = e^{-\beta(T-t)}$. From Eq. (24.4) we have

$$f(t, T) = -\frac{\sigma^2 \left(e^{-\beta(T-t)} - 1 \right)^2}{2\beta^2} - \mu(e^{-\beta(T-t)} - 1) + r(t) e^{-\beta(T-t)}.$$

As $E_t[r(T) | r(t)] = \mu + (r(t) - \mu) e^{-\beta(T-t)}$, the liquidity premium is $-\sigma^2 [e^{-\beta(T-t)} - 1]^2/(2\beta^2)$. Note that this premium is zero for $T = t$, as it should be. It is negative otherwise and converges to $-\sigma^2/(2\beta^2)$ as $T \to \infty$, which is the difference between the long rate and the long-term mean of the short rate, μ. See [855].

Exercise 25.1.3: Just plug in the following equations into Eq. (24.12):

$$\lambda(t, r) = 0,$$
$$\mu(r, t) = \beta(\mu - r),$$
$$\sigma(r, t) = \sigma,$$
$$\frac{\partial P/\partial T}{P} = -re^{-\beta(T-t)} + \mu e^{-\beta(T-t)} - \frac{\sigma^2}{\beta^2} e^{-\beta(T-t)} - \mu + \frac{\sigma^2}{2\beta^2} + \frac{\sigma^2}{2\beta^2} e^{-2\beta(T-t)},$$
$$\frac{\partial P/\partial r}{P} = -B(t, T) = \frac{e^{-\beta(T-t)} - 1}{\beta},$$
$$\frac{\partial^2 P/\partial r^2}{P} = B(t, T)^2 = \left[\frac{1 - e^{-\beta(T-t)}}{\beta}\right]^2.$$

The $\beta = 0$ case is simpler.

Exercise 25.1.4: Obtain the formula for $\partial P/\partial r$ from Exercise 25.1.3.

Exercise 25.1.5: From Exercise 25.1.4,

$$d \ln P = \left[r - \frac{1}{2} B(t, T)^2 \sigma^2\right] dt - B(t, T) \sigma \, dW.$$

Because $f(t, T) = -\partial \ln P(t, T)/\partial T$,

$$df = -\left[-B(t, T)\sigma^2 \frac{\partial B(t, T)}{\partial T}\right] dt + \sigma \frac{\partial B(t, T)}{\partial T} dW.$$

We can simplify the preceding equation by using $\partial B(t, T)/\partial T = e^{-\beta(T-t)}$ to

$$df = \left[\frac{1 - e^{-\beta(T-t)}}{\beta} e^{-\beta(T-t)} \sigma^2\right] dt + e^{-\beta(T-t)} \sigma \, dW.$$

See [16].

Exercise 25.1.6: Use $\ln P(t, T) = \ln A(t, T) - B(t, T) r(t)$ and Eq. (14.14) to prove that

$$\text{Var}[\ln P] = B(t, T)^2 \, \text{Var}[r(t)] = \frac{\sigma^2}{2\beta} \left[1 - e^{-2\beta(T-t)}\right].$$

Exercise 25.2.1: It approaches $2\beta\mu/(\beta + \gamma)$ by letting $T \to \infty$.

Exercise 25.2.4: The spot rate curve is $r(t, T) = -\frac{\ln P(t,T)}{T-t} = -\frac{\ln A(t,T)}{T-t} + \frac{B(t,T)r(t,T)}{T-t}$. So the spot rate volatility structure becomes $\frac{\partial r(t,T)}{\partial r} \sigma(r, t) = \sigma(r, t) B(t, T)/(T - t)$. See [731, p. 415].

Exercise 25.2.5: (2) It is $\sqrt{2(\phi_1\phi_2 - \phi_2^2)}$ [138].

Exercise 25.2.6: Equations (25.1) and (25.3) say that the spot rate is a linear function of the short rate under these two models, say $r(t, T) = a(T - t) + b(T - t) r(t)$. In particular, $r(T, T_1) = a' + b'r(T, T_2)$ for some a' and b'. Now

$$\max(0, r(T, T_1) - r(T, T_2)) = \max(0, a' + (b' - 1) r(T, T_2))$$
$$= (b' - 1) \times \max\left(0, r(T, T_2) - \frac{a'}{1 - b'}\right),$$

which means a portfolio of caplets. The case of floorlets is symmetric. See [616].

Exercise 25.2.9: Substitute the definitions of r^+, r^-, and $r = f(x)$ into Eq. (25.4) [268].

Exercise 25.2.11: See [868].

Exercise 25.2.12: Observe that all the rates on the same horizontal row are identical. Nodes with identical short rates generate identical term structures because the term structure depends solely on the prevailing short rate. As a result, we need to store only a vector of rates and a vector of probabilities. To answer (1), for example, we slide the two vectors backward in time as we perform the necessary computation on a third vector. See [405].

Exercise 25.2.13: Let $x \equiv f(y, t)$. By Ito's lemma,

$$dx = \left[\frac{\partial f}{\partial t} + \frac{\partial f}{\partial y} \alpha(y, t) + \frac{1}{2} \frac{\partial^2 f}{\partial y^2} \sigma(y, t)^2 \right] dt + \frac{\partial f}{\partial y} \sigma(y, t) \, dW.$$

The said transformation makes the diffusion term above equal one. See [696, 841].

Exercise 25.2.14: In asking if

$$\sigma(r, t)\sqrt{\Delta t} - \sigma(r_u, t)\sqrt{\Delta t} = -\sigma(r, t)\sqrt{\Delta t} + \sigma(r_d, t)\sqrt{\Delta t}, \tag{33.29}$$

we have $\sigma(r, t) = \sigma\sqrt{r}$ for the CIR model and $\sigma(r, t) = r\sigma$ for the geometric Brownian motion. Hence, for the CIR model,

$$r_u = r + \beta(\mu - r)\,\Delta t + \sigma\sqrt{r}\sqrt{\Delta t},$$
$$r_d = r + \beta(\mu - r)\,\Delta t - \sigma\sqrt{r}\sqrt{\Delta t},$$

whereas for the geometric Brownian motion,

$$r_u = r + r\mu\Delta t + r\sigma\sqrt{\Delta t},$$
$$r_d = r + r\mu\Delta t - r\sigma\sqrt{\Delta t},$$

It is easy to verify that Eq. (33.29) holds for the geometric Brownian motion but not for the CIR model.

Exercise 25.3.1: The tree should model $x(r) \equiv r^{1-\gamma}$ [475].

Exercise 25.5.1: Changes in slope are due to changes in a, changes in curvature are due to changes in b, and parallel moves are due to changes in r [239, 564].

CHAPTER 26

Exercise 26.1.1: The equilibrium model, as the no-arbitrage model is calibrated to the government bonds.

Exercise 26.2.3: Consider an n-period bond. Let $P_{ud}(t+2, t+n)$ denote its price when the short rate first rises and then declines. Similarly, let $P_{du}(t+2, t+n)$ denote its price when the short rate first declines and then rises. From Eq. (26.3),

$$
\begin{aligned}
P_{ud}(t+2, t+n) &= \frac{P_u(t+1, t+n)}{P_u(t+1, t+2)} \frac{2e^{v_3 + \cdots + v_n}}{1 + e^{v_3 + \cdots + v_n}} \\
&= \frac{P(t, t+n)}{P(t, t+1)} \frac{2}{1 + e^{v_2 + \cdots + v_n}} e^{r_2 + v_2} \frac{2e^{v_3 + \cdots + v_n}}{1 + e^{v_3 + \cdots + v_n}}, \\
P_{du}(t+2, t+n) &= \frac{P_d(t+1, t+n)}{P_d(t+1, t+2)} \frac{2}{1 + e^{v_3 + \cdots + v_n}} \\
&= \frac{P(t, t+n)}{P(t, t+1)} \frac{2e^{v_2 + \cdots + v_n}}{1 + e^{v_2 + \cdots + v_n}} e^{r_2} \frac{2}{1 + e^{v_3 + \cdots + v_n}}.
\end{aligned}
$$

These two formulas are indeed equal.

Exercise 26.2.5: The portfolio value is $P(t, t+T_1) + \beta P(t, t+T_2)$. It becomes $V_u = P_u(t+1, t+T_1) + \beta P_u(t+1, t+T_2)$ if the short rate rises and $V_d = P_d(t+1, t+T_1) + \beta P_d(t+1, t+T_2)$ if the short rate declines. To make it riskless, there must be no uncertainty, that is $V_u = V_d$. The implication is, by Eq. (26.3),

$$
\begin{aligned}
\beta &= -\frac{P_u(t+1, t+T_1) - P_d(t+1, t+T_1)}{P_u(t+1, t+T_2) - P_d(t+1, t+T_2)} \\
&= \frac{2 \frac{P(t, t+T_1)}{P(t, t+1)} \left(\frac{1 - \exp[v_2 + \cdots + v_{T_1}]}{1 + \exp[v_2 + \cdots + v_{T_1}]} \right)}{2 \frac{P(t, t+T_2)}{P(t, t+1)} \left(\frac{1 - \exp[v_2 + \cdots + v_{T_2}]}{1 + \exp[v_2 + \cdots + v_{T_2}]} \right)} \\
&= \frac{P(t, t+T_1)(1 - \exp[v_2 + \cdots + v_{T_1}])(1 + \exp[v_2 + \cdots + v_{T_2}])}{P(t, t+T_2)(1 - \exp[v_2 + \cdots + v_{T_2}])(1 + \exp[v_2 + \cdots + v_{T_1}])}.
\end{aligned}
$$

See [458].

Exercise 26.2.7: The return $r(t, t+n)$ is either $\ln P_d(t+1, t+n) - \ln P(t, t+n) = (n-1)v + C$ with probability $1 - p$ or $\ln P_u(t+1, t+n) - \ln P(t, t+n) = C$ with probability p for some constant C. The variance of $r(t, t+n)$ is hence $p(1-p)[(n-1)v]^2$. Equation (26.2) says that $\sigma^2 = p(1-p)v^2$. Hence the variance of $r(t, t+n)$ is $(n-1)^2\sigma^2$. As for the covariance, use Eq. (6.3). See [72].

Exercise 26.2.8: Observe that $\text{Var}[\xi_s] = (1/4)\ln\delta = v^2/4$ and $v = 2\sigma$.

Exercise 26.2.9:

$$-\ln\left(\frac{1+\delta^n}{1+\delta^{n+1}}\right) - \frac{1}{2}\ln\delta = \ln\left(\frac{\delta^{-1/2} + \delta^{n+1/2}}{1+\delta^n}\right)$$

$$= \ln\left(\frac{e^{-\sigma(\Delta t)^{1.5}} + e^{2\sigma\sqrt{\Delta t}\,[(T-t)+\Delta t/2]}}{1 + e^{2\sigma\sqrt{\Delta t}\,(T-t)}}\right)$$

$$\approx \ln\left(\frac{1 - \sigma(\Delta t)^{1.5} + e^{2\sigma\sqrt{\Delta t}\,[(T-t)+\Delta t/2]}}{1 + e^{2\sigma\sqrt{\Delta t}\,(T-t)}}\right)$$

$$\approx \frac{-\sigma(\Delta t)^{1.5} + e^{2\sigma\sqrt{\Delta t}\,[(T-t)+\Delta t/2]} - e^{2\sigma\sqrt{\Delta t}\,(T-t)}}{1 + e^{2\sigma\sqrt{\Delta t}\,(T-t)}}$$

$$\approx \frac{(1/2)(2\sigma\sqrt{\Delta t}\,[(T-t)+\Delta t/2])^2 - (1/2)[2\sigma\sqrt{\Delta t}\,(T-t)]^2}{2}$$

$$\approx \sigma^2(T-t)(\Delta t)^2.$$

Exercise 26.2.10: Following the proof of Exercise 26.2.9, we have

$$-\ln\left(\frac{1+\delta^{t-1}}{1+\delta^t}\right) - \frac{1}{2}\ln\delta = \ln\left(\frac{\delta^{-1/2} + \delta^{t-1/2}}{1+\delta^{t-1}}\right) \approx \sigma^2 t\,(\Delta t)^2.$$

Exercise 26.3.1: Rearrange Eqs. (26.7) and (26.8) as simultaneous equations:

$$f(P_u, P_d) \equiv P_u + P_d - \frac{2(1+r_1)}{(1+y)^i} = 0,$$

$$g(P_u, P_d) \equiv P_u^{-1/(i-1)} - 1 - e^{2\kappa}[P_d^{-1/(i-1)} - 1] = 0.$$

Because P_u and P_d are functions of r_i and v_i, $f(P_u, P_d)$ and $g(P_u, P_d)$ are also functions of r_i and v_i. Denote them by $F(r_i, v_i)$ and $G(r_i, v_i)$, respectively. For brevity, we use $f(r, v)$ instead of $f(r_i, v_i)$, $g(r, v)$ instead of $g(r_i, v_i)$, and so on. By the Newton–Raphson method, the $(k+1)$th approximation to (r_i, v_i) – denoted as $(r(k+1), v(k+1))$ – satisfies

$$\begin{bmatrix} \dfrac{\partial F(r(k), v(k))}{\partial r} & \dfrac{\partial F(r(k), v(k))}{\partial v} \\[2mm] \dfrac{\partial G(r(k), v(k))}{\partial r} & \dfrac{\partial G(r(k), v(k))}{\partial v} \end{bmatrix} \begin{bmatrix} \Delta r(k+1) \\[1mm] \Delta v(k+1) \end{bmatrix} = -\begin{bmatrix} F(r(k), v(k)) \\[1mm] G(r(k), v(k)) \end{bmatrix},$$

where $\Delta r(k+1) \equiv r(k+1) - r(k)$ and $\Delta v(k+1) \equiv v(k+1) - v(k)$. We need $\partial F/\partial r, \partial F/\partial v, \partial G/\partial r$, and $\partial G/\partial v$ to solve the preceding matrix for $(r(k+1), v(k+1))$. Obviously,

$$\frac{\partial F}{\partial r} = \frac{\partial P_u}{\partial r} + \frac{\partial P_d}{\partial r}, \qquad \frac{\partial F}{\partial v} = \frac{\partial P_u}{\partial v} + \frac{\partial P_d}{\partial v}.$$

By the chain rule,

$$\frac{\partial G}{\partial r} = \frac{\partial g}{\partial P_u}\frac{\partial P_u}{\partial r} + \frac{\partial g}{\partial P_d}\frac{\partial P_d}{\partial r}, \qquad \frac{\partial G}{\partial v} = \frac{\partial g}{\partial P_u}\frac{\partial P_u}{\partial v} + \frac{\partial g}{\partial P_d}\frac{\partial P_d}{\partial v}.$$

In the preceding four equations, the items we need to evaluate are

$$\frac{\partial P_u}{\partial r}, \quad \frac{\partial P_d}{\partial r}, \quad \frac{\partial P_u}{\partial v}, \quad \frac{\partial P_d}{\partial v}, \quad \frac{\partial g}{\partial P_u}, \quad \frac{\partial g}{\partial P_d}.$$

The differential tree method can compute them as follows. Working backward, the method ends up with $P_u, P_d, \partial P_u/\partial r, \partial P_d/\partial r, \partial P_u/\partial v$, and $\partial P_d/\partial v$. The remaining $\partial g/\partial P_u$ and $\partial g/\partial P_d$ can be computed directly from the definition of g:

$$\frac{\partial g}{\partial P_u} = -\frac{1}{i-1} P_u^{-i/(i-1)}, \qquad \frac{\partial g}{\partial P_d} = \frac{e^{2\kappa}}{i-1} P_d^{-i/(i-1)}.$$

Thus $\partial F/\partial r$, $\partial F/\partial v$, $\partial G/\partial r$, and $\partial G/\partial v$ can be computed.

This backward-induction algorithm runs in cubic time because the Newton–Raphson method takes only a few iterations to get to the desired accuracy. See [625].

Exercise 26.3.4: The proof is similar to Exercise 23.2.3. From the process,

$$
\begin{aligned}
\sigma(t)^2 \Delta t &= \text{Var}[\ln r(t+\Delta t) - \ln r(t)] \\
&= \text{Var}[\ln r(t)\{1 + \Delta r(t)/r(t)\} - \ln r(t)] \\
&= \text{Var}[\ln(1 + \Delta r(t)/r(t))] \\
&\approx \text{Var}[\Delta r(t)/r(t)] \\
&= \text{Var}[\Delta r(t)]/r(t)^2.
\end{aligned}
$$

So $\text{Var}[\Delta r(t)] \approx r(t)^2 \sigma(t)^2 \Delta t$. See [514, p. 482].

Exercise 26.4.1: The normal model does, for the following reason. It has fatter left tails and thinner right tails for the probability distribution of interest rates in the future (see Fig. 6.1). It therefore gives thinner left tails and fatter right tails for the probability distribution of bond prices. See [476].

Exercise 26.4.2: Apply the moment generating function of the normal distribution with mean $r_j + \mu_{i,j}\Delta t$ and variance $\sigma^2\Delta t$. In other words, use formula (6.8) with $t = -\Delta t$, $\mu = r_j + \mu_{i,j}\Delta t$, and $\sigma = \sigma\sqrt{\Delta t}$. The result is

$$e^{-r_j\Delta t - \mu_{i,j}(\Delta t)^2 + \sigma^2(\Delta t)^3/2} \approx e^{-r_j\Delta t}\left[1 - \mu_{i,j}(\Delta t)^2 + \frac{1}{2}\sigma^2(\Delta t)^3\right].$$

See [477].

Programming Assignment 26.4.3: Because the unreachable nodes' corresponding branch entries are useless, we can improve the algorithm by keeping track of the upper and the lower bounds of the reachable nodes in each column of branch and limiting the calculation to those nodes.

Programming Assignment 26.4.5: Apply the trinomial tree algorithm for the Hull–White model to $x \equiv \ln r$. See also [215, 848].

Exercise 26.5.1: It is $(\sigma^2/\kappa)(e^{-\kappa(T-t)} - e^{-2\kappa(T-t)})$ [746, p. 581].

Exercise 26.5.2: From Eq. (26.14), the process for $df(t, T)$ in Eq. (26.15) can be written as

$$df(t, T) = \mu(t, T)\,dt + \sigma(t, T)\,dW_t.$$

Now $r(t) = f(t, t) = f(0, t) + \int_0^t \mu(s, t)\,ds + \int_0^t \sigma(s, t)\,dW_s$. Differentiate it with respect to t to obtain

$$
\begin{aligned}
dr(t) &= \frac{\partial f(0, t)}{\partial t}\,dt + \left[\int_0^t \frac{\partial \mu(s, t)}{\partial t}\,ds\right]dt + \left[\int_0^t \frac{\partial \sigma(s, t)}{\partial t}\,dW_s\right]dt + \mu(t, t)\,dt + \sigma(t, t)\,dW_t \\
&= \left\{\frac{\partial}{\partial T}\left[f(0, T) + \int_0^t \mu(s, T)\,ds + \int_0^t \sigma(s, T)\,dW_s\right]\bigg|_{T=t}\right\}dt + \mu(t, t)\,dt + \sigma(t, t)\,dW_t \\
&= \left[\mu(t, t) + \frac{\partial f(t, T)}{\partial T}\bigg|_{T=t}\right]dt + \sigma(t, t)\,dW_t \\
&= \left[\sigma(t, t)\lambda(t) + \frac{\partial f(t, T)}{\partial T}\bigg|_{T=t}\right]dt + \sigma(t, t)\,dW_t.
\end{aligned}
$$

See [515, Appendix 1].

Exercise 26.5.4: From Eq. (26.21),

$$C_t = P(t, T)\,N(d_t) - XP(t, s)\,N(d_t - \sigma_t),$$

where $d_t \equiv (1/\sigma_t)\ln(P(t, T)/[XP(t, s)]) + (\sigma_t/2)$. (1) Let $\sigma_t \equiv \sigma(T-s)\sqrt{s-t}$. (2) Let $\sigma_t \equiv \beta(s-t, T-t)\sqrt{\phi(s-t)}$.

Exercise 26.5.5: From Example 26.5.3 it is $\frac{\partial r(t, T)}{\partial r(t)}\sigma = -\frac{1}{T-t}\frac{\partial \ln P(t, T)}{\partial r(t)}\sigma = \beta(t, T)\frac{\sigma}{T-t}$.

Exercise 26.5.7: If every node contains forward rates up to the maturity of the underlying bond, a term structure up to that maturity can be constructed. The tree then needs to extend to only the maturity of the claim.

Exercise 26.5.8: Use Eqs. (24.5), (26.23), and (26.24) to arrive at

$$
1 = \frac{1}{2} \times \exp\left[-\left\{ \Delta t \int_{t+\Delta t}^{T} \mu(t, u)\, du + \sqrt{\Delta t} \int_{t+\Delta t}^{T} \sigma(t, u)\, du \right\} \right]
$$
$$
+ \frac{1}{2} \times \exp\left[-\left\{ \Delta t \int_{t+\Delta t}^{T} \mu(t, u)\, du - \sqrt{\Delta t} \int_{t+\Delta t}^{T} \sigma(t, u)\, du \right\} \right].
$$

Exercise 26.5.9: Equation (26.25) can be applied iteratively to obtain

$$
(\Delta t)^2 \mu(t, T) \approx \ln \frac{e^x + e^{-x}}{e^y + e^{-y}}
$$

$$
\approx \frac{e^x - e^y + e^{-x} - e^{-y}}{e^y + e^{-y}}
$$

$$
\approx \frac{[e^{(\Delta t)^{1.5}\sigma(t,T)} - 1]e^y - [1 - e^{-(\Delta t)^{1.5}\sigma(t,T)}]e^{-y}}{e^y + e^{-y}}
$$

$$
\approx \frac{(\Delta t)^{1.5}\sigma(t, T)\, e^y - (\Delta t)^{1.5}\sigma(t, T)\, e^{-y}}{e^y + e^{-y}}
$$

$$
= (\Delta t)^{1.5}\sigma(t, T) \frac{e^y - e^{-y}}{e^y + e^{-y}}
$$

$$
\approx (\Delta t)^{1.5}\sigma(t, T) \frac{e^x - e^{-x}}{e^x + e^{-x}}
$$

$$
= (\Delta t)^{1.5}\sigma(t, T)\, \tanh(x),
$$

where $x \equiv \sqrt{\Delta t} \int_{t+\Delta t}^{T} \sigma(t, u)\, du$ and $y \equiv \sqrt{\Delta t} \int_{t+\Delta t}^{T-\Delta t} \sigma(t, u)\, du$.

Chapter 27

Exercise 27.3.1: Without loss of generality, assume that the par value of the bond is $1. Let the call price at time t be $C(t)$. Suppose the sinking-fund provision requires the issuer to retire the principal by F_i dollars at time t_i for $i = 1, 2, \ldots, n$, where $\sum_{i=1}^{n} F_i = 1$. As before, it is best to view the embedded option as n separate options. In particular, the ith option has the right to buy a face value F_i callable bond at par at maturity (time t_i) with the call schedule equal to $F_i C(t)$ for $0 \le t \le t_i$. The underlying bond is otherwise identical to the original bond. The option value can be calculated by backward induction. The desired price is that of the otherwise identical straight bond minus the n options. See [848].

Exercise 27.3.2: Suppose the bond were trading below its conversion value. Consider the following strategy: Buy one CB and sell short the stock with the number of shares equal to the conversion ratio, and cover the short position by conversion. Note that there is a positive initial cash flow, and the number of shares exactly covers the short position. See [325, p. 374].

Exercise 27.3.3: As bondholders convert, the price of the stock will decline because of dilution. The correct definition for the conversion value would use the stock price *after* conversion, taking into account the dilution issue. However, here we are assuming that the CBs in questions are the only securities to affect the number of outstanding shares, which is rarely true. See [221, Chapter 8] and [325, p. 373n].

Exercise 27.3.4: A CB's market value is at least its conversion value, but early conversion generates exactly the conversion value.

Exercise 27.3.5: If the current stock price is so high that the conversion option always expires in the money and is exercised, every node will be an exercise node by induction from the terminal nodes. As a result, every node's CB value equals the stock price.

Exercise 27.3.6: On a per-share basis, the desired relation is

$$
\text{CB value } (B) = \text{warrant value } (W) + \text{straight value } (s), \tag{33.30}
$$

where the warrant's strike price X equals the CB's conversion price P plus the coupon payment. Assume that the stock price of the issuer follows the binomial model and that there are n periods before maturity. Let c_i be the coupon payment at time, P be the par value of the option-free bond, R be the riskless gross return per period, and $s_i \equiv P R^{-(n-i)} + \sum_{j=i}^{n} c_j R^{-(j-i)}$, the straight value at time i. Identity (33.30) can be proved by induction as follows. At maturity, $W = \max(S - X, 0)$ and $B = \max(S, X) = W + (P + c_n) = W + s_n$. Inductively, at time i,

$$B = \frac{p B_u + (1 - p) B_d}{R} + c_i = \frac{p(W_u + s_{i+1}) + (1 - p)(W_d + s_{i+1})}{R} + c_i = W + s_i.$$

See [221, Chapter 8].

Exercise 27.4.1: It is identical to the valuation of riskless bonds by use of backward induction in a risk-neutral economy except that at every node at time i, the bond may branch with probability p_i into a default state and thus pay zero dollar at maturity. In other words, the risk-neutral probabilities at time i are those for the riskless bond multiplied by $1 - p_i$. See [514, p. 570] or [535] for the more general case in which the firm pays a fraction of the par value instead of zero.

Exercise 27.4.2: When interest rates rise, both the price of the noncallable bond component and the price of the embedded call option fall. The fall in price of the embedded option will offset some of the fall in price of the noncallable bond. See [325, p. 324].

Exercise 27.4.3: (1) From Eq. (27.1), $price_c = price_{nc} - call$ price. Differentiate both sides and then divide by the price of the callable bond, $price_c$, to obtain

$$-OAD_c = \frac{\partial(price_{nc})}{\partial y} \frac{1}{price_c} - \frac{\partial(call\ price)}{\partial y} \frac{1}{price_c}. \tag{33.31}$$

The first term on the right-hand side of Eq. (33.31) is

$$\frac{\partial(price_{nc})}{\partial y} \frac{1}{price_{nc}} \frac{price_{nc}}{price_c} = -duration_{nc} \times \frac{price_{nc}}{price_c}. \tag{33.32}$$

From the chain rule and delta's definition,

$$\frac{\partial(call\ price)}{\partial y} = \frac{\partial(call\ price)}{\partial(price_{nc})} \frac{\partial(price_{nc})}{\partial y} = \Delta \times \frac{\partial(price_{nc})}{\partial y}.$$

The second term on the right-hand side of Eq. (33.31) thus is

$$\frac{\partial(call\ price)}{\partial y} \frac{1}{price_{nc}} \frac{price_{nc}}{price_c} = \Delta \times \frac{\partial(price_{nc})}{\partial y} \frac{1}{price_{nc}} \frac{price_{nc}}{price_c}$$

$$= -\Delta \times duration_{nc} \times \frac{price_{nc}}{price_c}.$$

Combine Eq. (33.32) and the preceding equation to yield the desired result. See [325, pp. 332–334].

Exercise 27.4.4: See Exercise 23.3.1.

Exercise 27.4.5: Increased volatility makes the bond more likely to be called, reducing the investor's return.

Exercise 27.4.6: (1) The OAS increases. The price of course cannot drop below par. (2) The OAS decreases but will level off at zero after a certain point.

Exercise 27.4.7: Consider a T-period zero-coupon bond callable at time m with call price X. For ease of comparison, assume a binomial interest rate process in which $r_n(i)$ denotes the short rate for period i on the nth path. The true value is derived by backward induction thus:

$$P_b \equiv \frac{1}{2^T} \sum_{n_1=1}^{2^m} \times \min \left(\sum_{n_2=1}^{2^{T-m}} \frac{100}{\prod_{i=1}^{T}[1 + r_{n_1 n_2}(i) + OAS]}, \frac{X}{\prod_{i=1}^{m}[1 + r_{n_1 n_2}(i) + OAS]} \right),$$

whereas the Monte Carlo method calculates

$$P_{MC} \equiv \frac{1}{2^T} \sum_{n=1}^{2^T} \frac{1}{\prod_{i=1}^{m}[1+r_n(i)+\text{OAS}]} \times \min\left(\frac{100}{\prod_{i=m+1}^{T}[1+r_n(i)+\text{OAS}]}, X\right)$$

$$= \frac{1}{2^T} \sum_{n_1=1}^{2^m} \sum_{n_2=1}^{2^{T-m}} \frac{1}{\prod_{i=1}^{m}[1+r_{n_1 n_2}(i)+\text{OAS}]} \times \min\left(\frac{100}{\prod_{i=m+1}^{T}[1+r_{n_1 n_2}(i)+\text{OAS}]}, X\right)$$

$$= \frac{1}{2^T} \sum_{n_1=1}^{2^m} \sum_{n_2=1}^{2^{T-m}} \times \min\left(\frac{100}{\prod_{i=1}^{T}[1+r_{n_1 n_2}(i)+\text{OAS}]}, \frac{X}{\prod_{i=1}^{m}[1+r_{n_1 n_2}(i)+\text{OAS}]}\right).$$

From Jensen's inequality $\sum_{n=1}^{l} \min(x_n, c) \le \min(\sum_{n=1}^{l} x_n, c)$, we conclude that $P_{MC} \le P_b$. Similarly the Monte Carlo method tends to overestimate the value of putable bonds. See [349].

Exercise 27.4.11: In Fig. 27.7(d), (1) change 0.936 to 1.754 because $1.754 = (106.754 - 5) - 100$ and (2) change 0.482 to 0.873 because $0.873 = (1.754 + 0.071)/1.045$.

Exercise 27.4.12: We prove the claim for binomial interest rate trees with reference to Fig. 23.1. The same argument can be applied to any tree. Let s represent the OAS. Consider zero-coupon bonds first. Backward induction performs

$$P_A = \frac{P_B + P_C}{2e^{r+s}} = e^{-s}\frac{P_B + P_C}{2e^r}$$

at each node. It follows by mathematical induction that the n-period benchmark bond price $P_b(n)$ and the n-period nonbenchmark bond price $P_{nb}(n)$ are related by $P_b(n) = P_{nb}(n)\,e^{sn}$. Let $r_b(n)$ and $r_{nb}(n)$ denote their respective yields. Then $e^{-r_b(n)n} = e^{-r_{nb}(n)n}e^{sn}$, i.e., $r_b(n) = r_{nb}(n) - s$. By assumption, $r_b(n)$ are all identical. We can hence drop the dependence on the maturity n, and the identity becomes $r_b = r_{nb} - s$. Consider coupon bonds with cash flow $C_1, C_2, \ldots, C_n$. Then

$$P_{nb} \equiv \sum_{i=1}^{n} C_i\, P_{nb}(n) = \sum_{i=1}^{n} C_i e^{-r_{nb}i} = \sum_{i=1}^{n} C_i e^{-(r_b+s)i}.$$

The yield spread is hence s, as claimed.

CHAPTER 28

Exercise 28.5.1: It is because the servicing and guaranteeing fee is a percentage of the remaining principal, and the principal is paid down through time.

Exercise 28.6.1: The investor first has a capital gain. Now a pass-through trades at a discount because its coupon rate is lower than the current coupon rate of new issues. The prepayments therefore can be reinvested at a higher coupon rate. See [325, p. 254].

Exercise 28.6.2: Given the premise, there is a smaller incentive to refinance a loan with a lower remaining balance, other things being equal. Because 30-year loans amortize more slowly than 15-year loans, MBSs backed by 30-year mortgages should prepay faster. See [433].

Exercise 28.6.3: Even without the transactions costs, expression (28.2) says that refinancing does not make economic sense unless the old rate exceeds the new one.

Exercise 28.6.4: Expression (28.1) says that, at the first refinancing, the remaining balance is

$$C\frac{1 - (1+r)^{-n+a}}{r},$$

where C is the monthly payment. However, it is easy to show (or from Eq. (29.4)) that

$$C = \mathcal{O} \times \frac{r(1+r)^n}{(1+r)^n - 1}, \tag{33.33}$$

where $\mathcal{O}$ denotes the original balance. When the preceding expressions are combined, the remaining balance is

$$\mathcal{O}' = \mathcal{O} \times \frac{(1+r)^n}{(1+r)^n - 1}[1 - (1+r)^{-n+a}] = \mathcal{O} \times \frac{(1+r)^n - (1+r)^a}{(1+r)^n - 1}.$$

The new monthly payment is

$$C' = \mathcal{O}' \times \frac{r(1+r)^n}{(1+r)^n - 1} = \mathcal{O} \times \frac{(1+r)^n - (1+r)^a}{(1+r)^n - 1}\frac{r(1+r)^n}{(1+r)^n - 1}$$

by Eq. (33.33). After the preceding process is iterated, the monthly payment after the ith refinancing is

$$\mathcal{O} \times \left[\frac{(1+r)^n - (1+r)^a}{(1+r)^n - 1} \right]^i \frac{r(1+r)^n}{(1+r)^n - 1}.$$

Exercise 28.6.5: Although both scenarios have the same rate difference of 2%, the second deal is better because of a higher refinancing incentive, as prescribed by expression (28.3).

CHAPTER 29

Exercise 29.1.2: From Eq. (29.2) with $k = 1$, the remaining principal balance per $1 of original principal balance is

$$1 - \frac{x-1}{x^n - 1} = 1 - \frac{1}{x^{n-1} + x^{n-2} + \cdots + 1},$$

where $x \equiv 1 + r/m \geq 1$. The preceding equation is monotonically increasing in x. Hence different r's yield different balances as they give rise to different x's.

Programming Assignment 29.1.3: See Fig. 33.8.

Exercise 29.1.5: First, run the algorithm in Fig. 33.8 but on a short rate tree with $n + k - 1$ periods to obtain all the desired spot rates up to time n when the swap expires. Only the nodes up to time $n - 1$ are needed. Now use backward induction to derive at each node the price of a swap with $1 of notional principal initiated at that node and ending at time n. The swap's amortization amount follows the original swap's. Specifically, let a be the amortizing amount at the current node, which is determined uniquely by the node's k-period spot rate. Then the value at the node equals $C + (1-a)(p_u P_u + p_d P_d)/(1+r)$, where r is the node's short rate, C is its cash flow, p_u and p_d are the branching probabilities, and P_u and P_d are the values of the swaps as described above at the two successor nodes. Note the use of scaling. The running time is $O(kn^2)$.

Exercise 29.1.6:

Month	6	12	18	24	30	36
CPR	1.2	3.12	5.544	11.04	8.1	7.5

with the help of Eq. (29.5).

Exercise 29.1.7: No. See [323, p. 362] for an example.

Algorithm for generating spot rate dynamics:

```
input:    n, k, r[ n ][ n ];
real      s[ n − k + 1 ][ n − k + 1 ], P[ n ];
integer   i, j, l;
for (i = n − k down to 0) {
          // Backward induction to obtain discount factors.
          for (j = 0 to i + k − 1)
            P[ j ] := 1/(1 + r[ i + k − 1 ][ j ]);
          for (l = i + k − 2 down to i)
            for (j = 0 to l)
              P[ j ] := 0.5(P[ j ] + P[ j + 1 ])/(r[ l ][ j ] + 1);
          // Turn discount factors into spot rates.
          for (j = 0 to i) s[ i ][ j ] := P[ j ]^{−1/k} − 1;
}
return s[ ][ ];
```

Figure 33.8: Algorithm for generating spot rate dynamics. $r[i][j]$ is the $(j+1)$th short rate for period $i + 1$, the short rate tree covers n periods, and k is the maturity of the desired spot rates. The spot rates are stored in $s[\ ][\ ]$. Specifically, $s[i][j]$ refers to the desired spot rate at the same node as $r[i][j]$, where $0 \leq i \leq n - k$ and $0 \leq j \leq i$. All rates are measured by the period.

Exercise 29.1.8: Sum $\overline{P_i}$ of Eq. (29.6) and $\overline{I_i}$ of Eq. (29.7) with $\alpha = 0$.

Exercise 29.1.9: (1) Substitute Eq. (29.8) into the formula's B_i (2)

$$\overline{P_i} + \mathrm{PP}_i = b_{i-1} P_i + B_{i-1} \frac{\mathrm{RB}_i}{\mathrm{RB}_{i-1}} \times \mathrm{SMM}_i$$

$$= b_{i-1} P_i + \mathrm{RB}_{i-1} \times b_{i-1} \frac{\mathrm{RB}_i}{\mathrm{RB}_{i-1}} \times \mathrm{SMM}_i$$

$$= b_{i-1}(P_i + \mathrm{RB}_i \times \mathrm{SMM}_i).$$

Exercise 29.1.10: For Eq. (29.9), just observe that $B_i/\mathrm{Bal}_i = (B_{i-1}/\mathrm{Bal}_{i-1})(1 - \mathrm{SMM}_i)$. As for Eqs. (29.10), use Eqs. (29.2) and (29.3) to prove the formula for $\overline{P_i}$. The formula for $\overline{I_i}$ is due to amortization.

Exercise 29.1.12: Let s denote the SMM. For simplicity, assume that the original balance is \$1 and r is the period yield:

$$\mathrm{PO} = \sum_{i=1}^{n} \frac{\overline{P_i} + \mathrm{PP}_i}{(1+r)^i}$$

$$= \sum_{i=1}^{n} \frac{(1-s)^{i-1} \frac{r(1+r)^{i-1}}{(1+r)^n - 1} + B_i \frac{s}{1-s}}{(1+r)^i} \quad \text{from Eqs. (29.3) and (29.10) and Exercise 29.1.9(1)}$$

$$= \sum_{i=1}^{n} \frac{(1-s)^{i-1} \frac{r(1+r)^{i-1}}{(1+r)^n - 1} + \frac{(1+r)^n - (1+r)^i}{(1+r)^n - 1} s(1-s)^{i-1}}{(1+r)^i} \quad \text{from Eqs. (29.2) and (29.9)}$$

$$= \frac{1}{(1+r)^n - 1} \left\{ \frac{r}{1+r} \frac{1 - (1-s)^n}{s} + s \frac{(1+r)^n - (1-s)^n}{r+s} - [1 - (1-s)^n] \right\}.$$

Similarly,

$$\mathrm{IO} = \sum_{i=1}^{n} \frac{\overline{I_i}}{(1+r)^i} = \sum_{i=1}^{n} \frac{(1-s)^{i-1} \times \mathrm{RB}_{i-1} \times r}{(1+r)^i}$$

$$= r \sum_{i=1}^{n} \frac{(1-s)^{i-1} \frac{(1+r)^n - (1+r)^{i-1}}{(1+r)^n - 1}}{(1+r)^i} \quad \text{from (29.2)}$$

$$= \frac{r}{(1+r)^n - 1} \left[\frac{(1+r)^n - (1-s)^n}{r+s} - \frac{1}{1+r} \frac{1 - (1-s)^n}{s} \right].$$

Exercise 29.1.13: Let the principal payments be $P_1, P_2, \ldots, P_n$ and the interest payment at time i be $I_i \equiv r(P - \sum_{j=1}^{i-1} P_j)$ by the principle of amortization, where r is the period yield and $P = P_1 + P_2 + \cdots + P_n$ is the original principal amount. The FV of the combined cash flow $P_i + I_i$, $i = 1, 2, \ldots, n$, is

$$\sum_{i=1}^{n} \left[P_i + r \left(P - \sum_{j=1}^{i-1} P_j \right) \right] (1+r)^{n-i} = \sum_{i=1}^{n} \left(P_i + r \sum_{j=i}^{n} P_j \right) (1+r)^{n-i}$$

$$= \sum_{i=1}^{n} P_i (1+r)^{n-i} \left[1 + \sum_{j=1}^{i} r(1+r)^{i-j} \right]$$

$$= \sum_{i=1}^{n} P_i (1+r)^{n-i} (1+r)^i$$

$$= P(1+r)^n,$$

independent of how P is distributed among the P_is.

Exercise 29.1.15: They are \$125,618, \$115,131, \$67,975, \$34,612, \$20,870, and \$0 for months 1–6. See the formula in Exercise 29.1.9(1).

Programming Assignment 29.1.16: Validate your program by running it with zero original balance for tranche C (that is, $\mathcal{O}[3] = 0$) and comparing the output with those in Fig. 29.11.

Exercise 29.2.1: Because the fast refinancers will exit the pool at a faster rate, an increasingly larger proportion of the remaining population will be the slow refinancers. Consequently, the refinancing rate of the pool will move toward that of the slow refinancers. See [433].

Exercise 29.2.2: PO $\approx s/(r+s)$ and IO $\approx r/(r+s)$. Hence, $\frac{\partial(\mathrm{PO})/\partial s}{\mathrm{PO}} \approx \frac{r}{s(r+s)}$ and $\frac{\partial(\mathrm{IO})/\partial s}{\mathrm{IO}} \approx -\frac{1}{r+s}$.

Exercise 29.2.3: Like interests, servicing fees are a percentage of the principal [829, p. 105].

Exercise 29.3.1: A premium-priced MBS must have a coupon rate exceeding the market discount rate. This means that the prepaid principal is less than the PV of the future cash flow foregone by such a prepayment. Hence prepayment depresses the value of the cash flow. The argument for discount MBSs is symmetric.

Exercise 29.3.2: The formula is valid only if the cash flow is independent of yields. For MBSs, this does not hold. See [55, p. 132].

Exercise 29.3.3: Refer to Fig. 28.8. In a bull market, Treasury securities' prices go up, whereas the MBSs' prices go down. This hedger therefore has the worst of both worlds, losing money on both. See [54, p. 204].

Exercise 29.3.4: This model does not take into account the negative convexity of the security [304].

CHAPTER 30

Exercise 30.2.1:

LIBOR Change (Basis Points)	−300	−200	−100	0	+100	+200	+300
Conventional floater	4.5	5.5	6.5	7.5	8.5	9.5	10.5
Superfloater	2.0	3.5	5.0	6.5	8.0	9.5	11.0

CHAPTER 31

Exercise 31.1.1: Let $\omega \equiv [\omega_1, \omega_2, \ldots, \omega_n]^{\mathrm{T}}$ be the vector of portfolio weights, $\bar{r} \equiv [\bar{r}_1, \bar{r}_2, \ldots, \bar{r}_n]^{\mathrm{T}}$ be the vector of mean security returns, and Q be the covariance matrix of security returns. The efficient portfolio for a target return of $\bar{r}$ is determined by

$$
\begin{aligned}
\text{minimize} \quad & \min_{\omega} \omega^{\mathrm{T}} Q \omega, \\
\text{subject to} \quad & \omega^{\mathrm{T}} \bar{r} = \bar{r}, \\
& \omega^{\mathrm{T}} \mathbf{1} = 1,
\end{aligned}
$$

where $\mathbf{1} \equiv [1, 1, \ldots, 1]^{\mathrm{T}}$. See [3].

Exercise 31.1.2: The variance is $\omega_1^2 \sigma_1^2 + (1 - \omega_1)^2 \sigma_2^2 - 2\omega_1(1 - \omega_1)\sigma_1\sigma_2$. Pick $\omega_1 = \sigma_2/(\sigma_1 + \sigma_2)$, which lies between zero and one, to make the variance zero. See [317, p. 74].

Exercise 31.1.3: (1) The problem is

$$
\begin{aligned}
\text{minimize} \quad & -(1/2)\sum_i \sum_j \omega_i \omega_j \sigma_{ij}, \\
\text{subject to} \quad & \sum_i \omega_i = 1, \\
& \omega_i \geq 0 \quad \text{for all } i.
\end{aligned}
$$

Let i^* be such that $\sigma_{i^*i^*} = \max_i \sigma_{ii}$. It is not hard to show that $\sigma_{ii}\sigma_{jj} > \sigma_{ij}^2$ for all i, j [465, p. 398]; hence $\sigma_{i^*i^*} > \sigma_{i^*i}$ for all $i \neq i^*$. The **Kuhn–Tucker conditions** for the optimal solution say that there exist μ and $\nu_1, \nu_2, \ldots$ such that [721, pp. 522–523]:

$$
-\sum_j \sigma_{ij}\omega_j + \mu - \nu_i = 0 \quad \text{for all } i,
$$

$$
\sum_i \omega_i = 1,
$$

$$
\omega_i \geq 0 \quad \text{for all } i,
$$

$$
\nu_i \geq 0 \quad \text{for all } i,
$$

$$
\nu_i \omega_i = 0 \quad \text{for all } i.
$$

Observe that $\sum_i (\mu - v_i) \omega_i = \mu > 0$ is the desired value. The feasible solution in which $\omega_{i^*} = 1$ and $\omega_i = 0$ for $i \neq i^*$ satisfies the above by the selection of $\mu = \sigma_{i^* i^*}$ and $v_i = \sigma_{i^* i^*} - \sigma_{i^* i} \geq 0$ for all i.

(2) The problem is

$$\text{minimize} \quad (1/2) \sum_i \sum_j \omega_i \omega_j \sigma_{ij},$$

$$\text{subject to} \quad \sum_i \omega_i = 1,$$

$$\omega_i \geq 0 \quad \text{for all } i.$$

Use the Lagrange multiplier to show that the optimal solution satisfies

$$\sum_j \sigma_{ij} \omega_j + \mu = 0 \quad \text{for all } i,$$

$$\sum_i \omega_i = 1,$$

$$\omega_i \geq 0 \quad \text{for all } i.$$

Observe that $-\mu \sum_i \omega_i = -\mu > 0$ is the desired value. Let $\omega \equiv [\omega_1, \omega_2, \ldots, \omega_n]^T$ and $\mathbf{1} \equiv [1, 1, \ldots, 1]^T$. Hence, $\omega = -\mu C^{-1} \mathbf{1}$, which satisfies the conditions $\omega_i \geq 0$ for all i by the assumption. Combining this with $\mathbf{1}^T \omega = 1$, we have $\mathbf{1}^T (-\mu C^{-1} \mathbf{1}) = 1$, implying that $-\mu = 1/(\mathbf{1}^T C^{-1} \mathbf{1})$. If we let $C^{-1} \equiv [a_{ij}]$, then $-\mu = 1/(\sum_i \sum_j a_{ij})$.

Exercise 31.1.4: Note that $r(T) + 1$ is lognormally distributed and $r_c(T)$ is normally distributed. The mean and the variance of $r(T)$ and those of $r_c(T)$ are therefore related by

$$\mu(T) = e^{\mu_c(T) + 0.5\sigma_c^2(T)} - 1,$$

$$\sigma^2(T) = e^{2\mu_c(T) + 2\sigma_c^2(T)} - e^{2\mu_c(T) + \sigma_c^2(T)},$$

according to Eq. (6.11). Alternative formulations are

$$\mu_c(T) = 2\ln(1 + \mu(T)) - \frac{\ln(\sigma^2(T) + [1 + \mu(T)]^2)}{2},$$

$$\sigma_c^2(T) = \ln\left(1 + \left[\frac{\sigma(T)}{1 + \mu(T)}\right]^2\right).$$

See [646, p. 105].

Exercise 31.1.5: The return rate of each individual asset is $\mu - \sigma^2/2$ (see Exercise 13.3.8(2)). The variance of the portfolio is σ^2/n; the portfolio's return rate is hence $\mu - \sigma^2/(2n)$. Their difference is $\sigma^2/2 - \sigma^2/(2n) = (1 - 1/n)(\sigma^2/2)$. See [623, p. 429].

Exercise 31.1.6: Because all investors trade the same fund of risky assets, trading activity in each stock as a fraction of its shares outstanding is identical across all stocks [614].

Exercise 31.1.7: Replace every variable ω_i with $Y_i - Z_i$ and add the requirements $Y_i \geq 0$ and $Z_i \geq 0$. Under the new model, the equation about means remains a homogeneous linear form in the variables and the equation about covariances remains a homogeneous quadratic form. It is thus a general portfolio selection model. The new model is strictly equivalent to the original model for the following reasons. If $(Y_1, Z_1, Y_2, Z_2, \ldots, Y_n, Z_n)$ is a feasible portfolio for the new model, then $(\omega_1, \omega_2, \ldots, \omega_n)$ with $\omega_i \equiv Y_i - Z_i$ is feasible for the original model with the same mean and standard deviation. Conversely, if $(\omega_1, \omega_2, \ldots, \omega_n)$ is feasible for the original model, then $(Y_1, Z_1, Y_2, Z_2, \ldots, Y_n, Z_n)$ with $Y_i \equiv \max(\omega_i, 0)$ and $Z_i \equiv \max(-\omega_i, 0)$ is feasible for the new model, and with the same mean and standard deviation. See [642, p. 26].

Exercise 31.2.1: It has been shown in the text that the combination of minimum-variance portfolios results in a minimum-variance portfolio. In fact, the combination of efficient portfolios results in an efficient portfolio if the weights applied to the portfolios are all nonnegative. This is because the resulting portfolio is a minimum-variance portfolio with an expected rate of return at or above the MVP's, making it efficient by definition. Because investors hold only efficient portfolios, the market is a combination of efficient portfolios and thus is efficient as well.

Exercise 31.2.2: Let the market return be $r_M = \sum_i \omega_i' r_i$, where ω_i' are market proportions. The fact that every investor holds the same market portfolio implies that the market portfolio satisfies Eq. (31.1), or $\lambda \sum_i \sigma_{ij} \omega_i' = \bar{r}_j - r_f$. Note that λ is independent of the choice of j. Because

$$\text{Cov}[r_j, r_M] = E\left[(r_j - \bar{r}_j)\sum_i \omega_i'(r_i - \bar{r}_i)\right] = \sum_i \omega_i' E[(r_j - \bar{r}_j)(r_i - \bar{r}_i)] = \sum_i \omega_i' \sigma_{ij},$$

we recognize that $\lambda \text{Cov}[r_j, r_M] = \bar{r}_j - r_f$. For the market portfolio, $\lambda \text{Cov}[r_M, r_M] = \bar{r}_M - r_f$, implying $\lambda = (\bar{r}_M - r_f)/\sigma_M^2$. See [317, pp. 303–304], [623, p. 177], and [799, pp. 287–289].

Exercise 31.2.3: The risk of an asset that is uncorrelated with the market can be diversified away because purchasing many such assets that are also mutually uncorrelated results in a small total variance [623, p. 179].

Exercise 31.2.4: Such an asset reduces the overall portfolio risk when combined with the market. Investors must pay for this risk-reducing benefit. Another example is insurance. See [424, p. 213], [623, p. 179], and [799, p. 271].

Exercise 31.2.5: They all have the same level of systematic risk, thus beta. The part of the total risk that is specific will not be priced. See [88, p. 300].

Exercise 31.2.6: (1) We need to show that if $P_1 = \frac{\bar{Q}_1}{1+r_f+\beta_1(\bar{r}_M-r_f)}$ and $P_2 = \frac{\bar{Q}_2}{1+r_f+\beta_2(\bar{r}_M-r_f)}$, then

$$P_1 + P_2 = \frac{\bar{Q}_1 + \bar{Q}_2}{1 + r_f + \beta'(\bar{r}_M - r_f)},$$

where β' is the beta of the asset which is the sum of assets 1 and 2. Because both terms within the braces of Eq. (31.5) depend linearly on Q, the claim holds. (2) If otherwise, then we can buy the cheaper portfolio and sell the more expensive portfolio to earn arbitrage profits in a perfect market. See [623, p. 188].

Exercise 31.2.7:

Index Value in a Year	1200	1100	1000	900	800
Portfolio Value in a Year	1.36	1.16	0.96	0.76	0.56

Exercise 31.2.8: If the index is S, the portfolio will be worth $\$1,000 \times S$. However, the payoff of the options will be

$$10 \times \max(1,000 - S, 0) \times 100 \geq 1,000,000 - 1,000 \times S$$

dollars. (Recall that the size of a stock index option is \$100 times the index.) Add them up to obtain a lower bound of \$1,000,000.

Exercise 31.2.10: It is a bull call spread.

Exercise 31.2.11: Because the continuous compounded return $\ln S(t)/S(0)$ is a $(\mu - \sigma^2/2, \sigma)$ Brownian motion by Example 14.3.3, its mean is proportional to t, whereas its volatility or noise is proportional to $\sqrt{t}$. Shorter-term returns (i.e., small t's) are therefore dominated by noise.

Exercise 31.2.12: See [317, p. 359] or [424, pp. 132–135]. Whether riskless assets exist is not essential.

Exercise 31.3.1: $\text{Cov}[r_i, r_M] = b_i \times \text{Var}[r_M]$. The CAPM predicts that $\alpha_i = 0$. See [623, p. 205].

Exercise 31.3.2: Consider a k-factor model $r_i = a_i + b_{i1} f_1 + b_{i2} f_2 + \cdots + b_{ik} f_k + \epsilon_i$. Define $f \equiv [f_1, f_2, \ldots, f_k]^T$ and let C be the covariance matrix of the factors f. Assume that $E[f] = 0$ or apply the procedure below to $f - E[f]$ instead. By Eq. (19.1), there exists a real orthogonal matrix B such that $\Sigma \equiv B^T CB$ is a diagonal matrix. Because $[p_1, p_2, \ldots, p_k]^T \equiv B^T f$ has the covariance matrix Σ, $\{p_1, p_2, \ldots, p_k\}$ can be used as the desired set of uncorrelated factors.

Exercise 31.3.3: When the CAPM holds, Eq. (31.7) becomes $\bar{r} = r_f + b_1 \lambda_1$. Equate it with the security market line $\bar{r} - r_f = \beta(\bar{r}_M - r_f)$ in Theorem 31.2.1 to obtain $\lambda_1 = \beta(\bar{r}_M - r_f)/b_1$. Because $\beta = \text{Cov}[r, r_M]/\sigma_M^2 = b_1 \times \text{Cov}[f, r_M]/\sigma_M^2$, we obtain $\lambda_1 = (\text{Cov}[f, r_M]/\sigma_M^2)(\bar{r}_M - r_f)$. See [799, p. 335].

Exercise 31.3.4: Suppose we invest ω_i dollars in asset $i, i = 1, 2, \ldots, n$, in order to satisfy $\sum_{i=1}^{n} \omega_i = 0$, $\sum_{i=1}^{n} \omega_i b_{i1} = 0, \ldots, \sum_{i=1}^{n} \omega_i b_{im} = 0$. This portfolio requires zero net investment and has zero risk. Therefore its expected payoff must be zero, or $\sum_{i=1}^{n} \omega_i \bar{r}_i = 0$. Define $\omega \equiv (\omega_1, \omega_2, \ldots, \omega_n)^T$, $b_j \equiv (b_{1j}, b_{2j}, \ldots, b_{nj})^T$ for $j = 1, 2, \ldots, m$, $\mathbf{1} \equiv (1, 1, \ldots, 1)^T$, and $\bar{r} = (\bar{r}_1, \bar{r}_2, \ldots, \bar{r}_n)^T$. We can restate the conclusion as follows: For any ω satisfying $\omega^T \mathbf{1} = 0$, $\omega^T b_1 = 0, \ldots, \omega^T b_m = 0$, it holds that $\omega^T \bar{r} = 0$. That is, any ω orthogonal to $\mathbf{1}, b_1, \ldots, b_m$ is also orthogonal to $\bar{r}$. By a standard result in linear algebra, $\bar{r}$ must be a linear combination of the vectors $\mathbf{1}, b_1, \ldots, b_m$. Thus there are constants $\lambda_0, \lambda_1, \ldots, \lambda_m$ such that $\bar{r} = \lambda_0 \mathbf{1} + \lambda_1 b_1 + \cdots + \lambda_m b_m$.

Exercise 31.3.5: The proof of Exercise 31.3.4 goes through for the well-diversified portfolio as it is riskless.

Exercise 31.4.2: As $\ln S_\tau - \ln S \sim N((\mu - \sigma^2/2)\,\tau, \sigma^2\tau)$ (see Comment 14.4.1), with probability c, the return $\ln S_\tau/S$ is at least $(\mu - \sigma^2/2)\,\tau + N^{-1}(1 - c)\,\sigma\sqrt{\tau}$. The desired VaR thus equals $Se^{\mu\tau} - Se^{(\mu - \sigma^2/2)\tau + N^{-1}(1-c)\sigma\sqrt{\tau}}$. See [8].

Exercise 31.4.3: See Eq. (19.1) for (1) and (2). (3) It follows from

$$[\,dZ_1, dZ_2, \ldots, dZ_n\,]^T = \mathrm{diag}[\,\lambda_1^{-1}, \lambda_2^{-1}, \ldots, \lambda_n^{-1}\,]\,P^T\,[\,dW_1, dW_2, \ldots, dW_n\,]^T$$

and (2). (4) It results from the construction of the u_is.

NOTE

1. This can be verified formally as follows. Put in matrix terms, Eqs. (33.14) and (33.15) say that $Ak = 0$, where k is a column vector whose jth element is k_j and A is the $(n+1) \times (n+1)$ matrix whose ith row is $[\,\sigma_{i1} f_1, \sigma_{i2} f_2, \ldots, \sigma_{in} f_n\,]$ for $1 \le i \le n$ and whose $(n+1)$th row is $[\,f_1(\mu_1 - r), f_2(\mu_2 - r), \ldots, f_n(\mu_n - r)\,]$. For this set of equations to have a nontrivial solution, it is necessary that the determinant of A be zero. In particular, its last row must be a linear combination of the first n rows. This implies that $f_j(\mu_j - r) = \sum_i \lambda_i \sigma_{ij} f_j$, $j = 1, 2, \ldots, n+1$, for some $\lambda_1, \lambda_2, \ldots, \lambda_n$, which depend on only $S_1, S_2, \ldots, S_n$, and t.

Bibliography

[1] ABBEY, MICHAEL AND MICHAEL J. COREY. *Oracle8: A Beginner's Guide*. Berkeley, CA: Osborne/McGraw-Hill, 1997.

[2] ABKEN, PETER A. "An Empirical Evaluation of Value at Risk by Scenario Simulation." *The Journal of Derivatives*, 7, No. 4 (Summer 2000), 12–29.

[3] ABKEN, PETER A. AND MILIND M. SHRIKHANDE. "The Role of Currency Derivatives in Internationally Diversified Portfolios." *Economic Review* (3rd Quarter, 1997), 34–59.

[4] ABRAHAMS, STEVEN W. "The New View in Mortgage Prepayments: Insight from Analysis at the Loan-by-Loan Level." *The Journal of Fixed Income*, 7, No. 1 (June 1997), 8–21.

[5] ABRAMOWITZ, M. AND I.A. STEGUN, eds. *Handbook of Mathematical Functions*. New York: Dover, 1972.

[6] ACWORTH, PETER, MARK BROADIE, AND PAUL GLASSERMAN. "A Comparison of Some Monte Carlo and Quasi Monte Carlo Techniques for Option Pricing." In *Monte Carlo and Quasi-Monte Carlo Methods*. Edited by Harald Niederreiter et al. New York: Springer, 1998.

[7] ADAMS, K.J. AND D.R. VAN DEVENTER. "Fitting Yield Curves and Forward Rate Curves with Maximum Smoothness." *The Journal of Fixed Income*, 4, No. 1 (June 1994), 52–62.

[8] AHN, DONG-HYUN, JACOB BOUDOUKH, MATTHEW RICHARDSON, AND ROBERT F. WHITELAW. "Optimal Risk Management Using Options." *The Journal of Finance*, 54, No. 1 (February 1999), 359–375.

[9] AHN, DONG-HYUN, STEPHEN FIGLEWSKI, AND BIN GAO. "Pricing Discrete Barrier Options with an Adaptive Mesh Model." *The Journal of Derivatives*, 6, No. 4 (Summer 1999), 33–43.

[10] AINGWORTH, DONALD, RAJEEV MOTWANI, AND JEFFREY D. OLDHAM. "Accurate Approximations for Asian Options." In *Proceedings of the 11th Annual ACM-SIAM Symposium on Discrete Algorithms*, Philadelphia: Society for Industrial and Applied Mathematics, 2000.

[11] AÏT-SAHALIA, YACINE. "Testing Continuous-Time Models of the Spot Interest Rate." *The Review of Financial Studies*, 9, No. 2 (Spring 1996), 385–426.

[12] ALLEN, ARNOLD O. *Probability, Statistics, and Queueing Theory with Computer Science Applications*. 2nd ed. New York: Academic, 1990.

[13] ALSOP, STEWART. "E or Be Eaten." *Fortune* (November 8, 1999), 86–87.

[14] AMERMAN, DANIEL R. *Collateralized Mortgage Obligations*. New York: McGraw-Hill, 1996.

[15] AMES, WILLIAM F. *Numerical Methods for Partial Differential Equations*. 3rd ed. New York: Academic, 1992.

[16] AMIN, KAUSHIK I. "On the Computation of Continuous Time Option Pricing Using Discrete Approximations." *Journal of Financial and Quantitative Analysis*, 26, No. 4 (December 1991), 477–495.

[17] AMIN, KAUSHIK I. AND JAMES N. BODURTHA, JR. "Discrete-Time Valuation of American Options with Stochastic Interest Rates." *The Review of Financial Studies*, 8, No. 1 (1995), 193–234.

[18] AMIN, KAUSHIK AND AJAY KHANNA. "Convergence of American Option Values from Discrete to Continuous-Time Financial Models." *Mathematical Finance*, Vol. 4 (1994), 289–304.

[19] AMIN, KAUSHIK I. AND ANDREW J. MORTON. "Implied Volatility Functions in Arbitrage-Free Term Structure Models." *Journal of Financial Economics*, Vol. 35 (1994), 141–180.

[20] ANDERSEN, LEIF AND PHELIM P. BOYLE.

"Monte Carlo Methods for the Valuation of Interest Rate Securities." In [516, Chapter 13].

[21] ANDERSON, K. AND S. AMERO. "Scenario Analysis and the Use of Options in Total Return Portfolio Management." In [321, Chapter 7].

[22] ANDERSON, THEODORE W. *The Statistical Analysis of Time Series*. New York: Wiley, 1971.

[23] ANDERSON, THEODORE W. *An Introduction to Multivariate Statistical Analysis*. 2nd ed. New York: Wiley, 1984.

[24] ANDERSON, THOMAS E., DAVID E. CULLER, DAVID A. PATTERSON, AND THE NOW TEAM. "A Case for NOW (Networks of Workstations)." *IEEE Micro*, 15, No. 1 (February 1995), 54–64.

[25] ANDREASEN, JESPER. "The Pricing of Discretely Sampled Asian and Lookback Options: a Change of Numeraire Approach." *The Journal of Computational Finance*, 2, No. 1 (Fall 1998), 5–30.

[26] ANDREWS, GEORGE E. *The Theory of Partitions*. Reading, MA: Addison-Wesley, 1976.

[27] ANGEL, EDWARD AND RICHARD BELLMAN. *Dynamic Programming and Partial Differential Equations*. New York: Academic, 1972.

[28] AOYAMA, M. "Is MPT Applicable to Japan?" *The Journal of Portfolio Management*, 21, No. 1 (Fall 1994), 103–111.

[29] ARISTOTLE. *The Politics of Aristotle*. Translated by Ernest Barker. London: Oxford Univ. Press, 1971.

[30] ARNOLD, LUDWIG. *Stochastic Differential Equations: Theory and Applications*. New York: Wiley, 1973.

[31] ARROW, KENNETH J. AND F.H. HAHN. *General Competitive Analysis*. San Francisco: Holden-Day, 1971.

[32] ARTZNER, PHILIPPE AND FREDDY DELBAEN. "Term Structure of Interest Rates: The Martingale Approach." *Advances in Applied Mathematics*, 10, No. 1 (March 1989), 95–129.

[33] ASAY, M.R., F.-H. GUILLAUME, AND R.K. MATTU. "Duration and Convexity of Mortgage-Backed Securities: Some Hedging Implications from a Simple Prepayment-Linked Present-Value Model." In [320, pp. 103–125].

[34] ASNESS, C. "OAS Models, Expected Returns, and a Steep Yield Curve." *The Journal of Portfolio Management*, 19, No. 4 (Summer 1993), 85–93.

[35] ATKINSON, KENDALL E. *An Introduction to Numerical Analysis*. 2nd ed. New York: Wiley, 1989.

[36] AUDLEY, DAVID, RICHARD CHIN, SHRIKANT RAMAMURTHY, AND SUSAN VOLIN. "OAS and Effective Duration." In [328, Chapter 20].

[37] BAASE, S. *Computer Algorithms: Introduction to Design and Analysis*. Reading, MA: Addison-Wesley, 1988.

[38] BABBEL, DAVID F. AND CRAIG B. MERRILL. *Valuation of Interest-Sensitive Financial Instruments*. SOA Monograph M-FI96-1. Schaumburg, IL: The Society of Actuaries, 1996.

[39] BACKUS, DAVID, SILVERIO FORESI, ABON MOZUMDAR, AND LIUREN WU. "Predictable Changes in Yields and Forward Rates." To appear in *Journal of Financial Economics*, 35, No. 3.

[40] BACKUS, DAVID, SILVERIO FORESI, AND CHRIS TELMER. *Quantitative Models of Bond Pricing*. Manuscript, September 10, 1996. Available at www.stern.nyu.edu/~dbackus/bondnotes.htm.

[41] BACKUS, DAVID, SILVERIO FORESI, AND CHRIS TELMER. "Discrete-Time Models of Bond Pricing." In [516, Chapter 4].

[42] BACKUS, DAVID, SILVERIO FORESI, AND STANLEY ZIN. "Arbitrage Opportunities in Arbitrage-Free Models of Bond Pricing." *Journal of Business and Economic Statistics*, 16, No. 1 (January 1998), 13–26.

[43] BACKUS, DAVID K., ALLAN W. GREGORY, AND STANLEY E. ZIN. "Risk Premiums in the Term Structure: Evidence from Artificial Economies." *Journal of Monetary Economics*, Vol. 24 (1989), 371–399.

[44] BAKSHI, GURDIP, CHARLES CAO, AND ZHIWU CHEN. "Empirical Performance of Alternative Option Pricing Models." *The Journal of Finance*, 52, No. 5 (December 1997), 2003–2049.

[45] BAKSHI, GURDIP, CHARLES CAO, AND ZHIWU CHEN. "Do Call Prices and the Underlying Stock Always Move in the Same Direction?" *The Review of Financial Studies*, 13, No. 3 (Fall 2000), 549–584.

[46] BALDUZZI, PIERLUIGI, SANJIV RANJAN DAS, SILVERIO FORESI, AND RANGARAJAN SUNDARAM. "A Simple Approach to Three-Factor Affine Term Structure Models." *The Journal of Fixed Income*, 6, No. 3 (December 1996), 43–53.

[47] BALI, TURAN G. "Testing the Empirical Performance of Stochastic Volatility Models of the Short-Term Interest Rate." *Journal of Financial and Quantitative Analysis*, 35, No. 2 (June 2000), 191–215.

[48] BALL, CLIFFORD A. "Estimation Bias Induced by Discrete Security Prices." *The Journal of*

Finance, 43, No. 4 (September 1988), 841–865.

[49] BALL, CLIFFORD A. AND ANTONIO ROMA. "Stochastic Volatility Option Pricing." *Journal of Financial and Quantitative Analysis*, 29, No. 4 (December 1994), 589–607.

[50] BALL, CLIFFORD A. AND WALTER N. TOROUS. "Bond Price Dynamics and Options." *Journal of Financial and Quantitative Analysis*, 18, No. 4 (December 1983), 517–531.

[51] BANK FOR INTERNATIONAL SETTLEMENTS. "Central Bank Survey of Foreign Exchange and Derivatives Market Activity." Monetary and Economic Department, Bank for International Settlements, Basel, Switzerland, May 1999.

[52] BARBER, JOEL R. AND MARK L. COPPER. "Is Bond Convexity a Free Lunch?" *The Journal of Portfolio Management*, 24, No. 1 (Fall 1997), 113–119.

[53] BARRAQUAND, JÉRÔME AND THIERRY PUDET. "Pricing of American Path-Dependent Contingent Claims." *Mathematical Finance*, 6, No. 1 (January 1996), 17–51.

[54] BARTLETT, W.W. *Mortgage-Backed Securities*. New York: New York Institute of Finance, 1989.

[55] BARTLETT, W.W. *The Valuation of Mortgage-Backed Securities*. Burr Ridge, IL: Irwin, 1994.

[56] BATTIG, ROBERT J. AND ROBERT A. JARROW. "The Second Fundamental Theorem of Asset Pricing: A New Approach." *The Review of Financial Studies*, 12, No. 5 (Winter 1999), 1219–1235.

[57] BAUMOHL, BERNARD. "The Banks' Nuclear Secrets." *Time* (May 25, 1998).

[58] BAXTER, MARTIN. "General Interest-Rate Models and the Universality of HJM." In *Mathematics of Derivative Securities*. Edited by M. Dempster and S. Pliska. Cambridge, U.K.: Cambridge Univ. Press, 1997.

[59] BAXTER, MARTIN AND ANDREW RENNIE. *Financial Calculus: An Introduction to Derivative Pricing*. Cambridge, U.K.: Cambridge Univ. Press, 1998.

[60] BECKETTI, SEAN. "Are Derivatives Too Risky for Banks?" *Economic Review* (3rd Quarter, 1993), 27–42.

[61] BEDER, TANYA STYBLO. "VAR: Seductive but Dangerous." *Financial Analysts Journal*, 51, No. 5 (September–October 1995), 12–24.

[62] BENNINGA, S. *Numerical Techniques in Finance*. Cambridge, MA: MIT Press, 1989.

[63] BERNERS-LEE, TIM. "WWW: Past, Present, and Future." *Computer*, 29, No. 10 (October 1996), 69–77.

[64] BERNSTEIN, PETER L. *Capital Ideas: The Improbable Origins of Modern Wall Street*. New York: Free Press, 1992.

[65] BERNSTEIN, PETER L. *Against the Gods: The Remarkable Story of Risk*. New York: Wiley, 1996.

[66] BERTSEKAS, DIMITRI P. *Dynamic Programming: Deterministic and Stochastic Models*. Englewood Cliffs, NJ: Prentice-Hall, 1987.

[67] BERTSEKAS, DIMITRI P. AND JOHN N. TSITSIKLIS. *Parallel and Distributed Computation: Numerical Methods*. Englewood Cliffs, NJ: Prentice-Hall, 1989.

[68] BHAGAVATULA, RAVI S. AND PETER P. CARR. Valuing Double Barrier Options with Fourier Series." Manuscript, September 12, 1997.

[69] BHIDÉ, A. "Return to Judgment." *The Journal of Portfolio Management*, 20, No. 2 (Winter 1994), 19–25.

[70] BICK, AVI. AND WALTER WILLINGER. "Dynamic Programming without Probabilities." *Stochastic Processes and Their Applications*, 50, No. 2 (April 1994), 349–374.

[71] BIERWAG, GERALD O. "Immunization, Duration, and the Term Structure of Interest Rates." *Journal of Financial and Quantitative Analysis*, 12, No. 5 (December 1977), 725–742.

[72] BIERWAG, GERALD O. "The Ho–Lee Binomial Stochastic Process and Duration." *The Journal of Fixed Income*, 6, No. 2 (September 1996), 76–87.

[73] BILLINGSLEY, PATRICK. *Convergence of Probability Measures*. New York: Wiley, 1968.

[74] BJERKSUND, PETTER, AND GUNNAR STENSLAND. "Implementation of the Black–Derman–Toy Interest Rate Model." *The Journal of Fixed Income*, 6, No. 2 (September 1996), 67–75.

[75] BJÖRCK, ÅKE. *Numerical Methods for Least Squares Problems*. Philadelphia: Society for Industrial and Applied Mathematics, 1996.

[76] BJÖRK, TOMAS. "Interest Rate Theory." In *Financial Mathematics*, Lecture Notes in Mathematics 1676. Edited by W. Runggaldier. Berlin: Springer-Verlag, 1997.

[77] BJÖRK, TOMAS AND BENT JESPER CHRISTENSEN. "Interest Rate Dynamics and Consistent Forward Rate Curves." Working paper 209, Stockholm School of Economics, November 1997.

[78] BJÖRK, TOMAS, GIOVANNI DI MASI, YURI KABANOV, AND WOLFGANG RUNGGALDIER. "Towards a General Theory of Bond Markets." *Finance and Stochastics*, Vol. 1 (1997), 141–174.

[79] BLACK, FISCHER. "The Pricing of Commodity Contracts." *Journal of Financial Economics*, 3, Nos. 1/2 (January/March 1976), 167–179.

[80] BLACK, FISCHER. "Living Up to the Model." In [744, Chapter 1].

[81] BLACK, FISCHER. "Estimating Expected Return." *Financial Analysts Journal*, 49, No. 5 (September–October 1993), 36–38.

[82] BLACK, FISCHER. "Beta and Return." *The Journal of Portfolio Management*, 20, No. 1 (Fall 1993), 8–11.

[83] BLACK, FISCHER. "Interest Rates as Options." *The Journal of Finance*, 50, No. 5 (December 1995), 1371–1376.

[84] BLACK, FISCHER, EMANUEL DERMAN, AND WILLIAM TOY. "A One-Factor Model of Interest Rates and Its Application to Treasury Bond Options." *Financial Analysts Journal*, 46, No. 1 (January–February 1990), 33–39.

[85] BLACK, FISCHER AND PIOTR KARASINSKI. "Bond and Option Pricing When Short Rates Are Lognormal." *Financial Analysts Journal*, 47, No. 4 (July–August 1991), 52–59.

[86] BLACK, FISCHER AND MYRON SCHOLES. "The Valuation of Option Contracts and a Test of Market Efficiency." *The Journal of Finance*, Vol. 27 (1972), 399–418.

[87] BLACK, FISCHER AND MYRON SCHOLES. "The Pricing of Options and Corporate Liabilities." *Journal of Political Economy*, 81, No. 3 (May–June 1973), 637–654.

[88] BLAKE, DAVID. *Financial Market Analysis*. New York: McGraw-Hill, 1990.

[89] BLAKE, DAVID AND J. MICHAEL ORSZAG. "A Closed-Form Formula for Calculating Bond Convexity." *The Journal of Fixed Income*, 6, No. 1 (June 1996), 88–91.

[90] BLISS, ROBERT R. "Testing Term Structure Estimation Methods." Manuscript, November 1996. *Advances in Futures and Options Research*, Vol. 9 (1996), 197–231.

[91] BLISS, ROBERT R. "Movements in the Term Structure of Interest Rates." *Economic Review* (4th Quarter, 1997), 16–33.

[92] BLISS, ROBERT R. AND PETER RITCHKEN. "Empirical Tests of Two State-Variable Heath–Jarrow–Morton Models." *Journal of Money, Credit, and Banking*, 28, No. 3, Part 2 (August 1996), 452–476.

[93] BLISS, ROBERT AND DAVID SMITH. "The Elasticity of Interest Rate Volatility – Chan, Karolyi, Longstaff and Sanders Revisited." *The Journal of Risk*, 1, No. 1 (Fall 1998), 21–46.

[94] BLYTH, STEPHEN, AND JOHN UGLUM. "Rates of Skew." *Risk*, 12, No. 7 (July 1999), 61–63.

[95] BOARD OF TRADE OF THE CITY OF CHICAGO. *Commodity Trading Manual*. Chicago: Chicago Board of Trade, 1994.

[96] BODIE, ZVI, ALEX KANE, AND ALAN J. MARCUS. *Investments*. 4th ed. New York: McGraw-Hill, 1999.

[97] BODURTHA, JAMES N., JR. AND GEORGES R. COURTADON. "Tests of an American Option Pricing Model on the Foreign Currency Options Market." *Journal of Financial and Quantitative Analysis*, 22, No. 2 (June 1987), 153–167.

[98] BOGLE, JOHN C. *Bogle on Mutual Funds*. New York: Dell, 1994.

[99] BOLLERSLEV, TIM. "Generalized Autoregressive Conditional Heteroskedasticity." *Journal of Econometrics*, Vol. 31 (1986), 307–327.

[100] BOLLOBÁS, BÉLA. *Random Graphs*. New York: Academic, 1985.

[101] BOOCH, G. *Object Oriented Design with Applications*. Redwood City, CA: Benjamin/Cummings, 1991.

[102] BORODIN, ALLAN AND RAN EL-YANIV. *Online Computation and Competitive Analysis*. Cambridge, U.K.: Cambridge Univ. Press, 1998.

[103] BORODIN, ALLAN AND IAN MUNRO. *The Computational Complexity of Algebraic and Numeric Problems*. New York: American Elsevier, 1975.

[104] BORODIN, ANDREI N. AND PAAVO SALMINEN. *Handbook of Brownian Motion – Facts and Formulae*. Basel: Birkhäuser Verlag, 1996.

[105] BOUAZIZ, LAURENT, ERIC BRIYS, AND MICHEL CROUHY. "The Pricing of Forward-Starting Asian Options." *Journal of Banking & Finance*, Vol. 18 (1994), 823–839.

[106] BOYLE, PHELIM P. "Options: A Monte Carlo Approach." *Journal of Financial Economics*, Vol. 4 (1977), 323–338.

[107] BOYLE, PHELIM P. "A Lattice Framework for Option Pricing with Two State Variables." *Journal of Financial and Quantitative Analysis*, 23, No. 1 (March 1988), 1–12.

[108] BOYLE, PHELIM P., MARK BROADIE AND PAUL GLASSERMAN. "Monte Carlo Methods for Security Pricing." *Journal of Economic Dynamics & Control*, Vol. 21 (1997), 1267–1321.

[109] BOYLE, PHELIM P. AND DAVID EMANUEL. "Discretely Adjusted Option Hedges." *Journal of Financial Economics*, Vol. 8 (1980), 259–282.

[110] BOYLE, PHELIM, JEREMY EVNINE, AND STEPHEN GIBBS. "Numerical Evaluation of Multivariate Contingent Claims." *The Review of Financial Studies*, 2, No. 2 (1989), 241–250.

[111] BOYLE, PHELIM AND SOK HOON LAU.

"Bumping Up against the Barrier with the Binomial Method." *The Journal of Derivatives*, 1, No. 4 (Summer 1994), 6–14.

[112] BOYLE, PHELIM AND KEN SENG TAN. "Lure of the Linear." In *Over the Rainbow*. Edited by R. Jarrow. London: Risk Publications, 1995.

[113] BOYLE, PHELIM AND YISONG "SAM" TIAN. "Pricing Lookback and Barrier Options under the CEV Process." *Journal of Financial and Quantitative Analysis*, 34, No. 2 (June 1999), 241–264.

[114] BOYLE, PHELIM AND Y.K. TSE. "An Algorithm for Computing Values of Options on the Maximum or Minimum of Several Assets." *Journal of Financial and Quantitative Analysis*, 25, No. 2 (June 1990), 215–227.

[115] BOYLE, PHELIM AND TON VORST. "Option Replication in Discrete Time with Transaction Cost." *The Journal of Finance*, 47, No. 1 (March 1992), 271–293.

[116] BRACE, ALAN, DARIUSZ GĄTAREK AND MAREK MUSIELA. "The Market Model of Interest Rate Dynamics." *Mathematical Finance*, 7, No. 2 (April 1997), 127–155.

[117] BRAUER, J.S. AND L.S. GOODMAN. "Hedging with Options and Option Products." In [321, Chapter 9].

[118] BRAZIL, A.J. "Citicorp's Mortgage Valuation Model Option-Adjusted Spreads and Option-Based Durations." *Journal of Real Estate Finance and Economics*, 1, No. 2 (1988), 151–162.

[119] BRENNAN, MICHAEL J. "The Pricing of Contingent Claims in Discrete Time Models." *The Journal of Finance*, 34, No. 1 (March 1979), 53–68.

[120] BRENNAN, MICHAEL J. AND EDUARDO S. SCHWARTZ. "Convertible Bonds: Valuation and Optimal Strategies for Call and Conversion." *The Journal of Finance*, 32, No. 5 (December 1977), 1699–1715.

[121] BRENNAN, MICHAEL J. AND EDUARDO S. SCHWARTZ. "A Continuous Time Approach to the Pricing of Bonds." *Journal of Banking & Finance*, 3, No. 2 (July 1979), 133–155.

[122] BRENNAN, MICHAEL J. AND EDUARDO S. SCHWARTZ. "Analyzing Convertible Bonds." *Journal of Financial and Quantitative Analysis*, 15, No. 4 (November 1980), 907–929.

[123] BRENNAN, MICHAEL J. AND EDUARDO S. SCHWARTZ. "An Equilibrium Model of Bond Pricing and a Test of Market Efficiency." *Journal of Financial and Quantitative Analysis*, 17, No. 3 (September 1982), 301–329.

[124] BRENNAN, MICHAEL J. AND EDUARDO S. SCHWARTZ. "Determinants of GNMA Mortgage Prices." *Journal of the American Real Estate & Urban Economics Association*, 13, No. 3 (Fall 1985), 209–228.

[125] BRENNER, MENACHEM, GEORGES COURTADON, AND MARTI SUBRAHMANYAM. "Options on the Spot and Options on Futures." *The Journal of Finance*, 40, No. 5 (December 1985), 1303–1317.

[126] BRENNER, ROBIN J., RICHARD H. HARJES, AND KENNETH F. KRONER. "Another Look at Models of the Short-Term Interest Rate." *Journal of Financial and Quantitative Analysis*, 31, No. 1 (March 1996), 85–107.

[127] BROADIE, MARK AND JEROME DETEMPLE. "American Option Valuation: New Bounds, Approximations, and a Comparison of Existing Methods." *The Review of Financial Studies*, 9, No. 4 (Winter 1996), 1211–1250.

[128] BROADIE, MARK AND PAUL GLASSERMAN. "Estimating Security Price Derivatives Using Simulation." *Management Science*, 42, No. 2 (February 1996), 269–285.

[129] BROADIE, MARK AND PAUL GLASSERMAN. "Pricing American-Style Securities Using Simulation." *Journal of Economic Dynamics & Control*, Vol. 21 (1997), 1323–1352.

[130] BROADIE, MARK, PAUL GLASSERMAN, AND GAUTAM JAIN. "Enhanced Monte Carlo Estimates for American Option Prices." *The Journal of Derivatives*, 5, No. 1 (Fall 1997), 25–44.

[131] BROADIE, MARK, PAUL GLASSERMAN, AND STEVEN KOU. "A Continuity Correction for Discrete Barrier Options." *Mathematical Finance*, 7, No. 4 (October 1997), 325–349.

[132] BROCATO, J. AND P.R. CHANDY. "Does Market Timing Really Work in the Real World?" *The Journal of Portfolio Management*, 20, No. 2 (Winter 1994), 39–44.

[133] BROOKS, ROBERT AND DAVID YONG YAN. "London Inter-Bank Offer Rate (LIBOR) versus Treasury Rate: Evidence from the Parsimonious Term Structure Model." *The Journal of Fixed Income*, 9, No. 1 (June 1999), 71–83.

[134] BROWN, DAVID T. "The Determinants of Expected Returns on Mortgage-Backed Securities: An Empirical Analysis of Option-Adjusted Spreads." *The Journal of Fixed Income*, 9, No. 2 (September 1999), 8–18.

[135] BROWN, ROGER H. AND STEPHEN M. SCHAEFER. "The Term Structure of Real Interest Rates and the Cox, Ingersoll, and Ross Model." *Journal of Financial Economics*, 35, No. 1 (February 1994), 3–42.

[136] BROWN, ROGER H. AND STEPHEN M. SCHAEFER. "Interest Rate Volatility and the Shape of the Term Structure." *Philosophical*

Transactions of the Royal Society of London A, 347, No. 1684 (June 15, 1994), 563–576.

[137] BROWN, ROGER H. AND STEPHEN M. SCHAEFER. "Ten Years of the Real Term Structure: 1984–1994." *The Journal of Fixed Income* (March 1996), 6–22.

[138] BROWN, STEPHEN J. AND PHILIP H. DYBVIG. "The Empirical Implications of the Cox, Ingersoll, Ross Theory of the Term Structure of Interest Rates." *The Journal of Finance*, 41, No. 3 (July 1986), 617–632.

[139] BÜHLER, WOLFGANG, MARLIESE UHRIG-HOMBURG, ULRICH WALTER AND THOMAS WEBER. "An Empirical Comparison of Forward-Rate and Spot-Rate Models for Valuing Interest-Rate Options." *The Journal of Finance*, 54, No. 1 (February 1999), 269–305.

[140] BUSINESS WEEK. Special 1994 Bonus Issue: the Information Revolution. *Business Week*, 1994.

[141] BYKHOVSKY, MICHAEL AND LAKHBIR HAYRE. "Anatomy of PAC Bonds." *The Journal of Fixed Income*, 2, No. 1 (June 1992), 44–50.

[142] CAFLISCH, RUSSEL E. AND WILLIAM MOROKOFF. "Valuation of Mortgage Backed Securities Using the Quasi-Monte Carlo Method." Report 96-23, Department of Mathematics, University of California, Los Angeles, California, 1996.

[143] CAMPBELL, JOHN Y. "A Defense of Traditional Hypotheses about the Term Structure of Interest Rates." *The Journal of Finance*, 41, No. 1 (March 1986), 183–193.

[144] CAMPBELL, JOHN Y. "Some Lessons from the Yield Curve." *The Journal of Economic Perspectives*, 9, No. 3 (Summer 1995), 129–152.

[145] CAMPBELL, JOHN Y. "Understanding Risk and Return." *Journal of Political Economy*, 104, No. 2 (1996), 298–345.

[146] CAMPBELL, JOHN Y. AND JOHN AMMER. "What Moves the Stock and Bond Markets? A Variance Decomposition for Long-Term Asset Returns." *The Journal of Finance*, 48, No. 1 (March 1993), 3–37.

[147] CAMPBELL, JOHN Y., ANDREW W. LO AND A. CRAIG MACKINLAY. *The Econometrics of Financial Markets*. Princeton, NJ: Princeton Univ. Press, 1997.

[148] CAMPBELL, JOHN Y. AND ROBERT J. SHILLER. "Stock Prices, Earnings and Expected Dividends." *The Journal of Finance*, 43, No. 3 (July 1988), 661–676.

[149] CANABARRO, EDUARDO. "Where Do One-Factor Interest Rate Models Fail?" *The Journal of Fixed Income*, 5, No. 2 (September 1995), 31–52.

[150] CANINA, LINDA AND STEPHEN FIGLEWSKI. "The Informational Content of Implied Volatility." *The Review of Financial Studies*, 6, No. 3 (1993), 659–681.

[151] CAO, H. HENRY. "The Effect of Derivative Assets on Information Acquisition and Price Behavior in a Rational Expectations Equilibrium." *The Review of Financial Studies*, 12, No. 1 (Spring 1999), 131–163.

[152] CARAYANNOPOULOS, PETER. "The Mispricing of U.S. Treasury Callable Bonds." *The Journal of Futures Markets*, 15, No. 8 (1995), 861–879.

[153] CARLETON, W. AND I. COOPER. "Estimation and Uses of the Term Structure of Interest Rates." *The Journal of Finance*, 31, No. 4 (September 1976), 1067–1083.

[154] CARR, PETER. "Randomization and the American Put." *The Review of Financial Studies*, 11, No. 3 (1998), 597–626.

[155] CARR, PETER, REN-RAW CHEN, AND LOUIS O. SCOTT. "Valuing the Timing Option and the Quality Option in Treasury Bond Futures Contracts." Manuscript, January 1996. Available at www.rci.rutgers.edu/~rchen/papers.html.

[156] CARR, PETER AND MARC CHESNEY. "American Put Call Symmetry." Working Paper, Morgan Stanley. New York, November 1996.

[157] CARR, PETER AND ANDREW CHOU. "Breaking Barriers: Static Hedging of Barrier Securities." *Risk* (September 1997), 139–145.

[158] CARR, PETER AND ANDREW CHOU. "Hedging Complex Barrier Options." Working Paper, Morgan Stanley, New York, 1997.

[159] CARR, PETER, KATRINA ELLIS, AND VISHAL GUPTA. "Static Hedging of Exotic Options." *The Journal of Finance*, 53, No. 3 (1998), 1165–1190.

[160] CARR, PETER AND GUANG YANG. "Simulating American Bond Options in an HJM Framework." Working Paper, Morgan Stanley, New York, December 1996.

[161] CARRON, ANDREW S. "Understanding CMOs, REMICs, and Other Mortgage Derivatives." *The Journal of Fixed Income*, 2, No. 1 (June 1992), 25–43.

[162] CARRON, ANDREW S. "Collateralized Mortgage Obligations." In [328, Chapter 25].

[163] CARVERHILL, ANDREW. "When Is the Short Rate Markovian?" *Mathematical Finance*, 4, No. 4 (October 1994), 305–312.

[164] CARVERHILL, ANDREW. "A Note on the Models of Hull and White for Pricing

Options on the Term Structure." *The Journal of Fixed Income*, 5, No. 2 (September 1995), 89–96.

[165] CARVERHILL, ANDREW. "A Simplified Exposition of the Heath, Jarrow and Morton Model." *Stochastics and Stochastics Reports*, 53, Nos. 3+4 (1995), 227–240.

[166] CARVERHILL, ANDREW AND KIN PANG. "Efficient and Flexible Bond Option Valuation in the Heath, Jarrow, Morton Framework." *The Journal of Fixed Income*, 5, No. 2 (September 1995), 70–77.

[167] CASTRO, ELIZABETH. *HTML for the World Wide Web*. Berkeley, CA: Peachpit Press, 1998.

[168] CHALASANI, PRASAD, SOMESH JHA, FEYZUL-LAH EGRIBOYUN, AND ASHOK VARIKOOTY. "A Refined Binomial Lattice for Pricing American Asian Options." *Review of Derivatives Research*, Vol. 3 (1999), 85–105.

[169] CHALASANI, PRASAD, SOMESH JHA, AND ISAAC SAIAS. "Approximate Option Pricing." *Algorithmica*, Vol. 25 (1999), 2–21.

[170] CHALASANI, PRASAD, SOMESH JHA, AND ASHOK VARIKOOTY. "Accurate Approximations for European Asian Options." *The Journal of Computational Finance*, 1, No. 4 (Summer 1999), 11–29.

[171] CHAMBERS, DONALD R., WILLARD T. CARLETON AND DONALD W. WALDMAN. "A New Approach to Estimation of the Term Structure of Interest Rates." *Journal of Financial and Quantitative Analysis*, 19, No. 3 (September 1984), 233–252.

[172] CHAN, GEORGE WEI-TSO. *Theory and Practice of Option Pricing in Taiwan*. Master's Thesis. Department of Computer Science and Information Engineering, National Taiwan University, Taiwan, 1998.

[173] CHAN, K.C., G. ANDREW KAROLYI, FRANCIS A. LONGSTAFF AND ANTHONY B. SANDERS. "An Empirical Comparison of Alternative Models of the Short-Term Interest Rate." *The Journal of Finance*, 47, No. 3 (July 1992), 1209–1227.

[174] CHAN, L.K.C. AND J. LAKONISHOK. "Are the Reports of Beta's Death Premature?" *The Journal of Portfolio Management*, 19, No. 4 (Summer 1993), 51–62.

[175] CHANCE, DON M. AND DON RICH. "The Pricing of Equity Swaps and Swaptions." *The Journal of Derivatives*, 5, No. 4 (Summer 1998), 19–31.

[176] CHANG, CAROLYN W. AND JACK S.K. CHANG. "Forward and Futures Prices: Evidence from the Foreign Exchange Markets." *The Journal of Finance*, 45, No. 4 (September 1990), 1333–1336.

[177] CHANG, CAROLYN W., JACK S.K. CHANG AND KIAN-GUAN LIM. "Information-Time Option Pricing: Theory and Empirical Evidence." *Journal of Financial Economics*, 48, No. 2 (May 1998), 211–243.

[178] CHAPMAN, DAVIS AND JEFF HEATON. *SAMS Teach Yourself Visual C++ 6 in 21 Days*. Indianapolis, IN: SAMS, 1999.

[179] CHAO, KUN-YUAN. *Combinatorial Methods for Double-Barrier Option Pricing*. Master's Thesis. Department of Computer Science and Information Engineering, National Taiwan University, Taiwan, 1999.

[180] CHELO, NEIL R. "Can Volatility Be Your Friend?" *Indexes*, Issue 4 (April–June 2000), 22–25.

[181] CHEN, GEN-HUEY, MING-YANG KAO, YUH-DAUH LYUU, AND HSING-KUO WONG. "Optimal Buy-and-Hold Strategies for Financial Markets with Bounded Daily Returns." In *Proceedings of the 31st Annual ACM Symposium on the Theory of Computing*, New York: Association for Computing Machinery, 1999, pp. 119–128. To appear in *SIAM Journal on Computing*.

[182] CHEN, J. BRADLEY, YASUHIRO ENDO, KEE CHAN, DAVID MAZIÉRES, ANTONIO DIAS, MARGO SELTZER AND MICHAEL D. SMITH. "The Measured Performance of Personal Computer Operating Systems." *ACM Transactions on Computer Systems*, 14, No. 1 (February 1996), 3–40.

[183] CHEN, REN-RAW. *Understanding and Managing Interest Rate Risks*. Singapore: World Scientific, 1996.

[184] CHEN, REN-RAW AND LOUIS SCOTT. "Pricing Interest Rate Options in a Two-Factor Cox-Ingersoll-Ross Model of the Term Structure." *The Review of Financial Studies*, 4, No. 4 (1992), 613–636.

[185] CHEN, REN-RAW AND LOUIS SCOTT. "Maximum Likelihood Estimation for a Multifactor Equilibrium Model of the Term Structure of Interest Rates." *The Journal of Fixed Income*, Vol. 3 (December 1993), 14–31.

[186] CHEN, REN-RAW AND LOUIS SCOTT. "Interest Rate Options in Multifactor Cox-Ingersoll-Ross Models of the Term Structure." *The Journal of Derivatives*, 3, No. 2 (Winter 1995), 53–72.

[187] CHEN, REN-RAW AND TYLER T. YANG. "A Universal Lattice." *Review of Derivatives Research*, 3, No. 2 (1999), 115–133.

[188] CHEN, SI. "Understanding Option-Adjusted

Spreads: The Implied Prepayment Hypothesis." *The Journal of Portfolio Management* (Summer 1996), 104–113.

[189] CHEN, WEI-JUI. *Calibrating Interest Rate Models with Differential Tree Algorithms: The Case of Black–Derman–Toy Model.* Master's Thesis. Department of Computer Science and Information Engineering, National Taiwan University, Taiwan, 1997.

[190] CHEN, WEI-JUI AND YUH-DAUH LYUU. "Calibrating Interest Rate Models with Differential Tree Algorithms: The Case of the Black–Derman–Toy Model." In *Proceedings of the 1997 National Computer Symposium* (NCS'97), Tung-Hai University, Taiwan, December 1997, pp. A19–A24.

[191] CHEN, YUAN-WANG. *Towards Creating Taiwan's Put Market.* Master's Thesis. Department of Computer Science and Information Engineering, National Taiwan University, Taiwan, 1999.

[192] CHENG, CHIA-JEN. *On Hull–White Models: One and Two Factors.* Master's Thesis. Department of Computer Science and Information Engineering, National Taiwan University, Taiwan, 1998.

[193] CHENG, SUSAN T. "On the Feasibility of Arbitrage-Based Option Pricing When Stochastic Bond Price Processes Are Involved." *Journal of Economic Theory*, 5, No. 1 (February 1991), 185–198.

[194] CHERIAN, JOSEPH A. AND ROBERT A. JARROW. "Options Markets, Self-Fulfilling Prophesies, and Implied Volatilities." *Review of Derivatives Research*, 2, No. 1 (1998), 5–37.

[195] CHERNOFF, HERMAN AND LINCOLN E. MOSES. *Elementary Decision Theory.* New York: Dover, 1986.

[196] CHEUK, TERRY H.F. AND TON C.F. VORST. "Complex Barrier Options." *The Journal of Derivatives*, 4, No. 1 (Fall 1996), 8–22.

[197] CHEYETTE, OREN. "Term Structure Dynamics and Mortgage Valuation." *The Journal of Fixed Income* (March 1992), 28–41.

[198] CHEYETTE, OREN. "OAS Analysis for CMOs." *The Journal of Portfolio Management*, 20, No. 4 (Summer 1994), 53–66.

[199] CHEYETTE, OREN, SAM CHOI AND ELENA BLANTER. "The New BARRA Fixed Rate Prepayment Model." Spring 1996, BARRA.

[200] CHIEN, ANDREW, ET AL. "High Performance Virtual Machines (HPVM): Clusters with Supercomputing APIs and Performance." In *Proceedings of the Eighth SIAM Conference on Parallel Processing for Scientific Computing*, Philadelphia: Society for Industrial and Applied Mathematics, March 1997.

[201] CHO, D. CHINHYUNG AND EDWARD W. FREES. "Estimating the Volatility of Discrete Stock Prices." *The Journal of Finance*, 43, No. 2 (June 1988), 451–466.

[202] CHO, HE YOUN AND KI WOOK LEE. "An Extension of the Three-Jump Process Model for Contingent Claim Valuation." *The Journal of Derivatives*, Vol. 3 (Fall 1995), 102–108.

[203] CHOI, SAM AND MIKE SCHUMACHER. "GNMA II 30-Year Pass-Through MBS Prepayment Analysis." *The Journal of Fixed Income*, 6, No. 4 (March 1997), 99–104.

[204] CHOPRA, VIJAY K. AND WILLIAM T. ZIEMBA. "The Effect of Errors in Means, Variances, and Covariances on Optimal Portfolio Choice." *The Journal of Portfolio Management* (Winter 1993), 6–11.

[205] CHOW, Y.S., HERBERT ROBBINS AND DAVID SIEGMUND. *Great Expectations: The Theory of Optimal Stopping.* Boston: Houghton Mifflin, 1971.

[206] CHRISTENSEN, PETER E., FRANK J. FABOZZI AND ANTHONY LoFASO. "Bond Immunization: An Asset/Liability Optimization Strategy." In [328, Chapter 42].

[207] CHRISTENSEN, PETER OVE AND BJARNE G. SØRENSEN. "Duration, Convexity, and Time Value." *The Journal of Portfolio Management*, 20, No. 2 (Winter 1994), 51–60.

[208] CHRISTIE, ANDREW A. "The Stochastic Behavior of Common Stock Variances: Value, Leverage and Interest Rate Effects." *Journal of Financial Economics*, 10, No. 4 (December 1982), 407–432.

[209] CHUA, J. "A Closed-Form Formula for Calculating Bond Duration." *Financial Analysts Journal*, Vol. 40 (1984), 76–78.

[210] CHUNG, KAI LAI. *A Course in Probability Theory.* 2nd ed. New York: Academic, 1974.

[211] CHUNG, KAI LAI AND R.J. WILLIAMS. *Introduction to Stochastic Integration.* Boston: Birkhäuser, 1983.

[212] CHURCHILL, RUEL V. *Fourier Series and Boundary Value Problems.* 2nd ed. New York: McGraw-Hill, 1963.

[213] CHVÁTAL, VAŠEK. *Linear Programming.* New York: Freeman, 1983.

[214] CLEWLOW, LES AND ANDREW CARVERHILL. "On the Simulation of Contingent Claims." *The Journal of Derivatives* (Winter 1994), 66–74.

[215] CLEWLOW, LES AND CHRIS STRICKLAND. *Implementing Derivatives Models.* Chichester, U.K.: Wiley, 1998.

[216] CODD, EDGAR F. "A Relational Model of Data for Large Shared Data Banks." *Communications of the ACM*, 13, No. 6 (June 1970), 377–397.

[217] COHEN, J.B., E.D. ZINBARG, AND A. ZEIKEL. *Investment Analysis and Portfolio Management*. 4th ed. Burr Ridge, IL: Irwin, 1982.

[218] COHLER, GENE, MARK FELDMAN, AND BRIAN LANCASTER. "Price of Risk Constant (PORC): Going beyond OAS." *The Journal of Fixed Income*, 6, No. 4 (March 1997), 6–15.

[219] COLEMAN, THOMAS S., LAWRENCE FISHER, AND ROGER G. IBBOTSON. "Estimating the Term Structure of Interest Rates from Data That Include the Prices of Coupon Bonds." *The Journal of Fixed Income*, 2, No. 2 (September 1992), 85–116.

[220] COLLIN-DUFRESNE, P. AND JOHN P. HARDING. "A Closed Form Formula for Valuing Mortgages." *Journal of Real Estate Finance and Economics*, 19, No. 2 (September 1999), 133–146.

[221] CONNOLLY, KEVIN B. *Pricing Convertible Bonds*. New York: Wiley, 1998.

[222] CONNOR, G. "Hedging." In [307, pp. 164–171].

[223] CONSTANTINIDES, GEORGE M. "A Theory of the Nominal Term Structure of Interest Rates." *The Review of Financial Studies*, 5, No. 4 (1992), 531–552.

[224] CONTE, SAMUEL D. AND CARL DE BOOR. *Elementary Numerical Analysis: An Algorithmic Approach*. 3rd ed. New York: McGraw-Hill, 1980.

[225] CONZE, ANTOINE AND VISWANATHAN. "Path Dependent Options: the Case of Lookback Options." *The Journal of Finance*, 46, No. 5 (December 1991), 1893–1907.

[226] COOPERS & LYBRAND. *Interest Rate Swap*. Burr Ridge, IL: Irwin, 1992.

[227] CORMEN, THOMAS H., CHARLES E. LEISERSON AND RONALD L. RIVEST. *Introduction to Algorithms*. Cambridge, MA: MIT Press, 1992.

[228] CORRADO, CHARLES J. AND TIE SU. "Implied Volatility Skews and Stock Index Skewness and Kurtosis Implied by S&P 500 Index Option Prices." *The Journal of Derivatives*, 4, No. 4 (Summer 1997), 8–19.

[229] COURTADON, GEORGES. "An Introduction to Numerical Methods in Option Pricing." In [346, Chapter 14].

[230] COX, DAVID R. AND HILTON D. MILLER. *The Theory of Stochastic Processes*. London: Chapman & Hall, 1995.

[231] COX, JOHN C., JONATHAN E. INGERSOLL, JR., AND STEPHEN A. ROSS. "Duration and the Measurement of Basis Risk." *Journal of Business*, 52, No. 1 (1979), 51–61.

[232] COX, JOHN C., JONATHAN E. INGERSOLL, JR., AND STEPHEN A. ROSS. "A Re-Examination of Traditional Hypothesis about the Term Structure of Interest Rates." *The Journal of Finance*, 36, No. 4 (September 1981), 769–799.

[233] COX, JOHN C., JONATHAN E. INGERSOLL, JR. AND STEPHEN A. ROSS. "The Relationship between Forward Prices and Futures Prices." *Journal of Financial Economics*, Vol. 9 (December 1981), 321–346.

[234] COX, JOHN C., JONATHAN E. INGERSOLL, JR., AND STEPHEN A. ROSS. "A Theory of the Term Structure of Interest Rates." *Econometrica*, 53, No. 2 (March 1985), 385–407.

[235] COX, JOHN C., STEPHEN A. ROSS, AND MARK RUBINSTEIN. "Option Pricing: a Simplified Approach." *Journal of Financial Economics*, 7, No. 3 (September 1979), 229–263.

[236] COX, JOHN C. AND MARK RUBINSTEIN. *Options Markets*. Englewood Cliffs, NJ: Prentice-Hall, 1985.

[237] CRABBE, LELAND E. AND JOSEPH D. ARGILAGOS. "Anatomy of the Structured Note Market." *Journal of Applied Corporate Finance*, 7, No. 3 (Fall 1994), 85–98.

[238] CRABBE, LELAND E. AND PANOS NIKOULIS. "The Putable Bond Market: Structure, Historical Experience, and Strategies." *The Journal of Fixed Income*, 7, No. 3 (December 1997), 47–60.

[239] CRACK, TIMOTHY FALCON AND SANJAY K. NAWALKHA. "Interest Rate Sensitivities of Bond Risk Measures." *Financial Analysts Journal*, 56, No. 1 (January–February 2000), 34–43.

[240] CROWNOVER, RICHARD M. *Introduction to Fractals and Chaos*. Boston: Jones & Bartlett, 1995.

[241] CURRAN, MICHAEL. "Valuing Asian and Portfolio Options by Conditioning on the Geometric Mean Price." *Management Science*, 40, No. 12 (December 1994), 1705–1711.

[242] CURRAN, MICHAEL. "Accelerating American Option Pricing in Lattices." *The Journal of Derivatives*, 3, No. 2 (Winter 1995), 8–18.

[243] CUTLAND, NIGEL J., EKKEHARD KOPP, AND WALTER WILLINGER. "From Discrete to Continuous Financial Models: New Convergence Results for Option Pricing." *Mathematical Finance*, 3, No. 2 (April 1993), 101–123.

[244] DAHL, F. *The Random House Personal Investment Calculator*. New York: Random House, 1990.

[245] DAHL, HENRIK. "A Flexible Approach to Interest-Rate Risk Management." In [891, Chapter 8].

[246] DAHL, HENRIK, ALEXANDER MEERAUS AND STAVROS A. ZENIOS. "Some Financial Optimization Models: I Risk Management." In [891, Chapter 1].

[247] DAHL, HENRIK, ALEXANDER MEERAUS, AND STAVROS A. ZENIOS. "Some Financial Optimization Models: II Financial Engineering." In [891, Chapter 2].

[248] DAHLQUIST, MAGNUS. "On Alternative Interest Rate Processes." *Journal of Banking & Finance*, Vol. 20 (1996), 1093–1119.

[249] DAI, TIAN-SHYR. *Pricing Path-Dependent Derivatives*. Master's Thesis. Department of Computer Science and Information Engineering, National Taiwan University, Taiwan, 1999.

[250] DAI, TIAN-SHYR, GUAN-SHIENG HUANG, AND YUH-DAUH LYUU. "Accurate Approximation Algorithms for Asian Options." Manuscript, May 2001.

[251] DAI, TIAN-SHYR AND YUH-DAUH LYUU. "Efficient Algorithms for Average-Rate Option Pricing." In *Proceedings of the 1999 National Computer Symposium* (NCS'99), Tamkang University, Taiwan, December 1999, pp. A-359–A-366.

[252] DAI, TIAN-SHYR AND YUH-DAUH LYUU. "Efficient, Exact Algorithms for Asian Options with Multiresolution Lattices." Manuscript, January 2001.

[253] DAS, SANJIV RANJAN. "Credit Risk Derivatives." *The Journal of Derivatives* (Spring 1995), 7–23.

[254] DAS, SANJIV RANJAN AND RANGARAJAN K. SUNDARAM. "A Discrete-Time Approach to Arbitrage-Free Pricing of Credit Derivatives." *Management Science*, 46, No. 1 (January 2000), 46–62.

[255] DAS, SANJIV RANJAN AND RANGARAJAN K. SUNDARAM. "Of Smiles and Smirks: A Term Structure Perspective." *Journal of Financial and Quantitative Analysis*, 34 No. 2 (June 1999), 211–239.

[256] DATTATREYA, R.E. AND FRANK J. FABOZZI. "A Simplified Model for the Valuation of Debt Options." In [321, Chapter 4].

[257] DAVES, PHILLIP R. AND MICHAEL C. EHRHARDT. "Joint Cross-Section/Time-Series Maximum Likelihood Estimation for the Parameters of the Cox-Ingersoll-Ross Bond Pricing Model." *The Financial Review*, 28, No. 2 (May 1993), 203–237.

[258] DAVIDSON, ANDREW S. "Overview of Alternative Duration Measures for Mortgage-Backed Securities." In [320, pp. 67–79].

[259] DAVIDSON, ANDREW S. AND MICHAEL D. HERSKOVITZ. "Analyzing MBS: A Comparison of Methods for Analyzing Mortgage-Backed Securities." In [322, pp. 305–328].

[260] DAVIDSON, ANDREW S. AND MICHAEL D. HERSKOVITZ. *Mortgage-Backed Securities: Investment Analysis & Advanced Valuation Techniques*. Chicago: Probus, 1994.

[261] DAVIDSON, JAMES. *Stochastic Limit Theory: An Introduction for Econometricians*. London: Oxford Univ. Press, 1994.

[262] DAVIS, M.H.A. AND J.M.C. CLARK. "A Note on Super-Replicating Strategies." *Philosophical Transactions of the Royal Society of London A*, 347, No. 1684 (June 15, 1994), 485–494.

[263] DBMS. Special Report: Parallel Database Special. *DBMS*, March 1995, pp. A–X.

[264] DEITEL, HARVEY M. AND PAUL J. DEITEL. *Java How To Program: With an Introduction to Visual J++*. Upper Saddle River, NJ: Prentice-Hall, 1997.

[265] DEITEL, HARVEY M. AND PAUL J. DEITEL. *C++ How To Program*. 2nd ed. Upper Saddle River, NJ: Prentice-Hall, 1998.

[266] DELBAEN, FREDDY. "Consols in the CIR Model." *Mathematical Finance*, 3, No. 2 (April 1993), 125–134.

[267] DEMBO, RON S. "Scenario Immunization." In [891, Chapter 12].

[268] DENG, YONGHENG. "Mortgage Termination: An Empirical Hazard Model with a Stochastic Term Structure." *Journal of Real Estate Finance and Economics*, 14, No. 3 (1997), 309–331.

[269] DERMAN, EMANUEL AND IRAJ KANI. "Riding on a Smile." *Risk*, 7, No. 2 (February 1994), 32–39.

[270] DERMAN, EMANUEL, DENIZ ERGENER, AND IRAJ KANI. "Static Options Replication." *The Journal of Derivatives*, 2, No. 4 (Summer 1995), 78–95.

[271] DERMAN, EMANUEL, IRAJ KANI, DENIZ ERGENER, AND INDRAJIT BARDHAN. "Enhanced Numerical Methods for Options with Barriers." *Financial Analysts Journal*, 51, No. 6 (November–December 1995), 65–74.

[272] DEROSA, P., L. GOODMAN, AND M. ZAZZARINO. "Duration Estimates on Mortgage-Backed Securities." *The Journal of Portfolio Management*, 19, No. 2 (Winter 1993), 32–38.

[273] DEVORE, J.L. *Probability and Statistics for*

Engineering and the Sciences. Monterey, CA: Brooks/Cole, 1987.

[274] DEWYNNE, J.N., A.E. WHALLEY, AND P. WILMOTT. "Path-Dependent Options and Transaction Costs." *Philosophical Transactions of the Royal Society of London A*, 347, No. 1684 (June 15, 1994), 517–529.

[275] DEZHBAKHSH, HASHEM. "Foreign Exchange Forward and Futures Prices: Are They Equal?" *Journal of Financial and Quantitative Analysis*, 29, No. 1 (March 1994), 75–87.

[276] DHARAN, VENKAT G. "Pricing Path-Dependent Interest Rate Contingent Claims Using a Lattice." *The Journal of Fixed Income*, 6, No. 4 (March 1997), 40–49.

[277] DIMAND, ROBERT W. "The Case of Brownian Motion: A Note on Bachelier's Contribution." *British Journal for History of Science*, 26, Part 2, No. 89 (June 1993), 233–234.

[278] DIXIT, AVINASH K. *Optimization in Economic Theory.* 2nd ed. Oxford, U.K.: Oxford Univ. Press, 1995.

[279] DIXIT, A.K. AND R.S. PINDYCK. "The Options Approach to Capital Investment." *Harvard Business Review*, 73, No. 3 (May–June 1995), 105–115.

[280] DOOB, JUSTIN L. *Stochastic Processes.* New York: Wiley, 1953.

[281] DORFMAN, ROBERT, PAUL A. SAMUELSON AND ROBERT M. SOLOW. *Linear Programming and Economic Analysis.* New York: Dover, 1987.

[282] DOTHAN, M. "On the Term Structure of Interest Rates." *Journal of Financial Economics*, Vol. 7 (1978), 229–264.

[283] DOWD, KEVIN. "A Value at Risk Approach to Risk-Return Analysis." *The Journal of Portfolio Management*, 25, No. 4 (Summer 1999), 60–67.

[284] DOWNES, J. AND J.E. GOODMAN. *Dictionary of Finance and Investment Terms.* Hauppauge, NY: Barron's, 1987.

[285] DRAW, CHI-SHANG. *Path-Dependent Option Pricing.* Master's Thesis. Department of Computer Science and Information Engineering, National Taiwan University, Taiwan, 2000.

[286] DUAN, JIN-CHUAN. "The GARCH Option Pricing Model." *Mathematical Finance*, 5, No. 1 (January 1995), 13–32.

[287] DUAN, JIN-CHUAN. "A Unified Theory of Option Pricing under Stochastic Volatility—from GARCH to Diffusion." Manuscript, October 1996. Available at www.bm.ust.hk/~joduan/opm_sv.pdf.

[288] DUAN, JIN-CHUAN, GENEVIÈVE GAUTHIER AND JEAN-GUY SIMONATO. "An Analytical Approximation for the GARCH Option Pricing Model." *The Journal of Computational Finance*, 2, No. 4 (Summer 1999), 75–116.

[289] DUFFIE, DARRELL. *Security Markets: Stochastic Models.* New York: Academic, 1988.

[290] DUFFIE, DARRELL. *Dynamic Asset Pricing Theory.* 2nd ed. Princeton, NJ: Princeton Univ. Press, 1996.

[291] DUFFIE, DARRELL AND RUI KAN. "Multi-Factor Term Structure Models." *Philosophical Transactions of the Royal Society of London A*, 347, No. 1684 (June 15, 1994), 577–586.

[292] DUFFIE, DARRELL, JIN MA, AND JIONGMIN YONG. "Black's Consol Rate Conjecture." *The Annals of Applied Probability*, 5, No. 2 (May 1995), 356–382.

[293] DUFFIE, DARRELL AND JUN PAN. "An Overview of Value at Risk." *The Journal of Derivatives*, 4, No. 3 (Spring 1997), 7–49.

[294] DUFFIE, DARRELL AND PHILIP PROTTER. "From Discrete- to Continuous-Time Finance: Weak Convergence of the Financial Gain Process." *Mathematical Finance*, 2, No. 1 (January 1992), 1–15.

[295] DUMAS, BERNARD. "Partial Equilibrium versus General Equilibrium Models of the International Capital Market." In *The Handbook of International Macroeconomics.* Edited by Frederick van der Ploeg. Cambridge, MA: Blackwell, 1996.

[296] DUNN, KENNETH B. AND JOHN J. MCCONNELL. "A Comparison of Alternative Models for Pricing GNMA Mortgage-Backed Securities." *The Journal of Finance*, 36, No. 2 (May 1981), 471–484.

[297] DUNN, KENNETH B. AND JOHN J. MCCONNELL. "Valuation of GNMA Mortgage-Backed Securities." *The Journal of Finance*, 36, No. 3 (June 1981), 599–616.

[298] DUNN, KENNETH B. AND JOHN J. MCCONNELL. "Rate of Return Indexes for GNMA Securities." *The Journal of Portfolio Management* (Winter 1981), 65–74.

[299] DUPIRE, BRUNO. "Pricing with a Smile." *Risk*, 7, No. 1 (January 1994), 18–20.

[300] DYBVIG, PHILIP H. "Bond and Bond Option Pricing Based on the Current Term Structure." In *Mathematics of Derivative Securities.* Edited by M. Dempster and S. Pliska. Cambridge, U.K.: Cambridge Univ. Press, 1997.

[301] DYBVIG, PHILIP H., JONATHAN E. INGERSOLL, JR. AND STEPHEN A. ROSS. "Long Forward and Zero-Coupon Rates Can Never Fall." *Journal of Business*, 69, No. 1 (January 1996), 1–25.

[302] DYBVIG, PHILIP H. AND WILLIAM J. MARSHALL. "Pricing Long Bonds: Pitfalls and Opportunities." *Financial Analysts Journal*, 52, No. 1 (January–February 1996), 32–39.

[303] DYBVIG, PHILIP H. AND STEPHEN A. ROSS. "Arbitrage." In [307, pp. 57–71].

[304] DYER, LAWRENCE J. AND DAVID P. JACOB. "Guide to Fixed Income Option Pricing Models." In [321, Chapter 3].

[305] DYER, LAWRENCE J. AND DAVID P. JACOB. "An Overview of Fixed Income Option Pricing Models." In [328, Chapter 34].

[306] DYM, STEVEN I. "A Generalized Approach to Price and Duration of Non-Par Floating-Rate Notes." *The Journal of Portfolio Management*, 24, No. 4 (Summer 1998), 102–107.

[307] EATWELL, J., M. MILGATE, AND P. NEWMAN, eds. *The New Palgrave: Finance*. New York: Norton, 1987.

[308] ECONOMIST, THE. "The Risk Business." *The Economist* (October 17–23, 1998).

[309] ECONOMIST, THE. "The Trader's Lament." *The Economist* (October 16–22, 1999).

[310] EDWARDS, JERI. *3-Tier Client/Server at Work*. New York: Wiley, 1997.

[311] EDWARDS, JOHN. "Changing Database Market Hurts Major Vendors." *Computer*, 31, No. 3 (March 1998), 10–11.

[312] EHRBAR, A. "The Great Bond Market Massacre." *Fortune* (October 17, 1994), 77–92.

[313] EL BABSIRI, MOHAMED AND GERALD NOEL. "Simulating Path-Dependent Options: A New Approach." *The Journal of Derivatives*, 6, No. 2 (Winter 1998), 65–83.

[314] EL-JAHEL, LINA, WILLIAM PERRAUDIN AND PETER SELLIN. "Value at Risk for Derivatives." *The Journal of Derivatives*, 6, No. 3 (Spring 1999), 7–26.

[315] ELMASRI, R. AND S.B. NAVATHE. *Fundamentals of Database Systems*. Redwood City, CA: Benjamin/Cummings, 1989.

[316] ELMER, PETER J. AND ANTON E. HAIDORFER. "Prepayments of Multifamily Mortgage-Backed Securities." *The Journal of Fixed Income*, 6, No. 4 (March 1997), 50–63.

[317] ELTON, EDWIN J. AND MARTIN J. GRUBER. *Modern Portfolio Theory and Investment Analysis*. 5th ed. New York: Wiley, 1995.

[318] ELTON, EDWIN J., MARTIN J. GRUBER, AND RONI MICHAELY. "The Structure of Spot Rates and Immunization." *The Journal of Finance*, 45, No. 2 (June 1990), 629–642.

[319] ENGLE, ROBERT F. "Autoregressive Conditional Heteroskedasticity with Estimates of the Variance of UK Inflation." *Econometrica*, Vol. 50 (1982), 987–1008.

[320] FABOZZI, FRANK J., ed. *Mortgage-Backed Securities: New Strategies, Applications, and Research*. Chicago: Probus, 1987.

[321] FABOZZI, FRANK J., ed. *The Handbook of Fixed-Income Options: Pricing, Strategies & Applications*. Chicago: Probus, 1989.

[322] FABOZZI, FRANK J., ed. *Advances & Innovations in the Bond and Mortgage Markets*. Chicago: Probus, 1989.

[323] FABOZZI, FRANK J. *Fixed Income Mathematics: Analytical & Statistical Techniques*. Revised ed. Chicago: Probus, 1991.

[324] FABOZZI, FRANK J., ed. *The Handbook of Mortgage-Backed Securities*. 3rd ed. Chicago: Probus, 1992.

[325] FABOZZI, FRANK J., ed. *Bond Markets, Analysis and Strategies*. 2nd ed. Englewood Cliffs, NJ: Prentice-Hall, 1993.

[326] FABOZZI, FRANK J. "The Structure of Interest Rates." In [328, Chapter 6].

[327] FABOZZI, FRANK J. *Fixed Income Securities*. New Hope, PA: Frank J. Fabozzi Associates, 1997.

[328] FABOZZI, FRANK J. AND T. DESSA FABOZZI., eds. *The Handbook of Fixed Income Securities*. 4th ed. Burr Ridge, IL: Irwin, 1995.

[329] FABOZZI, FRANK J., ANDREW J. KALOTAY, AND GEORGE O. WILLIAMS. "Valuation of Bonds with Embedded Options." In [328, Chapter 28].

[330] FABOZZI, FRANK J. AND FRANCO MODIGLIANI. *Mortgage and Mortgage-Backed Securities Markets*. Boston: Harvard Business School, 1992.

[331] FABOZZI, FRANK J. AND DEXTER SENFT. "Introduction to Mortgages." In [324, Chapter 2].

[332] FAMA, EUGENE F. "Risk, Return and Equilibrium: Some Clarifying Comments." *The Journal of Finance*, 23, No. 1 (March 1968), 29–40.

[333] FAMA, EUGENE F. "Efficient Capital Markets: II." *The Journal of Finance*, 46, No. 5 (December 1991), 1575–1617.

[334] FAMA, EUGENE F. "Random Walks in Stock Market Prices." Reprinted in *Financial Analysts Journal*, 51, No. 1 (January–February 1995), 75–80.

[335] FAMA, EUGENE F. AND ROBERT R. BLISS. "The Information in Long-Maturity Forward Rates." *The American Economic Review*, 77, No. 4 (September 1987), 680–692.

[336] FAMA, EUGENE F. AND KENNETH R. FRENCH. "The Cross-Section of Expected Stock Returns." *The Journal of Finance*, 47, No. 2 (1992), 427–465.

[337] FANG, KAI-TAI AND YUAN WANG. *Number-Theoretic Methods in Statistics.* London: Chapman & Hall, 1994.

[338] FANG, MING, JAN STALLAERT, AND ANDREW B. WHINSTON. "The Internet and the Future of Financial Markets." *Communications of the ACM,* 43, No. 11 (November 2000), 83–88.

[339] FAREBROTHER, R.W. *Linear Least Squares Computations.* New York: Marcel Dekker, 1988.

[340] FELDMAN, AMY AND JOAN CAPLIN. "The Art of Managing Your Stock Options." www.money.com (December 16, 2000).

[341] FELLER, WILLIAM. "Two Singular Diffusion Problems." *Annals of Mathematics,* 54, No. 1 (July 1951), 173–182.

[342] FELLER, WILLIAM. *An Introduction to Probability Theory and Its Applications,* Vol. 1. 3rd ed. New York: Wiley, 1968.

[343] FELLER, WILLIAM. *An Introduction to Probability Theory and Its Applications,* Vol. 2. 2nd ed. New York: Wiley, 1971.

[344] FERGUSON, R. "Some Formulas for Evaluating Two Popular Option Strategies." *Financial Analysts Journal,* 49, No. 5 (September–October 1993), 71–76.

[345] FIGLEWSKI, STEPHEN. "Remembering Fischer Black." *The Journal of Derivatives,* 3, No. 2 (Winter 1995), 94–98.

[346] FIGLEWSKI, STEPHEN, WILLIAM L. SILBER, AND MARTI G. SUBRAHMANYAM. *Financial Options: From Theory to Practice.* Burr Ridge, IL: Irwin, 1990.

[347] FILIMON, RADU A. "COFI: An Index of Retail Interest Rates." *The Journal of Fixed Income,* 7, No. 3 (December 1997), 61–65.

[348] FINNERTY, JOHN D. "Measuring the Duration of Floating-Rate Debt Instruments." In [322, pp. 77–96].

[349] FINNERTY, JOHN D. AND MICHAEL ROSE. "Arbitrage-Free Spread: A Consistent Measure of Relative Value." *The Journal of Portfolio Management,* 17, No. 3 (Spring 1991), 65–77.

[350] FISHER, LAWRENCE AND ROMAN L. WEIL. "Coping with the Risk of Interest-Rate Fluctuations: Returns to Bondholders from Naive and Optimal Strategies." *Journal of Business,* Vol. 44 (1971), 408–431.

[351] FISHER, MARK AND CHRISTIAN GILLES. "Around and Around: The Expectations Hypothesis." *The Journal of Finance,* Vol. 53 (February 1998), 365–383.

[352] FISHER, MARK, DOUGLAS NYCHKA, AND DAVID ZERVOS. "Fitting the Term Structure of Interest Rates with Smoothing Splines." Manuscript, September 1994.

[353] FISHMAN, GEORGE S. *Monte Carlo: Concepts, Algorithms and Applications.* New York: Springer-Verlag, 1996.

[354] FISHMAN, VLADIMIR, PETER FITTON, AND YURI GALPERIN. "Hybrid Low-Discrepancy Sequences: Effective Path Reduction for Yield Curve Scenario Generation." *The Journal of Fixed Income,* 7, No. 1 (June 1997), 75–84.

[355] FITCH, T. *Dictionary of Banking Terms.* Hauppauge, NY: Barron's, 1990.

[356] FLANAGAN, DAVID. *Java in a Nutshell: A Desktop Quick Reference for Java Programmers.* 2nd ed. Sebastopol, CA: O'Reilly & Associates, 1997.

[357] FLANAGAN, DAVID. *JavaScript: The Definitive Guide.* 3rd ed. Sebastopol, CA: O'Reilly & Associates, 1998.

[358] FLEMING, MICHAEL J. AND ELI M. REMOLONA. "What Moves Bond Prices?" *The Journal of Portfolio Management,* 25, No. 4 (Summer 1999), 28–38.

[359] FLESAKER, BJORN. "Testing the Heath–Jarrow–Morton/Ho–Lee Model of Interest Rate Contingent Claims Pricing." *Journal of Financial and Quantitative Analysis,* 28, No. 4 (December 1993), 483–495.

[360] FOGLER, H.R. "A Modern Theory of Security Analysis." *The Journal of Portfolio Management,* 19, No. 3 (Spring 1993), 6–14.

[361] FOLLAIN, JAMES R., LOUIS O. SCOTT, AND TL TYLER YANG. "Microfunctions of a Mortgage Prepayment Function." *Journal of Real Estate Finance and Economics,* Vol. 5 (1992), 197–217.

[362] FONG, H. GIFFORD AND OLDRICH A. VASICEK. "Fixed-Income Volatility Management." *The Journal of Portfolio Management,* 17, No. 4 (Summer 1991), 41–46.

[363] FOSTER, CHESTER AND ROBERT VAN ORDER. "FHA Terminations: A Prelude to Rational Mortgage Pricing." *Journal of the American Real Estate & Urban Economics Association,* 13, No. 3 (Fall 1985), 273–291.

[364] FRIEDMAN, AVNER. *Stochastic Differential Equations and Applications,* Vol. I. New York: Academic, 1975.

[365] FROOT, K.A., D.S. SCHARFSTEIN AND J.C. STEIN. "A Framework for Risk Management." *Harvard Business Review,* 72, No. 6 (November–December 1994), 91–102.

[366] FU, MICHAEL C., DILIP B. MADAN, AND TONG WANG. "Pricing Continuous Asian Options: A Comparison of Monte Carlo and Laplace Transform Inversion Methods." *The Journal*

of Computational Finance, 2, No. 2 (Winter 1998/1999), 49–74.

[367] GAGNON, LOUIS AND LEWIS D. JOHNSON. "Dynamic Immunization under Stochastic Interest Rates." *The Journal of Portfolio Management*, 20, No. 3 (Spring 1994), 48–54.

[368] GALANTI, SILVIO AND ALAN JUNG. "Low-Discrepancy Sequences: Monte Carlo Simulation of Option Prices." *The Journal of Derivatives*, 5, No. 1 (Fall 1997), 63–83.

[369] GALITZ, LAWRENCE C. *Financial Engineering: Tools and Techniques To Manage Financial Risk*. Burr Ridge, IL: Irwin, 1995.

[370] GAO, BIN, JING-ZHI (JAY) HUANG, AND MARTI G. SUBRAHMANYAM. "An Analytical Approach to the Valuation of American Path-Dependent Options." Manuscript, October 4, 1996. Available at www.kenanflagler.unc.edu/faculty/directory/80.html.

[371] GARBADE, KENNETH D. "Invoice Prices, Special Redemption Features, Cash Flows, and Yields on Eurobonds." In [322, pp. 233–231].

[372] GARBADE, KENNETH D. "Managerial Discretion and Contingent Valuation of Corporate Securities." *The Journal of Derivatives*, 6, No. 4 (Summer 1999), 65–76.

[373] GARDINER, C.W. *Handbook of Stochastic Methods for Physics, Chemistry and the Natural Sciences*. 2nd ed. Berlin: Springer-Verlag, 1985.

[374] GARMAN, MARK B. AND MICHAEL J. KLASS. "On the Estimation of Security Price Volatilities from Historical Data." *Journal of Business*, 53, No. 1 (1980), 67–78.

[375] GARTLAND, W.J., T.W. RITCHFORD, AND N.C. LETICA. "Overview of Fixed-Income Options." In [321, Chapter 1].

[376] GASTINEAU, GARY L. "The Essentials of Financial Risk Management." *Financial Analysts Journal*, 49, No. 5 (September–October 1993), 17–21.

[377] GASTINEAU, GARY. "An Introduction to Special-Purpose Derivatives: Path-Dependent Options." *The Journal of Derivatives*, 1, No. 2 (Winter 1993), 78–86.

[378] GEMAN, HÉLYETE, NICOLE EL KAROUI, AND JEAN-CHARLES ROCHET. "Changes of Numéraire, Changes of Probability Measure and Option Pricing." *Journal of Applied Probability*, 32, No. 2 (June 1995), 443–458.

[379] GEMAN, HÉLYETE AND MARC YOR. "Bessel Processes, Asian Options, and Perpetuities." *Mathematical Finance*, 3, No. 4 (October 1993), 349–375.

[380] GEMAN, HÉLYETE AND MARC YOR. "Pricing and Hedging Double-Barrier Options: A Probabilistic Approach." *Mathematical Finance*, 6, No. 4 (October 1996), 365–378.

[381] GERALD, CURTIS F. AND PATRICK O. WHEATLEY. *Applied Numerical Analysis*. 5th ed. Reading, MA: Addison-Wesley, 1994.

[382] GERBER, ROBERT I. AND ANDREW S. CARRON. "The Option Feature in Mortgages." In [346, Chapter 10].

[383] GESKE, R. AND K. SHASTRI. "Valuation by Approximation: A Comparison of Alternative Option Valuation Techniques." *Journal of Financial and Quantitative Analysis*, Vol. 20 (March 1985), 45–71.

[384] GIBBONS, MICHAEL R. AND KRISHNA RAMASWAMY. "A Test of the Cox, Ingersoll, and Ross Model of the Term Structure." *The Review of Financial Studies*, 6, No. 3 (1993), 619–658.

[385] GILLES, CHRISTIAN, AND STEPHEN F. LEROY. "A Note on the Local Expectations Hypothesis: A Discrete-Time Exposition." *The Journal of Finance*, 41, No. 4 (September 1986), 975–979.

[386] GILLESPIE, D.T. *Markov Processes: An Introduction for Physical Scientists*. New York: Academic, 1992.

[387] GOLDMAN, D., D. HEATH, G. KENTWELL, AND E. PLATEN. "Valuation of Two-Factor Term Structure Models." *Advances in Futures and Options Research*, Vol. 8 (1995), 263–291.

[388] GOLDMAN, M. BARRY, HOWARD B. SOSIN, AND MARY ANN GATTO. "Path Dependent Options: 'Buy at the Low, Sell at the High.'" *The Journal of Finance*, 34, No. 5 (December 1979), 1111–1127.

[389] GOLLINGER, T.L. AND J.B. MORGAN. "Calculation of an Efficient Frontier for a Commercial Loan Portfolio." *The Journal of Portfolio Management*, 19, No. 2 (Winter 1993), 39–46.

[390] GOLUB, BENNETT W. AND LEO M. TILMAN. "Measuring Yield Curve Risk Using Principal Components Analysis, Value at Risk, and Key Rate Durations." *The Journal of Portfolio Management*, 23, No. 4 (Summer 1997), 72–84.

[391] GOLUB, GENE H. AND JAMES M. ORTEGA. *Scientific Computing and Differential Equations: An Introduction to Numerical Methods*. New York: Academic, 1992.

[392] GOLUB, GENE H. AND C.F. VAN LOAN. *Matrix Computations*. 2nd ed. Baltimore, MD: Johns Hopkins Univ. Press, 1989.

[393] GONÇALVES, FRANKLIN DE O. AND JOÃO VICTOR ISSLER. "Estimating the Term Structure of Volatility and Fixed-Income Derivative

Pricing." *The Journal of Fixed Income*, 6, No. 1 (June 1996), 32–39.

[394] GOODMAN, LAURIE S. AND JEFFREY HO. "Mortgage Hedge Ratios: Which One Works Best?" *The Journal of Fixed Income*, 7, No. 3 (December 1997), 23–33.

[395] GOODMAN, LAURIE S., JUDITH JONSON, AND ANDREW SILVER. "Trading and Investment Opportunities with Agency Securities." In [322, pp. 171–198].

[396] GOULDEN, IAN R. AND DAVID M. JACKSON. *Combinatorial Enumeration*. New York: Wiley, 1983.

[397] GRANNAN, L.E. AND S.L. NUTT. "Fixed Income Option Contracts." In [321, Chapter 2].

[398] GRANT, JEREMY. "Hidden Risks in Credit Derivatives." *The Financial Times* (November 2, 1998).

[399] GRAY, W.S. "Historical Returns, Inflation and Future Return Expectations." *Financial Analysts Journal*, 49, No. 4 (July–August 1993), 35–45.

[400] GREEN, T. CLIFTON AND STEPHEN FIGLEWSKI. "Market Risk and Model Risk for a Financial Institution Writing Options." *The Journal of Finance*, 54, No. 4 (August 1999), 1465–1499.

[401] GREIDER, W. *Secrets of the Temple: How the Federal Reserve Runs the Country*. New York: Simon & Schuster, 1987.

[402] GRINBLATT, MARK, AND NARASIMHAN JEGADEESH. "Futures vs. Forward Prices: Implications for Swap Pricing and Derivatives Valuation." In [516, Chaper 3].

[403] GRINOLD, R.C. "Is Beta Dead Again?" *Financial Analysts Journal*, 49, No. 4 (July–August 1993), 28–34.

[404] GUIDERA, JERRY. "Fannie and Freddie May Not Trim Cost of Home Financing." *The Wall Street Journal* (November 16, 2000), A2.

[405] GUO, JIA-HAU. *Option-Adjusted Spread of Mortgage-Backed Securities: A Client/Server System Based on Java and C++*. Master's Thesis. Department of Computer Science and Information Engineering, National Taiwan University, Taiwan, 1998.

[406] GUTTENTAG, JACK M. "The Evolution of Mortgage Yield Concepts." *Financial Analysts Journal*, 48, No. 1 (January–February 1992), 39–46.

[407] GYOURKO, J. AND D.B. KEIM. "Risk and Return in Real Estate: Evidence from a Real Estate Stock Index." *Financial Analysts Journal*, 49, No. 5 (September–October 1993), 39–46.

[408] HABERMAN, RICHARD. *Elementary Applied Partial Differential Equations with Fourier Series and Boundary Value Problems*. 2nd ed. Englewood Cliffs, NJ: Prentice-Hall, 1987.

[409] HACKING, IAN. *The Emergence of Probability: A Philosophical Study of Early Ideas about Probability, Induction, and Statistical Inference*. Cambridge, U.K.: Cambridge Univ. Press, 1975.

[410] HALD, ANDERS. *A History of Probability and Statistics and Their Applications before 1750*. New York: Wiley, 1990.

[411] HALL, ARDEN R. "Valuing the Mortgage Borrower's Prepayment Option." *Journal of the American Real Estate & Urban Economics Association*, 13, No. 3 (Fall 1985), 229–247.

[412] HALL, MARTY. *Core Web Programming*. Upper Saddle River, NJ: Prentice-Hall, 1998.

[413] HALL, P. AND C.C. HEYDE. *Martingale Limit Theory and Its Application*. New York: Academic, 1980.

[414] HAMILTON, ALEXANDER, JAMES MADISON, AND JOHN JAY. *The Federalist Papers*. First published in 1788. New York: New American Library, 1961.

[415] HAMILTON, JAMES D. *Time Series Analysis*. Princeton, NJ: Princeton Univ. Press, 1994.

[416] HAMILTON, MARC A. "Java and the Shift to Net-Centric Computing." *Computer*, 29, No. 8 (August 1996), 31–39.

[417] HAMMING, RICHARD W. *Numerical Methods for Scientists and Engineers*. 2nd ed. New York: Dover, 1986.

[418] HANSEN, LARS PETER. "Large Sample Properties of Generalized Method of Moments Estimators." *Econometrica*, 50, No. 4 (July 1982), 1029–1054.

[419] HARRISON, J. MICHAEL. *Brownian Motion and Stochastic Flow Systems*. New York: Wiley, 1985.

[420] HARRISON, J. MICHAEL AND STANLEY R. PLISKA. "Martingales and Stochastic Integrals in the Theory of Continuous Trading." *Stochastic Processes and Their Applications*, 11, No. 1 (March 1981), 215–260.

[421] HARTMANIS, JURIS. "Turing Award Lecture: On Computational Complexity and the Nature of Computer Science." *ACM Computing Surveys*, 27, No. 1 (March 1995), 7–16.

[422] HARVEY, ANDREW C. *The Econometric Analysis of Time Series*. 2nd ed. New York: Philip Allan, 1990.

[423] HAUG, ESPEN GAARDER. *The Complete Guide to Option Pricing Formulas*. New York: McGraw-Hill, 1998.

[424] HAUGEN, ROBERT A. *Modern Investment Theory*. 3rd ed. Englewood Cliffs, NJ: Prentice-Hall, 1993.

[425] HAWAWINI, G., ed. *Bond Duration and Immunization: Early Developments and Recent Contributions*. New York: Garland, 1982.

[426] HAYEK, FRIEDRICH A. VON. *New Studies in Philosophy, Politics, Economics and the History of Ideas*. Chicago: University of Chicago Press, 1978.

[427] HAYRE, LAKHBIR S. "Arbitrage-Free Spread: Response." *The Journal of Portfolio Management*, 17, No. 3 (Spring 1991), 78–79.

[428] HAYRE, LAKHBIR S. "Random Error in Prepayment Projections." *The Journal of Fixed Income*, 7, No. 2 (September 1997), 77–84.

[429] HAYRE, LAKHBIR S. AND HUBERT CHANG. "Effective and Empirical Durations of Mortgage Securities." *The Journal of Fixed Income*, 6, No. 4 (March 1997), 17–33.

[430] HAYRE, LAKHBIR S., SHARAD CHAUDHARY AND ROBERT A. YOUNG. "Anatomy of Prepayments." *The Journal of Fixed Income*, 10, No. 1 (June 2000), 19–49.

[431] HAYRE, LAKHBIR S., KENNETH LAUTERBACH AND CYRUS MOHEBBI. "Prepayment Models and Methodologies." In [322, pp. 329–350].

[432] HAYRE, LAKHBIR S., CYRUS MOHEBBI, AND THOMAS ZIMMERMAN. "Mortgage Pass-Through Securities." In [328, Chapter 24].

[433] HAYRE, LAKHBIR S. AND ARVIND RAJAN. "Anatomy of Prepayments: The Salomon Prepayment Model." Fixed-Income Research, Salomon Brothers, June 1995. Also in [516, Chapter 8] as "Anatomy of Prepayments: The Salomon Brothers Prepayment Model."

[434] HE, HUA. "Convergence of Discrete- to Continuous-Time Contingent Claims Prices." *The Review of Financial Studies*, 3, No. 4 (1990), 523–546.

[435] HEATH, DAVID, ROBERT JARROW, AND ANDREW MORTON. "Contingent Claim Valuation with a Random Evolution of Interest Rates." *The Review of Futures Markets*, 9, No. 1 (1990), 54–76.

[436] HEATH, DAVID, ROBERT JARROW, AND ANDREW MORTON. "Bond Pricing and the Term Structure of Interest Rates: A Discrete Time Approximation." *Journal of Financial and Quantitative Analysis*, 25, No. 4 (December 1990), 419–440.

[437] HEATH, DAVID, ROBERT JARROW, AND ANDREW MORTON. "Bond Pricing and the Term Structure of Interest Rates: A New Methodology for Contingent Claims Valuation." *Econometrica*, 60, No. 1 (January 1992), 77–105.

[438] HERSKOVITZ, MICHAEL D. "Option-Adjusted Spread Analysis for Mortgage-Backed Securities." In [321, Chapter 22].

[439] HESS, ALAN C. AND CLIFFORD W. SMITH, JR. "Elements of Mortgage Securitization." *Journal of Real Estate Finance and Economics*, Vol. 1 (1988), 331–346.

[440] HESTON, STEVEN L. "A Closed-Form Solution for Options with Stochastic Volatility with Applications to Bond and Currency Options." *The Review of Financial Studies*, 6, No. 2 (1993), 327–344.

[441] HESTON, STEVEN. "Discrete-Time Versions of Continuous-Time Interest Rate Models." *The Journal of Fixed Income*, 5, No. 2 (September 1995), 86–88.

[442] HESTON, STEVEN L. AND SAIKAT NANDI. "A Closed-Form GARCH Option Valuation Model." *The Review of Financial Studies*, 13, No. 3 (Fall 2000), 585–625.

[443] HEYNEN, RONALD C. AND HARRY M. KAT. "Discrete Partial Barrier Options with a Moving Barrier." *The Journal of Financial Engineering*, 5, No. 3 (1996), 199–209.

[444] HEYNEN, RONAND, ANGELIEN KEMNA, AND TON VORST. "Analysis of the Term Structure of Implied Volatilities." *Journal of Financial and Quantitative Analysis*, 29, No. 1 (March 1994), 31–57.

[445] HICKS, JOHN RICHARD. *Value and Capital*. 2nd ed. London: Oxford Univ. Press, 1946.

[446] HILDEBRAND, FRANCIS B. *Finite-Difference Equations and Simulations*. Englewood Cliffs, NJ: Prentice-Hall, 1968.

[447] HILDEBRAND, FRANCIS B. *Introduction to Numerical Analysis*. 2nd ed. New York: Dover, 1974.

[448] HILDEBRAND, FRANCIS B. *Advanced Calculus for Applications*. 2nd ed. Englewood Cliffs, NJ: Prentice-Hall, 1976.

[449] HILLIARD, JIMMY E., JAMES B. KAU, DONALD C. KEENAN AND WALTER J. MULLER, III. "Pricing a Class of American and European Path Dependent Securities." *Management Science*, 41, No. 12 (December 1995), 1892–1899.

[450] HILLIARD, JIMMY E. AND ADAM SCHWARTZ. "Binomial Option Pricing under Stochastic Volatility and Correlated State Variables." *The Journal of Derivatives*, 4, No. 1 (Fall 1996), 23–39.

[451] HILLIARD, JIMMY E., ADAM SCHWARTZ, AND ALAN L. TUCKER. "Bivariate Binomial Options Pricing with Generalized Interest Rate Processes." *The Journal of Financial Research*, 19, No. 4 (Winter 1996), 585–602.

[452] HO, TENG-SUAN, RICHARD C. STAPLETON,

AND MARTI G. SUBRAHMANYAM. "Multivariate Binomial Approximations for Asset Prices with Nonstationary Variance and Covariance Characteristics." *The Review of Financial Studies*, 8, No. 4 (Winter 1995), 1125–1152.

[453] HO, THOMAS S.Y. "Key Rate Durations: Measures of Interest Rate Risks." *The Journal of Fixed Income*, 2, No. 2 (September 1992), 29–44.

[454] HO, THOMAS S.Y. "CMO Yield Attribution and Option Spread." *The Journal of Portfolio Management*, 19, No. 3 (Spring 1993), 57–68.

[455] HO, THOMAS S.Y. "Primitive Securities: Portfolio Building Blocks." *The Journal of Derivatives*, 1, No. 2 (Winter 1993), 6–22.

[456] HO, THOMAS S.Y. "Evolution of Interest Rate Models: a Comparison." *The Journal of Derivatives*, 2, No. 4 (Summer 1995), 9–20.

[457] HO, THOMAS S.Y. AND ALLEN A. ABRAHAMSON. "Options on Interest Sensitive Securities." In [346, Chapter 8].

[458] HO, THOMAS S.Y. AND SANG-BIN LEE. "Term Structure Movements and Pricing Interest Rate Contingent Claims." *The Journal of Finance*, 41, No. 5 (December 1986), 1011–1029.

[459] HOCHBAUM, DORIT S., ed. *Approximation Algorithms for NP-Hard Problems*. Boston: PWS-Kent, 1997.

[460] HOFFMAN, W.C., T. KIGGINS, AND N.C. LETICA. "Covered Calls on Mortgage-Backed Securities." In [321, Chapter 11].

[461] HOFRI, MICHA. *Probabilistic Analysis of Algorithms*. New York: Springer-Verlag, 1987.

[462] HOGAN, MICHAEL. "Problems in Certain Two-State Term Structure Models." *Annals of Applied Probability*, 3, No. 2 (May 1993), 576–581.

[463] HOGG, ROBERT V. AND ALLEN T. CRAIG. *Introduction to Mathematical Statistics*. 4th ed. New York: Macmillan, 1978.

[464] HOPPER, GREG. "Value at Risk: A New Methodology for Measuring Portfolio Risk." *Business Review* (July/August 1996), 19–29.

[465] HORN, ROGER A. AND CHARLES R. JOHNSON. *Matrix Analysis*. Cambridge, U.K.: Cambridge Univ. Press, 1991.

[466] HORSTMANN, CAY S. *Mastering Object-Oriented Design in C++*. New York: Wiley, 1995.

[467] HORSTMANN, CAY S. AND GARY CORNELL. *Core Java 2, Vol. 1 – Fundamentals*. Mountain View, California: SunSoft Press, 1999.

[468] HOUTHAKKER, HENDRIK S. "Futures Trading." In [307, pp. 153–158].

[469] HU, JOSEPH. "Housing and the Mortgage Securities Markets: Review, Outlook, and Policy Recommendations." *Journal of Real Estate Finance and Economics*, 5, No. 2 (June 1992), 167–179.

[470] HULL, JOHN C. *Options, Futures and Other Derivatives*. 4th ed. Englewood Cliffs, NJ: Prentice-Hall, 1999.

[471] HULL, JOHN C. AND ALAN WHITE. "The Pricing of Options on Assets with Stochastic Volatilities." *The Journal of Finance*, 42, No. 2 (June 1987), 281–300.

[472] HULL, JOHN C. AND ALAN WHITE. "The Use of the Control Variate Technique in Option Pricing." *Journal of Financial and Quantitative Analysis*, 23, No. 3 (September 1988), 237–251.

[473] HULL, JOHN C. AND ALAN WHITE. "Valuing Derivative Securities Using the Explicit Finite Difference Method." *Journal of Financial and Quantitative Analysis*, 25, No. 1 (March 1990), 87–100.

[474] HULL, JOHN C. AND ALAN WHITE. "Pricing Interest-Rate-Derivative Securities." *The Review of Financial Studies*, 3, No. 4 (1990), 573–592.

[475] HULL, JOHN C. AND ALAN WHITE. "Root and Branch." In [744, Chapter 14].

[476] HULL, JOHN C. AND ALAN WHITE. "New Ways with the Yield Curve." In [744, Chapter 15].

[477] HULL, JOHN C. AND ALAN WHITE. "One-Factor Interest-Rate Models and the Valuation of Interest-Rate Derivative Securities." *Journal of Financial and Quantitative Analysis*, 28, No. 2 (June 1993), 235–254.

[478] HULL, JOHN C. AND ALAN WHITE. "Efficient Procedures for Valuing European and American Path-Dependent Options." *The Journal of Derivatives*, Vol. 1 (Fall 1993), 21–31.

[479] HULL, JOHN C. AND ALAN WHITE. "Numerical Procedures for Implementing Term Structure Models I: One-Factor Models." *The Journal of Derivatives*, 2, No. 1 (Fall 1994), 7–17.

[480] HULL, JOHN C. AND ALAN WHITE. "Numerical Procedures for Implementing Term Structure Models II: Two-Factor Models." *The Journal of Derivatives*, 2, No. 2 (Winter 1994), 37–48.

[481] HULL, JOHN C. AND ALAN WHITE. "The Impact of Default Risk on the Prices of Options and Other Derivatives. *Journal of Banking & Finance*, Vol. 19 (1995), 299–322.

[482] HULL, JOHN C. AND ALAN WHITE. "'A Note on the Models of Hull and White for Pricing Options on the Term Structure': Response." *The Journal of Fixed Income*, 5, No. 2 (September 1995), 97–103.

[483] HULL, JOHN C. AND ALAN WHITE. "Using Hull-White Interest Rate Trees." *The Journal of Derivatives*, 3, No. 3 (Spring 1996), 26–36.

[484] HULL, JOHN C. AND ALAN WHITE. "Value at Risk When Daily Changes in Market Variables Are Not Normally Distributed." *The Journal of Derivatives*, 5, No. 3 (Spring 1998), 9–19.

[485] HUNTER, WILLIAM C. AND DAVID W. STOWE. "Path-Dependent Options: Valuation and Applications." *Economic Review*, Vol. 77 (July/August 1992), 30–43.

[486] HUTCHINSON, JAMES M., ANDREW W. LO, AND TOMASO POGGIO. "A Nonparametric Approach to Pricing and Hedging Derivative Securities via Learning Networks." *The Journal of Finance*, 49, No. 3 (July 1994), 851–889.

[487] HWANG, KAI. AND ZHIWEI XU. *Scalable Parallel Computing: Technology, Architecture, Programming*. New York: McGraw-Hill, 2000.

[488] IANSITI, MARCO AND ALAN MACCORMACK. "Developing Products on Internet Time." *Harvard Business Review*, 75, No. 5 (September–October 1997), 108–117.

[489] ILMANEN, ANTTI. "Overview of Forward Rate Analysis (Understanding the Yield Curve: Part 1)." Fixed-Income Research, Salomon Brothers, May 1995.

[490] ILMANEN, ANTTI. "Convexity Bias and the Yield Curve." In [516, Chapter 2].

[491] INGERSOLL, JONATHAN E., JR. "A Contingent-Claims Valuation of Convertible Securities." *Journal of Financial Economics*, Vol. 4 (May 1977), 289–322.

[492] INGERSOLL, JONATHAN E., JR. *Theory of Financial Decision Making*. Savage, MD: Rowman & Littlefield, 1987.

[493] INGERSOLL, JONATHAN E., JR. "Interest Rates." In [307, pp. 172–179].

[494] INGERSOLL, JONATHAN E., JR. "Option Pricing Theory." In [307, pp. 199–212].

[495] INGERSOLL, JONATHAN E., JR. "Digital Contracts: Simple Tools for Pricing Complex Derivatives." *Journal of Business*, 73, No. 1 (January 2000), 67–88.

[496] INGERSOLL, JONATHAN E., JR., JEFFREY SKELTON, AND ROMAN L. WEIL. "Duration Forty Years Later." *Journal of Financial and Quantitative Analysis*, 13, No. 4 (November 1978), 627–650.

[497] INMON, WILLIAM H., CLAUDIA IMHOFF, AND RYAN SOUSA. *Corporate Information Factory*. New York: Wiley, 1998.

[498] INTERNETWEEK. "Transformation of the Enterprise 1999." *Internetweek* (October 28, 1999). Available at www.internetweek.com/trans/default.html.

[499] INUI, KOJI AND MASAAKI KIJIMA. "A Markovian Framework in Multi-Factor Heath–Jarrow–Morton Models." *Journal of Financial and Quantitative Analysis*, 33, No. 3 (September 1998), 423–440.

[500] ITO, KIYOSI. "Stochastic Integral." *Proceedings of the Imperial Academy of Tokyo*, Vol. 20 (1944), 519–524.

[501] ITO, KIYOSI. "On a Formula Concerning Stochastic Differentials." *Nagoya Mathematics Journal*, Vol. 3 (1951), 55–65.

[502] JACKWERTH, JENS CARSTEN. "Generalized Binomial Trees." *The Journal of Derivatives*, 5, No. 2 (Winter 1997), 7–17.

[503] JACKWERTH, JENS CARSTEN. "Option-Implied Risk-Neutral Distributions and Implied Binomial Trees: A Literature Review." *The Journal of Derivatives*, 7, No. 2 (Winter 1999), 66–82.

[504] JACOB, DAVID P., GRAHAM LOAD, AND JAMES A. TILLEY. "Price, Duration and Convexity of Mortgage-Backed Securities." In [320, pp. 81–101].

[505] JAMES, JESSICA AND NICK WEBBER. *Interest Rate Modelling*. New York: Wiley, 2000.

[506] JAMSHIDIAN, FARSHID. "An Exact Bond Option Formula." *The Journal of Finance*, 44, No. 1 (March 1989), 205–209.

[507] JAMSHIDIAN, FARSHID. "Bond and Option Evaluation in the Gaussian Interest Rate Model." *Research in Finance*, Vol. 9 (1991), 131–170.

[508] JAMSHIDIAN, FARSHID. "Forward Induction and Construction of Yield Curve Diffusion Models." *The Journal of Fixed Income*, Vol. 1 (June 1991), 62–74.

[509] JAMSHIDIAN, FARSHID AND YU ZHU. "Scenario Simulation: Theory and Methodology." *Finance and Stochastics*, Vol. 1 (1997), 43–67.

[510] JARROW, ROBERT A. *Modelling Fixed Income Securities and Interest Rate Options*. New York: McGraw-Hill, 1996.

[511] JARROW, ROBERT A. "The HJM Model: Its Past, Present and Future." *The Journal of Financial Engineering*, 6, No. 4 (1997), 269–279.

[512] JARROW, ROBERT A. "In Honor of the Nobel Laureates Robert C. Merton and Myron S. Scholes: a Partial Differential Equation That Changed the World." *The Journal of Economic Perspectives*, 13, No. 4 (Fall 1999), 229–248.

[513] JARROW, ROBERT A. AND STUART TURNBULL. "Pricing Derivatives on Financial Securities Subject to Credit Risk." *The Journal of Finance*, 50, No. 1 (March 1995), 53–85.

[514] JARROW, ROBERT A. AND STUART

TURNBULL. *Derivative Securities*. Cincinnati, OH: South-Western College Publishing, 1996.

[515] JEFFREY, ANDREW. "Single Factor Heath-Jarrow–Morton Term Structure Models Based on Markov Spot Interest Rate Dynamics." *Journal of Financial and Quantitative Analysis*, 30, No. 4 (December 1995), 619–642.

[516] JEGADEESH, NARASIMHAN AND BRUCE TUCKMAN. (Ed.) *Advanced Fixed-Income Valuation Tools*. New York: Wiley, 2000.

[517] JIANG, GEORGE J. AND JOHN L. KNIGHT. "Finite Sample Comparison of Alternative Estimators of Ito Diffusion Processes: A Monte Carlo Study." *The Journal of Computational Finance*, 2, No. 3 (Spring 1999), 5–38.

[518] JOHANSSON, FREDERIK, MICHAEL J. SEILER, AND MIKAEL TJARNBERG. "Measuring Downside Portfolio Risk." *The Journal of Portfolio Management*, 26, No. 1 (Fall 1999), 96–107.

[519] JOHNSON, HERB. "Options on the Maximum or the Minimum of Several Assets." *Journal of Financial and Quantitative Analysis*, 22, No. 3 (September 1987), 277–283.

[520] JOHNSON, HERB AND DAVID SHANNO. "Option Pricing When the Variance Is Changing." *Journal of Financial and Quantitative Analysis*, 22, No. 2 (June 1987), 143–151.

[521] JOHNSON, HERB AND RENÉ STULZ. "The Pricing of Options with Default Risk." *The Journal of Finance*, 42, No. 2 (June 1987), 267–280.

[522] JOHNSON, NORMAN LLOYD AND SAMUEL KOTZ. *Distributions in Statistics: Continuous Univariate Distributions*, Vol. 1. Boston: Houghton Mifflin, 1970.

[523] JOHNSON, RICHARD A. AND DEAN W. WICHERN. *Applied Multivariate Statistical Analysis*. 3rd ed. Englewood Cliffs, NJ: Prentice-Hall, 1992.

[524] JONES, F.J. AND A. JAIN. "Hedging Mortgage-Backed Securities." In [320, pp. 367–442].

[525] JORDAN, JAMES V. AND SATTAR A. MANSI. "How Well Do Constant-Maturity Treasuries Approximate the On-the-Run Term Structure?" *The Journal of Fixed Income*, 10, No. 2 (September 2000), 35–45.

[526] JORION, PHILIPPE. "Predicting Volatility in the Foreign Exchange Markets." *The Journal of Finance*, 50, No. 2 (June 1995), 507–528.

[527] JORION, PHILIPPE. "Orange County Case: Using Value at Risk To Control Financial Risk." 1997. Available at www.gsm.uci.edu/˜jorion/oc/case.html.

[528] JOUBERT, ADRIAAN. "Financial Applications and HPF." In *Proceedings of the Conference on High Performance Computing: Issues, Methods and Applications*, June 1994.

[529] JOUBERT, ADRIAAN AND L.C.G. ROGERS. "Fast, Accurate and Inelegant Valuation of American Options." In *Numerical Methods in Finance*. Edited by L.C.G. Rogers and Denis Talay. Cambridge, U.K.: Cambridge Univ. Press, 1997.

[530] JOY, CORWIN, PHELIM P. BOYLE, AND KEN SENG TAN. "Quasi-Monte Carlo Methods in Numerical Finance." *Management Science*, 42, No. 6 (June 1996), 926–938.

[531] JU, NENGJIU. "Pricing an American Option by Approximating Its Early Exercise Boundary as a Multipiece Exponential Function." *The Review of Financial Studies*, 11, No. 3 (Fall 1998), 627–646.

[532] JUNG, ALAN. "Improving the Performance of Low-Discrepancy Sequences." *The Journal of Derivatives*, 6, No. 2 (Winter 1998), 85–95.

[533] KAHNEMAN, DANIEL AND MARK W. RIEPE. "Aspects of Investor Psychology." *The Journal of Portfolio Management*, 24, No. 4 (Summer 1998), 52–65.

[534] KAHNEMAN, DANIEL AND AMOS TVERSKY. "Prospect Theory: An Analysis of Decision under Risk." *Econometrica*, Vol. 47 (1979), 263–291.

[535] KAJIMA, MASAAKI AND MATSUYA KOMORIBAYASHI. "A Markov Chain Model for Valuing Credit Risk Derivatives." *The Journal of Derivatives*, 6, No. 1 (Fall 1998), 97–108.

[536] KALOTAY, ANDREW J. AND LESLIE A. ABREO. "Putable/Callable/Reset Bonds: Intermarket Arbitrage with Unpleasant Side Effects." *The Journal of Derivatives*, 6, No. 3 (Spring 1999), 88–93.

[537] KAMAL, MICHAEL AND EMANUEL DERMAN. "Correcting Black–Scholes." *Risk*, 12, No. 1 (January 1999), 82–85.

[538] KAMPHOEFNER, J.E. AND R.J. MCKENDRY. "Trading and Arbitrage Strategies Using Debt Options." In [321, Chapter 8].

[539] KAMRAD, BARDIA AND PETER RITCHKEN. "Multinomial Approximating Models for Options with k State Variables." *Management Science*, 37, No. 12 (December 1991), 1640–1652.

[540] KANG, PAN AND STAVROS A. ZENIOS. "Complete Prepayment Models for Mortgage-Backed Securities." *Management Science*, 38, No. 11 (November 1992), 1665–1685.

[541] KANNAN, D. *An Introduction to Stochastic Processes*. New York: Elsevier North Holland, 1979.

[542] KARATZAS, IOANNIS. "Optimization Problems in the Theory of Continuous Trading." *SIAM Journal on Control and Optimization*, 27, No. 6 (November 1989), 1221–1259.

[543] KARLIN, SAMUEL AND HOWARD M. TAYLOR. *A First Course in Stochastic Processes*. 2nd ed. New York: Academic, 1975.

[544] KARLIN, SAMUEL AND HOWARD M. TAYLOR. *A Second Course in Stochastic Processes*. New York: Academic, 1981.

[545] KARLOFF, HOWARD. *Linear Programming*. Boston: Birkhäuser, 1991.

[546] KAU, JAMES B., DONALD C. KEENAN, WALTER J. MULLER III, AND JAMES F. EPPERSON. "Rational Pricing of Adjustable Rate Mortgages." *Journal of the American Real Estate & Urban Economics Association*, 13, No. 2 (Summer 1985), 117–127.

[547] KELLISON, S.G. *The Theory of Interest*. 2nd ed. Burr Ridge, IL: Irwin, 1991.

[548] KEMNA, A.G.Z. AND A.C.F. VORST. "A Pricing Method for Options Based on Average Asset Values." *Journal of Banking & Finance*, 14, No. 1 (March 1990), 113–129.

[549] KENDALL, WILFRID S. "Itovsn3: Doing Stochastic Calculus with *Mathematica*." In [854, Chapter 10].

[550] KENNEDY, P. *A Guide to Econometrics*. 3rd ed. Cambridge, MA: MIT Press, 1993.

[551] KIKUGAWA, T. AND K.J. SINGLETON. "Modeling the Term Structure of Interest Rates in Japan." *The Journal of Fixed Income*, 4, No. 2 (September 1994), 6–16.

[552] KIM, DONGCHEOL AND STANLEY J. KON. "Alternative Models for the Conditional Heteroskedasticity of Stock Returns." *Journal of Business*, 67, No. 4 (October 1994), 563–598.

[553] KING, TAO-HSIEN DOLLY AND DAVID C. MAUER. "Corporate Call Policy for Nonconvertible Bonds." *Journal of Business*, 73, No. 3 (July 2000), 403–444.

[554] KISHIMOTO, NAOKI. "Duration and Convexity of Coupon Bond Futures." *The Journal of Fixed Income*, 8, No. 1 (June 1998), 79–83.

[555] KLASSEN, TIMOTHY R. "Simple, Fast and Flexible Pricing of Asian Options." Manuscript, 1999. To appear in *The Journal of Computational Finance*.

[556] KLOEDEN, PETER E. AND ECKHARD PLATEN. *Numerical Solution of Stochastic Differential Equations*. Berlin: Springer-Verlag, 1992.

[557] KLOEDEN, PETER E., ECKHARD PLATEN, AND HENRI SCHURZ. *Numerical Solution of SDE through Computer Experiments*. Berlin: Springer-Verlag, 1994.

[558] KLOEDEN, P.E., E. PLATEN, H. SCHURZ, AND M. SØRENSEN. "On Effects of Discretization on Estimators of Drift Parameters for Diffusion Processes." *Journal of Applied Probability*, 33, No. 4 (December 1996), 1061–1076.

[559] KNECHT, LUKE AND MIKE MCCOWIN. "Valuing Convertible Securities." In [322, pp. 97–116].

[560] KNUTH, DONALD E. *The Art of Computer Programming, Vol. II: Seminumerical Algorithms*. 3rd ed. Reading, MA: Addison-Wesley, 1998.

[561] KON, STANLEY J. "Models of Stock Returns – A Comparison." *The Journal of Finance*, 39, No. 1 (March 1984), 147–165.

[562] KOPPRASCH, ROBERT W. "Option-Adjusted Spread Analysis: Going down the Wrong Path?" *Financial Analysts Journal*, 50, No. 3 (May–June 1994), 42–47.

[563] KOUTMOS, GREGORY. "Modeling Short-Term Interest Rate Volatility: Information Shocks versus Interest Rate Levels." *The Journal of Fixed Income*, 9, No. 4 (March 2000), 19–26.

[564] KRAUS, ALAN AND MAXWELL SMITH. "A Simple Multifactor Term Structure Model." *The Journal of Fixed Income*, Vol. 3 (June 1993), 19–23.

[565] KREMER, JOSEPH W. AND RODNEY L. ROENFELDT. "Warrant Pricing: Jump-Diffusion vs. Black–Scholes." *Journal of Financial and Quantitative Analysis*, 28, No. 2 (June 1992), 255ff.

[566] KRISHNAN, VENKATARAMA. *Nonlinear Filtering and Smoothing*. New York: Wiley, 1984.

[567] KRITZMAN, MARK. "Portfolio Insurance and Related Dynamic Trading Strategies." In [346, Chapter 11].

[568] KRITZMAN, MARK. "What Practitioners Need To Know about the Term Structure of Interest Rates." *Financial Analysts Journal*, 49, No. 4 (July–August 1993), 14–18.

[569] KRITZMAN, MARK. "What Practitioners Need To Know about Hedging." *Financial Analysts Journal*, 49, No. 5 (September–October 1993), 22–26.

[570] KRITZMAN, MARK. "What Practitioners Need To Know about Monte Carlo Simulation." *Financial Analysts Journal*, 49, No. 6 (November–December 1993), 17–20.

[571] KUPIEC, PAUL H. "Stress Testing in a Value at Risk Framework." *The Journal of Derivatives*, 6, No. 1 (Fall 1998), 7–24.

[572] KUSHNER, HAROLD J. *Probability Methods for Approximations in Stochastic Control and for Elliptic Equations*. New York: Academic, 1977.

[573] KUSHNER, HAROLD J. *Approximation and Weak Convergence Methods for Random Processes with Applications to Stochastic*

Systems Theory. Cambridge, MA: MIT Press, 1984.

[574] KWAN, THOMAS T., ROBERT E. MCGRATH, AND DANIEL A. REED. "NCSA's World Wide Web Server: Design and Performance." *Computer*, 28, No. 11 (November 1995), 68–74.

[575] KWOK, YUE-KUEN. *Mathematical Models of Financial Derivatives.* Singapore: Springer-Verlag, 1998.

[576] LAMBERTON, DAMIEN. "Convergence of the Critical Price in the Approximation of American Options." *Mathematical Finance*, 3, No. 2 (April 1993), 179–190.

[577] LAMLE, HUGH R. "Ginnie Mae: Age Equals Beauty." *The Journal of Portfolio Management* (Winter 1981), 75–79.

[578] LAMOUREUX, CHRISTOPHER G. AND WILLIAM D. LASTRAPES. "Heteroskedasticity in Stock Return Data: Volume versus GARCH Effects." *The Journal of Finance*, 45, No. 1 (March 1990), 221–229.

[579] LAMOUREUX, CHRISTOPHER G. AND WILLIAM D. LASTRAPES. "Forecasting Stock-Return Variance: Toward an Understanding of Stochastic Implied Volatility." *The Review of Financial Studies*, 6, No. 2 (1993), 293–326.

[580] LAMPORT, LESLIE. *LATEX: a Document Preparation System.* 2nd ed. Reading, MA: Addison-Wesley, 1994.

[581] LANCASTER, K. *Mathematical Economics.* New York: Dover, 1987.

[582] LANCZOS, C. *Applied Analysis.* First published in 1956. New York: Dover, 1988.

[583] LANDO, DAVID. "Some Elements of Rating-Based Credit Risk Modeling." In [516, Chapter 7].

[584] LAW, AVERILL M. AND W. DAVID KELTON. *Simulation Modeling and Analysis.* 2nd ed. New York: McGraw-Hill, 1991.

[585] LAWLER, GREGORY F. *Introduction to Stochastic Processes.* London: Chapman & Hall, 1995.

[586] LAWSON, CHARLES L. AND RICHARD J. HANSON. *Solving Least Squares Problems.* Philadelphia: Society for Applied and Industrial Mathematics, 1995.

[587] LEHMANN, BRUCE N. "Fads, Martingales, and Market Efficiency." *The Quarterly Journal of Economics*, 105, No. 1 (February 1990), 1–28.

[588] LEIGHTON, F. THOMSON. *Introduction to Parallel Algorithms and Architectures: Arrays, Trees, Hypercubes.* San Mateo, CA: Morgan Kaufmann, 1992.

[589] LEISEN, DIETMAR. "Pricing the American Put Option: A Detailed Convergence Analysis for Binomial Models." *Journal of Economic Dynamics & Control*, Vol. 22 (1998), 1419–1444.

[590] LEISEN, DIETMAR AND MATTHIAS REIMER. "Binomial Models for Option Valuation – Examining and Improving Convergence." *Applied Mathematical Finance*, Vol. 3 (1996), 319–346.

[591] LEKKOS, ILIAS. "A Critique of Factor Analysis of Interest Rates." *The Journal of Derivatives* (Fall 2000), 72–83.

[592] LELAND, HAYNE E. "Beyond Mean-Variance: Performance Measurement in a Nonsymmetrical World." *Financial Analysts Journal*, 55, No. 1 (January–February 1999), 27–36.

[593] LEONG, KENNETH. "Solving the Mystery." In [744, Chapter 11].

[594] LEROY, STEPHEN F. "Efficient Capital Markets and Martingales." *Journal of Economic Literature*, 27, No. 4 (December 1989), 1583–1621.

[595] LEROY, STEPHEN F. "Mortgage Valuation under Optimal Prepayment." *The Review of Financial Studies*, 9, No. 3 (Fall 1996), 817–844.

[596] LEVY, EDMOND. "Pricing European Average Rate Currency Options." *Journal of International Money and Finance*, Vol. 11 (1992), 474–491.

[597] LEVY, EDMOND AND FRANCQIS MANTION. "Discrete by Nature." *Risk*, 10, No. 1 (January 1997), 74–75.

[598] LEVY, EDMOND AND STUART TURNBULL. "Average Intelligence." In [744, Chapter 23].

[599] LEWIS, MICHAEL. "Wall Street Crashes Venture Caps Party." *Bloomberg News* (September 28, 2000).

[600] LI, ANLONG, PETER RITCHKEN AND L. SANKARASUBRAMANIAN. "Lattice Models for Pricing American Interest Rate Claims." *The Journal of Finance*, 50, No. 2 (June 1995), 719–737.

[601] LIAO, ANDY AND YUH-DAUH LYUU. "Computer Applications on Wall Street." Manuscript, Fixed Income Research, Citicorp Securities, New York, 1994.

[602] LIEPELT, MICHAEL AND KLAUS SCHITTKOWSKI. "Remark on Algorithm 746: New Features of PCOMP, a Fortran Code for Automatic Differentiation." *ACM Transactions on Mathematical Software*, 26, No. 3 (September 2000), 352–362.

[603] LINSMEIER, THOMAS J. AND NEIL D. PEARSON. "Value at Risk." *Financial Analysts Journal*, 56, No. 2 (March–April 2000), 47–67.

[604] LINT, J.H. VAN AND R.M. WILSON. *A Course in Combinatorics.* Cambridge, U.K.: Cambridge Univ. Press, 1994.

[605] LINTNER, JOHN. "The Valuation of Risk Assets and the Selection of Risky Investments in Stock Portfolios and Capital Budgets." *Review of Economics and Statistics*, Vol. 47 (February 1965), 13–37.

[606] LIONS, J.L. "Ariane 5 Flight 501 Failure: Report by the Inquiry Board." Paris, July 19, 1996. Available at www.esa.int/htdocs/tidc/Press/Press96/ariane5rep.html.

[607] LITTERMAN, ROBERT AND JOSÉ SCHEINKMAN. "Common Factors Affecting Bond Returns." *The Journal of Fixed Income*, Vol. 1 (June 1991), 54–61.

[608] LITZENBERGER, ROBERT H. "Swaps: Plain and Fanciful." *The Journal of Finance*, 47, No. 3 (July 1992), 831–850.

[609] LITZENBERGER, ROBERT H. AND JACQUES ROLFO. "An International Study of Tax Effects on Government Bonds." *The Journal of Finance*, 39, No. 1 (March 1984), 1–22.

[610] LIU, YU-HONG. *Barrier Options Pricing: Combinatorial Methods and Trinomial Tree Algorithms*. Master's Thesis. Department of Computer Science and Information Engineering, National Taiwan University, Taiwan, 1997.

[611] LO, ANDREW W. "Maximum Likelihood Estimation of Generalized Ito Processes with Discretely Sampled Data." *Econometric Theory*, 4, No. 2 (August 1988), 231–247.

[612] LO, ANDREW W. "Neural Networks and Other Nonparametric Techniques in Economics and Finance." In *Blending Quantitative and Traditional Equity Analysis*. Edited by H. Russell Fogler. Charlottesville, VA: Association for Investment Management and Research, 1994.

[613] LO, ANDREW W. AND JIANG WANG. "Implementing Option Pricing Models When Asset Returns Are Predictable." *The Journal of Finance*, 50, No. 1 (March 1995), 87–129.

[614] LO, ANDREW W. AND JIANG WANG. "Trading Volume: Definitions, Data Analysis, and Implementations of Portfolio Theory." *The Review of Financial Studies*, 13, No. 2 (Summer 2000), 257–300.

[615] LONG, D. MICHAEL AND DENNIS T. OFFICER. "The Relation between Option Mispricing and Volume in the Black–Scholes Option Model." *The Journal of Financial Research*, 20, No. 1 (Spring 1997), 1–12.

[616] LONGSTAFF, FRANCIS A. "The Valuation of Options on Yields." *Journal of Financial Economics*, Vol. 26 (1990), 97–121.

[617] LONGSTAFF, FRANCIS A. "Hedging Interest Rate Risk with Options on Average Interest Rates." *The Journal of Fixed Income*, 5, No. 2 (September 1995), 37–45.

[618] LONGSTAFF, FRANCIS A. AND EDUARDO S. SCHWARTZ. "Interest Rate Volatility and Bond Prices." *Financial Analysts Journal*, 49, No. 4 (July–August 1993), 70–74.

[619] LOWELL, L. "Mortgage Pass-Through Securities." In [324, Chapter 4] and [328, Chapter 27].

[620] LOWENSTEIN, LOUIS. *Sense and Nonsense in Corporate Finance*. Reading, MA: Addison-Wesley, 1991.

[621] LUBY, MICHAEL. *Pseudorandomness and Cryptographic Applications*. Princeton, NJ: Princeton Univ. Press, 1996.

[622] LUEHRMAN, TIMOTHY A. "Strategy as a Portfolio of Real Options." *Harvard Business Review* (September–October 1998), 89–99.

[623] LUENBERGER, DAVID G. *Investment Science*. New York: Oxford Univ. Press, 1998.

[624] LYUU, YUH-DAUH. "Very Fast Algorithms for Barrier Option Pricing and the Ballot Problem." *The Journal of Derivatives*, 5, No. 3 (Spring 1998), 68–79.

[625] LYUU, YUH-DAUH. "A General Computational Method for Calibration Based on Differential Trees." *The Journal of Derivatives*, 7, No. 1 (Fall 1999), 79–90.

[626] LYUU, YUH-DAUH AND CHEN-LEH WANG. "An Object-Oriented Framework for Financial Computation on Three-Tier Client/Server Architectures." In *Proceedings of the 13th Workshop on Combinatorial Mathematics and Computation Theory*, Providence University, Taiwan, 1996.

[627] MACAULAY, FREDERICK R. *Some Theoretical Problems Suggested by the Movements of Interest Rates, Bond Yields, and Stock Prices in the United States since 1856*. New York: National Bureau of Economic Research (NBER), 1938.

[628] MADAN, DILIP B., FRANK MILNE, AND HERSH SHEFRIN. "The Multinomial Option Pricing Model and Its Brownian and Poisson Limits." *The Review of Financial Studies*, 2, No. 2 (1989), 251–265.

[629] MADAN, DILIP AND HALUK UNAL. "A Two-Factor Hazard Rate Model for Pricing Risky Debt and the Term Structure of Credit Spread." *Journal of Financial and Quantitative Analysis*, 35, No. 1 (March 2000), 43–65.

[630] MAGHSOODI, YOOSEF. "Solution of the Extended CIR Term Structure and Bond Option Valuation." *Mathematical Finance*, 6, No. 1 (January 1996), 89–109.

[631] MAIOCCHI, ROBERTO. "The Case of Brownian Motion." *British Journal for History of Science*, 23, Part 3, No. 78 (September 1990), 257–283.

[632] MALINVAUD, E. *Statistical Methods of Econometrics.* 3rd revised ed. Amsterdam: North-Holland, 1980.

[633] MALKIEL, BURTON G. "Term Structure of Interest Rates." In [307, pp. 265–270].

[634] MALKIEL, BURTON G. "Returns from Investing in Equity Mutual Funds 1971 to 1991." *The Journal of Finance,* 50, No. 2 (June 1995), 549–572.

[635] MALKIEL, BURTON G. *A Random Walk Down Wall Street.* New York: Norton, 1999.

[636] MALKIEL, BURTON G. AND YEXIAO XU. "Risk and Return Revisited." *The Journal of Portfolio Management,* 23, No. 3 (Spring 1997), 9–14.

[637] MANDELBROT, B.B. "Louis Bachelier." In [307, pp. 86–88].

[638] MANES, S. AND P. ANDREWS. *Gates.* New York: Simon & Schuster, 1994.

[639] MANKIW, N. GREGORY AND JEFFREY A. MIRON. "The Changing Behavior of the Term Structure of Interest Rates." *The Quarterly Journal of Economics,* 101, No. 2 (May 1986), 211–228.

[640] MARGRABE, WILLIAM. "The Value of an Option To Exchange One Asset for Another." *The Journal of Finance,* 33, No. 1 (March 1978), 177–186.

[641] MARKOWITZ, HARRY M. "Portfolio Selection." *The Journal of Finance,* 7, No. 1 (1952), 77–91.

[642] MARKOWITZ, HARRY M. *Mean-Variance Analysis in Portfolio Choice and Capital Markets.* Oxford, U.K.: Blackwell, 1987.

[643] MARKOWITZ, HARRY M. "The General Mean-Variance Portfolio Selection Problem." *Philosophical Transactions of the Royal Society of London A,* 347, No. 1684 (June 15, 1994), 543–549.

[644] MARKOWITZ, HARRY M. "The Early History of Portfolio Theory: 1600–1960." *Financial Analysts Journal,* 55, No. 4 (July–August 1999), 5–16.

[645] MARSH, TERRY A. AND ERIC R. ROSENFELD. "Stochastic Processes for Interest Rates and Equilibrium Bond Prices." *The Journal of Finance,* 38, No. 2 (May 1983), 635–650.

[646] MARSHALL, JOHN F. AND VIPUL K. BANSAL. *Financial Engineering: A Complete Guide to Financial Innovation.* New York: New York Institute of Finance, 1992.

[647] MASON, SCOTT P., ROBERT C. MERTON, ANDRÉ F. PEROLD, AND PETER TUFANO. *Cases in Financial Engineering: Applied Studies of Financial Innovation.* Englewood Cliffs, NJ: Prentice-Hall, 1995.

[648] MAYR, ERNST. *The Growth of Biological Thought: Diversity, Evolution, and Inheritance.* Cambridge, MA: Harvard Univ. Press, 1982.

[649] MCCONNELL, JOHN J. AND MANOJ SINGH. "Rational Prepayments and the Valuation of Collateralized Mortgage Obligations." *The Journal of Finance,* 49, No. 3 (July 1994), 891–921.

[650] MCCONNELL, S. *Code Complete.* Redmond, WA: Microsoft Press, 1993.

[651] MCCOY, WILLIAM F. "Bond Dynamic Hedging and Return Attribution: Empirical Evidence." *The Journal of Portfolio Management* (Winter 1995), 93–101.

[652] MCCULLOCH, J. HUSTON. "The Tax-Adjusted Yield Curve." *The Journal of Finance,* 30, No. 3 (June 1975), 811–830.

[653] MCENALLY, RICHARD W. AND JAMES V. JORDAN. "The Term Structure of Interest Rates." In [328, Chapter 37].

[654] MEHRLING, PERRY. "Minsky and Modern Finance." *The Journal of Portfolio Management,* 26, No. 2 (Winter 2000), 81–88.

[655] MELEIS, HANAFY. "Toward the Information Network." *Computer,* 29, No. 10 (October 1996), 59–67.

[656] MELMAN, A. "Geometry and Convergence of Euler's and Halley's Methods." *SIAM Review,* 39, No. 4 (December 1997), 728–735.

[657] MENGER, CARL. *Investigations into the Method of the Social Sciences with Special Reference to Economics.* New York: New York Univ. Press, 1985.

[658] MERTON, ROBERT C. "Options." In [307, pp. 213–218].

[659] MERTON, ROBERT C. "Influence of Mathematical Models in Finance on Practice: Past, Present and Future." *Philosophical Transactions of the Royal Society of London A,* 347, No. 1684 (June 15, 1994), 451–463.

[660] MERTON, ROBERT C. *Continuous-Time Finance.* Revised ed. Cambridge, MA: Blackwell, 1994.

[661] MERTON, ROBERT C. "Finance Theory and Future Trends: The Shift to Integration." *Risk,* 12, No. 7 (July 1999), 48–51.

[662] MERTON, ROBERT C. AND MYRON S. SCHOLES. "Fischer Black." *The Journal of Finance,* 50, No. 5 (December 1995), 1359–1370.

[663] MILEVSKY, MOSHE ARYE AND STEVEN E. POSNER. "A Closed-Form Approximation for Valuing Basket Options." *The Journal of Derivatives,* 5, No. 4 (Summer 1998), 54–61.

[664] MILEVSKY, MOSHE ARYE AND STEVEN E. POSNER. "Asian Options, the Sum of

Lognormals, and the Reciprocal Gamma Distribution." *Journal of Financial and Quantitative Analysis*, 33, No. 3 (September 1998), 409–422.

[665] MILL, JOHN STUART. *Principles of Political Economy*. Books IV and V. First published in 1848. New York: Penguin, 1985.

[666] MILLER, MERTON H. "The History of Finance." *The Journal of Portfolio Management*, 25, No. 4 (Summer 1999), 95–101.

[667] MILLS, TERENCE C. *Time Series Techniques for Economists*. Cambridge, U.K.: Cambridge Univ. Press, 1990.

[668] MIL'SHTEIN, G.N. "Approximate Integration of Stochastic Differential Equations." *Theory of Probability and Its Applications*, 19, No. 2 (June 1974), 557–562.

[669] MIL'SHTEIN, G.N. "A Method of Second-Order Accuracy Integration of Stochastic Differential Equations." *SIAM Theory of Probability and Its Applications*, 23, No. 2 (June 1978), 396–401.

[670] MILTERSEN, KRISTIAN R. "An Arbitrage Theory of the Term Structure of Interest Rates." *Annals of Applied Probability*, 4, No. 4 (November 1994), 953–967.

[671] MISES, RICHARD VON. *Probability, Statistics, and Truth*. First published in 1928. New York: Dover, 1981.

[672] MOAD, J. "Time for a Fresh Approach to DOI." *Datamation*, 41, No. 3 (February 15, 1995), 57–59.

[673] MODIGLIANI, FRANCO AND LEAH MODIGLIANI. "Risk-Adjusted Performance." *The Journal of Portfolio Management*, 23, No. 2 (Winter 1997), 45–54.

[674] MODIGLIANI, FRANCO AND RICHARD SUTCH. "Innovations in Interest Rate Policy." *The American Economic Review*, Vol. 56 (May 1966), 178–197.

[675] MOHANTY, SRI GOPAL. *Lattice Path Counting and Applications*. New York: Academic, 1979.

[676] MONTESQUIEU, CHARLES DE. *The Spirit of Laws*. First published in 1748. Chicago: Encyclopaedia Britannica, 1952.

[677] MORALEDA, JUAN M. AND ANTOON PELSSER. "Forward versus Spot Interest Rate Models of the Term Structure: An Empirical Comparison." *The Journal of Derivatives*, 7, No. 3 (Spring 2000), 9–21.

[678] MORO, BORIS. "The Full Monte." *Risk*, 8, No. 2 (February 1995), 57–58.

[679] MOROKOFF, WILLIAM J. "Generating Quasi-Random Paths for Stochastic Processes." *SIAM Review*, 40, No. 4 (December 1998), 765–788.

[680] MOSSIN, JAN. "Equilibrium in a Capital Asset Market." *Econometrica*, Vol. 34 (October 1966), 768–783.

[681] MUSIELA, MAREK AND MAREK RUTKOWSKI. *Martingale Methods in Financial Modeling*. Berlin: Springer-Verlag, 1997.

[682] MUSIELA, MAREK, STUART M. TURNBULL, AND LEE M. WAKEMAN. "Interest Rate Risk Management." *Review of Futures Markets*, 12, No. 1 (1993), 221–261.

[683] MYNENI, RAVI. "The Pricing of the American Options." *Annals of Applied Probability*, 2, No. 1 (1992), 1–23.

[684] NABAR, P.G. AND J.E. TIERNEY. "Market Dynamics of Eleventh District COFI in a Post-RTC World." *The Journal of Fixed Income*, Vol. 3 (December 1993), 63–70.

[685] NAGOT, ISABELLE AND ROBERT TROMMSDORFF. "The Tree of Knowledge." *Risk*, 12, No. 8 (August 1999), 99–102.

[686] NARAYANA, T.V. *Lattice Path Combinatorics with Statistical Applications*. Toronto: University of Toronto Press, 1979.

[687] NASH, STEPHEN G. AND ARIELA SOFER. *Linear and Nonlinear Programming*. New York: McGraw-Hill, 1996.

[688] NATH, ALOKE. *The Guide to SQL Server*. 2nd ed. Reading, MA: Addison-Wesley, 1995.

[689] NAUSS, ROBERT M. "Bond Portfolio Analysis Using Integer Programming." In [891, Chapter 11].

[690] NAWALKHA, SANJAY K. AND DONALD R. CHAMBERS. "The Binomial Model and Risk Neutrality: Some Important Details." *The Financial Review*, 30, No. 3 (August 1995), 605–615.

[691] NEFTCI, SALIH N. "Value at Risk Calculation, Extreme Events and Tail Estimation." *The Journal of Derivatives*, 7, No. 3 (Spring 2000), 23–37.

[692] NEFTCI, SALIH N. *An Introduction to the Mathematics of Financial Derivatives*. 2nd ed. San Diego, CA: Academic, 2000.

[693] NEGRYCH, C. AND D. SENFT. "Portfolio Insurance Using Synthetic Puts – the Reasons, Rewards and Risks." In [321, Chapter 12].

[694] NELSON, C.R. *The Term Structure of Interest Rates*. New York: Basic Books, 1972.

[695] NELSON, CHARLES R. AND ANDREW F. SIEGEL. "Parsimonious Modeling of Yield Curves." *Journal of Business*, 60, No. 4 (1987), 473–489.

[696] NELSON, DANIEL B. AND KRISHNA RAMASWAMY. "Simple Binomial Processes as Diffusion Approximations in Financial

Models." *The Review of Financial Studies*, 3, No. 3 (1990), 393–430.

[697] NYBORG, K.G. "The Use and Pricing of Convertible Bonds." *Applied Mathematical Finance*, Vol. 3 (1996), 167–190.

[698] NEW YORK INSTITUTE OF FINANCE. *Stocks, Bonds, Options, Futures: Investments and Their Markets*. Englewood Cliffs, NJ: Prentice-Hall, 1987.

[699] NIEDERREITER, HARALD. *Random Number Generation and Quasi-Monte Carlo Methods*. Philadelphia: Society for Applied and Industrial Mathematics, 1992.

[700] NIELSEN, J.A. AND K. SANDMANN. "The Pricing of Asian Options under Stochastic Interest Rates." *Applied Mathematical Finance*, Vol. 3 (1996), 209–236.

[701] NOBLE, BEN. *Applied Linear Algebra*. Englewood Cliffs, NJ: Prentice-Hall, 1969.

[702] OGDEN, JOSEPH P. "An Analysis of Yield Curve Notes." *The Journal of Finance*, 42, No. 1 (March 1987), 99–110.

[703] OHMAE, KENICHI. "Five Strong Signals of Japan's Coming Crash." *The Straits Times* (July 6, 1998), 28.

[704] OMBERG, EDWARD. "A Note on the Convergence of Binomial-Pricing and Compound-Option Models." *The Journal of Finance*, 42, No. 2 (June 1987), 463–469.

[705] OREIZY, PEYMAN AND GAIL KAISER. "The Web as Enabling Technology for Software Development and Distribution." *Internet Computing*, 1, No. 6 (November/December 1997), 84–87.

[706] ORTEGA, J.M. *Introduction to Parallel and Vector Solution of Linear Systems*. New York: Plenum, 1988.

[707] ORTEGA, J.M. AND R.G. VOIGT. *Solution of Partial Differential Equations on Vector and Parallel Computers*. Philadelphia: Society for Industrial and Applied Mathematics, 1985.

[708] ORVIS, WILLIAM J. *Excel for Scientists and Engineers*. 2nd ed. San Francisco: SYBEX, 1995.

[709] OSBORNE, M. "Brownian Motion in the Stock Market." *Operations Research*, Vol. 7 (1959), 145–173.

[710] PAPADIMITRIOU, CHRISTOS H. *Computational Complexity*. Reading, MA: Addison-Wesley, 1995.

[711] PAPAGEORGIOU, A. AND J.F. TRAUB. "New Results on Deterministic Pricing of Financial Derivatives." Technical Report CUCS-028-96, Department of Computer Science, Columbia University, New York, 1996.

[712] PAQUETTE, L. AND P. VANKUDRE. "Citicorp's Rich/Cheap Model for Treasury Securities."

Research Series, Citicorp Fixed Income Research, New York, Vol. 3, No. 4, September 1988.

[713] PARKINSON, MICHAEL. "Option Pricing: The American Put." *Journal of Business*, 50, No. 1 (January 1977), 21–36.

[714] PARNAS, DAVID LORGE. "Why Software Jewels Are Rare." *Computer*, 29, No. 2 (February 1996), 57–60.

[715] PASKOV, SPASSIMIR H. "New Methodologies for Valuing Derivatives." Technical Report CUCS-029-96, Department of Computer Science, Columbia University, New York, 1996. Also in *Mathematics of Derivative Securities*. Edited by S. Pliska and M. Dempster. Cambridge, U.K.: Cambridge Univ. Press, 1996.

[716] PASKOV, SPASSIMIR H. AND J.F. TRAUB. "Faster Valuation of Financial Derivatives." Technical Report CUCS-030-96, Department of Computer Science, Columbia University, New York, 1996.

[717] PATTERSON, DAVID A. AND JOHN H. HENNESSY. *Computer Organization & Design: The Hardware/Software Interface*. 2nd ed. San Mateo, CA: Morgan Kaufmann, 1994.

[718] PELSSER, ANTOON AND TON VORST. "The Binomial Model and the Greeks." *The Journal of Derivatives*, 1, No. 3 (Spring 1994), 45–49.

[719] PETERSON, LARRY L. AND BRUCE S. DAVIE. *Computer Networks: A Systems Approach*. 2nd ed. San Mateo, CA: Morgan Kaufmann, 2000.

[720] PHELAN, MICHAEL J. "Probability and Statistics Applied to the Practice of Financial Risk Management: The Case of JP Morgan's RiskMetrics™." Working paper 95-19, Wharton School, University of Pennsylvania, Philadelphia.

[721] PHILLIPS, DON T., A. RAVINDRAN, AND J.J. SOLBERG. *Operations Research: Principles and Practice*. New York: Wiley, 1976.

[722] PHOA, WESLEY. "Can You Derive Market Volatility Forecasts from the Observed Yield Curve Convexity Bias?" *The Journal of Fixed Income*, 7, No. 1 (June 1997), 43–54.

[723] PIKOVSKY, IGOR AND IOANNIS KARATZAS. "Anticipative Portfolio Optimization." *Advances in Applied Probability*, Vol. 28 (1996), 1095–1122.

[724] PITTS, MARK AND FRANK J. FABOZZI. "Introduction to Interest-Rate Futures and Options Contracts." In [328, Chapter 50].

[725] PLISKA, STANLEY R. *Introduction to Mathematical Finance: Discrete-Time Models*. Malden, MA: Blackwell, 1997.

[726] PRECHELT, LUTZ. "An Empirical Comparison of Seven Programming Languages." *Computer*, 33, No. 10 (October 2000), 23–29.

[727] PRESS, WILLIAM H., SAUL A. TEUKOLSKY, WILLIAM T. VETTERLING, AND BRIAN P. FLANNERY. *Numerical Recipes in C: The Art of Scientific Computing*. 2nd ed. Cambridge, U.K.: Cambridge Univ. Press, 1992.

[728] PRITSKER, MATT. "Nonparametric Density Estimation and Tests of Continuous Time Interest Rate Models." *The Review of Financial Studies*, 11, No. 3 (Fall 1998), 449–487.

[729] PUBLIC SECURITIES ASSOCIATION. "Standard Formulas for the Analysis of Mortgage-Backed Securities and Other Related Securities." July 1, 1990.

[730] RAYMAR, STEVEN B., AND MICHAEL J. ZWECHER. "Monte Carlo Estimation of American Call Options on the Maximum of Several Stocks." *The Journal of Derivatives*, 5, No. 1 (Fall 1997), 7–23.

[731] REBONATO, RICCARDO. *Interest-Rate Option Models: Understanding, Analyzing, and Using Interest Rate Models for Exotic Interest-Rate Options*. 2nd ed. New York: Wiley, 1998.

[732] REDINGTON, F.M. "Review of the Principles of Life-Office Valuations." *Journal of the Institute of Actuaries*, 18, No. 3 (1952), 286–315.

[733] REGLI, WILLIAM. "Intranets." *Internet Computing*, 1, No. 5 (September/October 1997), 6–7.

[734] REINER, ERIC. "Quanto Mechanics." In [744, Chapter 22].

[735] REISMAN, HAIM. "A Binomial Tree for the Hull and White Model with Probabilities Independent of the Initial Term Structure." *The Journal of Fixed Income*, 6, No. 3 (December 1996), 92–96.

[736] RENAUD, P.E. *Introduction to Client/Server Systems: A Practical Guide for Systems Professionals*. New York: Wiley, 1993.

[737] RENDLEMAN, RICHARD J., JR. "Duration-Based Hedging with Treasury Bond Futures." *The Journal of Fixed Income*, 9, No. 1 (June 1999), 84–91.

[738] RENDLEMAN, RICHARD J., JR. AND BRIT J. BARTTER. "Two State Option Pricing." *The Journal of Finance*, Vol. 34 (1979), 1093–1110.

[739] RENDLEMAN, RICHARD J., JR. AND BRIT J. BARTTER. "The Pricing of Options on Debt Securities." *Journal of Financial and Quantitative Analysis*, 15, No. 1 (March 1980), 11–24.

[740] RICH, DON R. "The Mathematical Foundations of Barrier Option-Pricing Theory." *Advances in Futures and Options Research*, Vol. 7 (1994), 267–311.

[741] RICHARD, SCOTT F. "An Arbitrage Model of the Term Structure of Interest Rates." *Journal of Financial Economics*, 6, No. 1 (March 1978), 38–57.

[742] RICHARD, SCOTT F. AND RICHARD ROLL. "Prepayments on Fixed-Rate Mortgage-Backed Securities." *The Journal of Portfolio Management*, 15, No. 3 (Spring 1989), 73–82.

[743] RICHARDSON, MATTHEW AND TOM SMITH. "A Test for Multivariate Normality in Stock Returns." *Journal of Business*, 66, No. 2 (April 1993), 295–321.

[744] RISK MAGAZINE. *From Black-Scholes to Black Holes: New Frontiers in Options*. London: RISK/FINEX, 1992.

[745] RITCHKEN, PETER. "On Pricing Barrier Options." *The Journal of Derivatives*, 3, No. 2 (Winter 1995), 19–28.

[746] RITCHKEN, PETER. *Derivative Markets: Theory, Strategy, and Applications*. New York: HarperCollins, 1996.

[747] RITCHKEN, PETER AND L. SANKARASUBRAMANIAN. "Volatility Structures of Forward Rates and the Dynamics of the Term Structure." *Mathematical Finance*, 5, No. 1 (January 1995), 55–72.

[748] RITCHKEN, PETER, L. SANKARASUBRAMANIAN AND ANAND M. VIJH. "The Valuation of Path Dependent Contracts on the Average." *Management Science*, 39, No. 10 (October 1993), 1202–1213.

[749] RITCHKEN, PETER AND ROB TREVOR. "Pricing Options under Generalized GARCH and Stochastic Volatility Processes." *The Journal of Finance*, 54, No. 1 (February 1999), 377–402.

[750] RITTER, L.S. AND W.L. SILBER. *Principles of Money, Banking, and Financial Markets*. New York: Basic Books, 1991.

[751] ROBERDS, WILLIAM, DAVID RUNKLE AND CHARLES H. WHITEMAN. "A Daily View of Yield Spreads and Short-Term Interest Rate Movements." *Journal of Money, Credit, and Banking*, 28, No. 1 (February 1996), 34–53.

[752] ROBERTS, G.O. AND C.F. SHORTLAND. "Pricing Barrier Options with Time-Dependent Coefficients." *Mathematical Finance*, 7, No. 1 (January 1997), 83–93.

[753] ROGERS, L.C.G. "Equivalent Martingale Measures and No-Arbitrage." *Stochastics and Stochastics Reports*, Vol. 51 (1994), 41–49.

[754] ROGERS, L.C.G. "Which Model for Term-Structure of Interest Rates Should One Use?" In *Mathematical Finance*. Edited by Mark H.A. Davis, Darrell Duffie, Wendall H. Fleming, and Steven E. Shreve. Berlin: Springer-Verlag, 1995.

[755] ROGERS, L.C.G. AND Z. SHI. "The Value of an Asian Option." *Journal of Applied Probability*, 32, No. 4 (December 1995), 1077–1088.

[756] ROGERS, L.C.G. AND E.J. STAPLETON. "Fast Accurate Binomial Pricing." *Finance & Stochastics*, 2, No. 1 (1998).

[757] ROGERS, L.C.G AND O. ZANE. "Valuing Moving Barrier Options." *Journal of Computational Finance*, 1, No. 1 (1997).

[758] ROLL, RICHARD. "Collateralized Mortgage Obligations: Characteristics, History, Analysis." In [320, pp. 7–43].

[759] ROLL, RICHARD. "U.S. Treasury Inflation-Indexed Bonds: The Design of a New Security." *The Journal of Fixed Income*, 6, No. 3 (December 1996), 9–28.

[760] ROLL, RICHARD AND STEPHEN A. ROSS. "On the Cross-Sectional Relation between Expected Returns and Betas." *The Journal of Finance*, 49, No. 1 (March 1994), 101–121.

[761] ROSE, COLIN. "Bounded and Unbounded Stochastic Processes." In [854, Chapter 11].

[762] ROSE, M.E. "The Effective Cash Flow Method." *The Journal of Fixed Income*, 4, No. 2 (September 1994), 66–79.

[763] ROSS, SHELDON M. *Stochastic Processes*. New York: Wiley, 1983.

[764] ROSS, SHELDON M. *Introduction to Probability Models*. 5th ed. New York: Wiley, 1993.

[765] ROSS, STEPHEN A. "The Arbitrage Theory of Capital Asset Pricing." *Journal of Financial Economics*, Vol. 13 (1976), 341–360.

[766] ROSS, STEPHEN A. "Finance." In [307, pp. 1–34].

[767] ROSS, STEPHEN A., RANDOLPH W. WESTERFIELD AND JEFFREY F. JAFFE. *Corporate Finance*. Burr Ridge, IL: Irwin, 1993.

[768] RUBINSTEIN, MARK. "Guiding Force." In [744, Chapter 4].

[769] RUBINSTEIN, MARK. "One for Another." In [744, Chapter 29].

[770] RUBINSTEIN, MARK. "Implied Binomial Trees." *The Journal of Finance*, 49, No. 3 (July 1994), 771–818.

[771] RUBINSTEIN, MARK. "Edgeworth Binomial Trees." *The Journal of Derivatives*, 5, No. 3 (Spring 1998), 20–27.

[772] RUBINSTEIN, MARK AND HAYNE E. LELAND. "Replicating Options with Positions in Stock and Cash." *Financial Analysts Journal*, 37, No. 4 (July–August 1981), 63–72.

[773] RUBINSTEIN, REUVEN Y. *Simulation and the Monte Carlo Method*. New York: Wiley, 1981.

[774] RÜMELIN, W. "Numerical Treatment of Stochastic Differential Equations." *SIAM*

Journal on Numerical Analysis, 19, No. 3 (June 1982), 604–613.

[775] RUMSEY, JOHN. "Pricing Cross-Currency Options." *The Journal of Futures Markets*, 11, No. 1 (February 1991), 89–93.

[776] SAATY, T.L. *Modern Nonlinear Equations*. New York: Dover, 1981.

[777] SACK, BRIAN. "Deriving Inflation Expectations from Nominal and Inflation-Indexed Treasury Yields." *The Journal of Fixed Income*, 10, No. 2 (September 2000), 6–17.

[778] SAMUELSON, PAUL A. *Economics*. 11th ed. New York: McGraw-Hill, 1980.

[779] SANDMANN, KLAUS. "The Pricing of Options with an Uncertain Interest Rate: A Discrete-Time Approach." *Mathematical Finance*, 3, No. 2 (April 1993), 201–216.

[780] SANDMANN, KLAUS AND DIETER SONDERMANN. "A Term Structure Model and the Pricing of Interest Rate Derivatives." *The Review of Futures Studies*, 12, No. 2 (1993), 391–423.

[781] SANDMANN, KLAUS AND DIETER SONDERMANN. "A Note on the Stability of Lognormal Interest Rate Models and the Pricing of Eurodollar Futures." *Mathematical Finance*, 7, No. 2 (April 1994), 119–125.

[782] SANDMANN, KLAUS AND DIETER SONDERMANN. "Log-Normal Interest Rate Models: Stability and Methodology." Manuscript, January 10, 1997.

[783] SCHRODER, MARK. "Change of Numeraire for Pricing Futures, Forwards, and Options." *The Review of Financial Studies*, 12, No. 5 (Winter 1999), 1143–1163.

[784] SCHROEDER, M. *Fractals, Chaos, Power Laws: Minutes from an Infinite Paradise*. New York: Freeman, 1991.

[785] SCHUMPETER, JOSEPH ALOIS. *The Theory of Economic Development: An Inquiry into Profits, Capital, Credit, Interest, and the Business Cycle*. First published in 1911. Translated by R. Opie. Cambridge, MA: Harvard Univ. Press, 1934.

[786] SCHUMPETER, JOSEPH ALOIS. *History of Economic Analysis*. New York: Oxford Univ. Press, 1954.

[787] SCHUMPETER, JOSEPH ALOIS. *Essays on Entrepreneurs, Innovations, Business Cycles, and the Evolution of Capitalism*. Edited by K.V. Clemence. New Brunswick, NJ: Transaction Publishers, 1989.

[788] SCHWARTZ, EDUARDO S. AND WALTER N. TOROUS. "Prepayment and the Valuation of Mortgage-Backed Securities." *The Journal of Finance*, 44, No. 2 (June 1989), 375–392.

[789] SCHWARTZ, EDUARDO S. AND WALTER N. TOROUS. "Prepayment, Default, and the Valuation of Mortgage Pass-Through Securities." *Journal of Business*, 65, No. 2 (April 1992), 221–239.

[790] SCOTT, LOUIS O. "Option Pricing When the Variance Changes Randomly: Theory, Estimation, and an Application." *Journal of Financial and Quantitative Analysis*, 22, No. 4 (December 1987), 419–438.

[791] SEDGEWICK, R. *Algorithms*. 2nd ed. Reading, MA: Addison-Wesley, 1988.

[792] SEEMAN, B.S. "An Introduction to Index Amortizing Swaps (IASs)." *The Journal of Fixed Income*, 4, No. 2 (September 1994), 6–16.

[793] SELBY, MICHAEL J.P. AND CHRIS STRICKLAND. "Computing the Fong and Vasicek Pure Discount Bond Price Formula." *The Journal of Fixed Income*, 5, No. 2 (September 1995), 78–84.

[794] SENFT, DEXTER. "Parallel Processing on Wall Street." In *Proceedings of the Commercial Applications of Parallel Processing Systems Conference*, Austin, Texas: MCC, October 1993.

[795] SHARPE, WILLIAM F. "A Simplified Model for Portfolio Analysis." *Management Science*, 9, No. 1 (January 1963), 277–293.

[796] SHARPE, WILLIAM F. "Capital Asset Prices: A Theory of Market Equilibrium under Conditions of Risk." *The Journal of Finance*, 19, No. 3 (September 1964), 425–442.

[797] SHARPE, WILLIAM F. *Investments*. 3rd ed. Englewood Cliffs, NJ: Prentice-Hall, 1985.

[798] SHARPE, WILLIAM F. "The Sharpe Ratio." *The Journal of Portfolio Management*, 21, No. 1 (Fall 1994), 49–58.

[799] SHARPE, WILLIAM F., GORDON J. ALEXANDER AND JEFFERY V. BAILEY. *Investments*. 5th ed. Englewood Cliffs, NJ: Prentice-Hall, 1995.

[800] SHEA, GARY S. "Pitfalls in Smoothing Interest Rate Term Structure Data: Equilibrium Models and Spline Approximations." *Journal of Financial and Quantitative Analysis*, 19, No. 3 (September 1984), 253–269.

[801] SHEA, GARY S. "Interest Rate Term Structure Estimation with Exponential Splines: A Note." *The Journal of Finance*, 40, No. 1 (March 1985), 319–325.

[802] SILVEY, S.D. *Statistical Inference*. London: Chapman & Hall, 1988.

[803] SINGH, MANOJ K. "Estimation of Multifactor Cox, Ingersoll and Ross Term Structure Model: Evidence on Volatility Structure and Parameter Stability." *The Journal of Fixed Income*, 5, No. 2 (September 1995), 8–28.

[804] SINGH, MANOJ K. "Value at Risk Using Principal Components Analysis." *The Journal of Portfolio Management*, 24, No. 1 (Fall 1997), 101–112.

[805] SKIDELSKY, ROBERT. *John Maynard Keynes: Hope Betrayed, 1883–1920*. New York: Elisabeth Sifton, 1983.

[806] SKIDELSKY, ROBERT. *John Maynard Keynes: The Economist as Savior, 1920–1937*. New York: Penguin, 1992.

[807] SKIDELSKY, ROBERT. *John Maynard Keynes: Fighting for Britain, 1937–1946*. London: Macmillan, 2000.

[808] SMITH, ADAM. *Adam Smith's Moral and Political Philosophy*. Edited by H.W. Schneider. New York: Harper & Row, 1970.

[809] SMITH, CLIFFORD W., JR. "Corporate Risk Management: Theory and Practice." *The Journal of Derivatives*, 2, No. 4 (Summer 1995), 21–30.

[810] SMITH, G.D. *Numerical Solution of Partial Differential Equations: Finite Difference Methods*. 3rd ed. Oxford, U.K.: Oxford Univ. Press, 1985.

[811] SMITHSON, CHARLES. "Wonderful Life." In [744, Chapter 2].

[812] SMITHSON, CHARLES. "Path Dependency." *Risk*, 10, No. 4 (April 1997), 65–67.

[813] SMITHSON, CHARLES. "Does Risk Management Work?" *Risk*, 12, No. 7 (July 1999), 44–45.

[814] SONER, H.M., S.E. SHREVE AND J. CVITANIĆ. "There Is No Nontrivial Hedging Portfolio for Option Pricing with Transaction Costs." *The Annals of Applied Probability*, 5, No. 2 (May 1995), 327–355.

[815] SPARKS, ANDY AND FRANK FEIKEH SUNG. "Prepayment Convexity and Duration." *The Journal of Fixed Income*, 5, No. 2 (September 1995), 7–11.

[816] SPIEGEL, M.R. *Theory and Problems of Statistics*. Schaum's Outline Series. New York: McGraw-Hill, 1961.

[817] SPITZER, FRANK. *Principles of Random Walk*. Princeton, NJ: D. Van Nostrand, 1964.

[818] STANTON, RICHARD. "Rational Prepayment and the Valuation of Mortgage-Backed Securities." *The Review of Financial Studies*, 8, No. 3 (1995), 677–708.

[819] STANTON, RICHARD. "A Nonparametric Model of Term Structure Dynamics and the Market Price of Interest Rate Risk." *The Journal of Finance*, 52, No. 5 (December 1997), 1973–2002.

[820] STANTON, RICHARD AND NANCY WALLACE. "Anatomy of an ARM: The Interest-Rate Risk

of Adjustable-Rate Mortgages." *Journal of Real Estate Finance and Economics*, 19, No. 1 (1999), 49–67.

[821] STAVIS, R.M. AND V.J. HAGHANI. "Putable Swaps: Tools for Managing Callable Assets." In [321, Chapter 20].

[822] STEELE, J. MICHAEL AND ROBERT A. STINE. "*Mathematica* and Diffusions." In [854, Chapter 9].

[823] STEIN, ELIAS M. AND JEREMY C. STEIN. "Stock Price Distributions with Stochastic Volatility: An Analytic Approach." *The Review of Financial Studies*, 4, No. 4 (1991), 727–752.

[824] STEINER, MANFRED, MARTIN WELLMEIER, AND REINHOLD HAFNER. "Pricing Near the Barrier: The Case of Discrete Knock-Out Options." *The Journal of Computational Finance*, 3, No. 1 (Fall 1999), 69–90.

[825] STEWART, G.W. "On the Early History of the Singular Value Decomposition." *SIAM Review*, 35, No. 4 (December 1993), 551–566.

[826] STIGLER, GEORGE J. *Essays in the History of Economics*. First published in 1965. Midway Reprint. Chicago: University of Chicago Press, 1987.

[827] STIGUM, MARCIA. *Money Market Calculations: Yields, Break-Evens and Arbitrage*. Homewood, IL: Irwin, 1981.

[828] STIGUM, MARCIA. *The Money Market*. Newly revised. Homewood, IL: Irwin, 1983.

[829] STONE, CHARLES AUSTIN AND ANNE ZISSU. "The Risks of Mortgage Backed Securities and Their Derivatives." *Journal of Applied Corporate Finance*, 7, No. 3 (Fall 1994), 99–111.

[830] STRANG, GILBERT. *Linear Algebra and Its Applications*. 2nd ed. New York: Academic, 1980.

[831] STREETER, LYNN A., ROBERT E. KRAUT, HENRY C. LUCAS, JR., AND LAURENCE CABY. "How Open Data Networks Influence Business Performance and Market Structure." *Communications of the ACM*, 39, No. 7 (July 1996), 62–73.

[832] STROUSTRUP, BJARNE. *The C++ Programming Language*. 2nd ed. Reading, MA: Addison-Wesley, 1991.

[833] STULZ, RENÉ. "Options on the Minimum or the Maximum of Two Risky Assets." *Journal of Financial Economics*, 10, No. 2 (July 1982), 161–185.

[834] SULLIVAN, MICHAEL A. "Valuing American Put Options Using Gaussian Quadrature." *The Review of Financial Studies*, 13, No. 1 (Spring 2000), 75–95.

[835] SUN, TONG-SHENG. "Real and Nominal Interest Rates: A Discrete-Time Model and Its Continuous-Time Limit." *The Review of Financial Studies*, 5, No. 4 (1992), 581–611.

[836] SUNDARAM, RANGARAJAN K. "Equivalent Martingale Measures and Risk-Neutral Pricing: An Expository Note." *The Journal of Derivatives*, 5, No. 1 (Fall 1997), 85–98.

[837] SUNDARESAN, SURESH M. *Fixed Income Markets and Their Derivatives*. Cincinnati, OH: South-Western College Publishing, 1997.

[838] TAKÁCS, LAJOS. *Combinatorial Methods in the Theory of Stochastic Processes*. New York: Wiley, 1967.

[839] TAYLOR, STEPHEN. *Modelling Financial Time Series*. New York: Wiley, 1986.

[840] TAYLOR, STEPHEN J. "Modeling Stochastic Volatility: A Review and Comparative Study." *Mathematical Finance*, 4, No. 2 (April 1994), 183–204.

[841] TIAN, YISONG. "Pricing Complex Barrier Options under General Diffusion Processes." *The Journal of Derivatives*, 7, No. 2 (Winter 1999), 11–30.

[842] TILLEY, JAMES A. "Valuing American Options in a Path Simulation Model." *Transactions of the Society of Actuaries*, Vol. 45 (1992), 83–104.

[843] TOBIN, JAMES. "Financial Intermediaries." In [307, pp. 35–56].

[844] TOMPKINS, ROBERT. "Behind the Mirror." In [744, Chapter 19].

[845] TOMPKINS, ROBERT G. "Power Options: Hedging Nonlinear Risks." *The Journal of Risk*, 2, No. 2 (Winter 1999/2000), 29–45.

[846] TRIVEDI, KISHOR S. *Probability and Statistics with Reliability, Queueing and Computer Science Applications*. Englewood Cliffs, NJ: Prentice-Hall, 1982.

[847] TSIVERIOTIS, KOSTAS AND CHRIS FERNANDES. "Valuing Convertible Bonds with Credit Risk." *The Journal of Fixed Income*, 8, No. 2 (September 1998), 95–102.

[848] TUCKMAN, BRUCE. *Fixed Income Securities: Tools for Today's Markets*. New York: Wiley, 1995.

[849] TURNBULL, STUART M. "Swaps: A Zero Sum Game?" *Financial Management*, Vol. 16 (Spring 1987), 15–21.

[850] TURNBULL, STUART M. AND FRANK MILNE. "A Simple Approach to Interest-Rate Option Pricing." *The Review of Financial Studies*, 4, No. 1 (1991), 87–120.

[851] TURNBULL, STUART M. AND LEE MACDONALD WAKEMAN. "A Quick Algorithm for Pricing European Average Options." *Financial and Quantitative Analysis*, 26, No. 3 (September 1991), 377–389.

[852] U.S. DEPARTMENT OF COMMERCE. *The Emerging Digital Economy*. Springfield, VA: National Technical Information Service, 1998.

[853] VARIAN, HAL R. "The Arbitrage Principle in Financial Economics." *The Journal of Economic Perspectives*, 1, No. 2 (Fall 1987), 55–72.

[854] VARIAN, HAL R., ed. *Economic and Financial Modeling with* Mathematica. Berlin: Springer-Verlag, 1993.

[855] VASICEK, OLDRICH. "An Equilibrium Characterization of the Term Structure." *Journal of Financial Economics*, Vol. 5 (1977), 177–188.

[856] VASICEK, OLDRICH A. AND H. GIFFORD FONG. "Term Structure Modeling Using Exponential Splines." *The Journal of Finance*, 37, No. 2 (May 1982), 339–348.

[857] VENKATARAMAN, SUBU. "Value at Risk for a Mixture of Normal Distributions: The Use of Quasi-Bayesian Estimation Techniques." *Economic Perspectives*, 21, No. 2 (March/April 1997), 2–13.

[858] VETZAL, KENNETH R. "A Survey of Stochastic Continuous Time Models of the Term Structure of Interest Rates." *Insurance: Mathematics and Economics*, Vol. 14 (1994), 139–161.

[859] VETZAL, KENNETH R. "An Improved Finite Difference Approach to Fitting the Initial Term Structure." *The Journal of Fixed Income*, 7, No. 4 (March 1998), 62–81.

[860] VINER, JACOB. *Essays on the Intellectual History of Economics*. Edited by Douglas A. Irwin. Princeton, NJ: Princeton Univ. Press, 1991.

[861] VVEDENSKY, DIMITRI. *Partial Differential Equations with* Mathematica. Reading, MA: Addison-Wesley, 1993.

[862] WAGNER, W.H. "Ten Myths and Twenty Years of Betas." *The Journal of Portfolio Management*, 21, No. 1 (Fall 1994), 79–82.

[863] WAHBA, GRACE. *Spline Methods for Observational Data*. Philadelphia: Society for Industrial and Applied Mathematics, 1990.

[864] WALDMAN, M., A. SCHWALB, AND A.K. FEIGENBERG. "Prepayments of Fifteen-Year Mortgages." *The Journal of Fixed Income* (March 1993), 37–44.

[865] WALL STREET JOURNAL. *The Wall Street Journal Almanac 1998*. New York: Ballantine, 1997.

[866] WALLMAN, STEVEN M.H. "Technology Takes to Securities Trading." *IEEE Spectrum*, Vol. 34 (February 1997), 60–65.

[867] WANG, CHEN-LEH. *An Object-Oriented Framework Based on Three-Tier Client/Server Architecture for Financial Computation*. Master's Thesis. Department of Computer Science and Information Engineering, National Taiwan University, Taiwan, 1996.

[868] WANG, HUANG-WEN. *A Comparative Study of Numerical Algorithms for Interest Rate Models*. Master's Thesis. Department of Computer Science and Information Engineering, National Taiwan University, Taiwan, 1997.

[869] WARSH, DAVID. "Convergence? Not Exactly." *The Boston Globe* (September 29, 1998).

[870] WATKINS, DAVID S. *Fundamentals of Matrix Computations*. New York: Wiley, 1991.

[871] WATSON, RICHARD T. *Data Management: Databases and Organizations*. 2nd ed. New York: Wiley, 1999.

[872] WATSON, THOMAS J., JR. AND PETER PETRE. *Father, Son & Co.: My Life at IBM and Beyond*. New York: Bantam, 1990.

[873] WEI, JASON Z. "Valuing Differential Swaps." *The Journal of Derivatives*, 1, No. 3 (Spring 1994), 64–76.

[874] WEI, JASON Z. "Valuation of Discrete Barrier Options by Interpolations." *The Journal of Derivatives*, 6, No. 1 (Fall 1998), 51–73.

[875] WEINTRAUB, KEITH AND MICHAEL HOGAN. "The Lognormal Interest Rate Model and Eurodollar Futures." Global Research, Citicorp Securities, New York, May 1998.

[876] WILLARD, GREGORY A. "Calculating Prices and Sensitivities for Path-Independent Derivative Securities in Multifactor Models." *The Journal of Derivatives*, 5, No. 1 (Fall 1997), 45–61.

[877] WILLIAMS, DAVID. *Probability with Martingales*. Cambridge, U.K.: Cambridge Univ. Press, 1994.

[878] WILMOTT, PAUL. *Derivatives: The Theory and Practice of Financial Engineering*. New York: Wiley, 1998.

[879] WILMOTT, PAUL, SAM HOWISON AND JEFF DEWYNNE. *The Mathematics of Financial Derivatives: A Student Introduction*. Cambridge, U.K.: Cambridge Univ. Press, 1996.

[880] WINDAS, T. *An Introduction to Option-Adjusted Spread Analysis*. New York: Bloomberg Magazine Publication, 1993.

[881] WOLD, HERMAN. *A Study in the Analysis of Stationary Time Series*. 2nd ed. Uppsala, Sweden: Almqvist & Wiksells, 1954.

[882] WOLFRAM, STEPHEN. *Mathematica*. 2nd ed. Reading, MA: Addison-Wesley, 1994.

[883] WONG, SAMUEL S.M. *Computational Methods in Physics and Engineering*. Englewood Cliffs, NJ: Prentice-Hall, 1992.

[884] WOOLLEY, SCOTT. "Night Baseball without Lights." *Forbes* (August 11, 1997), 42–43.

[885] WU, CHAO-SHENG. *Numerical Methods for Model Calibration under Credit Risk*. Master's Thesis. Department of Computer Science and Information Engineering, National Taiwan University, Taiwan, 1999.

[886] XU, XINZHONG AND STEPHEN J. TAYLOR. "The Term Structure of Volatility Implied by Foreign Exchange Options." *Journal of Financial and Quantitative Analysis*, 29, No. 1 (March 1994), 57–74.

[887] YAMAMOTO, Y. AND STAVROS A. ZENIOS. "Predicting Prepayment Rates for Mortgages Using the Cascade-Correlation Learning Algorithm." *The Journal of Fixed Income*, 2, No. 4 (March 1993), 86–96.

[888] YARGER, RANDY JAY, GEORGE REESE, AND TIM KING. MYSQL AND MSQL. Sebastopol, CA: O'Reilly & Associates, 1999.

[889] YAWITZ, J.B. "Bonds with Embedded Options." In [321, Chapter 15].

[890] ZARETSKY, MICHAEL. "Generation of a Smooth Forward Curve for U.S. Treasuries."

The Journal of Fixed Income, 5, No. 2 (September 1995), 65–69.

[891] ZENIOS, STAVROS A. *Financial Optimization*. Cambridge, U.K.: Cambridge Univ. Press, 1993.

[892] ZENIOS, STAVROS A. "Parallel Monte Carlo Simulation of Mortgage-Backed Securities." In [891, Chapter 14].

[893] ZENIOS, STAVROS A. "A Model for Portfolio Management with Mortgage-Backed Securities." *Annals of Operations Research*, Vol. 43 (1993), 337–356.

[894] ZHANG, PETER G. *Exotic Options: A Guide to Second Generation Options*. Singapore: World Scientific, 1997.

[895] ZIMA, P. AND R.L. BROWN. *Theory and Problems of Contemporary Mathematics of Finance*. Schaum's Outline Series. New York: McGraw-Hill, 1984.

[896] ZIPKINS, PAUL. "Mortgages and Markov Chains: A Simplified Evaluation Model." In [891, Chapter 13].

[897] ZVAN, ROBERT, KENNETH VETZAL AND PETER FORSYTH. "Swing Low Swing High." *Risk*, Vol. 11 (March 1998), 71–75.

> This paper was a great failure, and I am ashamed of it.
> Charles Darwin (1809–1882), *Autobiography*

Glossary of Useful Notations

Index

Printed in the United States
By Bookmasters